秦仁昌教授

秦仁昌,字子农,江苏武进人。生于1898年,1914年入江苏省第一甲种农业学校,1919年考入金陵大学,1925年毕业。1923年任东南大学助教,1928年升为讲师,1929年任"中央研究院"自然历史博物院植物部技师,1930年春获中华教育文化基金会资助,前往丹麦等欧洲诸国访问研究。1932年回国,任静生生物调查所技师,兼植物标本室主任。1934年创建庐山森林植物园并任主任,1938年辗转至云南丽江,设立植物园丽江工作站。1945年任云南大学林学系教授,后任生物系教授兼系主任。1955年当选为学部委员,1956年任中国科学院植物研究所研究员兼植物分类与植物学地理学研究室主任。1986年在北京去世,享年八十有八。秦仁昌一生致力于蕨类植物分类研究,以创建秦仁昌系统而享誉国际,该项成果于1993年荣获中国国家自然科学一等奖。

Prof. Ren-Chang Ching

Ren-Chang Ching, whose courtesy name was Zi-Nong, was born in Wujin County, Jiangsu Province in 1898. He attended Jiangsu First Agriculture School in 1914. In 1919, he entered the University of Nanking and graduated in 1925. From 1923, he worked as an assistant in Southeast University, and was promoted to lecturer in 1928. In 1929, he became the Head of the Botany Section, Metropolitan Museum of Natural History, "Academia Sinica" at Nanjing. Under the support of China Foundation for the Promotion of Education and Culture, he visited Denmark and other European countries for fern study in spring of 1930, and returned back in 1932 to act as Head of the Herbarium of the Fan Memorial Institute of Biology at Beijing. In 1934, he founded and directed the Lushan Arboretum and Botanical Garden. In 1938, he transferred his team to Lijiang, Yunnan Province and founded the Lijiang Botanical Station of Lushan Arboretum and Botanical Garden. Later in 1945, he taught as a professor at the Forestry Department of Yunnan University and later as the Director of the Biology Department of Yunnan University. In 1955, he was elected to be a member of academic committee, the Chinese Academy of Sciences, soon he moved to Beijing and assumed as the Director of the Phytotaxonomy & Phytogeography Department of the Institute of Botany, Chinese Academy of Science in 1956. Professor Ching passed away in Beijing at the age of 88 in 1986.

Professor Ching dedicated his whole life to the taxonomical research of ferns, and was very well-known abroad and home by establishing the Ching's system, which was granted the National Natural Science Prizes (First Class) in 1993.

中国蕨类植物图谱

ICONES FILICUM SINICARUM

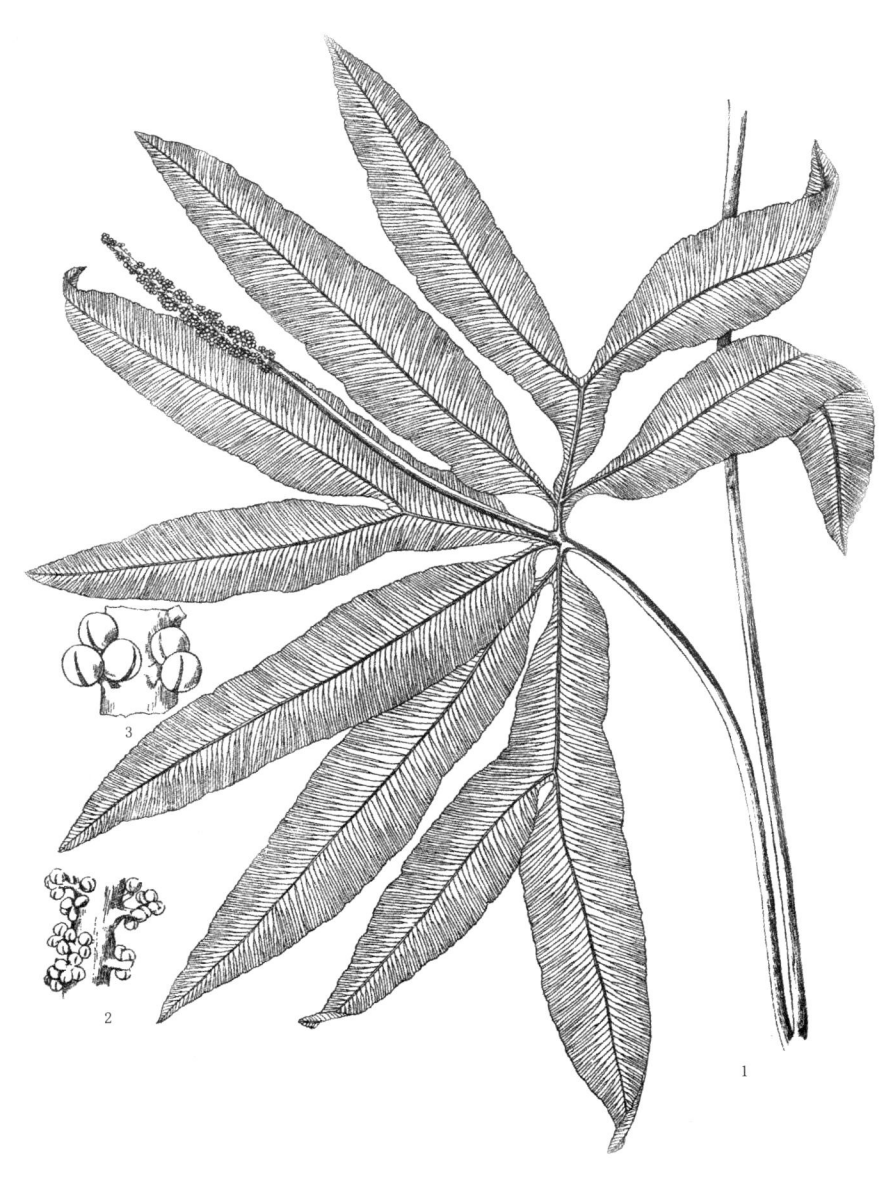

秦仁昌 编著
REN-CHANG CHING

图书在版编目(CIP)数据

中国蕨类植物图谱/秦仁昌编著. —北京：北京大学出版社，2011.10
ISBN 978-7-301-19475-1

Ⅰ.①中…　Ⅱ.①秦…　Ⅲ.①蕨类植物－中国－图谱　Ⅳ.①Q949.36-64

中国版本图书馆 CIP 数据核字(2011)第 185467 号

书　　　名：	中国蕨类植物图谱
著作责任者：	秦仁昌　编著
策 划 编 辑：	陈斌惠
责 任 编 辑：	陈斌惠
标 准 书 号：	ISBN 978-7-301-19475-1/N·0042
出 版 发 行：	北京大学出版社
地　　　址：	北京市海淀区成府路 205 号　100871
网　　　址：	http://www.pup.cn　电子信箱：zyjy@pup.cn
电　　　话：	邮购部 62752015　发行部 62750672　编辑部 62754934　出版部 62754962
印　刷　者：	北京中科印刷有限公司
经　销　者：	新华书店
	787 毫米×1092 毫米　8 开本　66.5 印张　936 千字
	2011 年 10 月第 1 版　2011 年 10 月第 1 次印刷
定　　　价：	280.00 元

未经许可，不得以任何方式复制或抄袭本书之部分或全部内容。
版权所有，侵权必究
举报电话：(010)62752024　电子信箱：fd@pup.pku.edu.cn

Preface to the Reprint of *Icones Filicum Sinicarum*

Icones Filicum Sinicarum (Fascicle 1-5) by Professor Ren-Chang Ching is well known as one of the early authoritative taxonomic works on Chinese ferns. Its publication spanned 28 years from the first Fascicle published in 1930, Fascicle II in 1934, Fascicle III in 1935, Fascile IV in 1937, to Fascicle V in 1958. The Icones included a total of 251 beautiful plates illustrating 252 taxonomically important Chinese fern species known at that time. The illustration plates depicted the fern species based on their natural sizes with amplified figures showing the taxonomically important characters. Bilingual descriptions of the taxonomy, morphology and distributions of each species, as well as information on synonymy, important publication sources and reference specimens, were provided also, with the English text more comprehensively written than the concise Chinese explanation.

This fern fascile is of great taxonomic value, not only for its scientific value, but also for the high standard and artistic merit of the illustrations. Due to the old date of publication and the limited printed copies, it is now very difficult to find this publication in local and overseas bookshops, although it had been printed for sale in India in the 20th century.

Thanks to the Peking University Press, this masterpiece of work is reprinted for the first time in China. The new reproduction of the Icones, with its historical and great taxonomic values in pteridophyte classification, hopes to continue serving the botanical community in the years to come.

I have examined all the illustrations of fern species in the Icones and re-assessed their taxonomy following mainly the classification adopted in *Flora Reipublicae Popularis Sinicae* and other recent studies. A table of comparison of the modern taxonomy with that used in the Icones is provided at the end of this newly printed book to reflect on the advancement of fern systematics today.

Xian-Chun Zhang[*],
May 2011, Xiangshan, Beijing

---

[*] Xian-Chun Zhang. Pteridologist, Research professor at the Institute of Botany, Chinese Academy of Sciences. Council member of International Association of Plant Taxonomy (IAPT), Member of Nomenclature Committee of Vascular Plants, Council member of International Association of Pteridologists (IAP), Honorary member of Indian Fern Society, Advisor of *Indian Fern Journal*, Editorial committee of *Flora of China*, Chair of the Fern Committee of China Flower Association, Chair of the Fern Committee of China Wild Plant Conservation Association, Member of the Society of Himalayan Botany.

再 版 序 言

《中国蕨类植物图谱》原书共五卷，由我国蕨类植物学家秦仁昌院士编著，是国际上有较大影响的关于中国蕨类植物分类的权威学术著作。该图谱一至五卷分别出版于1930年、1934年、1935年、1937年和1958年，共包括251张精美的图版，描述了中国252种重要的蕨类植物。图版是根据植物自然大小绘制，并附有放大的主要器官解剖图，对每种植物用中英文两种文字进行描述，但描述以英文为主，中文为简略介绍，包括形态特征、产地、地理分布、标本引证和主要文献及异名等内容。

《中国蕨类植物图谱》不仅有很高的分类学价值，其图版也有很高的艺术价值。由于图谱出版年代已久，发行量不大，目前国际上已难以见到，虽然也有印度出版商在20世纪复印销售过该书。

今北京大学出版社慧眼发现该套蕨类植物图谱的价值，汇集原图谱五卷之251张图版于一册，使该图谱继续发挥其在蕨类植物分类学研究上的价值，为植物学家和社会各行及植物学爱好者服务。

我主要根据已出版的《中国植物志》和有关研究对原图重新鉴定，在书末附上原图谱名称和目前接受名称对照表，以反映分类上的一些变化。此外，为了方便读者，再版时对书中主要地名的英文译法按照现在标准进行了修订。

张宪春[*]

2011年5月于香山

---

[*] 张宪春，中国科学院植物研究所研究员，博士生导师。蕨类植物学家，国际植物分类学会理事，国际维管植物命名委员会委员，国际蕨类植物学家协会理事，印度蕨类协会名誉会员，印度蕨类杂志编辑顾问，中国植物志编委，中国花卉协会蕨类植物分会主任，中国野生植物保护协会蕨类保育委员会主任，喜马拉雅植物学会会员。

目录 CONTENTS

第一卷　FASCICLE 1

1. Helminthostachys zeylanica，锡兰七指蕨　/ 2
2. Archangiopteris henryi，亨利马蹄蕨　/ 4
3. Trichomanes tereticaulum，圆柄石衣蕨　/ 6
4. Woodsia cinnamomea，赤色岩蕨　/ 8
5. Cystopteris moupinensis，滇冷蕨　/ 10
6. Dryopteris enneaphylla，宜昌金星蕨　/ 12
7. Aspidium longicrure，燕尾三叉蕨　/ 14
8. Aspidium ebenium，黑柄三叉蕨　/ 16
9. Polystichum duthiei，杜氏耳叶蕨　/ 18
10. Polystichum acanthophyllum，刺耳叶蕨　/ 20
11. Polystichum deltodon，对生耳叶蕨　/ 22
12. Polystichum hecatopterum，锯齿耳叶蕨　/ 24
13. Polystichum chunii，陈氏耳叶蕨　/ 26
14. Cyrtomium hemionitis，单叶贯众　/ 28
15. Cyrtomium nephrolepioides，低头贯众　/ 30
16. Cyrtomium lonchitoides，拟贯众　/ 32
17. Cyrtomium fraxinellum，槐叶贯众　/ 34
18. Microlepia chrysocarpa，黄胞鳞蕨　/ 36
19. Lindsaya chienii，钱氏假铁线蕨　/ 38
20. Lindsaya chinensis，华假铁线蕨　/ 40
21. Athyrium anisopterum，宿蹄盖蕨　/ 42
22. Asplenium finlaysonianum，网脉单盖蕨　/ 44
23. Diplazium pullingeri，波氏双盖蕨　/ 46
24. Diplazium pellucidum，膜叶双盖蕨　/ 48
25. Diplazium macrophyllum，大叶双盖蕨　/ 50
26. Phyllitis delavayi，荷叶对开盖蕨　/ 52
27. Phyllitis cardiophylla，琼崖对开盖蕨　/ 54
28. Blechnum eburneum，象牙乌毛蕨　/ 56
29. Gymnopteris vestita，金毛裸蕨　/ 58
30. Plagiogyria henryi，亨氏瘤足蕨　/ 60
31. (1-5) Adiantum greenii，格氏铁线蕨
    (6-9) Adiantum nanum，矮铁线蕨　/ 62
32. Pteris deltodon，岩凤尾蕨　/ 64
33. Pteris hui，胡氏凤尾蕨　/ 66
34. Pteris dimorpha，二形凤尾蕨　/ 68
35. Pteris actiniopteroides，猪鬣凤尾蕨　/ 70
36. Pteris paupercula，黄毛凤尾蕨　/ 72
37. Pteris excelsa，溪凤尾蕨　/ 74
38. Vittaria pauciareolata，阔叶书带蕨　/ 76
39. Vittaria nana，矮叶书带蕨　/ 78
40. Antrophyum petiolatum，长柄车前蕨　/ 80
41. Polypodium dareaeformioides，乌柄水龙骨　/ 82
42. Polypodium mengtzeense，蒙自水龙骨　/ 84
43. Polypodium oblongiosorum，珠带水龙骨　/ 86
44. Polypodium triglossum，三叶水龙骨　/ 88
45. Polypodium ellipticum，椭圆水龙骨　/ 90
46. Polypodium leveillei，莱氏水龙骨　/ 92
47. Loxogramme chinensis，华剑蕨　/ 94
48. Neocheiropteris palmatopedata，掌状扇蕨　/ 96
49. Elaphoglossum austro-sinicum，华南舌蕨　/ 98
50. Cheiropleuria bicuspis，二尖燕尾蕨　/ 100

第二卷　FASCICLE 2

51. Oleandra wallichii，高山蓧蕨　/ 104
52. Woodwardia harlandii，哈氏狗脊　/ 106
53. Asplenium adnatum，合生铁角蕨　/ 108
54. Taenitis blechnoides，竹叶蕨　/ 110
55. Pteris insignis，全缘凤尾蕨　/ 112
56. Adiantum faberi，峨眉铁线蕨　/ 114
57. Lepisorus lewisii，庐山瓦韦　/ 116
58. Lepisorus eilophyllus，高山瓦韦　/ 118
59. Lepisorus sublinearis，滇瓦韦　/ 120
60. Lepisorus pseudonudus，长瓦韦　/ 122
61. Lepisorus loriformis，带瓦韦　/ 124
62. Lepisorus macrosphaerus，大瓦韦　/ 126
63. Lepisorus macrosphaerus
    var. asterolepis，黄瓦韦　/ 128
64. Lepisorus kuchenensis，瑶山瓦韦　/ 130
65. Lepisorus bicolor，两色瓦韦　/ 132
66. Lepisorus obscure-venulosus，粤瓦韦　/ 134
67. Lepisorus clathratus，网眼瓦韦　/ 136
68. Lepisorus sordidus，黑鳞瓦韦　/ 138
69. Lepisorus suboligolepidus，拟鳞瓦韦　/ 140
70. Lepisorus oligolepidus，鳞瓦韦　/ 142
71. Lepisorus subconfluens，连珠瓦韦　/ 144
72. Lepisorus heterolepis，芨瓦韦　/ 146
73. Lepisorus angustus，狭叶瓦韦　/ 148
74. Lepisorus ussuriensis，乌苏里瓦韦　/ 150
75. Lepisorus contortus，扭瓦韦　/ 152
76. Lepisorus thunbergianus，瓦韦　/ 154
77. Lemmaphyllum subrostratum，骨牌蕨　/ 156
78. Lemmaphyllum adnascens，川石莲　/ 158
79. Lemmaphyllum drymoglossoides，抱石莲　/ 160
80. Lemmaphyllum microphyllum，抱树莲　/ 162
81. Cyclophorus calvatus，光石韦　/ 164
82. Elaphoglossum mcclurei，琼崖舌蕨　/ 166
83. Microsorium fortuni，福氏星蕨　/ 168
84. Microsorium hymenodes，滇星蕨　/ 170
85. Microsorium buergerianum，波氏星蕨　/ 172
86. Microsorium punctatum，星蕨　/ 174
87. Microsorium zippelii，戚氏星蕨　/ 176
88. Microsorium membranaceum，膜叶星蕨　/ 178
89. Neocheiropteris phyllomanes，单叶扇蕨　/ 180

90. Colysis bonii, 彭氏线蕨 / 182
91. Colysis henryi, 亨氏线蕨 / 184
92. Colysis longipes, 长柄线蕨 / 186
93. Christopteris tricuspis, 戟蕨 / 188
94. Polypodium manmeiense, 滇水龙骨 / 190
95. Polypodium lachnopus, 濑水龙骨 / 192
96. Polypodium microrhizoma, 栗柄水龙骨 / 194
97. Polypodium dielseanum, 川水龙骨 / 196
98. Polypodium niponicum, 水龙骨 / 198
99. Polypodium amoenum, 友水龙骨 / 200
100. Phymatodes nigrovenia, 黑鳞莐蕨 / 202
101. Microlepia tenera, 嫩毛蕨(Fas. 2) / 204

第三卷　FASCICLE 3

101. Woodsia macrochlaena, 大囊岩蕨（Fas. 3）/ 208
102. Woodsia manchuriensis, 满洲岩蕨 / 210
103. Davallia mariesii, 海州骨碎补 / 212
104. Davallia orientalis, 华南骨碎补 / 214
105. Humata assamica, 高山阴石蕨 / 216
106. Arthropteris obliterata, 藤蕨 / 218
107. Camptosorus sibiricus, 过山蕨 / 220
108. Athyrium sheareri, 庐山蹄盖蕨 / 222
109. Athyrium otophorum, 光蹄盖蕨 / 224
110. Athyrium goeringianum, 柯氏蹄盖蕨 / 226
111. Asplenium sarelii, 华中铁角蕨 / 228
112. Asplenium sampsoni, 岭南铁角蕨 / 230
113. Asplenium prolongatum, 长生铁角蕨 / 232
114. Asplenium crinicaule, 毛铁角蕨 / 234
115. Asplenium saxicola, 粤铁角蕨 / 236
116. Woodwardia orientalis, 东方狗脊 / 238
117. Dictyocline griffithii, 圣蕨 / 240
118. Quercifilix zeylanica, 地耳蕨 / 242
119. Bolbitis heteroclita, 长叶实蕨 / 244
120. Bolbitis subcordata, 海南实蕨 / 246
121. Hypodematium crenatum, 肿足蕨 / 248
122. Hypodematium fordii, 福氏肿足蕨 / 250
123. Hypodematium cystopteroides, 山东肿足蕨 / 252
124. Tectaria subtriphylla, 三叉蕨 / 254
125. Tectaria macrodonta, 高山三叉蕨 / 256
126. Cyrtomium fortunei, 贯众 / 258
127. Cyrtomium falcatum, 全缘贯众 / 260
128. Cyclopeltis crenata, 拟贯众 / 262
129. Polystichum fimbriatum, 瓦鳞耳蕨 / 264
130. Polystichum otophorum, 高山耳蕨 / 266
131. Polystichum omeiense, 峨眉耳蕨 / 268
132. Pellaea smithii, 史氏旱蕨 / 270
133. Cheilanthes duclouxii, 杜氏粉背蕨 / 272
134. Cheilanthes hancockii, 韩氏粉背蕨 / 274
135. Cheilanthes chusana, 舟山粉背蕨 / 276
136. Cheilanthes trichophylla, 毛粉背蕨 / 278
137. Adiantum refractum, 蜀铁线蕨 / 280
138. Adiantum davidi, 白背铁线蕨 / 282
139. Adiantum edgeworthii, 爱氏铁线蕨 / 284
140. Pteris fauriei, 傅氏凤尾蕨 / 286
141. Pteris dactylina, 掌凤尾蕨 / 288
142. Schizoloma ensifolium, 拟凤尾蕨 / 290
143. Coniogramme intermedia, 华凤了蕨 / 292
144. Vittaria forrestiana, 宽书带蕨 / 294
145. Antrophyum formosanum, 车前蕨 / 296
146. Saxiglossum taeniodes, 拟石苇 / 298
147. Pyrrosia sheareri, 庐山石苇 / 300
148. Pyrrosia drakeana, 毡毛石苇 / 302
149. Arthromeris himalayensis, 琉璃节肢蕨 / 304
150. Arthromeris lungtauensis, 粤节肢蕨 / 306

第四卷　FASCICLE 4

151. Gleichenia cantonensis, 广东里白 / 310
152. Gleichenia laevissima, 光里白 / 312
153. Gleichenia splendida, 硕里白 / 314
154. Dipteris chinensis, 双扇蕨 / 316
155. Plagiogyria assurgens, 峨眉瘤足蕨 / 318
156. Lindsaya lobbiana, 洛氏林蕨 / 320
157. Lindsaya decomposita, 网脉林蕨 / 322
158. Adiantum gravesii, 粤铁线蕨 / 324
159. Adiantum chienii, 钱氏铁线蕨 / 326
160. Adiantum roborowskii, 陇铁线蕨 / 328
161. Onychium contiguum, 高山乌蕨 / 330
162. Onychium moupinense, 木坪乌蕨 / 332
163. Onychium tenuifrons, 狭叶乌蕨 / 334
164. Onychium ipii, 叶氏乌蕨 / 336
165. Pleurosoriopsis makinoi, 睫毛蕨 / 338
166. Coniogramme fraxinea, 全缘凤了蕨 / 340
167. Coniogramme caudata, 毛叶凤了蕨 / 342
168. Coniogramme procera, 高山凤了蕨 / 344
169. Oleandra cumingii, 华南荼蕨 / 346
170. Oleandra whangii, 瑶山荼蕨 / 348
171. Oleandra undulata, 长柄荼蕨 / 350
172. Gymnocarpium remoti-pinnatum, 肢节蕨 / 352
173. Asplenium fugax, 阴地铁角蕨 / 354
174. Asplenium exiguum, 低头铁角蕨 / 356
175. Asplenium loriceum, 南海铁角蕨 / 358
176. Asplenium interjectum, 黔铁角蕨 / 360
177. Acrophorus stipellatus, 拟鳞毛蕨 / 362
178. Dryopteris serrato-dentata, 高山鳞毛蕨 / 364
179. Dryopteris scottii, 史氏鳞毛蕨 / 366
180. Dryopteris liankwangensis, 两广鳞毛蕨 / 368
181. Dryopteris championi, 张氏鳞毛蕨 / 370
182. Polystichum chingae, 滇耳蕨 / 372
183. Cyrtomium aequibasis, 滇贯众 / 374
184. Cyrtomium muticum, 大叶贯众 / 376
185. Hemigramma decurrens, 拟叉蕨 / 378
186. Leucostegia immersa, 膜盖蕨 / 380
187. Leucostegia hookeri, 霍氏膜盖蕨 / 382
188. Leucostegia multidentata, 毛膜盖蕨 / 384
189. Loxogramme grammitoides, 小叶剑蕨 / 386
190. Loxogramme salicifolia, 柳叶剑蕨 / 388

191. Loxogramme ensiformis，阔叶剑蕨 / 390
192. Drynaria fortunei，槲蕨 / 392
193. Drynaria sinica，华槲蕨 / 394
194. Colysis wui，吴氏线蕨 / 396
195. Colysis hemionitidea，断线蕨 / 398
196. Colysis wrightii，莱氏线蕨 / 400
197. Colysis hemitoma，胄叶线蕨 / 402
198. Colysis digitata，掌叶线蕨 / 404
199. Colysis pentaphylla，滇线蕨 / 406
200. Colysis morsei，马氏线蕨 / 408

第五卷　FASCICLE 5

201. Pseudodrynaria coronans，崖姜蕨 / 412
202. Angiopteris fokiensis，福建观音座莲 / 414
203. Archangiopteris bipinnata，二回原始观音座莲 / 416
204. Archangiopteris hokouensis，河口原始观音座莲 / 418
205. Archangiopteris somai，台湾原始观音座莲 / 420
206. Archangiopteris subrotundata，圆基原始观音座莲 / 422
207. Archangiopteris latipinna，阔叶原始观音座莲 / 424
208. Archangiopteris caudata，尾叶原始观音座莲 / 426
209. Archangiopteris tonkinensis，尖叶原始观音座莲 / 428
210. Adiantum caudatum，鞭叶铁线蕨 / 430
211. Adiantum sinicum，苍山铁线蕨 / 432
212. Adiantum juxtapositum，仙霞铁线蕨 / 434
213. Adiantum capillus-junonis，团叶铁线蕨 / 436
214. Adiantum philippense，半月形铁线蕨 / 438
215. Adiantum soboliferum，翅柄铁线蕨 / 440
216. Adiantum capillus-veneris，铁线蕨 / 442
217. Adiantum edentulum，月芽铁线蕨 / 444
218. Adiantum fengianum，冯氏铁线蕨 / 446
219. Adiantum smithianum，长盖铁线蕨 / 448
220. Adiantum venustum，细叶铁线蕨 / 450
221. Adiantum bonatianum，毛足铁线蕨 / 452
222. Adiantum muticum，鹤庆铁线蕨 / 454
223. Adiantum flabellulatum，扇叶铁线蕨 / 456
224. Adiantum induratum，海南铁线蕨 / 458
225. Adiantum diaphanum，长尾铁线蕨 / 460
226. Adiantum pedatum，掌叶铁线蕨 / 462
227. Adiantum myriosorum，灰背铁线蕨 / 464
228. Leptogramma mollissima，毛叶茯蕨 / 466
229. Leptogramma scallani，峨眉茯蕨 / 468
230. Leptogramma caudata，尾叶茯蕨 / 470
231. Stegnogramma cyrtomioides，波叶溪边蕨 / 472
232. Stegnogramma asplenioides，浅裂溪边蕨 / 474
233. Polystichum dielsii，圆顶耳蕨 / 476
234. Polystichum excellens，尖顶耳蕨 / 478
235. Polystichum yuanum，倒叶耳蕨 / 480
236. Polystichum kwangtungense，广东耳蕨 / 482
237. Polystichum consimile，刺叶耳蕨 / 484
238. Polystichum stenophyllum，芽胞耳蕨 / 486
239. Polystichum grossidentatum，粗齿耳蕨 / 488
240. Polystichum lanceolatum，亮叶耳蕨 / 490
241. Polystichum thomsoni，尾叶耳蕨 / 492
242. Polystichum bifidum，钳形耳蕨 / 494
243. Polystichum nepalense，软骨耳蕨 / 496
244. Polystichum xiphophyllum，革叶耳蕨 / 498
245. Polystichum tripteron，三叉耳蕨 / 500
246. Polystichum tsus-simense，马祖耳蕨 / 502
247. Cyrtomidictyum basipinnatum，单叶鞭叶蕨 / 504
248. Cyrtomidictyum conjunctum，卵形鞭叶蕨 / 506
249. Cyrtomidictyum lepidocaulon，鞭叶蕨 / 508
250. Cyrtomidictyum faberi，普陀鞭叶蕨 / 510

拉丁名索引 Index to Latin Names / 512

中文名索引 Index to Chinese Names / 515

分类和名称变化对照表 Taxonomic and Nomenclatural Changes / 517

中国蕨类植物图谱

ICONES FILICUM SINICARUM

BY

HSEN-HSU HU, D. S.

HEAD OF BOTANICAL DIVISION

FAN MEMORIAL INSTITUTE OF BIOLOGY

AND

REN-CHANG CHING, B. S.

BOTANIST OF

METROPOLITAN MUSEUM OF NATURAL HISTORY

"ACADEMIA SINICA"

FASCICLE I, PLATES 1-50

第一卷

TO

DR. CARL CHRISTENSEN

CURATOR OF THE UNIVERSITETETS BOTANISKE MUSEUM

COPENHAGEN, DENMARK

THROUGH WHOSE PAINS-TAKING LABOUR AND EFFORT, IN

HIS MONUMENTAL WORK OF INDEX FILICUM

THE PTERIDOLOGICAL KNOWLEDGE OF THE WORLD

HAS BEEN GREATLY ADVANCED AND SYSTEMATIZED

THIS FIRST FASCICLE OF THE ICONES FILICUM SINICARUM

IS RESPECTFULLY DEDICATED.

图版 1　锡兰七指蕨

根状茎粗壮，肥厚，匍匐；无鳞片；叶柄直立，长 20—30 cm，几绿色；无子囊群，叶掌状普通三裂，具短柄，再分叉或成二次羽状；最后裂片披针形，长 7—15 cm，宽 2—4 cm，渐尖头，略成齿牙状或几全缘，基部约成楔形；叶脉显明直立开张，羽状，通常分叉；叶体薄草质；孢子囊穗单生，自无子囊群叶片之基部生出，长 7—12 cm，宽 1 cm，柄长略同。

分布：澳洲，新喀里多尼亚，印度，斯里兰卡，菲律宾；中国：海南，台湾。

本种为亚洲之热带习见之蕨，多生于卑湿之地。

图注：1.本种全形(自然大)；2.孢子囊群之一部(放大 20 倍)；3.同前之具孢子囊者示其开裂之情形(放大 52 倍)。

HELMINTHOSTACHYS ZEYLANICA Hooker et Bauer

HELMINTHOSTACHYS ZEYLANICA Hooker et Bauer in Gen. Fil. t. 47 (1840); HK. in Gard. Ferns t. 28; Bedd. in Ferns S. Ind. t. 69; HK. in 2nd. Cent. Ferns t. 44; Christ in Farnkaut. Erde 365; Diels in Nat. Pfl. Fam. I. 4, 472; C. Chr. Ind. 344 (1906); v. A. v. R. in Malay. Ferns 777 (1908).

Osmunda zeylanica Linn. Sp. Pl. **2**: 1063 (1753).
Botrychium zeylanicum Sw. in Schrad. Jour. 1800, **2**: 111 (1801).
Helminthostachys dulcis Klf. in Enum. 28, t. 1. fig. 1 (1822).
Botryopteris mexicana in Rel. Haenk. **1**: 76, t. 12. fig. 1 (1825).
Helminthostachys mexicana Spring (1827).

Rhizome thick, fleshy, creeping, glabrous; *stipes* erect, 20-30 cm long, greenish; *sterile segment* palmately pinnate, usually in 3 principal divisions, which are shortly petiolate, forked or pinnate; the ultimate segments lanceolate, 7-15 cm long, 2-4 cm broad, acuminate, the edge slightly toothed or entire, the base more or less cuneate; *veins* distinct, erect-patent, pinnate, usually forked; *texture* thin herbaceous; fertile spike solitary, arising from the base of *sterile segment*, 7-12 cm long, 1 cm broad, on a peduncle of about the same length.

Distribution: Australia, New Caledonia, India, Sri Lanka, Philipines; China: Hainan, Taiwan.

A fairly common fern in Tropical Asia, inhabiting damp or swampy places.

Plate 1. Fig. 1. Habit sketch (natural size). 2. A portion of spike ($\times$ 20). 3. The same with sporangia, showing the manner of dehiscence ($\times$ 52).

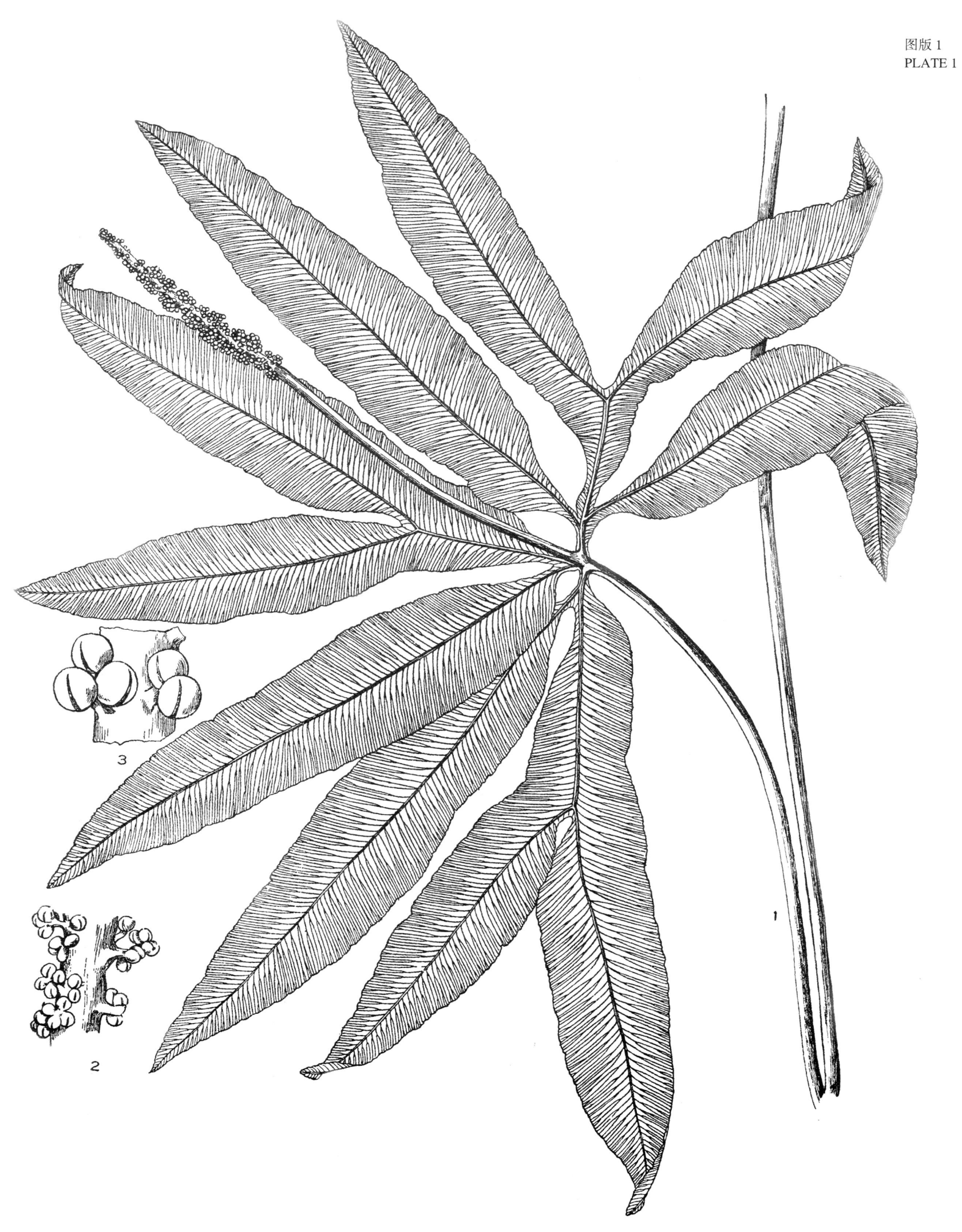

HELMINTHOSTACHYS ZEYLANICA (L.) Hooker
锡兰七指蕨

根状茎未详,叶柄长约 50 cm,粗如鹄翎管,圆筒形,暗绿色,与根状茎不成关节,覆以披针形之鳞片及毛,叶轴及小叶柄亦如之;叶体长 50 cm,宽 25 cm,卵圆形,小叶相距略远,互生,每侧有三五枚,与顶端小叶形态相同,长卵圆形,尖头,长至 25 cm,宽 6 cm,全缘,自中部以上略成小圆齿状,小叶柄长 1.5—2 cm,几黑色,中空,被有显明之长毛;叶体薄纸质,淡绿色,叶柄及小叶柄颇肥厚多汁;叶脉密生,每侧约 80 条,几平行开张,几达叶缘;子囊群居中部,与中肋或叶缘间成 0.5 cm 宽之空隙,红褐色;每侧有 80 个,长达 2 cm,宽达 1.5 mm,无子囊群盖。

分布:云南。

本种特殊蕨在系统学之地位在美洲之 *Danaea* 及亚洲热带之观音座莲(*Angiopteris*)之间;与后者差别在长念珠状生于中间之子囊群,由多数孢子囊集合而成,与其肥大之叶柄及一次羽状之叶体;此种为 A. Henry 氏最有价值之发现,在云南省外尚未发现也。

图注:1.本种全形(自然大);2.小叶之一部示叶脉及子囊群之一部(放大 16 倍);3.子囊群之横断面示孢子囊之附着及隔丝(放大 16 倍);4.两孢子囊(a,前面,b,侧面放大 52 倍)。

ARCHANGIOPTERIS HENRYI Christ et Giesenhagen

ARCHANGIOPTERIS HENRYI Christ et Giesenhagen in Flora Regensb. 73 (1899); Christ Bull. Boiss. **7**: (1899); Diels in Nat. Pfl. Fam. 1.4.439; C. Chr. Ind. 62 (1906); Matthew in Journ. Linn. Soc. **39**: 342 (1911).

Rhizome not seen; *stipe* about 50 cm long, thick as swan's quill, terete, obscurely green, not articulate to the rhizome, and like rachis and the petioles obscurely hairy with lanceolate scales; *frond* about 50 cm long, 25 cm broad, ovate, pinnae remote, alternate, 3-5 on each side, similar to the terminal one, oblong-ovate, acute, 25 cm long, 6 cm broad, entire, slightly crenulate above the middle, acutely serrate towards the apex, petiole 1.5-2 cm long, blackish, inflated, conspicuously pilose; *texture* thin chartaceous, light green, stipes and petioles rather succulent; *veins* dense, about 80 in each side, almost horizontally patent, almost extending to the margin; *sori* medial, with a broad free space about 0.5 cm broad from the margin and the costa, linear, reddish-brown, about 80 on each side, to 2 cm long, 1.5 mm broad, exindusiate.

Distribution: Yunnan.

This peculiar fern holds a systematic place between *Danaea* of America and *Angiopteris* of Tropical Asia, from the latter, it differs in elongate moniliform medial sori consisting of numerous sporangia, inflated petiole and simply pinnate frond. One of the most noteworthy discoveries made by A. Henry. A very rare plant, as not been known elsewhere in this country.

Plate 2. Fig. 1. Habit sketch (natural size). 2. A portion of pinna, showing venation and a portion of sorus ($\times$ 16). 3. A cross section of sorus, showing attachment of sporangia and paraphyses ($\times$ 46). 4. Two sporangia, a. front view; b, lateral view ($\times$ 52).

图版 2
PLATE 2

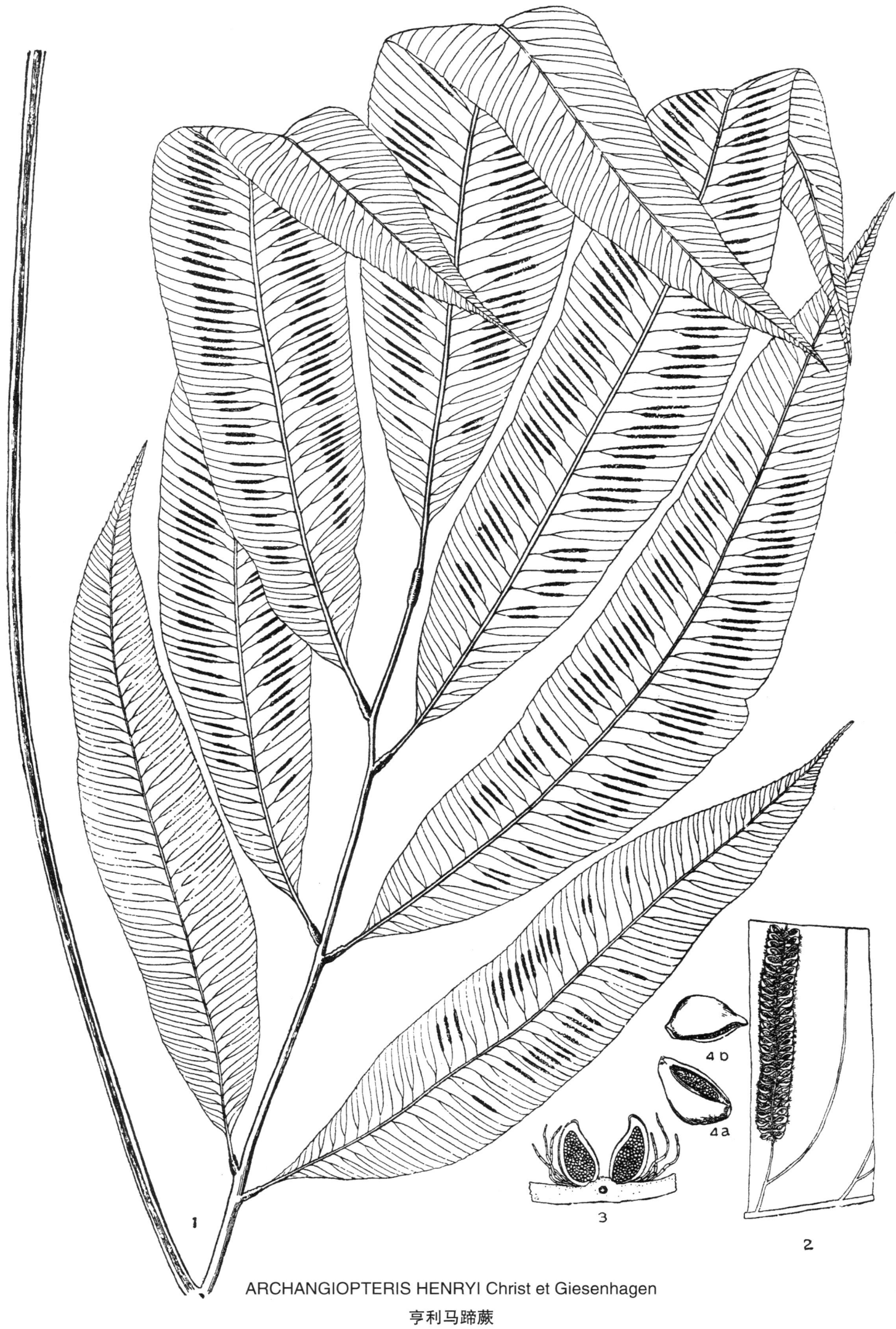

ARCHANGIOPTERIS HENRYI Christ et Giesenhagen
亨利马蹄蕨

根状茎短，裸露或几裸露，密织于铁丝状之纤维状根中；叶柄簇生而密，铁丝状，硬直；圆形，无翼或仅顶部有狭翼，通常长 4—6 cm，或更长，无鳞片；叶体无毛，几三角形至线长方形，长 4—6 cm，宽 2—2.5 cm，三次羽状分裂，叶轴全体具狭翼，一次小叶每侧 3—5 枚，其间相距颇远，直而开张，长方卵圆形，长 1—2 cm，上部者则渐短；二次小叶 2—4，全缘或二至三次羽状分裂，而成线形全缘之小裂片，叶薄膜质，半透明，暗绿色；叶脉清晰，每小叶片仅 1 条，无假脉；孢子囊群位于最后裂片之顶端通，常下陷，每一次小叶 2—3 枚，子囊群盖短，漏斗状，微张开，其口部成不显明波状，不二裂，孢子囊群之托长突出，甚粗大，暗棕色。

分布：广西。

图注 1. 本种全形（自然大）；2. 一次小叶之一部，示叶脉与孢子囊盖（放大 4 倍）。

TRICHOMANES TERETICAULUM Ching

TRICHOMANES TERETICAULUM Ching* in Sinensia **1**: 2 (1929)

Rhizome naked or nearly so, very short, interwoven in dense wiry fibrous roots; *stipes* densely caespitose-fasciculate, wiry, rigid, terete, not winged or only narrowly winged towards apex, 4-6 cm long or longer, naked; *fronds* glabrous, subdeltoid to linear; oblong, 4-6 cm long, 2-2.5 cm broad, 3-pinnatifid, rachis narrowly winged throughout; pinnae 3-5 on each side, remote, erect-patent, oblong-ovate, 1-2 cm long, the upper ones gradually shortened; *texture* thin-herbaceous, translucent, dull green; *veins* distinct, 1 to each segment, spurious veinlets wanting; *sori* terminal on ultimate segments, mostly immersed, 2-3 to each pinna, indusium short funnel-shaped, slightly dilated, obscurely undulate at mouth not bifid, receptacle long-exerted, stout, dark brown.

Distribution: Guangxi.

Plate 3. Fig. 1. Habit sketch (natural size). 2. A portion of pinna, showing venation and inducia ($\times$ 4).

---

* For the species by Ching, readers are referred to Sinensia 1: pp. 1—13 (1929), the Metropolitan Museum of Natural History, "Academia Sinica".

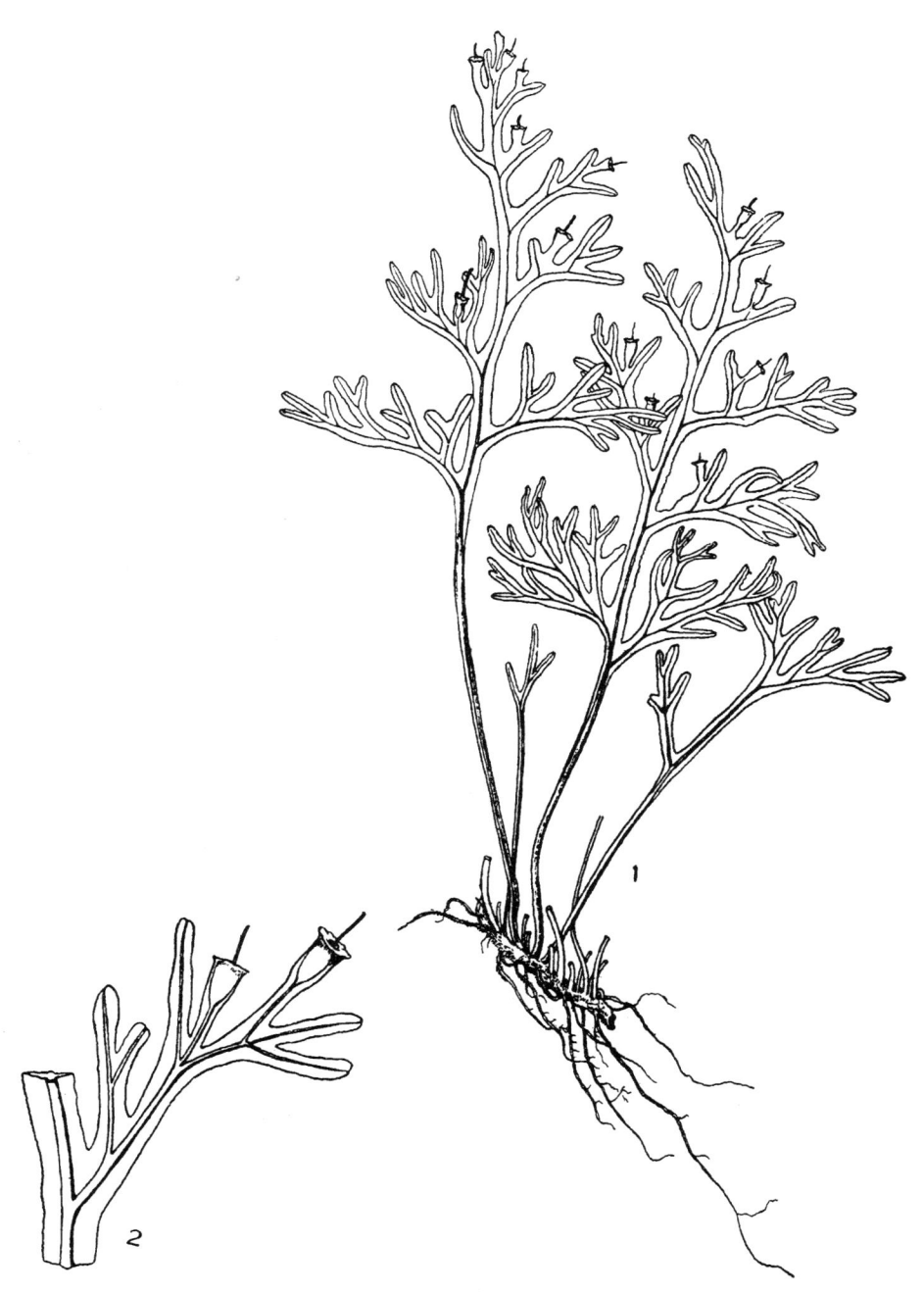

TRICHOMANES TERETICAULUM Ching
圆柄石衣蕨

图版 4　赤色岩蕨

根状茎肥厚而短,直立或斜卧,上覆鳞片,鳞片膜质,色淡红,披针形或线针形;叶柄5—15,簇生,长 2—5 cm,其不生孢子囊群者尤长,稻秆色,近基部处密覆同样之鳞片,他部则被锈色之薄毛;叶身通常椭圆披针形,长 6—12 cm,宽 1.8—2.5 cm;通常偶数羽状,鲜奇数羽状,向基部则略退化;小叶平展,无或几无叶柄,每侧 8—14 枚,对生或几对生,鲜互生,其间相距 6—8 mm,卵长方形,基部对称,圆形或略扩大,几截形或几心脏截形,边缘具波状锯齿,其在无孢子囊群之叶则为羽状深裂,长 8—12 mm,宽 3—5 mm,无孢子囊者则尤阔,叶基部短楔形,每侧具三至四羽状裂片,各裂片圆形,全缘或稍有小波齿;孢子囊群深藏于密毛之下,子囊群盖初为球形,破裂后则成具长纤毛之裂片,每孢子囊群中有 4—6 孢子囊,孢子囊暗褐色,柄甚短。

分布:四川,云南,贵州。

本种属于 Physematium 系,与 Woodsia rosthorniana Diels 相近,与之异者为较多小型有薄毛而较长之小叶及坚硬之叶轴。其外形在 W. polystichoides Eaton 与 W. rosthorniana Diels 之间。

图注:1.本种全形(自然大);2.除去绒毛之叶片,示脉与孢子囊群之分布(放大 5 倍)。3. 小叶片之一部,示孢子囊群与子囊群盖之排列(放大 15 倍);4.二子囊之毛(放大 108 倍);5.叶轴之毛(放大 108 倍);6.叶柄基部之鳞片(放大 35 倍)。

WOODSIA CINNAMOMEA Christ

WOODSIA CINNAMOMEA Christ in Bull. Geogr. Bot. Man. 122 (1906).

Rhizome thick, short, erect or oblique, clothed with palered lanceolate or linearsubulate entire membranaceous scales; *stipes* densely tufted (5-15 together), 2-5 cm long or much longer in barren leaves, stramineous, densely clothed near the base with scales similar to those on rhizome, ferruginously pilose in other parts, particularly the under surface; *frond* usually oblong-lanceolate, 6-12 cm long, 1.8-2.5 cm broad, simply and usually evenly pinnate, slightly reduced towards the base; pinnae sessile or subsessile, horizontally patent, 8-14 on each side, opposite or subopposite, sometimes alternate, 6-8 mm apart, ovate-oblong, base equal, rounded or somewhat dilated, subtruncate or subcordate-truncate, crenate-serrate or deeply pinnatifid in barren leaves, 8-12 mm long, 3-5 mm broad at the base, the barren ones much broader with shortly cuneate base and pinnatifid over half way down into 3-4 rounded, subentire or slightly crenulate lobes on each side; *sori* completely hidden in thick tomentum, indusium globose at first, ruptured at last into long-ciliate lacinae; sporangia 4-6 to each sorus, dark brown, very shortly stipitate.

Distributions: Sichuan, Yunnan, Guizhou.

This distinct fern belonging to § *Physematium* is allied to *Woodsia rosthorniana* Diels, but differs in small more numerous pubescent longer pinna, and stiff rachis; with a habit intermediate between *W. polystichoides* Eaton and *W. rosthorniana* Diels.

Plate 4. Fig. 1. Habit sketch (natural size). 2. A pinna with tomentum removed, showing venation and sori ($\times$ 5). 3. A portion of pinna, showing disposition of sori and inducia ($\times$ 15). 4. Two induciate hairs ($\times$ 108). 5. Hairs from the rachis ($\times$ 108). 6. Scales from the base of stipe ($\times$ 35).

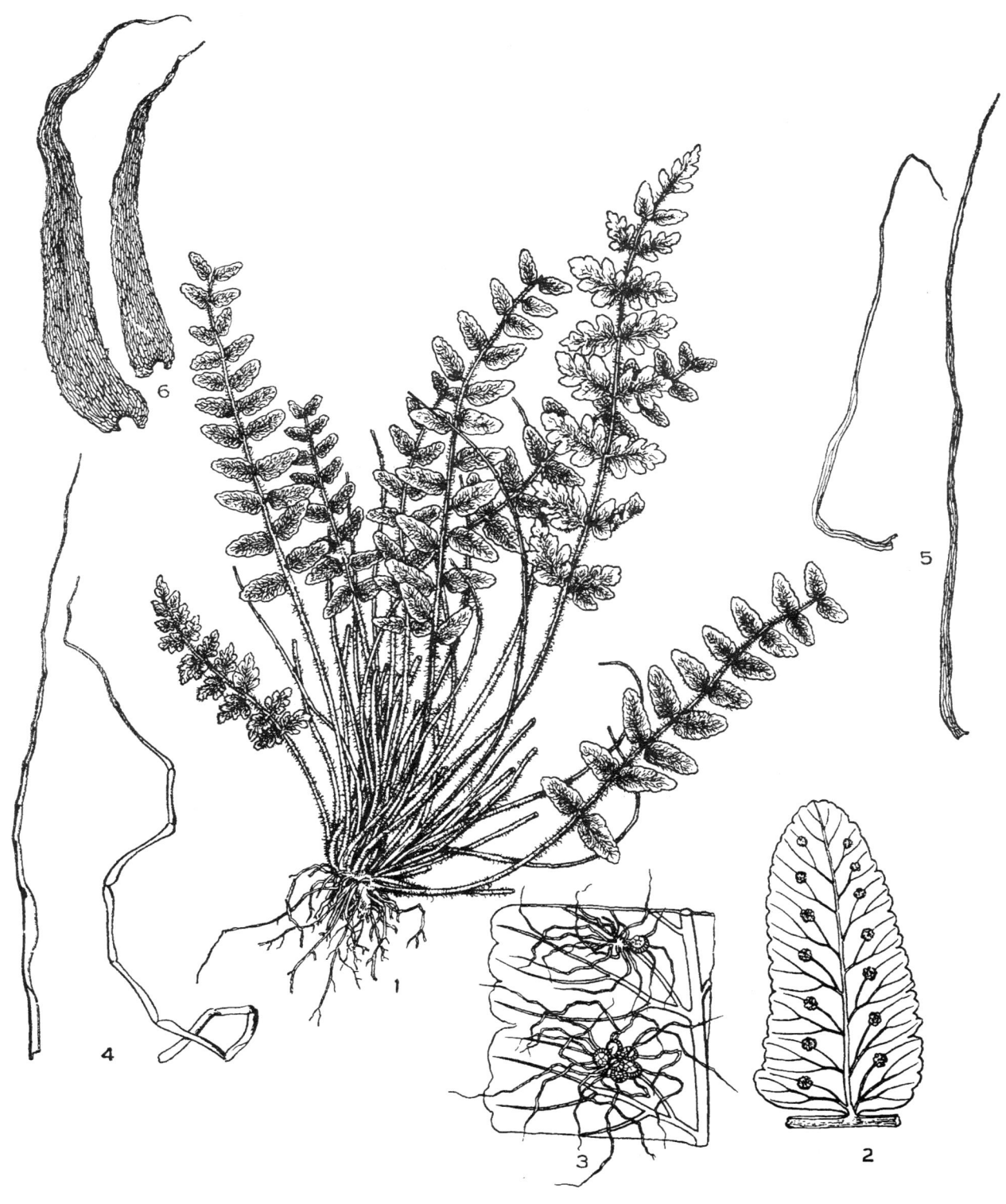

WOODSIA CINNAMOMEA Christ
赤色岩蕨

图版 5　滇冷蕨

　　根状茎微弱铁丝状，匍匐甚广，几光滑；叶柄细长，相距甚远，色微绿，无毛，长 10—15 cm，叶体三角卵圆形，几三次羽状分裂，长 10—17 cm，基部宽 6—10 cm，顶部由尖至细锐尖或尾状；一次小叶俱有小叶柄，直而开展，其间相距 2—3 cm，互生，在基部者呈披针形，长 4—7 cm，宽 1.5—2.2 cm，顶部尖锐，基部楔形；二次小叶几菱卵形或卵长方形，每侧 6—11 枚，上内基部修直，基部下部则割截至中肋，其在下部者每至叶基又呈羽状或深入之羽状分裂；裂片卵圆形，具圆形而有深刻之齿之顶部；叶薄草质；通常每二次小叶具孢子囊群 6—8 枚，有时仅 3 枚，位于细脉背面；子囊群盖无毛，薄膜质，终则失去。

　　分布：四川，云南，西藏。

　　此种为一显明特产之种，与 C. sudetica 相近，唯其子囊群盖无毛，叶较小，而裂片亦不同，Père David 最初发现于西藏，后 Delavay 采得于云南，Wilson 继得之于四川，本图则根据 Wilson 之标本而绘者也。

　　图注：1. 本种全形（自然大）；2. 小叶基部之二次小叶，示叶脉与孢子囊群（放大 16 倍）。

CYSTOPTERIS MOUPINENSIS Franchet

CYSTOPTERIS MOUPINENSIS Franch. in Nouv. Arch. Mus. 2. **10**: III (1887); C. Chr. Ind: Suppl. 46 (1913-17); Hand-Mzt. in Symb-Sinic. VI: 20 (1929).

Davalia triangularis Baker in Ann. Bot. **5**: 202 (1891).

Rhizome slender, wiry, wide. creeping, subglabrous; *stipes* slender, elongate, far apart, 10-15 cm long, greenish, glabrous; *frond* deltoid-ovate, subtripinnate, 10-17 cm long, 6-10 cm broad at the base, attenuate to the fine acuminate or caudate apex; pinnae all petiolate, erect-patent, 2-3 cm apart, alternate, the basal ones lanceolate, 4-7 cm long, 1.5-2.2 cm broad, acuminate, cuneate at the base; pinnules subrhombic-ovate-oblong, about 6-11 on each side, the upper inner base straight, the lower cut away to the costa, the lower ones pinnate or deeply pinnatifid towards the base; lobes ovate, with rounded inciso-dentate apex; *texture* thin herbaceous; *sori* mostly 6-8 sometimes only 3 to each pinnule, dorsal on the veinlets; indusium glabrous, thin membranaceous, at last evanescent.

Distribution: Sichuan, Yunnan, Tibet.

A fairly uniform and distinct endemic species, closely related to C. sudetica A. Br. et Milde, differs in glabrous indusium, smaller frond and cutting. First discovered by Père David in Tibet, later by Delavay in Yunnan and again by Wilson in W. Sichuan. Our figure is drawn from Wilson's No. 5311.

Plate 5. Fig. 1. Habit sketch (natural size). 2. A pinnule from basal part of the pinna showing venation and sori (× 16).

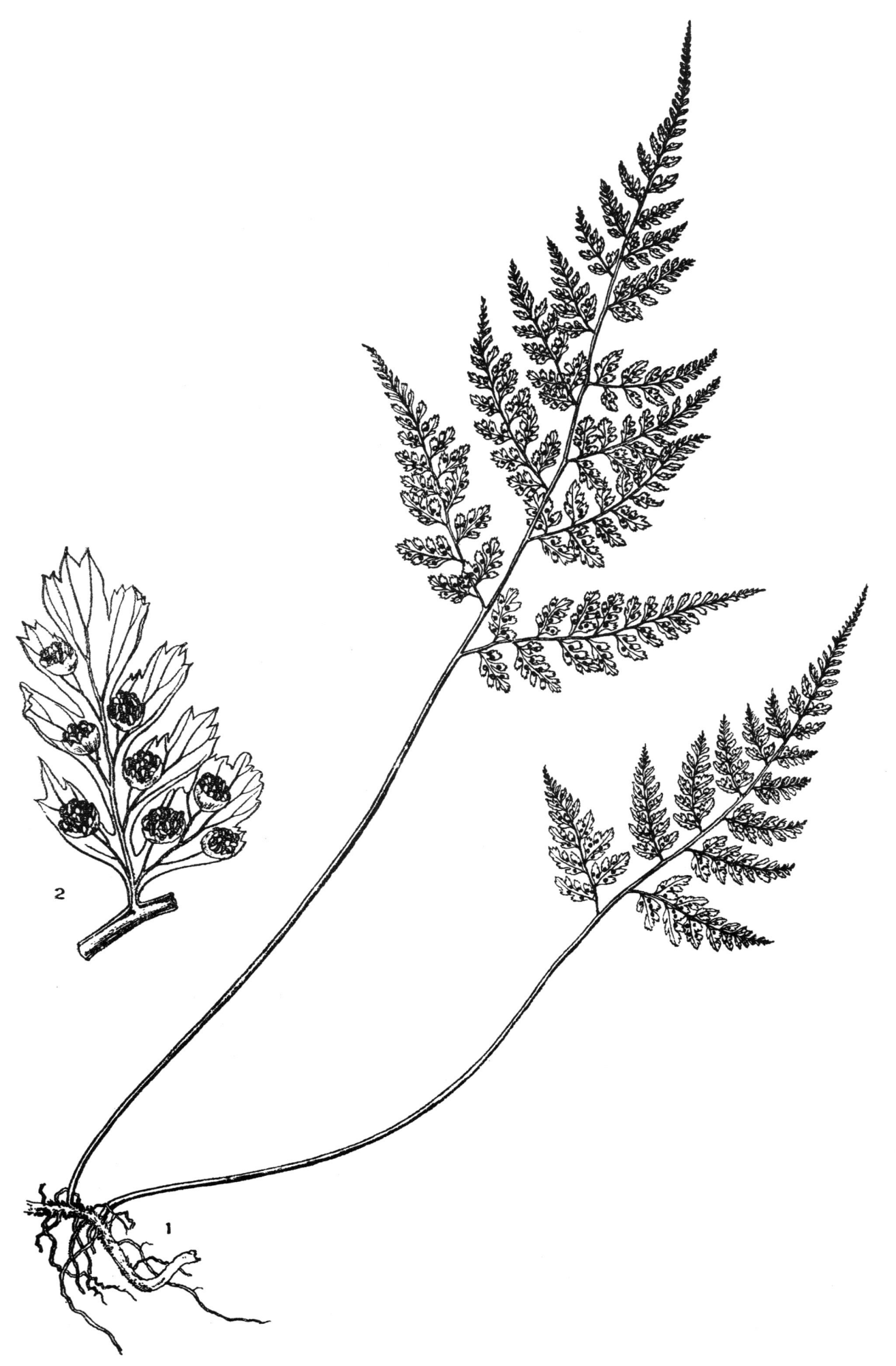

CYSTOPTERIS MOUPINENSIS Franchet
滇冷蕨

根状茎直立,大如拇指,密被鳞片,鳞片膜质,红褐色,披针形,上部尖锐,长 2 cm;叶柄簇生,稻秆色,下面呈圆形,上面与叶轴俱具深槽,长几 30 cm;近基部处密覆鳞片,渐上渐稀,其在叶柄基部者,与根状茎之鳞片相同;叶身三角长方形,长 20—30 cm,宽则较高为稍狭,一次羽状分裂,质颇坚固,上下两面俱绿色无毛;小叶 7—11 枚,顶生者与侧生者相似,具短柄,披针长方形,长 10—13 cm,宽 2.5 cm,缘具浅波状齿,基部截形,顶部钝形,叶脉分布成多数羽状而略展开之小群,小脉纤细而上弯,在侧脉之两侧各具 3—4 对;孢子囊群成 3—4 不整齐之行列,几全限于小叶中部之半,余有甚阔之边缘,位于细脉之中部,子囊群盖大,盾状,褐色无毛,中部略凹,几永存。

分布:湖北,福建。

本种为易于变易之一种,位于 D. podophylla (Hk.) 与 D. sieboldii (van Houtte) 之间,而尤与后种相近。A. Henry 始发现于鄂西之宜昌附近,Baker 之记述,盖根据其 32I7 号之标本也,小叶具深裂片,基部截形或几截形为其特征,而 7881 号亦采自宜昌附近,小叶颇大,(至 1.7 cm 宽 3.5 cm),基部圆形,仅有浅波齿之缘,孢子囊群分布为一至二不整齐之列,亦迄于小叶主肋至缘之半程,Dunn 在闽之中部亦采得此种标本,与 Baker 之原式颇似,与 7881 号较则仅波牙较细耳。

图注 1.本种全形(自然大);2.小叶之一部表示叶脉与孢子囊群(放大 2 倍);3.小叶之一部表示子囊群盖与子囊群及孢子囊附着于叶脉上之状(放大 6 倍);4.叶柄基部之鳞片(放大 8 倍)。

DRYOPTERIS ENNEAPHYLLA (Baker) C. Christensen

DRYOPTERIS ENNEAPHYLLA (Baker) C. Chr. Ind. 263 (1905); Matthew in Journ. Linn. Soc. **38**: 362 (1911).
Nephrodium enneaphyllum Baker in Journ. Bot. (1887) 170; Diels in Nat. Pfl. Fam. I. 4, 168.

Caudex erect, thick as a thumb, clothed in dense reddish-brown lanceolate acuminate membranaceous scales to 2 cm long; *stipes* tufted, stramineous, terete beneath, deeply grooved above and so is the rachis, nearly 30 cm long, densely scaly near the base, become rather sparsely so upward, scales at the baste similar to those only rhizome; *frond* deltoid-oblong, 20-30 cm long and a little less broad, simply impari-pinnate, moderately firm in *texture*, green and glabrous on both sides; pinnae 3-5 on each side, the terminal one similar to the lateral ones, shortly petiolate, lanceolate-oblong, 10-13 cm long, 2.5 cm broad, shallowly crenate, truncate or subtruncate at the base, blunt at the apex; *veins* in numerous pinnate subpatent groups, veinlets fine, ascending, 3-4 on each side of the lateral veins; *sori* large, in 3-4 irregular rows, almost restricted to the central half of the pinnae, leaving a broad free margin, medial on the veinlets, indusium large, peltate, brown, glabrous, with a slightly depressed centre, subpersistent.

Distribution: Hubei, Fujian.

Rather a variable species, evidently intermediate between D. *podophylla* (Hk.) and D. *sieboldii* (van Houtte) with a much stronger approach to the latter. It was discovered by A. Henry around Ichang, W. Hubei. Henry's No. 3217, which has rather deeply lobato-incised pinnae with truncate or subtruncate base, while Henry's No. 7881 from the same locality deviates from the type in much larger pinnae (to 17 cm long, 3.5 cm broad) with rounded base, only shallowly crenulato-serrate margin, and 1-2 irregularly seriate sori, confined to the inner half breadth of the pinnae. Dunn's specimen No. 3830, from Central Fujian (1905) agrees well with Baker's type except that it has still more finely crenato-serrate margin than Henry's No. 7881, and yet it should belong here.

Plate 6. Fig. 1. Habit sketch (natural size). 2. A portion of pinna showing venation and sori ($\times$ 2). 3. A portion of pinna showing indusium and sori in situ on the veinlet and the way of insertion of sporangia ($\times$ 6). 4. Scales on the base of stipe ($\times$ 8).

DRYOPTERIS ENNEAPHYLLA (Baker) C. Christensen
宜昌金星蕨

图版 7　燕尾三叉蕨

叶柄长达 30 cm,或较长,光泽无毛,褐红色;叶体薄膜质,鲜绿色,仅肋及脉上有短毛,三角长方形,长 40 cm,宽 35 cm,生在较下部者为二次羽状复叶至三次羽状分裂,肋轴与肋基俱光泽,淡黝色,上面具深槽,生于顶上者大而具小柄,深入三裂,三角形,中部裂片长方形,长 15—30 cm,阔 5—10 cm,缘有粗而不整齐之微裂,裂片下部并生,楔形修长,缘不整齐,侧生之一次小叶在每侧各 3 枚,对生,其间相距 12—16 cm,最下者 14 cm,其上者长 8 cm,具长柄,最上者几无柄,长方形,顶部尾状,基部钝戟形,二次小叶 2 枚,相距颇远,微有柄,基部戟形,或深裂,长 10—13 cm,缘具不整齐之波状齿,与一次小叶同具尖锐或长尾状之顶部,侧脉显明,几达于叶缘,孢子囊孔呈不整齐之四角形,成 2—3 列介于侧脉间,中肋至叶缘之间可容 6 枚,常无游离之细脉,孢子囊群颇多生边缘小脉上或背上,小而作点状,黯褐色,呈不整齐之排列,无子囊群盖。

分布:贵州,广西。

本种为 A. cicutarium 系中具二次羽状复叶之一种,最显明者在其一次小叶与下部之二次小叶俱长狭而有长柄,二次小叶之略分裂者与各裂片俱具长尖锐之顶部,叶柄叶轴俱赤黝色,有光泽;孢子囊群微小不整齐。此种 1904 年 Père Cavalerie 始发现于贵州,后 Père Esquirol 亦在贵州采得之,最近则见诸广西。

图注:1. 植物体之一部(原大);2. 叶体之一部,示脉与子囊群(放大 8 倍);3. 孢子囊(放大 106 倍)。

ASPIDIUM LONGICRURE Christ

ASPIDIUM LONGICRURE Christ in Bull. Geogr. Bot. Mans. (1909) Mém. xx. 169; C. Chr. Ind. Suppl. 8 (1906-13); Matthew in Journ. Linn. Soc. **39**: 343 (1911).

Sagenia longicrure Christ in Bull. Geogr. Bot. Mans. (1906) 250.

Stipe to 30 cm or more long, shining, reddish-brown, glabrous; *frond* deltoid-oblong, 40 cm long, 35 cm broad, bipinnate to tripinnatifid in the lower part, rachis and the base of costa shining, light ebeneous, deeply grooved above; terminal pinna large, petiolate, deeply trilobate, deltoid, the central lobe oblong, 15-30 cm long, 5-10 cm broad, margin coarsely repando-lobulate, with the base of lobes broadly adnate, cuneat-elongate, repand; lateral pinna 3 on each side, opposite, 12-16 cm apart, the lower ones 14 cm long, the uper ones 8 cm long, long-petiolate, the uppermost ones subsessile, oblong, caudate, with obtuse hastate base, the middle pinnae deeply 3-lobed down to the base, the lowest pinnate; pinnules 2, farapart, petiolulate, with hastate or deeply lobed base, the terminal pinnules deeply trilobate; all pinnae ascendingly patent, the uppermost pinnules to 10-13 cm long, 4 cm broad, margin repando-crenate, all pinnae and pinnules acuminate, or long caudate; lateral *veins* prominent, almost reaching the margin, areolae irregularly quadrangular, in 2-3 rows between main veins and about 6 between costa and margin, usually without free included veinlets; *sori* very numerous, campital or dorsal, small, almost punctate, dark brown, irregularly disposed, indusium destitute; *texture* thin herbaceous, lustrous green, surfaces naked but the costa and veins are shortly pubescent.

Distribution: Guizhou, Guangxi

A bipinnate species of the group of *A. cicutarium* Sw., very distinct in pinnae and lower pinnules being long-petiolate, narrowly elongate, few-lobed pinnules and segments with very long-acuminate apex, stipes and rachis reddish-ebeneous, polished, and very small irregular exindusiate sori. Collected for the first time by Père Cavalerie in Guizhou, (1904), later by J. Esquirol in the same locality and of very late from Guangxi.

Plate 7. Fig. 1. Habit sketch of a portion of frond (natural size). 2. A portion of segment, showing venation and sori ($\times$ 8). 3. Sporangium ($\times$ 106).

ASPIDIUM LONGICRURE Christ
燕尾三叉蕨

叶柄及叶轴深黑色有光泽，上面具短绒毛；叶体三角形颇大，长约 70 cm，底部宽约 60 cm，深绿色，薄膜质，主脉亦具绯色之短绒毛，此外则全部光滑，二次至三次羽状分裂，最下 2 对一次小叶有柄互生，相距约 15 cm，上部 2 小叶无柄，至最上之小叶则基部连合或连合其较上部分，中间联络之翼宽 1—3 cm；底部之一次小叶最大，羽状分裂至下部成无柄或合着之二次小叶，上部裂片 6—8 个，各裂片间有翼相连，翼宽 1—1.5 cm，底部 2 独立二次小叶长 10—17 cm，宽 5—6 cm，作长方披针形，渐尖头，底部圆楔形，羽状分裂至一半，中间裂片约长 10 cm，宽 3—5 cm，有宽裂片或羽状分裂，裂片略呈镰形，钝头有阔钝锯齿，长 1—1.5 cm，中部一次小叶与顶部下之裂片作羽状分裂与基部一次小叶略同；脉纹显明，网状，网孔作五或六角形；子囊群不甚规则，在最顶部之裂片上，成单行，延中肋而生，左右各有 3—7 个，中间留有甚阔之空隙；子囊群盖永存，盾状，深褐色。

分布：贵州。

本种为本属中最大之种，与 A. cicutarium Sw. 一种相近，而其异点则在黝黑光泽之叶轴与中肋，而下部二次小叶较其上者略短耳，本种最初为 Père J. Esquirol 于 1910 年采于贵州森林中。

图注：1. 全植物之一部（自然大）；2. 上部之叶片，示其脉络及子囊群。

ASPIDIUM EBENIUM C. Christensen

ASPIDIUM EBENIUM C. Chr. Bull. Geogr. Bot. Mans (1913) 138; Ind. Suppl. 5 (1913-16).

Stipes and rachis shining, ebeneous, very shortly tomentose on the upper side; *frond* ample, deltoid, about 70 cm long, 60 cm broad at the base, dark green; *texture* thin herbaceous, costa like the rachis very shortly reddish-tomentose on the upper side, otherwise the entire plant glabrous, bipinnate to tripinnatifid; the lower 2 pairs of pinnae petiolate, alternate, 15 cm apart, the upper 2 sessile but free, the uppermost broadly adnate or confluent upward, connected by a wing 1-3 cm broad, the basal pinnae much the largest, pinnate below into sessile or adnate pinnules, pinnatifid upward into 6-8 segments connected by a wing 1-1.5 cm broad, the basal 2 free pinnules 10-17 cm long, 5-6 cm broad, oblong-lanceolate, acuminate, cuneate-rounded at the base, pinnatifid half way down, the middle segments about 10 cm long, 3-5 cm broad, broadly lobed or pinnatifid; lobes subfalcate, obtuse, broadly crenate, 1-1.5 cm broad; the middle pinnae and the segments below the apex of the frond pinnatifid in the similar manner as the basal pinnae; *veins* distinct, reticulate, areolae 5-6-gonous, usually with included free veinlets; *sori* subirregular, on the ultimate lobes, uniseriate, 3-7 on each side of the costule, leaving a broad blank space in the middle, indusium corrugated, persistent, peltate, dark brown.

Distribution: Guizhou.

A large fern in the genus closely allied to A. cicutarium Sw., differs only in shining ebeneous stipe, rachis and costa, and in inferior pinnules shorter than those next above. Collected in Guizhou under forest by Père J. Esquirol, 1910.

Plate 8. Fig. 1. Habit sketch of a portion of frond (natural size). 2. A portion of ultimate segment, showing venation and sori.

ASPIDIUM EBENIUM C. Christensen
黑柄三叉蕨

根状茎肥大直立多根；叶丛生 4—7 枚成一簇；叶柄长 5 cm,密生鳞片,其形可别为二种,一种鳞片短纤维状,另一种则呈宽大卵形,锐尖头,长至 6 mm,红黄色；叶体长 12—15 cm,中部宽 1.5—2.5 cm,狭长披针形,两端渐窄,一次羽状分裂,叶轴覆以披针形具锥头之鳞片；一次小叶簇生,在下部者甚小,卵圆形,长 4 mm,中部者长至 12 mm,形长圆,基部相等,边缘反卷,波齿状,基部楔形或心脏形,叶体为厚革质,黄绿色,上部具疏松白色紧贴刚毛状之鳞片,下部唯中肋附近密生大形披针形褐色之鳞片,在脉络附近者则略小,作狭长披针形,膜质,长至 2 mm；脉纹内陷不显明,近基部处歧出或略成羽状；子囊中肋两侧各 3—4 枚,其形几圆,不显明,子囊群盖大,质薄,褐色,最易脱落。

分布：云南,四川,印度西北部亦有之。

注：本种经 Hope 氏采自印度西北部者其形比中国西部所产为小,本种性质与 P. lachensee (Hk.) Bedd. 相近,而本种之叶柄较短叶轴较硬,且上部密生大形鳞片及平铺之刚毛状之鳞片,叶片为厚革质,此其不同之点也。因 Père Delavay 氏于 1884 年采自云南冰川之麓,故 Christ 名之为 P. glaciale,后 Christensen 仍归纳之于 P. duthiei 云。

图注：1.本种全形（自然大）；2.叶体中部之小叶上面（放大 4 倍）；3.同上之背面（放大 3 倍）；4.同上示其托及子囊群附着状况；5.小叶之一部示子囊群之托及数鳞片（放大 15 倍）；6.叶背面之鳞片（放大 16 倍）；7.叶轴上之鳞片（放大 16 倍）；8.叶轴上之小形鳞片（放大 16 倍）；9.叶上面之鳞片（放大 16 倍）；10.叶柄下之鳞片（放大 16 倍）。

POLYSTICHUM DUTHIEI (Hope) C. Christensen

POLYSTICHUM DUTHIEI (Hope) C. Chr. Ind. 72 (1905), 581 (1906); Medd, Göteb. Bot. Trädg **1**: 94 (1924); Hand-Mzt. in Symb. Sine. vi. 26 (1929).

Aspidium duthiei Hope in Journ. Bomb. Nat. Hist. Soc. **12**: 532, t. 6 (899).

Polystichum glaciale Christ in Bull. Soc. Bot. Fr. 52. Mém. 1. 28 (1905).

Sorolepidium glaciale Christ in Bot. Gaz. **51**: 350 c. fig. (1911).

Rhizome thick, erect, densely rooted; leaves tufted (4-7 together), *stipes* 5 cm long, covered with dense dimorphic scales, the one short, fibrillose, the other broad, ovate, acuminate, 6 mm long, ochraceous-yellow; *frond* 12-25 cm long, 1.5-2.5 cm broad at the middle, linear-lanceolate, tapering to both ends, simply pinnate, rachis thickly clothed with lanceolate subulate scales; pinnae conferted, the lower ones much reduced, ovate, 4 mm long, the middle ones to 1.2 cm long, oblong, equal, margin strongly reflexed, crenate, truncate or cordate at the base; *texture* thick coriaceous, greenish-flavescent, villose with whitish adpressed setaceous scales above, the under surface clothed in dense large lanceolate brown scales from the costa and the smaller ones from the veins, the scales lanceolate-linear, scarious, 2 mm long; *veins* entirely hidden, forked or subpinnate near the base; *sori* 3-4 on each side of the costa, almost rounded, hidden, indusium large, thin, brown, very fugaceous.

Distribution: Yunnan, Sichuan, also N. W. India.

Hope's type from N. W. India was a much smaller plant than the plant from W. China. In general habit, this species resembles *P. lachenense* (HK.) Bedd., differs in much shorter stiff stipe and stouter rachis, the dense covering of disproportionally large scales, the adpressed white setae on the upper surface and thick coriaceous texture. Its first record from China was credited to Père Delavay, who found it in 1884 at the foot of a glacier in Sukiang, Yunnan, for which reason, Christ called the plant *P. glaciale* by referring to Hope's *P. duthiei*, to which it was finally reduced by Dr. Christensen.

Plate 9. Fig. 1. Habit sketch (natural size). 2. A pinna from middle part of the frond, upper side (× 4). 3. The same, under side (× 3). 4. The same, showing venation and attachment of sori. 5. A portion of the pinna, showing receptacles of sori with a few scales attached (× 15). 6. Scale from the under side of pinna (× 16). 7. Scale from rachis (× 16). 8. Smaller scale from the rachis (× 16). 9. Scale from the upper side of the pinna (× 16). 10. Scale from lower part of the stipe (× 16).

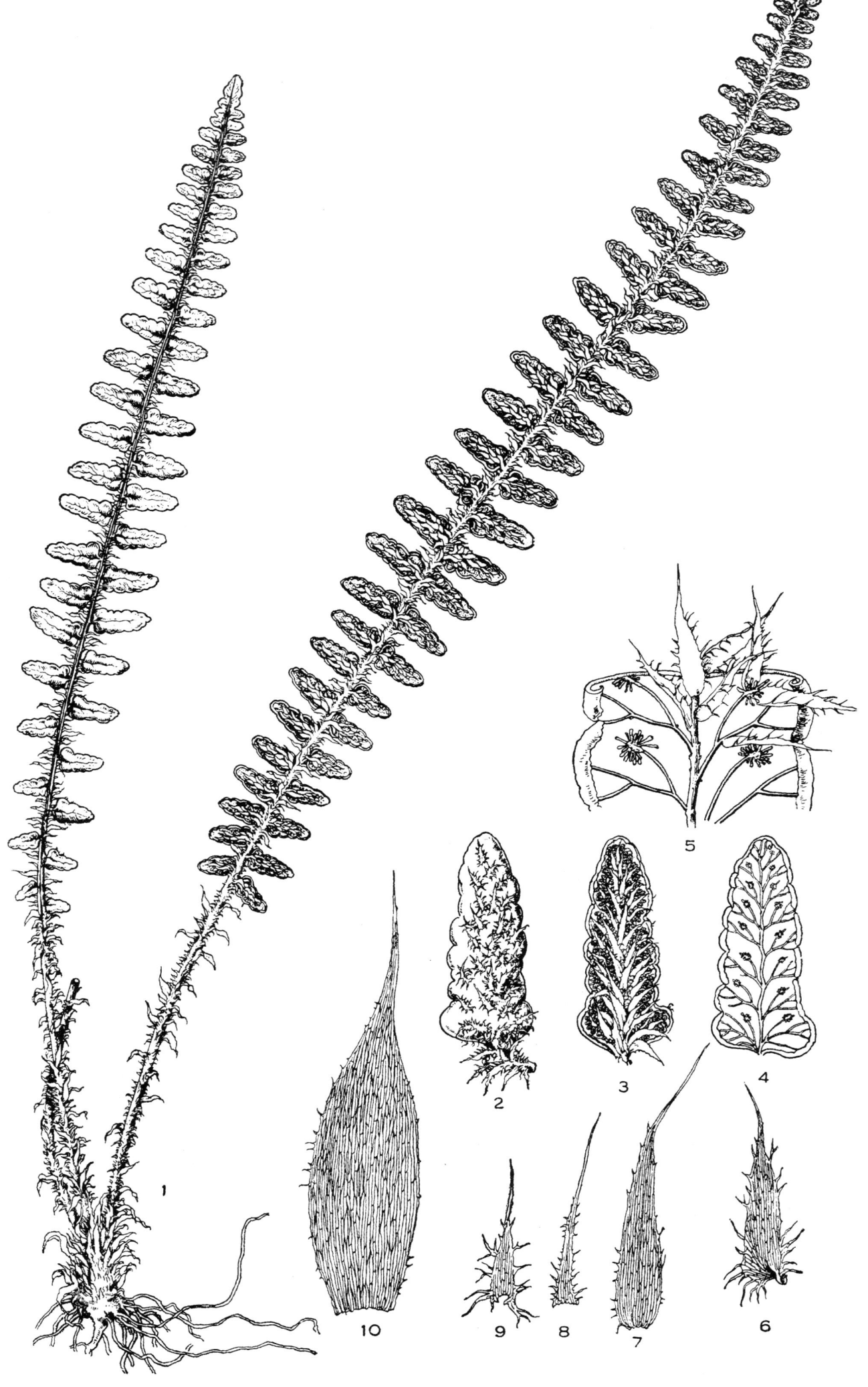

POLYSTICHUM DUTHIEI (Hope) C. Christensen
杜氏耳叶蕨

茎根部短而直立；叶柄簇生，长 3—8 cm，稻秆色，有棱，密生小枇糠状鳞片，鳞片黄色，俱一色，形式不一，一种细长线形，他种宽大卵形，突成尾状尖头，疏散包覆；叶体窄披针形，渐尖头，下部略缩小，长 10—20 cm，全部二次羽状分裂或上部二次下部简单羽状分裂，宽至 2.2 cm，叶轴密覆以鳞片，鳞片黄色线形刚毛状；一次小叶略具短柄或无柄，密集，互相掩覆，在下部则稍疏，硬革质，上面光泽绿色，下面略浅而平滑，卵圆披针形，宽 6—8 mm，略具羽状裂片；裂片卵圆形，每侧各具 2—4 枚，由中肋分裂，与叶轴平行，边缘较厚，向内卷，顶部呈刚刺状；子囊群成单列，生于支脉之上面，近小叶片之中肋或基部小裂片之肋深褐色，子囊群盖盾状，圆形，边缘微有缺刻。

分布：云南，四川。

本种之形态似 P. aculeatum Schott.，但小很多，而为革质，且叶之分裂比较简单，此其异点也，就此而言本种实具抗旱性之高山植物也。在云南及四川两省甚为普通，初为 Delavay 氏在 1882 年采得。

图注：1. 本种全形（自然大）；2. 一次小叶示叶脉及子囊群；3. a，子囊群盖之前面（放大 16 倍）b，同上侧面（放大 16 倍）；4. 叶轴上之鳞片（放大 27 倍）；5. 根状茎之鳞片（放大 27 倍）。

POLYSTICHUM ACANTHOPHYLLUM (Franch.) Christ

POLYSTICHUM ACANTHOPHYLLUM (Franch.) Christ in Bull. Soc. Bot. Fr. **52**: Mén Ⅰ. 30 (1905); C. Chr. Ind. 575 (1906); Matthew in Jour. Linn. Soc. **39**: 384 (1901).

Aspidium acanthophyllum Franch. in Bull. Soc. Bot. Fr. **32**: 28 (1885).

Polystichum aculeatum Schott. var. *acanthophyllum* Bedd. in Handb. Suppl. 43 (1892).

Caudex short, erect; *stipes* fasciculate, 3-8 cm long, stramineous, angular, densely paleaceous, paleae yellow, concolored, dimorphic, the one narrow, linear, the other broadly ovate, abruptly cuspidate, slightly imbricate; *frond* narrowly lanceolate, acuminate, slightly reduced downward, 10-20 cm long including the stipe, scarcely 2.2 cm broad, bipinnatifid or bipinnate below under the pinnatifid apex, rachis densely scaly, scales yellow, linear-setaceous; pinnae subsessile, close, imbricate, somewhat distant below, *texture* rigidly coriaceous, lustrous green above, pale below, glabrous, ovate, 2-4 on each side, cut straight on the inner side and parallel to the rachis, margin thickened, strongly revolute, apex rigidly spinescent; *sori* uniseriate, dorsal on lateral veinlets, closer to the costa of pinnae or costules on basal pinnules or lobes, dark brown, indusium peltate, orbicular, margin slightly erosed.

Distribution: Yunnan, Sichuan.

Habit of *P. aculeatum* Schott., but of much smaller size, thick coriaceous texture and simpler pinnae. Evidently an alpine plant of xerophytic habit. A fairly common fern in Yunnan and Sichuan, as it has been reported by many collectors from that region since its discovery by Delavay in 1882.

Plate 10. Fig. 1. Habit sketch (natural size). 2. A pinna, showing venation and sori. 3. Two inducia, a, front view; b, lateral view (× 16). 4. Scales from rachis, (× 27). 5. Scales from rhizome (× 27).

POLYSTICHUM ACANTHOPHYLLUM (Franch.) Christ
刺耳叶蕨

图版 11　对生耳叶蕨

　　根状茎短而直立,疏被鳞片,鳞片形小,暗褐色,披针形,渐尖头,薄膜质,多生基部,稀有延展至上部者;叶柄丛出,5—10枚,长5—15 cm,宽2.5 cm,下部略窄,光滑,质殊坚硬;叶轴几光滑,具少数紧贴三角形暗褐色之鳞片;一次小叶密接而生,无柄,两侧各20—25枚,略呈方卵圆形,内边直截,上有大耳,顶部圆形,具尾状尖头,上边有锯齿,下边通常全缘,脉纹不显,羽状,支脉一次分叉,子囊群生在中部之上半,形小,暗褐色,常限于小叶之外半邻,每小叶具3—6枚,子囊群盖小而光滑。

　　分布:四川,云南,湖北,广西,台湾。

　　本种最早由Maries氏采于离宜昌城二十里之宜昌峡,其性质与 P. auriculatum Presl相近,而本种之短小三角形之小叶具少数短而颇坚之齿则又与后者不同。

　　图注:1.植物全形之一部(自然大);2.一次小叶示其脉纹及子囊群(放大3倍);3.子囊群及子囊群盖之一部(放大16倍);4.孢子囊(放大52倍)。

POLYSTICHUM DELTODON (Baker) Diels

POLYSTICHUM DELTODON (Baker) Diels in Nat. Pfl. Fam. Ⅰ. 4. 191; C. Chr. Ind. 580 (1905); Matthew in Journ. Linn. Soc. **39**: 385 (1911).

Aspidium deltodon Baker in Gard. Chron. n. s. **14**: 494 (1880).

Hemesteum deltodon Léveille, Flore du Kouy-tscheou 496 (1915).

Rhizome short, erect, scales sparse, small, dark brown, lanceolate, acuminate, membranaceous, basal, rarely extending higher up; *stipes* tufted (5-10 together), 5-15 cm long, slender, naked, stramineous; *frond* linear, elongate, simply pinnate, 10-20 cm long, 2.5 cm broad, scarcely narrowed downward, glabrous, moderately firm in *texture*; rachis subglabrous with a few small adpressed deltoid dark brown scales; pinnae close, sessile, 20-25 on each side, quadrate-ovate, cut straight on the inner side and strongly auricled on the upper, apex rounded, cuspidate, denticulate on the upper edge, usually entire on the lower; *veins* inconspicuous, pinnate, once forked; *sori* supra-medial, small, dark brown, usually confined to the outer half of the pinnae, 3-6 to a pinna, indusium small, glabrous.

　　Distribution: Sichuan, Yunnan, Hubei, Guangxi, Taiwan.

　　This fern was first collected by Maries in Ichang Gorge, about 20 li to the west of the city of Ichang, Hubei. It is closely related to *P. auriculatum* Presl, differs only in short deltoid pinnae with few short rather stout teeth.

　　Plate 11. Fig. 1. Habit sketch (natural size). 2. A pinna showing venation and sori (× 3). 3. A sorus with indusium attached (× 16). 4. A sporangium (× 52).

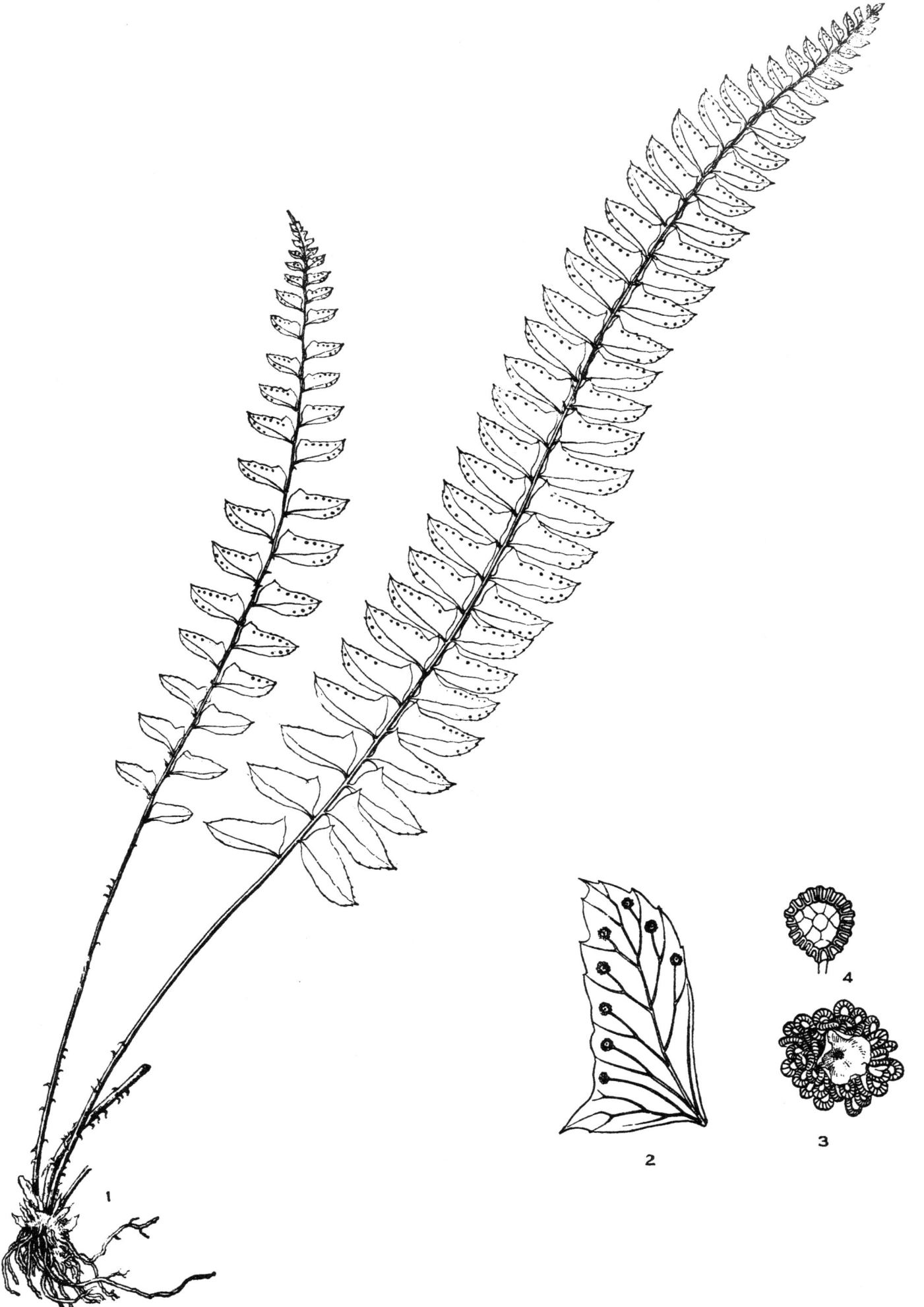

POLYSTICHUM DELTODON (Baker) Diels
对生耳叶蕨

图版 12　锯齿耳叶蕨

根状茎短，直立，宽如鸦喙，密覆以薄膜质褐色具两种形态之鳞片，鳞片一种宽披针形，他种尖线形，皆具须毛状顶端；叶柄3—4枚丛生，细瘦，下面颇圆，上面具两条深沟，长5—15 cm；叶体长线形两端渐尖，长20—35 cm，宽2—3 cm，一次羽状分裂；一次小叶每侧各约40—50个，小叶长12—20 mm，宽5 mm，密生或略相掩覆，略成长菱形，下部小叶渐向下弯，但上部者则平伸而出，小叶上方之内边与叶轴平行，形成一三角形之耳，其尖端具刺，下方至中肋而削去，小叶之尖端圆形，边缘具有刺头之小锯齿；叶脉甚细不显明，支脉羽状，作颇规则之一次分叉，但近顶部之支脉则为单行，上部之叶片具子囊群，子囊群生于中肋及叶缘之间，在中肋之上方有6—10个，下方有2—4个或全无，子囊群盖圆形，盾状，小于子囊群，中央附着点略陷入，深褐色。

分布：四川。

本种为最特殊之种，其短而有鳞片之叶柄密生栉状长圆形之小叶，每侧具八枚至十枚弯曲而有长刺之锯齿，及下部渐向下弯之小叶，皆为最显著之特性也。本种标本早经 E. Faber 氏采自峨眉，Baker 氏定之为 *Aspidium auriculatum* Sw. var. stenophyllum。至1900年 Von Rosthorn 氏又采之于南川，由 Diels 定今名，但当时尚不知与 Baker 氏之变种是一是二也，前著者于香港植物园中得见 Faber 氏所采之标本，其叶柄似较 Diels 之记载略长，本种之具宽叶者似与 *P. auriculatum* 之习性相近，但本种尖锐之齿及上方深刻之耳状裂片则又与后者不同，此种在四川分布甚广，尤多见诸峨眉山。——仁昌

图注：1. 本种全形（自然大）；2. 一生于中部之一次小叶示其脉络及子囊群（放大3倍）；3. 一次小叶之一部示脉纹，子囊群及鳞片（放大12倍）；4. 叶柄上之鳞片（放大31倍）；5. 叶轴上之鳞片（放大31倍）；6. 孢子囊（放大145倍）。

POLYSTICHUM HECATOPTERUM Diels

POLYSTICHUM HECATOPTERUM Diels in Engl. Jahrb. **29**: 193 (1900); C. Chr. Ind. 582 (1906); Mathew in Jour. Linn. Soc. **39**: 386 (1911).

Polystichum auriculatum Presl var. *stenophyllum* Baker in Jour. Bot. (1888) 227.

Hemesteum hecatopterum Léveille, Flore du Kouy-tscheou 496 (1915).

Rhizome short, erect, thick as raven's quill, clothed like the stipes with moderately dense membranaceous brown dimorphic scales, the one broad lanceolate, the other linear-subulate, both are hair-pointed; *stipes* tufted (3-4 together), slender, terete beneath, bisulcate above, 5-15 cm long; *frond* elongate linear, gradually tapering towards both ends, 20-35 cm long, 2-3 cm broad at the middle, pinnate; pinnae 40-50 on each side. 1. 2-2 cm long, 0. 5 cm broad, close or subimbricate, subrhombic-oblong, gradually deflexed downward, but the upper ones horizontally patent, the upper inner side straight and parallel to the rachis, much produced into a triangular auricle, provided at the apex with a bristle, the lower side cut away to the costa, apex rounded, margin spinuloso- or aristatato-denticulate; *veins* inconspicuous, fine, pinnate, regularly once forked, except those towards the apex which are simple, the upper pinnae fertile; *sori* between the costa and margin, mostly 6-10 on the upper side and none or 2-4 on the lower side of the costa, indusium rounded, peltate, smaller than the sori, attached by the centre, which is slightly depressed and blackish.

Distribution: Sichuan.

This is a uniquely distinct species, well marked by short scaly stipe and pectinately close oblong pinnae with 8-10 often incurved long-aristate teeth on each side, the lower pinnae considerably abbreviated and strongly deflexed. This pretty fern was first discovered by Rev. Earnst Faber on Omei Shan early in 1886 and described two years later by Baker as *Aspidium auriculatum* Sw. var. *stenophyllum* in following words: "Approaching to *Aspidium lonchitis* in habit, with a frond a foot long and scarcely above an inch broad at the middle, with subrhomboidal spinulose conspicuously auricled pinnae much cut away on the lower side of the midrib" (Jour. Bot. XXVI. 227, 1888). It was, however, not until 1900 when Herr von Rosthorn's plants collected in Tapao Shan, in Nanchuan, S. E. Sichuan (1891), was brought under the notice of Dr. Diels, who gave the name as it bears today, not quite sure then as to the identity of his species to Baker's var. *stenophyllum*. I have seen Faber's plant (No. 1305) in Hong Kong Herbarium and found it is typical of Diels' species except with longer stipe (—12 cm). Ample materials collected by Wilson in W. Sichuan show that fronds even from same rhizome may vary from 2-3-4 cm in width, and it is with the broadest frond that the species somewhat approaches *P. auriculatum* Presl in habit but for long-aristate teeth and more sharp auricle. Evidently a fairly common fern in Sichuan, particularly on Omei Shan. —R. C. C.

Plate 12. Fig. 1. Habit sketch (natural size). 2. A pinna from the middle part of the frond, showing venation and sori (× 3). 3. A portion of the pinna, showing venation, sori and scales attached (× 12). 4. Scales from the stipe. (× 31) 5. Scale from the rachis (× 31). 6. Sporangium (× 145).

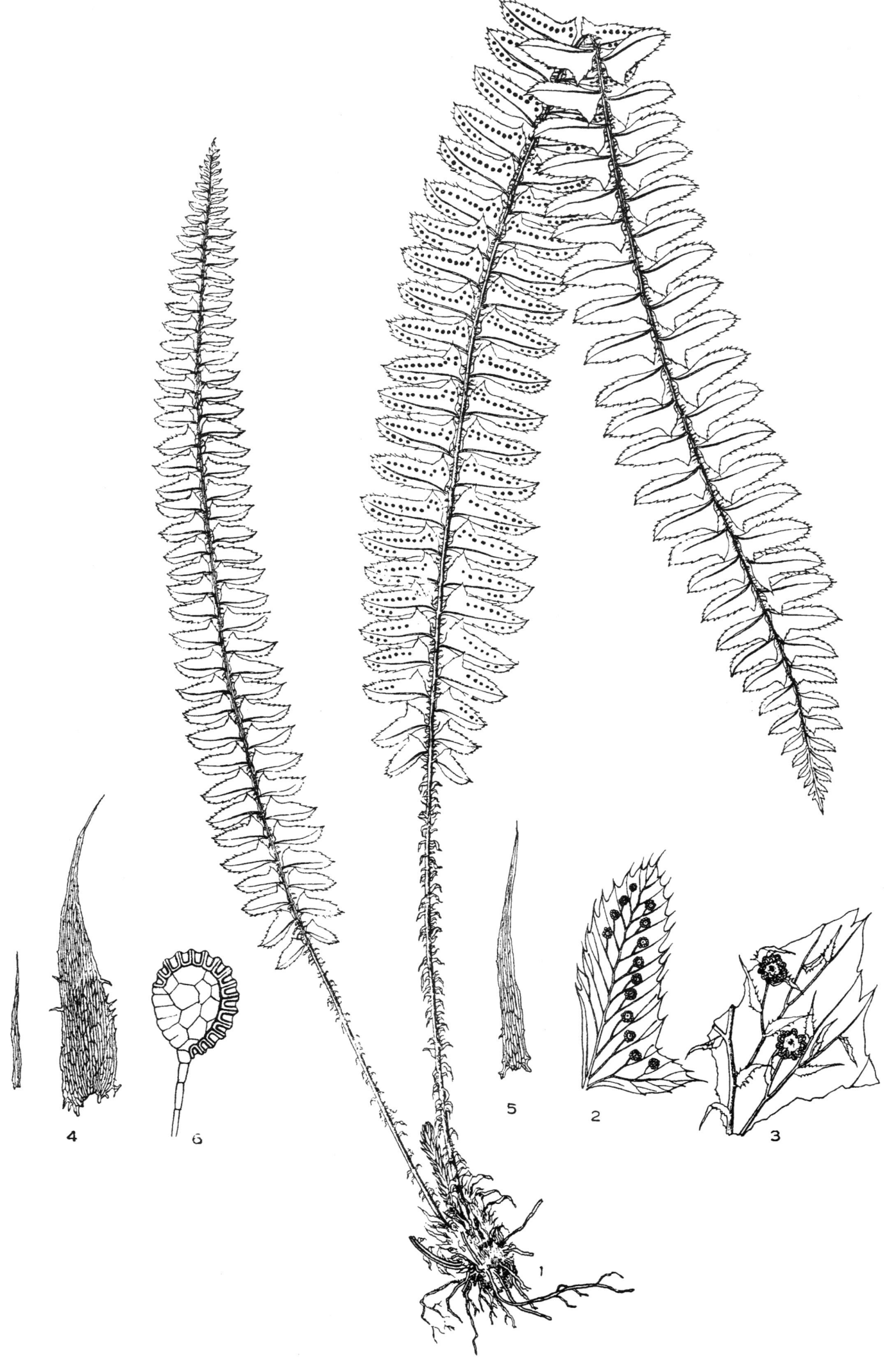

POLYSTICHUM HECATOPTERUM Diels
锯齿耳叶蕨

根状茎肥壮,木质,颇短,向上斜伸;鳞片大,密生,覆瓦状排列,膜质,锈赤色,阔卵形,长渐尖头,长 6—10 mm,阔 3—5 mm;叶柄簇生,成 5—10 丛,长 8—12 cm,粗 2 mm,红稻秆色,密被异形之鳞片,鳞片一种似根状茎者,一种较小,纤维针状或尖线状,延至叶轴之上;叶体线状披针形,连叶柄共长 35—42 cm,阔 4.2—5 cm,二次羽状,向基部缩小,向顶部狭长,在顶部或较下之处发芽生根,芽圆形,密生如根状茎上所生之鳞片;一次小叶几无柄,互生或几对生,颇密集,横开张,下部者向下弯,在叶轴每侧有 30—35 枚,上面光滑,下面被纤维状鳞片,在与下面叶轴连接处被有数个卵圆形尖头有纤毛之鳞片,长 2—2.5 cm,基部阔 1 cm,长圆披针形,下部成羽状复叶,顶端者仅有芒状波齿或锯齿,近顶部则羽状分裂,在前显成耳状;二次小叶 5—7 枚,密集,生基部者独立,无柄,生上部者基部附着,菱形卵圆形,全缘或具波齿,顶部圆而作芒状;叶体厚革质,上面光绿色,下面较浅;叶脉不显明,二次小叶上之支脉 3—5 对,分离,通常一次分叉;孢子囊群小,一列,在中肋两侧各 2—3 枚,生支脉背面之中部,子囊群盖小,革质,永存。

分布:广西。

图注:1.本种全形(自然大);2.一次小叶(放大 2 倍);3.一次小叶基部之二次小叶,示叶脉与孢子囊(放大 12 倍);4.叶柄上之鳞片(放大 5 倍);5.叶轴上之鳞片(放大 24 倍)。

POLYSTICHUM CHUNII Ching

POLYSTICHUM CHUNII Ching in Sinesia **1**: 2 (1929).

Rhizome thick strong, woody, rather short, oblique-ascending; scales large, dense, imbricate, membranaceous, ferruginous, broad ovate, long acuminate, 6-10 mm long, 3-5 mm broad; *stipes* fasciculate, 5-10-clustered, 8-12 cm long, 2 mm thick, rufostramineous, densely clad in dimorphic scales, the one similar to the rhizomatic ones, the other smaller, fibrillose-subulate or linear-subulate, extending throughout the rachis; *fronds* linear-lanceolate, 35-42 cm long (including stipes 9-11 cm long) 4.2-5 cm broad, bipinnate, abbreviate towards base, long attenuate towards apex, gemmiferous and radicant at tip or a considerable distance backward; gemmae globular, densely clad with imbricate scales similar to the rhizomatic ones; pinnae subsessile, alternate or sub-opposite, rather conferted, horizontally patent, with the lower ones deflexed, numerous, 30-35 on each side of the rachis, glabrous above, fibrillose beneath, clad at the point of insertion beneath with a few ovate, acuminate, ciliate scales, 2-2.5 cm long, 1 cm broad at base, oblong-lanceolate, pinnate (except the terminal ones which are only aristate-crenate, or -serrate) below the middle, pinnatifid towards apex, strongly auricled on the anterior side; i. e., the basal pinnule on the upper side of the costa much the largest with its inner side cut parallel to the rachis with aristato-acute apex; pinnules 5-7, conferted, basal ones free, sessile, upper ones adnate, rhomboidal-ovate, entire or crenate, rounded and aristate at apex; *texture* thick coriaceous, shining green above, pale below; *veins* inconspicous, lateral veinlets in the pinnules 3-5-jugate, free, mostly once-forked; *sori* small, uniseriate, 2-3 on each side of the costule, medial, dorsal on the veinlets, indusium small, coriaceous, persistent.

Distribution: Guangxi.

Plate 13. Fig. 1. Habit sketch(natural size). 2. A pinna (× 2). 3. A pinnule from the base of pinna, showing venation and sori (× 12). 4. Scales from stipe (× 24). 5. Scale from rachis (× 24).

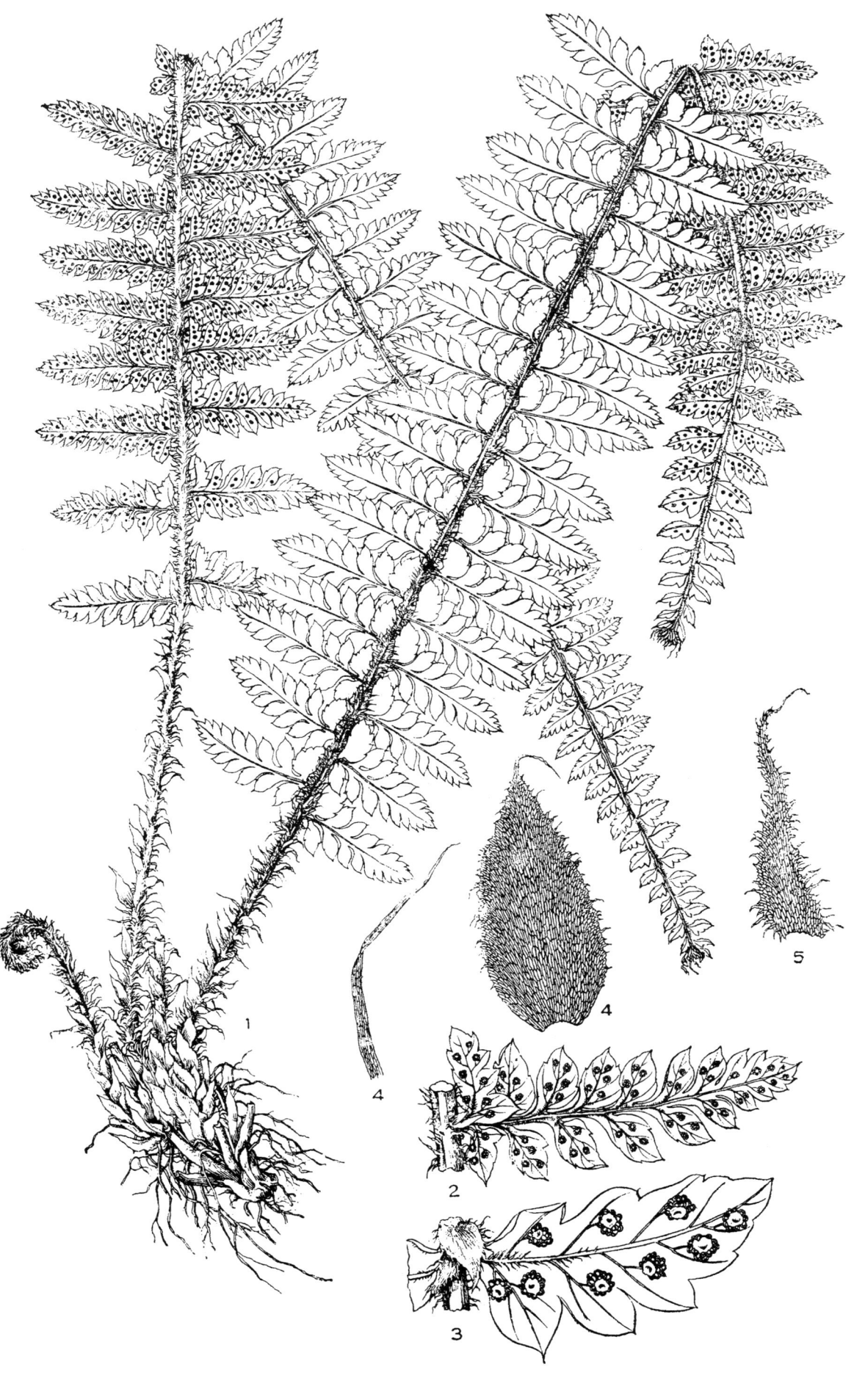

POLYSTICHUM CHUNII Ching
陈氏耳叶蕨

根状茎斜生或直立，粗如人指，黑色，被有叶柄之残余；叶 3—6 个簇生；叶柄细硬，稻秆色，基部微黑，长 10—15 cm 或 18 cm，基部被有卵圆披针形暗褐色全缘尖锐之鳞片，鳞片是 2/3 cm；其在叶体基部下之部分被有黑色刺毛，成熟之植物则几无毛；叶体单简，戟状三角形，基部深心脏形，具有浅或紧贴之凹处，两侧有宽耳，顶部短渐尖头，长阔各 8 cm，全缘；叶体硬革质，上面淡绿色，下面尤淡，不透明；叶脉显明，密生，扇状羽状，甚倾斜，联络成大形长网孔，包有 1—2 伸出至孢子囊群之支脉；孢子囊群甚多，褐色，不规则疏散于小叶下面全部，阔 1.5 mm，几圆形；子囊群盖盾状，较子囊群为大，硬，灰色，边缘有小齿，中部突起黑色，永存或久而脱落。

分布：贵州。

此种特殊之蕨与此属中各种皆异者在其简单全缘阔戟状心脏形之叶，外形与 *Phyllitis cardiophylla* (Hance) Ching 相似，最初为 Père J. Cavalerie 于 1908 年在贵州采得，后经 Père I. Esquirol 在原地采得。此种之采得可将此属发达由单叶至一次羽状复叶终至二次羽状复叶之程序完全表明。

图注：1. 本种全形（自然大）；2. 叶体之一部示叶脉与孢子囊群（自然大）；3. 同上（放大 3 倍）；4. 子囊群盖（放大 104 倍）；5. 二孢子囊（放大 160 倍）；6. 根状茎与叶柄基部上之鳞片（放大 5 倍）。

CYRTOMIUM HEMIONITIS Christ

CRYOTMIUM HEMIONITIS Christ in Bull. Geogr. Bot. Mans. (1900) 138; Chr. Ind. Suppl. 101 (1906-13).

Rhizome oblique or erect, thick as man's finger, black, covered with the vestiges of persistent stipes; leaves fasciculate (3-6 together); *stipes* slender, but rigid, stramineous, blackish at the base, 1.5 mm thick, 10-15-18 cm long, clothed near the base with ovate-lanceolate dark brown entire subulate scales 2/3 cm long, the parts near the base of frond clothed with black setae, the matured plant almost glabrous; *frond* simple, hastato-triangular, base deeply cordate with narrow or close sinus, broadly auriculate, on each side, shortly acuminate or acute at the apex, 8 cm long and broad, margin perfectly entire; *texture* rigid, coriaceous, color light green above, paler and opaque below; *veins* conspicuous, close, flabellato-pinnate, very oblique, anastomosing into elongate large areolae with 1-2 excurrent included soriferous veinlets; *sori* numerous, brown, sparsely and irregularly disposed all over the under surface, 1.5 mm broad, subrounded, indusium peltate, larger than sorus, turgid, gray, with denticulate margin and raised black centre, persistent or falling off at last.

Distribution: Guizhou.

This interesting fern differs from all known species in the genus by its simple entire, broadly hastato-cordate leaves, resembling our *Phyllitis cardiophylla* (Hance) Ching in outline. It was first discovered in Pinfa, Guizhou, by Père J. Cavalerie in 1908 and later in 1911 by Père J. Esquirol in the same locality. The addition of this species to the genus *Cyrtomium* has practically completed the uninterrupted line of development of all the types represented in the genus from the very simple form to pinnate and more ample bipinnate forms.

Plate 14. Fig. 1. Habit sketch (natural size). 2. A portion of the frond, showing venation and sori (natural size). 3. The same ($\times$ 3). 4. indusium ($\times$ 104). 5. Two sporangia ($\times$ 160). 6. Scales from rhizome and base of stipe ($\times$ 5).

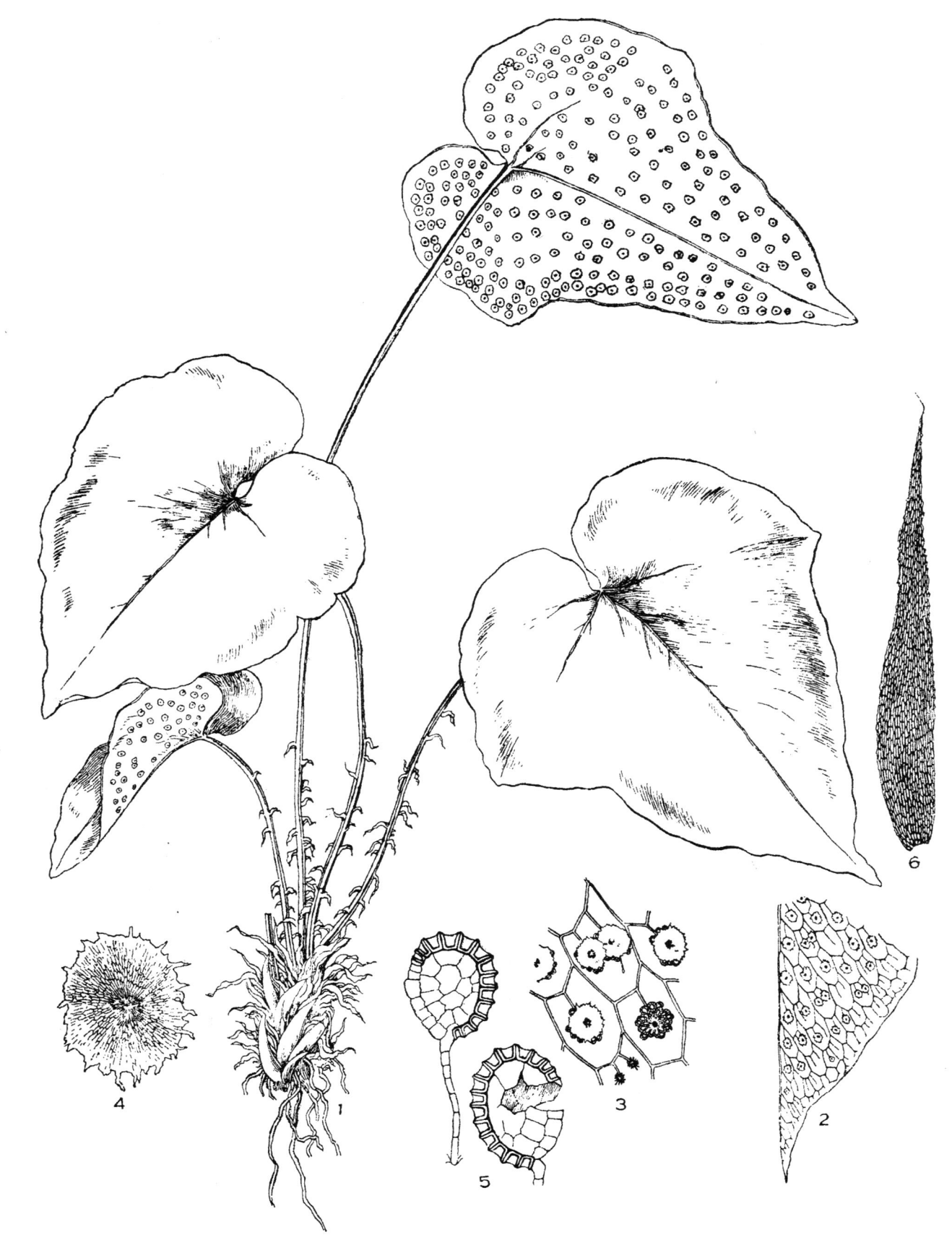

CYRTOMIUM HEMIONITIS Christ
单叶贯众

根状茎短而壮；叶柄通常簇生，长 3—9 cm，弯生或直立，密被赤褐色卵圆形透明长 3 mm 之鳞片，上部几光滑，下面圆筒形，上面有槽；叶体微俯，线披针形，长 5—28 cm，阔 3—4 cm，一次羽状，小叶在每侧 4—25 枚，掩覆密集，下部者略小，向下弯，几相称，卵圆形，基部显著心脏形，顶端甚钝头或圆形，有短柄，全缘，不透明，褐绿色，厚革质，光滑，上面平滑，下面皱折，边缘微卷，叶脉不显明，密集，联络成一行大形倾斜六角之网孔各有一内伸至子囊群之支脉，在边缘处游离具有加厚之顶端；孢子囊群成 1 列或在基部成 2 列，每侧 5—6 个，生于中部，阔 1.5 mm，圆形，子囊群盖小形，盾状，易脱落。

分布：贵州。

一稀有之种独生于贵州，为 Père J. Esquirol 在 1898 年所发现，其与他种异者为其叶体微俯之性质与其厚革质小卵圆心脏形全缘之小叶。

图注：1. 本种全形（自然大）；2. 小叶表示叶脉与孢子囊群（放大 3 倍）；3. 脱落之子囊群与子囊群盖（放大 52 倍）；4. 叶柄上之鳞片（放大 8 倍）。

CYRTOMIUM NEPHROLEPIOIDES（Christ）Ching

CYRTOMIUM NEPHROLEPIOIDES (Christ) Ching, comb. nov.

Polystichum nephrolepioides Christ in Bull. Geogr, Bot. Mans (1902) 258, c. fig; C. Chr. Ind. Ind. 582 (1906); Matthew in Journ. Linn. Soc. **39**: 387 (1911).

Rhizome short, strong; *stipes* mostly fasciculate, 3-9 cm long, curved or erect, densely clothed with reddish-brown, ovate diaphanous scales 3 mm long, subglabrous upward, terete below, sulcate above; *frond* nutant, linear-lanceolate, 5-28 cm long, 3-4 cm broad, simply pinnate under the terminal pinna; pinnae 4-25 on each side, imbricately conferted, the lower ones scarcely reduced, deflexed, subequal, ovate, strongly cordate at the base, very obtuse or rounded at the apex, shortly petiolate, entire, opaque, brownish-green, thick coriaceous, glabrous smooth above, corrugated beneath, margin narrowly reflexed; *viens* hidden, close, anastomosing in a row of large oblique hexagonal areolae each with 1 included soriferous veinlet, free towards margin with thickened apex; *sori* uniseriate or 2 at the base, 5-6 on each side, medial, 1.5 mm broad rounded, indusium small, peltate, deciduous.

Distribution: Guizhou.

A rare little fern endemic in the province of Guizhou, and its discovery is indebted to Père J. Esquirol, 1898. It differs from all its relatives by nodding character of its frond with rather small ovate-cordate entire pinnae of a thick coriaceous texture. Its anastomosed veins unmistakably indicate its proper genus; and a correction regarding venation in Christ's diagnosis must be made. —R. C. C.

Plate 15. Fig. 1. Habit sketch (natural size). 2. A pinna, showing venation and sori ($\times$ 3). 3. A detached sorus with indusium ($\times$ 52). 4. Scales from stipe ($\times$ 8).

CYRTOMIUM NEPHROLEPIOIDES (Christ) Ching
低头贯众

图版 16 拟贯众

根状茎短而直立,与叶柄同被有大形暗褐色披针形膜质之鳞片,鳞片长约 1 cm,阔约 3 mm,边缘完全成薄膜状,叶柄长 5—10 cm,暗稻秆色;叶体长 20—30 cm,阔 4 或 5—10 cm,长披针形,下部较窄,一次羽状,顶端小叶短,深羽状分裂,叶轴被有坚硬开张尖披针形淡褐色之鳞片;小叶每侧 16—20 个,具短柄或几无柄,相距甚远,几对生或互生,不等称,开张,下部者颇缩小,上部者具一有骤弯之顶之三角形耳,下部削去至中肋,长至 3 cm,阔 1.2 cm,全缘或圆波齿状;叶体薄草质;叶脉甚倾斜,密集,联络极复杂,每网孔中有一或二内伸至子囊群之小脉;孢子囊群小,多数,中肋每侧各三列,子囊群盖薄,盾状或肾状。

分布:四川,湖北,贵州,云南。

此种最初为 Père Bodinier 在贵州发现,与 *Cytomium falcatum* Presl var. *polypterum* (Diels) 相近,与之异者在长披针形几膜质之叶体与多数较小波齿状之小叶。

图注:1.本种全形(自然大);2.一小叶表示叶脉与子囊群(放大 2 倍);3.叶柄上之鳞片(放大 5 倍);4.叶轴上之鳞片(放大 5 倍);5.子囊群与子囊群盖(放大 46 倍)。

CYRTOMIUM LONCHITOIDES Christ

CYRTOMIUM LONCHITOIDES Christ in Bull. Geogr. Bot. Mans (1902) 264; C. Chr. Ind. Suppl. 101 (1906-13).

Polystichum lonchitoides Diels in Nat. Pfl. Fam. I. 4, 195; C. Chr. Ind. 585 (1906).

Aspidium lonchitoides Christ in Bull. Boiss. **7**: 16 (1899).

Rhizome short, erect, covered with large dark brown lanceolate, membranaceous scales and so is the stipe, scales about 1 cm long, 3 mm broad, with entire or scarious margin; *stipes* 5-10 cm long, dark stramineous; *frond* 20-30 cm long, 4-5 or 6 cm broad, elongate lanceolate, narrowed downward, pinnate under the deeply pinnatifid short apex; rachis covered with stiff spreading linear-subulate light brown scales; pinnae 16-20 on each side, shortly petiolate or subsessils, remote, subopposite or alternate, unequal, patent, the lower ones much reduced, triangular-auriculate with a shard deflexed apex on the upper side, the lower side cut away to the costa, to 3 cm long, 1.2 cm broad, entire or more or less sinuate-crenulate; *texture* thin herbaceous; *vein* very oblique, dense, copiously anastomosing usually with one rarely 2 excurrent included soriferous veinlets in each areola; *sori* very small, numerous, in 3 rows on each side of the costa, indusium very thin, peltate or reniform.

Distribution: Sichuan, Hubei, Guizhou, Yunnan.

This species was first discovered by Père Bodinier in Guizhou. It is most closely related to *Cyrtomium fortunei* J. Sm. var. *polypterum* Diels, differs in elongate lanceolate frond of almost membranaceons texture and numerous smaller undulatecrenate pinnae.

Plate 16. Fig. 1. Habit sketch (natural size). 2. A pinna, showing venation and sori ($\times$ 2). 3. A Scales from stipe ($\times$ 5). 4. Scales rachis ($\times$ 5). 5. A detached sorus with indusium ($\times$ 46).

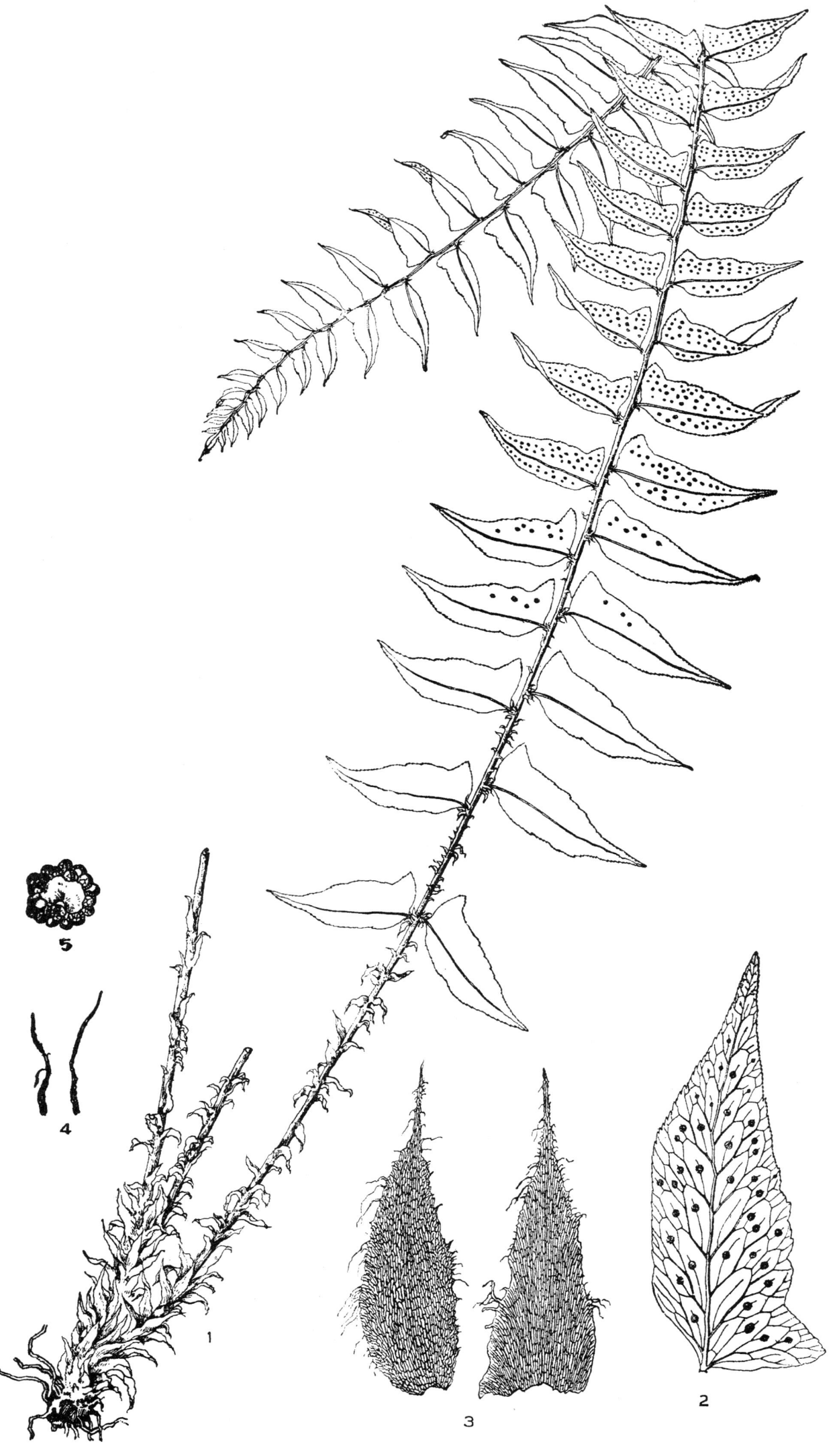

CYRTOMIUM LONCHITOIDES Christ
拟贯众

图版 17　槐叶贯众

根状茎短而粗壮,斜生,鳞片密生,阔形,渐尖头,膨大,赤褐色。叶柄成 2—6 簇,长 10—25 cm,基部暗褐色被有多数小形尖披针形暗褐色脱落之鳞片,上部则鳞片渐少而作淡灰色,叶体一次羽状,长卵圆形,长至 20 cm,阔 8 cm,基部略缩小,顶部有一全缘或羽状分裂之小叶;叶轴灰色,微被鳞片或光滑;小叶少,每侧通常 5—10 个,直立开张,相距甚远,互生,披针形,渐尖头,基部楔形,缩小成短叶柄,几全缘,近顶处则具波齿;叶体厚革质,光滑,上面暗光绿色,下面色甚淡而微皱折,光滑,边缘微卷;叶脉不显明,多而密,联络成 1 行狭长网孔几达于叶缘,无内伸游离之支脉,或不完全羽状,或一群中二三次分叉,孢子囊群直径 2—3 mm,圆形,黑褐色,成 1 列,生于网孔中之短脉上,每侧有 18—30 个,生于中肋与边缘之间。

分布:云南,贵州,四川,广西。

此特殊种先为 A. Henry 在云南发现,后在 1889 年 Père Bodinier 在贵州发现,最后 Wilson 在四川发现。在 1928 年作者在广西西北部发现,在该地此种甚为普通,生于石灰岩之裂缝中,此类蕨之外形颇似 C. falcatum Presl,其异点在黑褐色之一列孢子囊群,厚革质光绿色之叶与小叶楔形之基部,——仁昌。

图注:1. 本种全形(自然大);2. 一次小叶表示叶脉与孢子囊群(放大 2 倍);3. 脱离之子囊群盖(放大 52 倍);4. 叶柄基部之鳞片(放大 8 倍)。

CYRTOMIUM FRAXINELLUM Christ

CYRTOMIUM FRAXINELLUM Christ in Bull. Geogr. Bot. Mans (1902) 264; C. Chr. Ind. Suppl. 101 (1906-13).

Polystichum fraxinellum Diels in Nat. Pfl. Fam. I. 4, 494; C. Chr. Ind. 582 (1906); Matthew in Journ. Linn. Soc. **39**: 386 (1911).

Aspidium fraxinellum Christ in Bull. Boiss. **7**: 15 (1899).

Rhizome short, thick, oblique, strong, scales dense, broad, acuminate, inflated, reddish-brown; *stipes* 2-6 clustered, 10-25 cm long, dark brown at the base and covered with numerous small subulate-lanceolate dark brown deciduous scales, growing pale gray and sparse upward; *frond* simply pinnate, oval-oblong, to 20 cm long, 8 cm broad, scarcely reduced at the base, terminated at the apex sometimes by an entire of sometimes pinnatifid pinna; rachis gray, sparsely scaly or glabrous; pinnae few, usually 5-10 on each side, erect-patent, remote, alternate, lanceolate, acuminate, cuneate and attenuate towards the base on short petiole, the edge almost entire but distinctly crenate towards the apex; *texture* thick coriaceous, smooth, dark glossy green above very pale and somewhat corrugated (upon drying) below, glabrous, margin slightly reflexed; *veins* not prominent, numerous, close, anastomosing in a row of narrow areolae almost reaching the margin without included free veinlets or often imperfectly pinnate or rather 2-3-bifurcate in each group; *sori* 2-3 mm across, rounded, blackishbrown, uniseriate, terminating a short lateral veinlet in the areola, about 18-30 on each side, placed midway between the costa and margin.

Distribution: Yunnan, Guizhou, Sichuan and Guangxi.

This distinct species was first collected by A. Henry in Yunnan, on wooded clif, later (1889) by Père Bodinier in Guizhou, and still later by Wilson in Sichuan. It was of very late (1928) collected in N. W. Guangxi, where it is a fairly common fern, inhabiting the crevices and nitches in limestone cliff. In habit this fern is closely allied to some from of *C. fortunei* J. Sm., differs in blackish-brown uniseriate sori, thick leathery glossy green leaves and cuneate base of the pinnae.

Plate 17. Fig. 1. Habit sketch (natural size). 2. A pinna, showing venation and sori ($\times$ 2). 3. A detatched indusium ($\times$ 52). 4. Scales from the base of stipe ($\times$ 8).

CYRTOMIUM FRAXINELLUM Christ
槐叶贯众

根状茎匍匐甚广,粗 3—4 mm,密被硬毛,生根甚少;叶柄成一列,散生,相距 1.5—2 cm,壮而硬基部密被不整齐硬毛,向上则毛稀薄,有光泽,栗褐色,有槽,长 32—40 cm,粗与根状茎等或近基部者较粗;叶体阔长圆披针形,近基部处不短缩,长 35—42 cm,阔 10—12 cm,顶部渐尖头,三次羽状分裂或至下部小叶基部处成三次羽状;一次小叶具短柄,(叶柄长 3—4 mm)每侧 14—17 个,顶上具一深羽状分裂渐尖头之顶叶,下部小叶相距较上部者为远(相距 5—6 cm),直立开张,在中部密集下部小叶长 10—12 cm,阔 4—4.5 cm,长圆披针形,近基部微窄,羽状;二次小叶在一短深羽状分裂渐尖头之顶下每侧 14—17 个,密集,具短柄,长圆披针形,基部甚不等称,长 2—2.5 cm,阔 7—10 cm,羽状分裂几至中肋,裂片长圆,具圆顶,略具波齿状,裂片栉状分裂,在前面生基者最大,其内面边缘直,与小叶叶轴平行,长 5—7 mm,阔 3—4 mm,在锐尖头锯齿状之顶下每侧有 7 个,最小裂片上之叶脉显明,游离,羽状,成 4—5 对,每叶脉再分叉一次;孢子囊群暗褐色,密生于最小裂叶近叶缘处支脉之顶上,在基部裂片上有 5—6 个,在上部裂片上较少,在顶上裂片上 1 个;子囊群盖褐色,永存,几圆形;叶体厚革质,淡绿色,上面循中肋微有毛,在叶轴与小叶上面中肋上则密被硬毛,在下面则殊稀。

分布:广西。

图注:1. 本种全形(自然大);2. 小叶基部之二次小叶(放大 2 倍);3. 二次小叶基部之裂片表示叶脉与子囊群(放大 6 倍);4. 根状茎上之毛(放大 48 倍)。

MICROLEPIA CHRYSOCARPA Ching

MICROLEPIA CHRYSOCARPA Ching in Sinensia **1**: 3 (1929).

Rhizome wide-creeping, 3-4 mm thick, densely hirsute, sparsely rooted; *stipes* uniseriate, scattered, 1.5-2 cm apart, strong, rigid, densely shaggy hirsute near the base, sparsely hirtellous upward, light lustrous, castaneous, sulcate, 32-40 cm long, as thick as or slightly thicker near the base than the rhizome; *frond* broadly oblong-lanceolate, not abbreviate towards base, 35-42 cm long, 10-12 cm broad, gradually acuminate towards apex, tripinnatifid or almost tripinnate towards the base of the lower pinnae; pinnae shortly petiolate, (petiole 3-4 mm long), 14-17 on each side below a deeply pinnatifid acuminate apex, the lower ones much remoter than the upper (5-6 cm apart), erect-patent, confertad above the middle, the lower ones 10-12 cm long, 4-4.5 cm broad, oblong-lanceolate, slightly narrowed toward base, pinnate; pinnules 14-17 on each side below a short deeply pinnatifid acuminate apex, conferted, shortly petiolulate, oblong-lanceolate, strongly unequal at base, 2-2.5 cm long, 7-10 cm broad pinnatifid almost down to the costa into oblong, rounded, more or less crenate lobes; lobes pectinate, the basal ones on the anterior side much the largest with straight inner margin parallel to the rachis of pinnae, 5-7 mm long, 3-4 mm broad, about 7 on each side under the acute serrate apex; *veins* in the ultimate lobes distinct, free, pinnate, 4-5-jugate, each again once-forked; *sori* dark brown, dense, submarginal, ferminating the veinlets in the ultimate lobes, 5-6 on the basal lobes, fewer on the upper ones, unisorous on the uppermost lobes. indusium brown, persistent, almost rounded; *texture* thick, coriaceous, light green, sparsely hairy along costa above, densely hirtellous on the rachis, and upper side of the costa of the pinnae, much less so beneath.

Distribution: Guangxi.

Plate 18. Fig. 1. Habit sketch (natural size). 2. A pinnule from the basal part of pinna ($\times$ 2). 3. A segment from the basal part of the pinnule, showing venation and sori ($\times$ 6). 4. Hair from rhizome ($\times$ 48).

MICROLEPIA CHRYSOCARPA Ching
黄胞鳞蕨

根状茎匍匐颇广,斜向上伸,粗约 1.5—2 mm,薄被有小形紧贴针状锈色鳞片;叶柄几丛生,细瘦,长 15—25 cm,上面有槽,平滑,暗褐色,叶轴亦如之,稀基部微被鳞片,此外全部皆光滑;叶体长圆三角形,长 11—14 cm,基部阔约 7 cm,二次羽状,顶端有一羽状之顶;基部一次小叶最大,长 5 cm,阔 2 cm,羽状,上部者渐小渐简单,几无柄,互生,直立开张,下部 4—6 对,羽状长圆披针形,有一渐尖头羽状分裂之顶,二次叶轴下面栗色;二次小叶几无柄,斜菱形,基部者常作几圆形,楔形,外面上部分裂为数短阔截形生孢子囊群之裂片,下部与内面全缘,长 1—1.2 cm,阔 5 mm,互生,在基部一次小叶上有 5—6 对;顶端一次小叶之形状大小与侧生者相若;叶体薄草质,上面淡绿色,下面较淡,殊透明,中脉唯在较大之次小叶上显明;叶脉完全游离,扇状,半透明,一至二次分叉;孢子囊群在每二次小叶 5—7 个,短不相连,生近边缘处,横线长圆形,通常连接二叶脉之顶部,有时亦生于单脉之顶部;子囊盖膜质,窄,灰色。

分布:广西。

图注:1. 本种全形(自然大);2. 二次小叶表示叶脉与孢子囊(放大 4 倍);3. 根状茎上之鳞片(放大 16 倍)。

LINDSAYA CHIENII Ching

LINDSAYA CHIENII Ching in Sinensia **1**: 4 (1929)

Rhizome moderately wide-creeping, oblique-ascending, 1.5-2 mm thick, thinly covered by small adpressed subulate ferruginous scales; *stipes* subcaespitose, slender, 15-25 cm long, grooved above, polished, dark chestnut brown, and so is the rachis, rarely sparsely scaly near the base, otherwise glabrous in all parts; *fronds* oblong-deltoid, 11-14 cm long, about 7 cm broad at base, bipinnate, with a short simple pinnate apex; the basal pinnae much the largest, 5 cm long, 2 cm broad, pinnate, the upper ones gradually shortend and less compound, subsessile, alternate, erect-patent, the lower 4-6 pairs pinnate, oblong-lanceolate, with an acuminate pinnatifid apex, secondary rachis castaneous beneath; the pinnules subsessile, oblique rhomboid with basal ones often suborbicular, cuneate, cut on the outer and upper margin into few broad short truncate soriferous lobes, entire and straight on the lower and inner sides, 1-1.2 cm long, 5 mm broad, alternate, 5-6 pairs in the basal pinnae; the terminal pinnae are of same shape and size; *texture* thin herbaceous, light green above, paler below, rather pellucid, midrib distinct only in larger pinnules; *veins* all free, flabellate, transluscent, 1-2 forked; *sori* 5-7 on each pinnule, short, not confluent, submarginal, transversely linear-oblong, mostly uniting the apices of two veins, but not uncommon apical on single veins; indusium membranaceous, narrow, grayish.

Distribution: Guangxi.

Plate 19. Fig. 1. Habit sketch (natural size). 2. A pinnule, showing venation and sori (× 4). 3. Scales from rhizome (× 16).

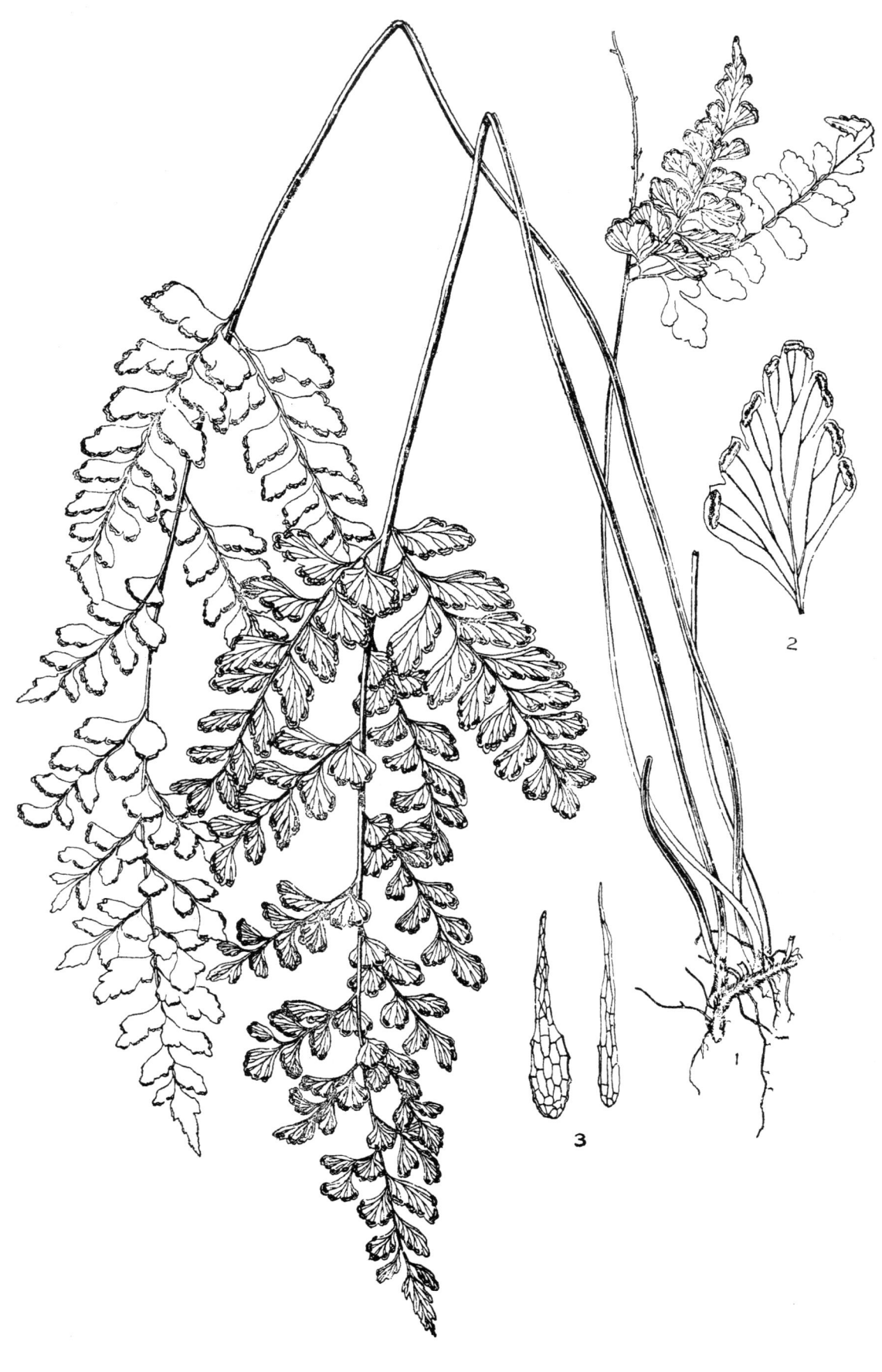

LINDSAYA CHIENII Ching
钱氏假铁线蕨

图版 20　华假铁线蕨

根状茎匍匐,粗约 3 mm,鳞片密生,开张,暗褐色,有光泽,线形,钝尖头,长 1—1.5 mm;叶柄相距 4—10 mm,细瘦,完全光滑,暗栗色,有光泽,渐上色渐淡,长至 9 cm,基部粗 1 mm,叶体三角披针形,长 10—16 cm,基部阔 55 mm,中部阔 12—18 mm,光滑,下部 1/3 二次羽状,上部 2/3 一次羽状,有时全体一次羽状,向顶部逐渐减小;一次小叶开张,有柄,互生,下部 3—6 对长圆披针形,长 2—3 cm,阔 6—10 mm,羽状,上有一羽状分裂钝尖头之顶;二次小叶 5—8 对,几等称三角形,有叶柄,基部楔形,自外缘深裂至 1/3 成为 2—4 个短阔裂片,裂片有钝而有缺刻截形之顶端,上部者常全缘,长 4 mm,阔如之;上部一次小叶有 22—30 对,肾状形,至扇状楔形,长 1 cm,阔如之,循圆形外缘分裂为 4—6 个长圆形之裂片,裂片有时再二裂,与下部一次小叶之二次小叶同形而较大;叶轴细瘦,圆筒形,上面微有槽,淡绿色,下部则暗稻秆褐色,光滑;叶脉细,游离,颇明显,扇状,通常一次分叉或简单,每裂片有 1—2 条;叶体透明草质,薄而硬;孢子囊群圆形或横长圆形,每裂片一个,生 1—2 叶脉之顶上;子囊盖膜质,淡绿色,几全线或二裂,距裂片边缘颇远。

分布:广西。

图注:1.本种全形(自然大);2.叶体上部之一小叶(放大 3 倍);3.根状茎上之鳞片(放大 24 倍)。

LINDSAYA CHINENSIS Ching

LINDSAYA CHINENSIS Ching in Sinensia **1**: 5 (1929).

Rhizome creeping, about 3 mm thick, scales dense, spreading, dark shining brown, linear with rather, a blunt apex, 1-1.5 mm long; *stipes* approximate, 4-10 mm apart, slender, glabrous throughout, dark shining chestnut brown and become lighter upward, to 9 cm long, and 1 mm thick at base; *fronds* deltoid-lanceolate, 10-16 cm long, about 5.5 cm broad at base, 1.2-1.8 cm broad in the middle, glabrous, bipinnate in the lower one-third, the remaining upper two-thirds simply pinnate, or sometimes the whole frond simply pinnate, gradually diminishing towards apex; pinnae spreading, petiolate, alternate, the lower 3-6 pairs oblong-lanceolate, 2-3 cm long, 6-10 mm broad, pinnate under a pinnatifid rather blunt apex; pinnules 5-8 pairs, almost equilateral triangular, petiolulate, cuneate at base, and cut on the rounded outer margin about way down into 2-4 short broad more or less dilated segments with a blunt or more or less erosed truncate appex, the upper ones often subentire, to 4 mm long, about as broad; the upper pinnae numerous (about 22-30 pairs), reniform to flabellato-cuneate, to 1 cm long and as broad, variously cut along the rounded outer edge into 4-6 oblong more or less dilated segments which are often again bifid and of the same shape as, but larger than, those in the pinnules of lower pinnae; rachis slender, terete, more or less channelled above, light green, except the lower part which is dark stramineous-brown, glabrous; *veins* slender, free, rather distinct, flabellate, mostly once-forked or simple, 1-2 (never 3) to each ultimate segment; *texture* pellucido-herbaceous, thin, but rigid; *sori* orbicular to transverse-oblong, one to each segment, terminating 1-2 veinlets, indusium membranaceous, pale green, margin subentire or bifid, falling quite a way short of the margin.

Distribution: Guangxi.

Plate 20. Fig. 1. Habit sketch (natural size). 2. A pinna from the upper part of frond ($\times$ 3). 3. Scales from rhizome ($\times$ 24).

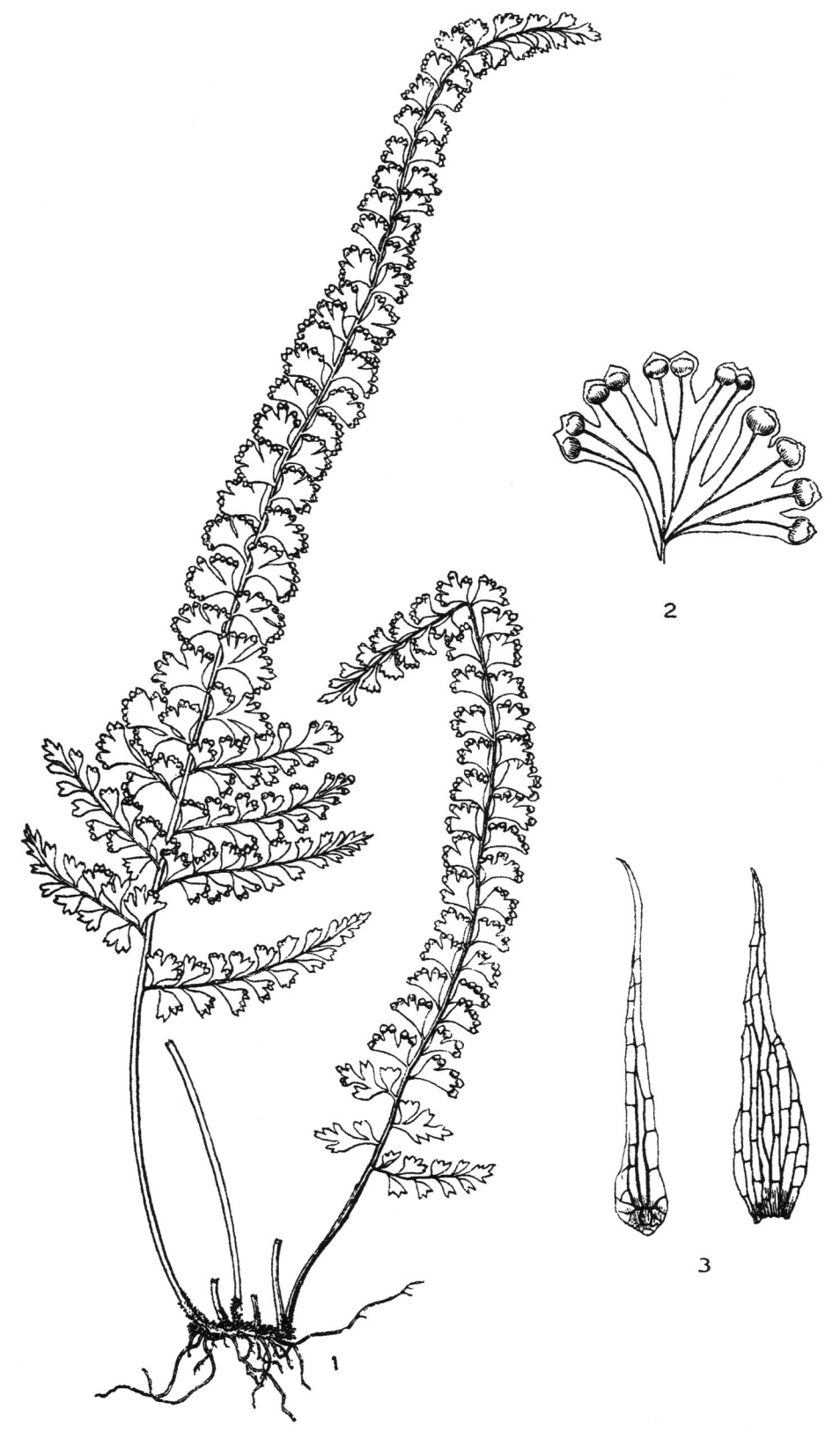

LINDSAYA CHINENSIS Ching
华假铁线蕨

根状茎粗短斜生；叶柄细瘦，丛生，3—4 根集合，基部较粗，暗褐色或淡黑色，密被褐色鳞片，上部作稻秆色，长 25 cm，其他部分光滑；叶体长至 28 cm，阔至 8 cm，向基部较阔，向顶部逐渐尖窄，长圆披针形，二次羽状分裂或近基部处微作二次羽状；一次小叶每侧约 17 个，互生，有柄，下部者横张，相距殊远，最下者最大，长圆三角形，极不等称，基部下面截形或楔形，上面有耳，顶部急尖头或短渐尖头，最大一次小叶长 5 cm，阔 2.5 cm，基部羽状，上部深羽状分裂，二次小叶或裂片斜长卵圆形，钝头，每侧约 7 枚，下部一次小叶所有之基部一对通常分离，上部者较之殊大，使一次小叶呈耳状，与叶轴相连，或掩覆其上，缺刻波齿状或深羽状分裂，其余裂片附着于有阔翼之中肋上，唯至近顶处有缺刻，上部小叶深缺刻分裂，裂片钝头，微有波齿；孢子囊群多数，大形，每裂片有 8、7、5 或 3 个，长 1.5—2 mm，通常马蹄状或卵圆形，子囊群盖永存，灰色，肾形；叶体薄草质，二面皆暗绿色。

分布：云南，湖北；菲律宾之吕宋岛亦有之。

此种蕨颇美观，其易辨别处为不等称在基部上面截形或楔形之小叶，与大形灰色马蹄状之子囊群与永存之子囊群盖，此种先为 A. Henry 在云南采得，后经 Père Ducloux 在同省采得，Tonglon 与 Copeland 复在吕宋采得。

图注：1. 本种全形（自然大）；2. 叶体基部之小叶（放大 2 倍）；3. 二次小叶示叶脉与子囊群盖（放大 5 倍）；4. 根状茎之鳞片（放大 30 倍）。

ATHYRIUM ANISOPTERUM Christ

ATHYRIUM ANISOPTERUM Christ in Bull. Boiss. **6**: 962 (1898), Bull. Geogr. Bot. Mans (1907) 133; C. Chr. Ind. 139 (1906); Matthew in Jour. Linn. Soc. **39**: 348 (1911).

Athyrium fauriei Mak. var. *elatius* Christ in Bull. Boiss **6**: 193 (1898).

Rhizome thick, short, oblique; *stipes* slender, fasciculate, but few (3-4) together, base incrassate, dark brown or blackish, clothed with moderately dense brown scales, stramineous upward, 25 cm long, glabrous in other parts; *frond* to 28 cm long, to 8 cm broad, somewhat broader towards the base, gradually attenuate towards the apex, oblong-lanceolate, bipinnatifid or sparsely bipinnate near the base; pinnae about 17 on each side, alternate, petiolate, the lower ones horizontally patent, remote, the lowest much the largest, oblong-deltoid, strongly unequal, truncate or cuneate on the under side at the base, auricled on the upper side, apex rather acute or shortly acuminate, the largest to 5 cm long, 2.5 cm broad, pinnate near the base, deeply pinnatifid above; pinnules or lobes oblique, oblong-ovate, obtuse, about 7 on each side, the basal pair in the lower pinnae usually free, the superior one much the largest, thus making the pinnae appearing auricled, contiguous to or somewhat imbricate on the rachis, inciso-crenate or deeply pinnatifid, the other lobes adnate to the broad winged costa and only deeply incised towards the apex, the upper pinnae deeply incisolobed, lobes obtuse, obscurely crenate; *sori* numerous, large, 8, 7, 5 or 3 to each lobe, 1.5-2 mm long, usually hippocrepiform or ovate, indusium persistent, gray, reniform; *texture* thin herbaceous, color obscurely green on both sides.

Distribution: Yunnan, Hubei, and also Luzon of the Philippine Islands.

A medium-sized fern of an elegant habit, easily recognized by its rather unequal pinnae, which are prominently auricled above and truncate or cuneate below and by large gray horseshoe-shaped sori with persistent indusia. It was first collected by A. Henry in Manmei, Yunnan, and later by Père Ducloux in the same province; and also recorded in Luzon, the Philippine Islands, by Tonglon and Copeland.

Plate 21. Fig. 1. Habit sketch (natural size). 2. A pinna from the basal part of frond (× 2). 3. A pinnule, showing venation and sori (× 5). 4. Scales from rhizome (× 30).

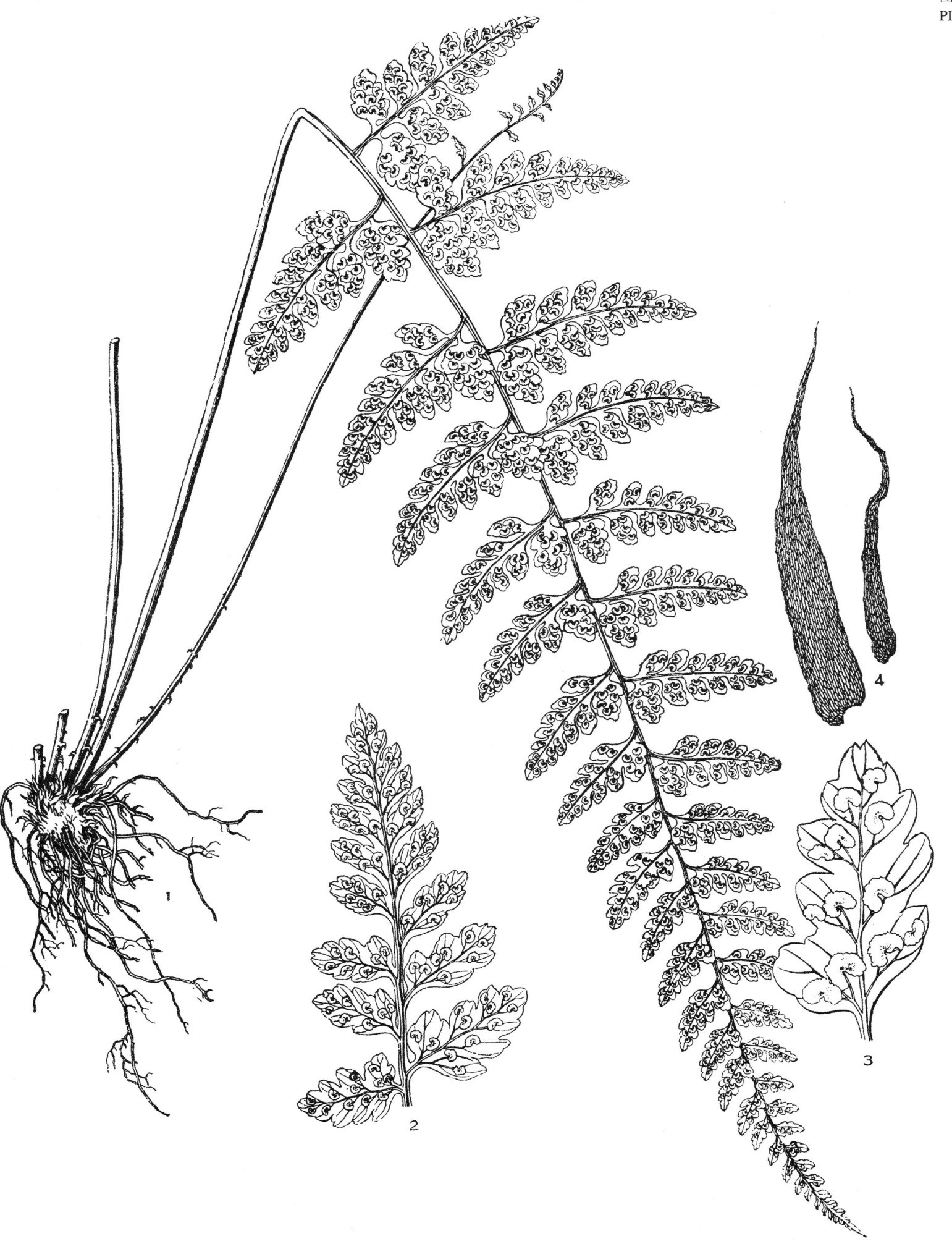

ATHYRIUM ANISOPTERUM Christ
宿蹄盖蕨

图版 22　网脉单盖蕨

　　根状茎短,粗壮,斜向上伸,密覆以坚硬略黑披针形之鳞片,叶柄长 15—23 cm,绿色,略扁,基部被以形似根状茎所有之鳞片,上部及叶轴微光滑或被以略黑色针状披针形之鳞片,叶体长 20—45 cm,宽 10—20 cm,一次羽状,顶部小叶甚大,全缘或稍有缺刻或成三裂片,两侧小叶 1—6 个,几对生或互生,下部之小叶较在次上部者略大,长 7—12 cm,宽 2—4 cm,尖端长渐尖头,基部相等或几相等,渐渐变窄而成叶柄,基部全缘或稍有缺刻;叶体几革质或纸质;叶脉略成扇状,甚斜,近叶缘处稍联络,有时连以不规则之缘内线;子囊群长 3.5—5 cm,疏散少规则,生近中肋,延至离叶缘甚远处而没。

　　分布:印度,马来半岛,在中国则云南有之。

　　本种在国内稀见,仅 A. Henry 氏在云南采得之。

　　图注:1.本种全形(自然大);2.一次小叶示其脉及子囊群(自然大);3.根状茎之鳞片(放大 30 倍)。

ASPLENIUM FINLAYSONIANUM Wallich

ASPLENIUM FINLAYSONIANUM Wall., HK. Ic. Pl. t. 937 (1854); HK. BK. Syn. Fil. 245; Christ Farnkraut. Erde. 199; Diels in Nat. Pfl. Fam. 1. 4, 239; C. Chr. Ind. Ⅲ (1906); v. A. v. R. in Malay. Ferns 476 (1908).

Rhizome short, thick, obliquely ascending, densely clothed with rigid blackish lanceolate scales; *stipes* 15-23 cm long, green, subcompressed, densely clothed near the base in the scales similar to those on the rhizome, the upper part and rachis subglabrous or clothed in blackish subulate-lanceolate scales; *fronds* 20-45 cm long, 10-20 cm broad, simply pinnate under an entire or subrepand or 3-lobed terminal pinna which is much the largest, lateral pinnae 1-6, subopposite or alternate, the lower ones somewhat larger than the ones next above, 7-12 cm long, 2-4 cm broad, the point very acuminate, the base equal or subequal, very gradually narrowed on both sides to a distinct petiole, the edge entire or subrepand; *texture* subcoriaceous or chartaceous; *veins* subflabellate, very oblique, slightly anastomosing towards the edge, sometimes bounded by an irregular intramarinal line; *sori* often 3.5-5 cm long, scattered irregularly, near the costa, but falling short far off from the margin.

　　Distribution: India, Malaya, China: Yunnan.

　　Evidently rather a rare plant in China, recorded so far only from Yunnan by A. Henry.

　　Plate 22. Fig. 1. Habit sketch (natural size). 2. A pinna, showing venation and sori (natural size). 3. Scale from rhizome (× 30).

ASPLENIUM FINLAYSONIANUM Wallich
网脉单盖蕨

图版 23　波氏双盖蕨

根茎肥壮,斜生,无鳞片,或有极小不显明之鳞片;叶柄密集,三五丛生,长 12—17 cm,基部暗黑色,向上作浅绿色,与主要叶轴俱被有粗糙坚硬之线状毛,叶体长 20—38 cm,中部阔 12—15 cm,卵长圆形,向顶部渐狭,向基部渐减缩,一次羽状,叶轴密生毛;小叶每侧 20—25 个,互生,颇密接,披针形,渐尖头,边缘波齿状,下部者无柄,中部者附生,最上者在羽状分裂具渐尖头之顶之下以阔翼相连,上半小叶横张,中部者甚向下弯,最大者长 7 cm,阔 1.5 cm,上面基部有耳,下面圆形或微削去;茎脉至叶缘皆明晰,半透明,一次或二次分叉,但在耳上则成扇状;孢子囊群线形,生于前面之支脉上,由中肋延至距叶缘处一半以上,微弯,长约 5 mm,稀类似 Diplazium 所有者;子囊群盖殊阔,永存,不生纤毛;叶体薄纸质,鲜绿色,两面皆密生软毛,尤以下面循中肋处为甚。

分布:广东,广西,台湾,槟榔屿亦有之。

此特殊蕨类最显著之性质为全部密生软毛与偶有类似 Diplazium 之孢子囊群,Baker 所以初称之为 Asplenium pullingeri 者因其记载完全本于一甚小之植物栽培于邱皇家植物园而或自 Pullinger 在 1871 年由香港寄去之根状茎产生者。朴氏所称甚短之叶柄显不真确,盖 Wilson 与 Tutcher 在香港所采者叶柄长至 30 cm,而 Matthew 在同地所采之二八〇号之为吾图所根据者,其叶柄亦长 20—17 cm 也。

图注:1. 本种全形(自然大);2. 小叶表示叶脉与孢子囊群;3. 叶轴上之鳞片(放大 48 倍)。

DIPLAZIUM PULLINGERI (Baker) J. Smith

DIPLAZIUM PULLINGERI (Baker) J. Sm. in Ferns Brit. and For. ed. II. 315 (1877); C. Chr. Ind. 238 (1906); Matthew in Jour. Linn. Soc. **39**: 358 (1911).

Asplenium pullingeri Baker in Gard. Chron. n. s. **4**: 484 (1875).

Asplenium bireme Wright in Kew Bull. (1908) 182.

Asplenium chlorophyllum Baker in Jour. Bot. (1885) 104; Bedd. in Handb. Suppl. 39 (1892).

Caudex rather thick, oblique; scales none or very small and obscure; *stipes* densely tufted (3-5 together), 12-17 cm long, sordid blackish at the base, growing greenish upward, densely clothed like the entire main rachis with shaggy rather stiff, brown glandular hairs; *frond* 20-38 cm long, 12-15 cm broad at the middle, which is the broadest part, ovate-oblong, narrowed very gradually towards the apex and reduced towards the base, simply pinnate, rachis densely pilose; pinnae 20-25 on each side, alternate, moderately close, lanceolate, acuminate, crenate, the lower ones sessile, the middle ones adnate, the uppermost ones connected by a broad wing under the pinnatifid acuminate apex, those in the upper half of the frond horizontally patent, the basal ones strongly deflexed, the largest to 7 cm long, 1.5 cm broad, distinctly auricled on the upper side at the base, rounded or slightly cut away below; *veins* distinct to the margin, transluscent, forked once or twice, but pinnate in the auricles; *sori* linear, borne on the anterior veinlets and extending from the costa to more than half way to the margin, slightly curved, about 5 mm long, rarely diplazioid, indusium moderately broad, persistent, not ciliate; *texture* thin papery, bright green, densely villose on both surfaces, and especially along the costae beneath.

Distribution: Guangdong, Guangxi, Taiwan, also Penang.

A medium-sized and very distinct fern, characterized by all parts being densely villose and rarely diplazioid sori. Baker's diagnosis for his *Asplenium pullingeri* was evidently based upon a rather smaller plant then grown in Kew Gardens perhaps from the rootstocks sent to that institution by Mr. Pullinger from Hong Kong, 1871. His statement, "its very short stipes" is certainly incorrect, as the specimens collected in Taimoshan, Hong Kong New Territory, where most likely be the type locality for Baker's species, by E. H. Wilson and W. T. Tutcher, upon which Wright based the description for his *Asplenium bireme*, has stipes to 30 cm long, and Matthew's plant, No. 280 (February 4th. 1907), from Ma-on-shan in the same locality, upon which our present figure is based, has stipes from 12 to 17 cm long.

Plate 23. Fig. 1. Habit sketch (natural size). 2. A pinna, showing venation and sori. 3. Scales from rachis (× 48).

DIPLAZIUM PULLINGERI (Baker) J. Smith
波氏双盖蕨

图版 24　膜叶双盖蕨

根状茎短而壮,斜向上伸;叶柄密集丛生,疏被以膜质褐色尖线形有纤毛开张之鳞片,长 20—23 cm,下半部几黑色,上部稻秆色,上面有深槽;叶体长圆卵形,长 30—32 cm,中部 17—20 cm,近基部微缩小,一次羽状,叶轴稻秆色,有鳞片;小叶在一阔羽状分裂渐尖头之顶下有 10—12 对,中部者长 9—11 cm,近基部阔 1.6—1.8 cm,披针形,横张,基部者略短(8—9 cm)而与之等阔微向下弯,互生,殊密集,下部者无柄,上部者附生,基部几截形或圆形,在两面俱扩张,顶部渐尖头,中部以下疏生不整齐波状齿,近顶部则显明具整齐之锯齿,中肋细瘦,侧脉半透明,显明,游离,相距 4—5 mm,成 2 对,基部者则多成 3 对,直至叶缘;叶体质薄,透明纸质,上面淡绿色,有光泽,下面较淡,孢子囊线形,长 3—5 mm,微弯,斜列,延生于每群中最前面之支脉上,生于中肋与叶缘之间,甚稀类 Diplacium 属所有者,子囊盖颇阔,永存。

分布:广西。

图注:1.本种全形(自然大);2.小叶之一部表示叶脉与孢子囊(放大 7 倍);3.根状茎上之鳞片(放大 24 倍)。

DIPLAZIUM PELLUCIDUM Ching

DIPLAZIUM PELLUCIDUM Ching in Sinensia **1**: 7 (1929).

Rhizome short, thick, obliquely erect; *stipes* densely tufted, clad rather sparsely in membranaceous brown subulate ciliate spreading scales, 20-23 cm long, blackish on the lower half, dull stramineous upward, deeply channelled above; *fronds* oblong-ovate, 30-32 cm long, 17-20 cm broad in the middle, which is much the broadest, slightly abbreviate towards base, simply pinnate, rachis stramineous, scaly; pinnae 10-12 pairs under a broad pinnatifid acuminate apex, the middle ones 9-11 cm long, 1.6-1.8 cm broad near the base, lanceolate, subhorizontally patent, the basal ones shorter (8-9 cm) and as broad, only slightly deflexed, alternate, rather approximate, the lower ones sessile, the upper ones broadly adnate, subtruncate, or rounded at base, dilated on both sides, apex acuminate, remotely and rather irregularly crenate below the middle, but distinctly and regularly serrate towards apex, *costa* slender, lateral veins transluscent, distinct, free, 4-5 mm apart, 2-jugate, except the basal pair which is mostly 3-jugate, very ascending, extending to the margin; *texture* thin, pellucido-chartaceous, lustrous, light green above, paler below (when living); *sori* linear, 3-5 mm long, slightly curved, oblique, following the anterior veinlet of the group, placed midway between costa and margin, very rarely diplazioid, indusium moderate broad, persistent.

Distribution: Guangxi.

Plate 24. Fig. 1. Habit sketch (natural size). 2. A portion of pinna, showing venation and sori ($\times$ 7). 3. Scales from rhizome ($\times$ 24).

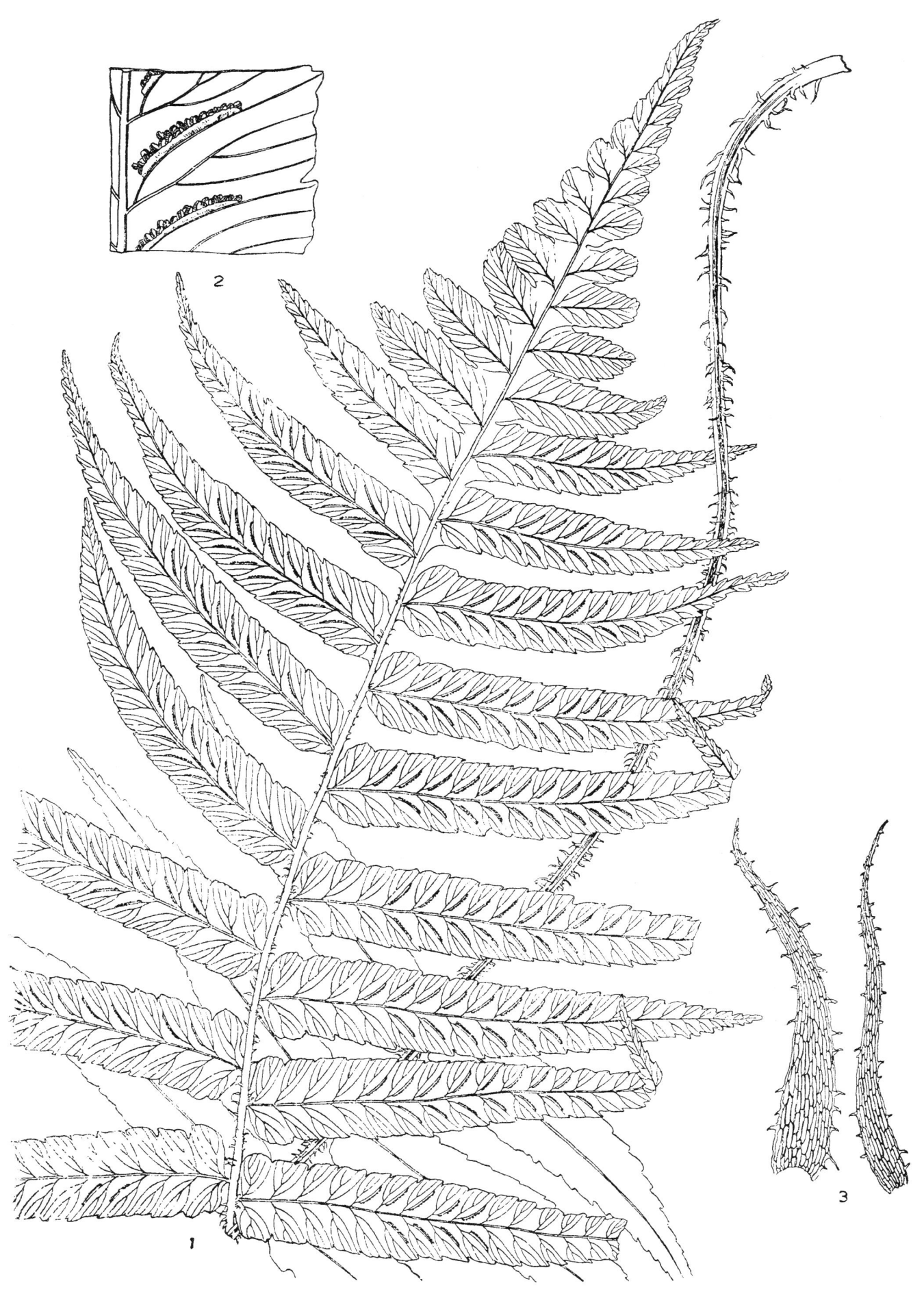

DIPLAZIUM PELLUCIDUM Ching
膜叶双盖蕨

图版 25　大叶双盖蕨

根状茎粗壮而短,木质,直立;鳞片密生,覆瓦状排列,线披针形,至尖线形,长 1.5—2 cm,顶端尖长有毛,每二裂,几黑色,有纤毛;叶柄强壮,长 50—85 cm,粗 1 cm,暗褐色,基部有粃糠状鳞片,上部几绿色,干燥后则成暗稻秆色,上面有阔槽;鳞片与根状茎上者相似,疏生于基部,上部则光滑;叶体大,长 40—60 cm,阔 20—26 cm,一次羽状,阔长圆形,基部微缩小,两面皆光滑;叶体草质,颇坚实,上面光绿色,下面色较淡;小叶在阔形浅羽状分裂渐尖头之顶下每侧 7—9 个,相距 2—3 cm,长圆披针形,有柄长至 5 mm,最上者附着,长 17—19 cm,阔 3—4 cm,基部几截形或几圆形,顶端渐尖头,边缘远裂为阔圆形短裂片;中脉甚显著,侧脉开张,相距约 7 mm,支脉 5—6 对,游离,上伸;孢子囊群密生,终至连合,长 5—7 mm,循每群之下部三四支脉而生,距边缘颇远;子囊群盖阔。

分布:广西。

图注:1.本种全形(自然大);2.小叶之一部表示叶脉与孢子囊(放大 2 倍);3.叶柄基部之鳞片(放大 24 倍)。

DIPLAZIUM MACROPHYLLUM Ching

DIPLAZIUM MACROPHYLLUM Ching in Sinensia **1**: 6 (1929).

Rhizome thick, short, woody, erect, scales dense, imbricate, linear-lanceolate to linear-subulate, 1.5-2 cm long, with long hair-pointed frizzy apices, often 2-cleft, blackish, ciliate; *stipes* strong, 50-85 cm long, about 1 cm thick, dark brown, and paleaceous near the base, greenish upward, turning dark stramineous upon drying, broadly channelled above, scales similar to the rhizomatic ones, sparse on the basal part of the stipe, glabrous upward; *fronds* ample, 40-60 cm long, 20-26 cm broad, simply pinnate, broadly oblong, slightly abbreviate towards base, glabrous on both surfaces; *texture* herbaceous, moderately firm, glossy green above, paler below (when living); pinnae 7-9 on each side below a broad and shallowly pinnatifid acuminate apex, 2-3 cm apart, oblong-lanceolate, petiolate (petiole 5 mm long), the uppermost ones adnate, 17-19 cm long, 3-4 cm broad, base subtruncate to subrotund, apex acuminate, margin remotely lobato-crenate into broad rounded short lobes; *midrib* prominent, lateral veins spreading about 7 mm apart, veinlets 5-6-jugate, free, very ascending; *sori* dense, confluent at last, oblong, 5-7 mm long, following the lower 2-4 veinlets of the group, falling considerably short from the margin, indusium broad, spurious.

Distribution: Guangxi.

Plate 25. Fig. 1. Habit sketch (natural size). 2. A portion of pinna, showing venation and sori ($\times$ 2). 3. Scale from the base of stipe ($\times$ 24).

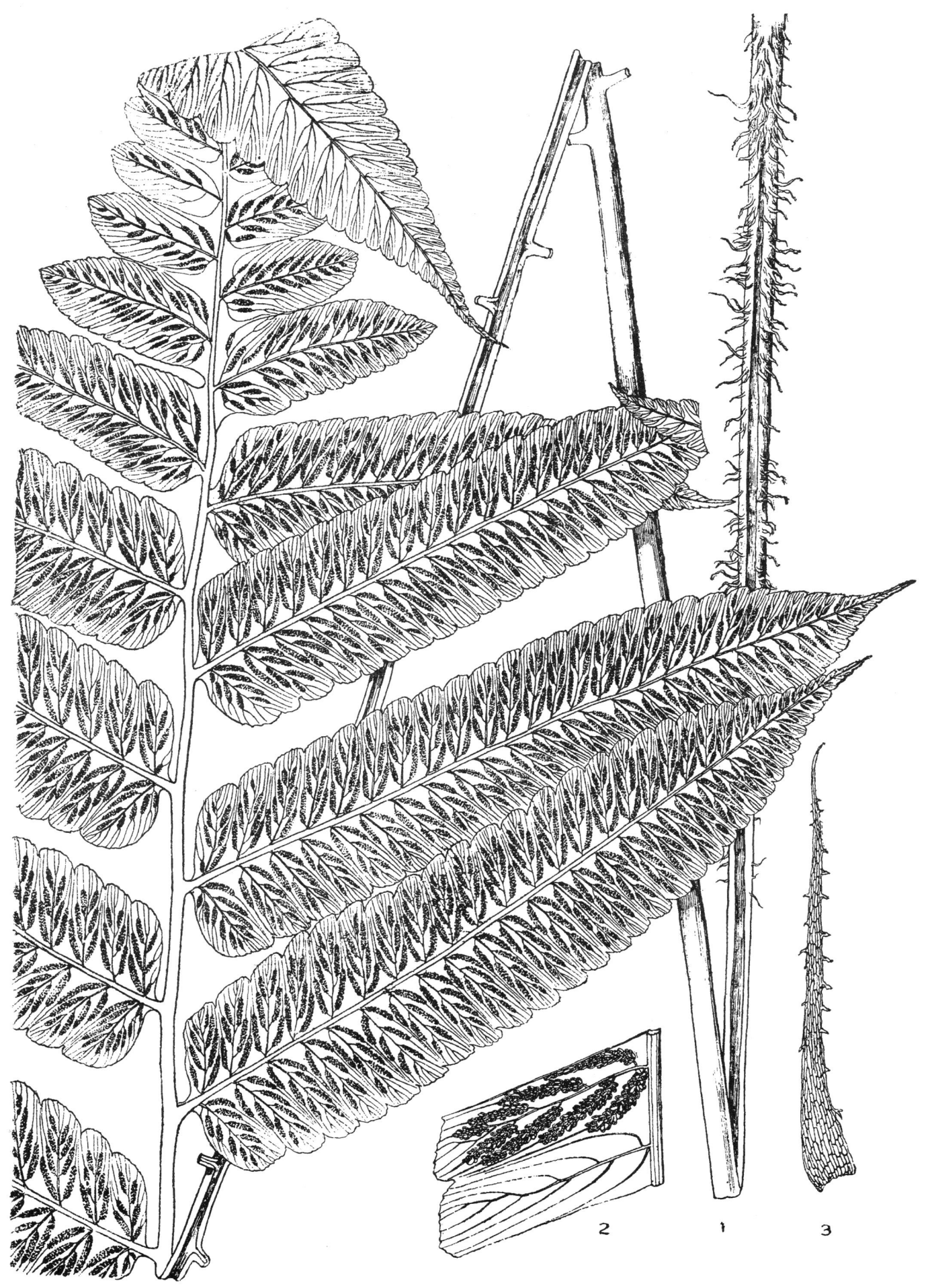

DIPLAZIUM MACROPHYLLUM Ching
大叶双盖蕨

图版 26　荷叶对开盖蕨

　　根状茎短小直立，被有黑色尖披针形鳞片；叶柄丛生，光黑色，光滑，细瘦，上面有槽，长 7—12 cm；叶体简单，全缘，圆形或几圆形，不显明波状，基部心脏形，凹处阔或殊窄，直径 3—6 cm，透明，膜质浅绿色；叶脉扇状，多次歧出，游离；孢子囊群生于第二次或第三次支脉上，长形；子囊群盖薄膜质，永存相对开。

　　分布：云南，四川，贵州；缅甸亦有之。

　　此种之特性为其圆形具深心脏形基部之单叶，形状颇似睡莲，其叶之大小与叶柄之长短变易甚大，叶体有小至 1.5 cm 直径，而有颇长之叶柄，或叶柄有长止 1 cm 而叶体颇大者；此种在中国西部颇普通，喜生有树木山峡之阴湿处。

　　图注：1. 本种全形（自然大）；2. 叶体之一部表示叶与孢子囊群（自然大）；3. 根状茎上之鳞片（放大 20 倍）。

PHYLLITIS DELAVAYI (Franch.) C. Christensen

PHYLLITIS DELAVAYI (Franch.) C. Chr. Ind. 492 (1906); Matthew in Journ. Linn. Soc. **39**: 376 (1911).

　　Scolopendrium delavayi Franch. in Bull. Soc. Bot Fr. 32: (1885); Christ in Farnkraut. Erde 213; Diels in Nat. Pfl. Fam. I. 4. 233.

　　Rhizome short, small, erect, clothed up to the collum with blackish lanceolate subulate scales; *stipes* tufted, shining ebeneous, glabrous, slender, channelled above, 7-12 cm long; *frond* simple, entire, orbicular or suborbicular, obscurely undulate, deeply cordate at the base with broad or rarely close sinus, 3-6 cm across, *texture* pellucid, light green; *veins* flabellate, repeatedly dichotomous, free; *sori* borne on the secondary or tertiary veins, elongate, not attingent at the ends, indusium thin membranaceous, persistent, opening towards each other.

　　Distribution: Yunnan, Sichuan, Guizhou, also Burma.

　　Distinct for its simple orbicular deeply cordate leaves, resembling those of our water lilies. The size of leaves and the length of stipes vary great deal, some plants having leaves as small as 1.5 cm across, provided with rather a long stalk, others having stalks not exceeding 1 cm in length but with rather large leaves. A fairly common fern in W. China, thriving best in moist shades in wooded gorges and ravines.

　　Plate 26. Fig. 1. Habit sketch (natural size). 2. A portion of frond, showing venation and sori (natural size). 3. Scales from rhizome (× 20).

PHYLLITIS DELAVAYI (Franch.) C. Christensen
荷叶对开盖蕨

根状茎管状，匍匐甚广，直径约 3—4 mm，密覆以小形光泽褐色开张之刺毛状鳞片；叶柄相距 2—5 cm，疏生，细瘦，柔韧，基部略粗，黑色有光泽，上面具三槽，中间者最宽，皆由基部直达尖端，下面隆起，长达 20 cm，有时更长，稀有基部宽达 2 mm 者，与根状茎不成关节；叶体简单深心脏卵圆形，渐尖头，基部两侧形成扁圆形有时截圆形之耳，长 10—15 cm，宽 7—10 cm，全缘，通常成不显明之浅波状，稀有成深缺刻者，基部凹处宽 5—10 mm，深 2—3 cm，通常开张，有时为基部之耳所遮掩；叶体纸质，光滑，透明，上面浅绿色，下面尤淡，中肋在下面甚显明，近顶部则不显明，下半部颜色与叶柄相同，在上面不显明，侧脉甚细，在下面几不能见，但持之向日光则立现，斜行开张，相离 1—2.5 mm，平行，三四次歧出，至边缘处常二股相合，形成一行有时两行窄线形之网孔，叶脉梢端通常略粗，简单，有时两股相交，不及叶缘而止；子囊群线形，长 2—3 cm，稀有更长者，生中肋及叶缘之间，距叶缘 1—1.5 cm，距中肋不过三五下部子囊群形似 Scolopendrium 而上部者则通常似 Asplenium 所有者，子囊群盖之形状一如子囊群，薄膜质，灰白色，全缘。

分布：海南岛原产，台湾省亦有之。

图注：1. 本种全形（自然大）；2. 叶体之下部表示叶脉与子囊样（放大 2.5 倍）；3. 孢子囊（放大 116 倍）；4. 根状茎之鳞片（放大 40 倍）。

本种初为 B. C. Henry 氏采自海南岛，由 Hance 氏定为 *Micropodium cardiophyllum*，其后 Baker 氏改定为 *Asplenium cardiophyllum*，但作者在审慎研究多数海南标本之后，确定本种当隶于 *Phyllitis* 一属，因叶体下半部之子囊群通常类似 *Scolopendrium*，逐渐向上始变为 *Asplenium* 之模式也。作者研究标本中三分之一皆具此特性，且 *Scolopendrium* 模式之子囊群皆生于叶基 2/3 之部分，或 Hance 氏检定之标本之定名为 *Micropodium cardiophyllum* 者，实本种多数仅具 *Asplenium* 模式子囊群者，故其系统关系在当时不能决定之，本种性质介乎 *Asplenium* 与多数 *Phyllitis* 之间，与后者之亚属 *Antigramme* 之性质最接近，但其异处则在匍匐甚广之根状茎与深心脏卵圆形纸质之叶体，产于日本之 *Phyllitis ikenoi* (Mak.) C. Christensen 一种亦有具两种子囊群之特性，习性亦同，或当归纳于本种也——仁昌

PHYLLITIS CARDIOPHYLLA (Hance) Ching

PHYLLITIS CARDIOPHYLLA (Hance) Ching, Comb. nov.
Micropidium cardiophyllum Hance in Journ. Bot. (1883) 268.
Asplenium cardiophyllum Baker in Ann. Bot. **5**: 311 (1891); C. Chr. Ind. (1906); Matthew in Journ. Linn. Soc. **39**: 344 (1911).

Rhizome terete, wide-creeping, 3-4 mm across, clad in dense, minute shining brown, spreading setae; *stipes* 2-5 cm apart, uniseriate, slender, flexible, somewhat incrassate at the base, shining ebeneous, 3-sulcate on the upper side with the middle groove the broadest, all extending from the base up to the apex, broadly keeled on the under side, to 20 cm or more long, scarcely 2 mm broad at the base, not articulate to the rhizome; *frond* simple, deeply cordato-ovate in outline, acuminate, with broad rounded, sometimes rotundo-truncate auricle on each side at the base, 10-15 cm long, 7-10 cm broad, margin entire, usually somewhat obscurely undulate, very rarely sinuate, basal sinus 5-10 mm broad, 2-3 cm deep, usually open, sometimes close by the imbrications of the auricles; *texture* chartaceous, glabrous, pellucid, light green above, pale below, *costa* rather prominent below except towards the apex, the lower half similar to the stipe in coloration, inconspicuous above, lateral veins very fine, scarcely seen on the under surface, but very distinct when held up against light, obliquely patent, 1-2.5 mm apart, parallel, 3-4-dichotomously branched and meeting once or twice near the margin forming mostly one sometimes two rows of narrowly linear areolae between the costa and margin; with somewhat thickened mostly free or sometimes connivent apex ended submarginally; *sori* linear, 2-3 cm rarely more long, nearer to the costa than the margin, about 1-1.5 cm from the margin, 3-5 mm or farther from the costa, the lower ones often scolopendroid, the upper ones usually asplenioid, inducia of the same shape as the sori, membranaceous, grayish-white, margin entire.

Distribution: Endemic in the Island of Hainan, Taiwan.

Plate 27. Fig. 1. Habit sketch (natural size). 2. A lower portion of frond, showing venation and sori (×2.5). 3. Sporangia (× 116). 4. Scale from rhizome (× 40).

This interesting little fern was first collected in the Island of Hainan by Rev. B. C. Henry, November 20, 1882 and named by Dr. Hance as *Micropodium cardiophyllum* which was changed later to *Asplenium* by Baker and has remained as such ever since. A recent careful examination of an ample stock of materials from Hainan* has vividly convinced me of considering this species most fit as a legitimate member of *Phyllitis* by the presence of its scolopendrioid sori on the lower part of the frond, although they all gradually become asplenioid as go upward. Of about 50 plants examined by me, I found that about one-third, which are in full fruitification, exhibits the character peculiar to the genus, and, moreover, that the scolopendrioid sori seem mostly confined to the lower one-third of the frond; while those in the upper 2-3 are almost all strictly of asplenioid nature. It is most likely that the type specimens for *Micropodium cardiophyllum* Hance are plants with sori in asplenioid condition as I have seen most of the plants are in, that perhaps chiefly accounts for the uncertainty of systematic position for the present species. Besides the asplenioid character of the sori on the upper part of the frond, it is very interesting, however, to note that while some of the sori are strictly of scolopenrioid nature, i. e. they are in opposite pairs, with their indusia directly opening towards each other, there are good many others, which are separated by one or more intervenning veins of the same group and still some others which are not in connnivent pairs, opening face to face, but either opening downward or upward singly. In spite of this deviation the reason for which was quite explicitly explained by W. J. Hooker (cf Sp. Fil. IV. p. 1), it seems most fit to incoporate this plant in *Phyllitis* or probably better still to remove into a distinct genus intermediate between *Asplenium* and *Phyllitis* as the nature of its inducia indicate. In view of the presence of distinct midrib and the anastomozination of the veinlets towards the margin, the present species naturally falls in ANTIGRAMME Presl, Diels in Engl, and Prantl. Nat. Pfl. Fam. 1: 4, 232 f. 124, E., a section of about 4 or 5 species endemic in Brasil. The present species differs from its far-off relatives by wide-creeping rhizome, deeply cordate-ovate frond of pellucid chartaceous texture, etc. The Japanese *P. ikenoi* (Mak.) C. Chr. exhibits, among other characters identical to the present species, the same irregularity in the nature of sori, and might perhaps be well regarded as identical to our species. —R. C. C.

---

* I wish to tender my heartiest thanks to Prof. F. A. McClure, the College of Science, Lingnan University, Canton, for his courtesy in handing over to me for examination all the material at his disposal. —*R.C.C.*

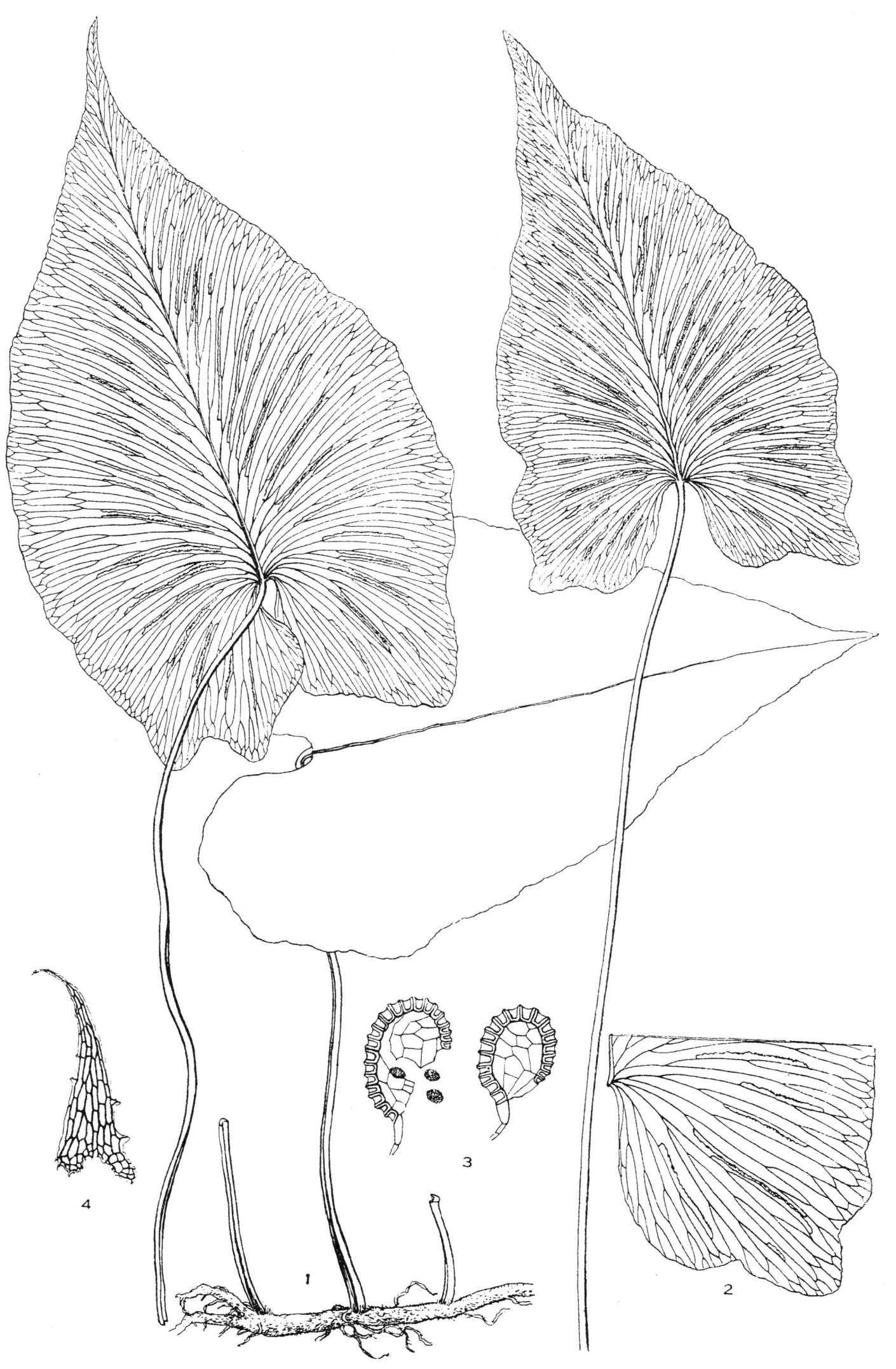

PHYLLITIS CARDIOPHYLLA (Hance) Ching
琼崖对开盖蕨

根状茎颇长,径约 4 mm,斜行,被以暗褐色披针状针形之鳞片,鳞片长 5 mm,其余部分则完全光滑,叶柄簇生,在无孢子囊叶叶柄长 48 cm,有孢子囊叶叶柄略长,坚硬,粗似鸦翎管,红稻秆色;无孢子囊叶叶体长 20—30 cm,宽 2.5—4 cm,长披针形,渐尖头,至尾部渐窄缩,一次羽状;叶轴上面有显明之槽,下面则圆柱状,小叶每侧有 30—58 个,在下部者短缩成耳状,在上部者则密凑成栉齿状,开张,其凹处窄而锐,小叶线形披针形,镰状弯曲,基部宽而贴附但不向上延展,上侧基脚扩张,长约 18—25 mm,阔 4 mm,全缘,锐尖头,叶体厚革质,下面几白色,上面淡绿色,边缘反卷,在叶体与叶轴相连处生一暗褐色易脱落之鳞片,叶脉完全不显,有孢子囊叶略长,基部颇瘦削,一次羽状,小叶多数,疏生,近基部者相距至 1 cm 左右,近顶部者则较密集,无柄,但附着部分较窄,其凹处钝圆而宽,小叶线形,形如荚,尖头,长 15 mm,阔 2.5 mm,上面扁平,下面隆起,子囊群盖永存,绿色有光泽,凸圆形,完全包覆子囊群。

分布:四川,湖北,贵州。

本种为一特殊之种类,最早在 1897 年 Père J. Bodinier 采于贵阳附近溪畔石上,同时 Rev. E. Faber 又采得于四川之峨眉山,后者经 Baker 氏定为 *Blechnum spicant* (L.) Wither,但经 Christensen 氏证明凡产自中国西部而前人定为 *B. spicant* 之种皆为 *B. eburneum* Christ.

图注:1. 植物全形之一部(自然大);2. 一有子囊叶叶体之小叶(放大 6 倍);3. 同上横切面示其子囊盖及子囊群附着之情况(放大 16 倍);4. 孢子囊及孢子(放大 75 倍);5. 根状茎之鳞片(放大 30 倍)。

BLECHNUM EBURNEUM Christ

BLECHNUM EBURNEUM Christ in Bull. Geogr. Bot. Mans (1902) 233, c. fig.; C. Chr. Ind. 153 (1906); Matthew in Journ. Linn. Soc. **39**: 351 (1911).

Rhizome rather long, 4 mm thick, oblique, clothed with dark brown, lanceolate subulate scales 5 mm long, entirely glabrous in other parts; *stipes* fasciculate, those of the sterile fronds 48 cm long, those of the fertile somewhat longer, rigid, thick as raven's quill, reddish straminous; the sterile *frond* 20-30 cm long, 2.5-4 cm broad, elongate lanceolate, acuminate, gradually attenuate towards the base, pinnate; rachis quite distinctly sulcate above, terete underneath; pinnae 30-58 on each side, the lower ones strongly reduced into auricles, the upper ones closely pectinato-patent with very narrow acute sinuses, linear-lanceolate, falcate, broadly adnate but not decurrent and dilated on the upper side at the base, 1.8-2.5 cm long, 4 mm broad, very entire and acute, *texture* toughly coriaceous, whitish below and pale green above, margin strongly reflexed, provided at the point of insertion with the rachis with a dark brown deciduous linear scale; *veins* hidden completely; fertile *frond* as long as or slightly longer than the sterile, strongly attenuate towards the base, pinnate, pinnae numerous, remote, the lower ones about 1 cm apart and much closer upward, adnate but contracted above the base with round broad sinuses, linear, pod-like, apiculate, 1.5 cm long, 2.5 mm broad, plane above, inflated below; indusium persistent, shining green, convex, covering the entire sorus.

Distribution: Sichuan, Hubei, Guizhou.

To characterize this very distinct species, it deems fit to quote Christ's states ment, "Espèce tres particuliere par son tissu durement coriace, sa couleur claire, sepinnae fertiles enflees par dès sores cylindrique a indusie persistent". This unique fern was first collected by Père J. Bodinier, Nov. 4. 1897, in the vicinity of Kweiyang, Guizhou, on rocks by the stream, and about simultaneously by Rev. E. Faber in Omei Shan, W. Sichuan, the latter specimen was identified by Baker at Kew as *B. spicant* (L.) Wither. It is most likely as Dr. Christensen remarked that all *B. spicant* reported from W. China by earlier collectors is *B. eburneum* Christ.

Plate 28. Fig. 1. Habit sketch (natural size). 2. A fertile pinna (× 6). 3. The same, a cross section, showing indusium and attachment of sporangia (× 16). 4. Two sporangia with spores (× 75). 5. Scales from rhizome (× 30).

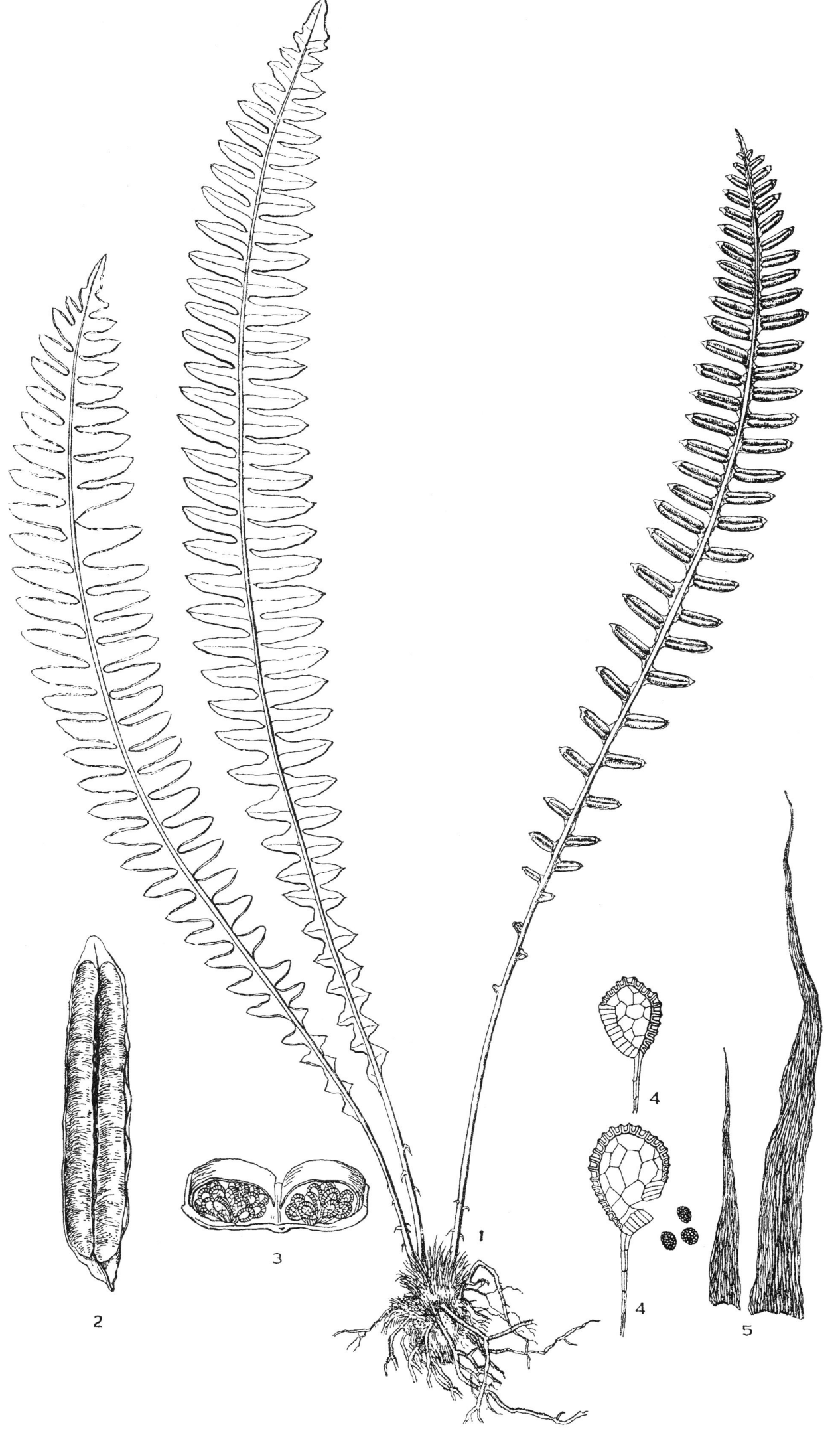

BLECHNUM EBURNEUM Christ
象牙乌毛蕨

图版 29　金毛裸蕨

　　根状茎短颇肥壮,斜向上伸密覆以黄色丝光长尖披针形之鳞片;叶柄长 7—15 cm,铁丝状,甚坚硬,黑色有光泽,密被粃糠状鳞片,基部之鳞片一如根状茎上所生者;叶体长 15—30 cm,宽 3—4 cm,一次羽状,顶部小叶最大,两侧小叶各 4—7 个,几对生或互生,卵圆形、心脏卵圆形或长圆形,全缘,在上面基部具一耳,钝头,上部小叶有显明之柄,下部者则无柄,中部小叶约 20—25 mm,质厚而软,两面皆被绒状锈色之毛,叶脉近边缘处作扇状,子囊群分叉,生于脉上,几全为茸毛包覆,无子囊群盖。

　　分布：自喜马拉雅山区至云南、四川、陕西、甘肃、河北皆有之。

　　图注：1.本种全形(自然大);2.小叶示其叶脉及子囊群之位置(放大 3 倍);3.子囊群(放大 149 倍);4.叶柄基部之鳞片(放大 27 倍);5.叶体下面之茸毛(放大 27 倍);6.叶体上面之茸毛(放大 27 倍)。

GYMNOPTERIS VESTITA (Wall.) Underwood

GYMNOPTERIS VESTITA (Wall.) Underw. in Bull. Torr. Club. **29**: 627 (1902); C. Chr. Ind. 342 (1906); Matthew in Journ. Linn. Soc. **39**: 369 (1911).

Grammitis vestita Wall., List 12 (1828).

Gymnogramma vestita Presl, Tent. 218 (1838); HK. Ic, Pl. t. 115; HK. BK. Syn. Fil. 379; Christ Farnkraut, Erde 66.

Syngramma vestita Moore, Ind. LX (1857).

Neurogramme vestita Diels in Nat. Pfl. Fam. I. 4. 262

　　Rhizome rather thick, short, obliquely ascending, clothed with dense silky yellow linear-subulate scales; *stipes* 7-15 cm long, wiry, but rigid, shining ebeneous, more or less furfuraceous, the scales at the base are dense and similar to those on rhizome; *fronds* 15-30 cm long, 3-4 cm broad, simply pinnate under a larger terminal pinna, pinnae 7-14 on each side, in distinct supposte or alternate pairs, ovate or cordate-ovate or oblong, entire, often with an auricle at the superior base, obtusish, the lower ones distinctly stalked, the upper subsessile, the middle 2-2.5 cm long, 7-12 mm broad; *texture* thick, but flaccid, both sides, especially the under, densely coated with fine velvety ferruginous hairs; *veins* flabellate towards the edge; *sori* forked, on the veins, almost completely hidden in hairs, indusium destitute.

　　Distribution: Himalayas to Yunnan, Sichuan, Shaanxi, Gansu, Hebei.

　　Plate 29. Fig. 1. Habit sketch (natural size). 2. A pinna, showing venation and disposition of sori (× 3). 3. Sporangium (× 149). 4. Scale from the base of stipe (× 27). 5. Hair from the under side of pinna (× 27). 6. Hair from the upper side of pinna (× 27).

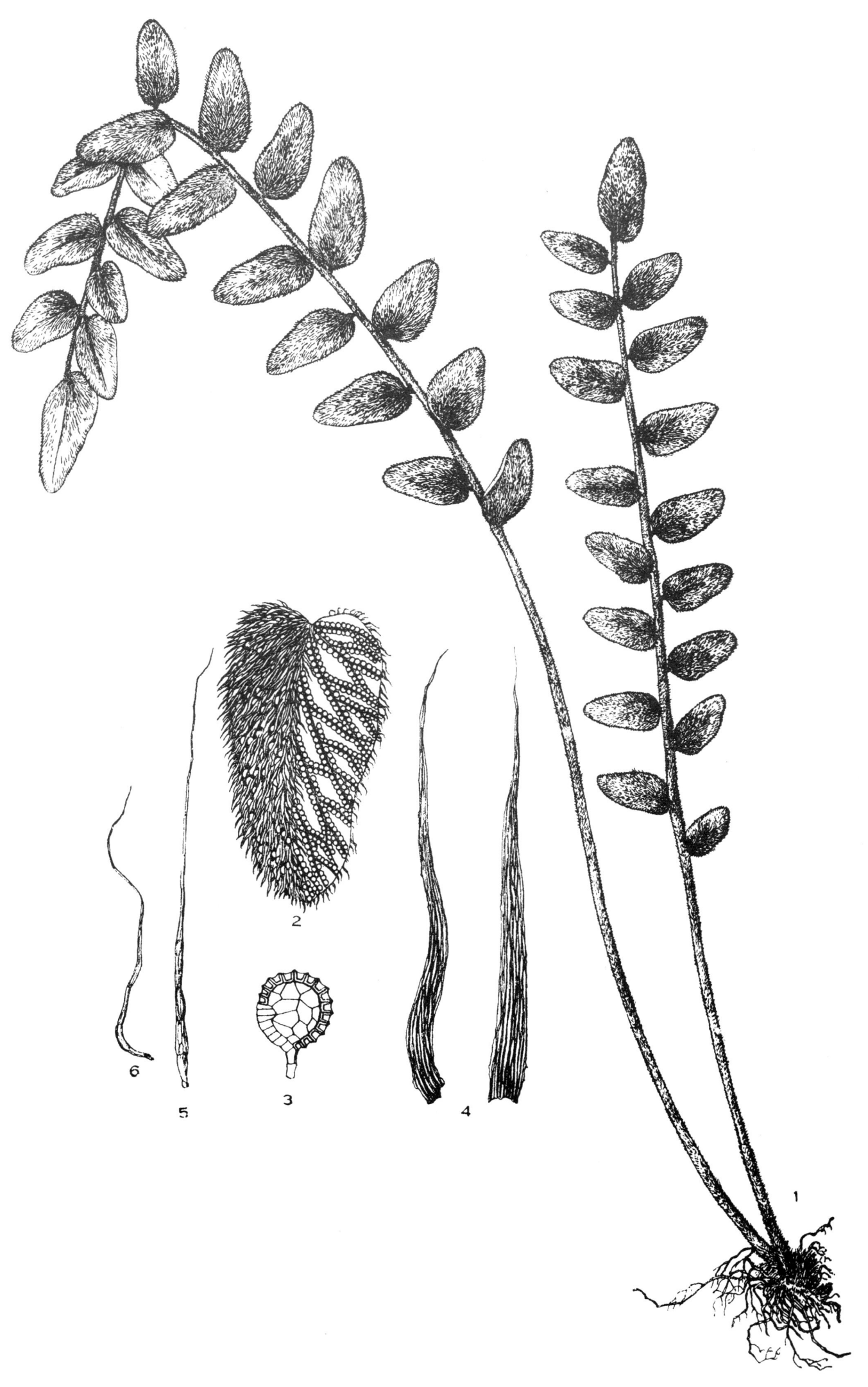

GYMNOPTERIS VESTITA (Wall.) Underwood

金毛裸蕨

图版 30　亨氏瘤足蕨

茎部直立,裸露无鳞,叶柄簇生,栗褐色,基部粗壮,三棱,无子囊叶叶柄长 10—13 cm,有子囊叶叶柄长达 30—37 cm,无子囊叶叶体披针形,薄膜质,一次羽状,长 37—45 cm,中部阔 6—10 cm,近基部渐窄缩,最低三对小叶有显明宽形附着之耳,中部小叶披针形,其间凹处窄而锐,长 3—5 cm,基部阔 8—12 mm,无柄,近顶部边缘锐锯齿状,下部不显明圆齿状;叶脉颇稀,显明,不分叉或一次分叉;有子囊叶叶体约宽 7 cm,小叶相距较远,长 4 cm,阔 3—4 mm,基部窄瘦,线形,有柄,尖端喙状。

分布:云南,安徽,浙江,在越南亦有之。

本种特有之性质为长披针形之叶体渐向下窄瘦,至基部小叶则直成耳状,有子囊叶之尖端成喙状。

图注:1.植物全形之一部(自然大);2.无子囊叶之小叶示其叶脉(放大 2 倍);3.有子囊叶之横断面示其子囊群着生状况(放大 27 倍);4.孢子囊(放大 76 倍);5.两孢子(放大 145 倍)。

PLAGIOGYRIA HENRYI Christ

PLAGIOGYRIA HENRYI Christ in Bull. Boiss. **7**: 8 (1899); Diels in Nat. Pfl. Fam, I. 4, 282; C. Chr. Ind. 496 (1906). *Lomaria decurrens* Baker in Kew Bull. (1906) 9.

Caudex erect, naked; *stipes* tufted, castaneous, base incrassate, sharply trigonous, those of the sterile frond 10-13 cm long, those of the fertile frond 30-37 cm long; sterile *frond* oblanceolate, thin membranaceous, simply pinnate, 37-45 cm long, 6-10 cm broad at the middle, gradually attenuate towards the base with the lowest 3 pairs reduced to distinct and broadly adnate auricles, the central pinnae lanceolate, conferted with very narrow sharp sinuses, 3-5 cm long, 8-12 mm broad at the base, which is broadly adnate, margin prominently serrate towards the apex and obscurely crenate below; *veins* lax, distinct, mostly once-forked or simple; fertile frond about 7 cm broad, pinnae far apart, 4 cm long, 3-4 mm broad, reduced towards the base, linear, distinctly stalked and beaked at the apices.

Distribution: Yunnan, Anhui, Zhejiang, and also Vietnam.

Distinguished by its long oblanceolate frond, gradually attenuate downward with the lower pinnae reduced to distinct auricles and by its stalked fertile pinnae distinctly beaked at the apex.

Plate 30. Fig. 1 Habit sketch (natural size). 2. A sterile pinna, showing venation, (× 2) 3. A cross section of fertile pinna, showing attachment of sori (× 27) 4. A sporangium (× 76). 5. Two spores (× 145).

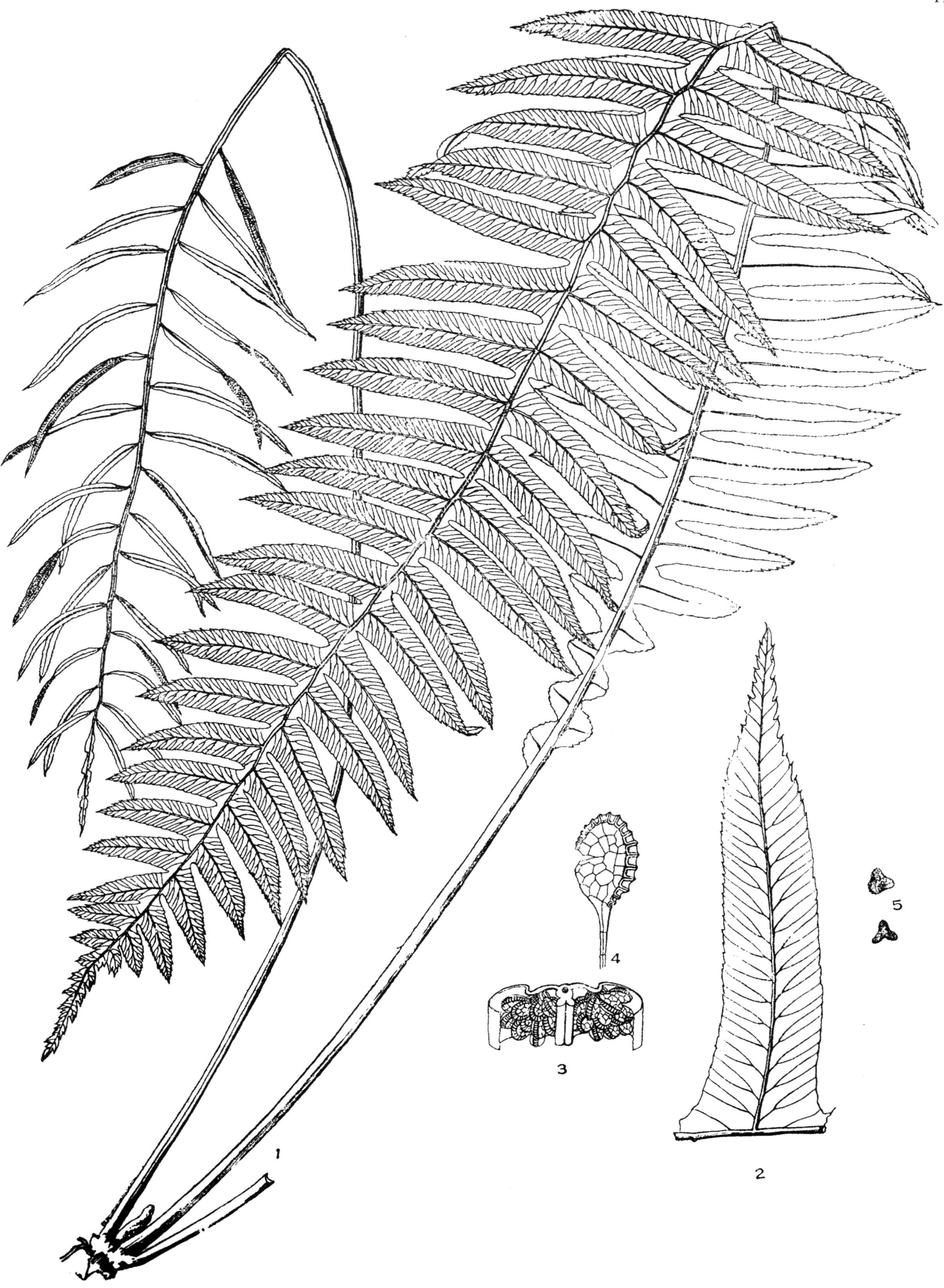

PLAGIOGYRIA HENRYI Christ
亨氏瘤足蕨

图版 31(1—5) 格氏铁线蕨 (6—9) 矮铁线蕨

格氏铁线蕨

根状茎短，多须根，鳞片黑色，密生向上，线形，尖端细针状；叶柄长 2—3 cm，簇生如草丛，深黑色，无鳞片，纤细如丝状，叶轴亦如之；叶体长 15—35 mm，稀有略长者，宽 20 mm，一次羽状；小叶三四个，有时仅两个，相距约 6—8 mm，有柄，柄长 1.5 mm，直立舒展，顶部小叶长 5—7 mm，基部耳状，脱落时叶柄残留叶轴上，宽短三角形，稀有长倒卵形者，基部圆楔形全缘，顶部截形，隐约成波状，长 6—9 mm，宽 7—10 mm，主脉 4—5 条从基部展开成扇状，每条再三次或五次分叉，支脉细微不显明，子囊群大，单生于顶部，子囊盖扁长肾状，长 4—6 mm，宽 1.3 mm，暗褐色，着生于宽而浅之边缘缺刻上，略带革质，稍透明，两面皆呈蓝色。

分布：广西。

图注：1.本种全形(自然大)；2.小叶一片示其脉及孢子囊群(放大 3 倍)；3.根状茎上之鳞片(放大 48 倍)；4.孢子囊(放大 100 倍)；5.子囊群盖之背面，表示子囊群附着之状(放大 10 倍)。

矮铁线蕨

根状茎甚短小，鳞片刚硬，密生向上，针状，微黑；叶柄多数，簇生，纤细如丝，坚硬，深黑色，无鳞片；叶轴亦如之，长 1 cm 弱；叶体长 10—15 mm，宽 7 mm，一次羽状，光滑，上面光深绿色，下面稍带蓝色；小叶三五个，稀更多，互生，有柄(柄长 1—2 mm，永存)，向外横舒，圆形至略成三角形，基部圆形或短楔形，全缘，外缘稍成波皱，长 2—4 mm，稀更长；厚革质，脉纹甚细，具两条粗脉，自小叶柄顶部射出，直行向上，至边际再一次或二次分叉，稍透明；孢子囊群单生，每小叶上仅一枚，子囊盖圆形，全缘，暗褐色，革质，附着于圆齿之阔而浅之缺刻上，由叶脉及薄膜组织生出，直达小叶之中部。

分布：广西。

图注：6.本种全形(自然大)；7.一小叶示其脉及子囊群(放大 8 倍)；8.根状茎之鳞片(放大 48 倍)；9.孢子囊(放大 100 倍)。

ADIANTUM GREENII Ching

ADIANTUM GREENII Ching in Sinensia **1**: 8 (1929).

Rhizome short, profusely fibrous-rooted, scales blackish, dense, ascending, linear-subulate; *stipes* 2-3 cm long, caespitose-fasciculate, dark ebeneous, naked, capillaceous and so is the rachis; *fronds* 1.5-3.5 cm long, rarely longer, about 2 cm broad, simply pinnate; pinnae 3-4, sometimes 2 only, 6-8 mm apart, petiolate (petioles 1.5 mm long), erect-patent, the terminal one with a petiole 5-7 mm long, articulate at base, leaving a persistent petiole upon falling, broadly short-triangular, rarely oblong-obovate, rotundo-cuneate and entire at base, truncate and very obscurely undulate on the broad truncate apex, 6-9 mm long, 7-10 mm broad; *veins* 4-5, flabellate from the base, each again 3-5 forked, veinlets fine, distinct; *sori* large, solitary, placed at the apex, indusium transversely oblong-reniform, 4-6 mm long, 1.3 mm broad, dark brown, attached to a broad shallow sinus; *texture* subcoriaceous, translucent, bluish on both sides.

Distribution: Guangxi.

Plate 31. Fig. 1. Habit sketch (natural size). 2. A pinna, showing venation and sorus (× 3). 3. Scales from rhizome (× 48). 4. Sporangia (× 100). 5. Dorsalside of indusium, showing attachment of sori (× 10).

ADIANTUM NANUM Ching

ADIANTUM NANUM Ching in Sinensia **1**: 9 (1929).

Rhizome very short, small, scales stiff, ascending, dense, subulate, blackish; *stipes* numerous, fasciculato-rosulate; capillaceous, but stiff, dark ebeneous, naked, so is the rachis, 1 cm or less long; *fronds* 1-1.5 cm long 7 mm broad, simply pinnate, glabrous, deep shining green above, bluish beneath; pinnae 3-5, rarely more, alternate, petiolate (petioles 1-2 mm long, persistent), spreading orbicular to subtriangular, rounded or shortly cuneate at base, entire, obscure undulate on the outer margin, 2-4 mm long, rarely longer; *texture* thick, stiff, coriaceous; *veins* fine, free, principal veins 2, radiating from the apex of petiole, erect-ascending, each again 1-2-forked towards outer margin, translucent; *sori* solitary, one to each pinna, small, indusium orbicular, entire, dark brown, coriaceous, attached to a rounded shallow sinus of the crenature, springing from veins and parenchyma, and reaching down as far as the centre of the pinnae.

Distribution: Guangxi.

Plate 31. Fig. 6. Habit sketch (natural size). 7. A pinna, showing venation and sorus (× 8). 8. Scales from rhizome (× 48). 9. Sporangium (× 100).

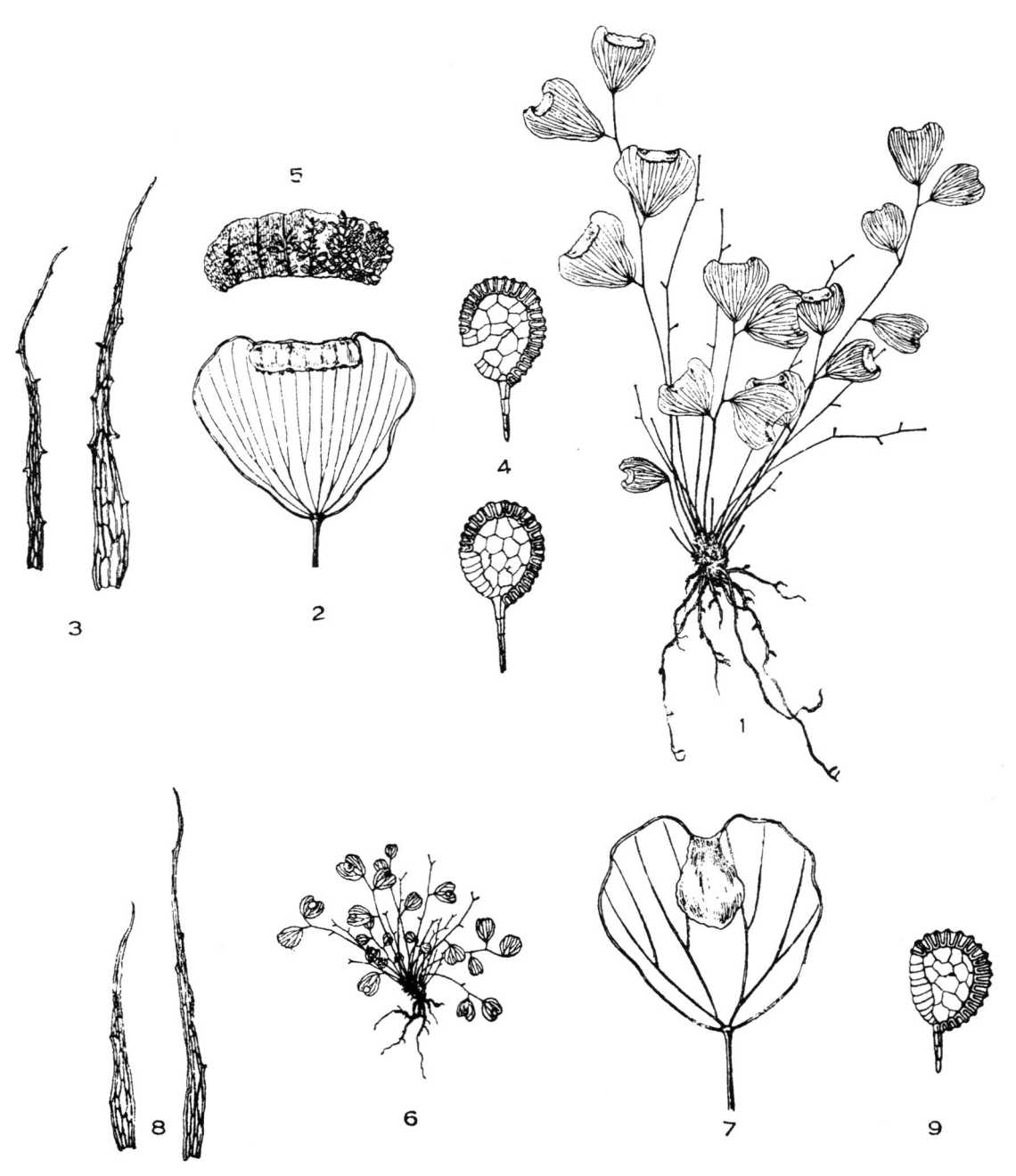

ADIANTUM GREENII Ching (1–5)
格氏铁线蕨
ADIANTUM NANUM Ching (6–9)
矮铁线蕨

图版 32　岩凤尾蕨

根状茎短，直立，密生须状根；叶柄簇生（4—7 枚），细瘦，无鳞片，有光泽，稻秆色，长 10—30 cm；叶体三角卵圆形，一次羽状，厚革质，长 10—20 cm，两面皆绿色光滑；小叶 3—5 个，顶端一个异常宽大，长卵圆形至广披针形，长 5—15 cm，中部宽 12—25 mm，两侧小叶均无柄，顶端小叶楔形具短柄，其顶端渐尖头，无子囊群，缺刻齿牙状，两侧小叶顶端急尖头，余相似；叶脉殊疏，由中肋分出成 75°斜角，单股或分叉，甚显明；子囊群由基部延展至小叶顶端之下，子囊群盖窄，光滑，暗褐色。

分布：四川，贵州，云南，广东。

本种与 *Pteris cretica* L. 相近，其异点为小叶数较少，基部简单，叶脉较稀，至于叶顶缺刻形状亦不同。叶体通常俱生 3 小叶，但大者亦有时生 5 叶，基部一对小叶与中部小叶相离至 1—2 cm。产自云南之 *Pteris nana* Christ 与 Baker 氏采自四川峨眉之标本完全相同，仅小叶略宽耳。香港博物院中采自广东之第八三六六号标本亦同此种，但叶体全缘而无子囊群。*Pteris nana* Christ var. quiquefoliata Copel.（福建产）比 Christensen 之原种稍大，若非 P. cretica L. 之强健品种，当改定名为 *Pteris deltodon* Baker var. quiquefoliata (Copel.) Ching.——仁昌。

图注：1. 本种全形（自然大）；2. 有子囊叶小叶之一部（背部）……（放大 2 倍）；3. 有子囊叶小叶（腹部）……（放大 2 倍）。

PTERIS DELTODON Baker

PTERIS DELTODON Baker in Journ. Bot. **26**: 226 (1888); C. Chr. Ind. 596 (1906); Matthew in Journ. Linn. Soc. **39**: 389 (1911).

Pteris nana Christ in C. Chr. Ind. 603 (1906).

Pteris trifoliata Christ in Bull. Boiss. **7**: 7 (1899); Diels in Nat. Pfl. Fam. 1. 4. 292 (non Fée 1857).

Rhizome short, erect, densely fibrous-rooted; *stipes* tufted (4-7), slender, naked, shining straminous, 10-30 cm long; *frond* deltoid-ovate, simply pinnate, of thick coriaceous *texture*, 10-20 cm long, green and glabrous on both sides; pinnae 3-5, with the terminal one much the largest, oblong-ovate to broad lanceolate, 5-15 cm long, 1.2-2.5 cm broad in the middle, sessile except the terminal one, which is cuneate at base and usually shortly petiolate, conspicuously inciso-dentate at the sterile tips, which is acuminate in the terminal pinna, and usually acute in the lateral ones; *veins* lax, branching at angles of 75° from the costa, simple or forked, very distinct; *sori* continuous from the base up to within a short distance of the tip of pinnae, indusium narrow, glabrous, dark-brown.

Distribution: Sichuan, Guizhou, Yunnan, Guangdong.

This species is nearest to *Pteris cretica* L. from which differs in but fewer pinnae, of which the lowest are simple, the serrations in sterile parts are also different and further differs by laxer veins. The frond is generally trifoliate but larger forms often have five leaflets with the lower pair separate from the one next above by a wingless rachis about 1-2 cm long. *Pteris nana* Christ from Yunnan is perfectly identical with Baker's type from Omei Shan, Sichuan, differing only in somewhat broader pinnae. A specimen, No. 8366, in Hong Kong Herbarium from Guangdong (without precise locality) appears almost congeneric with the present species, differing only by perfect entire sterile pinnae. *Pteris nana* Christ var. *quiquefoliata* Copel. from Fujian, a plant larger than the type having lateral pinnae cleft at the base, should be called *P. deltodon* Baker var. *quiquefoliata* (Copel.) Ching, if it does not represent a mere vigorous form of *P. cretica* L. —R. C. C.

Plate 32. Fig. 1. Habit sketch (natural size). 2. A portion of fertile pinna, dorsal side (× 2). 3. A fertile pinna, ventral side (× 2).

PTERIS DELTODON Baker
岩凤尾蕨

根状茎甚短，直立；鳞片线形，尖刺头，暗褐色，颇厚，长 2—3 mm；叶柄 4—8 枚簇生，无鳞片，稻秆色，干时扁平，上面有槽，长 6—17 cm，阔 2 mm；叶体短三角形，两次羽状，长 5—10 cm，稍窄，两面皆光滑；两侧小叶 1—2 对，对生，有柄；基部一对颇大，叶柄长 5—10 mm。小叶再分裂成 2—3 个二次小叶，二次小叶长圆披针形，至长圆卵形，长 2.5—6 cm，中部宽 1.5—2 cm，具短柄或几无柄，顶端无子囊群，显然成缺刻齿牙状；中部一对小叶简单，具短柄或几无柄，与基部一对形态修广各相同；顶部小叶永为简单，形态亦相类，但柄较长，达 9 cm；叶轴稻秆色，无鳞片，亦无翼；叶体厚革质，坚硬，绿色，有光泽，中肋在下面隆起，上面平滑；支脉颇多，上面显明，隆起，开张，相离 1—1.5 mm，直达叶缘锯齿之基部而止，从不突出成刺头；子囊群暗褐色，线形，由下部直走至叶中部或近顶处而没，有时仅在近基部处子囊群盖窄而颇厚，坚实，暗灰色，连续。

分布：广西。

图注：1. 本种全形（自然大）。

PTERIS HUI Ching

PTERIS HUI Ching in *Sinensia* **1**：9 (1929).

Rhizome very short, upright; scales linear-subulate, dark brown, rather thick, 2-3mm long; *stipes* tufted, 4-8 together, naked, straminous, flattend upon drying, broadly channelled above, 6-17 cm long, 2 mm broad ; *fronds* short deltoid, uniform, bipinnate, 5-10 cm long, a little less broad, glabrous on both sides ; lateral pinnae 1-2 pairs, opposite, petiolate; the lower pair much the largest, with petioles 5-10 mm long, 2-3-partite, pinnules oblong-lanceolate to oblong-ovate, 2.5-6 cm long, 1.5-2 cm broad at the middle, shortly petiolate or subsessile, distinctly inciso-dentate on the upper sterile portion; the middle pair (if there any) simple, shortly petiolate or subsessile, of same shape and size as the lower ones; the terminal pinna always simple, of same shape, but much longer (-9 cm) petiolate; rachis staminous, naked, wingless; *texture* thick, stiff, coriaceous, shining light green (when living); *midrib* prominently raised below, rather flattened above, lateral veins numerous, distinctly raised above, patent, 1-1.5 mm apart, ended near the base of the teeth, never in the mucron; *sori* dark brown, linear, continuous from the base to within a short distance from the tip, or only up to the middle of the margin, or sometimes only a short way upward from the base, indusium narrow, rather thick, rigid, dark gray, continuous.

Distribution：Guangxi.

Plate 33. Fig. 1. Habit sketch (natural size).

PTERIS HUI Ching
胡氏凤尾蕨

图版 34　二形凤尾蕨

　　根状茎短，上升，径约 4 mm，具密生须根，顶部密被小型栗色针形之鳞片，叶柄有光泽，近基部略带红色，上部暗稻秆色，顶部有窄翼；叶体显然具两种形态，无子囊叶长 10—15 cm，有子囊叶长 25—30 cm，无子囊叶长三角形，三出，光滑，纸质，上面浅绿有光泽，侧生小叶镰状卵圆形，长 5 cm，有深缺刻，裂片每侧 5 枚，长圆形，钝头，顶端一片最大长 2 cm，锐锯齿状或几全缘，中央小叶最大，长 10—15 cm，宽 3—4 cm，广披针形，渐尖头，每侧有 6—10 裂片，基部裂片窄缩，循叶轴向下延展；有子囊叶长 10—15 cm，有时仅长 5 cm，羽状一如 *Pteris ensiformis* Burm.，顶端小叶之下每侧各具 1—2 枚小叶，基部小叶分叉，或简单，宽达 6 mm，皆向下延长，形成窄翼，顶端小叶最大，长达 10 cm，通常不规则羽状分裂；子囊群盖狭窄连续达齿状之钝圆叶端，暗灰色。

　　分布：广东。

　　本种为 Charles G. Matthew 博士发现于广东东南部之山中，其后陈焕镛教授又采之于北江附近，作者详为审定与 Matthew 之标本无异，但其无子囊叶及有子囊叶之差别较少耳，本种位置当介乎 *Pteris multifida* Poir. 与 *Pteris ensiformis* Burm. 之间 ——仁昌。

　　图注：1. 本种全体（自然大）；2. 无子囊叶之裂片示其叶脉及锯齿（放大 4 倍）；3. 有子囊叶之一部示其叶脉及子囊群盖（放大 4 倍）。

PTERIS DIMORPHA Copeland

PTERIS DIMORPHA Copel. in Phil. Journ. Sci. Bot. 3 C: 282 (1908); C. Chr. Ind, Suppl. 66 (1906-12); Matthew in Journ. Linn, Soc. **39**: 389 (1911).

Rhizome short, ascending, about 4 mm thick, densely fibrous-rooted, clad at the apex in minute subulate castaneous scales; *stipes* shining, reddish towards the base, dark straminous upward, narrowly winged on the uppermost part; *frond* strongly dimorphous, the sterile ones 10-15 cm long, the fertile 25-30 cm long, the sterile oblong-deltoid, tripartite, glabrous, of chartaceous *texture*, shining light green on both surfaces, lateral pinnae ovate, falcate, to 5 cm long, deeply incised, segments about 5 on each side, close, oblong, obtuse, the terminal one much the largest, to 2 cm long, sharply serrate or subentire, the middle pinna the largest, to 10-15 cm long and 3-4 cm broad, broad lanceolate, gradually acuminate, with 6-10 segments on each side, the lowest ones reduced and narrowly decurrent along the rachis; fertile frond 10-15 cm long (sometimes only 5 cm long), pinnate like *Pteris ensiformis* Burm. with 1-2 pinnae on each side under the terminal one, the lowest ones forked or simple, 6 mm broad, all decurrent in a narow wing, the terminal pinna the largest, to 10 cm long, often irregularly pinnatifid; *indusium* narrow, continuous to the blunt serrate apex, dark gray.

Distribution: Guangdong.

This distinct and rather uniform species was first discovered by Dr. Charles G. Matthew of the British Navy in the mountains in southeastern Guangdong, and later in 1927 was collected in the mountains on North River by Prof. W. Y. Chun, of Botanical Laboratory, Canton, whose copious materials have recently been received by me. The specimens match the type very well, except that it show somewhat less dimorphism and in this respect it may be well considered as an intermediate between *P. multifida* Poir and *P. ensiformis* Burm. —R. C. C.

Plate 34. Fig. 1. Habit sketch (natural size). 2. A portion of sterile segment, showing venation and serration ($\times$ 4). 3. A portion of fertile segment, showing venation and indusium ($\times$ 4).

PTERIS DIMORPHA Copeland
二形凤尾蕨

图版 35 猪鬣凤尾蕨

根状茎短；叶柄簇生，直立或开张，细瘦有棱，长 3—6 cm，红稻秆色，或浅黑色，平滑；叶体具两种形态，无子囊叶较有子囊叶短，具 5 个小叶，有子囊叶长 10—15 cm，广三角形，二次羽状，顶端小叶三出分裂，以下侧生小叶每侧各有 2—3 个，在基部者具短柄，二或三出，中部者简单或分叉，最上者皆简单，小叶裂片等长，线形，长 4—6 cm，宽 1—2 mm，全缘，近梢端则作锐锯齿状，纸质，有光泽，完全光滑，叶轴红色；孑囊群盖甚宽，约 0.5 mm，白色，延叶缘附着，止于裂片之齿状顶端。

分布：云南，四川，湖北。

本种习性介乎 *Pteris multifida* Poir. 及 *Pteris dactylina* HK. 之间，但异点则为形态较小，无子囊裂片较短，及叶柄叶轴作栗色，此种首由 A. Henry 发现于云南，按 Christensen 之记载，产于四川西部者高至 40 cm，上着 6—7 对小叶，但钟观光教授由云南采得之标本（第二千〇六十五号）与 Christensen 氏之记录几全相同，在云南殊为习见之种也。

图注：1. 本种全形（自然大）；2. 小叶最上裂片示其脉纹及子囊群（放大 3 倍）；3. 有子囊叶之横剖图示其子囊群及子囊群盖（放大 5 倍）。

PTERIS ACTINIOPTEROIDES Christ

PTERIS ACTINIOPTEROIDES Christ in Bull. Boiss. **7**：6(1899); Diels in Nat. Pfl. Fam. 1. 4. 292 ; C. Chr. Ind. 591 (1906) ; Matther in Journ. Linna. Soc. **39**：388 (1911) ; C. Chr. in Medd. Göteb. Bot. Trädg. Ⅰ. 96 (1924).

Rhizome short; *stipes* in dense tuft, upright or spreading, slender, angular, 3-6 cm long, reddish-straminous or light ebeneous, polished; *frond* dimorhpous, the sterile one much shorter than the fertile, quiquefoliate, the 7-15 cm long, broadly deltoid, bipinnate, pinnae 2-3 on each side under the tri-foliolulate terminal one, the basal ones shortly petiolate, usually 2-3-forked, the middle ones forked or simple, the uppermost ones always simple, segments equal, linear, 4-6 cm long, 1-2 mm broad, margin entire except the apex, which is usually sharply serrate, *texture* chartaceous, glossy, entirely glabrous, rachis reddish; *indusium* rather broad (0.5 mm broad), whitish, continuous with entire margin, not extending to the serrated apex of the segment.

Distribution : Yunnan, Sichuan, Hubei.

A very interesting little fern, which is, in habit, somewhat intermediate between *P. multifida* Poir. and *P. dactylina* Hk, from both of which differs, by smaller size, very narrow fertile segments and castaneous stipes and rachis. In was first collected in Mengtze, Yunnan, by A. Henry and reported ever since from Sichuan and Hubei. According to Christensen the plant from W. Sichuan grows to 40 cm tall including the stipe, with 6-7 pairs of lateral pinnae under a fan-shaped apex consisting 7 segments; but all the specimens, No. 2965, collected by Prof. K. K. Tsoong (1921) on Tai Hwa Shan, Yunnan, perfectly agree with Christ's type in dimension, and it is evidently a common fern, growing in the crevices of exposed rocky cliff as his notes indicate.—R. C. C.

Plate 35. Fig. 1. Habit sketch (natural size). 2. An ultimate segment, showing venation and and inducia ($\times$ 3). 3. A cross section of fertile segment, showing inducia and sori ($\times$ 5).

PTERIS ACTINIOPTEROIDES Christ
猪鬣凤尾蕨

图版 36　黄毛凤尾蕨

根状茎肥短，密覆以褐色尖披针形之鳞片，叶柄如之，簇生，多数，直立，红稻秆色，管状，坚硬，密被以针状有刺头短小开张褐色之鳞片，鳞片至距基部 2—3 cm 处则脱落；无子囊叶柄长 6 cm，有子囊叶达 15 cm；无子囊叶体长 6 cm，宽 3 cm，长卵圆形，二次羽状，小叶远离，每侧有 5 个，近基部者有柄，削去至中肋，而下部之小裂片更分叉，上部裂片附着向下延长，简单或分叉，裂片钝头，修广各约 0.5 cm，边复波齿状；有子囊叶长 11 cm，宽 6 cm，一次羽状，小叶远离，每侧 5—6 个，最下一个再裂成 5—6 个二次小叶，在上部者则简单，二次小叶线形，长 4—5 cm，宽 2—3 mm，锐头，全缘；叶脉不显明，二三次分叉；叶体淡绿色，光滑，膜质，微透明；子囊群连续及顶部，宽 0.5 mm，子囊群盖薄膜质，狭窄，不规则流苏状分裂。

分布：四川。

本种在 1903 年为 E. H. Wilson 氏发现于四川，其习性与 *Pteris ensiformis* Burm. 相近，但较小，有子囊叶与无子囊叶之差异较少，尤显著之异点为密生红褐鳞片叶柄及叶轴与无子囊叶之复圆齿状之裂片。

图注：1. 本种全形（自然大）；2. 有子囊叶裂片之一部正面（放大 20 倍）；3. 叶轴上部之鳞片（放大 108 倍）；4. 叶柄基部之鳞片（放大 30 倍）；5. 叶脉及子囊群之附着状况（放大 40 倍）。

PTERIS PAUPERCULA Christ

PTERIS PAUPERCULA Christ in Bull. Geogr. Bot. Mans (1906) 131 ; C. Chr. Ind. Supl. 67 (1906-13); Matthew in Journ. Linn. Soc. **39**: 389 (1911).

Rhizome thick, short, densely clothed in brown lanceolate subulate scales and so is the stipe; leaves densely fasciculate, numerous; *stipes* erect, reddish-straminous, terete, rigid and like rachis densely asperous with subulate and setaceous short spreading brown slender scales, falling off at distance of 2-3 cm above the base; stipes of sterile leaves 6 cm long, those of the fertile 15 cm long, the sterile *frond* 6 cm long, 3 cm broad, oval-oblong, bipinnatifid; pinnae remote, about 5 on each side, the lowest shortly petiolate, cut down nearly to the costa with lower segments forked again, the upper ones adnately decurrent, simple or forked, segments obtuse, 0.5 cm long and broad, margin double crenulate; the fertile frond 11 cm long, 6 cm broad, bipinnate; pinna remote, 5-6 on each side, the lower ones pinnate into 5-6 pinnules, the upper ones simple, pinnules linear, 4-5 cm long, 2-3 mm broad, acute, entire; *veins* hidden, twice or thrice forked; color light green, surfaces glabrous, opaque, *texture* herbaceous, somewhat pellucid; *sori* continuous to the apex, 0.5 mm broad, indusium membranaceous, narrow, lacerato-fimbriate.

Distribution: Sichuan.

A remarkable fern discovered by E. H. Willson in W. China, 1973. In habit it closely allied to *P. ensiformis* Burm. but differs in smaller dimension, much less pronounced dimorphism between the fertile and sterile leaves and, particularly, by densely scaly reddish-brown stipes and rachis, and double crenulate pinnules of the sterile leaves.

Plate 36. Fig. 1. Habit sketch (natural size). 2. A portion of fertile segment, dorsal side (× 20). 3. Scale from the upper part of rachis (× 108). 4. Scales from the base of stipe (× 30). 5. Venation and attachment of sori (× 40).

PTERIS PAUPERCULA Christ
黄毛凤尾蕨

图版 37　溪凤尾蕨

　　根状茎短,粗壮,直立,密覆以狭披针形褐色鳞片;叶柄簇生,粗壮,长达 90 cm,径 6—10 mm,直立,裸露,有光泽,绿色或稻秆色或几褐色,鳞片甚稀,褐色;叶体长 175—200 cm,广三角形,顶端小叶达 30 cm,或稍长,阔 7.5 cm,削至中肋形成多数密集镰状披针形之裂片长约 4 cm,宽 6—10 mm,扩张,窄斜向下延长,在无子囊叶略成锯齿状,侧生小叶每侧各 5—10 个,形与顶端小叶几同,在基部之小叶最大长越 30 cm,相距 15 cm,长圆披针形,顶端简单披针尾状,顶端以下羽状分裂直至叶轴,有时基部小叶分叉;叶体纸质,上面浅绿色,下面色较淡或有光,无毛,叶脉一次分叉,脉基相距 3 mm;子囊群不达到裂片之顶端,子囊群盖灰色,薄膜质,全缘。

　　分布:印度,菲律宾群岛,马来西亚,夏威夷群岛,朝鲜半岛,与中国四川,云南,贵州,湖北,广西等处。

　　图注:1.本种全形之一部(自然大);2.小叶两顶端裂片示其叶脉及子囊群(放大)。

PTERIS EXCELSA Gaudichaud-Beaupré

PTERIS EXCELSA Gaud. in Freyc. Voy. Bot. 388 (1824); HK. Sp. Fil. Ⅱ. t. 136; HK. BK. Syn. Fil. 159; Bedd. Ferns S. Ind. t. 218, Handb. 114 (1892); Christ, Farnkraut. Erde. 167; Diels in Nat. Pfl. Fam. I. 4, 292; C. Chr. Ind. 597; v. A. v. R. Malay. Ferns 368 (1908); Matthew in Journ. Linn. Soc. **39**: 389 (1911).

Pteris terminalis Wall. List 101 (1828); Ag. Rec. 20 (1839).

Rhizome short, thick, erect, densely clothed in narrow lanceolate brown scales; *stipes* tufted, stout, to 90 cm long, 6-10 mm thick, erect, naked, glossy, green or straminous or brownish, scales sparse and brown; *frond* 175—200 cm long, broadly deltoid, terminal pinna to 30 cm long or longer, 7.5 cm broad, cut down to the costa into numerous rather closely placed falcate lanceolate segments on each side 4 cm long, 6-10 mm broad, dilated, narrowly oblique decurrent, slightly serrated on the barren parts, lateral pinnae 5-10 on each side, similar to the terminal one, the lowest much the largest, over 30 cm long, 15 cm apart, oblong-lanceolate, pinnatifid down to the rachis under the entire lanceolate caudate apex, sometimes forked in the basal pinnae; *texture* chartaceous, light green above, pale or glaucescent beneath, naked on both surfaces; *veins* once-forked, 3 mm apart at the base; *sori* not reaching the apex of the segments, indusium gray, membranaceous, entire.

Distribution: India, the Philippine Islands, Malayasia, Hawaii Islands, Korean peninsula, China: Sichuan, Yunnan, Guizhou, Hubei, Guangxi.

Plate 37. Fig. 1. A portion of the plant (natural size). 2. Two ultimate segments, showing venation and sori, (enlarged).

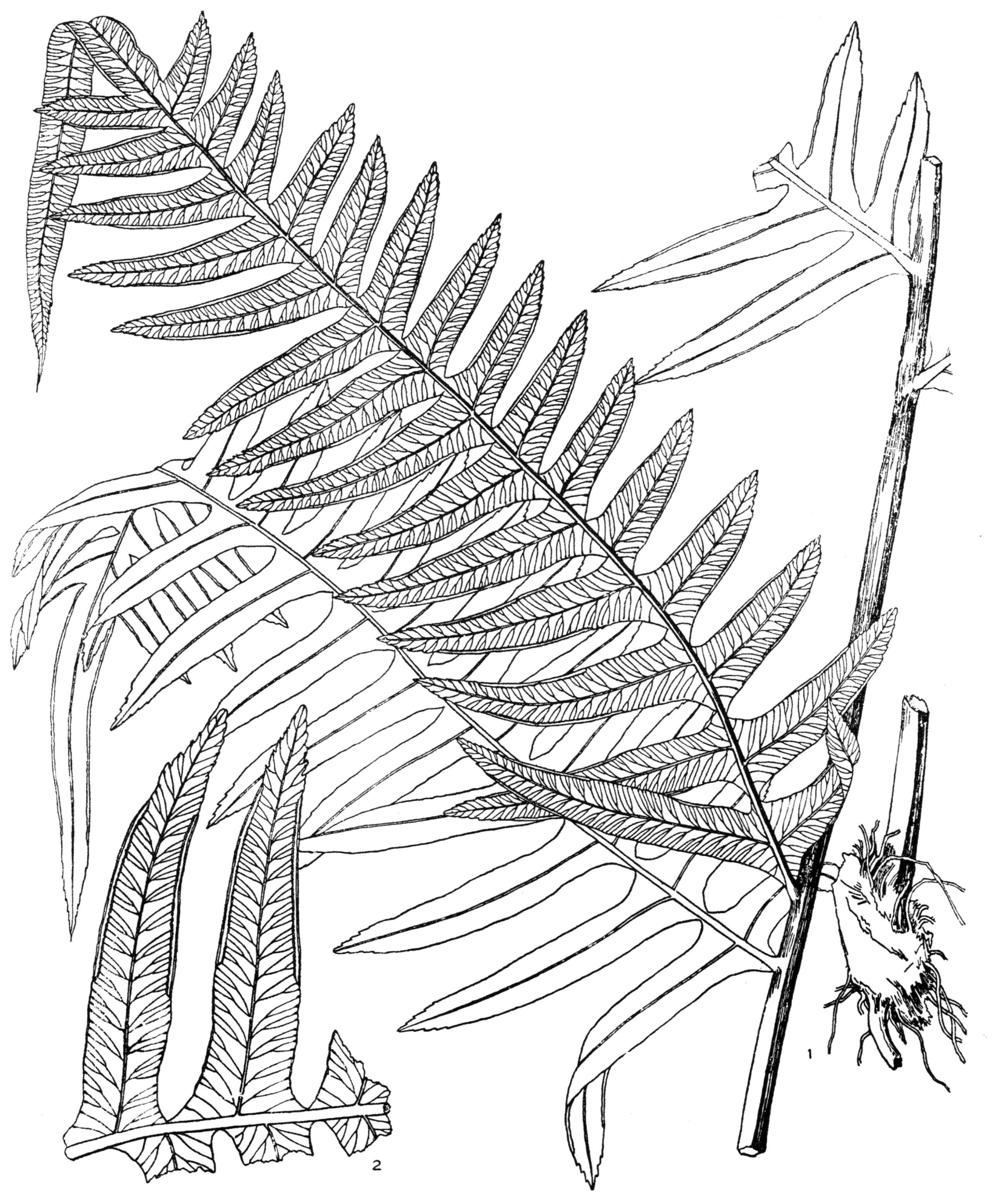

PTERIS EXCELSA Gaudichaud–Beaupré
溪凤尾蕨

根状茎匍匐，颇粗壮，鳞片密生，几黑色，披针形，尖端细长，引展成长须，脉络显明网状，长 5 mm，叶体远离，禾草状，长 90—130 cm（包括长 10—14 cm 之叶柄，叶柄圆管状，无鳞片），宽 13—15 mm，线形，下垂，由距基脚 2/3 部分渐向下窄缩形成叶柄，但上部稍扩张成广圆形或钝形尖端，几革质，略透明，干皱，中肋扁平，在两面皆不显明，侧脉细，3—4 条，与中肋几平行，不甚透明，但清晰可辨，由近叶缘之横脉相连贯，有时分叉，近叶梢处通常相并成少数窄线形之网孔，子囊群细长，从叶基直达顶端，陷入向外卷折之槽中，深 1 mm。

分布：广西，海南。

图注：1.本种全形（自然大）；2.叶之横断面（放大 7 倍）；3.叶缘有子囊群部分之横断面示子囊群盖（放大 50 倍）；4.根状茎上之鳞片（放大 36 倍）。

VITTARIA PAUCIAREOLATA Ching

VITTARIA PAUCIAREOLATA Ching in Sinensia **1**: 11 (1929).

Rhizome creeping, rather thick; scales dense, blackish; lanceolate with a long hair-pointed apex, distinctly reticulate, 5 mm long; *fronds* remote, grass-like, 90-130 cm long (including stipes, which are 10-14 cm long, terete, naked), 1.3-1.5 cm broad, linear, tape-like, pendulous, gradually attenuate from 2/3 way downward into the stipe, but not or very slightly narrowed towards a broad rounded or blunt apex; *texture* subcoriaceous, translucent, flaccid, *midrib* flattened and indistinct on both surfaces, lateral veins slender, 3-4 subparallel to the midrib, translucently distinct and connected by a transverse intramarginal vein, not rarely forked and often towards the apex of the frond scantily anastomosing into narrow linear meshes; *sori* long, reaching from near the base to the very apex, sunk in an extrorse marginal groove, 1 mm deep.

Distribution: Guangxi, Hainan.

Plate 38. Fig. 1. Habit sketch (natural size). 2. A cross section of frond (× 7). 3. A cross section of soriferous margin, showing indusium (× 50). 4. Scales from rhizome (× 36).

VITTARIA PAUCIAREOLATA Ching
阔叶书带蕨

图版 39 矮叶书带蕨

根状茎匍匐,细瘦,殊密覆以暗色,短形,栗色,尖线形,略成短齿状之鳞片;叶体多数,簇生,直立,长 3—7 cm,宽 1—1.5 mm,线形,渐向下窄缩形成短柄,向下渐窄,成钝形稀至渐尖头之顶端,边缘反卷,中肋上面成深槽,下面隆起,叶体几革质,软弱,光滑;叶缘反卷,其后展舒,叶脉不显;子囊群殊短,仅生于上部 2/3 部分至近极端而止,往往仅一边有之,生在中肋与边缘之间。

分布:广西。

图注 1. 本种全形(自然大);2. 叶体示子囊群盖及叶缘展开之情形(放大);3. 叶体横断面示子囊群之位置(放大 30 倍);4. 根状茎之鳞片(放大 20 倍)。

VITTARIA NANA Ching

VITTARIA NANA Ching in Sinensia **1**: 11 (1929).

Rhizome creeping, slender, clad in moderately dense dark short, castaneous, linear-subulate scales with obscurely denticulate margin; *fronds* numerous, caespitose, erect, 3-7 cm long, 1-1.5 mm broad, linear, gradually narrowed into the base of a short stipe, slightly narrowed towards a blunt, rarely acuminate apex, margin revolute, midrib deeply sulcate above, broadly keeled below; *texture* subcoriaceous, flaccid, glabrous; *veins* hidden; *sori* short, confined to upper 1/3, and quite a way short from the apex, often one-sided, and placed midway between costa and margin, which is at first rather broadly revolute, and finally relaxed.

Distribution: Guangxi.

Plate 39. Fig. 1. Habit sketch (natural size). 2. Fronds, showing different stages of opening induciate margin (enlarged). 3. A cross section of frond, showing disposition of sori ($\times$ 30). 4. Scales from rhizome ($\times$ 20).

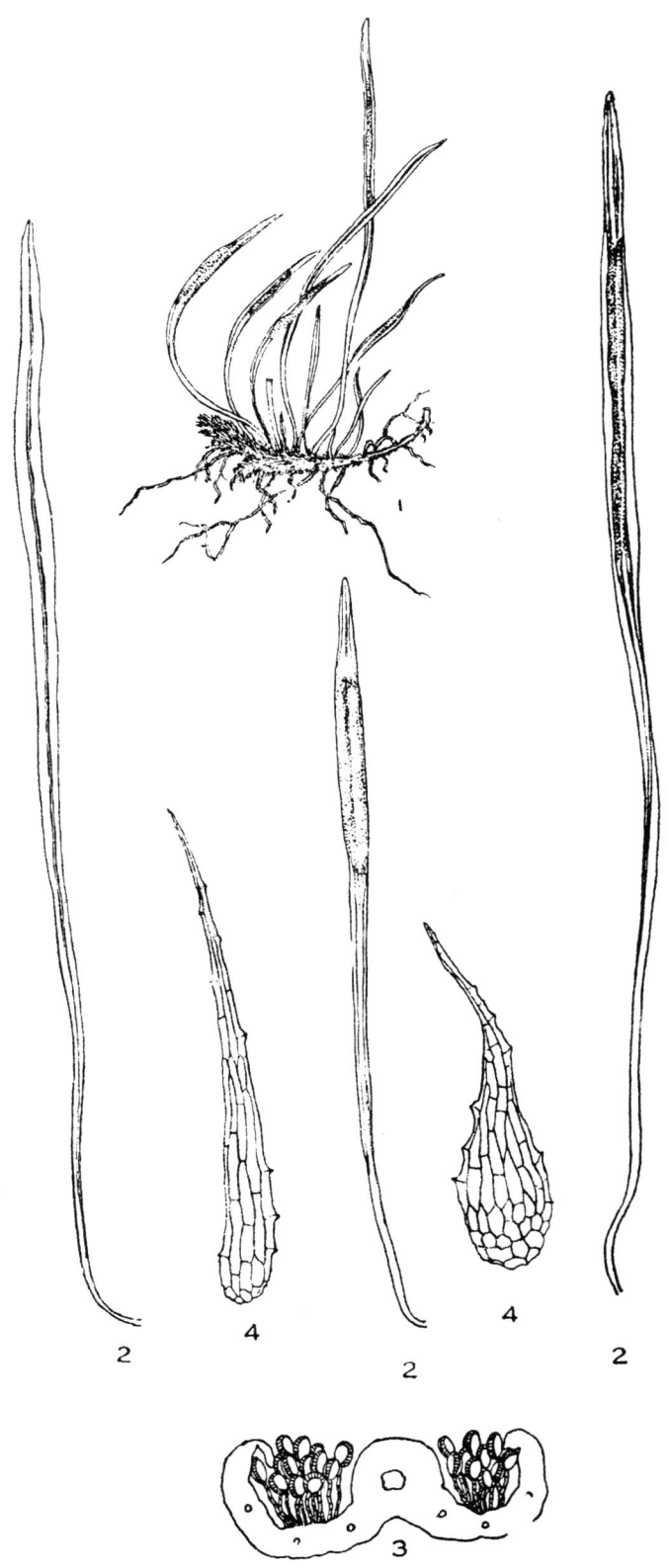

VITTARIA NANA Ching
矮叶书带蕨

根状茎短，直立；鳞片在近基部，暗褐色，尖线形，颇密，膜质；叶体数个丛集，叶柄无鳞片，绿色扁圆，长 7—14 cm；叶体倒卵楔形，顶端长尾状，长 10—13 cm，宽 5—6.5 cm，顶端 2/3 部分或更近顶部处最宽，肥厚革质，两面皆绿色，光滑；脉纹殊显明，直行，稀有联络者，网纹窄长，长约 2 cm，宽 3 mm；子囊群颇多，微陷入宽浅之槽中，有时顶端相连。

分布：云南，贵州，广西。

本种与 *Antrophyum plantagineum* Klf. 相近，但异点在较长之叶柄，宽倒卵楔形顶端长尾状之叶体，子囊较密集微陷入宽而浅之槽中。此种最初 A. Henry 采自云南，其后 Père Bodinier 又采之于贵州，今在广西各地亦发现之，最喜生山峡两侧荫翳之石上。

图注：1. 本种全形（自然大）；2. 叶体之一部分示脉纹及子囊群（放大 2 倍）；3. 孢子囊（放大 76 倍）；4. 根上之毛（放大 76 倍）；5. 根状茎之鳞片（放大 30 倍）。

ANTROPHYUM PETIOLATUM Baker

ANTROPHYUM PETIOLATUM Baker mss. Christ in Bull. Geogr. Bot. Mans: (1902) 202; Baker in Kew Bull. (1906) 14; Matthew in Journ. Linn. Soc. **39**: 342 (1911).

Rhizome short, erect; scales basal, dark brown, linear-subulate, rather dense, membranaceous; leaves several together, *stipes* naked, green, compressed, 7-14 cm long; *frond* obovate-cuneate, long-cuspidate, 10-13 cm long, 5-6.5 cm broad, broadest at the upper 2/3 or still higher up; *texture* carnoso-coriaceous, green and glabrous on both surfaces; *vein* rather distinct, vertical, rarely anastomosing, areolae narrow, above 2 cm long, 3 mm broad; *sori* copious, slightly immersed in broad shallow grooves, sometimes jointed at the ends.

Distribution: Yunnan, Guizhou, Guangxi.

Nearest to *A. plantagineum* Klf., yet distinct for its much longer stipe, broadly obovate-cuneate frond with cuspidate apex and much broader, closer sori only superficially immersed in broad shallow grooves. First collected in Yunnan by A. Henry, later in 1898 by Père Bodinier in Guizhou and recently known from several localities in Guangxi. As seen in its natural habitat, it generally grows on dripping or shaded rocky cliff by mountain torrents.

Plate 40. Fig. 1. Habit sketch (natural size). 2. A portion of the frond, showing venation and sori ($\times$ 2). 3. Sporangium ($\times$ 76). 4. Hairs from the roots ($\times$ 76). 5. Scales from rhizome ($\times$ 30).

ANTROPHYUM PETIOLATUM Baker
长柄车前蕨

图版 41　乌柄水龙骨

根状茎匍匐甚广，粗壮，密覆以覆瓦状，有光泽、褐色之鳞片；鳞片长达四种，基部广圆形，渐向上窄瘦，成线形刺头边缘有纤毛之顶端，在叶柄基部与根状茎附着点作暗褐色；叶柄远离，坚硬，长达 18 cm，无鳞片，有光泽，栗色，叶轴亦如之，与根状茎相接处有关节，叶体三角形，短渐尖头，长 22 mm，基部宽 15 mm，四次羽状；一次叶每侧约有十个，在基部者最大，有柄，直立开张，长 8—9 cm，宽 6—7 cm，三角形，二次叶轴无鳞片，栗色有光泽，长 3—4 cm，宽 12—15 mm，上部者则渐小，羽状，三次小叶长圆形，生于有翼之柄上，其中肋基部栗色，羽状，裂片线长圆形，有爪，长 2—3 mm，通常二裂为不等长之短裂片；叶体草质，主脉分叉，支脉简单，每裂片有一条，不显明，半透明，子囊群位于主脉分叉之附近部位，形小而圆，宽约当裂片三分之一。

分布：广西。

图注：1. 植物全形（自然大）；2. 一小叶片（放大 2 倍）；3. 二次小叶示其及脉纹子囊群（放大 4 倍）；4. 根状茎上之鳞片（放大 24 倍）。

POLYPODIUM DAREAEFORMIOIDES Ching

POLYPODIUM DAREAEFORMIOIDES Ching in Sinensia **1**: 12 (1929).

Rhizome wide-creeping, thick, densely covered with imbricate shining brown scales, which are to 4 mm long, broadly rounded at base, gradually narrowed upward to a linear-subulate apex with ciliate margin, attached to the rhizome by a dark brown centre of the broad base; *stipes* remote, rigid, to 18 cm long, naked, lustrous castaneous and so is the entire rachis, articulated to the rhizome; *fronds* deltoid, shortly acuminate, 22 cm long, 15 cm broad at base, quadripinnate; pinnae about 10 on each side, the lower ones much the largest, petiolate, erecto-patent, 8-9 cm long, 6-7 cm broad, deltoid, secondary rachis naked, shining castaneous, 3-4 cm long, 1.2-1.5 cm broad, the upper ones gradually smaller, pinnate, the pinnules of third order oblong, on a winged petiole, costule shaded castaneous near base, pinnate, segments linear-oblong, clavate, 2-3 mm long, often bifid into short lobes of unequal length; *texture* thin herbaceous; *veins* forked, veinlets simple, 1 to each segment, obscure, diaphanous; *sori* placed mostly some distance below the branching of veins, medial, small, round, about 1/3 as broad as the segment.

Distribution: Guangxi.

Plate 41. 1. Habit sketch (natural size.) 2. A pinna ($\times$ 2). 3. A pinnule, showing venation and sori ($\times$ 4). A scale from rhizome ($\times$ 24).

POLYPODIUM DAREAEFORMIOIDES Ching
乌柄水龙骨

根状茎匍匐生于地上,屈折,径 4 mm,覆以殊密三角形刺头暗褐色筛孔状之鳞片;叶柄相距 1.4—2 cm,细瘦,无鳞片,有光泽,淡稻秆色,长 9—20 cm;叶体披针长圆形,一次羽状,长 26—35 cm,宽 12—15 cm,几薄膜质,绿色,两面光滑,叶轴细瘦无鳞片,每侧小叶 6—10 个,生于长尖具深缺刻圆齿之顶端小叶下,几对生,相距 2—2.5 cm,披针形,渐尖头,不显著缺刻圆齿状,基部斜截形,略成心脏形,两侧皆具扩大之耳,耳圆形,在下侧者较在上侧者稍发达,下部小叶与上部小叶长短相同,横列开张,无柄,独立,下侧之耳常掩覆叶轴,长 6—8 cm,近基部宽 8—14 mm,上部小叶多少附着,最近顶端之一对侧生小叶基部附着甚阔,向叶轴延展甚长,顶端小叶最大,长 9—12 cm,近基部作深缺刻齿牙状;脉纹虽细而清晰,中肋两面隆起,网孔颇大,中肋之两侧各有一行,包括一具子囊群之细脉,稀具两条或近顶处又分叉,其余支脉游离,支脉至近叶缘处而止,皆具粗肥顶端;子囊群形小而圆,生于表面,疏散于中肋及叶缘之间。

分布:云南。

此特殊美丽之种与 *Polypodium subauricultum* Blume. 相近,但本种形小,细瘦,质薄,小叶亦短而少,叶脉网孔大,单行排列,子囊群亦较小,在中肋与叶缘之间甚疏,此种最初为 A. Henry 发现于云南,生于树干或林中积石上,*Polypodium aspersum* Baker 与 Christ 氏之原型种相符合。

图注:1.本种全形(自然大);2.一小叶示其叶脉及子囊群(放大 2 倍);3.根状茎上之鳞片(放大 76 倍)。

POLYPODIUM MENGTZEENSE Christ

POLYPODIUM MENGTZEENSE Christ in Bull. Boiss. **6**: 869 (1898); C. Chr. Ind. 544 (1906) Matthew in Journ. Linn. Soc. **39**: 381 (1911).

Polypodium aspersum Baker in Kew Bull. (1898) 231 (non *P. adspersum* Schrad. 1818, nec Blume 1828).

Polypodium argutum Wall. var. *khasianum* Clarke in Ferns N. Ind. (1880).

Rhizome creeping, epigaeous, flexuose, 4 mm thick, clad in moderately dense deltoid-subulate dark brown clathrate scales; *stipes* 1.4-2 cm apart, slender, naked, shining, pale straminous, 9—20 cm long; *frond* lanceolate-oblong, simply pinnate, 26-35 cm long, 12-15 cm broad; *texture* almost membranaceous, green and glabrous on both surfaces, rachis slender and naked; pinnae 6-10 on each side under the long attenuate deeply inciso-crenate apex, subopposite, 2-2.5 cm apart, lanceolate, acuminate, obscurely inciso-crenate, base obliquely truncate, somewhat cordate, dilatato-auricled on both sides, auricles rounded, the lower one more developed than the upper, the lower pinnae as long as the upper ones, horizontally patent, free, sessile, with the lower auricle usually imbricating the rachis, 6-8 cm long, 8-14 mm broad above the base, the upper ones more or less adnate, the uppermost one pair below the apex broadly adnate and decurrent some distance along the rachis, the terminal pinna the largest, 9-12 cm long, and generally deeply inciso-crenate near the base; *veins* slender but distinct, costa raised on both surfaces, areolae large, uniseriate along both sides of the costa usually with 1 included soriferous veinlet, or very rarely 2, or forked at the apex, the other veinlets free falling some distance short from the margin, all with clavate apex; *sori* small, globose, superficial, sparsely disposed midway between the costa and margin.

Distribution: Yunnan.

A very distinct, elegant fern and evidently a close ally to *P. subauriculatum* Blume, from which differs in slender habit, smaller size, fewer and shorter pinnae of still thinner texture, uniseriate large areolae along the both sides of the midrib and a few sparse smaller and remote sori midway between the midrib and margin. It was discoverd by A. Henry in Mengtze, Yunnan, on wooded rocks or on the trunk of trees. *P. aspersum* Baker is conspecific with Christ's type.

Plate 42. Fig. 1. Habit sketch (natural size). 2. A pinna, showing venation and sori ($\times$ 2). 3. A scale from the rhizome ($\times$ 76).

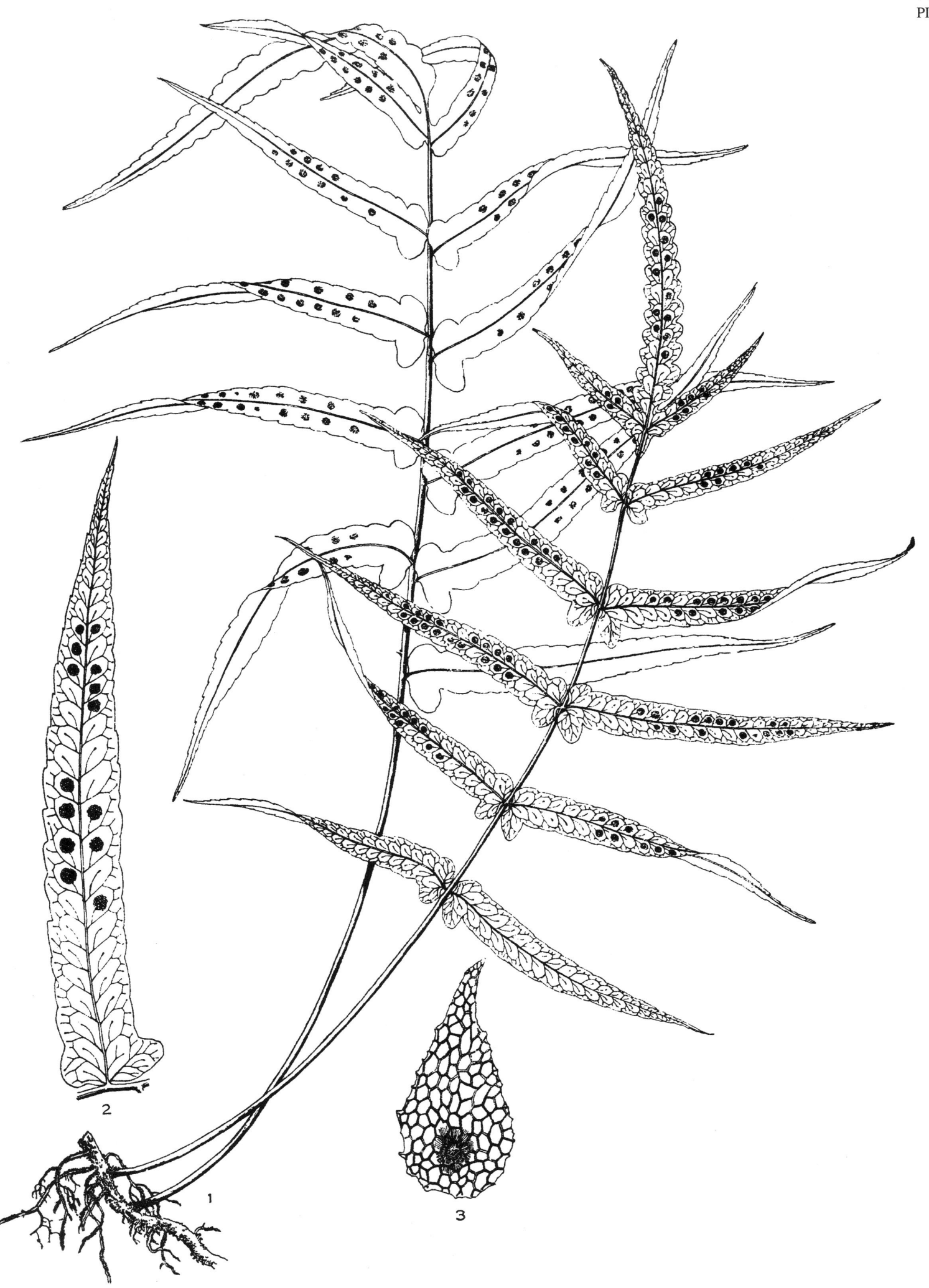

POLYPODIUM MENGTZEENSE Christ
蒙自水龙骨

图版 43　珠带水龙骨

根状茎生于地上，匍匐甚广，密覆以广披针形几黑色膜质筛孔状之鳞片，叶体相离 1—2 cm，窄线形，革质，长 25—40 cm，中部宽 3—8 mm，渐向下窄缩，形成短柄（柄长 2—4 cm），长渐尖头，正面绿色光滑，下面疏生短小盾状褐色脱落之鳞片，边缘成不显著波纹状，反卷殊窄；叶脉通常皆隐没，中肋则显明，下面尤甚，子囊群宽大，生于边缘，相离 1—2 cm，几对生或互生，长圆形，生于表面，泡肿状，长 4 mm，成熟时突出叶缘，殊显明，因而无子囊群之部分乃成反卷状。

分布：云南。

本种最初为 A. Henry 在云南发现，生于树干上，其大型子囊群突出于长线形叶体之外，最易在野外识别也，本种习性与 *Polypodium eilophyllum* Diels 最相近，但叶体较窄，反卷较少，子囊群突出叶缘之外。

图注：1. 本种全形（自然大）；2. 根状茎之鳞片（放大 30 倍）；3. 叶体之一部示其叶脉（放大 30 倍）。

POLYPODIUM OBLONGIOSORUM C. Christensen

POLYPODIUM OBLONGIOSORUM C. Chr. Ind. 549 (1906); Matthew in Journ. Linn. Soc. **39**: 381 (1911).

Polypodium subintegrum Baker in Kew Bull. (1898) 231 (non 1877).

Rhizome epigaeous, wide-creeping, clothed with broad lanceolate blackish membranaceous clathrate dense scales; *frond* 1-2 cm apart, narrowly linear, coriaceous, 25-40 cm long, 3-8 mm broad at the middle, gradually attenuate towards the base into a short stipe (2-4 cm long), apex longacuminate, green and glabrous above, the under side sparsely covered with minute peltate brown deciduous scales, margin obscurely repandulous, narrowly revolute; *veins* obscurely immersed except costa, which is prominent, particularly underneath; *sori* large, marginal, 1-2 cm apart, subopposite or alternate, oblong, superficial, bullate, short oblong, 4 mm long, prominently projecting beyond the margin of the frond when mature, in consequence of the free parts of the margin becoming revolute.

Distribution: Yunnan.

A very remarkable plant, first discovered by A. Henry in Manmei, Yunnan, epiphytic on the trunk of forest trees, and easily distinguished by its relatively large bead-like sori projecting well beyond the margin of long linear frond. From *P. eilophyllum* Diels to which the present species is closely allied, it differs in narrower, less revolute frond and the projecting marginal sori.

Plate 43. Fig. 1. Habit sketch (natural size). 2. Scales from rhizome (× 30). 3. A portion of frond, showing venation (× 30).

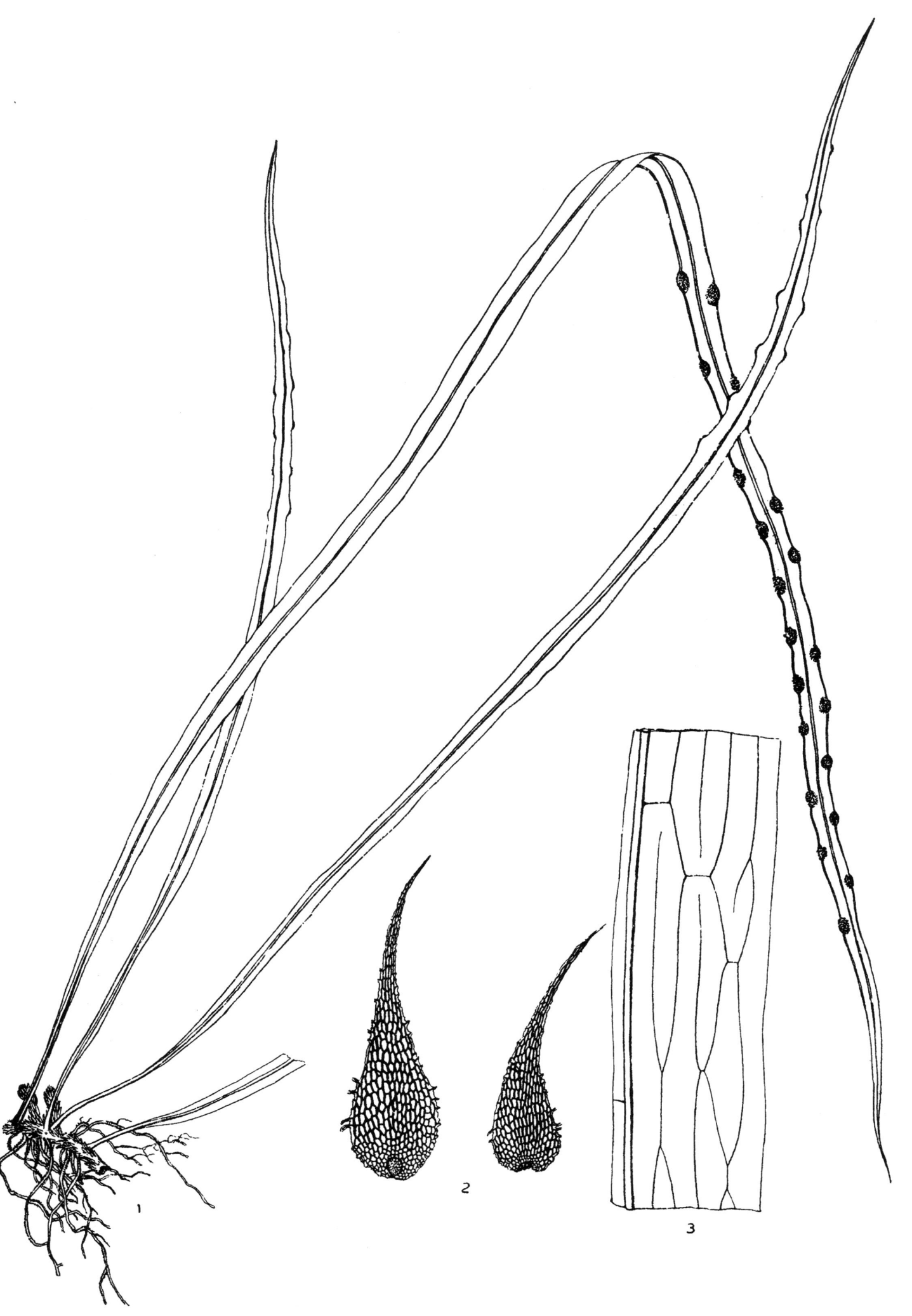

POLYPODIUM OBLONGIOSORUM C. Christensen
珠带水龙骨

图版 44　三叶水龙骨

根状茎匍匐,生于地上,覆以小披针形膜质锈色密集之鳞片;叶柄几无鳞片,细瘦,褐色,长 30—45 cm;叶体三角形,长 30—40 cm,三裂成 3 个向上披针形之裂片,中央裂片最大,长达 40 cm,两侧者略短,宽 4—5.5 cm,两端渐窄缩,中间凹处圆形,削入至距叶基 2—3 cm 处,全缘或不显着波纹状;叶体纸质,上面平滑,绿色,下面淡色,中肋疏被以紧贴膜质暗褐色盾状或卵圆形刺头之鳞片;脉纹甚显明,侧脉平行直伸向外开张,几达叶缘,中间细脉联络成小六角形之网孔,包有内伸游离之细脉;子囊群形小,长圆形,生于表面,有鳞片,一列或近中肋处成二列。

分布:云南。

本种最初为 A. Henry 博士发现于云南山地森林中,但至今尚无采之于他地者;本种习性与 *Polypodium pteropus* Blume 之壮大者相近,但异处则为殊大之子囊群,叶下面生长之卵圆披针形,具纤毛暗褐色之鳞片,及子囊群中间深齿牙状,显然成筛状斑纹盾状隔丝;本种之子囊群比同属他种所有者皆窄,在基部者为长圆形,在顶端乃成几圆形,短线形老亦几或有之,多在基部近中肋处,与中肋几成平列,以是本种当视为 *Sellignea* 及 *Pleopeltis* 两系之中间种。

图注:1. 本种全形(自然大);2. 小叶之一部示叶脉及孢子囊群(放大 2.5 倍);3. 根状茎之上鳞片(放大 30 倍);4. 示叶之下面叶脉上的鳞片(放大 46 倍);5. 子囊群上之鳞片(放大 46 倍)。

POLYPODIUM TRIGLOSSUM Baker

POLYPODIUM TRIGLOSSUM Baker in Kew Bull. (1898) 232; C. Chr. Ind. 571 (1906);
　　Matthew in Journ. Linn. Soe. **39**: 383 (1911).
　　Selliguea tripylla Christ in Bull. Boiss. **6**: 878 (1899).

Rhizome creeping, epigaeous, clothed with small lanceolate, membraneceous, ferruginous dense scales; *stipes* subnaked, slender, brown, 30-45 cm long; *frond* deltoid, 30-40 cm long, trisect into 3 ascending lanceolate segments, the middle one the largest, to 40 cm long, the two lateral ones somewhat shorter, 4-5.5 cm broad, gradually tapering towards both ends, sinuses rounded, cleft down to 2-3 cm from the base of the frond, margin entire or obscurely undulate; *texture* chartaceous, green and glabrous above, pale and sparsely clothed near the costa beneath with adpressed membraneceous dark-brown peltate or ovate-cuspidate scales; *venations* prominent, the lateral veins parallel, erect-patent, extending almost to the margin, intermediate veinlets anastomosing copiously into small hexagonal areolae with free included veinlets; *sori* small, oblong, superficial, scaly, uniseriate or tending to be biseriate along the costa between main veins.

Distribution: Yunnan.

The discovery of this beautiful species was indebted to Dr. A. Henry, who found it in the mountain forest in the district of Mile, Yunnan, and so far not yet known elsewhere. It is nearest to the large trifid forms of *P. pteropus* Blume, differs in larger sori, the presence of ovate-lanceolate dentate-ciliate dark brown scales on the under surface, the strongly dentate and conspicuously clathrate peltate scales in the sori and by the veinlets not forming a distinct series of large areolae. It is perhaps most interesting to note that this species is distinct from most of the members in § *Selliguea* in rather very short sori, which vary from oblong to almost globular towards the apex of the segments; a few short liner ones are, however, occasionally present near the basal part and very close and parallel or subparallel to the costae. In this respect it is not perhaps without reason to consider it as a linking species between § *Selliguea* and § *Pleopeltis*, which generally has sori not more than three times as long as broad.

Plate 44. Fig. 1. Habit sketch (natural size). 2. A portion of pinna, showing venation and sori (× 2.5). 3. Scales from rhizome (× 30). 4. Scales from the veinlet on the under surface of the pinna (× 46). 5. Scales from the sori (× 46).

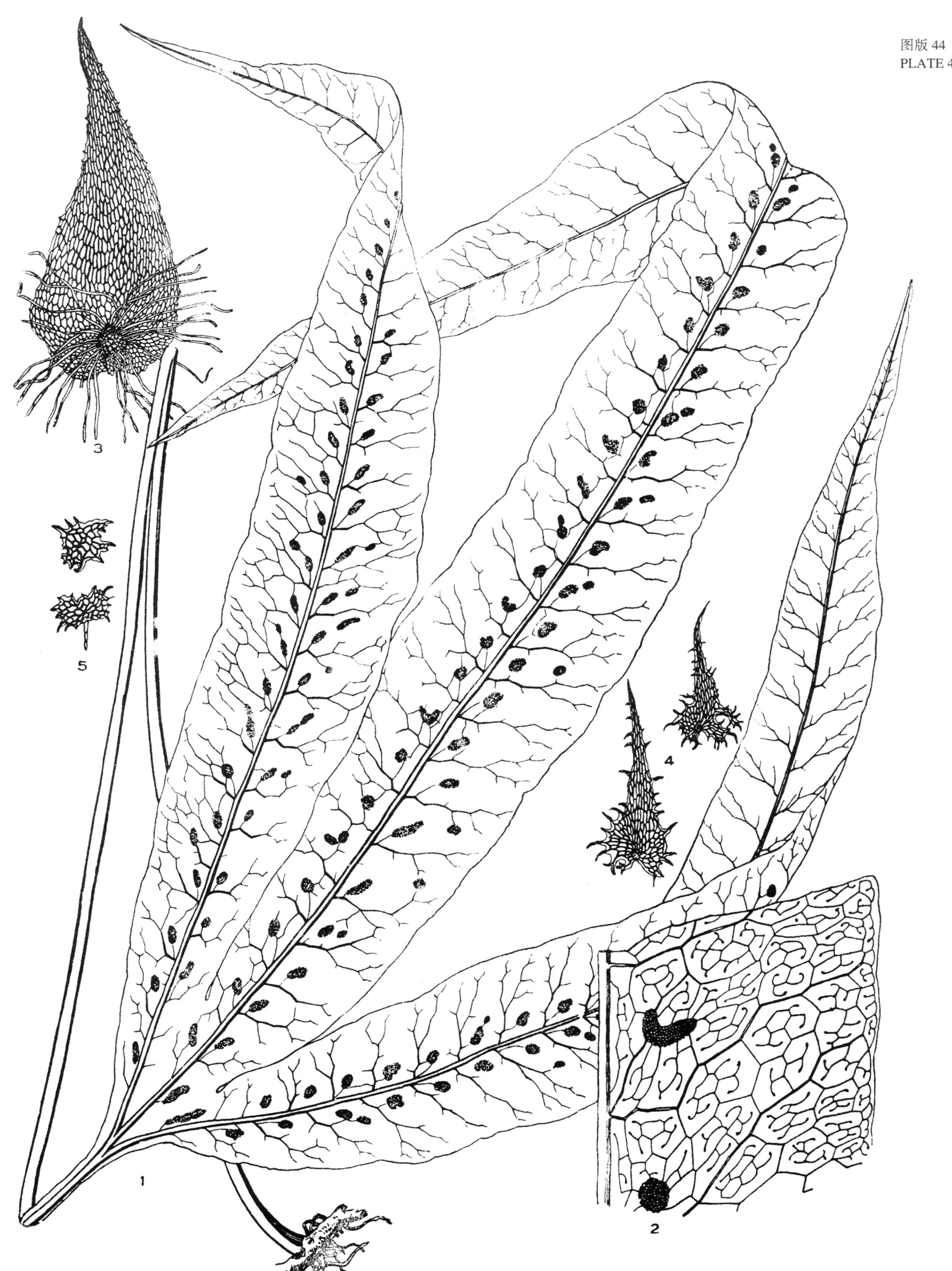

POLYPODIUM TRIGLOSSUM Baker
三叶水龙骨

图版 45　椭圆水龙骨

根状茎粗壮，木质，匍匐甚广，几黑色，密覆以披针形渐尖头几黑色筛孔状贴附之鳞片，叶柄疏生长 20—40 cm，坚实，直立无鳞片，淡黄或稻秆色，叶体长 20—25 cm，宽 13—25 cm，或略宽，长卵圆形，羽状深裂至中肋，每侧约有小叶 4—10 个，顶端之小叶与两侧相同；小叶细披针形或披针长圆形，渐尖头，宽 1—3 cm，长 5—20 cm，平行开张，近基部者形体稍较短小，多数以斜向下延之基部相连，形成叶轴之翼，小叶全缘，或成不显著波纹或皱波状；叶体草质，殊光滑，主脉细，至近叶处不显明。中间细脉联络成一行大形网孔，中含支脉作双叉状，在近边缘处相连，子囊群线形，斜生，几达中肋但距叶缘则较远。

分布：亚洲热带及中国：广东，广西，云南，四川，湖北，江西，安徽，浙江，江苏，福建；朝鲜半岛及日本亦有之。

本种在中国南部及中部分布甚广，形体几微之变异亦多，高者达 80 cm，矮者仅 15 cm，侧生小叶自二至十个，长 5—25 cm，宽乃由 8 mm—4 cm，*Polypodium flexilobium* Christ 及 *Polypodium fauriei* (Christ) Nakai. 当认为本种之形态稍异者，此由作者比较多数标本而知之。——仁昌

图注：1. 本种全形（自然大）；2. 叶之下面一部分，示叶脉之分布（放大 2 倍）；3. 根茎上之鳞片（放大 16 倍）。

POLYPODIUM ELLIPTICUM Thunberg

POLYPODIUM ELLIPTICUM Thunb., Fl. Jap. 335 (1784); Christ, Farnkraut. Erde 107; Diels in Nat. Pfl. Fam. 1. 4. 318; v. A. v. R., Malay. Ferns 677 (1908).

Gymnogramme elliptica Baker in Syn. Fil. 388.

Selliguea elliptica Bedd., Handb. 392

Gymnogramme decurrens Hook., Spec. Fil. 5; 161

Selliguea decurrens Bedd., Ferns Brit. Ind. t. 150.

Grammitis decurrens Wall., HK. et Grev. in Ic. Fil. t. 6.

Gymnogramme pentaphylla Baker in Kew Bull. (1898) 233.

Rhizome thick, woody, wide-creeping, blackish, clad in dense lanceolate acuminate, blackish, clathrate more or less adpressed scales; *stipes* scattered, 20-40 cm long, firm, erect, naked, pale or straw-coloured; *fronds* 20-50 cm long, 13-25 cm or more broad, oblong-ovate, pinnatifid down to the rachis into 4-10 rarely more pinnae on each side under the terminal segment similar to the lateral ones; pinnae linear-lonceolate or lanceolate-oblong, acuminate, 1-3 cm broad, 5-20 cm long, horizontally patent, the lower ones scarcely reduced, mostly connected by decurrent oblique bases, which form a wing to the rachis, margin usually entire, sometimes obscurely undulate or repand; *texture* herbaceous, quite glabrous; main *veins* slender, not distinct to the edge, intermediate veinlets anastomosing copiously with a row of large costal areolae with included bifid veinlets, all jointed near the margin; *sori* linear, oblique, almost reaching the midrib, but not the margin.

Distribution：Tropical Asia; China: Guangdong, Guangxi, Yunnan, Sichuan, Hubei, Jiangxi, Anhui, Zhejiang, Jiangsu, Fujian, also in Korea penninsula and Japan.

A Common but most variable fern in South and Central China. It often varies from 15-80 cm tall with 2-10 lateral pinna 5-25 cm long, 0.8-4 cm broad. In extreme forms, the pinnae are only about 5 mm broad. *P. flexilobium* Christ and *P. fauriei* (Christ) Nakai are better regarded as reduced forms of the present species, as the ample materials at my disposal show all gradations from Christ's *P. flexilobium* upward. — R. C. C.

Plate 45. Fig. 1. Habit sketch (natural size). 2. A portion of frond, showing venation ($\times$ 2). 3. Scales from rhizome ($\times$ 16).

POLYPODIUM ELLIPTICUM Thunberg
椭圆水龙骨

图版 46　莱氏水龙骨

根状茎匍匐甚广,宽 3 mm,黑色,密覆以黑色疏生坚硬刚毛状之鳞片;叶柄稻秆色,独生,相距 1—3 cm,细瘦,下面有棱,长 10—15 cm,叶体基部渐向下延展形成窄长之翼;叶体长 20—30 cm,中部宽 2.2—4 mm,或稍宽,窄披针形,渐尖头,基部突然窄缩,形成叶柄之翼,边缘卷皱,依稀成圆齿状,中肋两面皆明稻秆色,侧脉斜向开张,伸至叶缘与之成锐角,直出或几弯屈,每侧有 30 条,中部者约相距 0.5 cm,甚细,横显脉中间之网孔约 5 个,长圆扁方形,包括一简单棒形内卷之细脉,子囊群生侧脉上,从中肋至叶缘,甚细,褐色;叶体草质,柔软,淡绿色,两面光滑。

分布:云南,贵州,广西。

本种与 *Polypodium wightii*（HK.）Mett. 相近,但后者之叶比本种窄瘦,下部亦不缩成叶柄之翼状体,且细脉分叉较多,此种最初为 Père Cavalerie 于 1904 年采自贵州之溪畔。

图注:1.本种全形(自然大);2.叶体之一部示其脉纹及孢子囊群(放大 2 倍);3.孢子囊(放大 145 倍);4.根状茎上之鳞片(放大 30 倍)。

POLYPODIUM LEVEILLEI（Christ）C. Christensen

POLYPODIUM LEVEILLEI (Christ) C. Chr. Ind. Suppl. 60 (1906-13).

Selliguea léveillei Christ in Bull. Geogr. Bot. Mans (1906) 236.

Rhizome wide-creeping, 3 mm across, black, clothed with black sparse rigid setaceous scales; *stipe* stramineous, solitary, 1-3 cm apart, slender, sharply angular below, 10-15 cm long, with the base of frond gradually decurrent along it into a long narrow wing on each side; *frond* about 20-30 cm long, 2.2-4 cm broad, or broader at the middle, narrowly lanceolate, acuminate, suddenly narrowed near the base and long decurrent along the stipe, margin crispato-undulate and obscurely crenate; *casta* prominent on both sides, stramineous, lateral veins obliquely patent, extending to the margin at acute angles, straight or subflexuose, about 30 on each side, the middle ones 0.5 cm apart, very slender, areolae between transverse veins about 5, oblong-rectangular with simple clavate retrorsed included veinlets; *sori* running along the lateral veins from the costa to the margin, very slender, brown *texture* flaccid, herbaceous, light green, and glabrous on both surfaces.

Distribution: Yunnan, Guizhou, Guangxi.

A close ally to *P. wrightii* (HK.) Mett. which, however, has much narrower frond, not abruptly narrowed towards the base and numerous copiously forked, included veinlets. A pretty fern, discovered by Père Cavalerie in Guizhou, 1904, growing on moist banks of stream.

Plate 46. Fig. 1. Habit sketch (natural size). 2. A portion of frond, showing venation and sori ($\times$ 2). 3. Sporangium ($\times$ 145). 4. Scales from rhizome ($\times$ 30).

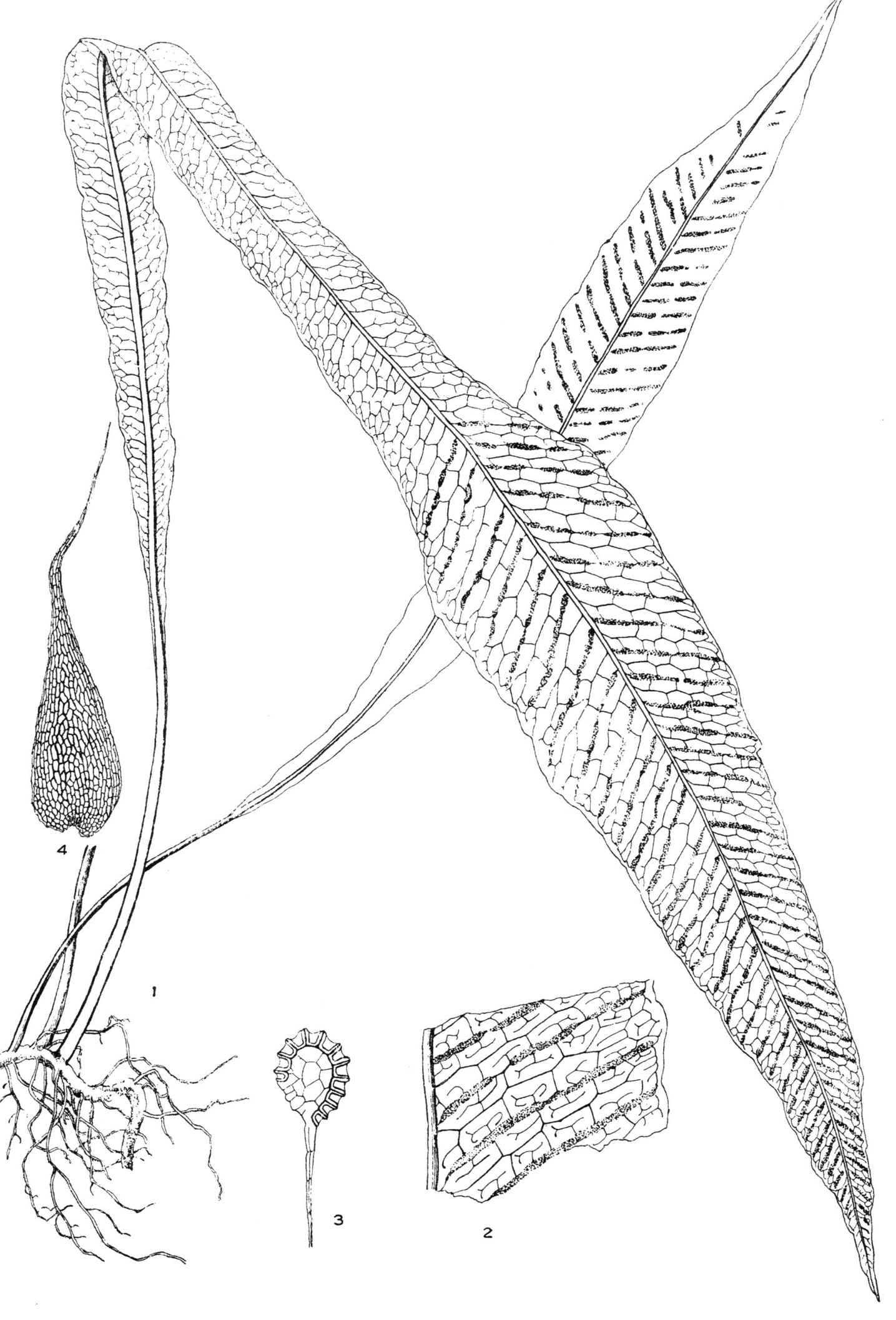

POLYPODIUM LEVEILLEI (Christ) C. Christensen
莱氏水龙骨

根状茎匍匐甚广，粗 2—3 mm，鳞片暗褐色，密生，贴附或几开张，基部殊宽，渐向顶端窄缩成长尾状，显明筛孔状；叶柄无或殊短；叶体几密接，淡绿色，光滑，形态约分二种，长 5—14 cm，最宽部分宽达 7 mm，无子囊群叶略宽（达 9 mm）而较短，自上部 1/3 处最宽，渐向下窄缩形成短柄，基约宽 2 mm，顶端尖形或钝形，全缘，反卷，几微成波状；中肋细，上面隆起，下面平或微有槽，侧脉及支脉不显明；叶体革质，柔软；子囊群斜生或几平行，通常倾向于合并，不生于中肋上，但几达叶缘，长 3—8 mm，仅限于叶体上部 1/3。

分布：广西。

图注：1.本种全形(自然大)；2.叶体之一部示叶脉及子囊群(放大 4 倍)；3.根状茎之鳞片(放大 24 倍)。

LOXOGRAMME CHINENSIS Ching

LOXOGRAMME CHINENSIS Ching in Sinensia **1**: 13 (1929).

Rhizome wide-creeping, 2-3 mm thick, scales dark brown, dense, adpressed or subpatent, broad at base, gradually attenuate to a long subulate apex, distinctly clathrate; *stipes* none or very short; *frond* subapproximate, pale green, glabrous throughout, somewhat dimorphic, 5-14 cm long, to 7 mm broad in the broadest part, the sterile ones broader (9 mm) and generally shorter, broadest in the upper one-third, slowly narrowed downward to the base or a short stipe, about 2 mm broad near the base, apex acute to blunt, margin entire, revolute, somewhat undulate; *midrib* slender, raised above, flattened or slightly grooved below, veins and veinlets hidden; *texture* coriaceous, but flaccid; *sori* oblique to subparallel, more often tending to be confluent, not costal, but well extending to the margin, 3-8 mm long, confined to the upper one-third portion of the frond.

Distribution: Guangxi.

Plate 47. 1. Habit sketch (natural size). 2. A portion of frond, showing venation and sori (× 4). 3. Scales from rhizome (× 24).

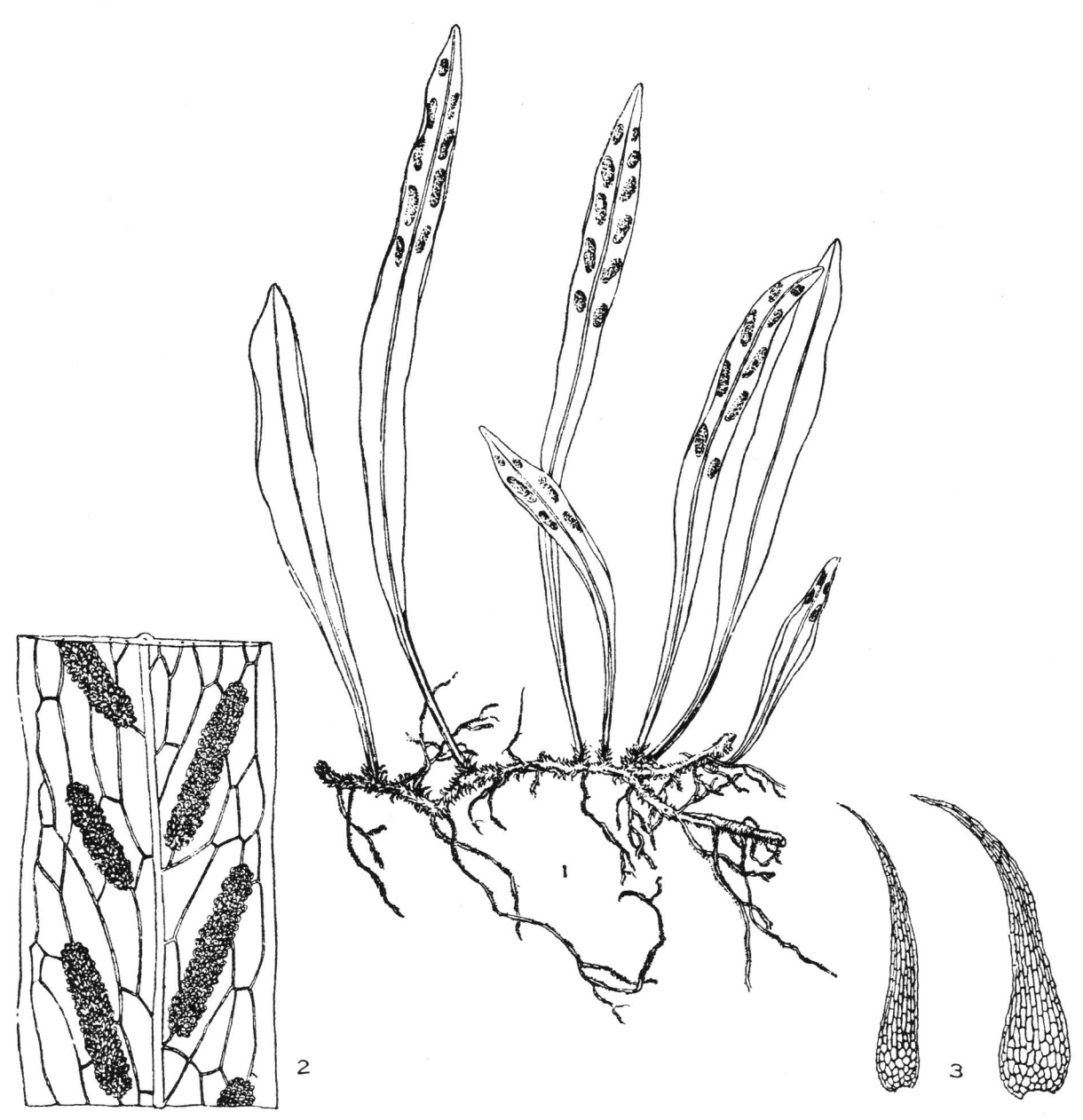

LOXOGRAMME CHINENSIS Ching
华剑蕨

图版 48　掌状扇蕨

　　根状茎生地面上,匍匐甚广,宽约 7 mm,密被鳞片,鳞片大,披针形,有尾状尖端,薄膜质,暗褐色,覆瓦状排列,有纤毛;叶柄相距甚远,长 30—45 cm,坚硬,全体裸露,稻秆色,底部管状,上部具 3 条深纹,叶体方形扇状,二裂掌状分裂,基部楔形,长阔各 25—30 cm 之间,或略宽,纸质绿色,上面平滑,下面具疏散之短小褐色之鳞片,叶片长 10—12 cm,长披针形,略相掩覆,中央之裂片长 17—20 cm,宽 2.5—3.5 cm,两侧之裂片则较小,全缘,中肋甚显明而略隆起,下部尤甚;脉络交错成六角形之网孔;子囊群生于表面,常集合于叶裂片之下部,外部裂片往往无子囊群或略有数枚生于中肋附近;子囊群在下部者形长方或长线状,在上部者圆形。

　　分布:云南,四川。

　　本种在 1883 年已为 Père Delavay 采自云南,但其后十五年 A. Henry 氏又在该省采得,始为植物学者注意。Baker 氏定为新种,名曰 *Polypodium palmatopedata*,至 1905 年,Christ 氏发现前者之标本于巴黎自然博物馆始定今名。本种性质介乎 *Dipteris* 及 *Polypodium* 之 *Pleopeltis* 系之间,此三者皆产云南,相杂而生,但本种生于中肋两侧之子囊群,及其宽而坚实由背面脉络分化而成之子囊托,则为最显著之特性,易与其他区别者也。

　　图注:1.本种全形(自然大);2.叶片一部示其脉络与子囊群(放大 3 倍);3.孢子囊(放大 90 倍);4.根状茎之鳞片(放大 40 倍)。

NEOCHEIROPTERIS PALMATOPEDATA (Baker) Christ

NEOCHEIROPTERIS PALMATOPEDATA (Baker) Christ in Bull. Soc. Bot. Fr. 52; Mém. I. 21 (1905) C. Chr. Ind. 432 (1906).

Polypodium palmatopedatum Baker in Kew Bull. (1895) 232.

Cheiropleris henryi Christ in Bull. Boiss, 6; 876 (1898); Diels in Nat. Pfl. Fam. 1. 4. 189. fig. 98 (1902).

Cheiropteris palmatopedata Christ in Bull. Boiss. 7; 21. t. 1 (1899);

Rhizome epigenous, wide-creeping, about 7 mm thick, clothed in large lanceolate, cuspidate membranaceous sordid brown imbricate ciliate dense scales; *stipes* distant, 30-45 cm long, stiff, naked throughout, stramineous, teret underneath, deeply 3-striate above; *frond* flabellate-quadrate pedate-palmatifid, base cuneate, 25-30 cm long and broad or broader; *texture* chartaceous, green and glabrous above, sparsely clothed with minute brown scales beneath; segments 10-20 cm long, upright, lanceolate, slightly imbricate, the middle ones 17-20 cm long, 2.5-3 cm broad, the lateral ones much smaller, margin entire; *costa* prominently raised below and less so above, veins anastomosing into small hexagonal areolae; *sori* superficial, usually confined to the lower half of the segments, the outermost segments sterile or at best sparsely soriferous, close to the costa, the lower ones oblong or linear-oblong, the upper ones globose.

Distribution: Yunnan, Sichuan.

This unique fern was first discovered by Père Delavay in the mountains in Tapingtze, Yunnan, September 4th. 1883, and by A Henry at Mile in the same province about fifteen years later, but it was the latter's specimens which received the first attention from Baker, who called it *Polypodium palmatopedatum*, while the former's specimen remained unkown in the Museum d'histoire naturelle de Paris until about 1905 when it was identified by Christ, who changed his *Cheiropteris palmatopedata* to the present name[1]. This fern must be considered as one of the most outstanding additions to our Chinese flora, discovered by early European botanical explorer in this country. As already quite thoroughly discussed by Christ[2], the present fern is entitled a systematic place between *Dipteris* and *Polypodium* § *Pleopeltis*, both of which are represented in Yunnan by their respective members growing side by side with our species there. However, the plant differs from either of its relatives in large elongated costal sori and, particularly, in the broad and incrassate receptacle, formed by the dilatation by the dorsal part of the costal viens, on which the sorus is borne.

Plate 48. Fig. 1. Habit sketch (natural size). 2. A portion of segment, showing venation and sori (× 3). 3. Sporangium (× 90). 4. Scales from rhizome (× 40).

---

[1] *Cheiropteris* was preoccupied in Kurr, Gen. Palaentologique.
[2] Bull. Boiss. 7; 21. 2. 1899.

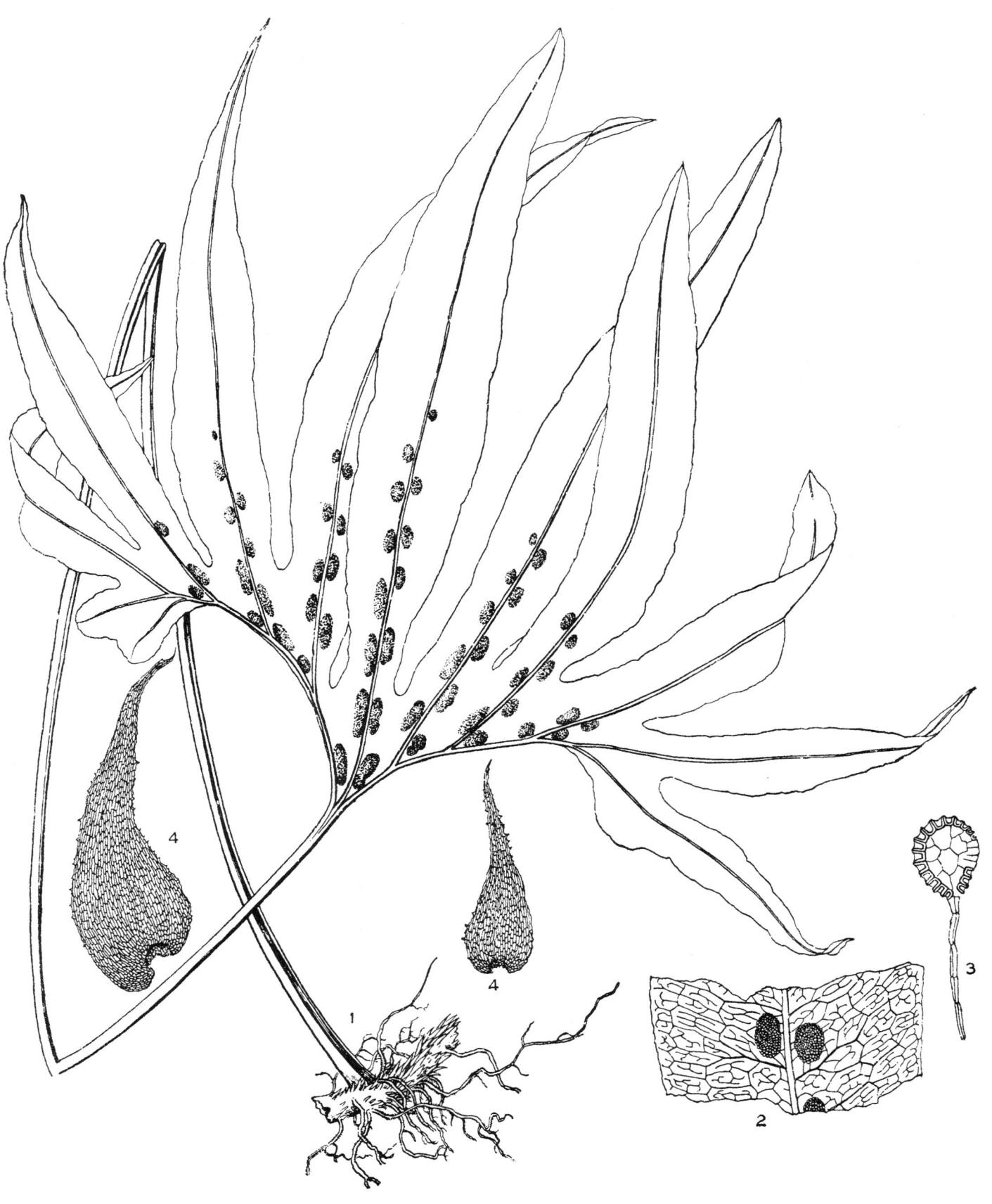

NEOCHEIROPTERIS PALMATOPEDATA (Baker) Christ
掌状扇蕨

根状茎短,粗壮,有多根;叶柄4—6个簇生,叶柄粗壮,丛生,径达3 mm,淡绿色,上面有深槽下面圆形,长3—6 cm,密覆以宽卵圆形,薄膜质,全缘,淡褐色,钝头之鳞片,鳞片长约5 mm;无子囊叶长15—30 cm,中间宽3—3.5 cm,刀锋披针形,自中部向上窄缩成渐尖而钝头之顶端,基部向下延长,叶缘完全,加厚,密覆以几黑色碎裂扩张贴附之小鳞片,下面尤甚;中肋甚阔而平,宽2—3 mm,淡绿色;侧脉甚斜,不显明,相距2 mm,直达叶缘,通常多单枝,亦有分叉者;叶体肥厚革质,上面浅绿色,下面淡绿色;有子囊叶略窄而短,叶柄长7—10 cm;子囊群暗褐色,密生于叶体下面,仅中肋附近及缘内无之。

分布:广东,广西,福建。

本种为一特殊之蕨,其特征在平阔之中肋,广阔之厚边缘及其肥厚革质之叶体,并下面密生之碎裂褐色鳞片,此种最初在1907年为Matthew采自香港附近阴翳之岩石上,近年曾在广西发现数次。其模式标本具三四粗厚无子囊叶,实为本种中之特殊情形。

按Dunn采自福建中部之标本3821号之 *Elaphoglossum parvulum* Copel. 实与本种无别,但形态略小叶略宽耳,但作者所鉴定之多数广西标本皆与Matthew氏之模式大小相同。——仁昌

图注:1.本种全形(自然大);2.两个孢子囊(放大106倍);3.叶柄基部之鳞片(放大10倍)。

ELAPHOGLOSSUM AUSTRO-SINICUM Matthew et Christ

ELAPHOGLOSSUM AUSTRO-SINICUM Matthew et Christ in Lecomte Not. Syst. **1**: 57 (1909); C. Chr. Ind. Suppl. 41 (1906-13); Matthew in Journ. Linn. Soc. **39**: 368 (1911).

Arostichum austro-sinicum Tutcher in Fl. Kwant. and Hongk. 355 (1912).

Elaphoglossum parvulum Copel. in Phil. Journ. Sci. Bot. C **11**: 40 (1916).

Rhizome short, thick, densely rooted; leaves 4-6 together, *stipes* tufted, thick, stout, 3 mm across, greenish, deeply sulcate above, terete below, 3-6 cm long, copiously clothed in broad, ovate, membranaceous, entire, scarious light brown, obtuse scales 5 mm long; sterile *frond* 15-30 cm long, 3-3.5 cm broad at the middle, cultrato-lanceolate, from the middle upward gradually acuminate in an obtuse apex, long decurrent towards the base, margin entire, thickened, surfaces, particularly the under, densely clothed with small blackish lacerato-dilated appressed scales; *costa* very broad, plane, 2-3 mm broad, pale green; lateral *veins* very oblique, inconspicuous, 2 mm apart, extending to the thickened margin, mostly simple or forked; texture carnoso-coriaceous, thick, light green above, pale green beneath; fertile fronds somewhat narrower, much shorter than the sterile, stipes 7-10 cm long; *sori* dark brown, covering the entire under surface except the midrib and a narrow free margin.

Distribution: Guangdong, Guangxi, Fujian.

A very distinct species, characterized by a broad flat midrib, broad thickened margin, thick carnoso-subcoriaceous texture and dense coating of lacerate-dilated brown scales underneath. It was first found by Dr. Matthew in Tai-mo-shan, New Territory, opposite Hong Kong, in 1907, on shaded granite cliff, and of late has been reported from several localities from Guangxi. The type specimen of the present species consisting of 3-4 robust sterile leaves represents rather an extreme form.

Elaphoglossum parvulun Copel. based upon Dunn's specimen, No. 3821, from Central Fujian (1905), is not specifically different from the type except of smaller size with somewhat broader leaves, only about half as long as the type. All the specimens from Guangxi I have examined are mostly approaching the type in dimension. —R. C. C.

Plate 49. Fig. 1. Habit sketch (natural size). 2. Venation (× 2). 3. Two sporangia (× 106). 4. Scales from the base of stipe (× 10).

ELAPHOGLOSSUM AUSTRO–SINICUM Matthew et Christ
华南舌蕨

图版 50　二尖燕尾蕨

　　根状茎木质,粗如手指,短而匍匐,密覆以几黄色有光泽之毛;叶柄数个簇生,无子囊叶之柄长 30 cm,或尤长,下面圆形,上面具宽槽,暗稻秆色,有光泽;无子囊叶体长 10—15 cm,稍窄,卵圆形,基部圆形,叶上部成两广三角形渐尖头之裂片,凹处宽阔圆形,全缘;叶体厚革质,上面绿色有光泽,下面稍淡两面皆无毛;主脉掌状,由叶柄顶端向叶缘辐射,其间有多列网孔,有子囊叶长 15—25 cm,宽 1—1.2 cm,舌状披针形,两端窄缩,简单,具三条主脉;子囊群暗褐色,密布于叶体下面,仅在中肋附近及叶缘内留一窄隙,叶柄长 45 cm。

　　分布:爪哇,菲律宾,苏门答腊,日本;中国:广西,台湾。

　　Cheiropleura 一属在中国发现实以 1928 年"中央研究院"广西采集团为嚆矢,本种由作者采自广西北部与贵州交界之石灰岩上,此次之标本皆具两裂片之叶体与 Blume 氏之模式标本无不符合,但台湾之变种 integrifolia Eat. 之具全叶者尚未见之于中国也。——仁昌

　　图注:1.本种全形(自然大);2.无孢子囊群之叶体之一部表示叶脉(放大 6 倍);3.有孢子囊群之叶体之横切面表示孢子囊群与隔丝之位置(放大 25 倍);4.根状茎上之鳞片(放大 25 倍);5.孢子囊群(放大 100 倍);6.隔丝(放大 40 倍)。

CHEIROPLEURIA BICUSPIS (Bl.) Presl

CHEIROPLEURIA BICUSPIS (Bl.) Presl, Epim. 189 (1839); Christ, Farnkraut. Erde 128, fig. 360; Diels in Nat. Pfl. Fam. 1. 4. 336, fig. 175; C. Chr. Ind. 181 (1905).

Polypodium bicuspe Bl., Enum. 125 (1828), Fl. Jav. Fil. t. 78 B.

Anapausia bicuspis Moore. Ind. XXI (1857).

Acrostichum bicuspe HK. Sp. Fil. 5: 271, Syn. Fil. 421.

Gymnopteris vespertilia HK. in Lond. Journ. Bot, 5: 193, t. 7-8 (1846).

Cheiropleuria vespertilia Presl; Epim. 190 (1849).

　　Rhizome woody, thick as man's small finger, short-creeping, densely clothed with yellowish silky hairs; *stipes* several together, those of barren frond 30 cm or more long terete below, broadly grooved above, dark stramineous, shining; barren *frond* 10-15 cm long and a little less broad in the entire portion, ovate, rounded at the base, the upper part consisting of 2 broad divaricated deltoid acuminate lobes with a broad rounded sinus between, margin entire; *texture* thick coriaceous, shining green above, pale below, glabrous on both surfaces; main *veins* palmate, radiating from the apex of the stipe to the upper edge, with copious areolae between them; fertile fronds 15-25 cm long, 1-1.2 cm broad, ligulate lanceolate, tapering on both ends, simple, with 3 prominent ribs; *sori* dark brown, densely covering the whole under surface except the midrib and a very narrow free margin, the stipe 45 cm or longer.

　　Distribution: Java, Philippines, Sumatra, Japan, China: Guangxi, Taiwan.

　　It is interesting to note here that the genus *Cheiropleuria* was not known from China prior to the "Academia Sinica" Kwangsi Expedition, 1928, when it was discovered by me for the first time in N. Guangxi on the border of Guizhou, in the chink of rather exposed limestone cliff. A very rare fern in the region. Our Chinese plant matches Blume's type very well in having bilobed frond. The var. *integrifolia* Eat. with simple entire ovate-acute or short acuminate barren leaves, commnon in Taiwan, is not yet known in this country. —R. C. C.

　　Plate 50. Fig. 1. Habit sketch (natural size). 2. A portion of sterile frond, showing venation ($\times$ 6). 3. A portion of cross section of fertile frond, showing disposition of sori and paraphyses ($\times$ 25). 4. Hair from rhizome ($\times$ 25). 5. A sporangium ($\times$ 100). 6. Paraphyses ($\times$ 40).

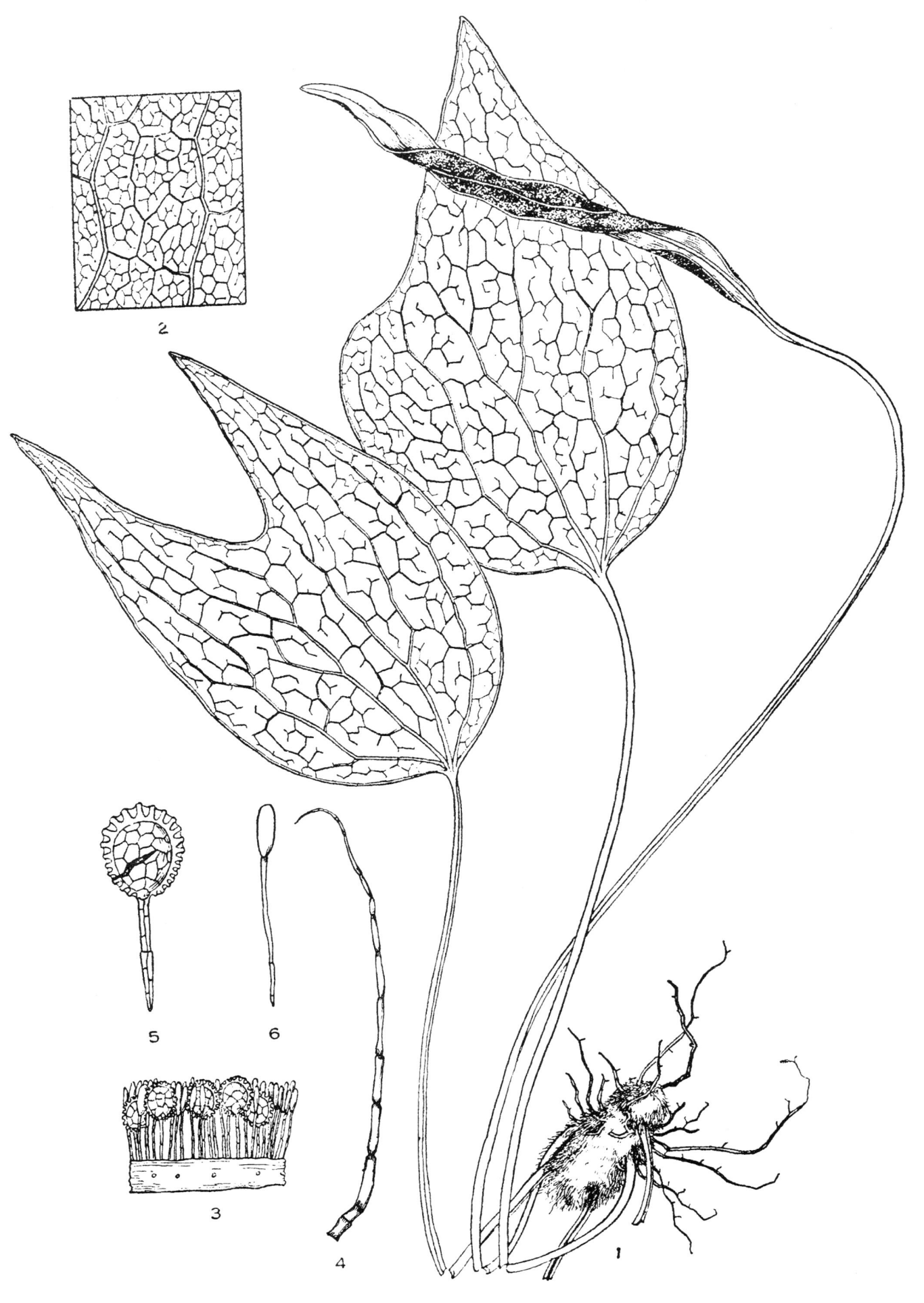

CHEIROPLEURIA BICUSPIS (Bl.) Presl
二尖燕尾蕨

中国蕨类植物图谱

ICONES FILICUM SINICARUM

BY

REN-CHANG CHING, B. S.

CURATOR OF HERBARIUM

FAN MEMORIAL INSTITUTE OF BIOLOGY

FASCICLE 2, PLATES 51-101

第二卷

TO

PROFESSOR HSEN-HSU HU, D. Sc.

DIRECTOR OF THE FAN MEMORIAL INSTITUTE OF BIOLOGY

AS ONE OF THE OUTSTANDING BOTANISTS OF PRESENT-DAY CHINA

IN RECOGNITION OF

HIS LEADERSHIP IN PROMOTING THE STUDY OF SYSTEMATIC BOTANY

IN CHINA AND

HIS SUPPORT IN VARIOUS MANNERS FOR THE ICONOGRAPHY OF CHINESE

FERNS FROM ITS INCIPIENCE AS ONE OF THE MOST RELIABLE

MEANS OF DISSEMINATING CHINESE PTERIDOPHYTIC KNOWLEDGE

THIS SECOND FASCICLE OF ICONES FILICUM SINICARUM

IS RESPECTFULLY DEDICATED

图版 51 高山篆蕨

根状茎广匍匐于岩石上；鳞片松密，黄色，开张，狭披针形，边缘具茸毛，腹部着生。叶疏生或颇密生，叶柄长仅 1.5—5 cm，被长毛，近基部有明显之骨节，由此脱落后，留一短柄于茎上，几为鳞片所掩。叶体长 15—30 cm，宽 2—3.5 cm，叶端急渐尖头，基部亚圆或短楔形，两边平行，略呈波状，薄纸质，中肋下面鳞片颇密；叶脉明显，两面均具短毛；子囊群位于中肋两边，为不规则之一列；子囊群盖圆肾形，膜质，被刚毛，以缺刻着生于叶脉。

分布：云南，四川，西藏；缅甸，锡金-喜马拉雅山区及马来半岛。

图注：1. 本种全形（自然大）；2. 叶体之一部，表明叶脉及子囊群之位置（放大 2 倍）；3. 叶面上之毛（放大 50 倍）；4. 根状茎上之鳞片（放大 50 倍）。

OLEANDRA WALLICHII (Hooker) Presl

OLEANDRA WALLICHII (Hooker) Presl, Tent. Pterid. 78 (1836); Hk. et Bak. Syn. Fil. 303 (1868); C. Chr. Ind. Fil. 467 (1906).

Aspidium wallichii Hooker, Exot. Fl. **1**: t. 5 (1823); Bedd. Ferns Brit. Ind. t. 265 (1868).

Neuronia asplenioides Don, Prod. Fl. Nepal. 6 (1825).

Oleandra wallichii var. *lepidota* Christ, Bull. Acad. Géogr. Bot. (1906) 140.

Rhizome horizontally creeping, densely shaggy scaly; *scales* ferruginous, spreading, linear-subulate, fimbriate, dorsally affixed; stipe distant or approximate, 1.5-5 cm long, ferruginously hairy, jointed close to the base, so that the very short, lower articulation is quite concealed among the scales; *lamina* 15-30 cm long, 2-3.5 cm broad, apex suddenly and sharply acuminate; base subrounded or rotundo-cuneate, both edges parallel, somewhat wavy; *texture* papyraceo-herbaceous, midrib densely scaly beneath, rather densely hairy on veins on both sides and densely ciliate along the edges; *venation* distinct, veins patent, fine, closely parallel, generally forked from near, or above the base; *sori* rather irregularly uniseriate on each side, close and parallel to the midrib; *indusium* orbicular-reniform, hispidato-ciliate, attached by a deep sinus and opening outwardly.

Yunnan: Shweli-Salwin Divide, *G. Forrest 11799*, *18581*, *24738*. Sichuan occid. *E. H. Wilson 5246*. Tibet: Adung-seninghku junction, *Capt. Kingdon Ward* (1926).

Burma, Sikkim-Himalayas, Malyan Peninsula.

A well marked species of the genus, common in the Sikkim-Himalayas and the southwestern part of Yunnan. In habit, it resembles O. cumingii J. Sm. from the Philippines and South China, which differs from the present species in rhizome scales being imbricatingly addressed, usually much higher articulation of the stipe and leaf-margin not densely ciliate.

Plate 51. Fig. 1. habit sketch (natural size). 2. portion of a lamina, showing venation and position of sori (× 2). 3. hairs from lamina (× 50). 4. scales from rhizome (× 50).

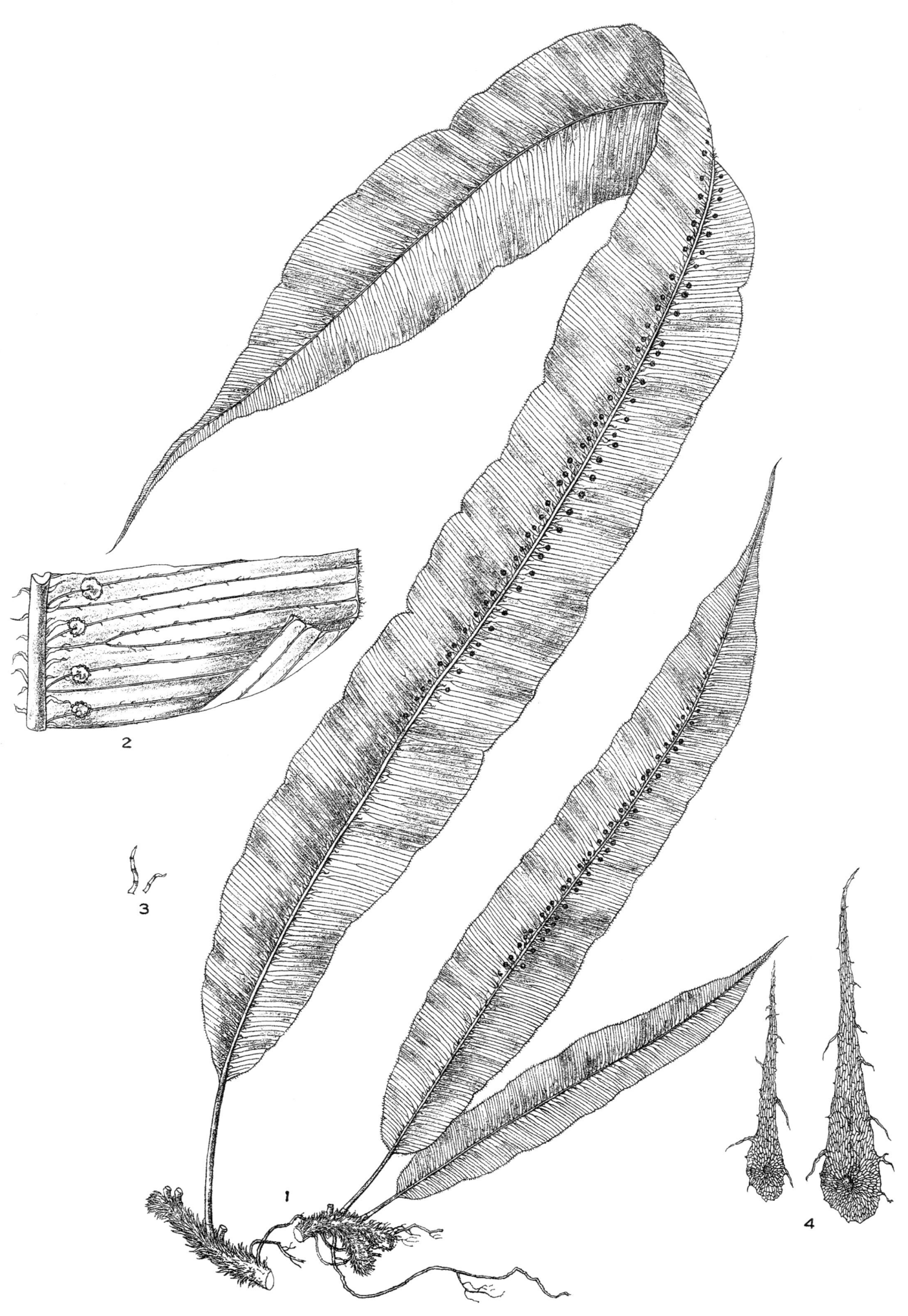

OLEANDRA WALLICHII (Hooker) Presl
高山蓧蕨

根状茎肥大，横生于土中；鳞片疏生，宽披针形，锈黄色，全缘，叶为一形或稍为二形，叶间距离大，柄长短不一，其在着生子囊群之叶者，长可 40 cm 以上，其在不生子囊群之叶者，长约半之，稻秆色，光滑，或下部鳞片疏生，叶体不一其形：或为披针形之单叶，或为三裂，或为羽状分裂，甚至为不完全的二次羽状分裂；其在羽状分裂者，小叶 1—3 对，长约 20 cm，宽约 2—3 cm，基部为具翅之中肋连着，顶部之小叶尤长；亚革质，边缘具小锯齿，叶脉不甚显明，于主脉之两旁成 3 列之斜长六角形网眼，近边缘分开；子囊群形长，生于主脉之两侧，具膜质全缘之盖向主脉开离。

分布：香港，广东，海南，广西，台湾，近发现于越南。

本种为本属特殊之一种，其异于其他产于亚洲之种者，为其全形或三裂或仅一位羽状分裂之叶体。W. kempii Cop. 者仅为本种之变形，以其两侧小叶有再行羽状分裂之倾向也。

图注：1.本种全形（自然大）；2.子囊群叶体之一部，表示其叶脉及子囊群之位置（放大 3 倍）；3.根状茎上之鳞片（放大 50 倍）。

WOODWARDIA HARLANDII Hooker

WOODWARDIA HARLANDII Hooker, Fil. Fxot. t. 7 (1857); C. Chr. Ind, Fil. 658 (1906); Nakai, Tokio Bot. Mag. **39**: 102 (1925); Ching, Bull. Fan Mem. Inst. Biol. **1**: 148 (1930); ibid. **2**: 2 (1931); Wu, Polyp. Yaoshan. in Bull. Dept. Biol. Sun Yatsen Univ. n. **3**: 280 t. 95 (1932); Ogata, Ic. Fil. Jap. **5**: t. 250 (1933).

Lorinseria harlandii J. Sm. Hist. Fil. 311 (1675).

Woodwardia kempii Cop. Phil. Journ. Sci. Bot.. **3**: 280 (1908).

Rhizome wide-creeping, hypogaeous, thick; scales broadly lanceolate, acuminate, brown, entire; *frond* uniform or subdimorphous, far apart, stipe variable in length, in fertile frond about 40 cm long, in sterile frond about half as long, straminous, naked or sparingly scaly below; lamina variable, from simple to trifid or regularly pinnate, or even irregularly bipinnatifid (*W. kempii*), in pinnate forms ovate-detoid, rounded at base, about 30 cm long, 20 cm broad, pinnae 2-3 on each side, more or less connected by a narrow wing along rachis, linear-lanceolate, acuminate, about 20 cm long, 2-3 cm broad, the terminal segment much the longest, margin sinuate, rather irregularly serrulate; *texture* subcoriaceous; *veins* quite visible in young state but obscure in fully grown leaves, regularly anastomosing in 3 rows of elongate, oblique, large, subhexagonal areolae; *sori* linear-oblong, rather short, in pairs, close to the midrib, *indusium* membranaceous, vaulted, opening towards midrib.

Hong Kong. *Dr. Harland* (type); *Lorraine*; Victoria Peak, C. *Wilford 15*; *Matthew 359, 360, 361* (1904). Guangdong: Tai-mo Shan, *Bodinier 1191*; Loh-fau Shan, *C. Ford* (1883); Swatow, Thaiyong, *Dr. Dalziel*, July, 1901; Lockchong, *N. K. Chun*, 42688; Yao Shan, *S. P. Ko 51973*. Guangxi: Yao Shan, Peniu, *S. S. Sin & K. K. Wang 177* (1928). Hainan: Ng Chi Leng, Fanyah, *C. L. Tso & N. K. Chun 44180*, Oct. 26, 1932, in forest ravine. Taiwan.

Also Vietnam.

An unique species of the genus, differing from all other known Asiatic species in simple or trifid or simple pinnate frond, with 2-3 pairs of linear-lanceolate, long segments. *W. kempii* Cop. is only a form with pinnae being more or less lobed.

Plate 52. Fig. 1. habit sketch (natural size). 2. section of a fertile pinna, showing venation and position of sori ($\times$ 3). 3. scales from rhizome ($\times$ 50).

WOODWARDIA HARLANDII Hooker
哈氏狗脊

图版 53　合生铁角蕨

根状茎短肥,斜生;鳞片密生,长约 5 mm,狭披针形,全缘,黝色,叶簇生,柄长 10—15 cm,黝色,鳞片密生,久则脱落,叶体长 15—20 cm,宽约 3 cm,线披针形,一回羽状深裂,中肋绿色,上面鳞片疏生,小叶无柄,但合生于中肋,长椭圆形,开张,边缘具钝锯齿,上面光滑,下面具针形鳞片,亚革质,淡绿色,叶脉分开,不显明;子囊群每小叶 3—5 个,斜出,长约 5 mm,盖线形,全缘。

分布:广东之天马山。

本种为特殊之种,其一回羽状分裂之线披针形之叶,与合生之小叶,易与他种分别。

图注:1.本种全形(自然大);2.小叶,表明其叶脉及子囊群(放大 4 倍);3.叶柄基部之鳞片(放大 50 倍);4.叶轴上之鳞片(放大 50 倍)。

ASPLENIUM ADNATUM Copeland

ASPLENIUM ADNATUM Copeland, Phil. Journ. Sci. Bot. **3**: 284 (1909); C. Chr. Ind. Fil. Suppl. 10 (1906-12).

Rhizome suberect, 5 mm thick, densely scaly; *scales* 5 mm long, linear-lanceolate, entire, blackish; *fronds* caespitose, several together, stipe 10-15 cm long, obscurely blackish, densely clothed in similar but smaller and subdeciduous scales; lamina 15-20 cm long, about 3 cm broad, linear-lanceolate, simple pinnate below pinnatifid apex, rachis green, unisulcate above, sparsely scaly beneath; *pinna* broadly adnate, the basal ones somewhat shortened, rachis wingless, those above the middle are connected by a narrow wing along the rachis, oblong, subacute, patent, margin obscurely dentate above the entire, constricted base, glabrous above, fibrillose beneath; *texture* subcoriaceous, color pale green; *veins* subflabellate, inconspicuous but caniculate, extending well into the teeth; *sori* 3-5 to each pinna, linear, oblique, about 5 mm long, *indusium* linear, broad, entire.

Guangdong: Taimo Shan, opposite Hong Kong, *C. G. Matthew* (type), November 7, 1907.

An unique species, characterized by narrowly lanceolate, simple pinnate leaves with short, adnate pinnae and dense, black, fibrillose scales on the stipe and rachis beneath. However, it may possibly prove to be a young form of the variable *Asplenium crinicaule* Hance, common in the locality.

Plate 53. Fig. 1. habit sketch (natural size). 2. pinna, showing venation and position of sori ($\times$ 4). 3. scales from the base of stipe ($\times$ 50). 4. scales from rachis ($\times$ 50).

ASPLENIUM ADNATUM Copeland
合生铁角蕨

图版 54 竹叶蕨

根状茎横生于土中,被红棕色之开张针状粗毛;叶疏生,柄长 26—50 cm,稻秆色,光滑,仅基部具同样之红棕毛。叶体略与柄等长或过之,一回单数羽状分裂;小叶长披针形,2—7 对或较多,下者亚对生,上者亚互生,具短柄,形体略等,渐尖头,基部楔形,亚革质,深绿色,上面光泽,全缘,具角质透明之狭平边,叶脉网状,网眼大,斜长方形,中无小脉;子囊群无盖,线形,一列,位于中肋及叶边之间,或稍贴近叶边,唯具无数深红棕色锤形之线状体。

分布:亚洲热带各地;最近发现于海南。

图注:1.本种全形(自然大);2.小叶一部,表示叶脉及子囊群之位置(放大 2 倍);3.叶柄横切面;4.根状茎上之毛(放大 50 倍);5.子囊群内之隔丝(放大 50 倍);6.孢子(放大 150 倍)。

TAENITIS BLECHNOIDES (Willdenow) Swartz

TAENITIS BLECHNOIDES (Willdenow) Swartz, Syn. Fil. 24, 220 (1806); Hk. et Bak. Syn. Fil. 397 (1868); Christ, Farnkr. d. Erde 130 (1897); Diels, Nat. Pfl. Fam. **1**: 4. 305 (1899); C. Chr. Ind. Fil. 630 (1906).

Pteris blechnoides Willd. Phytographia 13 t. 9, f. 3 (1794).

Taenitis chinensis Desv. Berl. Mag. **5**: 308 (1811).

Rhizome thick, creeping, densely beset with atropurple, spreading, stout, setose *hairs*, which extend upward to some distance above the base of stipe; *frond* distant, large, stipe 26-50 cm long, firm, naked, glossy, straminous, continuous to the rhizome, lamina as long as the stipe, oblong-ovate, impari-pinnate, *pinnae* 2-7-jugate, lanceolate, acuminate, margin entire, more or less wavy, cartilaginous, 15-25 cm long, 1.5-2 cm broad, oblique, lower ones subopposite, not shortened, generally broader and sterile, the upper ones subalternate, fertile, the terminal one as large as the lateral ones, shortly petiolate, naked in all parts; *texture* subcoriaceous, color lustrous green above; *veins* reticulated in 2-3 rows of oblong, oblique, hexagonal areolae without including veinlets; *sori* linear, continuous or rarely interrupted, transversing the reticulated veins between the margin and costa, or often nearer to the margin; *paraphyses* dense, atropurple, clavate, multiseptate and higher than sporangia.

Hainan: On the way from Dung Ka to Win Fa Shi, *C. L. Tso & N. K. Chun* 43462, 43708, August 15, 1932; Mo Shan, *C. L. Tso & N. K. Chun* 52278, April 30, 1932; on wet rocks under woods.

Malaysia, Polynesia, Sri Lanka, and Indo-China.

This interesting fern, though seems to be common in other parts of tropical Asia, is known for the first time in the island Hainan, the range of distribution being thus extended farther north. According to Copeland, the affinity of this fern to *Syngramma* and *Schizoloma* is unusually clear, though both von Goebel and Bower have regarded it as one of the "genera incertae sedis."

Plate 54. Fig. 1. habit sketch (natural size). 2. a section of pinna, showing venation and position of sori ($\times$ 2). 3. cross section of the lower part of a stipe. 4. setose hair from the base of stipe ($\times$ 50). 5. paraphysis from a sorus ($\times$ 50). 6. spore ($\times$ 150).

TAENITIS BLECHNOIDES (Willdenow) Swartz
竹叶蕨

根状茎粗短,木质,斜生,具黄褐色狭披针形之鳞片;叶单生,柄长 40—60 cm,粗健,稻秆色,光泽。叶体长 45—80 cm,宽 20—30 cm,卵椭圆形,一回单数羽状分裂,小叶 6—13 对,具柄,线披针形,全缘,顶生小叶具柄,形体同于侧生小叶,厚纸质,光滑,绿色,叶脉显明,开张,分叉,子囊群缘边着生,盖膜质,灰白色,全缘。

分布:香港,广东,贵州,广西,江西,福建,云南。

本种为特殊之种,形似凤尾蕨(*Pteris cretica* L.)然较大,基部一对小叶不分叉,叶边全缘,故易于识别。

图注:1.本种全形(自然大);2.子囊群叶之一部(放大 3 倍)。

PTERIS INSIGNIS Mettenius

PTERIS INSIGNIS Mettenius; Kuhn, Journ. Bot. (1868) 269; C. Chr. Ind. Fil. 599 (1906).

Rhizome oblique, thick, woody, blackish, clothed in ferruginous, subulate *scales*; *frond* solitary, uniform, stipe 40-60 cm long, rather thick near the base, terete, narrowly sulcate above, straminous, naked, lamina 45-80 cm long, 20-30 cm broad at base, ovate or ovate-oblong, pinnate, *pinnae* 6-13 pairs, petiolate (petioles in lower pinnae 1.2 cm long), to 20 cm long, 2-2.6 cm broad, the uppermost ones generally smaller and also free, similar to the lower ones, margin very entire throughout, cartilaginous; *texture* thickly chartaceous, naked; *venation* distinct, lateral veins patent, forked; *sori* narrow, confined to the upper half of the margin, leaving a considerable portion of the entire apex sterile, *indusium* narrow, gray, entire.

Hong Kong: *Hance 187* (type). Guangdong: Taimo Shan, *C. G. Matthew*, Nov. 7, 1907; Swatow, Thaiyong, *Dr. Dalziel*, July, 1901; Yingtak, *C. L. Tso 21980*; Lockchong, *C. L. Tso 7559*; *N. K. Chun 41995*, *42877*; Kochow, *Y. Tsiang 20782*; Lungtau Shan, *Y. K. Wang 31645*, *31545*. Guizhou: Chenfeng, *Y. Tsiang 4346*; Pinfa, *Cavalerie 3394*. Guangxi: Yao Shan, *S. S. Sin & K. K. Wang 183B*, in bamboo grove; San-fang, Lu-chen, *R. C. Ching 6239*; Lin Yin, *R. C. Ching 7037*. Jiangxi: Tsoongjen, Lepeichaio, *Y. Tsiang 10244* (1932). Fujian: Amoy, *Swinhoe* (1870). Yunnan: The Red River, *Hancock 179*.

One of the most distinct of the group of *Pteris cretica* L. from which it differs in decidedly larger size, simple basal pinnae, the lower few pairs of pinnae remaining sterile, and in the very entire margin throughout. Our plate represents a rather small form.

Plate 55: Fig. 1. habit sketch (natural size). 2. portion of a fertile pinna (× 3).

PTERIS INSIGNIS Mettenius
全缘凤尾蕨

根状茎短,斜生,叶簇生,叶柄细长,栗褐色,光泽,基部具褐色硬鳞片;叶体卵椭圆形,三回分裂,长 12—15 cm,厚纸质,光滑,一次基部小叶卵形,具柄,末次小叶圆倒卵形,具柄,长宽 3—7 mm,非子囊群小叶,顶边略呈波状,叶脉显明,扇形分叉,不达叶边。子囊群 1—2 个或 3 个,不大,圆形,盖同形,黑色,质坚,生于深圆缺刻内。

分布:四川峨眉山。

本种为稀见之种,仅采得一次;其形体极似甘肃产之 *Adiantum Roborowskii* Maxim. 所不同者,为其较大之倒卵圆形末回小叶,各具 1—2 个子囊群及黑色角质之盖。

图注:1.本种全形(自然大);2.末回小叶,表明叶脉及子囊群(放大 8 倍);3.根状茎上之鳞片(放大 50 倍)。

ADIANTUM FABERI Baker

ADIANTUM FABERI Baker, Journ. Bot. (1888) 225; Diels, Nat. Pfl. Fam. **1**: 4. 284 (1899); C. Chr. Ind. Fil. 26 (1905).

Rhizome short, oblique, densely clothed at the base of stipe in linear-acuminate, dark chestnut-brown, lustrous, subentire, spreading *scales*; *fronds* caespitose, stipe wiry, naked, shining, castaneous, 10-15 cm long; *lamina* ovate-oblong, tripinnate, 12-15 cm long, 5-8 cm broad, firm, green, glabrous, lower *pinna* ovate-oblong or deltoid, petiolate, ultimate pinnules orbicular, petiolulate, 0.3-0.7 cm broad, sometimes obscurely crenate on the upper margin of the sterile ones; *veins* distinct, free, flabellately branched, falling short from the narrowly cartilaginous margin; *sori* 1-2, or rarely 3, to each ultimate pinnule, small, globose, deeply indented on the upper margin, *indusium* orbicular, black, much smaller than that in *A. monochlamys* and curled-up at last.

Sichuan: Mt. Omei, *E. Faber 1033* (1887), 3,000 ft. alt.

A rare fern, being collected only once in the locality. It is a close ally to *A. roborowskii* Maxim. from Gansu, differs chiefly in much broader frond, larger orbicular ultimate pinnules each generally with 1-2 smaller sori and smaller black indusia.

Plate 56. Fig. 1. habit sketch (natural size). 2. ultimate pinnule, showing venation and sori (× 8). 3. scales from rhizome (× 50).

ADIANTUM FABERI Baker
峨眉铁线蕨

根状茎颇肥,横生;鳞片密生,色黑,质厚,形体及组织一如瓦苇;叶颇密生,长 6—9 cm,鲜有过此者,宽约 1.5 mm,线形,边缘强度反卷及于中肋,呈念珠状,坚革质,淡绿色;子囊群椭圆形,生于叶之上半部,幼时全为叶边所掩,终则半露于外。

分布:江西庐山,浙江,安徽黄山,广西北部,广东北部。

本种为本属最小之种,其叶反卷达于中肋,呈念珠状,故易与他种识别;本种常生于藓苔生之阴湿岩石上。

图注:1.本种全形(自然大);2.着生于子囊群上之盾状隔丝(放大 50 倍);3.根状茎上之鳞片(放大 50 倍)。

LEPISORUS LEWISII (Baker) Ching

LEPISORUS LEWISII (Baker) Ching, Bull. Fan Mem. Inst. Biol. **4**: 65 (1933).
　Polypodium lewisii Baker, Journ. Bot. (1875) 201; C. Chr. Ind. Fil. 529 (1906); Takeda, Notes, R. Bot. Gard. Edinb. **8**: 275 (1915).

Rhizome rather thick, creeping, epigaeous, densely scaly; *scales* atratous. rigid, lanceolate-subulate from an ovate base, margin slightly erosed, luminae opaque, very narrow with thick walls, dark-colored, except the narrow margin, which is hyaline with clear elongate meshes; *frond* approximate, 6-9 cm long, or rarely longer, about 1.5 mm broad, rigid, linear, with strongly revoluted margin throughout the entire length of the lamina; *texture* rigid, color pale green; *sori* oblong-ovate, confined to the upper part, wholly or partly covered by the revoluted leaf-margin, which appears bead-like, due to the expanding sori.

Jiangxi: LuShan, *Dr. Shearer* (type), 1873. Zhejiang: Siakan, Fanchiao, *R. C. Ching 3684*, October, 1927. Anhui: Whang Shan, *K. K. Tsoong 3172*. Guangxi: Sanfeng, Chufeng Shan, *R. C. Ching 5926* (1928). Guangdong: Lungtau Shan, Kook Kiang, *C. Wang 31652*, Dec. 19, 1931.

A peculiarly interesting little rock-dwelling fern with short, narrowly linear and strongly revoluted leaves, which appear bead-like due to the more or less concealed but expanding sori. All the specimens cited above are very uniform in all respects. It is found inhabiting the moss-clad cliffs in deep ravines or under dense forests, and proves to be the smallest known species of the genus.

Plate 57. Fig. 1. habit sketch (natural size). 2. peltate paraphysis from a sorus (× 50). 3. scale from rhizome (× 50).

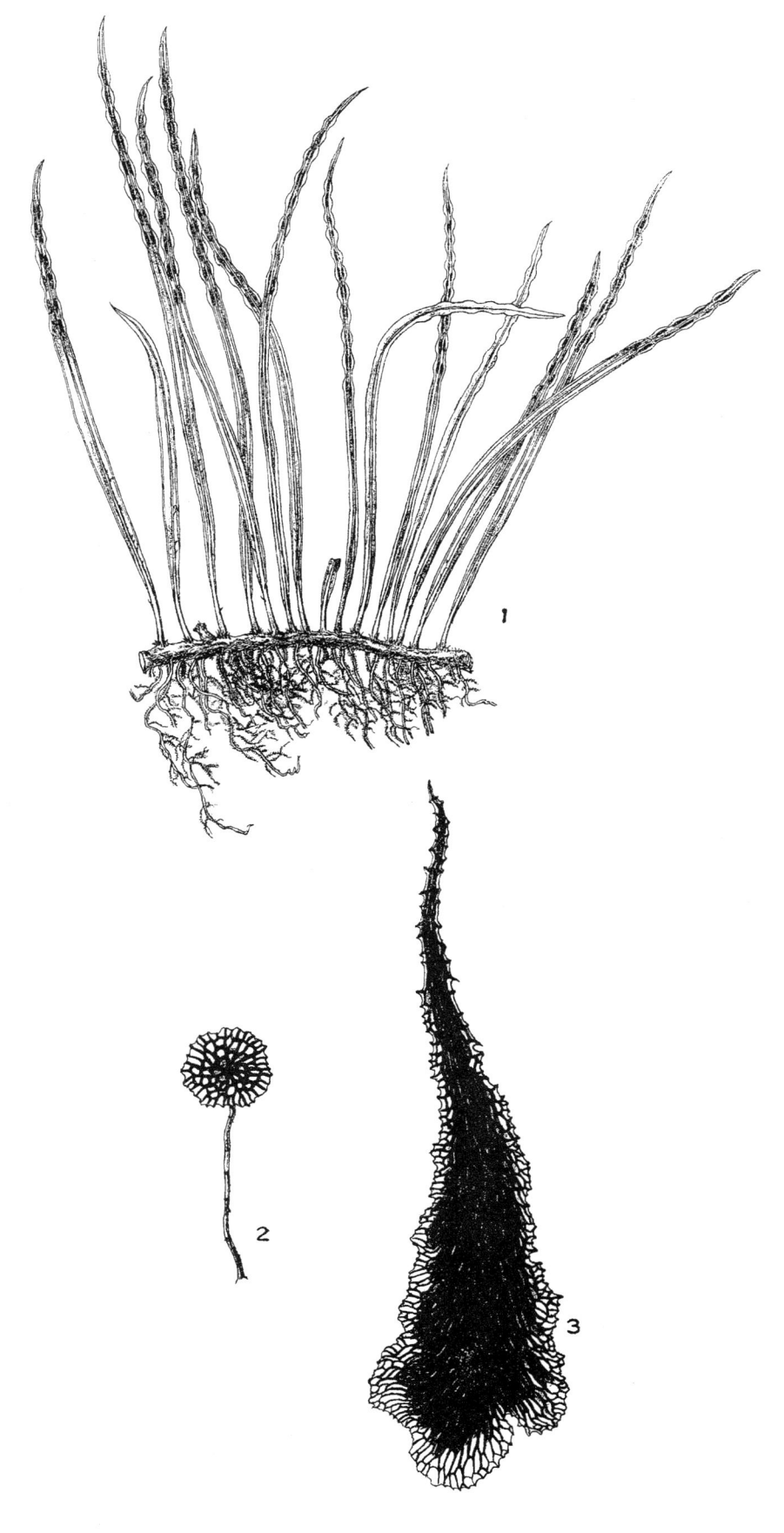

LEPISORUS LEWISII (Baker) Ching
庐山瓦苇

根状茎横生,颇肥;鳞片密生,卵形,渐尖头,开张,具小齿,色黑,质薄,网眼颇大而透明;叶颇密生,狭线披针形,叶边强度反卷,几达中肋,长 25—35 cm,宽约 5 mm,向基部渐狭,几无柄,厚革质,淡绿色,下面幼时稍被鳞片;子囊群卵形,紧位于中肋与反卷叶边之间,常半为后者所掩。

分布:湖北宜昌,甘肃南部,云南,四川,陕西及川边。

本种与庐山瓦苇(*Lepisorus lewisii*)同为本属特殊之种,因其狭线形之长叶具强度反卷之边,常蔽覆卵形之子囊群也。

图注:1. 本种全形(自然大);2. 叶体之一部,表示叶脉及子囊群之位置(放大 7 倍),3a—b. 着生于子囊群上之盾状隔丝(放大 50 倍);4. 根状茎上之鳞片(放大 50 倍);5. 叶体下面之鳞片(放大 50 倍)。

LEPISORUS EILOPHYLLUS (Diels) Ching

LEPISORUS EILOPHYLLUS (Diels) Ching, Bull. Fan Mem. Inst. Biol. **4**: 65 (1933).

Polypodium eilophyllum Diels in Engl. Jahrb. **29**: 204 (1901); C. Chr. Ind. Fil. 524 (1906); Acta Horti Gotob. **1**: 100 (1924); Dansk Bot. Archiv. **6**: 53 pl. Ⅷ. f. 3 (1929).

Polypodium involutum Baker (non Desv. 1811, nec Mett. 1856), Journ. Bot. (1889) 177; Diels Nat. Pfl. Fam. **1**: 4. 315 (1899).

Polypodium lewisii Christ (non Baker, 1875), Nuovo Giron. Bot. Soc. Ital. n. s. **4**: 97 pl. 1. f. 1 (1807).

Rhizome thick, creeping, densely scaly; *scales* shaggy, ovate-acuminate, denticulate, fuscous, concolorous, finely clathrate with medium-sized but clear, uniform luminae; *frond* rather approximate, narrowly linear-elongate, margin strongly revoluted, 25-35 cm long, about 5 mm broad, gradually decurrent to the base; *texture* rigidly coriaceous, color greenish, sparsely scaly beneath; *sori* ovate, closely packed between the midrib and the revoluted margin, which completely conceals the sori, when young, and becomes somewhat dilated due to the expanding sori.

Hubei: Ichang, *A. Henry 6859* (type); *E. H. Wilson 2636*. Sichuan: Tchenkoutin, *R. Farges*; Lower Tebbu Country, *J. F. Rock 14810*, Sept, 2, 1926; Luting Hsien, *W. P. Fang 3734* (1928); Tatsienlu, *Souliè 23*; Moupin, *David*, April, 1869. Yunnan: Mengtze, *A. Henry 9194B*. Shaanxi: Thaepei Shan, *Giraldi*, August, 1896; ibid., *Giraldi*, Sept. 1897; *Purdom 90* (1910).

One of the most distinct fern of the genus *Lepisorus*, being unlike any other known species, except *L. lewisii* (Baker), which is also provided with narrowly linear fronds with strongly revoluted margin and more or less concealed sori, but differs in decidedly smaller size, and entirely different type of scales on the rhizome.

Plate 58. Fig. 1. habit sketch (natural size). 2. section of a lamina, showing venation and position of sori ($\times$ 7). 3a-b. peltate paraphyses from a sorus ($\times$ 50). 4. scale from rhizome ($\times$ 50). 5. scale from underside of a lamina ($\times$ 50).

LEPISORUS EILOPHYLLUS (Diels) Ching
高山瓦韦

根状茎颇肥,着土横生;鳞片密生,卵形,渐尖头,色黑,质薄,纲眼大而透明,等形,边缘具粗锯齿;叶颇密生,长 25—35 cm,宽 2—2.5 cm,阔披针形,下部 1/3 处最宽,柄短,叶边平,厚纸质或亚革质,黄色,叶脉隐约;子囊群中大,形圆,位于叶边与中肋之间。

分布:云南;缅甸。

本种为本属之一大种,其茎上鳞片之组织,一如带瓦苇,然其叶不为细长线披针形,基部具较长之柄,子囊群圆形,位于叶边与中肋之间,故易于识别。

图注:1.本种全形(自然大);2.着生于子囊群上之盾状隔丝(放大 50 倍);3.根状茎上之鳞片(放大 50 倍)。

LEPISORUS SUBLINEARIS (Baker) Ching

LEPISORUS SUBLINEARIS (Baker) Ching, Bull. Fan Mem. Inst. Biol. **4**: 78 (1933).

Polypodium sublineare Baker; Takeda, Notes, R. Bot. Gard. Edinb. **8**: 276 (1915); C. Chr. Ind. Fil. Suppl. 28 (1913-16); Hand-Mzt. Symb. Sinic. **6**: 43 (1929); C. Chr. Contr. U. S. Nat. Herb. **26**: 320 pl. 22 (1931).

Rhizome rather thick, creeping, hypogaeous, densely scaly; *scales* ovate-acuminate, fuscous, coarsely clathrate with large, clear, almost isodiametrical luminae, concolorous, margin with short, stout, protruding teeth; *frond* rather approximate, 25-35 cm long, 2-2.5 cm broad, broadly lanceolate, broadest at the lower one-third, attenuate to a short stipe, margin plane; *texture* thick chartaceous, or subcoriaceous, color brown; *venation* obscure; *sori* medium-sized, rounded, medial, extending halfway downward.

Yunnan: Tengchwan Lin Mt., *A. Henry 9062A* (type); Mengtze, *Henry 11827*, *11827A*, *11827B*, *11828*; *Hancock 83*; Mekong, *Gebauer*; Yuangchang, *Henry 13603*. Sichuan: Moupin, *David* (1870).

Burma: Between Sadon and Yunnan border, *J. F. Rock 7424*, *7507*.

A large fern in the genus *Lepisorus*, and a close ally to *L. loriformis* (Wall.) in scale-character, differs in stipitate leaves and rounded medial sori.

Plate 59. Fig. 1. habit sketch (natural size). 2. peltate paraphysis from a sorus (× 50). 3. scale from rhizome (× 50).

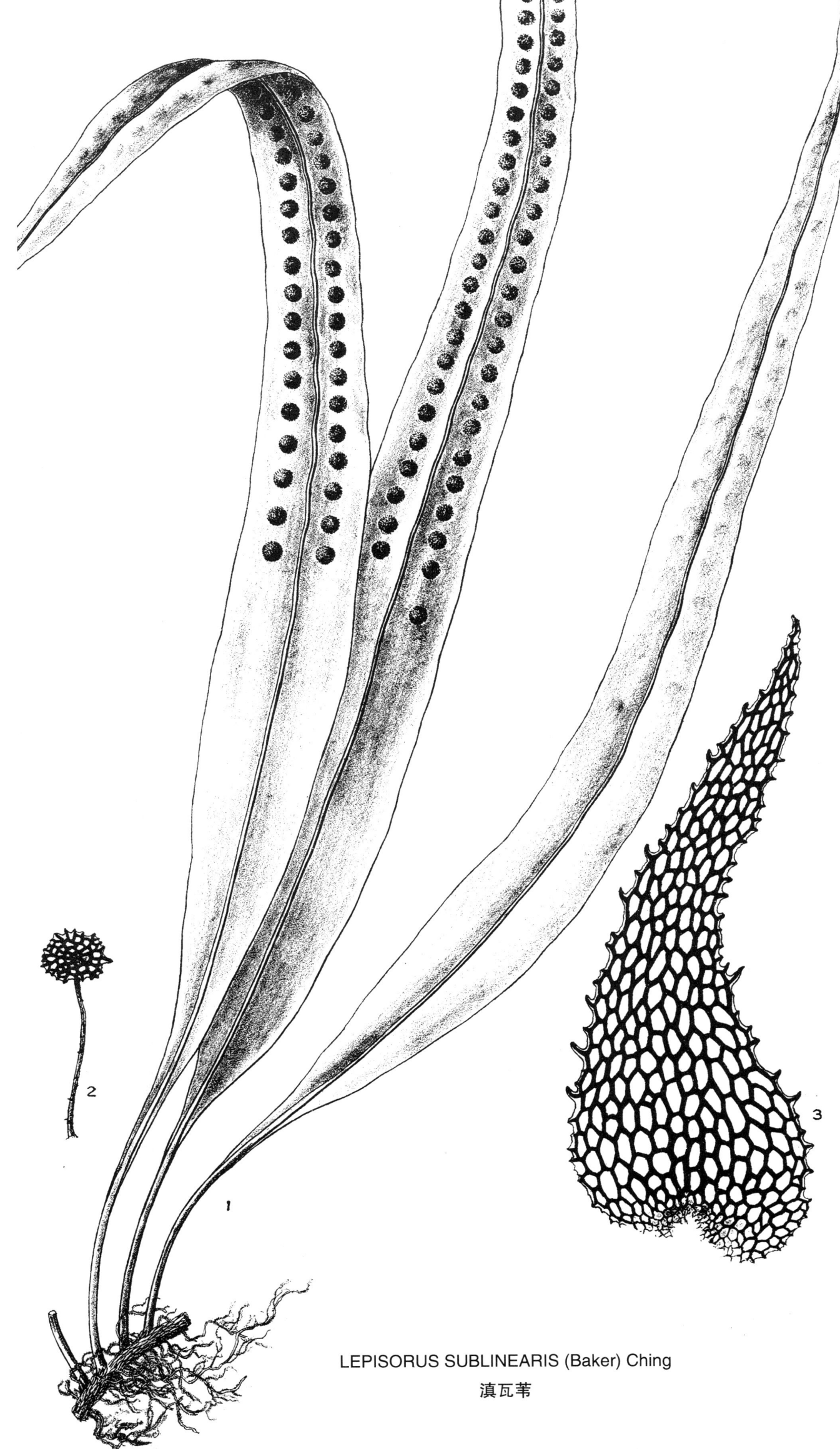

LEPISORUS SUBLINEARIS (Baker) Ching
滇瓦苇

根状茎蔓生;鳞片密生,卵状披针形,渐尖头,基部圆形,边缘具齿,色黑,网眼大而透明,一色,等形;叶间距离约1 cm,体长 20—25 cm,宽1.5 cm,长倒披针形,柄短,向叶端急尖,两面光滑,亚革质,叶脉不显,子囊群中大,圆形,远离,位于叶边与中肋之间。

分布:四川西部高山,云南;缅甸。

本种为本属颇为特殊之一种,形体略似光瓦苇(Lepisorus nudus),然其茎上之鳞片则大异,且叶端为急尖头,故易于识别;与本种更为类似者,为带瓦苇,然其叶为倒披针形,具较长之柄,叶端为急尖头,故亦易于识别。

图注:1.本种全形(自然大);2.着生于子囊群上之盾状隔丝(放大50倍);3.根状茎上之鳞片(放大50倍)。

LEPISORUS PSEUDONUDUS Ching

LEPISORUS PSEUDONUDUS Ching, Bull. Fan Mem. Inst. Biol. **4**: 83 (1933).

Rhizome wide-creeping, densely scaly; *scales* ovate-lanceolate, acuminate with rounded base, margin with protruding teeth, concolorous, fuscous, clathrate with large, clear, rather uniform luminae; *frond* about 1 cm apart, 20-25 cm long, about 1.5 cm broad, elongate-oblanceolate, gradually attenuate to a short but distinct stipe, apex caudato-acuminate, quite glabrous on both sides; *texture* subcoriaceous *venation* completely hidden; *sori* medium-sized, rounded, medial, extending halfway downward.

Sichuan austro-occid: Kanyeon Hsien, *W. P. Fang 6875* (type), on rock, 3,000 ft. alt., Oct. 12, 1030, common; O-shan, *W. P. Fang 6742*; Dongrergo, *H. Smith 3667* (in Herb. C. Chr.); Mt. Omei, *W. P. Fang 2984, 2995*, August 15, 1928; Mapien Hsien, *W. P. Fang 470, 1571* (f. major); ibid., *W. P. Fang 1586*; Honya Hsien, *W. P. Fang 8058* (f. reducta). Yunnan: Mengtze, *W. Hancock 104*; Lickiang Snow Range, *J. F. Rock 5400*; Lickiang, *J. F. Rock 3387* (f. parva), *5400*, May 30-June 6, 1922.

Burma: Between Sadon and the Yunnan border, *J. F. Rock 7498*.

A well marked fern, somewhat resembles *L. nudus* (Hooker) in general habit, differs, above all, in dense, spread, fuscous, lanceolate, long-acuminate scales with dentate margin, clear, quite uniform luminae, and in rather suddenly caudato-acuminate leaf-apex; from *L. loriformis* (Wall.) it differs in oblanceolate, stipitate leaves with suddenly caudato-acuminate apices and rounded medial sori.

This is a common fern in the mountains in western Sichuan and Yunnan and generally considered by previous authors as *Polypodium lineare* Thbg., or some other closely related species.

Plate 60. Fig. 1. habit sketch (natural size). 2. paraphysis from a sori ($\times$ 50). 3. scale from rhizome ($\times$ 50).

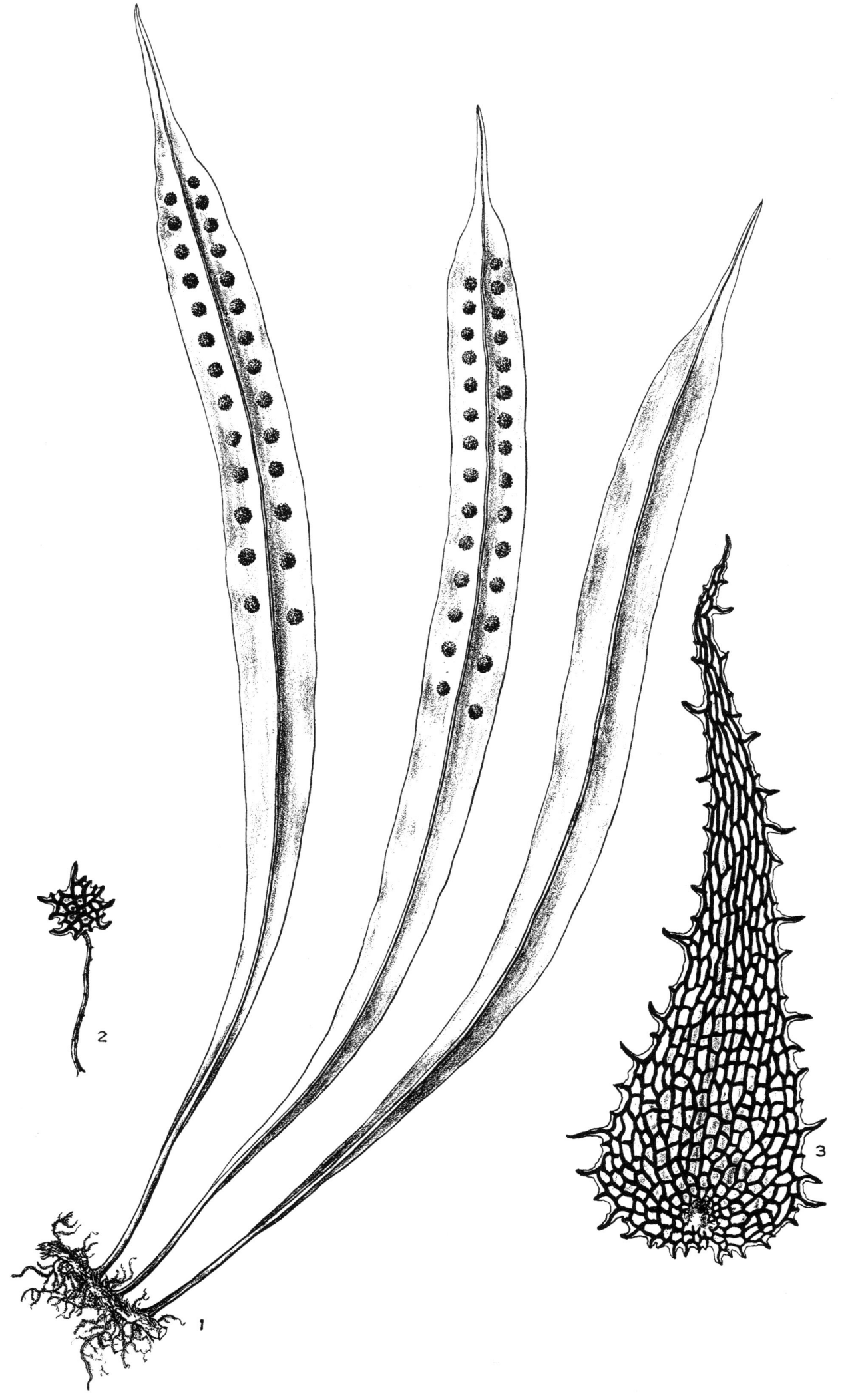

LEPISORUS PSEUDONUDUS Ching
长瓦韦

图版 61 带瓦韦

根状茎横生,较短,鳞片密生,色黑,质薄,卵形渐尖头,边缘具齿,网眼大而明,等形,为六角形,一色。叶间距离甚小,体长 30—45 cm,宽 6—13 mm,狭长,两边几平行,两端渐尖,基部具短柄,长 1—3 cm,全缘,干则稍反卷呈波形,亚革质,色黄褐,中肋显著,侧脉及细脉不显,仅于强光下隐约可见。子囊群中大,通常为椭圆形,贴近叶边,群间距离约 1.2 cm,幼时密覆不规则星芒形之盾状隔丝。

分布:印度北部,缅甸,尼泊尔;中国云南。

按本种为瓦韦属之最奇特之一种,惜自 Hooker 氏及 Mettenius 氏以来,几无人能识之矣。凡治印度蕨类者,如 Clarke、Beddome 及 Baker 诸氏,无一不以瓦韦之异名同物视之;直至 1931 年,Christensen 氏始将其由瓦韦分出,另立一种,并详为记载,以示此种与瓦韦不同之点。

图注:1. 本种全形(自然大);2. 子囊群上盾状隔丝(放大 50 倍);3. 根状茎上鳞片(放大 50 倍);4. 子囊全形(放大 150 倍);5. 孢子(放大 150 倍);6. 叶体之一部,表示叶脉及子囊群之位置(放大 4 倍)。

LEPISORUS LORIFORMIS (Wall.) Ching

LEPISORUS LORIFORMIS (Wall.) Ching, Bull. Fan Mem. Inst. Biol. **4**: 81. 1933.

  Polypodium loriforme Wall. Cat. n. 271. 1828 (nom. nud.); Mett. Polyp. 92 (exel. t. 1. f. 49-50). 1857; Hooker, Gard. Ferns pl. 14. 1862 (excl. syn.); C. Chr. Contr. U. S. Nat. Herb. **20**: 217, 1931.

  Drynaria loriforma J. Sm. Journ. Bot. (1841) 61.

  Polypodium xiphiopteris Baker, Kew Bull. (1906) 13.

  Polypodium excavatum var. *loriforme* C. Chr. Ind. Fil. 541. 1906.

  Polypodium subimmersum Baker, Kew Bull. (1895) 55; Takeda, Notes R. Bot. Gard. Edinb. **8**: 275. 1915.

  Polypodium mengtzeanum Baker, Kew Bull. (1906) 14.

  Polypodium lineare var. *loriforme* Takeda, Notes R. Bot. Gard. Edinb. **8**: 275. 1915; Acta Horti Goteb. **1**: 100. 1924.

  Polypodium subimmersum f. *mengtzeana* Takeda, Notes R. Bot. Gard. Edinb. **8**: 276. 1915; Hand-Mzt, Symb. Sinic. **6**: 43. 1929.

Rhizome rather short-creeping, densely scaly, scales fuscous, ovate-acuminate, denticulate, coarsely clathrate with large clear uniform hexagonal luminae, concolorous; *frond* rather approximate, 30-45 cm long, 6-13 mm broad, linear-elongate, gradually attenuate to short stipe (1-3 cm long), with long-acuminate apex, margin entire or slightly repando-undulate, somewhat revoluted in dry state; *texture* subcoriaceous, color brown; *costa* prominent, *lateral veins* and *veinlets* completely immersed but transluscent under strong light; *sori* medium-sized, roundish, or more often oblong, submarginal, 1.2 cm apart, densely covered when young with rather small, irregularly stellato-peltate scales.

Yunnan: Mengtze, *A. Henry* 11827B (pro parte), 11926A; *Hancock* 92; Szemeo, *Henry* 1339, 13249, 9134; on route to Tengyueh, *J. F. Rock* 7020; Between Tengyueh and Burmese border, *J. F. Rock* 7303; Salween, *Gebauer*. Tibetan border: *Capt. Kingdon Ward* 591; July 28, 1910.

Upper Burma: *G. Forrest* 26131.

Type from Nepal (*Wallich* 271).

The fern before us proves to be one of the most distinct of the genus but, unfortunately, was entirely forgotten since the time of Hooker and Mettenius over half a century ago. It is almost inconceivable when the authors of Indian ferns, like C. B. Clarke, R. H. Beddome and J. G. Baker, should have entirely suppressed it as a synonym of *Polypodium lineare* Thunberg. In 1915, H. Takeda took a sounder return of ideals of this fern by raising it to a varietal rank of Thunberg's species, but it is not until over two years ago that this fern, as a very distinct species, has happily been revived through the effort of Dr. C. Christensen, who having explicitly redescribed it concluded that it was sufficiently different from *Polypodium lineare* Thunberg, or other related species in general habit, position of sori and structure of scales on rhizome.

Plate 61. Fig. 1. Habit sketch (natural size). 2. Peltate paraphysis from a sorus ($\times$ 50). 3. Scale from rhizome ($\times$ 50). 4. Sporangium ($\times$ 150). 5. Spores ($\times$ 150). 6. Portion of a frond, showing venation and position of sori ($\times$ 4).

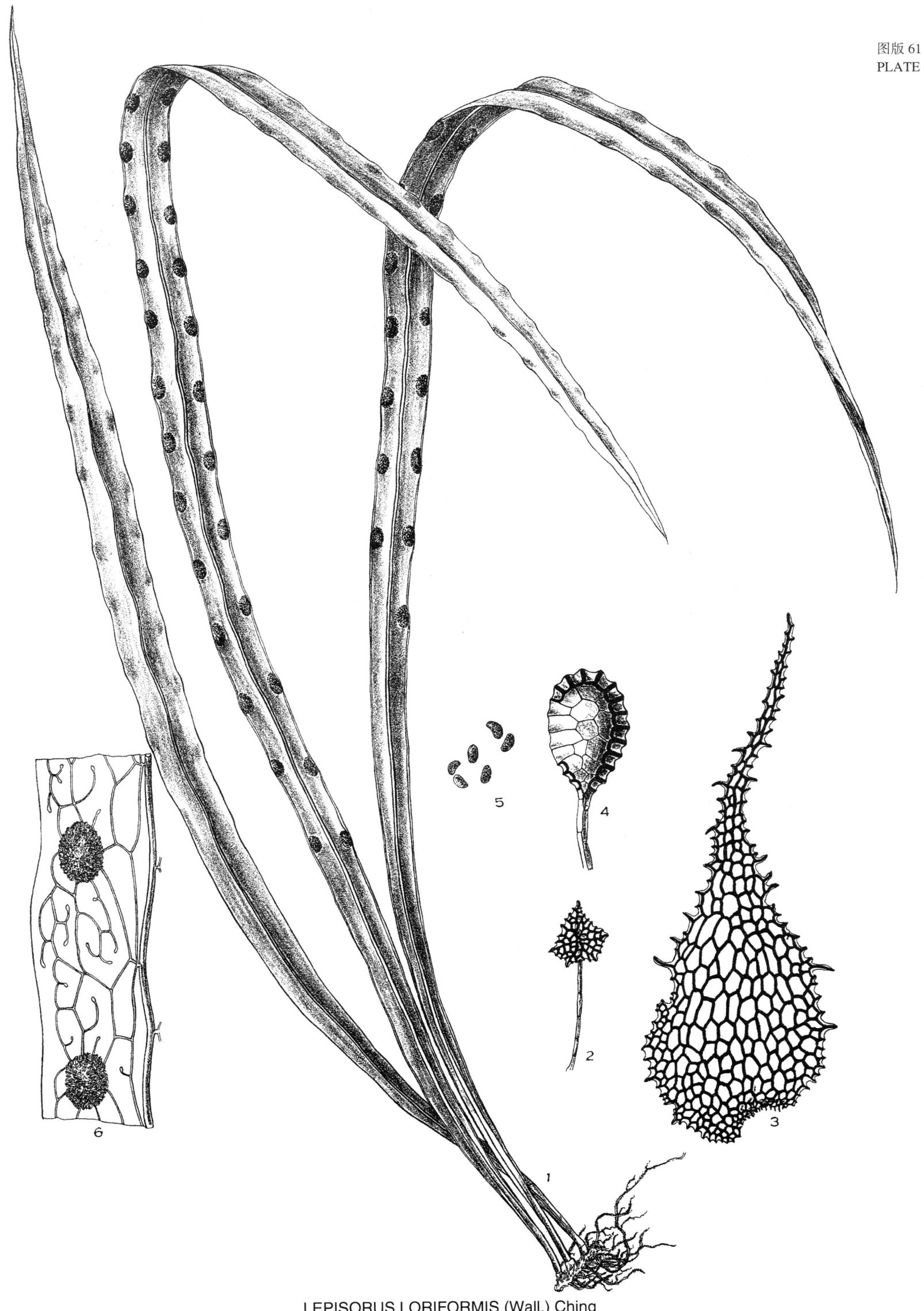

LEPISORUS LORIFORMIS (Wall.) Ching
带瓦苇

根状茎横生，颇肥，光滑；仅于叶之基部及茎之幼部鳞片疏生，黄色，卵形，圆头，全缘，网眼等大，形颇大而透明，久则脱落；叶间距离甚大，叶长 30—50 cm，宽约 3 cm，叶柄亦甚长，叶面光滑，绿色，或于下面着疏生之鳞，革质，叶脉颇明显；子囊群形大，椭圆形，近边着生，然有时离边颇远。

分布：云南，四川，贵州，广西，浙江（天目山），湖南及喜马拉雅山区。

本种为本属特殊之一种，常生于藓苔所生之石上。

图注：1. 本种全形（自然大）；2. 叶体之一部，表示叶脉及子囊群之位置（放大 4 倍）；3. a—b. 子囊群上之盾状隔丝（放大 50 倍）；4. 叶体下面之鳞片（放大 50 倍）；5. 根状茎上之鳞片（放大 50 倍）。

LEPISORUS MACROSPHAERUS (Baker) Ching

LEPISORUS MACROSPHAERUS (Baker) Ching, Bull. Fan Mem. Inst. Biol. **4**: 73 (1933).

Polypodium macrosphaerum Baker, Kew Bull. (1895) 55; Christ, Bull. Soc. Bot. France **52**: Mém. 1. 15 (1905); C. Chr. Ind. Fil. 542 (1906); Takeda, Notes, R. Bot. Gard. Edinb. **8**: 283 (1915); C. Chr. Contr. U. S. Nat. Herb. **26**: 319 (1931).

Polypodium intramarginale Baker; Christ, Bull. Herb. Boiss. ser. 2. **3**: 509 (1903); Baker, Kew Bull. (1906) 13.

Rhizome rather thick, creeping, epigaeous, naked, except the growing tip and the base of stipe, which both are sparcely scaly; *scales* brown, thin, concolorous, ovate, obtuse, dorsally affixed, margin entire, luminae medium-sized, uniform, clear, deciduous; *frond* rather distant, 30-50 cm long, to 3 cm broad, wingless stipe long, terete, naked, glabrous, or more often scaly underneath, when young; *texture* coriaceous, color greenish; *venation* prominent; *sori* large, oblong, intramarginal, or sometimes some distance below the margin.

Yunnan: Mengtze, *Hancock 49* (type); *A. Henry 10042A*, *13129 13363*; west of Tali, *J. F. Rock 6869*; between Tengyueh and Lungling, *J. F. Rock 7103*, *335*; Tsekou, *Ducloux 438*, *186*, *20*, *1384. 955* (1908); Tahwa Shan, *K. K. Tsoong 2991*; Salween, *Capt. Kingdon Ward*, Jan. 1, 1924; près Tapintze, *Delavay 4358*; *Forrest 16195*, *16219*; between Tengyueh and Burmese border, *J. F. Rock 7283*; between Kambaiti and Tengyueh, *J. F. Rock 7538*. Guizhou: Houangtsaopa, *Cavalerie 7293*; Kweiting, *Y. Tsiang 5319*; Tuyum, *Y. Tsiang 5707*; Paichai, *Y. Tsiang 6128*. Guangxi: Lin Yin, Tsinglung Shan, *R. C. Ching 7002*. Zhejiang: Tienmo Shan, *S. S. Chien 49*, *159* (1929). Sichuan occid: Mapien, *W. P. Fang 3961*; O-shan, *W. P. Fang 6664*, *6672*; Opien, *W. P. Fang 7253*.

Himalaya: Khasia, *C. B. Clarke 43627A*.

A very distinct species endemic in Central and West China, creeping over mossclad but rather exposed cliffs, or on rocks in thickets.

Plate 62. Fig. 1. habit sketch (natural size). 2. section of a lamina, showing venation and position of sori ($\times$ 4). 3a-b. peltate paraphyses ($\times$50). 4. scale from underside of a lamina ($\times$ 50). 5. scale from rhizome ($\times$ 50).

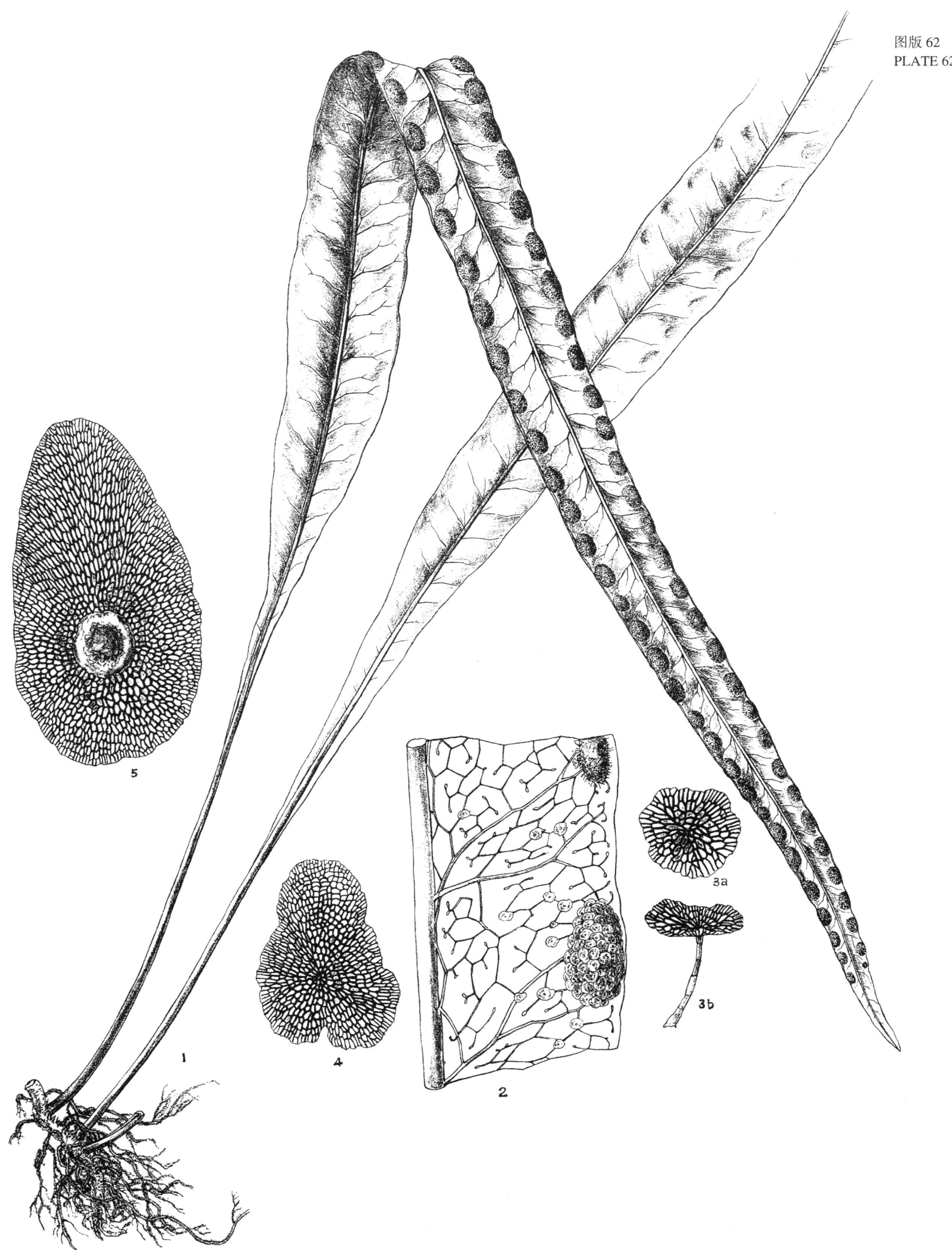

LEPISORUS MACROSPHAERUS (Baker) Ching
大瓦韦

本亚种之异于原种者，为其较小之叶体，厚革质，呈鲜橙黄色，及子囊群位于叶边与中肋之中央；茎上之鳞片为急尖头，然此不尽然者。

分布：同于原种，北向直达陕西南部，南及皖南之黄山与江西之庐山。

本亚种为长江上下游各省普通产，Baker 氏初名之曰 *Polypodium asterolepis*，似为一不同之种，然其后由各地所得标本渐丰，始证明其不过为大瓦苇之一亚种，其形体之不同，概由于生产地点湿度之差异有以致之耳；换言之，其生于干燥地点者，则叶体小，子囊群位于叶之中央。

图注：1.本种全形(自然大)；2.叶体之一部，表示叶脉及子囊群之位置(放大4倍)；3.着生于子囊群上之盾状隔丝(放大50倍)；4.叶体下面之鳞片(放大50倍)；5.根状茎上之鳞片(放大50倍)。

LEPISORUS MACROSPHAERUS (Bak.) var. ASTEROLEPIS (Bak.) Ching

LEPISORUS MACROSPHAERUS (Bak.) var. **ASTEROLEPIS** (Bak.) Ching, Bull. Fan Mem. Inst. Biol. **4**: 74 (1933).
  Polypodium asterolepis Baker, Journ. Bot. (1888) 330; Christ, Bull. Acad. Géogr. Bot. (1906) 105.
  Polypodium aspidiolepis Baker, Ann. Bot. **5**: 474 (1891); Diels in Engl. Jahrb. **29**: 204 (1901).
  Polypodium excavatum var. *asterolepis* C. Chr, Ind. Fil. 511 (1906).
  Polypodium macrosphaerum var. *asterolepis* C. Chr. Acta Horti Gotob. **1**: 101 (1024).

The present variety differs from the type in much smaller lanceolate frond of a more coriaceous texture, often clear flavescent color, more or less, oblong, oblique, medial sori and completely hidden venation.

Sichuan: Mt. Omei, *E. Faber 1063* (type); *Brown 76*; *Wilson 5221*; *W. P. Fang 3057*; Tchenkoutin, *Farges 95*; Nanchuan, *W. P. Fang 1080*, *1424*; Kuan Hsien, *W. P. Fang 2162*; Hanynang Hsien, *W. P. Fang 3769*; Rating, *W. P. Fang 3682*; Honya Hsien, *W. P. Fang 8419*, *8423*, *8424*; Opien, *T. F. Lou 159*, *49*, Sept. 15, 1929. Shaanxi: Tsinglin, *David*, March, 1837; Thaepei Shan, *Giraldi*, Sept. 1397; *Purdom 95*. Guizhou: Kweiyang, *Bodinier 23*; Ganchow, *Cavalerie 3734*; Vanching Shan, *Y. Tsiang 7585*; Tsingai, *Bodinier 2088*. Guangxi: Lungchow, *Morse 24*. Anhui: Wu-yuan, *R. C. Ching 8912*; Whang Shan, *R. C. Ching 8552*. Jiangxi: Lushan, *A. N. Steward 2674*; *N. K. Up 1829*. Hubei: Patung, *Henry 3653*, *1723*, *2556*; *E. H. Wilson 2625*; Shinshan Hisen, *S. S. Chien 8329*. Yunnan: Haut Mekong, *Soulié 1609*; *E. E. Maire 1388* (1913); environs de Yunnansen, *Ducloux 185*; *1384*; Chetchotze, *Delavay 10*; Tsang Shan, *Delavay 4215*, March, 1890; Mengtze, *Leduc* (1891).

The type specimen seems to be very distinct from *Lepisorus macrosphaerus* (Baker), but in the presence of ample material from different localities and diverse habitats, it is gradually led up by intermediate forms to the preceding species. The differences in characters referred to above are due largely to the drier conditions under which this fern happens to grow. Consequently, specimens from southern part of Shaanxi and Eastern China are more referrable to this variety than to the typical form of *L. macrosphaerus*, because ecological conditions in these localities are usually much drier than those in the southwestern part of the country.

Plate 63. Fig. 1. habit sketch (natural size). 2. section a lamina, showing venation and position of sorus ($\times$ 4). 3. peltate paraphysis from a sorus ($\times$ 50). 4. scale from underside of a lamina ($\times$ 50). 5. scale from rhizome ($\times$ 50).

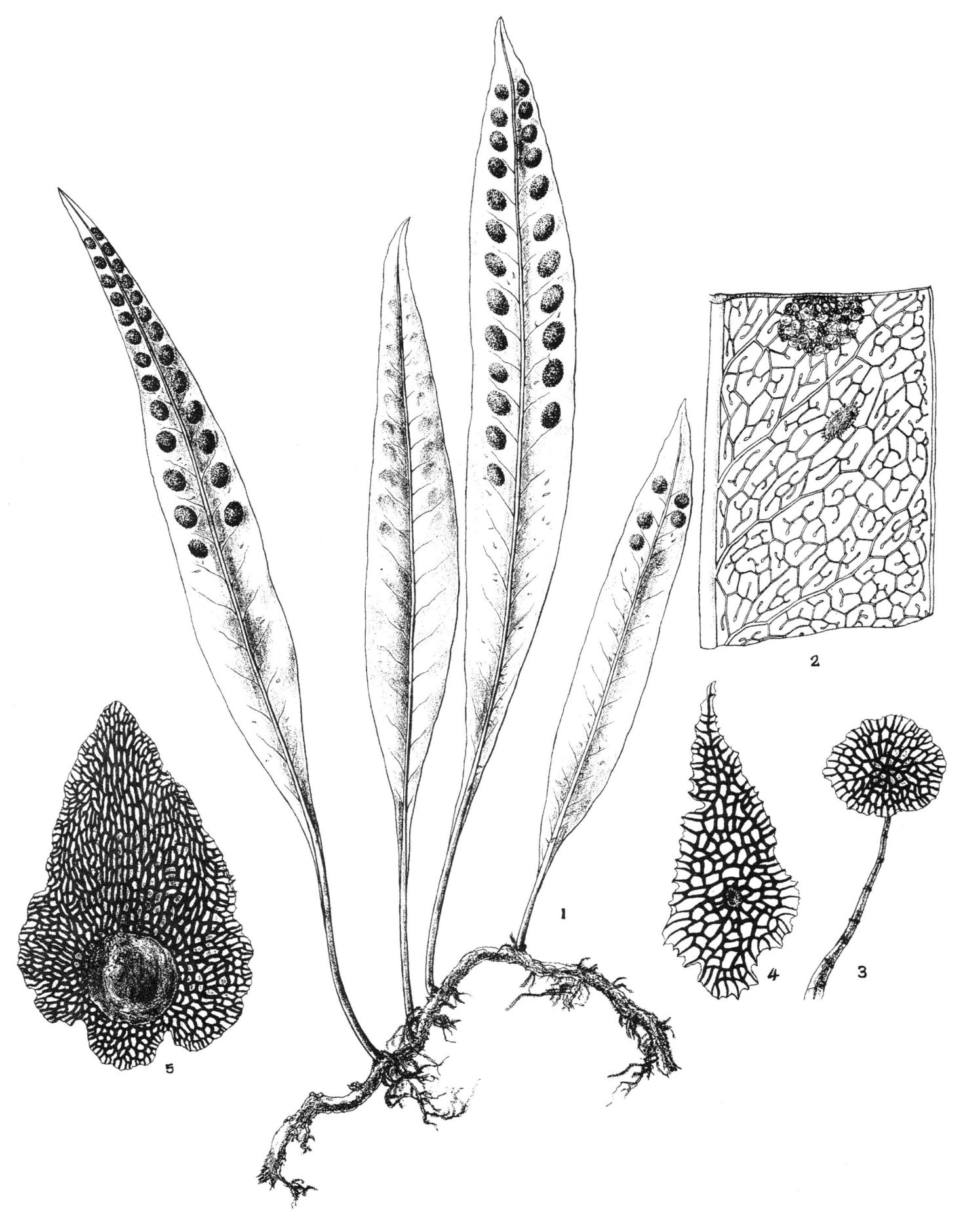

LEPISORUS MACROSPHAERUS (Bak.) var. ASTEROLEPIS (Bak.) Ching
黄瓦苇

根状茎颇肥，匍匐于岩石上；鳞片密生如瓦覆，长卵椭圆形，淡褐色，质薄，网眼透明，一色，全缘，腹部着生；叶疏生，甚大，长 30 cm，宽 4—5 cm，阔披针形，渐尖头，柄极短，叶边平而薄，软纸质，淡绿色，光滑，叶脉颇明显；子囊群大，略圆，稍贴近中肋。

分布：广西瑶山；模式标本在广州国立中山大学生物系。

本种为本属特殊之一种，为广西瑶山特产，其形体甚类凹瓦苇(*Lepisorus excavatus*)，然其叶体宽而质薄，子囊群大而叶上面不呈凹形。

图注：1.本种全形（自然大）；2.叶之一部，表示叶脉及子囊群之位置（放大 2 倍）；3.着生于子囊群上之盾状隔丝（放大 50 倍）；4.根状茎上之鳞片（放大 50 倍）。

LEPISORUS KUCHENENSIS (Wu) Ching

LEPISORUS KUCHENENSIS (Wu) Ching, Bull. Fan Mem. Inst. Biol. **4**: 69 (1932).

Polypodium kuchenense Wu, Bull. Dept. Biol. Sun Yatsen Univ. n. **3**: 276 pl. 129 (1932); Ching, ibid., n. **6**: 29. 1933.

Rhizome thick, creeping over rocks, densely imbricatingly scaly; *scales* oblong-ovate, thin, light-brown, entire, dorsally affixed, finely reticulated, luminae mediumsized, clear, dimorphous, concolorous; *frond* far apart, about 30 cm long, 4-5 cm broad, broadly lanceolate, acuminate, very gradually attenuate to a thick, short stipe, margin thin, plane; *texture* herbaceous, color pale green, naked; *venation* quite distinct; *sori* large, roundish, somewhat nearer to the midrib.

Guangxi: Yao Shan, Kushen, *S. S. Sin & K. K. Whang 2121*, *8980*, alt. 4,500 feet, on rocks.

An unique species, endemic in Yao Shan, in the eastern part of Guangxi Province. Closely allied to *L. excavatus* (Bory) and particularly the form from the Himalayas, differs chiefly in herbaceous (not chartaceous) leaves to 5 cm broad at the middle, and large, submedial, superficial sori, not punctate on the upperside.

Plate 64. Fig. 1. habit sketch (natural size). 2. portion of a lamina, showing venation and position of sori (× 2). 3. peltate paraphysis from a sorus (× 50). 4. scale from rhizome (× 50).

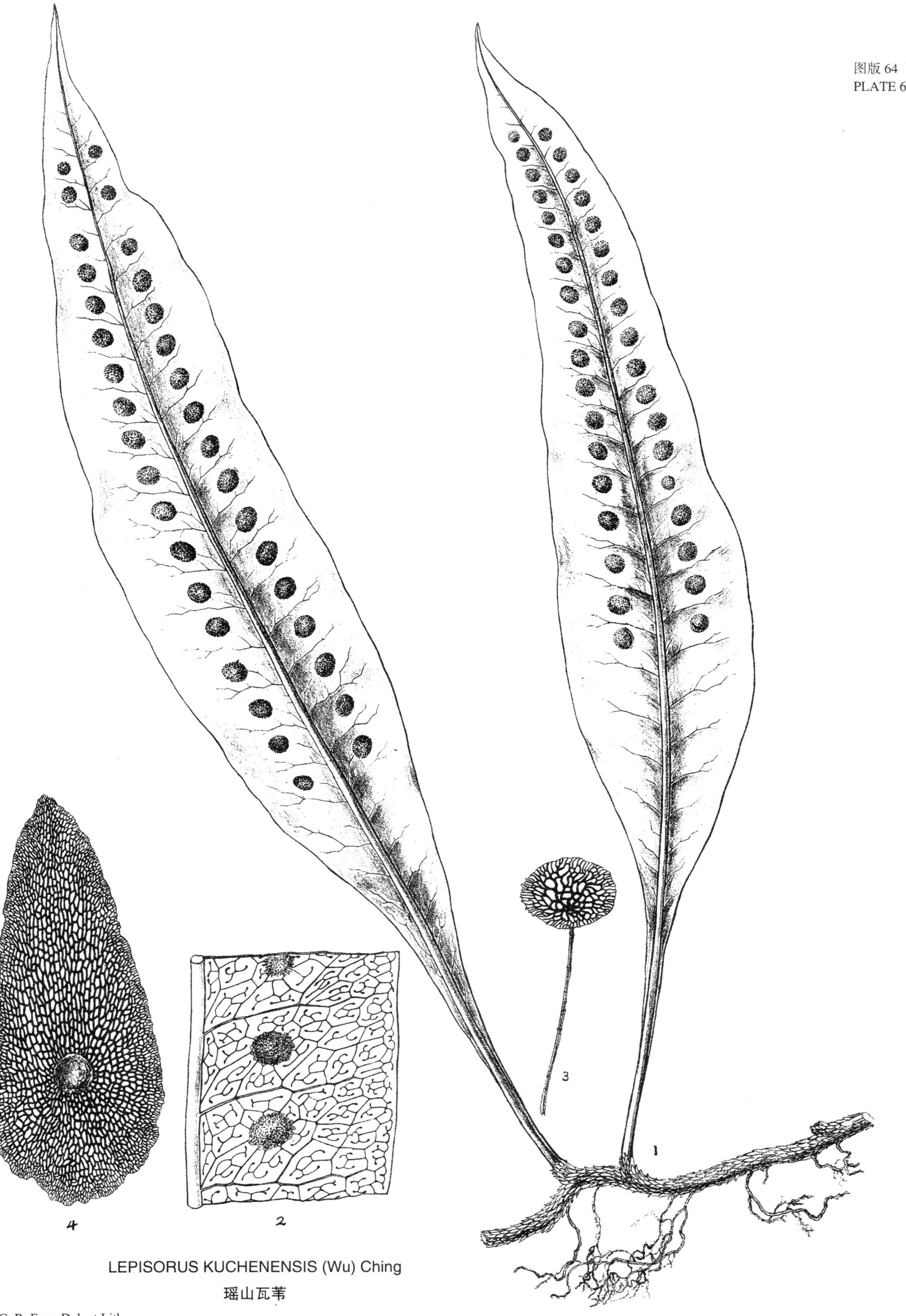

LEPISORUS KUCHENENSIS (Wu) Ching
瑶山瓦苇

根状茎伏石壁或树干横生，颇肥；鳞片紧覆，圆卵形，短尖头，边缘具不规则之齿，二色，中部为不甚透明之深褐色，边缘质薄而透明；叶间距离约 1 cm，柄短，叶体披针形，长 20 cm，宽约 1.5—2 cm，下部 2/3 处最宽，边平，厚纸质，色绿而常呈黄色，柄及下面常被多少之黑色鳞片，尤以中肋为甚；叶脉隐约可识；子囊群形圆，稍贴近中肋。

　　分布：云南及四川西部，湖北宜昌附近，陕西南部；印度之锡金亦产之，唯极稀少耳。

　　本种形体酷似凹瓦苇(*Lepisorus excavatus*)，昔之学者多以此名之，泊乎一九一五年，Takeda 氏始认为亚种，以其茎上鳞片呈二色故也；据余最近之研究，见其茎上鳞片，非特呈二色，且紧覆如瓦，其形体及大小亦大异，故认为新种，其地理分布，概限于中国西部高山，且甚普通，而与其相近之凹瓦苇则于此区域内甚为罕见。

　　图注：1. 本种全形（自然大）；2. 子囊群上之盾状隔丝（放大 50 倍）；3. 根状茎上之鳞片（放大 50 倍）。

LEPISORUS BICOLOR Ching

LEPISORUS BICOLOR Ching, Bull. Fan Mem. Inst. Biol. **4**: 66 (1933).
　　Polypodium excavatum var. *bicolor* Takeda, Notes, R. Bot. Gard. Edinb. **8**: 279 (1915); Hand-Mzt. Symb. Sinic. **6**: 43 (1929); C. Chr. Contr. U. S. Nat. Herb. **26**: 319 (1931).

　　Rhizome rather thick, epigaeous, densely imbricatingly scaly; *scales* adpressed, broadly ovate, short-acuminate with erosed margin, discolorous, luminae in central band small, narrow, dark-colored and thick-walled, but quite clear with flexous, thin cell-walls towards the light-colored hyaline margin; *frond* about 1 cm apart, shortly but distinctly stipitate, stipe covered more or less in similar scales, lamina linear-lanceolate, attenuate to both ends, 15-20 cm long, 1.5-2 cm broad at the lower one-third, being the broadest part, margin plane; *texture* chartaceous, greenish, tinged brown, the underside and particularly along the midrib more or less scaly; *veins* obscure, but distinct under light; *sori* rounded, nearer to the midrib than to the margin.

　　Yunnan: Mengtze, *A. Henry 10088*; Toudza *Ducloux 2813*; Tong Chow, *E. E. Maire 6918B*, *6061* (partim); Laohouy Shan, près Tali, *Ducloux 5043*; Lickiang, *C. Schnesider 2956, 2117, 2808, 3125*; *G. Forrest 8083*; *J. F Rock 5887*; Salween, *Forrest 15091, 27076*; Laokou Chow, *Ducloux 955*; Maoku Chang, *Delavay*, July 7, 1883; Mengtze, *Hancock 39* (Kew No.); between Yungbei and Boloti, *Handel-Mazzetti 3329*; near Lickiang, *Handel-Mazzetti 3588*. Hubei: Ichang, *A. Henry 2465*; Patung, *Henry 1739, 5315*. Shaanxi: Thaepei Shan, *Purdom 96*. Sichuan occid: *Wilson 5317, 5317A, 2637, 5315*; Tatsienlu, *Wilson 2637*; Tenghsiangying, *H. Smith 1893A*; Tahsiangling, *H. Smith 2088*; Moupin, *David*.

　　India: Kumaon, *Strachey and Winterbottom 2*; Darjeeling, *Griffith*.

　　Almost without exception, this fern has generally been identified as *Polypodium excavatum* by authors on the Northern Indian and western Chinese ferns, and more recently as var. *bicolor* by Takeda. In fact, it differs from the Himalayan *Polypodium excavatum* particullarly in scales, which are not only discolorous as first observed by Takeda but also much smaller than, and of an entirely different outline from, those in the other species. The extant herbarium material has also shown that this fern occupies rather a distinct though contiguous geographical area, namely, the western China, whence *P. excavatum* is so far unknown.

　　Plate 65. Fig. 1. habit sketch (natural size). 2. paraphysis from a sorus ($\times$ 50). 3. scales from rhizome ($\times$ 50).

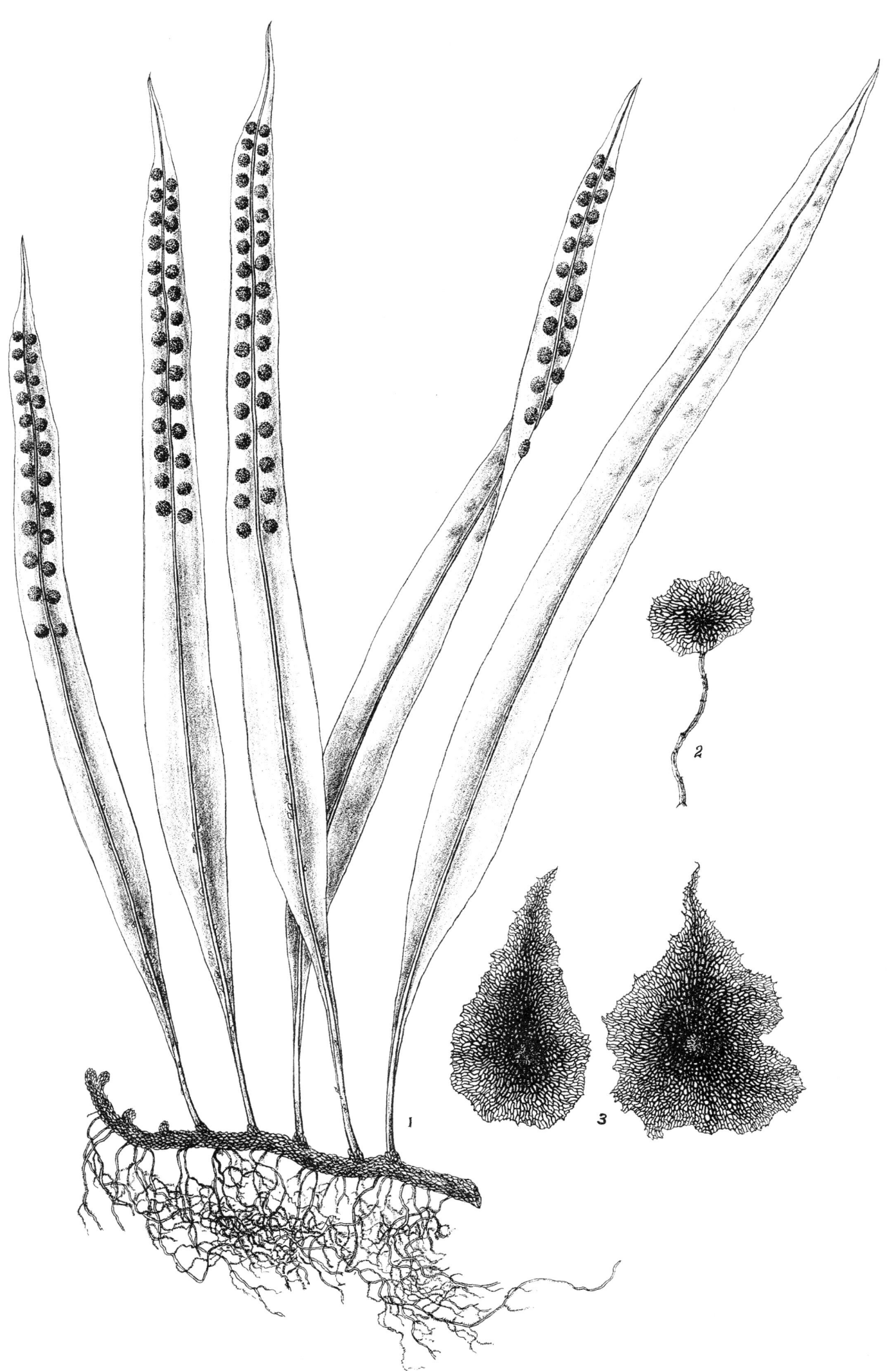

LEPISORUS BICOLOR Ching
两色瓦韦

根状茎横生,黑色,幼部鳞片密生,余半光滑;鳞片长披针形,渐尖头,膜质,淡黄色,边缘具短齿,网眼细而透明,唯中央呈深褐色;叶长 25—30 cm 或过之,宽 2 cm,叶端嘴形,上部渐狭,柄颇长,色黝而光滑,亚革质或厚纸质,绿色,叶脉隐约,上面水孔明显;子囊群中大,位于中肋及叶边中央,幼时遍覆黄色短柄盾状隔丝。

分布:贵州,广东,海南,广西,台湾。在广东尤为普通。

本种颇类印度及斯里兰卡产之光瓦苇(*Lepisorus nudus*),然其叶柄为黝靛青色,叶体较宽,色绿,其最宽处在下部 1/3,顶部嘴形,故易于识别。

图注:1.本种全形(自然大);2.着生于子囊群上之盾状隔丝(放大 50 倍);3.根状茎上之鳞片(放大 50 倍)。

LEPISORUS OBSCURE-VENULOSUS (Hayata) Ching

LEPISORUS OBSCURE-VENULOSUS (Hayata) Ching, Bull. Fan Mem. Inst Biol. **4**: 76 (1933).

Polypodium obscure-venulosum Hayata, Ic. Pl. Form. **5**: 322 (1915); C. Chr. Ind. Fil. Suppl. 27 (1913-16); Ogata, Ic. Fil. Jap. **3**: pl. 141 (1930).

Polypodium suprapunctatum Ching, nom. in herb.

Rhizome creeping, epigaeous, atratous, subnaked but densely scaly at the tip; *scales* lanceolate, acuminate, thin, brown, margin erosed, the central part above the base dark-colored, the base and margin light-colored, with irregular hyaline luminae with thin flexuous walls, shrivelled, at last fallen off; *frond* 25-30 cm long, or longer, 2 cm broad, or broader, apex rostrate, gradually attenuate downward into the long, blackish, naked stipe, margin plane; *texture* subcoriaceous, or thick chartaceous, greenish; *venation* obscure with distinctly punctate hydathodes on the upper surface; *sori* medium-sized, medial, covered when young with brown, peltate, shortly stalked paraphyses.

Guizhou: Keengtze, *Y. Tsiang 5000*; Vanching Shan, *Y. Tsiang 7082 7903* (1930). Guangdong: Sunyi, *P. S. Ko 51363*; *Y. K. Wang 31106*; Lohfau Shan, *C. O. Levine 1486*; *Levine & McClure 6819*; *N. K. Chun 41573*; Kochow, *Y. Tsiang 3208*; Lockchong, *N. C. Chun 43024*; Lungtau Shan, *To & Tsang 12513*, *12051* (1923). Hainan: Five Finger Mt., *F. A. McClure 9504*. Guangxi: Baimarchen, near Lin Yin, *R. C. Ching 6670* (1928); Tsing-Lung Shan, north of Lin Yin, *R. C. Ching 6889*. Hong Kong: Lantao, *Faber*, Aug. 1888; *Warburg*.

Closely related to *L. nudus* (Hooker) from India and Sri Lanka, differs in obscurely indigo-bluish stipe, broader lamina of green color, which is always broadened towards the base and rostrate at apex.

Plate 66. Fig. 1. habit sketch (natural size). 2. peltate paraphysis from a sorus (× 50). 3. scale from rhizome (× 50).

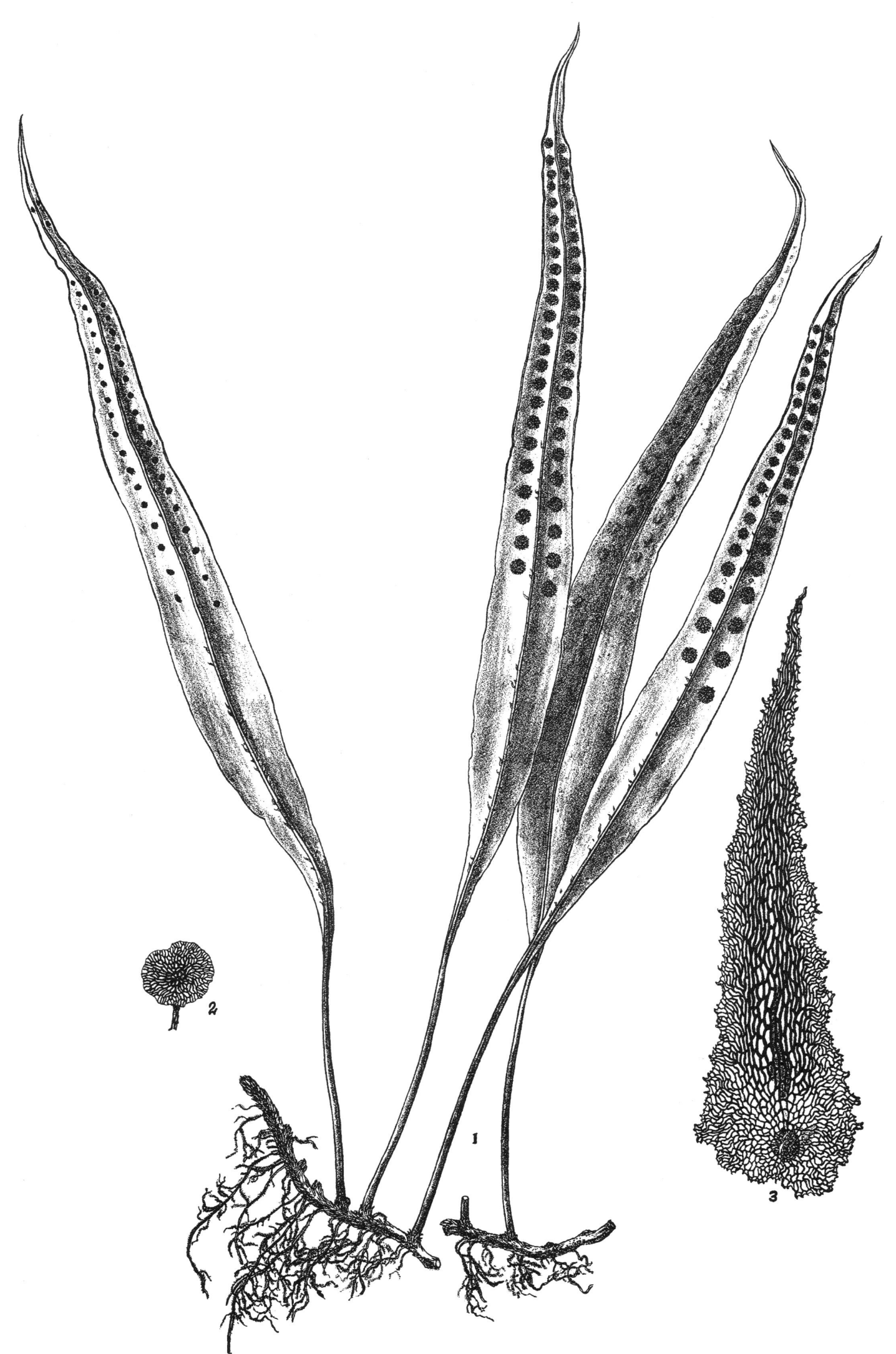

LEPISORUS OBSCURE–VENULOSUS (Hayata) Ching
粤瓦韦

根状茎横生土中,颇肥;鳞片密生,卵形,渐尖,色黑,质薄,边缘具长齿,网眼大而透明;叶狭披针形,下部渐狭,钝头或短尖头,形体大小不一,普通10—15 cm,宽1—1.3 cm,生于干燥地者,长仅4—5 cm,质薄,色绿,叶脉颇显著,边平,下面常覆黑色鳞片;子囊群形中大,圆形,位于中肋与叶边之中央。

分布:云南,山西,陕西,察哈尔,河北(北平附近),四川,甘肃,新疆;西伯利亚,日本亦产之。

本种为亚洲温带山地普通产,虽其形体大小不一,然因其绿色薄纸质之叶体及其茎部着生黑色薄质大网眼之鳞片,易与他种识别。

图注:1.本种全形(自然大);2.叶体之一部,表示叶脉及子囊群之位置(放大2倍);3.根状茎上之鳞片(放大50倍)。

LEPISORUS CLATHRATUS (Clarke) Ching

LEPISORUS CLATHRATUS (Clarke) Ching, Bull. Fan Mem. Inst. Biol. **4**: 71 (1933).

Polypodium clathratum Clarke, Ferns N. Ind. in Trans. Linn. Soc. ser. 2. Bot. **1**: 559 pl. 82 f. **1** (1880); C. Chr. Ind. Fil. 517 (1906).

Polypodium alberti Regel, Acta Horti Petr. **7**: 622 (1881).

Pleopeltis clathrata Bedd. Handb. Fern Brit. Ind. 348 (1883).

Polypodium uchiyamae Makino, Tokio Bot. Mag. **20**: 30 (1906); Ogata, Ic. Fil. Jap. **2**: pl. 91 (1929), teste Takeda.

Polypodium soulieanum Christ, Bull. Soc. Bot. France **52**: Mém. 1. 15 (1905).

Polypodium lineare var. *abbreviatum* Christ, Bull. Acad. Géogr. Bot. (1902) 153.

Rhizome quite thick, creeping, hypogaeous, densely scaly; *scales* ovate-acuminate fuscous, thin, with some long-protruding teeth on the margin, coarsely clathrate with large, uniform, isodiametrical and more or less iridescent luminae; *frond* linear-lanceolate, generally 10-15 cm long, 1-1.3 cm broad, or sometimes much reduced, gradually attenuate towards the slender, pale stramineous, and rather long stipe, leaf-apex blunt or subacute, or rarely subacuminate, margin entire, plane, or sometimes slightly wavy; *texture* thin herbaceous, color light green, underside with a few similar scales at first; *venation* distinct; *sori* medium-sized, rounded, medial.

Yunnan: Hokin, *Delavay 1716*. Gansu: Choni, *Purdom 1965*; near Pinfan, R. C. *Ching 571*, July 20, 1923. Shaanxi: Thae-pei Shan, *Giraldi*, August, 1890; *Purdom 96*, *1065*; K. S. *Hao 4414*, *4470*, Sept. 20, 1932. Hebei: Peiping Mountains, *Hancock 9*, *17* (1876); Siawutai Shan, C. T. *Li 2484*, August, 1932; ibid. H. W. *Kung 1201*, Sept. 9, 1930. Shanxi: Koti Shan, W. Y. *Hsia 1751*, Sept. 7, 1929; Tawutai Shan, T. H. *Serre 2335* (1929); Suiyuen, Halachingkow, Tatsin Shan, W. Y. *Hsia 2855*, July 24, 1931. Sichuan: Tatsienlu, *Soulié 759* (1893); Sungpan, H. *Smith 4082*, *2536*, *2726*; Chenchiangkuan, H. *Smith 2609*; Dongrergo, H. *Smith 3615*; Drogochi, H. *Smith 4470*.

Turkestan: *Fetosson*.

Distr: Turkestan, Afghanistan, Kashmir, Nepal to W. and N. W. China, Siberia, Japan.

Although quite variable in size, this fern can always be distinguished from others by its hypogaeous rhizome, thin herbaceous, green leaves with rather distinct venation, and very characteristic fuscous, clathrate, iridescent scales on rhizome and sori.

Plate 67. Fig. 1. habit sketch (natural size). 2. portion of a lamina, showing venation and position of sori ($\times$ 2). 3. scale from rhizome ($\times$ 50).

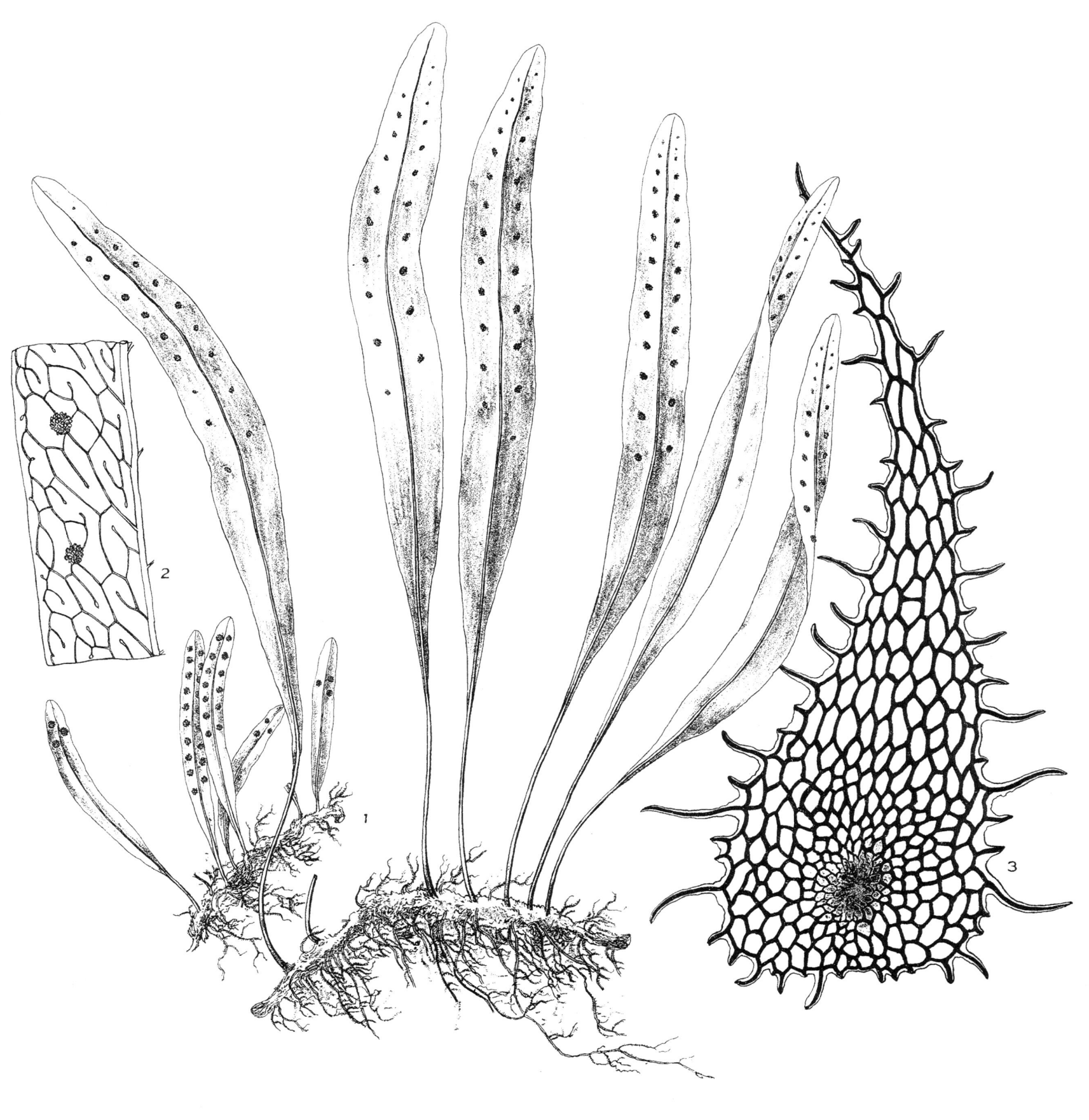

LEPISORUS CLATHRATUS (Clarke) Ching
网眼瓦韦

根状茎颇短,横生,黑色,幼部鳞片密生,炭黑色,质厚,不透明,披针形,渐尖头,亚全缘或具较小疏锯齿;叶颇密生,长30 cm,宽3—4 cm,披针形,柄颇长,淡绿色,然呈炭灰状,革质,叶脉不显;子囊群大,圆形,贴近叶边,位于叶之上半部。

分布:云南,四川西南部。

本种形似滇瓦苇,然其茎上鳞片为炭黑色,质厚不透明,叶柄亦较长,故易于识别。

图注:1. 本种全形(自然大);2. 着生于子囊群上之盾状隔丝(放大50倍);3. 根状茎上之鳞片(放大50倍)。

LEPISORUS SORDIDUS (C. Chr.) Ching

LEPISORUS SORDIDUS (C. Chr.) Ching, Bull. Fan Mem. Inst. Biol. **4**: 78 (1933).

Polypodium sordidum C. Chr. Contr. U. S. Nat. Herb. **26**: 320 (1931).

Rhizome rather short-creeping, hypogaeous, 3 mm thick, blackish, young parts densely scaly; *scales* black, entirely opaque, rigid, lanceolate, acuminate, subentire or sparingly short-toothed, concolorous, luminae very narrow with thick, dark-colored cell-walls; *frond* approximate, 30 cm or longer, 3-4 cm broad, lanceolate, attenuated to both ends, stipe rather long, greenish-straminous, smocky; *texture* coriaceous, pale green, smocky on both sides; *venation* hidden; *sori* large, rounded, supramedial, superficial, in the upper part of lamina.

Yunnan: Between Man Lo and Lungling, *J. F. Rock 7160* (type); between Kambaiti and Tengyueh, *J. F. Rock 7537*. Sichuan austro-occid: Honya Hsien, *W. P. Fang 8493*, August 21, 1930.

A large fern of the genus, habitally resembles *L. macrosphaerus* (Baker), or *L. sublinearis* (Baker), differing from both in entirely black and nearly opaque rhizome scales and in smocky appearance of leaves. From *L. sublinearis* (Baker) it further differs in its relatively long stipe and supramedial sori.

Plate 68. Fig. 1. habit sketch (natural size). 2. peltate paraphysis from a sorus ($\times$ 50). 3. scale from rhizome ($\times$ 50).

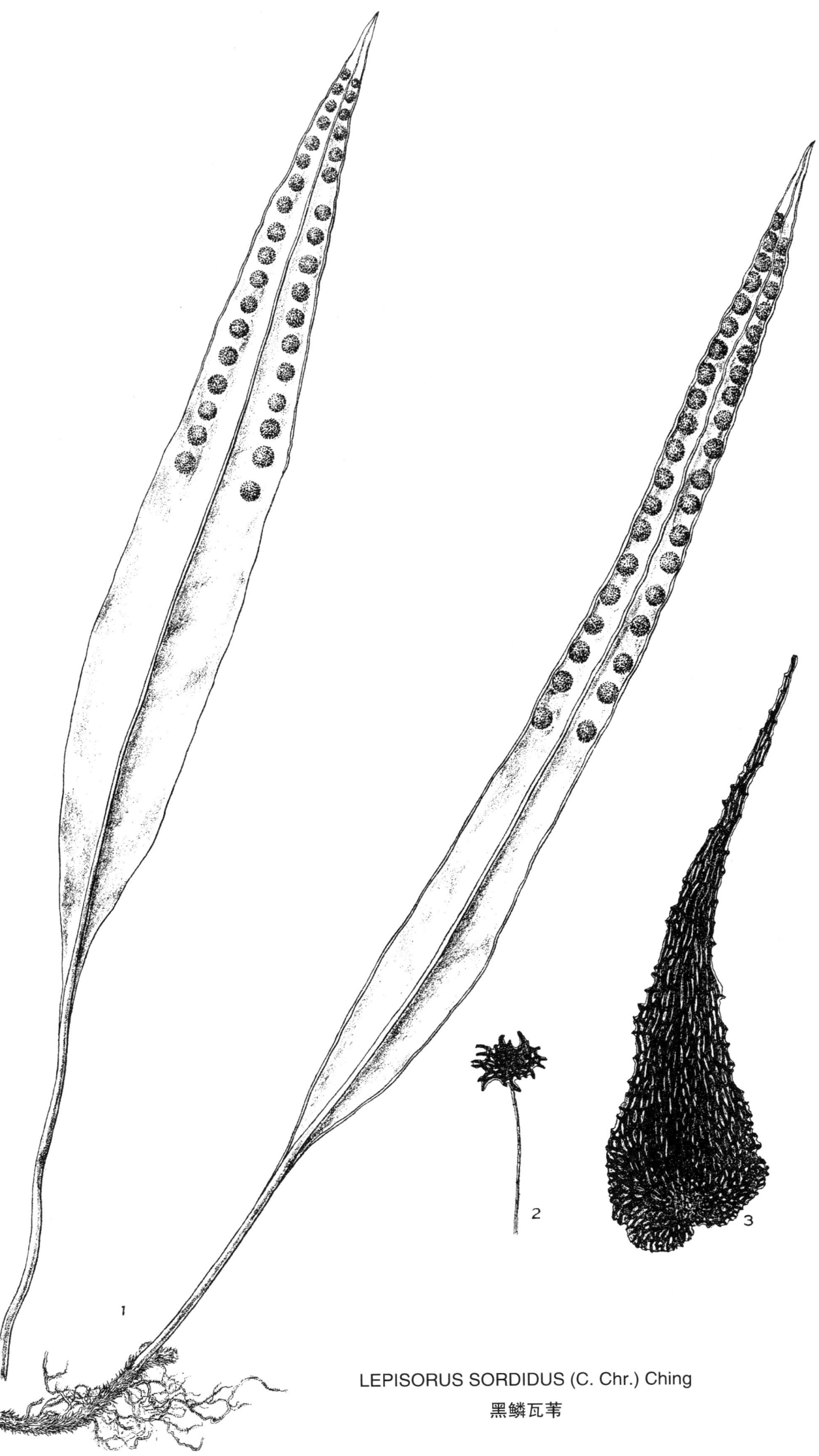

LEPISORUS SORDIDUS (C. Chr.) Ching
黑鳞瓦韦

图版 69　拟鳞瓦韦

　　根状茎横生,颇肥;鳞片密生,其形体及组织一如瓦韦;叶间距离颇大,体长 20—30 cm,宽 2—3 cm,披针形,向端部细长,下部 1/3 处最宽,亚革质,上面光滑,下面黑色鳞片密生;子囊群中大,生于叶之上半部,位于中央。

　　分布:云南。

　　本种颇为特殊,形体颇似鳞瓦韦(*L. oligolepidus*),然其叶质较薄,叶之上部细长,其下部 1/3 处最宽,子囊群较小而生于中央,不贴近中肋,故易识别。

　　图注:1.本种全形(自然大);2.叶体之一部,表明叶脉及子囊群之位置(放大 3 倍);3a—b.着生于子囊群上之盾状隔丝(放大 50 倍);4a—b.叶体下部之鳞片(放大 50 倍);5.根状茎上之鳞片(放大 50 倍)。

LEPISORUS SUBOLIGOLEPIDUS Ching

LEPISORUS SUBOLIGOLEPIDUS Ching, Bull. Fan Mem. Inst. Biol. **4**: 77 (1933).

　　Polypodium neurodioides C. Chr. var. **2**. C. Chr. Contr. U. S. Nat. Herb. **26**: 319 (1931).

　　Rhizome creeping, thick, densely scaly; *scales* as in *L. thunbergianus* (Kaulf.); *frond* rather distant, 20-30 cm long, 2-3 cm broad, lanceolate, long-attenuate towards apex, the broadest part being at the lower one-third, densely scaly on the under side and stipe; *texture* rigidly chartaceous or subcoriaceous; *sori* medium-sized, medial, confined to the upper part of lamina.

　　Yunnan: Mile District, *A. Henry 10088A* (type); between Tengyueh and Lungling, *J. F. Rock 7113*, on tree.

　　Quite a distinct species, habitally resembling *L. oligolepidus* (Baker), differing in thinner texture, long-attenuate leaf-apex and the broadest part of the lamina being at, the lower one-third, as well as in much smaller medial sori.

　　Plate 69. Fig. 1. habit sketch (natural size). 2. portion of a leaf, showing venation and position of sori (× 3). 3a-b. peltate paraphyses from a sorus (× 50). 4a-b. scales from underside of lamina (× 50). 5. scale from rhizome (× 50).

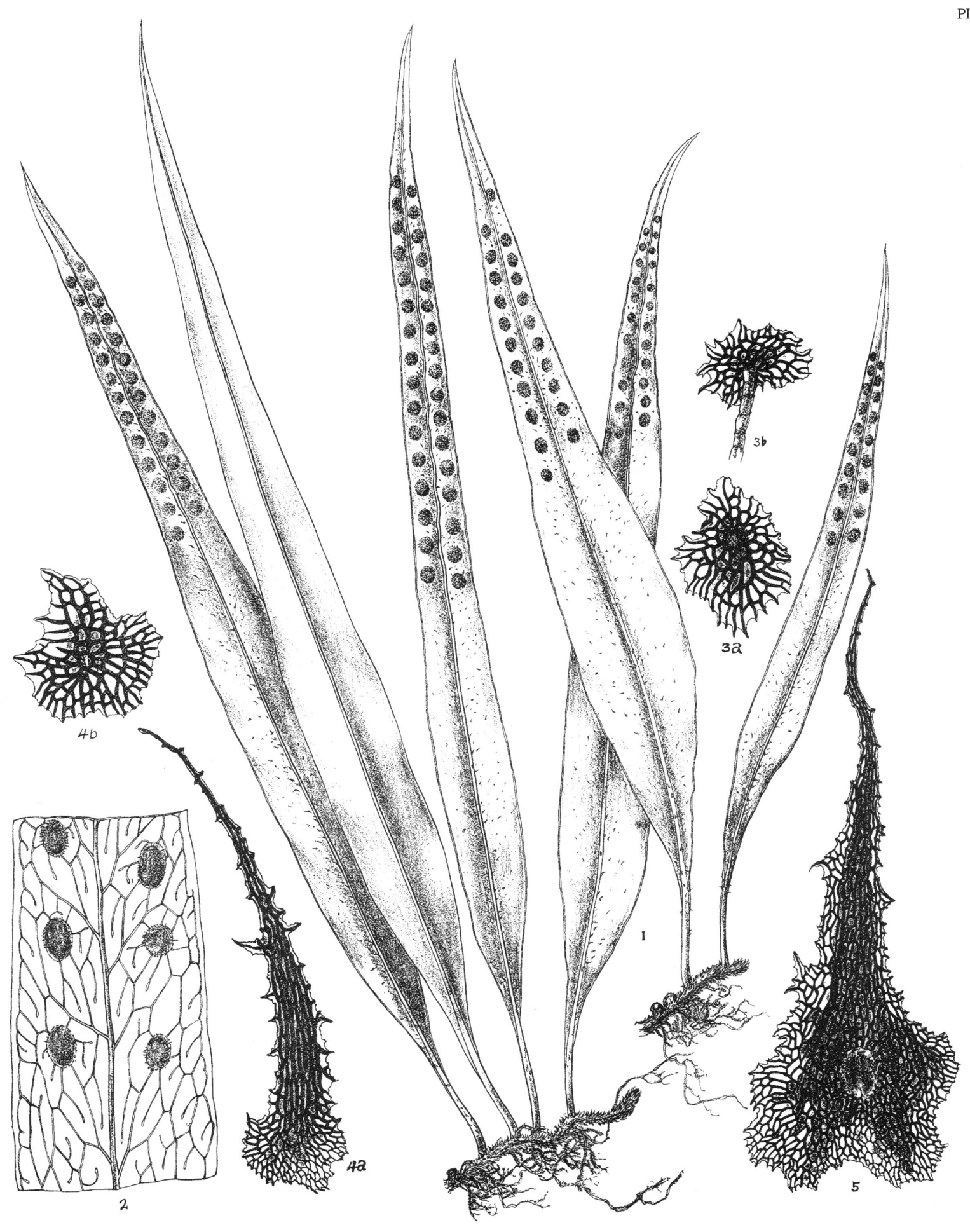

LEPISORUS SUBOLIGOLEPIDUS Ching
拟鳞瓦韦

根状茎较肥,横生;鳞片密生,色黑,质厚,线披针形,其构造同于瓦苇;叶披针形,长 15—30 cm,宽 1.5—2 cm,中部最宽,具短柄,革质,略带黑色,下面黑色鳞片密生,久则脱落,上面光滑;子囊群形大而圆,贴近中肋。

分布:江西庐山,湖北,贵州,云南,广东。

本种形体极似瓦苇,然形体较宽,子囊群极大,叶之下面有黑色鳞片密生,故易与瓦苇分别。

图注:1.本种全形(自然大);2.着生于子囊群上之盾状隔丝(放大 50 倍);3.根状茎上之鳞片(放大 50 倍)。

LEPISORUS OLIGOLEPIDUS (Baker) Ching

LEPISORUS OLIGOLEPIDUS (Baker) Ching, Bull. Fan Mem. Inst. Biol. **4**: 80 (1933).

Polypodium oligolepidum Baker, Gard. Chron. n. s. **14**: 494 (1880); Takeda, Notes, R. Bot. Gard. Edinb. **8**: 276 (1915); C. Chr. Contr. U. S. Nat. Herb. **26**: 316 (1931), partim.

Polypodium lineare var. *oligolepidum* Christ, Bull. Soc. Bot. France **52**: Mém. 1. 15 (1905); C. Chr. Ind. 540 (1906)

Polypodium trabeculatum Cop. Phil. Journ. Soc. Bot. **3**: 283 (1908).

Polypodium lineare Takeda, l. c. 269 (Henry 7795).

Rhizome rather thick, creeping, densely scaly; *scales* black, lanceolate-subulate from a broad ovate base, rigid, structurally similar to those in *L. thunbergianus* (Kaulf.); *froud* lanceolate, 15-30 cm long, 1.5-2 cm broad at the middle, shortly stipitate; *texture* coriaceous; *veins* hidden, color obscurely blackish, underside densely clad in small, ovate, cuspidate, adpressed, blackish scales, glabrous and punctate on the upper side; *sori* large, rounded, closer to the midrib than to the margin.

Jiangxi: Lu-shan, *Maires* (type); *Dr. Shearer* (1873). Hubei: Ichang, Nanto, *A. Henry 3049*. Guangdong: Lienchow, *C. G. Matthew*, December 4, 1907. Guizhou: Ganchow, *Cavalerie 7795*, *155*; Pinfa, *Cavalerie 1096*, *3748*, *1597*; Ganpin, *Bodinier 2017*, Yinkiang, *Y. Tsiang 7848*, December 20, 1930; Kwangcheng, *Y. Tsiang 8607*. Yunnan: Tsekou (Haut Mekong), *Soulié 1060* (partim); Salween, *Capt. Kingdon Ward*, Jan. 2, 1924; Tongchow, *E. E. Maire*; *Esquirol 3750*; Mengtze, *A. Henry 9062*; Hancock 104 (1894-95); Tchongshan, *Ducloux 39*.

Burma: Sadon, *J. F. Rock 7494*.

Plate 70. Fig. 1. habit sketch (natural size). 2. peltate paraphysis from a sorus (× 50). 3. scales from rhizome (× 50).

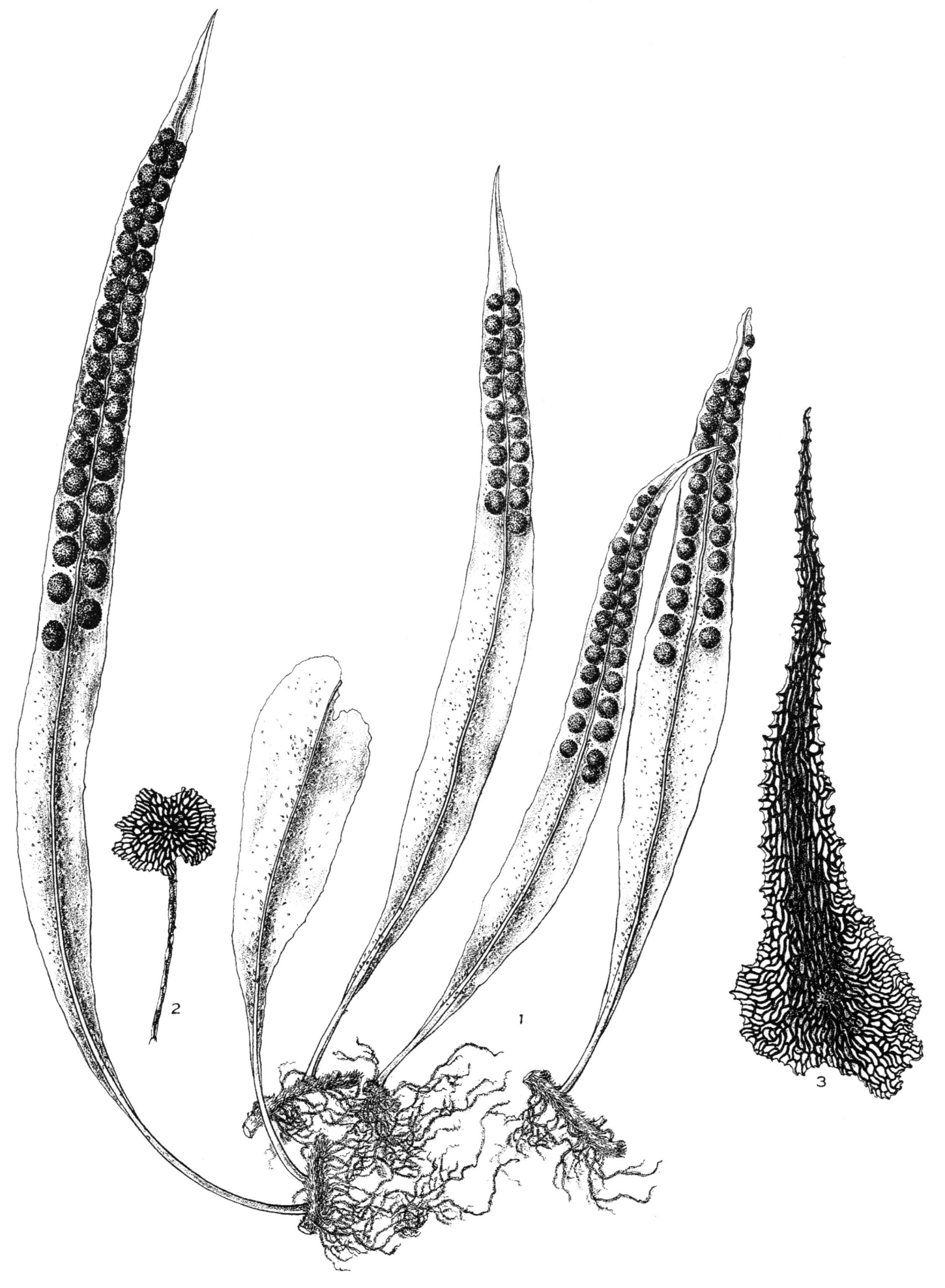

LEPISORUS OLIGOLEPIDUS (Baker) Ching
鳞瓦韦

根状茎横生；鳞片密生，细长，黑色，边缘具齿，网眼细长，黝色，不透明；基部圆形具透明阔边；叶颇密生，长 20—30 cm，宽 5 mm，细长线形，边缘稍反卷；革质，淡绿色，子囊群长椭圆形，接近，向叶端常接合。

分布：云南西部高山。

本种为高山产，其形体极似书带瓦苇（*Lepisolus vittarioides* Ching），然其叶狭长，子囊群向叶端常接合呈连珠状，故易识别。

图注：1. 本种全形（自然大）；2. 着生于子囊群上之盾状隔丝（放大 50 倍）；3. 根状茎上之鳞片（放大 50 倍）。

LEPISORUS SUBCONFLUENS Ching

LEPISORUS SUBCONFLUENS Ching, Bull. Fan Mem. Inst. Biol. **4**: 83 (1933).

Polypodium neurodioides C. Chr. var. **4**. C. Chr. Contr. U. S. Nat. Herb. **26**: 319 (1931).

Rhizome creeping, densely scaly; *scales* long, linear-subulate from an ovate base, atratous, dentate, luminae narrow, elongate, opaque with dark-colored cell-walls, lower part provided with broad hyaline margin, consisting narrow clear luminae with thin flexous walls; *frond* rather close together, 20-30 cm long, 5 mm broad, linear-elongate, gradually acuminate towards apex, margin somewhat narrowly revoluted; *texture* coriaceous, color greenish; *sori* elongate-oblong, approximate, or more or less fused towards the leaf-apex.

Yunnan: Alpine Regions of Silo, above Tseku, *J. F. Rock 8727*. Type in the Botanical Museum, Copenhagen, cotype in U. S. National Herbarium, Washington, D. C, and Herbarium of the Metropolitan Museum of Natural History, Nanking.

An alpine species of the genus, closely related to *L. vittarioides* Ching, differs in more elongate fronds and subconfluent upper sori.

Plate 71. Fig. 1. habit sketch (natural size). 2. peltate paraphysis from a sorus ($\times$ 50). 3. scale from rhizome ($\times$ 50).

LEPISORUS SUBCONFLUENS Ching
连珠瓦苇

图版 72 苋瓦苇

　　根状茎横生；鳞片密生，狭长披针形，基部卵形，黑色，边缘透明，具不规则之锯齿；叶颇密生，狭长线形，长 20—27 cm，宽达 5 mm，边稍反卷，革质，淡绿色，叶脉不显；子囊群椭圆，群间距离大，紧位于中肋及波形叶边之间。

　　分布：云南，台湾，喜马拉雅山区。

　　本种形体极似叶体狭长之带瓦苇，Baker 氏名之曰 *Polypodium subintegrum*，然其茎上鳞片之组织则大异，本种与连珠瓦苇十分相似，所不同者，为其彼此远离之子囊群及波形之叶边耳。

　　图注：1. 本种全形（自然大）；2. 着生于子囊群上之盾状隔丝（放大 50 倍）；3. 根状茎上之鳞片（放大 50 倍）。

LEPISORUS HETEROLEPIS (Rosenstock) Ching

LEPISORUS HETEROLEPIS (Rosenstock) Ching, Bull. Fan Mem. Inst. Biol. **4**: 86 (1933).

Polypodium lineare var. *heterolepis* Rosenstock in Fedde, Repert. Sp. Nov. **12**: 247 (1913).

Rhizome creeping, epigaeous, densely scaly; *scales* lanceolate-subulate from a broadly ovate base, atratous, margin hyaline with numerous irregularly protruding teeth, luminae opaque, very narrow with thick, dark-colored walls; *frond* rather approximate, linear-elongate, 20-27 cm long, to 5 mm broad, margin narrowly reflexed; *texture* coriaceous, color greenish; *venation* entirely hidden; *sori* oblong, far apart, closely packed between the midrib and margin, which is more or less undulated due to the expanding sori.

Yunnan: Between Tengyueh and Lungling, *J. F. Rock 7168*; below Kuyung, *J. F. Rock 7558*. Taiwan, Himalayas.

In habit this fern is exactly identical with the narrow-leaved form of *L. loriformis* (Wall.), called *Polypodium subintegrum* Baker, which differs in fuscous, ovate-acuminate, concolorous and coarsely clathrate scales with large, clear, almost isodiame-trical luminae. The affinity with *L. subconfluens* Ching is very great, which differs in close, or subconfluent, longer sori and hardly wavy leaf-margin.

Plate 72. Fig. 1. habit sketch (natural size). 2. peltate paraphysis from a sorus (× 50). 3. scale from rhizome (× 50).

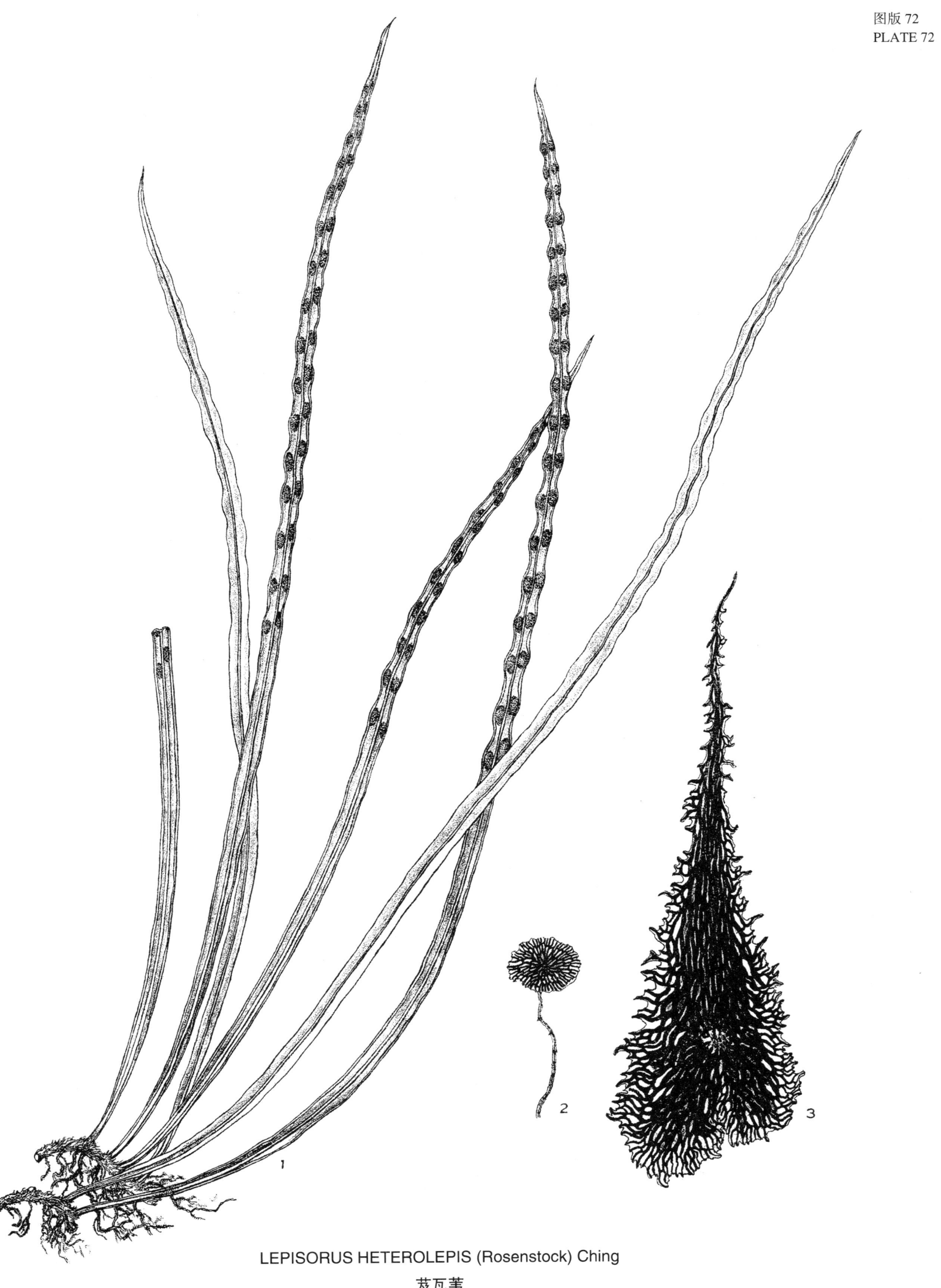

LEPISORUS HETEROLEPIS (Rosenstock) Ching
芨瓦苇

根状茎颇肥,横生;鳞片密生,其形体及组织同于瓦苇;叶间距离较大,体为狭线形,长 20 cm 或过之,宽达 5 mm,边缘稍反卷,软革质,淡绿色;子囊群长椭圆形,紧位于中肋及叶边之间。

分布:湖北西部,云南,浙江天目山,甘肃西部(舟曲及文县),陕西(周至)。

本种为本属细狭之一种,昔人认为瓦苇之变种,然其狭长线形之叶及绿色较软之质,易于识别。

图注:1. 本种全形(自然大);2. 着生于子囊群上之盾状隔丝(放大 50 倍);3. 根状茎上之鳞片(放大 50 倍)。

LEPISORUS ANGUSTUS Ching

LEPISORUS ANGUSTUS Ching, Bull. Fan Mem. Inst. Biol. **4**: 86 (1933).

Polypodium lineare var. *thunbergianum* f. *caudato-attenuata* Takeda, Notes, R. Bot. Gard. Edinb. **8**: 269 (1915).

Rhizome rather thick, wide-creeping, densely scaly; *scales* as in *L. thunbergianus* (Kaulf.); *frond* distant, narrowly linear, 20 cm or longer, about 0.5 cm broad, margin narrowly linear, 20 cm or longer, about 0.5 cm broad, margin narrowly revoluted; *texture* soft, subcoriaceous, color pale green; *sori* oblong-elongate, closely placed between the midrib and margin.

Hubei: *E. H. Wilson 2642*. Yunnan: Szemeo, *Henry 10062A*; Mengtze, *Henry 9194B*. Sichuan austro-occid: Opien, *T. F. Lou 171*, *107* (type), Sept. 21, 1929. Zhejiang: Tienmo Shan, *S. S. Chien* 170 (1920).

The affinity of this species to *L. thunbergianus* (Kaulf.) is sufficiently clear, to which it was referred as forma *caudato-attenuata* by Takeda, from which it differs in extraordinarily slender habit with caudato-atteunate leaf-apex, and in elongate sori.

Plate 73. Fig. 1. habit sketch (natural size). 2. peltate paraphysis from a sorus (× 50). 3. scale from rhizome (× 50).

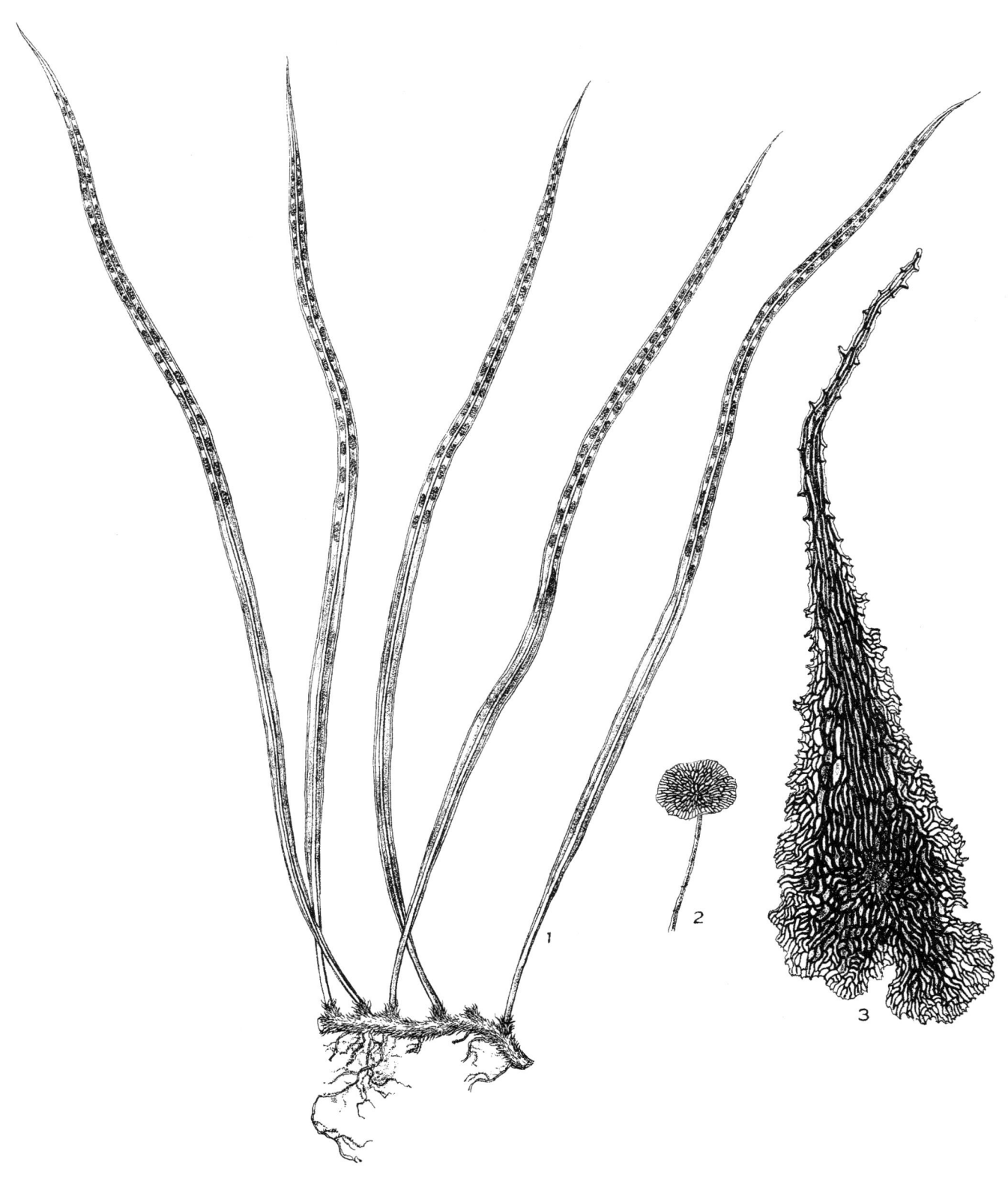

LEPISORUS ANGUSTUS Ching
狭叶瓦苇

根状茎细长,匍匐甚广;鳞片疏生,极似瓦苇;叶疏生,细长,狭线披针形,柄颇细长,叶端尖长,革质,绿色;子囊群中大,位于中肋与叶边之间,相距甚远。

分布:东三省,热河,河北,山东;朝鲜半岛,日本北部及西伯利亚东部。

本种在其分布区域极为普通,昔人认为与瓦苇同,实则不然,因其根状茎及叶体均甚细长,子囊群形体小而距离大,且其分布区域,位于瓦苇之北,实有显著之差异也。

图注:1.本种全形(自然大);2.着生于子囊群上之盾状隔丝(放大50倍);3.根状茎上之鳞片(放大50倍)。

LEPISORUS USSURIENSIS (Regel et Maack) Ching

LEPISORUS USSURIENSIS (Regel et Maack) Ching, Bull. Fan Mem. Inst. Biol. **4**: 91 (1933).

Pleopeltis ussuriensis Regel et Maack, Mém. Acad. Sci. Petersb. VII. **4**: 4. 175 (1861).

Polypodium ussuriense Regel, Acta Horti Petr. **7**: 663 (1881); Ogata, Ic. Fil. Jap. **2**: pl. 92 (1929).

Polypodium lineare var. *distans* Makino, Tokio Bot. Mag. **15**: 60 (1901).

Polypodium distans Makino, Tokio Bot. Mag. **20**: 33 (1906).

Polypodium lineare var. *ussuriense* C. Chr. Ind. Fil. 572 (1906); Takeda, Notes, R. Bot. Gard. Edinb. **8**: 271 (1915).

Polypodium lineare var. *coraiense* Christ in Fedde, Repert. Sp. Nov. **5**: 10 (1908).

Polypodium annuifrons var. *distans* Nakai, Fl. Kor. **2**: 44 (1911).

Rhizome rather slender, wide-creeping, sparsely scaly; *scales* appressed, atratous, rigid, ovate-acuminate, dentate, luminae opaque, small, with thick, dark-colored cell-walls; *frond* distant, slender, narrowly linear-lanceolate, rather long-stipitate, apex long-attenuate; *texture* coriaceous, color greenish; *sori* medium-sized, medial, far apart.

Northeast China: Er Tiengtien Sze, *P. H. Dorsett 3048* (1925); Henhtao Hotze, *P. H. Dorsett 3244*; Mefuv, *P. H. Dorsett 4216*; Kirin, *Komarov 46*; from Mukden to Kirin, *James*, May-August, 1886; Sching-king, *J. Ross 4*. Hebei: Shanhaikwan, Kiaoshan, *R. P. Licent 1525* (1915); Tungliang, *C. F. Li & W. K. Hsia 2331*. Shandong: Taishan, *Jacob 1346*; Tsingtao, Laushan, *Reymand 226*, Aug. 1911; *A. Engler 7038*, June, 1913.

Korea peninsula: Quelpaert, *Taquet 83*; Diamont Mt., *Faurie 134*, *112*. Also Japan and eastern Siberia.

A common fern in the regions noted. This is generally considered by authors as identical with, or as a variety of, *Polypodium lineare* Thunberg from warmer regions farther south, but differs in much slender habit, long-stipitate, narrower and distant leaves on much thinner rhizome, and in more widely separated, smaller, medial sori.

Plate 74. Fig. 1. habit sketch (natural size). 2. peltate paraphysis from a sorus ($\times 50$). 3. scale from rhizome ($\times 50$).

LEPISORUS USSURIENSIS (Regel et Maack) Ching
乌苏里瓦苇

根状茎匍匐甚广；端部鳞片密生，紧覆，卵形渐尖头，红褐色，边缘具齿，网眼狭长，透明，唯位于中央者不透明；叶疏生，长 15—40 cm，宽约 1 cm，狭披针形，光滑，革质，绿色；子囊群卵形，位于中肋及叶边之间。

分布：陕西，四川，湖北，云南及西藏。

本种为中国西部及西北部极普通之蕨种，昔人概认为瓦苇，然其茎鳞呈红褐色，短而宽，网眼唯位于中央者不甚透明；叶疏，绿色，故易于识别。

图注：1. 本种全形（自然大）；2. 着生于子囊群上之盾状隔丝（放大 50 倍）；3. 根状茎上之鳞片（放大 50 倍）。

LEPISORUS CONTORTUS (Christ) Ching

LEPISORUS CONTORTUS (Christ) Ching, Bull. Fan Mem. Inst. Biol. **4**: 90 (1933).

Polypodium contortum Christ, Bot. Gaz. **51**. 347 (1911).

Polypodium lineare var. *contortum* Christ, Nuovo Giorn. Bot. Soc. Ital. n. s. **4**: 98 pl. 1. f. 3 (1897); Diels in Engl. Jahrb. **29**: 204 (1901).

Polypodium lineare var. *Thunbergianum* f. *contorta* Takeda, Notes, R. Bot. Gard. Edinb. **8**: 270 (1915).

Polypodium onoei Franch. Pl. David. in Nouv. Archiv. II. **7**: 355 (1883, non 1879).

Polypodium lineare Franch. (non Thunberg) l. c. 354; Christ, Bull. Herb. Boiss. **6**: 860 (1898); Bull. Acad. Géogr. Bot. (1906) 105; Bot. Gaz. **51**: 374 (1911).

Rhizome thick, wide-creeping, densely scaly on the young parts; *scales* ovate-acuminate, rufo-brown, appressed, dentate, luminae elongate, rather clear except the central ones being darker colored with thick walls; *frond* distant, 15-40 cm long, about 1 cm broad, linear, naked; *texture* coriaceous, color pale green; *sori* ovate, large, medial, but oblong and oblique when young.

Shaanxi: Mt. Kulu, *Giraldi*, August, 1894 (type); Kuantou San, *Giraldi*, Nov. 5, 1896; Ngoshan, *Giraldi*, Oct. 1894; Rinsan, *Giraldi*, July 23, 1897; Tsinglin, *David*, April, 1874; ibid. *David*, March, 1873; *Purdom 92*, *93*. Sichuan: Tatsienlu, *E. H. Wilson 2633*, *2634*, June, 1908; Fang Hsien, *Wilson 2636*, *2635*, August, 1908; Mt. Omei, *W. P. Fang 3054*, *2995*, *2984*, *3055*; *Faber 1069*; Hanynon Hsien, *P. W. Fang 4779*; Hon-ya, O-shan, *W. P. Fang 6445*, *6665*. Hubei: *Henry 5654*. Yunnan: Lickiang, *C. Schneider 2685*; *G. Forrest 15556*; *Henry 9893*, *10087A Hancock 93*; *Maire 2779*; Sungwei, *G. Forrest 295*; *Delavay 10* (1883). Tibet: Yatung, *Hobson* (1897),

A common fern in the above localities and generally considered by authors as *Polypodium lineare*, from which it differs in rufo-brown, ovate-acuminate rhizome scales, which are reticulate with rather clear, large luminae except the centre being dark colored, and in thick coriaceous leaves of a pale green color and attached to the wide-creeping rhizome at greater distance.

Plate 75. Fig. 1. habit sketch (natural size). 2. peltate paraphysis from a sorus ($\times$ 50). 3. scales from rhizome ($\times$ 50).

LEPISORUS CONTORTUS (Christ) Ching
扭瓦苇

根状茎颇肥,横生;鳞片密生,黑色,质厚,狭披针形,基部卵圆形,网眼不透明,唯向边则为透明,具锯齿;叶颇密生,长 10—18 cm,宽 1—1.8 cm,狭披针形,柄极短或无,革质,色黝黄;子囊群大而圆,接近。

分布:中部及东南部各省,台湾省;日本,菲律宾群岛亦产之。

本种最初在日本发现,现则中国东南各省均产之;其产于中国北部,东三省,朝鲜半岛及西伯利亚东部,昔人认为即为本种者,当为另一种,即乌苏里瓦苇是也。

图注:1. 本种全形(自然大);2. 着生于子囊群上之盾状隔丝(放大 50 倍);3. 根状茎上之鳞片(放大 50 倍)。

LEPISORUS THUNBERGIANUS (Kaulfuss) Ching

LEPISORUS THUNBERGIANUS (Kaulfuss) Ching, Bull. Fan Mem. Inst. Biol. **4**: 88 (1933).

Pleopeltis thunbergiana Kaulfuss, Wesen. d. Farrnke. 113 (1827).

Polypodium lineare Thunberg (non Burm. 1768, nec. Houtt. 1783), Fl. Jap. 335 (1784), Iconogr. Fl. Jap. 11. t. 9 (1794); Baker in Hooker, Syn. Fil. 354 (1868), pro parte; C. Chr. Ind. Fil. 540 (1906), pro parte; Ogata, Ic. Fil. Jap. **2**: pl. 86 (1929).

Pleopeitis linearis Moore, Ind. Fil. 346 (1826); Bedd. Handb. Ferns Brit. Ind. 346 (1883).

Pleopeitis nuda Hooker, Journ. Bot. (1875) 355 (non 1823).

Rhizome rather thick, creeping, densely scaly; *scales* atratous, rigid, opaque, linear-subulate from a broadly ovate base, dark-colored, except the broad margin of the base being hyaline with clear, elongate luminae with thin walls; *frond* rather approximate, 10-18 cm long, 1-1.5 cm broad, linear-lanceolate, acuminate, gradually attenuate to a short stipe; *texture* coriaceous, color obscure; *sori* large, rounded, nearer to the costa than to the margin, close or almost contiguous at maturity.

Jiangxi: Kiukiang, *David 837* (1867). Guizhou: Vanching Shan, *Y. Tsiang 7848*; Pinfa, *Cavalerie 2179*, Fujian: Amoy, *Swinhoe* (1870); Foochow *Alexander* (pro parte); Kushan, *H. H. Chung 3772*. Zhejiang: Tientai Shan, *H. H. Hu 282* (1920); *C Y. Chiao 14391, 14203* (1927); Hangchow, Fan Chiao, *R. C. Ching 3865*. Guangdong: Lungtau Shan, *C. C. C. 12051*; Tsiangliang Shan, *McClure 6722*; Lockchong, *N. K. Chun 43006, 42557*; Lohfau Shan, *140 ex Herb. Hong Kong*; Swatow, Thaiyong, *Dalziel*, July, 1910; Guangxi: Yao Shan, *Sin & Whang 190*, on recks; Sanfang, Tangloo, *R. C. Ching 5675* (1928); Schfang Dar Shan, *R. C. Ching 7865*. Sichuan: Mapien Hsien, *W. P. Fang 206*; ibid. *W. P. Fang 1769*. Hainan: Hancock 111 (Kew No.). Jiangsu: Fenwhang Shan, *C. G. Matthew*, June 1, 1904; *Carles 416*, Oct. 1881; Wusih, Weishan, *R. C. Ching 3436* (1927); Ishing, *Y. L. Keng 2444*. Anhui: Whang Shan *A. N. Steward 7194*; *S. S. Chien 1067*; *R. C. Ching 8532*; *K. K. Chow* (1925).

Type from Japan. A common fern in the eastern and southeastern part of China as well as Japan, Quelpaert and the Philippines. Its previous report by authors from Northeast China, northern Korea and North China proves to be a mistake for *L. ussuriensis* (Kegel et Maack). The specimens so-called *Polypodium lineare* from western and northwestern China should for the most part be referred to *L. contortus* (Christ).

Plate 76. Fig. 1. habit sketch (natural size). 2. peltate paraphysis from a sorus (× 50). 3. scale from rhizome (× 50).

LEPISORUS THUNBERGIANUS (Kaulfuss) Ching
瓦苇

图版 77　骨牌蕨

根状茎细长,匍匐于树干上,绿色;鳞片疏生,深褐色,基部圆形具粗大锯齿,上部狭披针形,全缘,叶疏生,长 6—10 cm,宽 2—2.5 cm,阔披针形,柄短,叶端呈嘴形,有时亚二形,肉质,光滑,淡绿色,叶脉不显;子囊群中大,分离或向叶端为不规则之接合。

分布:香港,广东,贵州,广西,云南,台湾;越南,缅甸,喜马拉雅山区。

本种为附生于老林中树干上之小形蕨种,在其分布区内,颇为普通。

图注:1.本种全形(自然大);2.着生于子囊群上之盾状隔丝(放大 50 倍);3.根状茎上之鳞片(放大 50 倍)。

LEMMAPHYLLUM SUBROSTRATUM（C. Chr.）Ching

LEMMAPHYLLUM SUBROSTRATUM (C. Chr.) Ching, Bull. Fan Mem. Inst. Biol. **4**: 97 (1933).

Polypodium subrostratum C. Chr. Ind. Fil. 567 (1906); Takeda, Notes, R. Bot. Gard. Edinb. **8**: 311 (1915).

Polypodium rostratum Hooker (non Burm. 1768, nec Cav. 1802), Ic. Pl. t. 953 (1854); A Cent Ferns t. 53 (1854); Hayata, Ic. Pl. Form. **4**: 253 f. 177 (1914).

Pleopeltis rostratum Bedd. Ferns Brit. Ind. t. 159 (1866); Handb. Ferns Brit. Ind. 345 (1883).

Rhizome slender, wide-creeping generally on the trunk of trees, greenish, sparcely scaly; *scales* thin, clathrate, fusco-brown, lanceolate-subulate from a rounded base provided with a few large teeth or protruding arms; *frond* far apart, short-lanceolate, 6-10 cm long, 2-2.5 cm broad, apex rostrate, gradually attenuate to the short stipe, often subdimorphous, fertile one being longer and narrower; *texture* carnoso-subcoriaceous, naked, color pale green; *venation* more or less distinct at least under light; *sori* medium-sized, distinct or rarely two imperfectly fused.

Guangdong: Lockchong, *Y. K. Wang 31446*, *42270*; Lohfau Shan, *83* ex Herb. Hong Kong (1883); *S. P. Ko 50074*; Yingtak, *C. G. Matthew*, April 7, 1907; Lungtau Shan, *C. C. C. 12531*; North River, *C. L. Tso 21114*. Yunnan: Mengtze, Henry *10448*, *13302*; *Hancock 181*; Keng Hung, *J. F. Rock 2690*; Kulongtchang, *E. E. Maire*, Guizhou: Chengfeng, *Y. Tsiang 4793*; Yaoren Shan, Sanhoa, *Y. Tsiang 6305*; Yingkiang, Tushan, *Y. Tsiang 6716* (f. typica). Guangxi: *Sin & Whang 46* (1926); Yao Shan, *Sin & Whang 7* (1928); Yaomar Shan, Lin Yen, *R. C. Ching 7006*, *7104*; Chu-feng Shan, Sanfang, *R. C. Ching 5807* (f. typica).

A common fern, generally epiphytic on the trunk of trees in deep ravines under forests. A transitional species from *Lepisorus* to *Lemmaphyllum*.

Plate 77. Fig. 1. habit sketch (natural size). 2. peltate paraphysis from a sorus ($\times$ 50). 3. scale from rhizome ($\times$ 50).

LEMMAPHYLLUM SUBROSTRATUM (C. Chr.) Ching
骨牌蕨

根状茎细长横生；鳞片疏生，卵形，渐尖头，黝色，具小锯齿；叶疏生，叶间距离大，二形，其非子囊群叶长 8—11 cm，宽不及 2 cm，舌形或倒披针状舌形，钝头，厚肉质，淡绿色，遍体被鳞片，叶脉不显；子囊群叶为细长线形，长达 15 cm 或过之，宽不及 1 cm，下面密生鳞片，子囊群大，椭圆形，常突出于叶边外。

分布：四川中部之灌县，模式标本在静生生物调查所，副本标本在中国科学社生物研究所。

本种为特殊之一种，其形体同于抱石莲，然各部之大，则远过之，且叶体鳞片密生，子囊群为大椭圆形，常突出于叶边外。按川省植物，在最近五十年内，几经采集，今复发现如此特殊新种，则其富藏之未尽也明矣。

图注：1.本种全体（自然大）；2a.子囊群叶之一部，表示其叶脉及子囊群之位置（放大 4 倍）；2b.非子囊群叶之一部，表示其叶脉（放大 2 倍）；3a—b.着生于子囊群上之盾状隔丝（放大 50 倍）；4a—b.叶体下面之鳞片（放大 50 倍）；5.根状茎上之鳞片（放大 50 倍）。

LEMMAPHYLLUM ADNASCENS Ching

LEMMAPHYLLUM ADNASCENS Ching, Bull. Fan Mem. Inst. Biol. **4**: 101 (1933).

Rhizome slender, wide-creeping, 1 mm thick, sparcely scaly; *scales* ovate, acuminate, fuscous, coarsely clathrate, denticulate; *frond* distant, strongly dimorphous, the sterile 8-11 cm long, below 2 cm broad, ligulate or oblanceolate-ligulate, obtuse, gradually attenuate to a short stipe; *texture* thickly carnose, pale green, scaly on both surfaces; *venation* completely hidden; *fertile frond* much contracted, narrowly linear-elongate, to 10 cm or longer, below 1 cm broad, underside densely scaly; *sori* large, oblong-ovate, filling up the space between the midrib and margin, which latter sometimes appears wavy, due to the protruding sori.

Sichuan merid: Kuan Hsien, *F. T. Wang 20691*. Type in Herbarium of the Fan Memorial Institute of Biology and, of the Biological Laboratory of the Science Society of China.

A remarkable fern and perhaps the largest of the genus, not expected at all at such a late date from a region, where the botanical exploration has supposedly been carried out intensively over half a century. Evidently a rare fern in the locality, only two good specimens being collected. Its affinity to L. *drymoglossoides* (Baker) is clear enough, differs chiefly in much larger size, thicker and much harder texture, densely scaly under surface, in rhizome scales of different outline and sori being of much larger size, completely filling up the space between the midrib and margin.

Plate 78. Fig. 1. habit sketch (natural size). 2a. portion of a fertile leaf, showing venation ($\times$ 2). 3a-b. peltate paraphyses from a sorus ($\times$ 50). 4a-b. scales from underside of a leaf ($\times$ 50). 5. scale from rhizome ($\times$ 50).

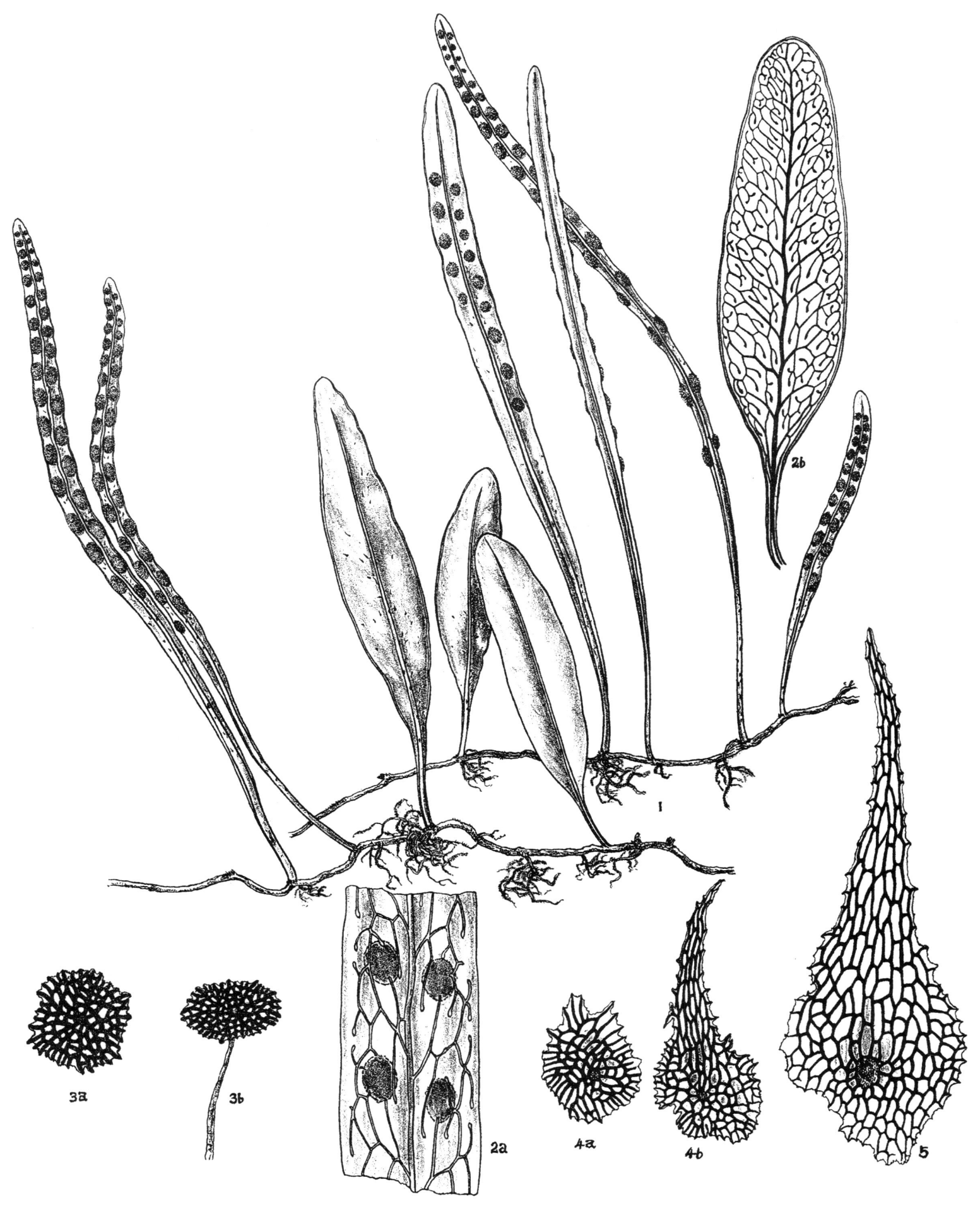

LEMMAPHYLLUM ADNASCENS Ching
川石莲

根状茎细长如铜丝,附生于石上,绿色;鳞片疏生,黄褐色,质薄,细长,基部宽而不规则分叉;叶疏生,亚二形,非子囊群叶为卵圆形或卵椭圆形,长约 1—2 cm,色淡绿,肉质,下面鳞片疏生,叶脉不显,子囊群叶细长如舌形或匙形,长 3—6 cm,宽不及 1 cm;子囊群中大,分离或向叶端为不规则之接合。

分布:湖北,安徽,四川,浙江,江苏,贵州,河南,生于阴湿石壁上。

本种之叶形及子囊群变异甚大,其在极端情形下者,颇类南方温热带之抱树莲。

图注:1. 本种全形(自然大);2a. 子囊群叶(放大 2 倍),2b. 非子囊群叶(放大 3 倍);3. 着生于子囊群上之盾状隔丝(放大 50 倍);4. 根状茎上之鳞片(放大 50 倍)。

LEMMAPHYLLUM DRYMOGLOSSOIDES (Baker) Ching

LEMMAPHYLLUM DRYMOGLOSSOIDES (Baker) Ching, Bull. Fan Mem. Inst. Biol. **4**: 100 (1933).

Polypodium drymoglossoides Baker, Journ. Bot. (1887) 170; Diels in Engl. Jahrb. **29**: 204 (1901); Christ, Bull. Acad. Géogr. Bot. (1902) 206 c. fig; C. Chr. Ind. Fil. 523 (1906); Takeda, Notes, R. Bot. Gard. Edinb. **8**: 302 (1915); Hand-Mzt. Symb. Sinic. **6**: 42 (1929); C. Chr. Dansk Bot. Archiv. **6**: 47 pl. V. f. 8-9 (1929).

Polypodium moupinense Franchet, Nouv. Archiv. Mus. II. **10**: 121 (1887); Christ, Bull. Acad. Géogr. Bot. (1906) 105.

Polypodium cyclophyllum Baker, Ann. Bot. **5**: 473 (1891); C. Chr. Ind. Fil. 520 (1906).

Rhizome slender, wide-creeping, greenish, sparcely scaly; *scales* light-brown, thin, coarsely clathrate, subulate from a broad, irregularly branched base; *frond* subdimorphous, distant, the sterile orbicular-ovate to oblong-ovate, 1-2 cm long, pale green, carnose, furnished with a few appressed scales on the underside; *venation* obscure even under light; *fertile frond* generally spathulate, 3-6 cm long, below 1 cm broad; but not infrequently almost similar to the sterile in outline; *sori* medium-sized, distinct, or not infrequently the upper ones imperfectly fused.

Hubei: Ichang, *A. Henry 1576* (type); Nanto, *Henry 2965, 4892, 5963*; Changyang, *E. H. Wilson 1450, 2647*. Anhui: Suinin, *K. K. Tsoong 4530*, 117. Sichuan: Moupin, *David*; *Henry 7532A*; Mt. Omei, *Faber 1046*; *E. H. Wilson 5314*; *W. P. Fang 5735*; Oshan, *W. P. Fang 8000* (1930); Honya, *W. P. Fang 8747* (1930). Zhejiang: Tientai Shan, *C. Y. Chiao 14719*, August, 1927; Bunji islank, *C. G. Matthew*, Sept. 30, 1907; Ningpo, *Hancock 32* (1877). Jiangsu: Soo-chow, Linyin Shan, *C. G. Matthew*, June 4, 1904; Wusih, Wei Shan, *R. C. Ching 3430* (1927); Ishing, *R. C. Ching & C. L. Tso 544*; Nanking, Tsihsia Shan, *Chen 196*. Guizhou: Pinfa, *Cavalerie 627*; Gaotsha, *Handel-Mazzetti 10261*; Vanching Shan, *Y. Tsiang 7619, 7748, 7952*; Anlung, *Y. Tsiang 9346, 7409*; Tungtze, *Y. Tsiang 5174, 4936*; Houang-tsao-pa, *Cavalerie 7051*, May, 1919. Hunan: Hsikwang Shan, *Handel-Mazzetti*.

A very unstable fern as to the shape of leaves and conditions of sori, which both in some extreme forms appear hardly distinguishable from its relative, *L. microphyllum* Presl, species inhabiting a more southern area.

Plate 79. Fig. 1. habit sketch (natural size). 2a. fertile leaf ($\times$ 2). 2b. sterile leaf ($\times$ 3). 3. peltate paraphysis from a sorus ($\times$ 50). 4. scale from rhizome ($\times$ 50).

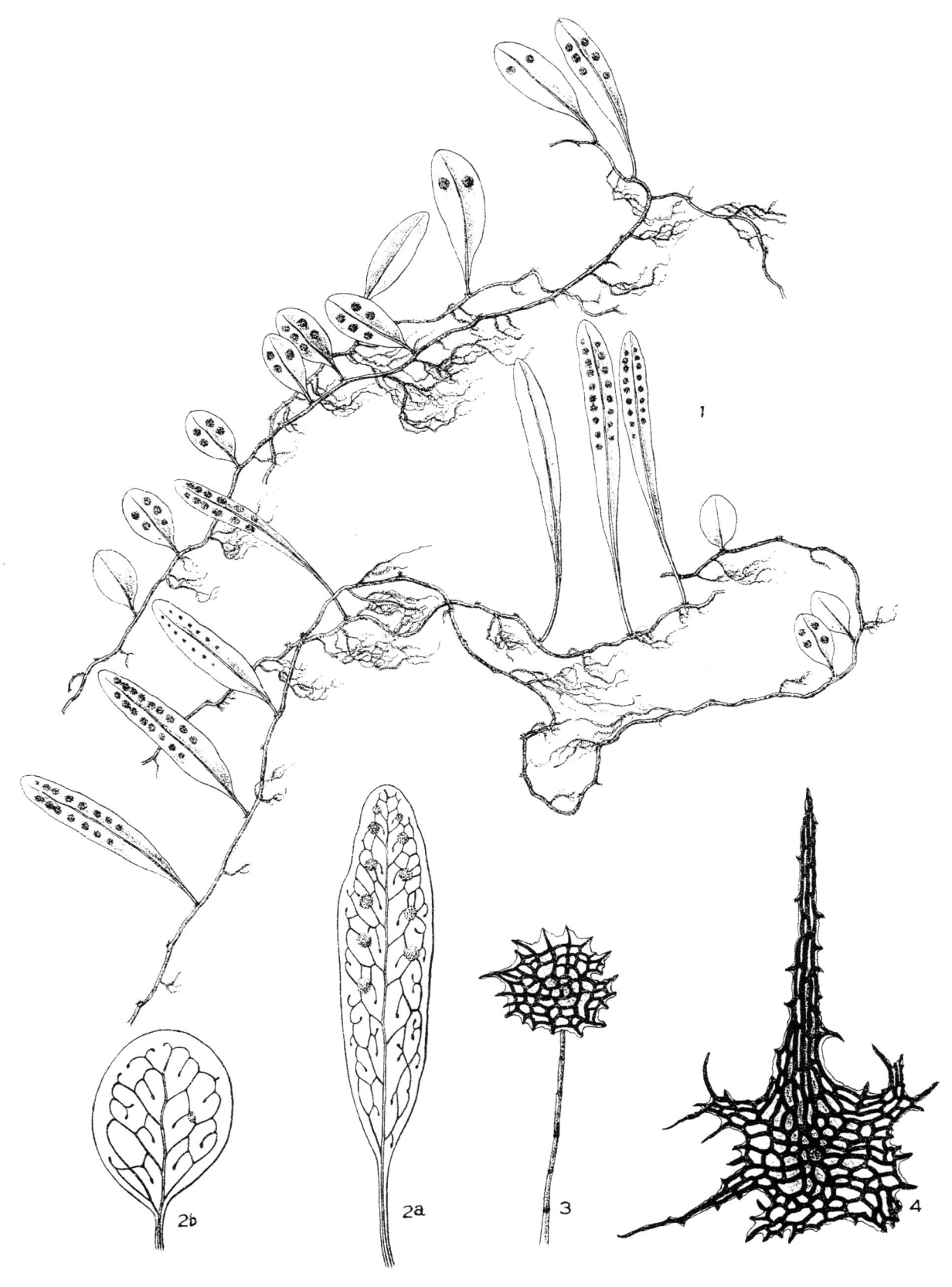

LEMMAPHYLLUM DRYMOGLOSSOIDES (Baker) Ching
抱石莲

根状茎细长，匍匐于树干上，绿色；鳞片疏生，黄褐色，基部圆形，不规则分叉，上部细长，全缘；叶二形，非子囊群叶为圆形或卵圆形，基部圆形，或心脏形或楔形，肉质，淡绿色，全体光滑或鳞片疏生，叶脉不显，在强光下可见，为有规则之网状；子囊群叶细长，舌形，长 3—4 cm，宽约 2—3 mm；子囊线形，一列，位于叶边与中肋之间，成熟后张开。

分布：福建，广东，广西，香港，湖北，台湾；日本及朝鲜半岛南部亦产之。

本种最初发现于日本，其叶形变异甚大，在极端情形之下，几与抱石莲无异。

图注：1. 本种全形（自然大）；2. 着生于子囊群上之盾状隔丝（放大 50 倍）；3. 根状茎上之鳞片（放大 50 倍）；4. 非子囊群叶，表示其叶脉（放大 3 倍）。

LEMMAPHYLLUM MICROPHYLLUM Presl

LEMMAPHYLLUM MICROPHYLLUM Presl, Epim. Bot. 263 (1849); C. Chr. Dansk Bot. Archiv. **6**: 46 pl. V. f. 1-4 (1929); Ching, Bull. Fan Mem. Inst. Biol. **4**: 102 (1933).

Drymoglossum carnosum Hooker (non J. Sm.), Kew Journ. Bot. **9**: 358 (1857); Benth. Fl. Hongk. 444 (1861).

Pteris piloselloides Thumberg (non Linnaeus), Fl. Jap. 331 (1784).

Drymoglossum carnosum var. *obovatum* Harr. Journ. Linn. Soc. **16**: 33 (1867).

Drymoglossum microphyllum C. Chr. Ind. Fil. 246 (1906); Ogata, Ic. Fil. Jap. **4**: pl. 164. (*f. typica*); I. c. pl. 166 (var. *obovatum*) (1931).

Drymoglossum carnosum var. *subcordatum* Baker in Hooker, Syn. Fil. 397 (1868).

Drymoglossum obovatum Christ, Journ. de Bot. France **19**: 73 (1905).

Rhizome wiry, wide-creeping generally on the trunk of trees, greenish, sparcely scaly; *scales* thin, clathrate, light-brown, entire, subulate from a broad, lacerato-fimbriate base; *frond* dimorphous, the *sterile* mostly orbicular with rounded or subcordate base (f. *typica*), or ovate or broadly obovate with shortly cuneate base (var. *obovatum*), 5-15 mm broad, surfaces naked or with a few brown, ovate, clathrate scales; *texture* carnose (in fully grown leaves), or submembranaceous (when young), color pale green, but obscurely brown or even blackish when dried; *venation* not distinct, rather regularly reticulate and looped some way below the thin, plane margin, meshes provided each with one simple, recurrent, clavate veinlet; *fertile frond* strongly contracted, ligulate, 2-3 mm broad, 3-4 cm long including the short stipe; *sori* linear, medial, densely scaly, laterally broadened at maturity; *spores* subreniform. pellucide, smooth.

Fujian: Kuliang Hill near Foochow, *J. B. Norton 1085*; Yuanshan, *H. H. Chung 1202*. Guangdong: Lockchong, *C. L. Tso 20371*; Canton, *Hance 31*; Ting Woo Shan, *Karl Buswell 6332*. Guangxi: Yao Shan, *S. S. Sin & K. K. Whang 1512B*; Lungchow, *Morse 68*. Hong Kong: *C. Wright*, *Bowring*, *C. Wilford 332*. Hubei: Ichang Gorge, *Maires* (1880). Taiwan.

Also Japan and South Korea.

Type from Japan, first collected by Thunberg and named *Pteris piloselloides* by himself. A very variable fern as to the shape of sterile leaves; in some extreme forms it approaches *L. drymoglossoides* (Baker) so closely that it can be distinguished from the latter species with difficulty, as already pointed out by Dr. Christensen in his recent critical work on the genus. In typical form, the sterile fronds orbicular, rounded or cordate at base; in var *obovatum* (Harr.), they are obovate or ovate, shortly cuneate at base on rather long petioles. *Drymoglossum Nobukoanum* Makino, Journ. Jap. Bot. **7**: 8 (1931), now to be called **LEMMAPHYLLUM NOBUKOANUM** (Makino) Ching, comb. nov. from Taiwan seems to be specifically distinct from var. *obovatum* by slender, small, spothulate sterile fronds and almost linear fertile ones on long, wiry stipes (cf. Ogata, Ic. Fil. Jap. pl. 165. 1931).

Plate 80. Fig. 1. habit sketch (natural size). 2. peltate paraphysis from a sorus ($\times$ 50). 3. scale from rhizome ($\times$ 50). 4. sterile frond, showing venation ($\times$ 3).

LEMMAPHYLLUM MICROPHYLLUM Presl
抱树莲

根状茎肥大,短而横生;鳞片密着,锈黄色,线披针形,边缘具细长齿,上部开张;叶簇生,柄长 5—9 cm,幼时被细长节状星芒形之松毛,渐次脱落,终变为光滑,叶体披针形,长 20—30 cm,宽 2—2.5 cm;基部渐狭,端部渐尖,全缘,厚革质,光绿色,上面多小孔且具疏生之星状细毛,久则脱落,下面初具灰白色之星状绒毛,终变为光滑,侧脉颇显明;子囊群着生于叶体之端部,圆形,于两侧脉之间成 3—4 列,明白可识,幼时具灰白色星状细绒毛,终为光滑。

分布:广东北江,贵州,四川,云南,湖北,江西。

本种为本属特殊之种,易于识别,与此相近的蕨为绒毛石韦(C. subfurfuraceus C. Chr.),产于锡金喜马拉雅山区及云南,然其体形较大,叶下面之厚绒毛终不脱落,故异于本种。

图注:1.本种全形(自然大);2.叶体之一部,表示其叶脉(放大 2 倍);3.叶柄基部之鳞片(放大 50 倍);4.叶柄上之星芒状松毛(放大 30 倍);5.子囊(放大 50 倍)。

CYCLOPHORUS CALVATUS (Baker) C. Christensen

CYCLOPHORUS CALVATUS (Baker) C. Christensen, Ind. Fil. 198 (1905).

Polypodium calvatum Baker, Journ. Bot. (1879) 304.

Niphobolus calvatus Diels in Engl. Jahrb. **29**: 207 (1901).

Polypodium stigmosum Christ (non Swartz) in Warburg, Monsunia **1**: 62 (1900).

Rhizome thick, oblique, rather short-creeping, densely scaly; *scales* ferruginous, spreading, linear-lanceolate, fimbriate; *frond* caespitose, several together, stipe 5-9 cm long, thick, when young densely clothed in loose, long-armed articulated but deciduous fine hairs, lamina lanceolate, 20-30 cm long, 2-2.5 cm broad, gradually narrowed downward, apex acuminate, margin entire; *texture* thick coriaceous, color clear light-green, pitted above and with a few fine but deciduous stellate hairs, thinly subfurfuraceous with gray, wooly, stellate hairs beneath, but becomes perfectly glabrous at last; lateral *main veins* rather distinct in the lower part; *sori* confined to the upper one-third, sparcely intermixed with wooly hairs, rounded, distinct, or sometimes imperfectly fused; arranged in 3-5 distinct rows between lateral main veins.

Guangdong: North River, *C. Ford* (type), May, 1875; Lien Hsien, *C. G. Matthew*, December 2, 1907; *S. P. Ko 51035*; *W. Y. Chun 5563*. Guizhou: Tungtze, *Y. Tsiang 5099*; Tsehheng, *Y. Tsiang 9249*; Chengfeng, *Y. Tsiang 4523*; Tinfa, *Cavalerie*. Sichuan: Mt. Omei, *E. Faber 1049*; *A. Henry 5570*; Nanchuan, *von Rosthorn 1721*, *1723*, *1708*; Moupin, *David*. Hubei: Nanto, *A. Henry 4024*; *E. H. Wilson 2671*. Yunnan: Mengtze, *A. Henry 9834*; *Hancock 52*; Paeulgai, *E. E. Maire*, Sept. 1916.

A pretty and distinct fern, with which C. *subfurfuraceus* (Hook.) from the Himalayas is perhaps the only close ally, which differs in much larger dimension, persistent, thicker, gray wooly hairs on the underside of lamina and in frond of a different color.

Plate 81. Fig. 1. habit sketch (natural size). 2. portion of a lamina, showing venation ($\times$ 2). 3. scales from the base of a stipe ($\times$ 50). 4. branched hair from stipe ($\times$ 30). 5. sporangium ($\times$ 50).

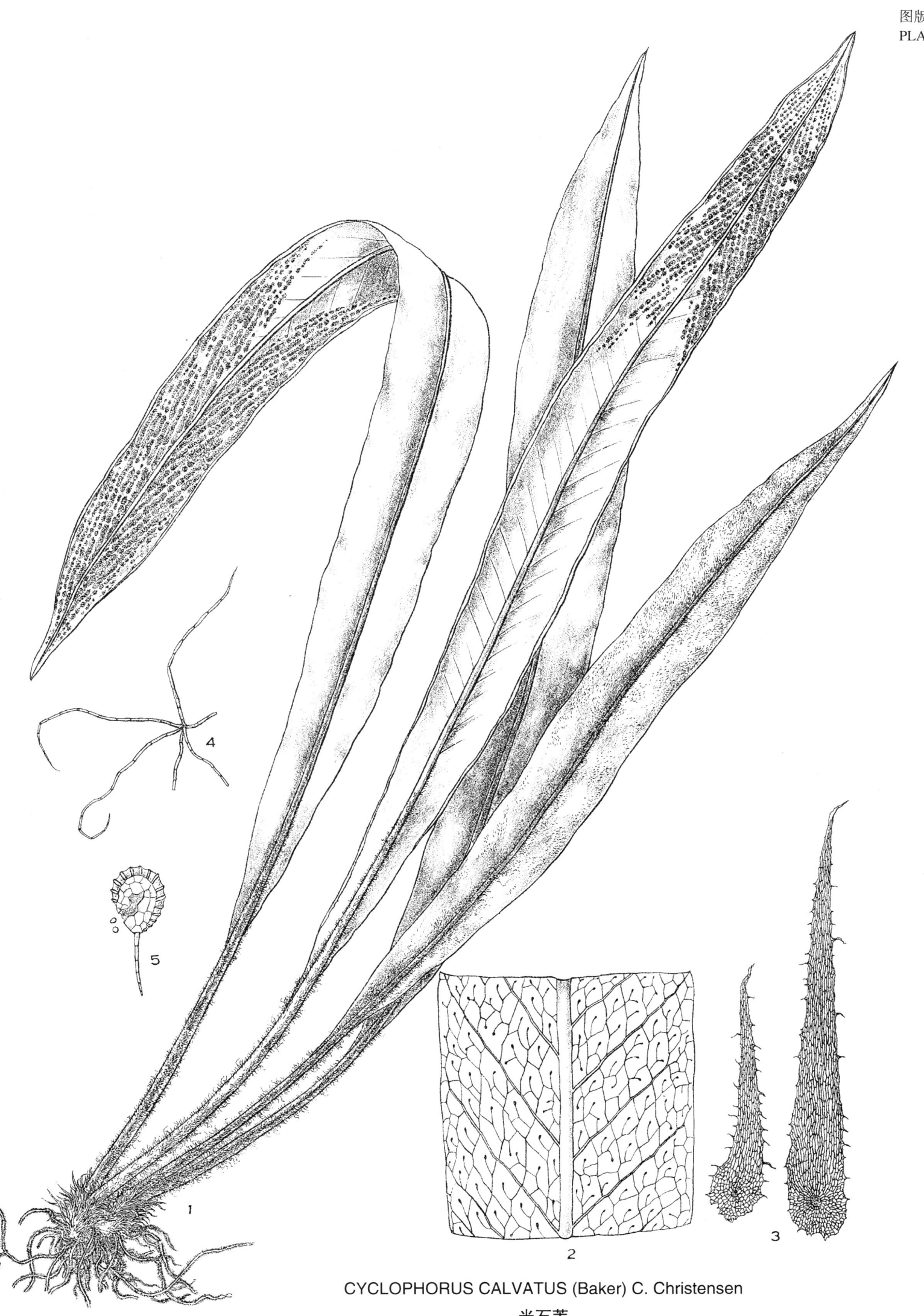

CYCLOPHORUS CALVATUS (Baker) C. Christensen 光石苇

根状茎短,半直立;鳞片密生,黄褐色,阔披针形,渐尖头,边缘具睫毛;叶簇生,二形,非子囊群叶每簇3—5个,柄长约4 cm,稻秆色,同样之鳞片密生,叶体披针状舌形,钝头,下部渐狭,长10—14 cm,宽达3 cm,坚革质,淡绿色,下面遍被黄褐色之鳞片,全缘平薄,角质,透明,侧脉不见,中肋下面扁平,被鳞片;子囊群叶稍狭,柄长过10 cm,叶体长约10 cm,宽仅1.5 cm,披针形,急尖头,边缘稍反卷。子囊群密布叶之下面,仅中肋可见耳。

分布:海南。

本种为海南特产,其形体颇似广东本部所产之 E. austrosinicum Matthew et Christ,然其形体较小,叶质较坚,子囊群叶较非子囊群叶为长,非子囊群叶舌形,钝头,基部楔形,柄长而瘦,得以识别。

图注:1.本种全形(自然大);2.非子囊群叶之一部,表示其叶脉(自然大);3.根状茎上之鳞片(放大30倍),4.叶下面之线状鳞片(放大30倍);5.孢子囊(放大150倍)。

ELAPHOGLOSSUM MCCLUREI Ching

ELAPHOGLOSSUM MCCLUREI Ching, *Sinensia* **1**: 55(1930).

Rhizome short, oblique, densely scaly; *scales* brown, broadly lanceolate, acuminate, margin sparcely fimbriate, texture rather thick; *frond* subcaespitose, several together, dimorphous, the fertile ones somewhat longer than the sterile, stipe of sterile frond 4 cm long, that of the fertile over 10 cm long, straminous, densely scaly; lamina of sterile frond 10-14 cm long, to 3 cm broad, lanceolate-ligulate, apex obtuse, base shortly atftenuate; *texture* thick coriaceous, color pale green, underside rather densely clothed with brown, appressed, branched hairs, upperside naked; margin plane, narrowly hyaline, midrib broadened, slightly raise above; *veins* completely hidden; *fertile* lamina to 10 cm long, 1.5 cm broad, lanceolate, apex acute, margin narrowly reflexed; *sori* obscure brown, dense, covering the entire underside except the midrib; *spores* opaque, ovoid, coarsely crested.

Hainan: Hung Mo Tung, *F. A. McClure 18257*, August 22, 1929, summit of highest peak, on mossy trunk of forest tree, 6,000 ft. alt.; on route from Dung Ka to Win Fa Shi, *C. L. Tso & N. K. Chun 43729*, August 26, 1932, on rocks in thickets, 1,700 ft, alt.

This is closely related to *E. austro-sinicum* Matthew et Christ from Guangdong, differs in much smaller size, rigid texture, ligulate sterile frond with obtuse apex and short-attenuate base provided with longer and thinner stipe, and fertile frond being longer than the sterile.

Plate 82. Fig. 1. habit sketch (natural size). 2. portion of a sterile frond, showing venation (natural size). 3. scale from rhizome ($\times$ 30). 4. branched hairs from the underside of a frond ($\times$ 80). 5. sporangium ($\times$ 250).

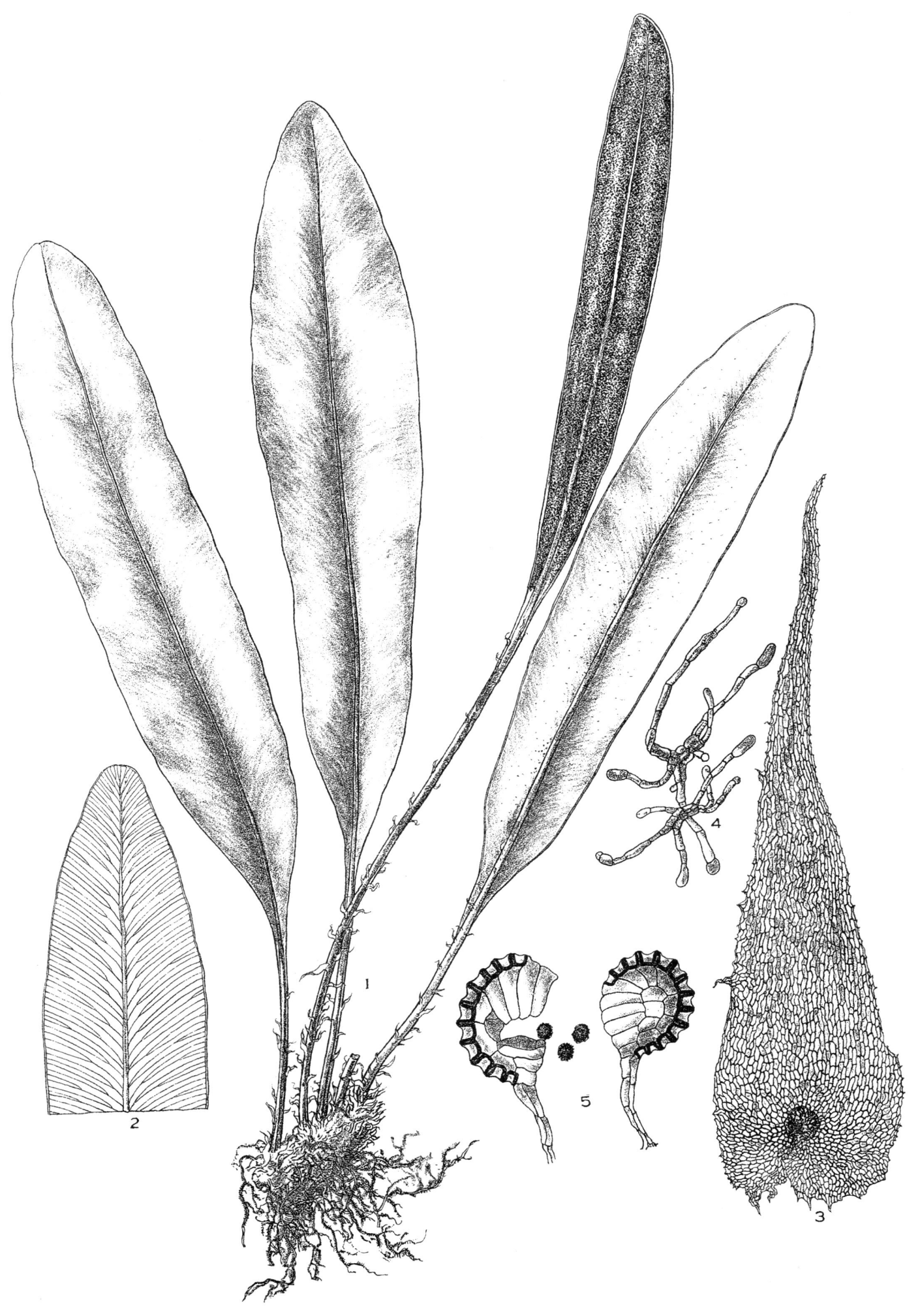

ELAPHOGLOSSUM MCCLUREI Ching
琼崖舌蕨

图版 83　福氏星蕨

根状茎长，匍匐或攀悬，绿色，半光滑；鳞片仅疏生于叶之基部或茎之幼部，黄褐色，卵形，急尖头，全缘，久则脱落；叶疏生，叶间距离约 1.5 cm，带状披针形或阔带状披针形，宽 1.5—5 cm，长 30—45 cm，渐尖头，基部渐狭，延入无翅之短柄，全缘，亚革质，淡绿色，干则两面起皱，叶脉不见；子囊群形圆而大，橙色，在狭叶上者一列，距离甚大，在宽叶上者为不规则之二列，稍贴近中肋。

分布：长江各省均产之，南达广东北部，东至台湾省及日本。

模式种产于舟山群岛，其叶细狭，子囊群一列，形体极类瓦苇，然其茎细长，常攀悬于树干，其子囊群上不被盾状透明隔丝；通常叶阔达 4—5 cm，子囊群为不规则之二列。

图注：1. 本种全形（自然大）；1a. 模式形，1b. 阔叶形；2. 叶之一部，表示其叶脉及子囊群之位置（放大 3 倍）；3. 根状茎上之鳞片（放大 50 倍）。

MICROSORIUM FORTUNI (Moore) Ching

MICROSORIUM FORTUNI (Moore) Ching, Bull. Fan Mem. Inst. Biol. **4**: 304 (1933).

Drynaria fortuni Moore, Gard. Chron. (1885) 708.

Polypodium fortunii Lowe, Ferns Brit. & Exot. **1**: t. 42B (1856); Takeda, Notes. R. Bot. Gard. Edinb. **8**: 284 (1915), quoad plant. ex China et Taiwan; C. Chr. Ind Fil. Suppl. 54 (1913-17).

Polypodium chinense Mett. apud Kuhn in Seemann's Journ. Bot. **6**: 270 (1868); Christ, Bull. Herb. Boiss. ser. 2. **4**: 611 (1904).

Polypodium henryi Christ, Bull. Herb. Boiss. **6**: 873 (1898).

Polypodium austro-sinicum Christ in C. Chr, Ind. Fil. 512 (1906, non Christ, Bull. Acad. Géogr. Bot. 1906).

Polypodium mengtzeanum Baker, Kew Pull. (1906) 14; C. Chr. Ind. Fil. Suppl. 60 (1906-12).

Polypodium simplex var. *esquirolii* Christ, Bull. Acad, Géogr. Bot. (1906) 247.

Polypodium normale Christ (non Don, 1825), Bull. Soc. Bot. France **52**: Mém. 1. 16 (1905); Dunn et Tutcher, Fl. Hongk. & Kwangt. 352 (1912); Cgata, Ic. Fil. Jap, **2**: pl. 89 (1929).

Polypodium normale var. *polysorum* Baker, Journ. Bot. (1875) 202.

Polypodium lineare Hk. et Bak. Syn. Fil. 354 (1868), partim.

Rhizome scandent, greenish, subglabrous; *scales* rather sparce at the base of stipe, or on the growing tip, brown, ovate, acute, erosed, concolorous, thin, fallen off at last; *frond* uniseriate, about 1.4 cm apart, erect, narrowly linear-lanceolate to broadly linear-lanceolate, 1.5-5 cm broad, 30-45 cm long, including the short, wingless stipe, gradually attenuate to both ends, entire; *texture* subcoriaceous, light green, more or less wrinked when dried; *veins* obscure; *sori* large, orange-yellow, uniseriate and far apart in the narrow-leaved forms (f. *typica*), but irregularly 1-2-seriate in the broader forms, closer to the costa, leaving a broad sterile margin, naked from the beginning.

Common in the Yangtze Valley and extending southwardly to Guangdong and Taiwan.

Type from the island Chusan (leg. Fortune) and represents a very narrow-leaved form, which appears so similar in general outline to the large form of *Polypodium lineare* Thunberg, that it was subsequently reduced to that species by Hooker and Baker. *P. chinense* Mett. is partly based upon Fortune's plant and partly upon the plant collected by De Griji from Foochow, Fujian, which latter represents a broad leaved form common in South and Southwestern China, known hitherto under half dozen other names as cited above.

Generically, this fern differs from *Lepisorus* (J. Sm.) Ching in tall scandent habit, the inherent tendency for the sori to be irregularly disposed and in the absence of the characteristic peltate scales on the sori. For over half a century, this fern has always been mistaken by authors for *Polypodium normale* Don from the Himalayas and Southwestern China, which differs, above all, in the presence on the back of the rhizome scales a large tuft of red setiform bristles.

Plate 83. Fig. 1. habit sketch (natural size). 1a. typical form; 1b. broad-leaved form. 2. portion of a frond, showing venation and sori ($\times$ 3). 3. scales from rhizome ($\times$ 30).

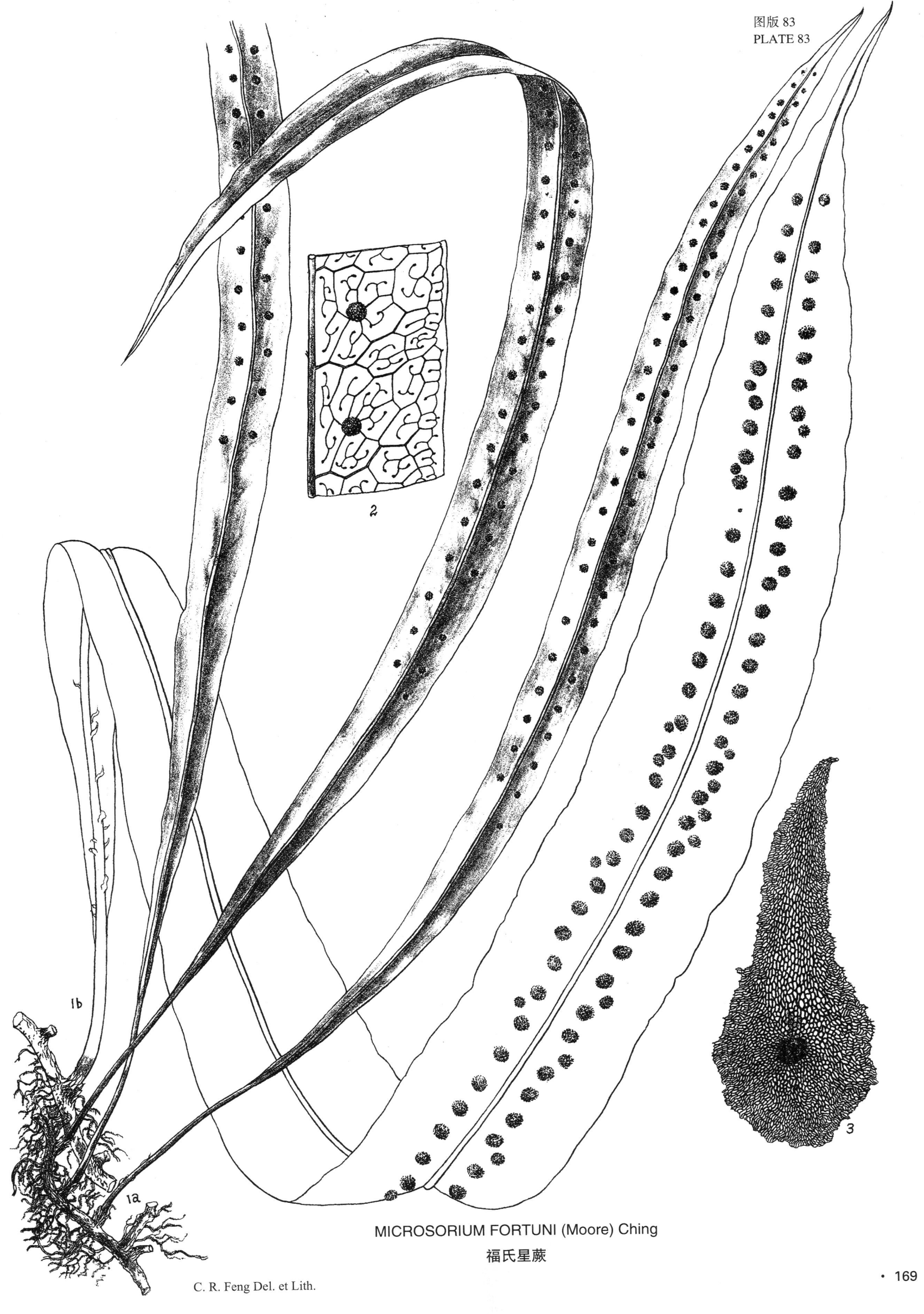

MICROSORIUM FORTUNI (Moore) Ching
福氏星蕨

图版 84　滇星蕨

　　根状茎攀悬甚高，略呈扁平；鳞片稀生，开张，深褐色，卵形，渐尖头；叶间距离甚大，形体大小不一，长 15—30 cm，宽 1.7—3 cm，向两端渐尖，柄短，无翅，光滑，全缘或间呈波形，纸质，光滑，干则呈黄绿色。叶脉颇明显，侧脉不发达；子囊群形大，体圆或椭圆，光滑，星散于叶之下面，或间贴近边缘而留一阔空白带于中肋两旁。

　　分布：云南，贵州，四川，广西；印度北部，越南。

　　本种形体大小极类波氏星蕨（*M. buergerianum*），然其叶质较薄，叶脉颇明显，体叶下部渐狭与较大之子囊群，故易于识别。

　　图注：1. 本种全形（自然大）；2. 根状茎上之鳞片（放大 50 倍）。

MICROSORIUM HYMENODES (Kunze) Ching

MICROSORIUM HYMENODES (Kunze) Ching, Bull. Fan Mem. Inst. Biol. **4**: 301 (1933).

　　Polypodium hymenodes Kunze, Linnaea **23**: 279 (1850); Mett. Fil. Hort. Lips. 37 t. 25, f. 40-41 (1856); Takeda, Notes, R. Bot. Gard. Edinb. **8**: 287 (1915); C. Chr. Ind. Fil. Suppl. 54 (1913-17).

　　Polypodium superficiale var. *semilinearis* Clarke, Ferns N. Ind. in Trans. Linn. Soc. ser. 2. Bot. **1**. 558 (1880).

　　Polypodium subhemionitideum Christ, Bull. Herb. Boiss. **7**: 5 (1899).

　　Rhizome scandent, more or less compressed, spareely scaly; *scales* spreading, fusco-brown, ovate, acuminate, clathrate; *frond* far apart, variable in size, 15-30 cm long, 1.7-3 cm broad, gradually attenuate to both ends, stipe short, 3-5 cm long, margin entire, more or less wavy; *texture* herbaceous, or papyraceous, naked, color brownish-green; *veins* rather distinct; *sori* large, rounded to oblong, scattered over the entire under surface (var. *sparsiorum* Takeda), or sometimes placed near to the margin, leaving a broad sterile space on each side of costa (var. *marginale* Takeda).

　　Yunnan: Mengtze, *A. Henry 9265B*; *Hancock 98*; Tengyueh, *Forrest 9453*; Longkang, *Delavay 4928*, Guizhou: Kweiyang, *Bodinier 2088*. Sichuan occid: *E. H. Wilson 2633*; Kuan Hsien, *W. P. Fang 2100*; Mt. Omei, *W. P. Fang 2493*, Guangxi: Sanfang, Chufang Shan, *R. C. Ching 5749*.

　　A tall scandent fern, clinging to the trunk of trees up to 20 feet high; habit and size are exactly similar to M. buergerianum (Miq.), differs in frond of thinner texture with quite visible venation, gradually attenuate base, and much larger sori. Geographically, this ferns is known only from Northern India and West China as well as Vietnam.

　　Plate 84. Fig. 1. habit sketch (natural size). 2. scale from rhizome ($\times$ 50).

MICROSORIUM HYMENODES (Kunze) Ching
滇星蕨

图版 85　波氏星蕨

根状茎攀悬或匍匐于石上，略呈扁平，黝色；鳞片深褐色，卵圆形，渐尖头，久则脱落；叶颇远生，披针形，长10—20 cm，宽2—3 cm，向端渐尖，下部急狭成翅，循叶边下延甚长，或达基部，全缘，常略呈波状，厚纸质，黄褐色，叶脉不显明，于强光下可见；子囊群形小，体椭圆，光滑，星散于叶之下面。

分布：香港，广东，福建，浙江，广西，四川，江西庐山及台湾；日本亦产之。

本种在广东极为普通，自来学者概误识为南洋群岛所产之 M. superficiale，实则大不相类，与本种最相似者，莫过于前种，然以其叶质较厚，叶脉不甚显明，叶之下部急狭，与子囊群较小而密，故易识别。

图注：1. 本种全形（自然大）；2. 根状茎上之鳞片（放大50倍）。

MICROSORIUM BUERGERIANUM (Miquel) Ching

MICROSORIUM BUERGERIANUM (Miquel) Ching, Bull. Fan Mem. Inst. Biol. **4**: 302 (1933).

  Polypodium buergerianum Miquel, Ann. Lugd. Bat. **3**: 170 (1867); C. Chr. Ind. Fil. 514 (1906); Takeda, Notes, R. Bot. Gard. Edinb. **8**: 290 (1915).

  Polypodium superficiale Hooker, Sp. Fil. **5**: 71 (1864); Syn. Fil. 355 (1868), quoad pl. ex Hong Kong et Tsushima (non Blume); Ogata, Ic. Fil. Jap. **3**: pl. 145 (1930).

  Polypodium hymenodes Benth. (non Wallich, 1828) Fl. Hongk. 548 (1861).

  Polypodium brachylepis Baker, Gard. Chron. n. s. **14**: 494 (1880).

  Polypodium ningpoense Baker, Ann. Bot. **5**: 474 (1891).

  Polypodium buergerianum var. *ningpoense* Takeda, Notes, R. Bot. Gard. Edinb. **8**: 290 (1915).

  Polypodium superficiale var. *attenuata* Rosenst. in Fedde, Repert. Sp. Nov. **13**: 314 (1914).

  Polypodium superficiale var. *anguinum* Christ, Bull. Soc. Bot. France **52**: Mém 1. 16 (1905).

Rhizome scandent, 2-4 mm broad, compressed, atratous, scaly; *scales* rufo-brown, thin, clathrate, ovate, acuminate, fallen off at last; *frond* 2-3 cm apart, lanceolate, 10-20 cm long, 2-3 cm broad, gradually attenuate upwards, suddenly narrowed at base, thence decurrent either nearly down to the base of the short stipe, or leaving a rather distinct wingless stipe to 3 cm long, margin often slightly wavy; *texture* firmly chartaceous, color brownish; *veins* hardly visible but distinct under strong light; *sori* small, scattered over the entire under surface, naked.

Hong Kong: Mt. Gaugh, *C. Wilford 38* (1858); Victoria Peak, *J. Lament 7303* in Herb. Hance; ibid. *R. C. Ching*, Dec. 15, 1929. Guangdong: North River, *C. L. Tso 21105, 20542*; Lockchong, *Y. Tsiang 1243*; Winglar, Lungtau Shan, *To & Tsang 12529* ex Herb. Lingnan Univ. Jiangxi: Lu Shan, *Maires* (1880); *Dr. Shearer* (1873). Fujian: Foochow, *H. H. Chung 1259, 3714*; Yuenfu, *Warburg* (1885-89). Zhejiang: Ningpo, *Hancock 24* (1877); Pootu island, *K. K. Tsoong 96*. Guizhou: Ganchow, *Cavalerie 1472, 4009*; Tuyun: *Y. Tsiang 5884, 5994*. Guangxi: Tsinlung Shan, Lin Yen, *R. C. Ching 6898*. Hubei: Ceinscian Hsien, *Silvestri 65*. Sichuan: Opien; *T. F. Lou 131* (1929). Taiwan.

Also Vietnam and Japan.

A very common fern in South China, particularly in the province Guangdong; generally epiphytic on the trunk of trees or over rocks under forests. Authors on Chinese and Japanese ferns have generally taken this for Malayan *Polypodium superficiale* Blume, with which it has in fact little in common, except the general outline. To *M. hymenodes* (Kze.) this species is very closely related, differs but in suddenly narrowed leaf-base, thicker texture and consequently less visible venation, and in smaller, denser sori.

Plate 85. Fig. 1. habit sketch (natural size). 2. scale from rhizome ($\times$ 50).

MICROSORIUM BUERGERIANUM (Miquel) Ching
波氏星蕨

图版 86　星蕨

根状茎肥短，横生，半光滑，附生于树干或岩石上，常略被白粉；鳞片疏生于叶柄基部，阔卵形，渐尖头，腹部着生，久则脱落；叶近生，直立，长 35—55 cm，宽 3—5 cm，带状披针形，渐尖头，下部渐狭，具短肥柄，或无柄，基部呈圆楔形或亚耳形，全缘或间呈不规则之深波形，坚纸质或为薄纸质，两面光滑，淡绿色，侧脉不甚发达而甚显明，小脉亦明白可见；子囊群中大，密生，不规则分布于叶之下面，上部淡黄色，光滑，常为不规则之连合。

分布：自非洲至马来群岛，菲律宾，全印度，越南而达中国之台湾，海南，广东，广西，贵州。

本种之模式标本原产于广东，Linné 氏初名之为 *Acrostichum* 属之一种，后人改隶水龙骨属，Copeland 氏于 1929 年始改为今名；本种叶体基部变异甚大，然除此则颇为一致。

图注：1. 本种全形（自然大）；2. 根状茎上之鳞片（放大 50 倍）。

MICROSORIUM PUNCTATUM (L.) Copeland

MICROSORIUM PUNCTATUM (L.) Copeland, Orient. Gen. Polyp. in Univ. Calif. Publ. Bot. **16**: 111 (1929); Ching, Bull. Fan Mem. Inst. Biol. **4**: 309 (1933).

Acrostichum punctatum L. Sp. Pl. ed. II. 1524 (1763).

Polypodium punctatum Sw. in Schrad. Journ. **1800^2**: 21 (1801); Christ, Farnkr. d. Erde 106 (1897); C. Chr. Ind. Fil. 557 (1906).

Polypodium iridioides Poiret in Encycl. Bot. **5**: 513 (1804); Hooker et Grev. Ic. Fil. t. 125 (1831); Dunn et Tutcher, Fl. Hongk. & Kwangt. 352 (1912); Ogata, Ic. Fil. Jap. **2**: pl. 136 (1930).

Microsorium irregulare Link, Hort. Berol. **2**: 110 (1833); Fée, Gen. Fil. 268 t. 20B. f. (1850-52).

Pleopeltis punctata Bedd. Ferns Brit. Ind. Suppl. 22 (1867); Handb. Ferns Brit. Ind. 357 (1883).

Pleopeltis irridioides Moore, Ind. Fil. LXXVII (1857); Bedd. Ferns S. Ind. t. 178 (1864).

Rizome short-creeping, thick, subglabrous, epiphytic on the trunk of trees, or on rocks; *scales* sparce at the base of stipe, fusco-brown, clathrate, broadly ovate, acuminate, dorsally affixed, fallen off at last; *frond* subcaespitose or close together, erect, 35-55 cm long, 3-5 cm broad, broadly linear-lanceolate, acuminate, gradually narrowed downward, base long-attenuate, or often rotundo-cuneate or subauriculate, stipe thick, short or almost none, leaf-margin entire, or sometimes irregularly undulate; *texture* thin herbaceeus, or chartaceous when dried, perfectly naked, pale green on both sides; lateral *veins* and veinlets quite distinct, copiously anastomosing with divaricating included veinlets; *sori* medium-sized, dense, irregular, covering the whole of under surface in the upper part of frond, lemon-yellow, naked, generally apical on the included veinlets, irregularly fused.

A widely dispersed epiphytic fern, ranging from Africa, Polynesia, Malaysia, over India generally, Indo-China to South China. In China, specimens have been seen from Taiwan, Guangdong, Hainan, Guangxi and the southern part of Guizhou.

This fern was first described from plant from Canton by Linné under *Acrostichum*, presumably on the ground of the dense, naked, irregularly subconfluent sori much like those in *Acrostichum*. It is a very variable fern, particularly as to the shape of leaf-base even in the the specimens from the same region, which is sometimes gradually attenuate to quite a distinct, thick stipe, or sometimes only slightly narrowed quite down to the base into a broadly rotundo-cuneate or subauricled shape.

Plate 86. Fig. 1. habit sketch (natural size). 2. scale from rhizome ($\times$ 50).

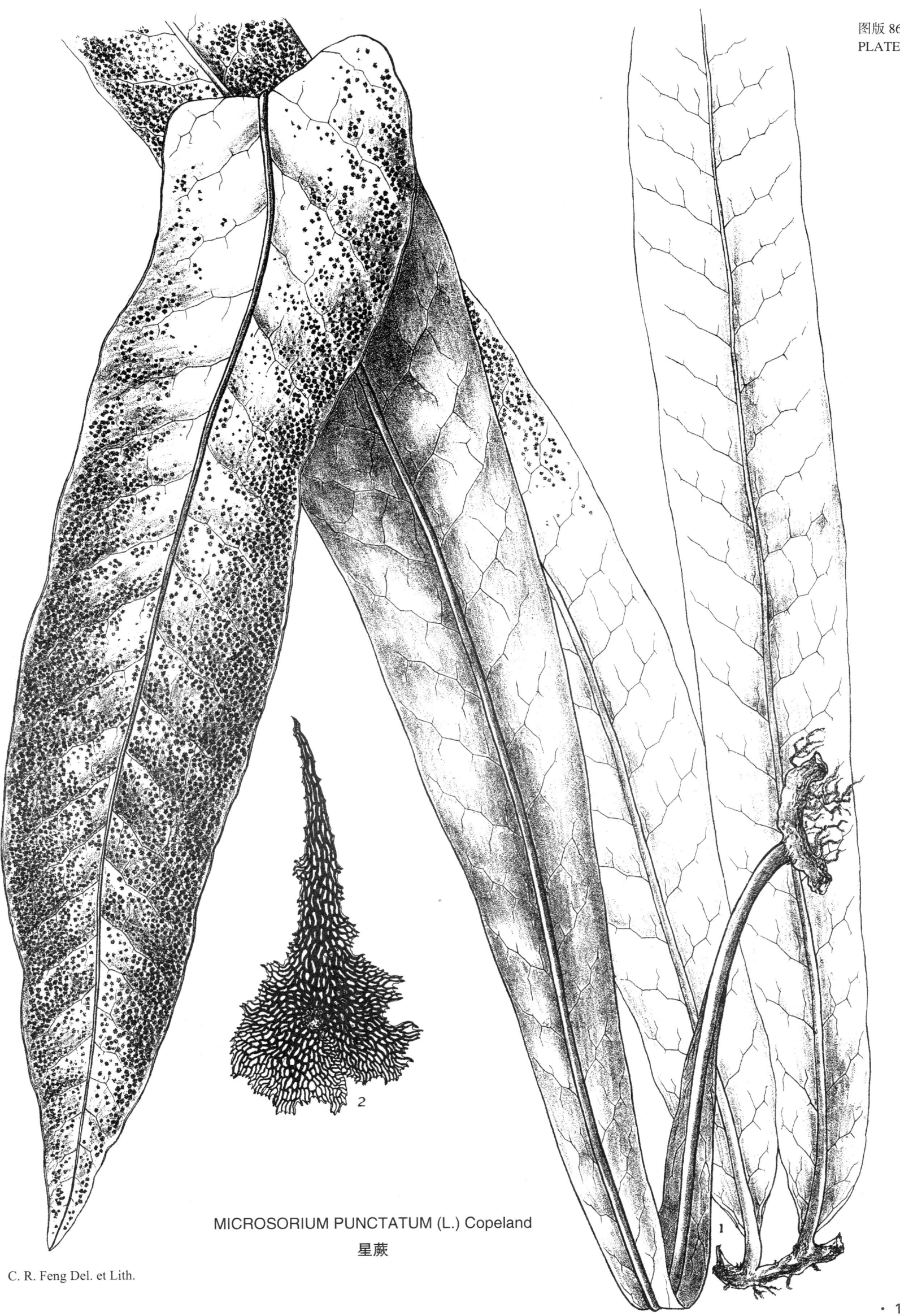

MICROSORIUM PUNCTATUM (L.) Copeland
星蕨

图版 87 戚氏星蕨

根状茎横生,颇肥,土生,黑色;鳞片深褐色,卵形,短尖头;叶颇近生,阔披针形,长 50—70 cm,下部渐狭,柄无翅,圆形,全缘,间呈波形,亚革质或厚纸质,深绿色,光滑,侧脉显著,粗强,几达边缘,网脉隐约;子囊群形大体圆,概为有规则之二列,位于两侧脉之间。

分布:海南,广西瑶山,马来群岛,菲律宾,全印度。

本种形体,极似膜叶星蕨(*M. membranaceum*),然其叶体通常较小,质厚而色深,子囊群较大且为颇有规则之二列,位于两侧脉之间,故易于识别。

图注:1.本种全形(自然大);2.叶之一部,表示其叶脉及子囊群之位置(放大 2 倍);3.根状茎上之鳞片(放大 50 倍)。

MICROSORIUM ZIPPELII (Blume) Ching

MICROSORIUM ZIPPELII (Blume) Ching, Bull. Fan Mem. Inst. Biol. **4**: 308 (1933).

Polypodium zippelii Blume, Fl. Javae. Fil. 172 t. 80 (1829); Mett. 1 olyp. n. 224 (1857); C. Chr. Ind. Fil. 575 (1906).

Polypodium heterocarpa Bedd. (non Mettenius) Ferns Brit. Ind. t. 324 (1870).

Pleopeltis zippelii Moore, Ind. Fil. 348 (1862); Bedd. Handb. Ferns Brit, Ind. 357 (1883).

Polypodium heterocarpum var. *zippelii* Hk. et Bak. Syn. Fil. 360 (1868).

Rhizome rather thick, short-creeping, hypogaeous, atratous; *scales* fusco-brown, rather firm, ovate, acute, appressed; *frond* rather approximate, broadly lanceolate, 50-70 cm long, including the short, but wingless, terete stipe, gradually attenuate along the upper part of stipe in a narrow wing, apex rather subcaudate, margin entire, but somewhat wavy; *texture* subcoriaceous, or thick chartaceous, color dark green, naked on both sides; lateral main *veins* distinct near to the margin, oblique, 1.2 cm apart, intervening veinlets obscure; *sori* large, rounded, naked, rather regularly two-rowed between main veins, and about 5-6 between the costa and margin.

Hainan: Ng Chi Leng, *F. A. McClure 8648*; Khio Tswi, *Eryl Smith 1544*; Fan Yah, *C. L. Tso & N. K. Chun 44027*, Oct. 7, 1932. Guangxi: Yao Shan, *S. S. Sin & K. K. Whang 698*, on stream side.

Distr. Malaysia, the Philippines, Indo-China, India generally and South China.

Habit of *M. membranaceum* (Don), differs in generally smaller size, subcoriaceous frond with short, but wingless, terete stipe, and larger biseriate sori between lateral main veins.

Plate 87. Fig. 1. habit sketch (natural size). 2. portion of a frond, showing venation and position of sori ($\times$ 2). 3. scale from rhizome ($\times$ 50).

MICROSORIUM ZIPPELII (Blume) Ching
戚氏星蕨

图版 88　膜叶星蕨

根状茎横生，肥肉质，土生或上生；鳞片密生，深褐色，阔卵形，渐尖头，全缘；叶近生或稍远生，阔披针形，长 30—50 cm，宽 6—14 cm，渐尖头，下部渐狭，延及基部，柄具狭翅，横切面呈三角形，因下面具棱角也。薄纸质，绿色，全缘，两面光滑，侧脉明显，不达叶边，网脉可见；子囊群形小体圆或椭圆，不规则散生，常不规则连合，光滑。

分布：云南，贵州，广西，广东，台湾；斯里兰卡，全印度，越南及菲律宾群岛均产之。

图注：1.本种全形（自然大）；2.叶之一部，表示其叶脉及子囊群之位置（放大 3 倍）；3.根状茎上之鳞片（放大 50 倍）；4.叶柄下部之横切面（放大 20 倍）。

MICROSORIUM MEMBRANACEUM (Don) Ching

MICROSORIUM MEMBRANACEUM (Don) Ching, Bull. Fan Mem. Inst. Biol. **4**：309 (1933).
　　Polypodium membranaceum Don, Prod. Fl. Nepal. 2 (1825); Mett. Polyp. n. 235 (1857); Hk. et Bak. Syn. Fil. 360
　　　　(1868); Christ, Farnkr. d. Erde 105 (1897); C. Chr. Ind. Fil. 544(1906); Ogata, Ic. Fil. Jap. **2**：pl. 88 (1929).
　　Pleopeltis grandifolia Bedd. Ferns S. Ind. t. 177 (1864); Christ, Bull. Herb. Boiss. **6**：874(1898).
　　Pleopeltis membranacea Mcore, Ind. Fil. 191 (1857); Bedd. Handb. Ferns Brit. Ind. 355 (1883).

Rhizome creeping, thick, fleshy, over 6 mm across, subhypogaeous or epiphytic, densely scaly; *scales* fusco-brown, thin, clathrate, broadly lanceolate, acuminate, suben-tire; *frond* sutcaespitcse or approximate, very broadly lanceolate, 6-14 cm broad, 30-50 cm long, acnminate, gradually attenuate down to the base of the winged and rather sharply carinate, herbaceous stipe, margin entire or somewhat wavy; *texture* thin herbaceous, green on both sides, rachis sharply ridged on lower part underneath, lateral main *veins* wide apart, distinct near to the margin, intervening veinlets conspicuous, areolae large with divaricated including veinlets; sori small, irregularly disposed between main veins, mostly compital, naked.

In China, numerous specimens have been examined from provinces Yunnan, Guizhou, Guangxi, Guangdong, Taiwan. It ranges from Sri Lanka, whole of India, Indo-China, Southwestern China to the Philippines.

Plate 88. Fig. 1. habit sketch (natural size). 2. portion of a frond, showing venation and position of sori ($\times$ 3). scale from rhizome ($\times$ 50). 4. cross section of the lower part of a stipe ($\times$ 20).

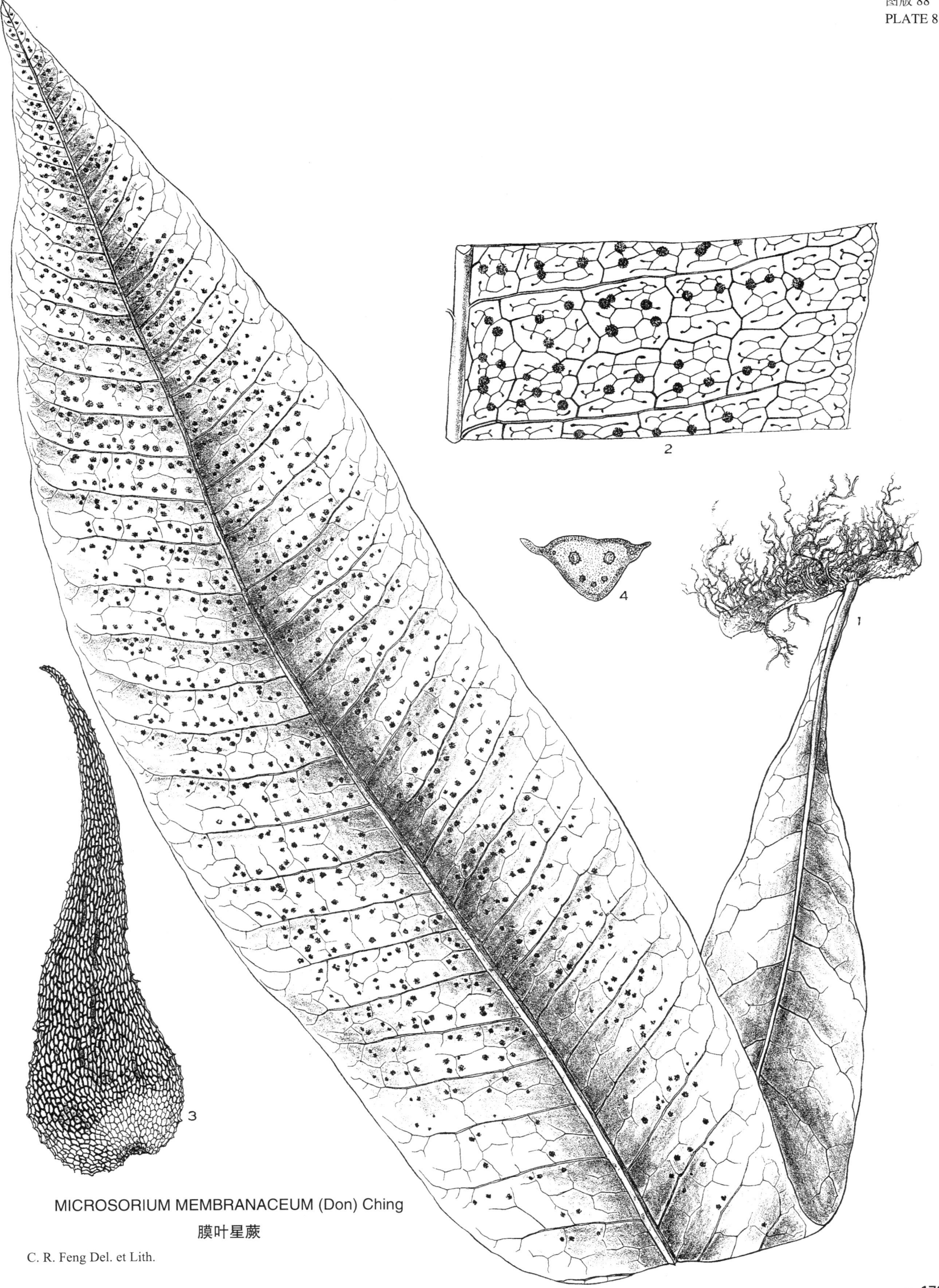

MICROSORIUM MEMBRANACEUM (Don) Ching
膜叶星蕨

图版 89　单叶扇蕨

根状茎匍匐甚广,半土生;鳞片密生,卵形,渐尖头;叶远生,直立,柄长 10—17 cm,或过之,黝色,下部具细长鳞片,叶体长卵形,长 13—23 cm,宽 7—12 cm,基部最阔,截形,或圆楔形,或间为楔形,全缘,或下部具 1—2 对裂片,或一至二次羽状分裂,厚纸质,淡绿色,上面光滑,下面略具鳞片,侧脉明显,开展,几达叶边;子囊群形圆体大,不规则排列或向叶端为一列,幼时密覆盾状隔丝,一如瓦苇然。

分布:云南,四川,贵州,广东,广西,湖北,安徽,江西,浙江,江苏,福建及台湾;喜马拉雅山区亦产之。

本种变异甚大,非特大小不一,即形体亦由卵形单叶至不规则之二次羽状分裂,有时单叶之形体极类原产于日本之 *Neoch. ensata* (Thunberg),难于识别。

图注:1a. 本种模式全形(自然大),1b. 本种变形之一,1c. 本种变形之二(1/2 大);2. 叶体之一部,表示其叶脉及子囊群之位置(放大 2 倍);3. 根状茎上之鳞片(放大 30 倍);4. 子囊群上之盾状隔丝(放大 50 倍);5. 叶体下面之鳞片(放大 50 倍);6. 叶柄上之鳞片(放大 50 倍);7. 茎之横切面,表示其维管束之排列(放大 10 倍)。

NEOCHEIROPTERIS PHYLLOMANES (Christ) Ching

NEOCHEIROPTERIS PHYLLOMANES (Christ) Ching, Bull. Fan Mem. Inst. Biol. **4**: 110 (1933).
　　Polypodium phyllomanes Christ, Bull. Acad. Géogr. Bot. (1902) 210. c. fig.; C. Chr. Ind. Fil. 553 (1906).
　　Polypodium ovatum Wall. (non Burm. 1768) Hooker et Grev. Ic. Fil. t. 41 (1827); Takeda, Notes, R. Bot. Gard. Edinb. **8**: 289 (1915).
　　Pleopeltis ovata Bedd. Ferns Brit. Ind. pl. 157 (1866); Handb. Ferns Brit. Ind. 354 (1883).
　　Polypodium ensatum Diels (non Thunberg) in Engl. Jahrb. **29**: 203 (1901); Hayata, Ic. Pl. Form. **5**: 312 (1915).

Rhizome wide-creeping, subhypogaeous, rather thick, densely scaly; *scales* fusco-brown, clathrate, ovate, acuminate; *frond* far apart, erect, stipe 10-17 cm long or longer, obscurely blackish and rather densely, fibrillose-scaly, particularly, on the lower part, lamina ovate-oblong, acuminate, 13-23 cm long, 7-12 cm broad at base, which is subtruncate, or rotundo-cuneate, or sometimes cuneate, margin entire (f. *typica*), or provided with one or two broadly short lanceolate lobes at base (f. *deltoidea*), or not infrequently copiously laciniato-lobed with basal lobes again divided on both sides (f. *doryopteris*); *texture* firm, thickly chartaceous, color light green, glabrous above, more or less scaly underneath; lateral main *veins* distinct nearly to the margin, patent, nearly 1 cm apart, straight, but in lobed forms often looped at base; intervening veinlets rather obscure, copiously anastomosing with divaricating included, clavate veinlets; *sori* large, dark-brown, globose, irregularly multiseriate on each side of costa or sometimes subuniseriate, especially, towards apex, clothed when young with fuscous, clathrate, peltate scales as in *Lepisorus*.

A common fern in Yunnan, Sichuan, Guizhou, Guangdong and eastwardly to the provinces Anhui, Zhejiang and Taiwan; also recently known in Jiangsu (Ishing, leg, C. Y. Luh n. 462). It is also found abundant in the Himalayas.

Very variable in the outline of lamina as very well illustrated by Christ (l. c.), and in some extreme forms it differs little from *N. ensata* (*Polypodium ensatum* Thunberg), with which it has often been confused.

Plate 89. Fig. 1a habit sketch (natural size). 1b. forma *deltoidea* (natural size). 1c. forma *doryopteris* ($\times$ 1/2). 2. portion of a frond, showing venation and position of sori ($\times$ 2). 3. scale from rhizome ($\times$ 30). 4. peltate paraphyses from a sorus ($\times$ 50). 5. scales from underside of a frond ($\times$ 50). 6. scale from a stipe ($\times$ 50). 7. cross section of a rhizome, showing perforated type of steles ($\times$ 10).

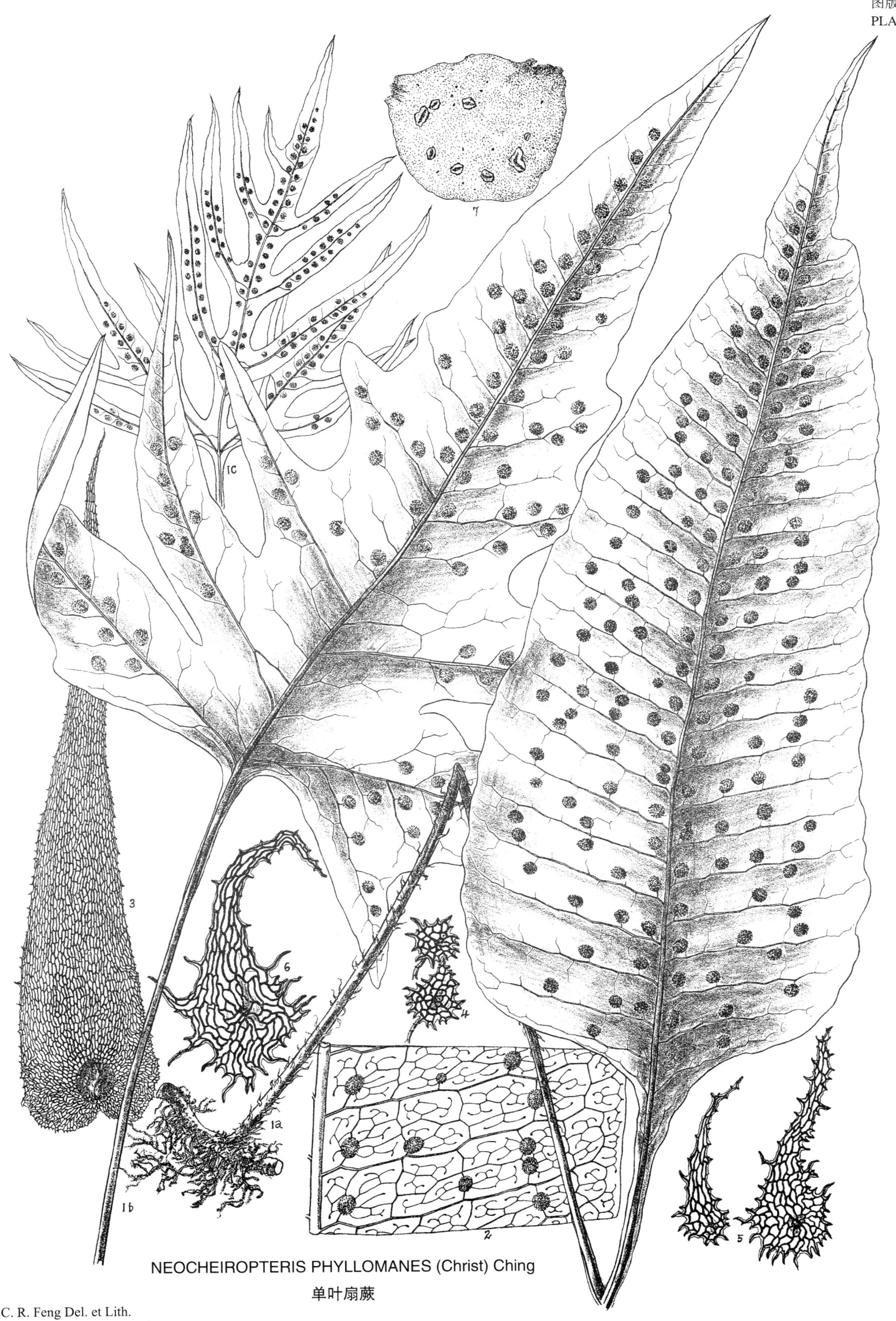

NEOCHEIROPTERIS PHYLLOMANES (Christ) Ching
单叶扇蕨

根状茎匍匐生于土中；鳞片深褐色，卵形，端狭长，开张；叶颇近生，柄长 8—18 cm，稻秆色，光滑，叶体长 16—27 cm，宽 3.5—5 cm，间甚狭，披针形，渐尖头，基部圆楔形，稍下延，全缘而呈细波状反卷，厚纸质，淡绿色，侧脉及小脉颇明显；子囊群叶之柄较长，叶体与非子囊群叶略同，子囊群线形，细长，略呈曲折，由中肋斜出达于叶边，群间距离约 4 mm。

分布：海南；越南。

本种颇类马来岛之 C. membranacea (Bl.) 与中国南部之 C. Wrightii (Hook.)，然其叶质较厚，无翅叶柄较长，易于识别，又后者之叶为黝褐色，本种则为淡绿色。

图注：1. 本种全形（自然大）；2. 根状茎上之鳞片（放大 50 倍）。

COLYSIS BONII Ching

COLYSIS BONII Ching, Bull. Fan Mem. Inst. Biol. **4**: 322 (1933).

Rhizome creeping, 3 mm thick, hypogaeous; *scales* fusco-brown, thin, ovate, long-acuminate, spreading; *frond* subdimorphous, 1 cm apart, erect, stipe 8-18 cm long, straminous, sterile lamina 16-27 cm long, 3.5-5 cm broad; or sometimes much narrower, lanceolate, acuminate, base rotundo-cuneate, narrowly decurrent for some distance downward, margin entire but crispo-repand; *texture* firm, thickly chartaceous, color light green; lateral main *veins* and veinlets quite distinct; *fertile* lamina on longer stipe, not much different from the sterile but in somewhat narrower lamina with more undulato-repand margin; *sori* linear, continuous, flexuose, very oblique, from costa extending to the margin, 4 mm apart.

Hainan: Nam Kao, *Eryl Smith 1516*, *1517*, January 19, 1923; Dai Land, Dung Ka, *C. L. Tso & N. K. Chun 43945*, Sept. 25, 1932, on moist rocks by streams.

Vietnam occid: *Bon 2395* (type).

The present fern seems to be a close ally to the Malayan *C. membranacea* (Bl.) Presl and *C. wrightii* (Hooker) from South China, differs from both in much thicker texture, and long-stalked frond with broad and shortly decurrent base. From the latter it further differs in light-green leaves instead of invariably turning obscurely blackish upon drying.

Plate 90. Fig. 1. habit sketch (natural size). 2. scale from rhizome ($\times$ 50).

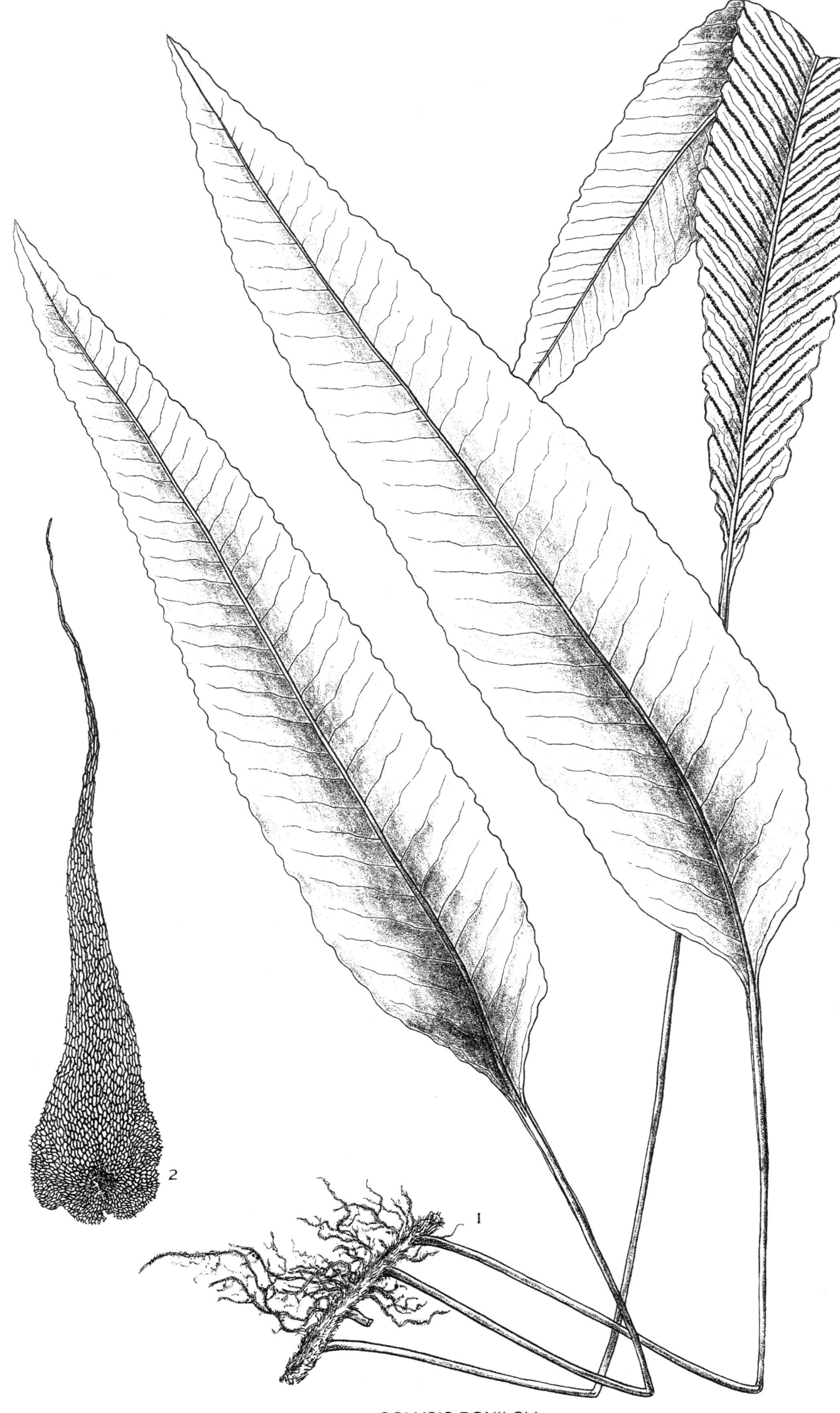

COLYSIS BONII Ching
彭氏线蕨

根状茎匍匐甚广，半土生。鳞片密生，深褐色，卵形，渐尖头；叶间距离约 3 cm，柄长 10—15 cm，无翅，光滑，略呈酒红色；叶体广披针形，长 24—30 cm，中部宽 5—7 cm，下部急狭成翅下延，短尖头，纸质，淡黄色而略呈酒红色，两面光滑，全缘，稍呈细波状反卷，侧脉细而颇明显，斜出，小脉亦颇显明；子囊群线形，一列位于侧脉之间，由中肋斜出，直达叶边，群间距离约 5 mm。

分布：湖北，四川，广西，安徽。

本种在峨眉山甚为普通，其形体颇似马来岛所产之 C. macrophylla (Blume)，然其叶之下部急狭，叶色略呈黄红色，故易识别。

图注：1. 本种全形（自然大）；2. 叶体之一部，表示叶脉及其子囊群之位置（放大 3 倍）；3. 根状茎上之鳞片（放大 50 倍）。

COLYSIS HENRYI (Baker) Ching

COLYSIS HENRYI (Baker) Ching, Bull. Fan Mem. Inst. Biol. **4**：325 (1933).
   Gymnogramme henryi Baker, Journ. Bot. (1887) 171.
   Gymnogramme macrophylla Baker, Journ. Bot. (1888) 230 (non Hook.).
   Selliguea henryi Christ, Bull. Herb. Boiss. **6**：879 (1898); Bull. Soc. Bot. France **52**：Mém. 1. 25 (1905).
   Polypodium henryi C. Chr. Ind. Fil. 512 (1906).
   Selliguea cochlearis Christ (non Reinw.), Bull. Acad. Géogr. Bot. (1907) 142.
   Polypodium mon-changense C. Chr. Ind. Fil. Suppl. 60 (1906-1912).

Rhizome wide-creeping, subhypogaeous, densely scaly; *scales* fusco-brown, thin, ovate, acuminate, clearly reticulate; *frond* far apart, erect, wingless stipe 10-15 cm long, rufo-stramineous, naked, lamina broadly lanceolate, 24-30 cm long, 5-7 cm broad at the middle, which is the broadest part, thence downward suddenly narrowed and then gradually attenuate in a narrow wing on each side of stipe, apex short-acuminate; *texture* herbaceous, brownish green, tinged wind-reddish and so is the midrib, naked on both sides, margin entire, but slightly repando-undulate; lateral main *veins* fine, but quite distinct, oblique, intervening veinlets also visible; *sori* linear, normally continuous, one between each pair of lateral veins, from midrib extending nearly to the margin, about 5 mm apart.

Hubei: Nanto near Ichang, *Henry 2114* (type); Ichang, *Henry 7880*, Guizhou: Kweiyang, *Bodinier 1572*; Meitan, *Y. Tsiang 8039*; Lopie, *Bodinier 1965*. Sichuan: Mt. Omei, *E. Faber 1002*; *W. P. Fang 2586*, Aug. 4, 1928, in thickets. Guangxi: Tien Lian Shan, *R. C. Ching 5357, 5322*. Anhui: Suinin Hsien, Lantien, *K. K. Tsoong 1950*.

A common, pretty fern in West China, particularly on the sacred Mount Omei. It is closely related to the Malayan *C. macrophylla* (Bl.), to which Rev. Faber's plant was referred by Baker, but from which it differs, apart from geographical distinction, in frond being suddenly narrowed below the middle and in different color.

Plate 91. Fig. 1. habit sketch (natural size). 2. portion of a frond, showing venation and position of sori ($\times$ 3). 3. scale from rhizome ($\times$ 50).

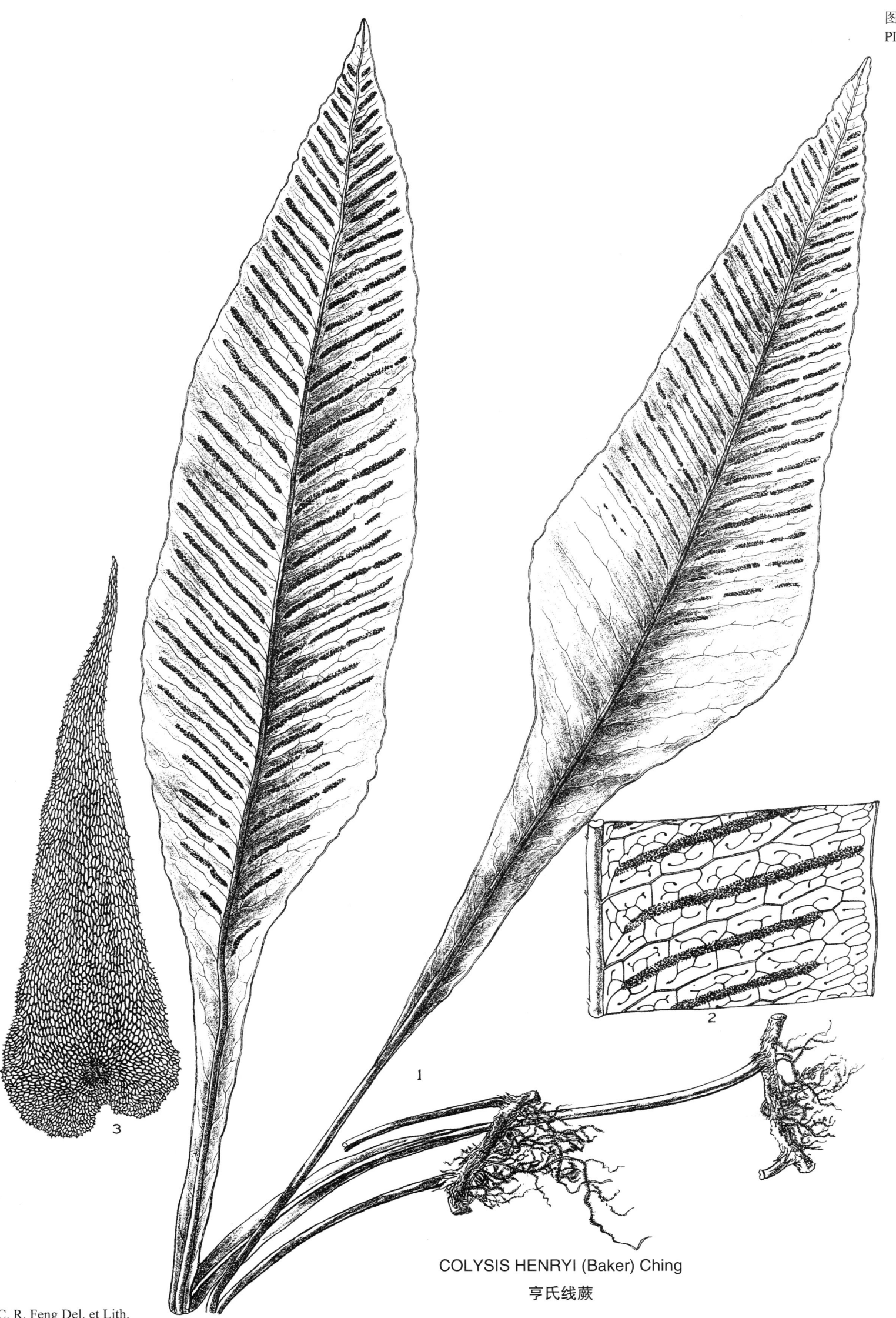

COLYSIS HENRYI (Baker) Ching
亨氏线蕨

根状茎肥,匍匐甚广,黝绿色;鳞片密生,深褐色,卵形,短尖头。叶颇远生,一形,柄长 15—20 cm,稻秆色,光滑;叶体三角卵形,长约 6—8 cm,宽如之,一回羽状分裂,基部圆截形,稍下延,硬革质,淡绿色,光滑,小叶 2—3 对,披针形,短尖头,向基部稍狭,顶生叶片形略同于侧生各对;叶脉不显,子囊群细长,斜出。

分布:海南。

本种为本属特殊之种,与 C. elliptica (Thunberg) 相近,然其极长之柄与极短阔硬革质之叶体,仅具 2—3 对之小叶,易别于他种。

图注:1.本种全形(自然大);2.根状茎上之鳞片(放大 50 倍)。

COLYSIS LONGIPES Ching

COLYSIS LONGIPES Ching, Bull. Fan Mem. Inst. Biol. **4**: 332 (1933).

Rhizome wide-creeping, thick, 4 mm across, greenish, densely scaly; *scales* fusco brown, clathrate, ovate, shortly acuminate, subentire; *frond* monomorphous, about 2 cm apart, stands on the same plane as does the rhizome, stipe 15-20 cm long, slender, naked, straminous, base articulate, lamina 6-8 cm long, as broad or broader, ovate-deltoid, deeply pauci-pinnatifid, base rotundo-truncate, slightly decurrent; *pinnae* generally 3, or rarely more, on each side, the terminal one similar to the lateral, lanceolate, to 7 cm long, 1.4 cm broad at the middle, shortly acuminate, somewhat narrowed towards the base, margin entire, thin, plane; *texture* firm, coriaceous, color pale green, naked on both sides; *veins* hidden; *sori* rather few-paired to each segment, oblique, linear, far apart.

Hainan: En route from Dung Ka to Win Fa Shi, *C. L. Tso & N. K. Chun 43697*, August 24, 1932, in thickets.

Endemic in the island Hainan. A very distinct fern, of the group of *C. elliptica* (Thunberg), differs above all, in short and broadly deltoid, pauci-pinnatifid frond of hard, coriaceous texture, provided with unusually long stipe, which is twice as long as the lamina.

Plate 92. Fig. 1. habit sketch (natural size). 2. scale from rhizome ($\times$ 50).

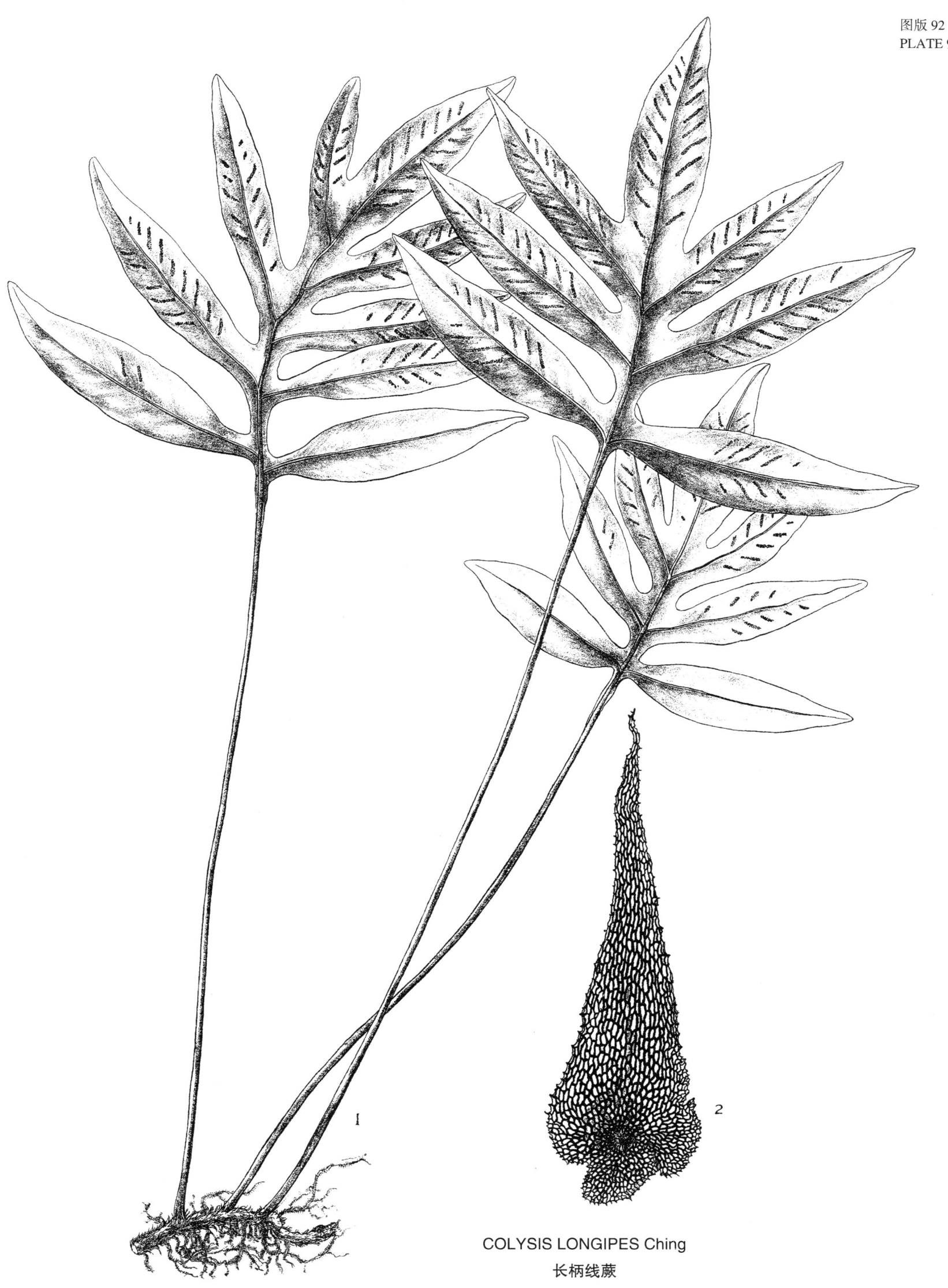

COLYSIS LONGIPES Ching
长柄线蕨

图版 93 戟蕨

根状茎匍匐甚广，附生于树干上；鳞片密生，深红棕色，上部细长如针，开张；叶二形，叶柄基部具骨节，长约 30 cm，子囊群叶之柄长约二倍之，稻秆色，光滑，非子囊群叶体三裂或间为四裂，基部楔形，侧生裂片宽 4 cm，长达 30 cm，其中央裂片较长，全缘，渐尖头，亚革质，淡绿色，两面光滑，叶脉不显明，网眼内具分叉之小脉；子囊群叶亦为三裂，裂片仅基部连合，狭长，宽不及 1 cm，带形，子囊群遍生于下面，仅中肋可见。

分布：海南之五指岭；锡金，缅甸，越南。

本种昔日仅见于锡金，缅甸，今发现于海南，实为分布上一大开展；越南产之 C. Eberhardtii Christ，具掌状五裂之非子囊群叶体，实与本种无他异。

图注：1. 本种全形（自然大）；2. 非子囊群叶之一部，表示其叶脉（放大 3 倍）；3. 根状茎上之鳞片（放大 50 倍）；4. 子囊群叶之横切面，表示其二层叶脉之分布情形及分叉之线形体；5. 孢子（放大）。

CHRISTOPTERIS TRICUSPIS (Hooker) Christ

CHRISTOPTERIS TRICUSPIS (Hooker) Christ, Journ. de Bot. France **21**: 273 (1908); Bower, Ferns III. 213 (1928).

Acrostichum tricuspe Hooker, Sp. Fil. **5**: 272 t. 304 (1864); Syn. Fil. 422 (1868).

Gymnopteris tricuspis Bedd. Ferns Brit. Ind. t. 53 (1866); Handb. Ferns Brit. Ind. 435 (1883); Christ, Farnkr. d. Erde 49 (1897); Diels, Nat. Pfl. Fam. **1**: 4. 299 (1899).

Cheiropleuria tricuspis, J. Sm. Hist. Fil. 139 (1875).

Leptochilus tricuspis C. Chr. Ind. Fil. 187 (1906).

Christopteris Eberhardtii Christ, I. c.

Rhizome wide-creeping, rather stout, epiphytic on the trunk of trees, densely scaly; *scales* atropurple, ovate-lanceolate, with very long, spreading hair-point; *frond* close, strongly dimorphous, stipe articulated to a prominent pseudopodium on the rhizome, about 30 cm long, that of fertile frond about twice as long, rather slender, straminous, naked, lamina of sterile frond deeply trilobed or 4-lobed, base cuneate, lateral lobes to 30 cm long, the middle one much longer, about 4 cm broad, entire, acuminate, *texture* subcoriaceous, rather soft, color light green, naked on both sides; *veins* obscure, but distinct against light, veinlets between the fine, wide-apart, lateral main veins copiously anastomosing with divaricating included veinlets; *fertile* lamina much elongate, tripartite nearly to the base with much contracted segments below 1 cm broad, and as long as the sterile ones, linear, strap-shaped, acuminate, ascendingly oblique; *texture* subcoriaceous; *sori* covering the entire underside except the costa and the very tip; *paraphyses* rather sparce, short, septate and branched; *spores* oblong-reniform, pellucide, greenish-yellow, smooth.

Hainan: Ng Chi Leng, Ah Ping, *C. L. Tso & N. K. Chun* 44147, October 24, 1932.

Sikkim: Darjeeling, 1,500 ft. alt. Also Burma and Vietnam.

This remarkable fern, known previously only from Sikkim, has recently received a thorough anatomical study from Prof. Bower, who is of the opinion that it has its systematic relation to the genus *Cheiropleuria*. *Christopteris Eberhardtii* Christ from Vietnam appears by no means specifically different from the present species, except the sterile lamina being palmately divided, i. e. with five lobes.

Plate 93. Fig. 1. habit sketch (natural size). 2. portion of a sterile segment, showing venation ($\times$ 2). 3. scale from rhizome ($\times$ 50). 4. part of a cross section of a fertile segment, showing the diplodesmic venation and branched paraphyses associated with sporangia ($\times$ 16), after Bower. 5. spores (magnified).

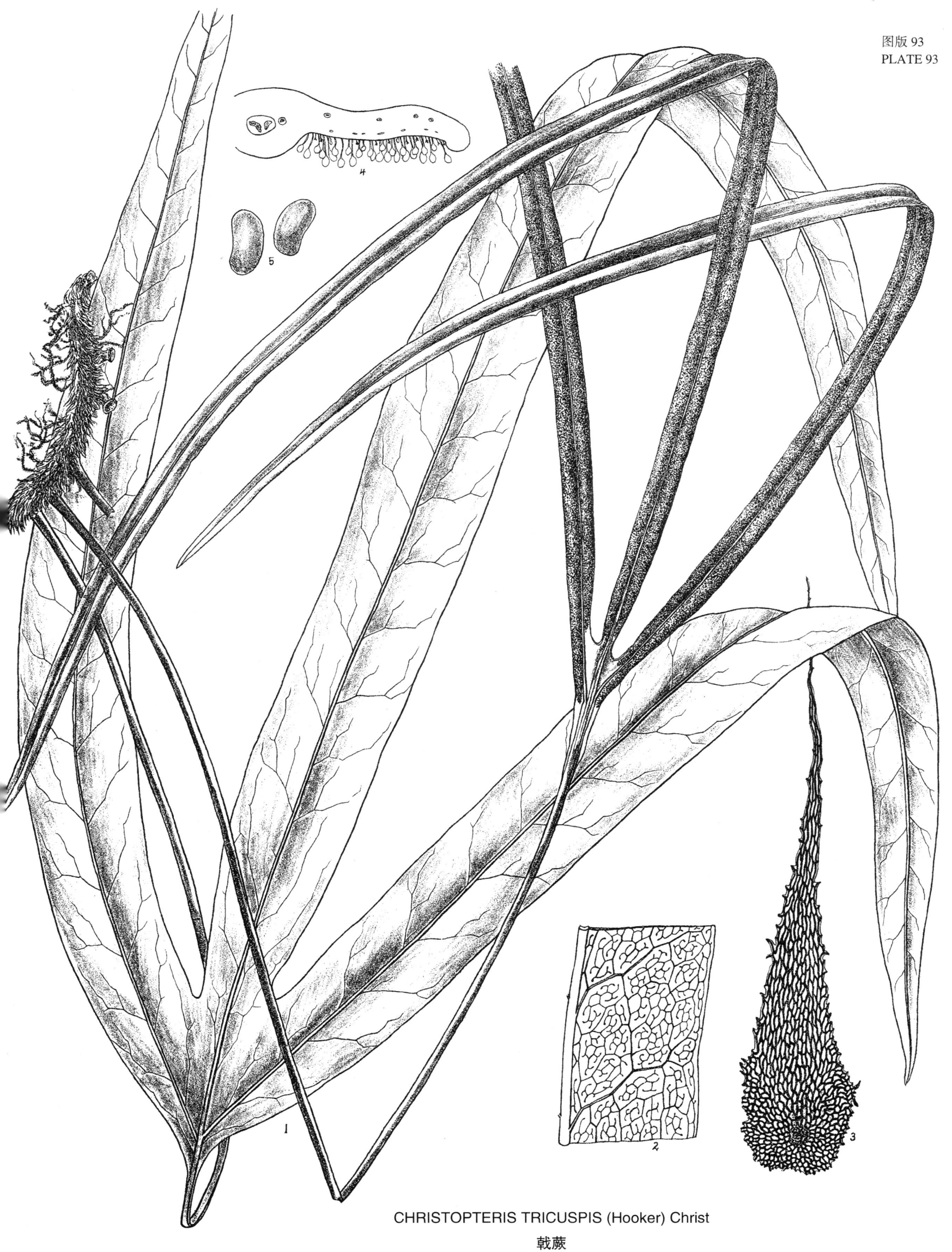

CHRISTOPTERIS TRICUSPIS (Hooker) Christ
戟蕨

图版 94　滇水龙骨

根状茎细长，匍匐甚广；鳞片颇密，披针形，长尖头，网眼透明；叶疏生，直立，柄细长，光滑，淡稻秆色，长 10—15 cm，叶体阔披针形，光滑，宽仅 5—6 cm，基部截形，顶部尾形而具粗锯齿，纸质，淡绿色，各部光滑，一次羽状分裂，小叶数十对，开张，长约 3 cm，宽仅 4 mm，狭披针形，基部一对较短，边缘具疏矮锯齿；叶脉分开，不成网状；子囊群形小而圆，位于主脉与边缘之间。

分布：云南，四川；越南及泰国。

本种形体稍似栗柄水龙骨（*P. microrhizoma* Clarke），其不同处为叶脉分开，不成网状，叶体狭而质较厚，叶柄短而不呈光亮之淡栗褐色。

图注：1. 本种全形（自然大）；2. 小叶，表示叶脉及其子囊群之位置（放大 3 倍）；3. 根状茎上之鳞片（放大 50 倍）。

POLYPODIUM MANMEIENSE Christ

POLYPODIUM MANMEIENSE Christ, Bull. Herb. Boiss. **6**: 870 (1898); Bull. Acad. Géogr. Bot. (1906) 104; Bull. Soc. Bot. France **52**: Mém. 1. 13 (1905); C. Chr. Ind. Fil. 543 (1906).

Polypodium simulans Baker, Kew Bull. (1906) 13.

Rhizome rather slender, wide-creeping, epigaeous, clothed in moderately dense, lanceolate, long-acuminate, sparcely denticulate, sorbid-brown *scales*; *frond* distant, erect, stipe rather slender, naked, pale straminous, 10-15 cm long, lamina lanceolate, simple pinnate, 20-34 cm long, 5-6 cm broad, base truncate, apex caudate with gross serration, *texture* herbaceous, pale green, naked in all parts; *pinnae* numerous, horizontally patent, 3-3.5 cm long, 4 mm broad, linear-lanceolate, decurrent, or rarely with the lower few pairs adnate to the wingless rachis, the basal ones generally slightly abbreviated with low and remote teeth; *veins* free or with open areolae; *sori* small, globose, superficial, placed midway between costa and margin.

Yunnan: Mengtze, *A. Henry 10081* (type), *13036*; *Hancock 152*; Tengyueh, *G. Forrest 26715*; Pautiho, *Forrest 27030*; Ta Hwa Shan, *K. K. Tsoong 4631*. Sichuan occid: *Wilson 5330*, *2649*; *David* (1870).

Sikkim: Shillong, ex herb. *H. Z, Darrah* (1888). Thailand: Doi Chiang Dap, *Kerr 1868*, *6573*.

In habit, this species appears exactly like P. *microrhizoma* Clarke from the same locality and the neighbourhood, differs only in open venation, shorter, pale-colored stipe and shorter pinnae of somewhat thicker texture; from P. *vulgare* L. it differs chiefly in slender habit, thinner texture, narrower segments with much broader interspaces, and having veinlets of contiguous groups with a greater tendency to join into a row of somewhat open costal areolae.

Plate 94. Fig. 1. habit sketch (natural size). 2. pinna, showing venation and position of sori (× 3). 5. scale from rhizome (× 50).

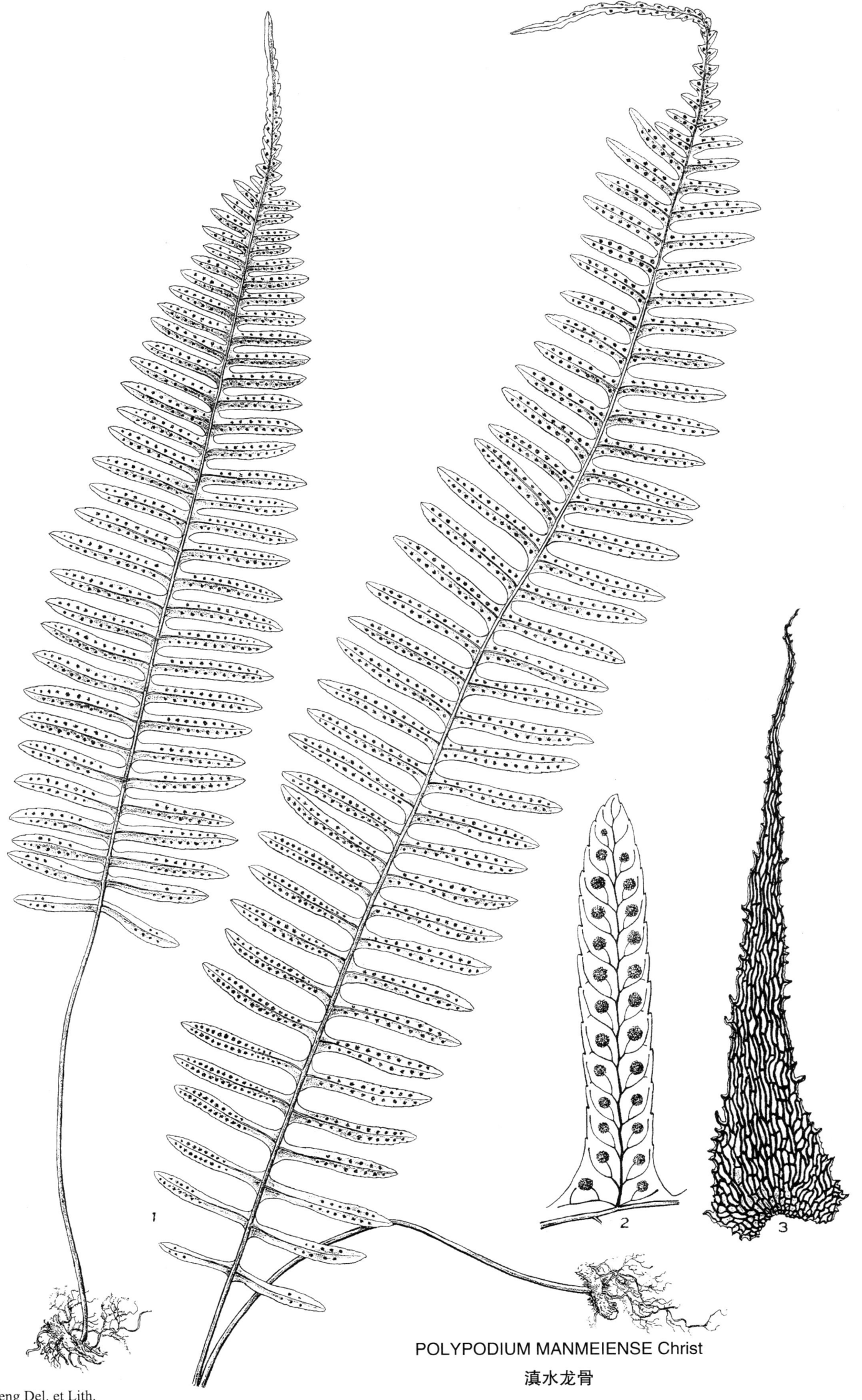

POLYPODIUM MANMEIENSE Christ
滇水龙骨

根状茎上生,匍匐甚广;鳞片密生,深褐色,基部卵形,上部急狭,细长如毛,开张,叶近生,柄短,通常 6—10 cm,光滑,或幼时鳞片与细毛疏生;叶体狭长,宽仅达 7 cm,长达 50 cm,一次羽状分裂,小叶多数,开张,近生,具疏钝锯齿,亚渐尖头,薄纸质,淡黄绿色,上面光滑,下面除中肋及主脉具鳞片甚密外,并有细毛疏生,久则变光滑;叶脉在光下十分显明,网脉一列;子囊群中大,幼时深褐色,小鳞片密覆。

分布:云南蒙自;锡金,喜马拉雅山区。

本种为本属颇特殊之一种,其异于被毛之 P. amoenum Wall. 者,为其茎上鳞片,较短之叶柄,甚细长之叶体与薄纸质之叶质是。

图注:1.本种全形(自然大);2.小叶,表示叶脉及其子囊群之位置(放大 3 倍);3.根状茎上之鳞片(放大 50 倍);4.子囊群上之鳞片(放大 50 倍);5.叶体下面之毛(放大 50 倍)。

POLYPODIUM LACHNOPUS Wallich

POLYPODIUM LACHNOPUS Wallich, Catal. n. 310 (1828); Hooker, Ic. Pl. t. 952 (1854); A Cent. Ferns t. 52 (1854); Mett. Polyp. n. 113 (1857); Hk. et Bak. Syn. Fil. 342 (1868); Diels, Nat. Pfl. Fam. **1**: 4. 312 (1899); C. Chr. Ind. Fil. 536 (1906).

Goniophlebium lachnopus J. Sm. in Hooker, Gen. Fil. ad t. 51 (1840); Bedd. Ferns Brit. Ind. t. 163 (1866); Handb. Ferns Brit. Ind. 319 (1883).

Schellolepis lachnopus J. Sm. Hist. Fil. 93 (1875).

Polypodium yunnanense Christ (non Franchet, 1885), Bull, Herb. Boiss. **6**: 869 (1898).

Rhizome epigaeous, wide-creeping, densely shaggy scaly; *scales* rufo-brown, clath-rate, iridescent, broadly ovate, cuspidate or gradually attenuate with long subulate and spreading apices, margin with long or short teeth; *frond* rather close, stipe short, generally 6-10 cm long, naked, straminous, lamina narrowly lanceolate, 30-50 cm long, about 7 cm broad, pinnatifid down quite to the rachis into numerous, spreading, close, remotely crenato-serrate, subacuminate *pinna*; *texture* papyraceo-herbaceous, glabrous above, densely scaly and sparcely hairy along rachis and costa underneath; *veins* hardly visible but distinct against light, areolae uniseriate; *sori* in a single row on each side of costa, densely scaly when young.

Yunnan: Mengtze, *A. Henry 9748A*; *Hancock 89, 91* (1893); Chungtien, *H. Handel-Mazzetti 6852*; Tengyueh, *G. Forrest 26753*. Sichuan: Tchen-kou-tin, *Farges 311*.

A very slender fern, endemic in the Himalayas, characterized by relatively short stipe and narrowly lanceolate, elongate lamina, dark-colored almost orbicular or ovate dense, finely reticulate scales having a long cuspidate or setiform spreading point, and consisting of large, clear, iridescent luminae. The sparcely pubescent underside of leaves may be a cause of confusion with the pilose form of *P. amoenum*, common in Yunnan and other parts of West China, which differs, however, in much broader leaves, decidedly stouter habit and thicker texture.

Plate 95. Fig. 1. habit sketch (natural size). 2. segment. showing venation and position of sori ($\times$ 3). 3. scale from rhizome ($\times$ 50). 4. scale from a sorus ($\times$ 50). 5. hairs from underside of a lamina ($\times$ 50).

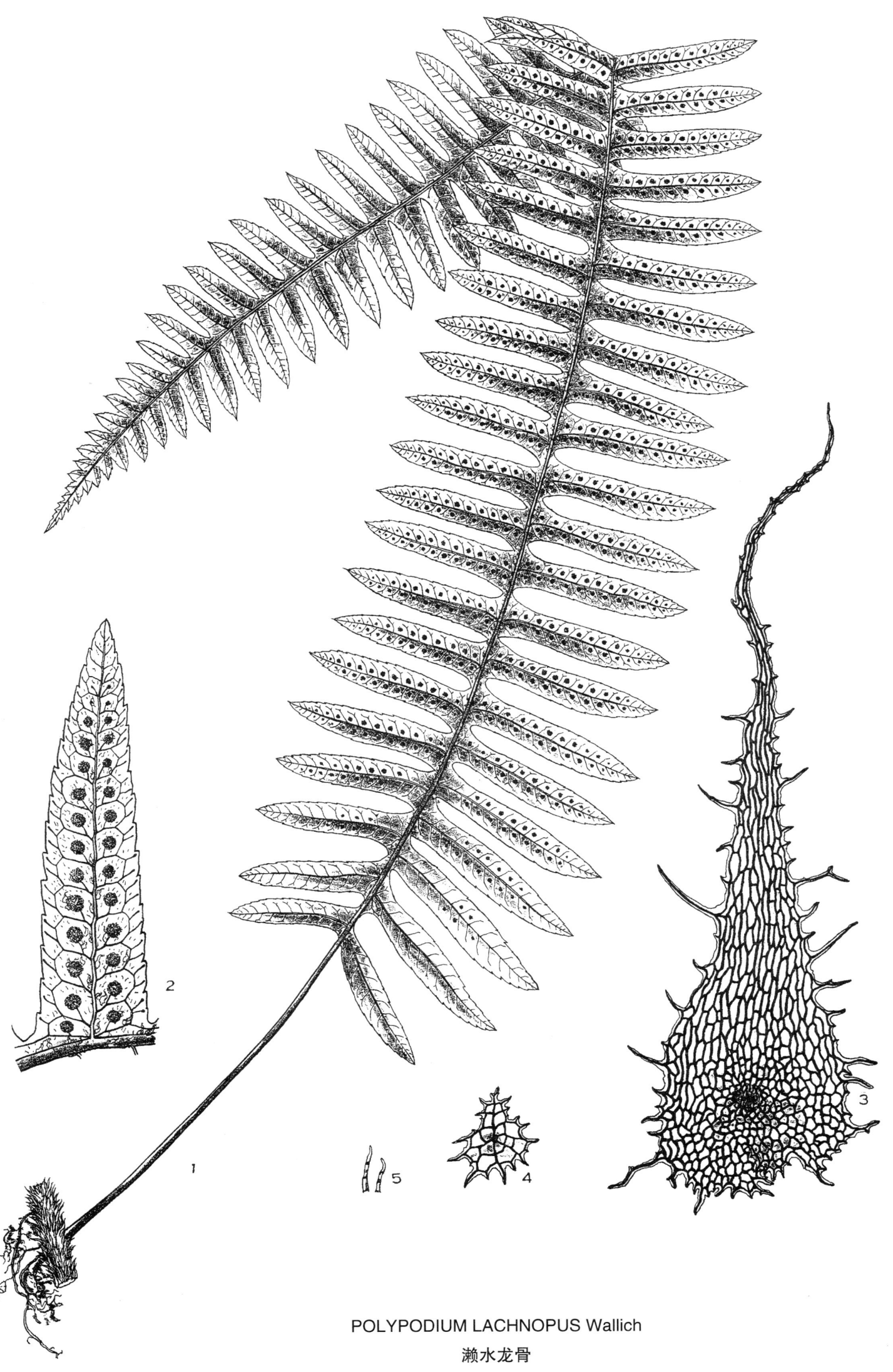

POLYPODIUM LACHNOPUS Wallich
濑水龙骨

图版 96　栗柄水龙骨

根状茎细长,土生;鳞片疏生,黝色,披针形,渐尖头,网眼透明;叶疏生,直立,柄长 10—14 cm,光滑,淡栗褐色,有光泽,中肋下面亦然,叶体长 22—30 cm,宽 7—10 cm,一次羽状分裂,小叶 20—30 对,开展,披针形,急尖头,钝锯齿疏生,基部一对与上部各对等长,唯其中肋不如上部之具有夹翅耳,薄纸质,色绿,具光泽,两面光滑,叶脉明显,网脉沿主脉一列;子囊群小而圆,稍贴近主脉。

分布:云南,四川西南部及喜马拉雅山区之克什米尔。

图注:1.本种全形(自然大);2.根状茎上之鳞片(放大 50 倍);3.小叶,表示叶脉及其子囊群之位置(放大 2 倍)。

POLYPODIUM MICRORHIZOMA Clarke

POLYPODIUM MICRORHIZOMA Clarke, Baker in Hooker Syn. Fil. 511 (1874); Clarke, Ferns N. Ind. in Trans. Linn. Soc. ser. 2. Bot. **1**. 551 (1880); C. Chr. Ind. Fil. 545 (1906).

Goniophlebium microrhizoma Bedd. Ferns Brit. Ind. Suppl, 22 t. 386 (1876).

Polypodium taliense Christ, Bull. Soc. Bot. France **52**; Mém. 1. 14 (1905); C. Chr. Ind. Fil. 569 (1906).

Rhizome hypogaeous, 2 mm thick, wide-creeping; *scales* sparce, dark brown, ovate, acuminate; *frond* distant, erect, stipe 10-14 cm long, naked, shining castaneous on the lower side and so is the rachis, lamina lanceolate, 22-30 cm long, 7-10 cm broad, simple pinnate; *pinnae* 20-30-jugate, patent, lanceolate, acute, distantly inciso-serrate, basal pinnae hardly abbreviated, but free from those next above, which are connected by a narrow wing along the rachis; *texture* membranaceous, green, lucide, naked on both surfaces; *veins* distinct on both sides, areolae uniseriate, generally close, but some are open-particularly towards apex; *sori* small, oblong, medial, far apart.

Yunnan: Mengtze, *A. Henry 10168*; Hokintien, *Delavay 27*, July 24, 1883; Longpon, *E. E. Maire 6047*; Pautiho, *C. Forrest 27029*. Sichuan: Yienpin Hsien, *T. T. Yü 1701* (1932).

A pretty species endemic in the Himalayas. It differs from other species with goniophlebioid venation in almost membranaceous, lucide, naked leaves with castaneous, shining stipe, and the remotely inciso-serrate pinnae. The venation in this species is generally of goniophlebioid type, but some of the areolae are open, particularly towards the apex of pinnae.

Plate 96. Fig. 1. habit sketch (natural size). 2. scale from rhizome ($\times$ 50). 3. pinna, showing venation and position of sori ($\times$ 2).

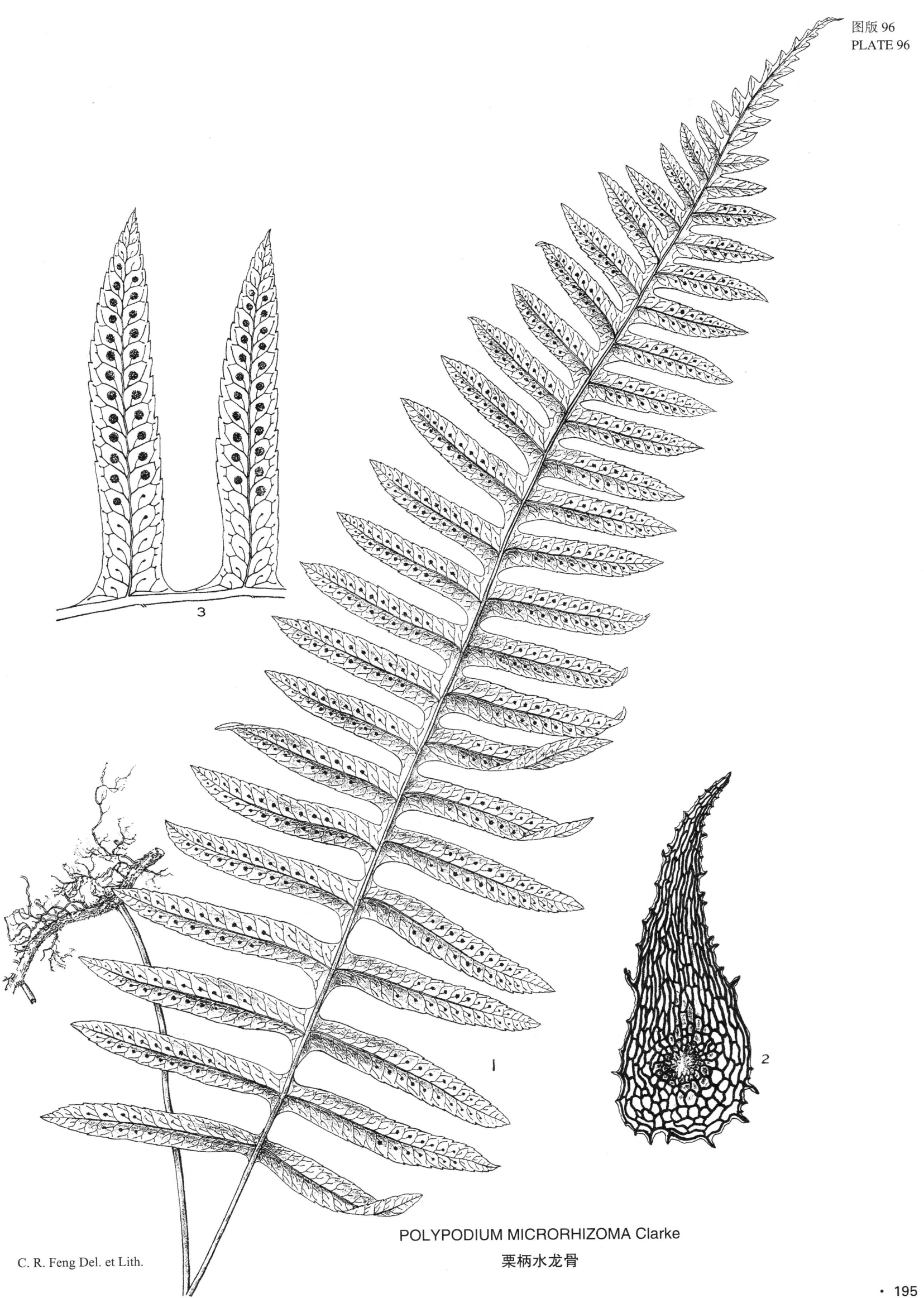

POLYPODIUM MICRORHIZOMA Clarke
栗柄水龙骨

图版 97　川水龙骨

根状茎长,蔓生,颇肥;鳞片瓦覆,深棕色,卵披针形,边缘具小齿,网眼大而明,膜壁薄;叶间距离大,柄长 8—12 cm,稻秆色,光滑;叶体长椭圆卵形,长 25—35 cm,基部截形,宽可 15 cm,一次羽状分裂,叶端呈尾形,小叶 12—20 对,开张,披针线形,钝头,边缘具钝锯齿,基部膨大而合生于中肋上,纸质,绿色,光滑,叶脉明显,网脉仅一列;子囊群一列,光滑。

分布:四川之南川,峨眉山及峨边等地。

本种形体极似友水龙骨(*Polypodium amoenum* Wall.),所不同者为小叶细狭,远隔,小叶间之中肋不具翅,基部膨大;又近于耳形水龙骨(*Polypodium subauriculatum* Bl.),其异点为根状茎不为附生而为土生,且不呈白粉状,亦不为半光滑,无鳞片盖覆。

图注:1.本种全形(自然大);2.侧生小叶,表示叶脉及其子囊群之位置(放大 3 倍);3.根状茎上之鳞片(放大 50 倍)。

POLYPODIUM DIELSEANUM C. Christensen

POLYPODIUM DIELSEANUM C. Chr. Ind. Fil. 522 (1906).
　　Polypodium leuconeuron Diels (non Christ, 1900) in Engl. Jahrb. **29**: 203 (1901).
　　Polypodium wilsonii Christ, Bull. Acad. Géogr. Bot. (1906) 104.

Rhizome wide-creeping, rather thick, densely scaly; *scales* fusco-brown, imbricate, ovate-lanceolate, minutely denticulate, with large clear luminae and thin walls; *frond* distant, stipe 8-12 cm long, straminous or light castaneous, glabrous, lamina oblong-ovate, 25-35 cm long, to 15 cm broad at base, pinnate, truncate at base, caudate at apex, *pinnae* 12-20 pairs, patent, linear, apex obtuse, margin crenate-serrate, adnate and somewhat dilated above base, far apart, the greater part of rachis between pinnae wingless; *texture* chartaceous, color green, moderately villose on both sides of rachis and upper side of lamina; *venation* distinct, veins in pinnae anastomosing in one row of areolae on each side of costa, beyond which veinlets are free; *sori* small, one-rowed, naked.

Sichuan: Nanchuan, Tanchiawan, *von Rosthorn 398* (type); Mt. Omei, *E. H. Wilson 5336*; *P. W. Fang 2736*, *2745*; O-pien Hsien, *P. W. Fang 8066*, *8474*, *8502*, *8703*, *6673*.

Closely allied to *P. amoenum* Wallich, differs by narrowly linear pinnae with peculiarly dilated base being widely separated by wingless rachis. In habit it resembles P. *subauriculatum* Blume, but differs in rhizome being not epigaeous, nor glaucous, nor subnaked, in much smaller size and more or less adnate pinnae.

Plate 97. Fig. 1. habit sketch (natural size). 2. lateral pinna, showing venation and position of sori ($\times$ 3). 3. scale from rhizome ($\times$ 50).

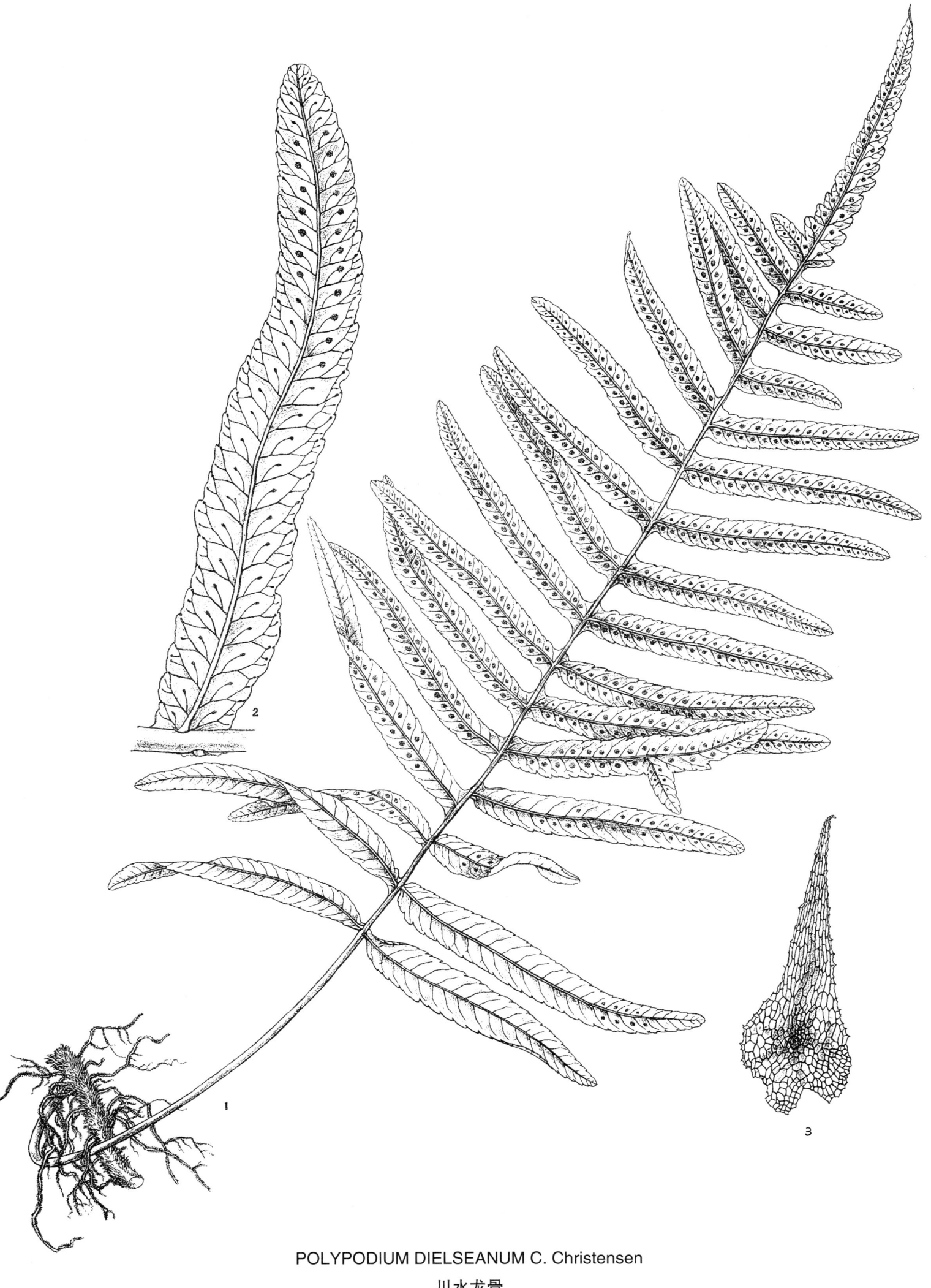

POLYPODIUM DIELSEANUM C. Christensen
川水龙骨

图版 98　水龙骨

根状茎颇肥,匍匐于石上甚广,光滑或鳞片疏生,黑色而被白粉;鳞片通常仅疏生于叶柄基部或茎之幼部,深褐色,狭长,网眼透明;叶疏生,直立,柄稻秆色,光滑,长 15—20 cm,叶体一次羽状分裂,小叶约 20 对,开展,披针形,钝头,全缘,基部一对通常较短而下向,纸质,两面具密绒毛,色稍黑,叶脉不显,网眼于中肋及主脉两旁一列;子囊群形圆,颇大,贴近于主脉。

分布:广东,广西,浙江,江苏,安徽,江西,湖北,云南,四川,贵州;日本亦产之。

本种为本属特殊之一种,其黑色半光滑之根状茎横卧于石上,常呈粉白色,叶体两面密被短绒毛。

图注:1. 本种全形(自然大);2. 小叶,表明叶脉及其生绒毛之面(放大 3 倍);3. 根状茎上之鳞片(放大 50 倍)。

POLYPODIUM NIPONICUM Mettenius

POLYPODIUM NIPONICUM Mettenius, Ann. Lugd. Bat. **2**: 222 (1866); Hk. et Bak. Syn. Fil. 341 (1868); Christ, Farnkr. d. Erde 93 (1897); Diels, Nat. Pfl. Fam. **1**: 4. 311 (1899); C. Chr. Ind. Fil. 564 (1906); Ogata, Ic. Fil. Jap. **3**: pl. 140 (1931).

Polypodium bodinieri Christ, Bull. Acad. Géogr. Bot. (1902) 203.

Polypodium silvestrii Christ, in Lecomte, Not. Syst. **1**: 58 (1909).

Polypodium longkyense Rosenst, in Fedde, Repert, Sp. Nov. **13**: 134 (1914).

Rhizome rather thick, fleshy, long-creeping, epigaeous, black, subglabrous, covered by whitish blooms; *scales* sparce at the base of stipe, or on the growing tip, fusco-brown, thin, clathrate, linear-subulate from an ovate base, margin subentire from base upwards; *frond* far apart, erect, stipe straminous, naked, 15-20 cm long, lamina pinnatifid nearly to the rachis into about 20 pairs or more, spreading; linear-oblong, entire, acute, close *pinnae*, the lowest often subfree, slightly reduced and deflexed; *texture* papyraceous, both sides densely pubescent with short, whitish, 2-3-celled, soft hairs; color obscurely blackish; *veins* anastomosing in one row of areolae along rachis and costa of pinnae; *sori* rather large, rounded, naked, nearer to the costa than to the margin, which is finely ciliate.

Guangdong: Lokchong, *N. K. Chun 42660*, *42963*. Jiangxi: Ruling, *Dr. Shearer* (1873); *Forbes 752* (1874); *H. H. Hu 2309*. Hubei: Nanto near Ichang, *A. Henry 2660*; *3196*; Ma Kia Keou, *Silvestri 62*, *3283*. Zhejiang: Ningpo, *Hancock 19*. Guizhou: Pinfa, *Cavalerie 7239*, *7012*; Shihtsien, *Y. Tsiang 4115*; Tushan, *Y. Tsiang 6922*; Tsingay, *Bodinier 2031*, March, 1898. Yunnan: Longky, *E. E. Maire*. Guangxi: Miu Shan, Bin-long, *R. C. Ching 5961*. Sichuan: Tchenkoutin, *R. P. Farges 2031*; Mapien Hsien, *W. P. Fang 6455*. Jiangsu: Ishing, Hufu, *C. Y. Luh 456*; *R. C. Ching & C. L. Tso* (1926).

Type from Japan. A fairly common fern in East China, and generally growing over rocks. This is one of the most distinct species of the genus by its rather thick, fleshy subglabrous, epigaeous rhizome covered by whitish blooms, and by its broadly lanceolate, herbaceous lamina, which is densely but shortly pubescent on both sides. As to all these essential characters, this is a very uniform species, although differs more or less in the dimension of leaves. The Chinese plant described twice by Christ and once by Rosen-stock as distinct species is perfectly identical with the Japanese type.

Plate 98. Fig. 1. habit sketch (natural size). 2. segment, showing venation and pubescent surface ($\times$ 3). 3. scale from rhizome ($\times$ 50).

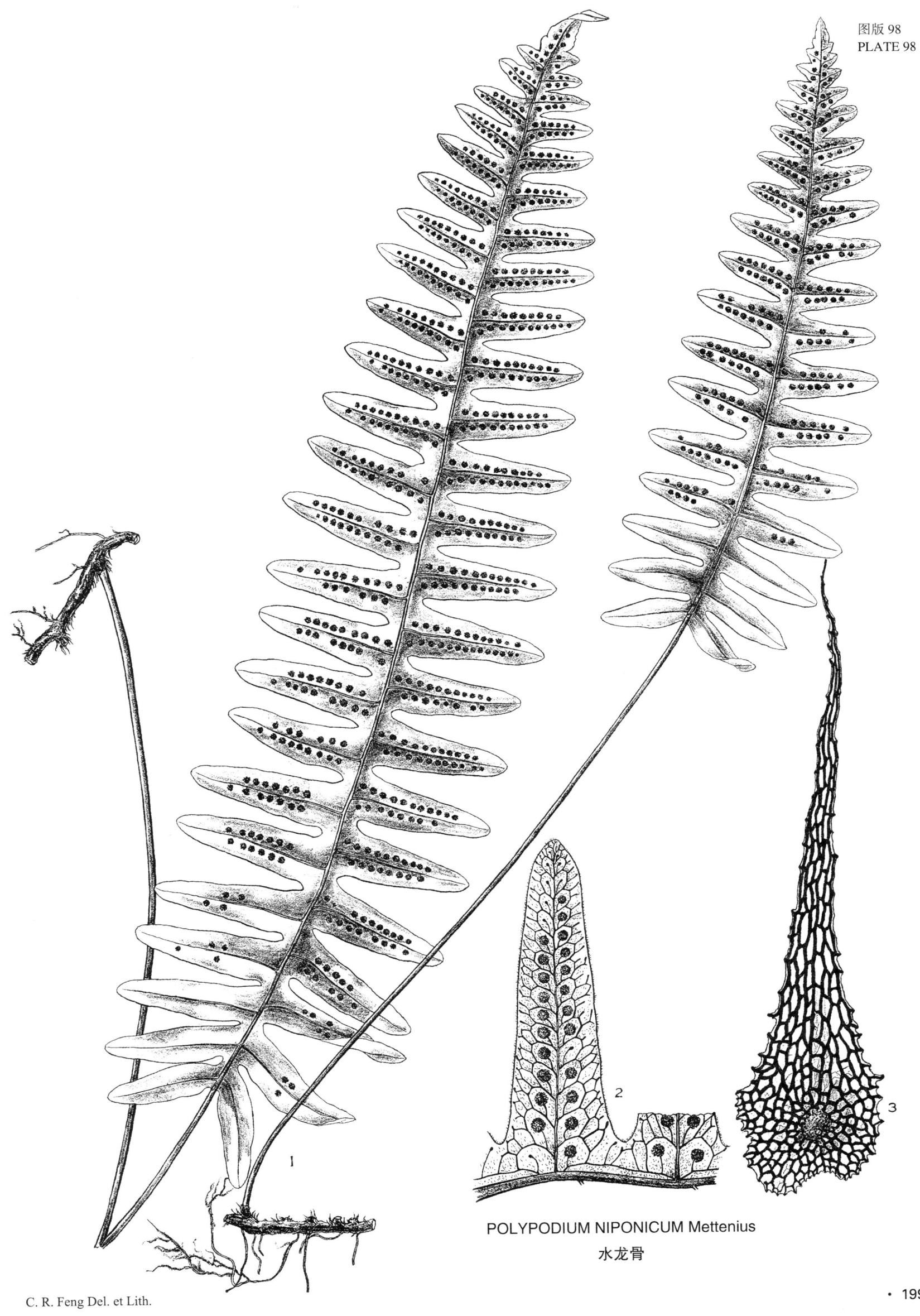

POLYPODIUM NIPONICUM Mettenius
水龙骨

根状茎匍匐甚广，颇肥，土生或亚土生；鳞片疏生，卵形，渐尖头，深褐色；叶疏生，直立，柄长 12—30 cm，光滑，稻秆色，叶体长 25—40 cm，宽 10—20 cm，一次羽状分裂，小叶 15 对或过之，开张，渐尖，边缘多少具齿，基部一对通常最长，开张或稍下向，厚纸质，光绿色，中肋下面鳞片疏生，两面光滑或短毛疏生，叶脉明显，网脉于中肋及主脉两侧一列，间为不完全之两列；子囊群中大，疏生，贴近主脉，光滑。

分布：原产锡金喜马拉雅山区，然中国中部，西南部及南各省均产之，近于江苏宜兴亦采得之。

本种形体大小不一，其产于陕西南部及中国中部者，形体大概较小，Clarke 氏初名之曰 P. subamoenum Clarke var. chinese，有时叶体之一面或二面具甚密短毛，P. yunnanense France；P. amoenum f. pilosa Rosenst 等属之，然与光滑无毛，形体甚大之模式种无显著之异同可分。

图注：1. 本种全形（自然大）；2. 根状茎上之鳞片（放大 50 倍）。

POLYPODIUM AMOENUM Wallich

POLYPODIUM AMOENUM Wallich, Catal. n. 290 (1828); Mett. Polyp. 80, n. 131 (1875); Hk. et Bak. Syn. Fil. 341 (1868); Diels, Nat. Pfl. Fam. **1**: 4, 311 (1899); C. Chr. Ind. 508 (1906).

Goniophlebium amoena J. Sm. in Hooker, Gen. Fil. ad t. 50 (1840); Bedd. Ferns Brit. Ind. t. 5 (8165); Handb. Ferns Brit. Ind. 317 (1883).

Marginaria amoena Presl, Tent. Pterid. 188 (1816).

Polypodium valdealatum Christ, Bull. Herb. Boiss **7**: 4 (1899).

Polypodium amoenum var. *latedeltoideum* Christ, Bull. Acad. Géogr. Bot. (1907) 142.

Polypodium subamoenum var. *chinensis* Christ, Nuovo Giorn. Bot. Soc. Ital. n. s. **4**: 99 (1897).

Polypodium bonatianum Brause, Hedwigia **54**: 207 pl. 4. f. L (1914).

Polypodium yunnanense Franch. Bull. Soc. Bot. France **32**: 29 (1885).

Polypodium duclouxii Christ in Lecomte, Not. Syst. **1**: 14 (1909).

Polypodium amoenum var. *pilosa* Clarke, Journ Linn. Soc. **24**: 417. (1888).

Polypodium amoenum f. *pilosa* Rosenst. in Fedde, Report. Sp. Nov. **13**: 134 (1914).

Rhizome wide-creeping, rather thick, hypogaeous or subepigaeous; *scales* fusco-brown, thin, clathrate, broadly ovate-acuminate; *frond* far apart, stipe 12-30 cm long, firm, erect, naked, straminous, lamina 25-40 cm long, 10-20 cm broad, pinnatifid nearly to the rachis into 15, or more, pairs of spreading acuminate, close segments, with more or less distantly serrate margin; the lowest not shortened, slightly deflexed; *texture* charta-ceous, lustrous green, scaly on rachis beneath, naked, or often more or lesss shortly pilose on both or one side; *veins* distinct on both surfaces, areolae uniseriate along the costa, or rarely imperfectly biseriate in broader segments; *sori* medium-sized, far apart, nearer to the costa than to the margin, naked.

Type from Nepal (leg. Wallich). A very common fern in the Himalayas as well as in Southwestern and Central China, extending northwardly to the southern part of Shaanxi. P. *subamoenum* var. *chinense* Christ is a reduced form common in North China and the Lower Yangtze Valley; P. *yunnanense* Franch. is the same form with lamina more or less pilose on one or, not infrequently, on both sides, and to this P. *Bonatianum* Brause and P. *amoenum* f. *pilosa* Rosenst. should be referred as direct synonyms. *Polypodium valdealatum* Christ, based upon Aug. Henry n. 11315 from Mengtze, Yunnan, differs from the type only in segments provided with gross teeth; P. *Duclouxii* Christ differs in stipe and lamina being tinged wine-red when dried.

Plate 99. Fig. 1. habit sketch (natural size). 2. scale from rhizome ($\times$ 50).

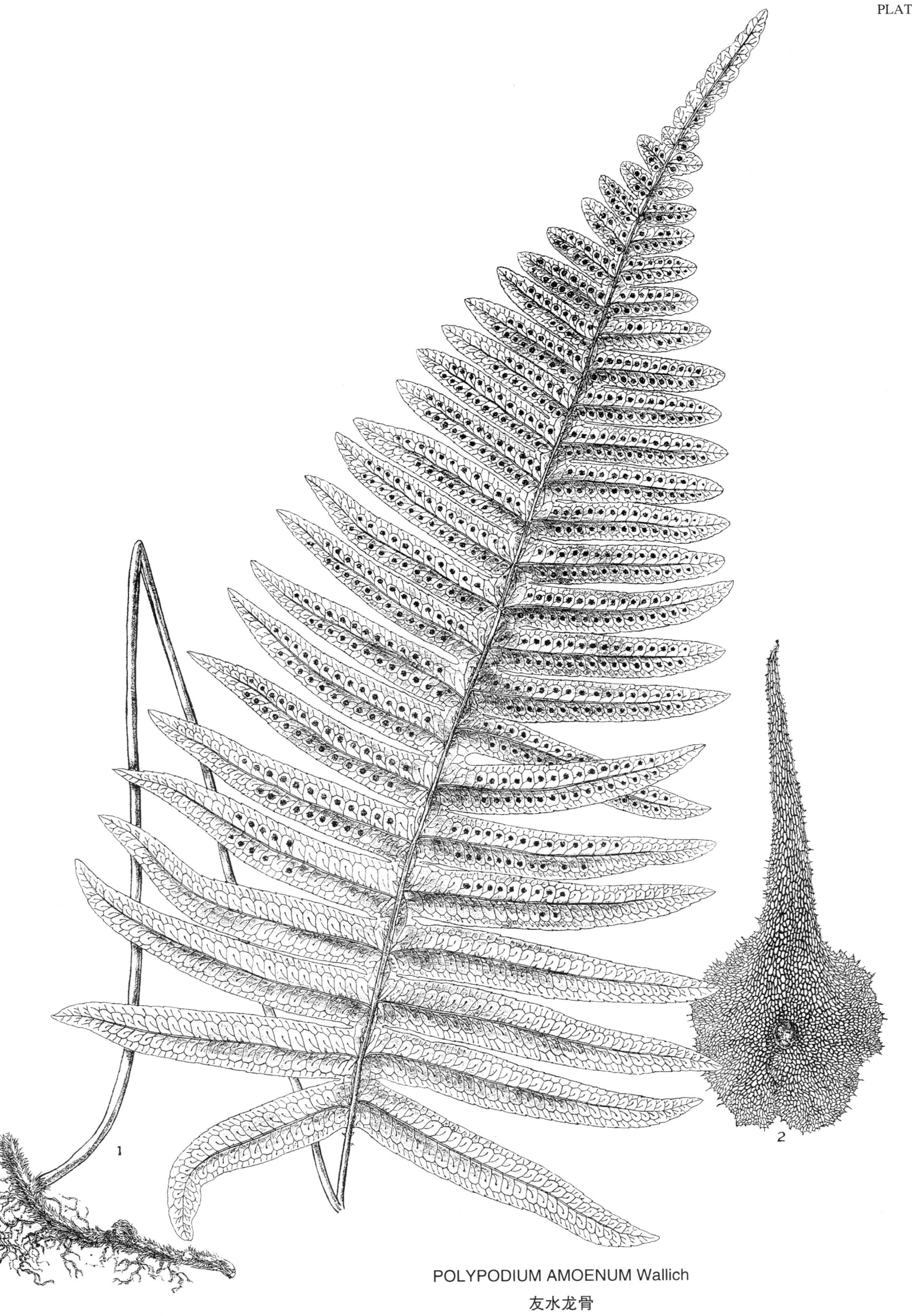

POLYPODIUM AMOENUM Wallich
友水龙骨

根状茎细长，蔓生，直径约 1 mm，鳞片密覆，卵披针形，色黑，质硬，边缘具小齿；叶间距离大，柄长 3—6 cm，稻秆色，光滑，叶身卵椭圆形，长 5—8 cm，宽 4—5 cm，基部心脏形，一回羽状分裂，小叶 2—4 对，基部一对最长且强度下向，其上各对渐短，开张或上向，边缘具疏生小锯齿，纸质，绿色，上面具短毛，尤以中肋及主脉为甚，下面光滑，淡绿，叶脉明显，于主脉两旁成二排有规则之网脉，网眼内具分叉小脉一或二，子囊群形小，一列，光滑。

分布：湖北及四川，生于荫湿之石壁上或林中树干上。

本种形体极类费氏茀蕨（*Phymatodes veitchii*），其不同之点为：黑色之茎鳞及叶体上面被短毛。

图注：1. 本种全形（自然大）；2. 叶体之一部，表示其上面所生之短毛（放大 4 倍）；3. 叶体之一部，表示其光滑下面及子囊群之位置（放大 4 倍）；4. 叶体上面之毛之构造（放大）；5. 根状茎上之鳞片（放大 50 倍）。

PHYMATODES NIGROVENIA (Christ) Ching

PHYMATODES NIGROVENIA (Christ) Ching, Contr. Bot. Inst., Nat. Acad. Peiping **2**: 79 (1933).

Polypodium nigrovenium Ching, Bull. Fan Mem. Inst. Biol. **1**: 150 (1930).

Polypodium Shaanxiense Christ var. *nigrovenium* Christ, Bull. Acad. Géogr. Bot. (1906) 106.

Polypodium veitchii Baker var. *nigrovenium* Takeda, Notes, R. Bot, Gard. Edinb. **8**: 296 (1915); C. Chr. Acta Horti Gotob. **1**: 102 (1924).

Rhizome rather slender, wide-creeping, about 1 mm thick, densely scaly; *scales* ovate-lanceolate, black, firm, denticulate; *frond* far apart, stipe 3-6 cm long, straminous, naked, lamina oblong-ovate, 5-8 cm long, 4-5 cm broad, cordate at base, simple pinnate *pinna* 2-4 pairs, the basal pair strongly deflexed and much the longest, the upper ones gradually decrescent, patent or oblique, margin remotely serrulate; *texture* chartaceous, color green, more or less shortly glandular pubescent especially along the midrib and costa above, pale green and glabrous underneath; *veins* distinct, rather regularly anastomosing in two rows of areolae on each side of costa, each provided with one, or two, forked including veinlets; *sori* small, uniseriate, naked.

Hubei: Changyang Hsien, *E. H. Wilson 5341A* (type). Sichuan: Mt. Omei, *P. W. Fang 2966*, *2985* (1928), on cliff and on the trunk of trees.

Habit exactly similar to *Ph. veitchii* (Baker), differs in black, firm rhizome scales and shortly glandular pubescence on the upper surface of lamina. *Polypodium erythrocarpum* Mettenius from Sikkim-Himalayas appears to be a still closer ally to the present fern, differs in leaves being more densely pubescent on both sides, and scales of a lighter color.

Plate 100. Fig. 1. habit sketch (natural size). 2. a portion of lamina, showing pubescent upperside ($\times$ 5). 3. a portion of lamina, showing glabrous underside and position of sori ($\times$ 5). 4. articulated hair from upperside of lamina (enlarged). 5. scales from rhizome ($\times$ 50).

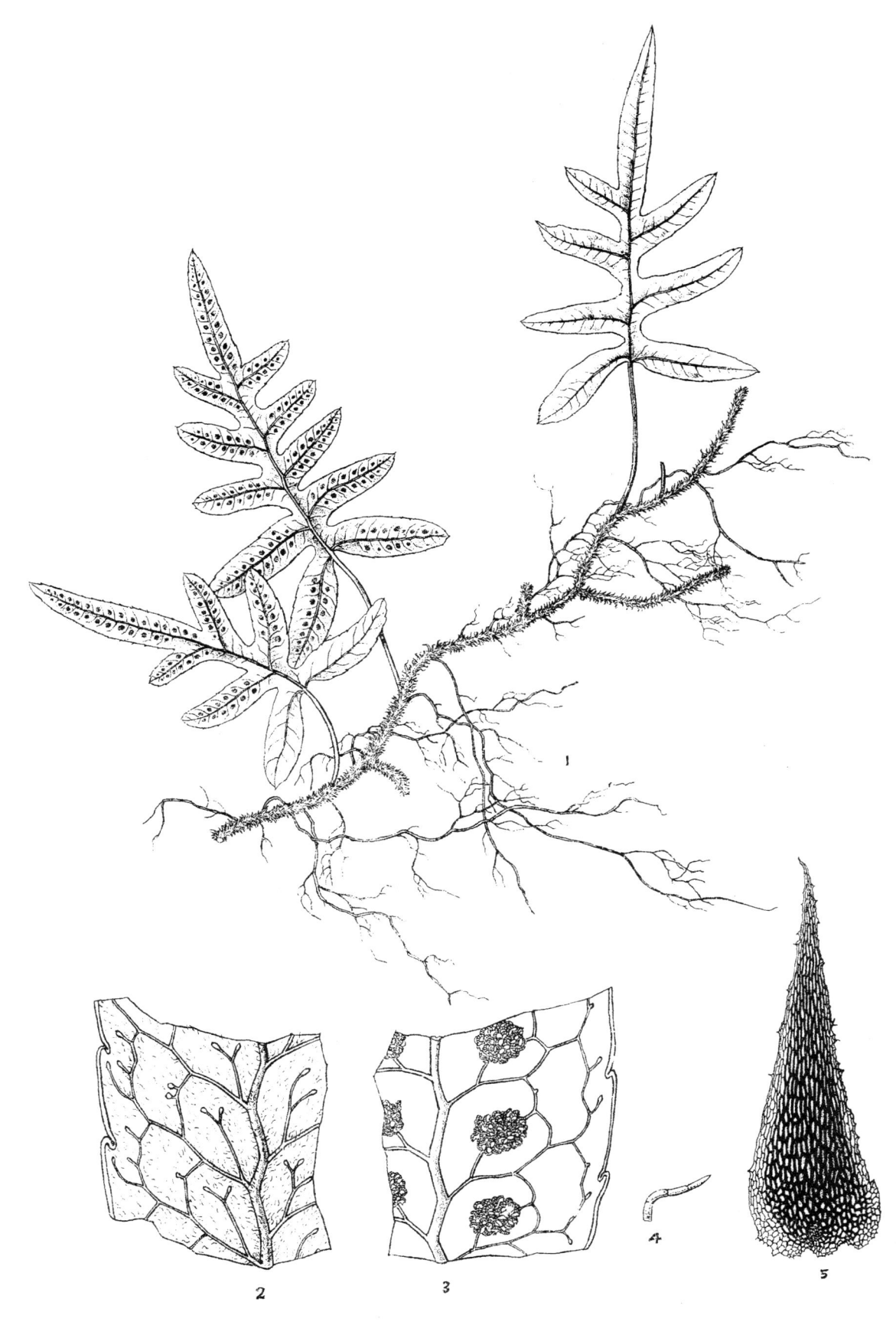

PHYMATODES NIGROVENIA (Christ) Ching
黑鳞茀蕨

图版 101(Fas.2) 嫩毛蕨

嫩毛蕨

根状茎细长,匍匐甚广,粗约 1 mm,颇光滑,叶疏生,柄长 25—35 cm,细圆,红稻秆色,叶体阔披针形,长 30—40 cm,基部宽 13—15 cm,渐尖头,不呈楔形,三回分裂,一回裂片约 15 对,位于下部者疏生,长 7—8 cm,宽 3—4 cm,基部等形,具极短之柄,中肋,主脉,侧脉两边均被疏毛,叶质薄,色绿;子囊群顶生于叶脉,形小,通常每裂片一个,盖膜质,绿色,肾形以缺刻着生,边缘分离。

分布:云南,贵州。

本种之形态及其子囊群盖绝类产于同地之 *Dryopteris flaccida* (Bl.) O. Ktze.,然其茎内具单筒形之维管束及其子囊群顶生于叶脉,易于识别。

图注:1.本种全形(自然大);2.三次裂片(放大 12 倍)。

MICROLEPIA TENERA Christ

MICROLEPIA TENERA Christ in Lecomte, Not. Syst. **1**: 53 (1909); C. Chr. Ind. Fil. Suppl. 51 (1906-1912).

Rhizome slender, wide-creeping, about 1 mm thick, subglabrous; *frond* far apart, stipe 25-35 cm long, slender, terete, glabrescent, rufo-straminous; *lamina* oblong-lanceolate or broadly lanceolate, 30-40 cm long, 13-15 cm broad at base, acuminate or caudato-acuminate, base not attenuate, deeply tripinnatifid, *lateral pinna* about 15-jugate, alternate, the lower ones far apart (about 5 cm apart), the basal ones as long as, or slightly longer than, those next above, 7-8 cm long, 3-4 cm broad at base, patent, the upper middle ones close together, oblique-patent, subfalcate, about as long as the lower ones, those above the upper two-thirds are gradually shortened below the pinnate apex, all very shortly petiolate (petiole about 2 mm long), oblong-lanceolate, acuminate, base equal, *secondary pinnules* in the lower pinna about 10-jugate, patent, the basal ones of equal length, as long as those next above, 2-2.5 cm long, 1 cm broad, base subequal, i. e, the posterior side cut away, oblong, obtusish, sessile but never adnate; *ultimate pinnules* in the lower pinna about 6-8-jugate, close, oblique, 5-7 mm long, decurrent, apex roundish, 3-4-lobulato-incised, or entire on the lower side, 2-3 bluntly toothed on the upperside; *veins* in the ultimate pinnule pinnate, one to each lobe or tooth, veinlet simple, fine, but distinct; *rachis* and *rachilets* moderately strigose-hairy, both sides sparsely hirsute on costules, veins and veinlets; *texture* thin herbaceous, color light green; *sori* small, one to each ultimate pinnule, or to each segment in lower pinna, borne on the anterior veinlet, falling short from the margin, *indusium* small, green, sub-glabrous, membranaceous, of the *lastreoid* type, i. e. reniform, attached by a deep sinus and free all round.

Yunnan: Szemeo, *Henry 13155A* (type), *13155*; Mengtze, *Hancock 75*. Guizhou: Houangtsapa, *Cavalerie 7288*.

One of the most distinct species of the genus, endemic in Southwestern China, The slender habit, thin herbaceous texture only sparsely strigose on both surfaces, the widely separated and very shortly petiolate lower pinna with almost equal base, and the small, thin, green indusium of a lastreoid type are the outstanding characters of this species. The general habit, thin herbaceous texture and the small, greenish, membranaceous indusium of a lastreoid type tend to make this fern appearing so alike *Dryopteris flaccida* (Blume) O. Ktze that it would have been strongly suspected to be a form of that species also known in the locality, had it not been for its primitive solenostelic vascular structure in the rhizome and its terminal sori.

Plate 101. Fig. 1. habit sketch (natural size). 2. an ultimate pinnule from the lower pinna ($\times$ 12).

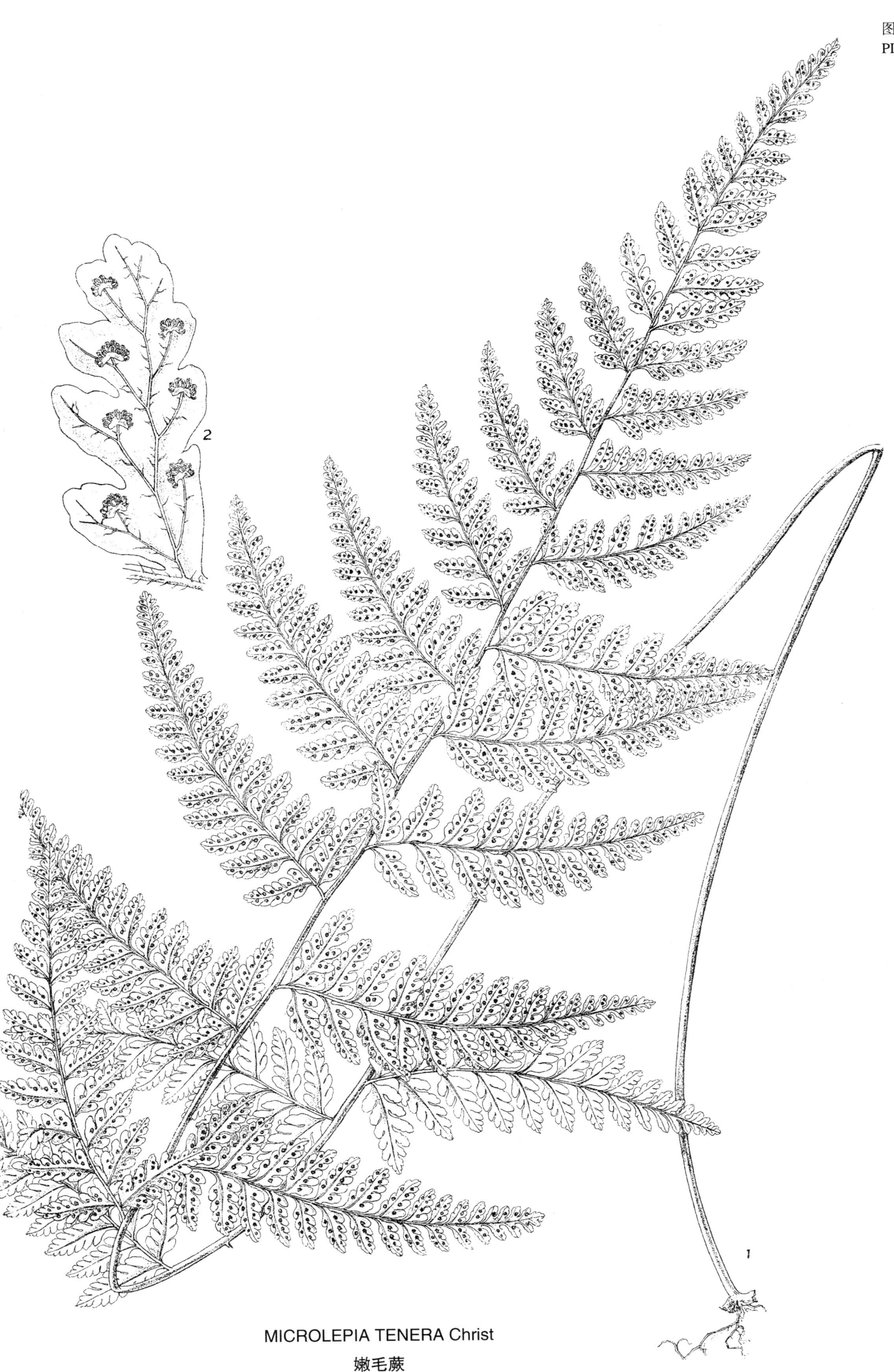

MICROLEPIA TENERA Christ 嫩毛蕨

中国蕨类植物图谱

ICONES FILICUM SINICARUM

BY

REN-CHANG CHING, B. S.

KEEPER

THE LU-SHAN ARBORETUM AND BOTANICAL GARDEN OF

THE FAN MEMORIAL INSTITUTE OF BIOLOGY AND

KIANGSI PROVINCIAL AGRICULTURAL INSTITUTE

FASCICLE 3, PLATES 101—150

第三卷

TO

PROFESSOR SUNG-SHU CHIEN

HEAD OF THE BOTANICAL DIVISION, BIOLOGICAL LABORATORY OF

THE SCIENCE SOCIETY OF CHINA

AS ONE OF THE FOREMOST MODERN CHINESE BOTANISTS

IN RECOGNITION OF

HIS EXEMPLARY WORK AS A TEACHER

UNDER WHOSE DELIGHTFUL GUIDANCE

THE AUTHOR FIRST BEGAN OVER TWENTY YEARS AGO

TO LEARN BOTANY IN ITS VARIOUS PHASES

THIS THIRD FASCICLE OF ICONES FILICUM SINICARUM

IS RESPECTFULLY DEDICATED

大囊岩蕨

根状茎短而直立，顶部具棕黄色之披针形密鳞片；叶簇生，柄长 2—10 cm，淡黄棕色，基部具鳞片，上部遍被刚毛，叶体长 5—12 cm，宽约 2—4 cm，一回羽状分裂，下部小叶几无柄，基部一对着生于叶柄顶部之节上，较小，下向，上部小叶与中肋合生，长约 1.2—2 cm，椭圆卵形，波状分裂，或深裂，中肋及两面被短刚毛，不具鳞片，纸质；叶脉羽状分裂；子囊群形圆，着生于侧脉之顶，盖为圆杯形，膜质，被刚毛，顶部开裂，口具 4—6 数锯齿。

分布：山东北部，河北，陕西，东三省；朝鲜半岛及日本亦产之。

本种为亚洲东北部特有之种，最初在山东烟台发现，现为华北常见之品。

图注：1. 本种全形（自然大）；2. 中部小叶，表明其叶脉及子囊群着生状（放大 6 倍）；3. 子囊群与盖（放大 20 倍）；4. 同上，垂直切开（放大 20 倍）；5. 叶柄基部之鳞片及刚毛（放大 16 倍）；6—7. 叶体上之刚毛（放大 30 倍）；8. 基部一对小叶着生于叶柄之节上（放大 20 倍）；9. 子囊（高倍放大）；10. 孢子（高倍放大）。

WOODSIA MACROCHLAENA Mettenius

WOODSIA MACROCHLAENA Mettenius ex Kuhn, Journ. Bot. (1868) 270; Linnaea **36**: 126 (1869); C. Chr. Ind. Fil. 657 (1905); Suppl. Ⅲ. 195 (1934); Fomin, Fl. Sib. et Orient. Extr. **5**: 10 c. fig. (1930); Ching, Sinensia **3**: 143. 1932.

Woodsia insularis Hk. et Bak. (non Hance, 1861) Syn. Fil. 47 (1867).

Woodsia brandtii Franch. et Sav. Enum. Pl. Jap. **2**: 205, 616 (1879).

Woodsia japonica Makino, Bot. Mag. Tokio **18**: 134 (1904); Ogata, Ic. Fil. Jap. **3**: t. 149 (1930).

Woodsia sinuata Makino, Bot. Mag. Tokio **11**: 64 (1897, non 1904); Christ, Bull. Acad. Géogr. Bot. **20**: 157 (1909, non 1902).

Woodsia frondosa Christ in Fedde, Repert. Sp. Nov. **5**: 12 (1908).

Rhizome short, erect, densely radicose, apex densely clothed in bright brown, lanceolate, acuminate, sparcely fimbriate *scales*; *fronds* fasciculated, stipe 2-10 cm long, straminous or brownish, scaly at base, hirsute upwards, lamina 5-12 cm long, 2-4 cm broad, pinnate under pinnatifid acutish apex, *pinnae* 5-12-jugate, the lower ones free, subsessile, the upper ones adnate, the basal pair springs from the prominent articulation of stipe, always somewhat smaller than those above, deflexed, ovate-deltoid, with more or less auriculated anterior base, the middle pinna 1.2-2 cm long, oblong-ovate, adnate with wingless rachis between contiguous pairs, lobato-sinuate or pinnatifid into oblong sinuate segments, rachis and both sides moderately clothed in ferruginous needle-like, articulated appressed hairs, no trace of scales; *texture* herbaceous, soft; *veins* in segments pinnate; *sori* large, terminal on veinlets, 1-2 to each lobe, or 4-6 to each segment, *indusium* cup-shaped, large, membranaceous, setose hairy, irregularly 4-6 erosed at top; *sporangia* shortly stipitate.

Shandong: Chefoo, *Schottmuller* (type), August, 1861; B. *Drug 285*, Sept. 17, 1907; *A. C. Maingay 13*, August-Sept. 1862; *Cowdry 644* (1920); *Forbes 2233* (1881); Weihaiwei, *C. G. Matthew*, Sept. 1906. Shanxi: *E. Licent* (without locality). Northeast China: *Swinhae 6516* in herb. Hance. Port Arthur: *Miss Möller*, Oct. 28, 1928.

Korean peninsula: Ouen San, *U. Faurie 101* (type of *W. frondosa*). Quelpaert: *U. Faurie 5610, 2172* (f. typica).

Japan: Prov. Chochiou, *Kramer 1569* (type of *W. Brandtii*); *C. G. Matthew*, July **23, 1906**.

For further comment and discussion about this distinct but hitherto much neglected fern, readers are referred to my paper on the genus *Woodsia* in *Sinensia* **3**: pp. **143**.

Plate 101. Fig. 1. Habit sketch, f. typica (natural size). 2. Upper middle pinna, showing venation and sori (× 6). 3. Sorus with indusium (× 20). 4. the same, cut vertically (× 20). 5. Scales and hair from the base of stipe (× 16). 6-7. Hairs from rachis and leaf surfaces (× 30). 8. Basal pinna attached to the articulation of stipe (× 20). 9. Sporangium (greatly enlarged). 10. Spores (greatly enlarged).

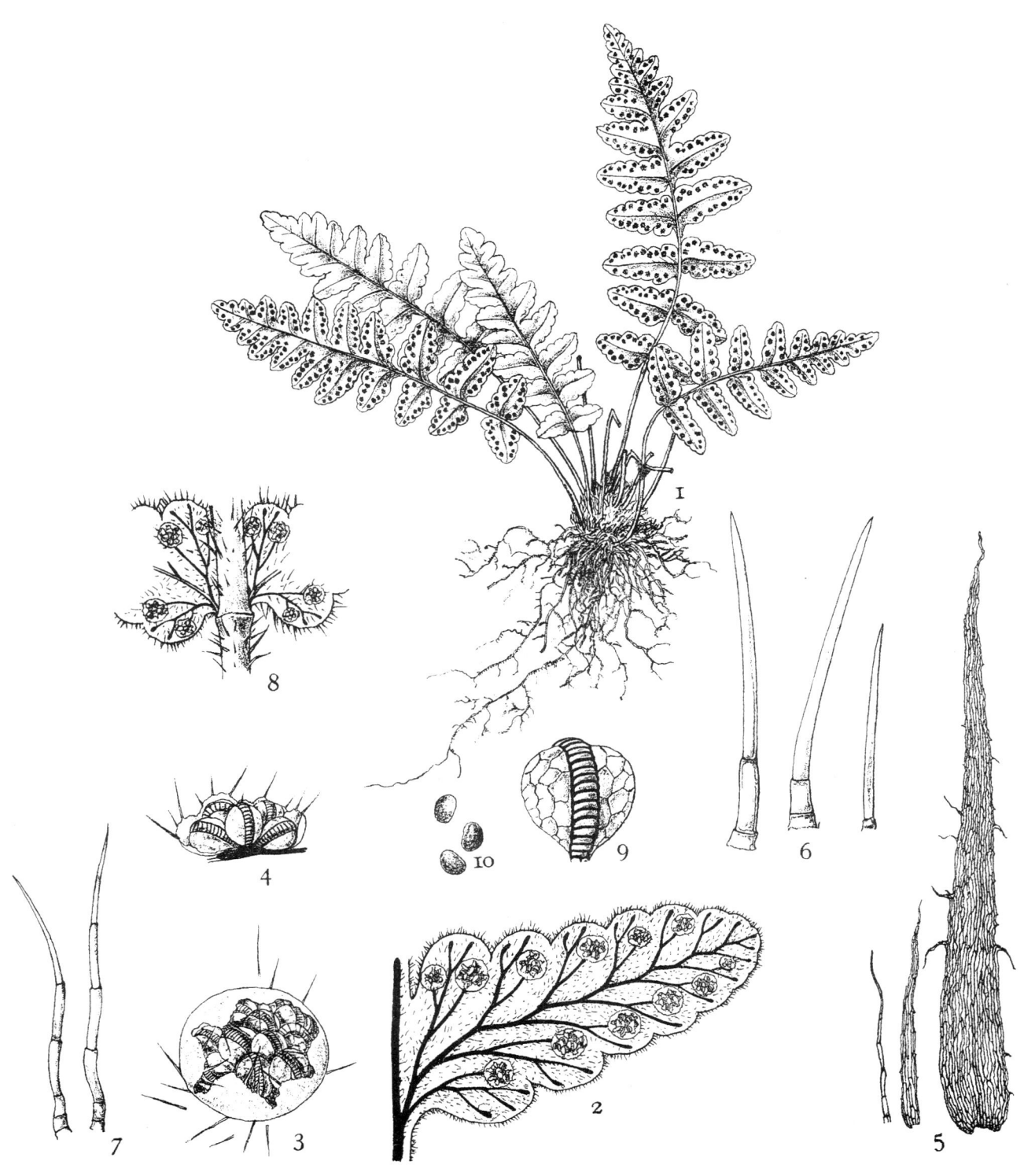

WOODSIA MACROCHLAENA Mettenius
大囊岩蕨

根状茎短而直立,端具棕黄色卵披针形之鳞片;叶簇生,柄长 2—5 cm,呈稻秆色,有光泽,近基部略具鳞片,叶体披针形,长 10—30 cm,宽 2—4 cm,或较宽,薄纸质,淡绿色,两面光滑,二回羽状深裂,小叶 15—40 对,斜出,无柄,下部数对较小,对生,上部小叶长 1—3 cm,互生,椭圆形,钝头,基部亚截形,羽状深裂,裂片 5—8 对,斜出,卵椭圆形,钝头,全绿,或呈波形;叶脉羽状分叉,2—3 对,顶稍膨大,不达叶缘;子囊群大,着生于上方基部叶脉,盖为圆球形,薄膜质,顶部开一小口。

分布:山东,河北,东三省;朝鲜半岛及日本亦产之。

本种之特点为其光绿色薄纸质之叶体及其球圆形膜质之子囊群盖。

图注:1. 本种全形(自然大);2. 小叶(放大 5 倍);3. 子囊群及其盖(放大 50 倍);4. 孢子(高倍放大);5. 根状茎上之鳞片(放大 30 倍)。

WOODSIA MANCHURIENSIS Hooker

WOODSIA MANCHURIENSIS Hooker, 2nd. Cent. Ferns t. 98 (1861); Syn. Fil. 48 (1867); Milde, Fil. Europ. et Atlant. 168 (1867); Christ, Farnkr. d. Erde 283 (1897); Diels in Engl. u. Prantl: Nat. Pflanzenfam. **1**: 4. 161 (1899); C. Chr. Ind. Fil. 657 (1905); Fomin, Fl. Sib. et Orient, Extr. **5**: 9 (1930); Ogata, Ic. Fil. Jap. **5**: t. 249 (1933).

Diacalpe manchuriensis Trev. Nuov. Giorn. Bot. Soc. Ital. **7**: 160 (1875).

Physematium manchuriensis Nakai, Bot. Mag. Tokio **39**: 176 (1925).

Woodsia insularis Hance, Ann. Sci. Nat. IV. **15**: 228 (1861); Diels in Engl. u. Prantl: Nat. Pflanzenfam. **1**: 4. 161 (1899).

Rhizome short, erect; *scales* light brown, thin, ovate-lanceolate, acuminate, clothing the apex; *fronds* caespitose, stipe short, terete, 2-5 cm long, glossy, strami-nous, sparcely scaly, lamina lanceolate, 10-30 cm long, 2-4 cm broad, or even broader, membranaceous, clear light green, glabrous on both sides, deeply bipinnatifid under pinnatifid acuminate apex, rachis glabrous, deeply sulcate above; *pinnae* 15-40-jugate, oblique, sessile. , the lower ones much abbreviated and opposite, the upper ones alternate, 1-3 cm long, oblong-obtuse, base truncate, deeply pinnatifid; *segments* 5-8-jugate, oblique, oblong-ovate, obtuse, entire or slightly sinuate; *veins* pinnate, 2-3-jugate, veinlets fine, oblique, fall short of the margin with enlarged apex; *sori* large, on the anterior basal veinlets of each segment, below the sinus, *indusium* large, globose, gray, thin membranaceous, opening with a toothed edge of the contracted mouth at the top.

Shandong: Chefoo, *Hancock 11*; *W. R. Carles*, Sept. 1889; *E. Faber 1060*, Oct. 1898; Tai Shan, *Jacob 39*, July 11, 1923; *K. S. Hao 1730*, July 10, 1931 (partly). Northeast China: *C. Wilford 1094* (type), July-August, 1859; *H. E. M. James*, May-August, 1886; *Komarov 2*; Chienshan, *J. Ross 594* (1877); Kirin, *F. H. Chen 201* (1931). Hebei: Tsang-ho, *M. S. Clemens 6038A*, August 9, 1913; Changli, *M. S. Clemens*, Oct. 13, 1913.

Also Korean peninsula and Japan.

One of the most distinct species of the subgenus *Physematium*, characterized by light green, membranaceous glabrous leaves and large, membranaceous sac-like indusium opening by the contracted mouth at the top. Its occurence in North China is said to be by no means common. *W. insularis* Hance, based upon Clarke's plant from Sagalien, only represents a small form of this species, evidently growing in a dry situation.

Plate 102. Fig. 1. Habit sketch (natural size). 2. Pinna, showing venation and position of sori ($\times$ 5). 3. Sorus with one side of the indusium removed ($\times$ 50). 4. Spores (greatly enlarged). 5. Scale from rhizome ($\times$ 30).

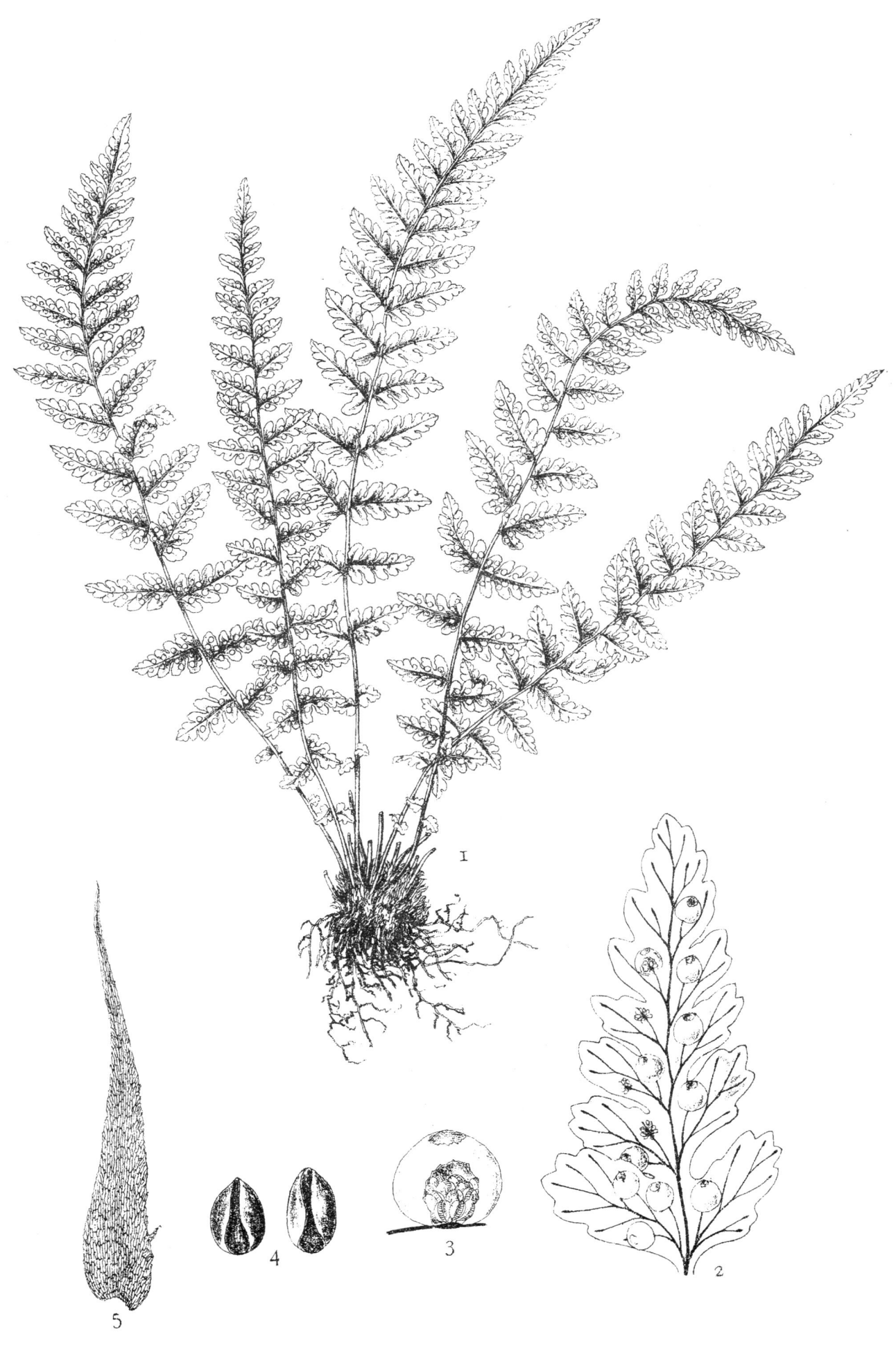

WOODSIA MANCHURIENSIS Hooker
满洲岩蕨

图版 103　海州骨碎补

　　根状茎肥大,肉质,蔓生,横行,具披针形之密鳞片,长达 1 cm 余,灰白色或白锈黄色;叶散生,叶柄长 6—10 cm,稻秆色,光滑,叶体呈五角形,长宽各 8—14 cm,尖头,三回分裂,一回小叶 6—7 对,具短柄,基部一对最大,长宽各 5—7 cm,二回小叶 6—7 对,具短柄,中肋具狭翅,基部一对较大,卵椭圆形,长 2.5—4 cm,宽 1—1.5 cm,自基部渐狭,深裂,裂片约五数,线形,钝头,或二裂,具一叶脉,革质,两面光滑;子囊群近于叶缘,盖为管状,革质,黄色,长二倍于宽,端部具小角状叶质。

　　分布:山东及江苏北部沿海及台湾;日本,朝鲜半岛亦产之。

　　此为亚洲东北部特有之种,与印度北部产之 *Davallia bullata* Wall. 相似,然叶体形小而分裂较甚,茎上鳞片亦较长,且呈瓦覆状,故得区别。

　　图注:1. 本种全形(自然大);2. 末回小叶,表示其叶脉及子囊群盖形状(放大 10 倍);3. 根状茎上之鳞片(放大 10 倍)。

DAVALLIA MARIESII Moore

DAVALLIA MARIESII Moore ex Baker, Ann. Bot. **5**: 201 (1891); C. Chr. Ind. Fil. 212 (1905): Nakai, Bot, Mag. Tokio **39**: 120 (1925).

　　Davallia bullata Franch. et Sav. Enum. Pl. Jap. **2**: 208 (1877); Christ in Warburg, Monsunia **1**: 86 (1900).

　　Rhizome thick, **7** mm across, long-creeping, densely scaly; *scales* large, imbricate, lanceolate from broadly ovate base, fimbriate, ferruginous on the growing tip, but whitish on the old part; *frond* far apart, articulated at base, stipe 6-10 cm long, straminous, naked, terete underneath, grooved on the upper side, lamina deltoid-pentagonous, 8-14 cm long and broad, shortly acuminate, tripinnate; *pinnae* **6**-7 -jugate, rather shortly petiolate, the basal ones much the largest, deltoid, **5-7** cm long and broad; *pinnules* **6**-7-jugate, shortly petiolulate on narrowly winged rahilet, the lower basal ones the largest, oblong-ovate, **2.5-4** cm long, **1-1.5** cm broad, ovate-oblong, base cuneate-decurrent, deeply pinnatifid into **5** linear, blunt or bifid uni-nerved segments; *texture* coriaceous, glabrous on both sides; *indusium* brown, tubular, coriaceous, twice as long as broad, overtopped by two horns of generally unequal length.

　　Shandong: Chefoo mountains, *Faber 58*; *Carles* (1888); *R. Zimmermann 545*, May 5, 1902; *Maingay 5* (1862); *Hancock 5*; Tsingtau, Lao Shan, *F. H. Sha 528* in herb. Shandong Univ., July 18, 1932; Yuchiaho, Mooping, *T. N. Liou 905*, May 27, 1930. Jiangsu: Haichow, Haishan, *Jacob 7*, *86*, *890*. Taiwan: *Hancock 115*, *119*.

　　Japan: Ex Hort. Veitch (type); Nagasaki, *Maximowicz 116* (1863).

　　Korean peninsula: *Taquet 2330*; Söul, *Warburg*.

　　This fern is a native of the northeastern Asia and a close relative of *D. bullata* Wall. of Eastern India, which differs, however, in larger and less divided frond, shorter, rufo-brown and very shaggy scales with scarious margin and in more shortly petiolate basal pinnae. The difference in the scale between this and the Indian species was first noted by Moore himself, as shown by an enlarged sketch on the type sheet in the herbarium at Kew.

　　Plate 103. Fig. 1. Habit sketch (natural size). 2. Ultimate pinnule, showing venation and shape of indusium ($\times$ 10). 3. Scale from rhizome ($\times$ 10).

DAVALLIA MARIESII Moore
海州骨碎补

图版 104 华南骨碎补

根状茎大如小指,肉质,蔓生,横行,具光亮锈黄色之披针形密鳞片,长逾 1 cm;叶散生,柄长 30—60 cm,自基部以上光滑,黄褐色,叶体大,长宽约各 60—90 cm,呈三角形,渐尖头,四回羽状分裂,一回小叶约 10 对,互生,具长柄,基部一对最大,长三角形,长 20—30 cm,宽 12—18 cm,渐尖头,上部各对渐小,中部一对为宽披针形,长 15 cm,宽 6 cm,具二回小叶约 10 对,三回小叶长约 1.5 cm,无柄,羽状,深裂,末回裂片细长,有锯齿;叶系亚革质,光滑,叶脉细密,分离,每一数至一锯齿;子囊群近于叶缘,每锯齿一个,盖管状,长二倍于宽,端呈截形,其上具小角形之叶质。

分布:广东,广西,香港,海南及云南南部;越南亦产之。

本种为亚洲南部大陆特有之种,形体极类马来群岛所产之 Davallia divaricata Bl. 所不同者,为其较薄之叶质,子囊群盖较狭长,子囊群着生之叶脉亦较长而几为向上。

图注:1.本种全形(自然大);2.二回小叶(放大 8 倍);3.子囊群及一部分之盖(放大 25 倍);4.根状茎上之鳞片(放大 12 倍)。

DAVALLIA ORIENTALIS C. Christensen

DAVALLIA ORIENTALIS C. Christensen ex Wu, Bull. Dept. Biol. Sun Yatsen Univ. No. **3**: 104 t. **43** (**1932**); No. **6**: 4 (**1933**); Ind. Fil. Suppl. Ⅲ. 68 (**1934**).

Davallia elegans Hk. (non Sw, 1801) Florul. Hongk. in Journ. Bot. (1857) 333; Syn. Fil. 95 (1867); Benth. Fl. Hongk. 461 (1861).

Davallia divaricata Christ (non Bl. 1828), Bull. Soc. Fr. et Belg. **28**: 260 (1898); Dunn & Tutch. Fl. Kwangt. & Hongk. 337 (1912); Merr. Enum. Hainan Pl. in Lingnan Sci. Journ. 5: 11 (1927).

Davallia denticulata Merr. l. c. (non Mett. 1867).

Rhizome thick as a small finger, fleshy, wide-creeping, densely clothed in bright brown lanceolate *scales* to 1 cm long or longer; *frond* far apart, stipe glabrous, brown, terete below, deeply bisulcate above, 30-60 cm long, 4 mm thick at the base, lamina ample, 60-90 cm each way, deltoid, 4-pinnate or 5-pinnatifid; *pinnae* about 10-jugate, alternate, long-petiolate, the basal ones much the largest, deltoid, 20-30 cm long, 12-18 cm broad, acuminate, the upper ones gradually smaller, the middle ones broadly lanceolate, 15 cm long, 6 cm broad, with about 10 pairs of *pinnules*, which are petiolulate, the anterior basal one much the largest, deltoid, acuminate, the others smaller with unequal base; *pinnules* of 2nd. order sessile, 1-1.5 cm long, pinnatifid into elongate denticulate *segments*; *texture* rigi-dulously coriaceous, glossy; *veinlets* in the segments oblique, one to each tooth; sori submarginal, one to each tooth, *indusium* tubular, twice as long as broad, apex truncate, with projecting teeth above.

Hainan: Yih Tsoh Mao, *F. A. McClure 9717* (type); Nodoa, Lin Fa Ling, *W. Y. Chun 1799*; *McClure 7919*; Five Finger Mt., *McClure 8463, 8678*; Hung Mo Shan, *W. T. Tsang 17771*. Guangdong: Man Sei Ho, *Tutcher 10505*; Lohfau Shan, *C. Ford* (1883); *N. K. Chun 41276*; Swatow, *Dalziel*, Sept. 1899. Hong Kong: *Champion 552*; *Matthew 72* (1905); May 18, 1904; April 12, 1907; *C. Wright* (1853-1856); Lantao Island, *C. L. Tso* (1929); *Borther* (1856); *Alexander*. Guangxi: Yao Shan, *S. S. Sin 3698*, Sept. 26. 1928, Linyen Hsien, outside of the south city gate, *R. C. Ching 6665, 6761*; Seh Fang Dar Shan, on the border of Vietnam, *R. C. Ching 1860*. Yunnan: Szemeo, *Henry 13141*.

Vietnam: Mt. Bana, *Clemens 3855*; environs de Chobo, *Pételot*, April 14, 1926; Chapa. *Pételot 3306*; Cao Bang, *Pételot 2740*; *Billet 4182*; *Bon 3257*.

The present species has hitherto been generally considered as identical with the Malaysia-Polynesian *D. divaricata* Bl. from which it differs in thinner leaves, greenish color, even when dried, much longer and more ascending soriferous veinlets and in longer, narrower indusium with exerted sterile margin as high as the sorus itself.

Swartz cited Canton as the type locality for his *D. elegans* (Syn. Fil. pp. 132, 347), but this is certainly wrong, for his species, actually based upon Thunberg's plant from Java, agrees exactly with *D. denticulata* of Malaysia-Polynesia, not known from China. The same mistake was since repeated by Hooker, Bentham and other authors on the flora of South China.

Plate 104. Fig. 1. Habit sketch (natural size). 2. pinnule of 2nd. order ($\times$ 8). 3. Sorus, with part of indusium removed ($\times$ 25). 4. Scale from rhizome ($\times$ 12).

DAVALLIA ORIENTALIS C. Christensen
华南骨碎补

图版 105　高山阴石蕨

　　根状茎肥厚,肉质,具银灰色披针形之大鳞片;叶散生,柄细长,长 5—10 cm,光滑,叶体披针形,长 15—28 cm,宽 5—7 cm,渐尖头,基部与上部几等宽,三回羽状深裂,一回小叶 20—30 对,互生,镰状披针形,具短柄,基部不等形,长达 4 cm,基部宽 2.5 mm,渐尖头,中肋上部具狭翅,二回小叶 7—10 对,无柄,卵椭圆形,长 1.2 cm,宽 5—8 mm,其位于下部者羽状深裂,裂片 5—7 数,具锯齿 2—3 数;叶脉不甚明显,羽状分叉;革质,有光泽,中肋下部下面具卵圆形之褐色大鳞片,余均光滑;子囊群近于叶缘,盖为半圆形,光亮,仅基部着生。

　　分布:云南及印度北部。

　　本种原产于印度北部,最近在云南西部发现,其异于 *Humata Griffithiana* Hk. 者,为其长披针形之叶体及其一回小叶中肋下部下面所具之卵圆形褐色大鳞片是也。

　　图注:1. 本种全形(自然大);2. 二回小叶(放大 8 倍);3. 根状茎上之鳞片(放大 14 倍);4. 一回小叶中肋下部下面所具之鳞片(放大 14 倍)。

HUMATA ASSAMICA (Bedd.) C. Christensen

HUMATA ASSAMICA (Bedd.) C. Christensen, Contr. U. S. Nat. Herb. **26**: 293 (1931); Ind. Fil. Suppl. III. 112 (1934).

　　Acrophorus assamicus Bedd. Ferns Brit. Ind. t. 94 (1866).

　　Leucostegia assamica J. Sm. Hist. Fil. 84 (1875); Bedd. Handb. Ferns, Brit. Ind. etc. 51 (1883); Suppl. 13 (1892).

　　Davallia assamica Baker in Hk. Syn. Fil. ed. 1. 452 (1868); ed. 2. 467 (1873); Clarke, Trans. Linn. Soc. II. Bot. **1**: 445 (1880).

　　Davallia micans Mett. ex Baker in Hk. Syn. Fil. ed. 1. 95 (1867).

　　Humata micans Diels in Engl. u. Prantl; Nat. Pflanzenfam. **1**: 4. 209 (1899).

　　Rhizome thick, wide-creeping, densely scaly; *scales* large, lanceolate, acuminate, hair-pointed, spread, silvery-brown, denticulate; *frond* far apart, stipe slender, firm, naked, rufo-stramineous, 5-10 cm long, lamina lanceolate, 15-28 cm long 5-7 cm broad, acuminate, base hardly narrowed, tripinnatifid; *pinnae* 20-30-jugate. alternate, lanceolate-falcate, shortly petiolate, unequal at base, the basal ones deltoid-lanceolate, with subequal and cordate base, to 4 cm long, 2.5 cm broad at the base, acuminate, rachis narrowly winged from the middle upward; *pinnules* 7-10-jugate, sessile, ovate-oblong, 1.2 cm long, 5-8 mm broad, the lower ones deeply pinnatifid with 5-7 *segments*, of which the lowest are pinnatifid, the upper ones 3-2-dentate, rachilets with a few large, brown, broadly ovate appressed scales underneath, surfaces naked, glossy in living state; *texture* coriaceous; *veinlets* hardly distinct, pinnate or bifurcate in segment; *sori* submarginal in ultimate lobes, mostly with a horn above, *indusium* suborbicular, broader than deep, glossy, rounded and free all around, except the broad base which is attached to the leaf tissue.

　　Yunnan: Shweli-Salween divide, *G. Forrest 24500*; South of Tengyueh, *G. Forrest. 26681*, June, **1925**; between Tengyueh and the Burmese border, *J. F. Rock 7312*; between Kambaiti and Tengyueh, *J. F. Rock 7580*; Salwin, *H. Handel-Mazzetti 9564*.

　　Bothan: Mishee, *Griffith*; *Jordon* (type).

　　Munipore: *Clarke*; and also Upper Burma: *Forrest 26604*.

　　This distinct species which was previously known only from Bothan and Munipore of Northwestern India, is very near *H. Griffithiana* of the same region, differing mainly in its lanceolate and shortly petiolate fronds and in the presence of a few large broadly ovate scales on the lower part of rachilets underneath. Our plate based upon Forrest No. 26681.

　　The genus *Humata* is closely related to both *Davallia* and *Leucostegia*, from the former it differs chiefly in thick leathery, shining, pale-colored reniform or suborbicular indusium attached only by its broad base, generally less divided frond of more rigid texture and the pale-colored scales on the rhizome, from the latter, in somewhat dimorphous and less divided frond of more thick texture and glossy leathery indusium. In fact, its affinity to *Leucostegia* seems to be too close to warrant a generic separation, particularly when the present species is taken into consideration.

　　Plate 105. Fig. 1. Habit sketch (natural size). 2. Pinnule from the middle pinna ($\times$ 8). 3. Scale from rhizome ($\times$ 14). 4. Scale from the lower part of rachilet underneath ($\times$ 14).

HUMATA ASSAMICA (Bedd.) C. Christensen 高山阴石蕨

根状茎蔓生,缘树干上升,高可达十数尺;鳞片疏生;叶散生,叶柄长 1—2 cm,节形着生于蔓茎上,叶体披针形,长 20—40 cm,宽 4—6 cm,一回奇数羽状分裂,小叶数十对,长 3—4 cm,宽约 1 cm,顶部一叶与下部数对较小,椭圆披针形,上基部耳形突出,下基部呈斜形,钝头或圆头,叶缘呈波状,几无柄,节状着生于具密腺毛之中肋,两面光滑,唯主脉上面略具腺毛;叶脉明显,分叉,不达于叶缘,纸质;子囊群圆形,着生于每一上基部叶脉之端,盖肾形,红棕色。

分布:广布于越南,马来群岛及南洋群岛,菲律宾,澳大利亚及非洲热带;最近发现于我国之海南及广西西南部之八角山,然不多见也。

本种为本属在亚洲热带大陆仅见之种,其特异点为其藤形之地上茎缘树干或悬崖上升,叶与小叶均节状着生,易于脱落是也。

图注:1.本种全形(自然大);2.小叶,表明其叶脉及子囊群之位置(放大 2 倍);3.子囊群(放大 16 倍);4.蔓生茎上之鳞片(放大 40 倍)。

ARTHROPTERIS OBLITERATA (R. Br.) J. Smith

ARTHROPTERIS OBLITERATA (R. Br.) J. Smith, Cat. Cult. Ferns **62** (1827); C. Chr. Ind. Fil. 63 (1905); v. A. v. R. Handb. Mal. Ferns 155 (1908); Bonap. Notes Pterid. Pt. **14**: 103 (1923); Merr. Enum. Hainan Pl. in Lingnan Sci. Journ. **5**: 11 (1927).

Nephrodium obliteratum R. Br. Prod. Fl. Novae-Holl. 148 (1810).

Aspidium obliteratum Spr. Syst. Veg. **4**: 99 (1827).

Nephrolepis obliterata J. Sm. Journ. Bot. **4**: 197 (1841); Hk. Sp. Fil. **4**: 154 (1862); Bedd. Ferns S. Ind. t. 251 (1864).

Nephrolepis ramosa Moore, Ind. Fil. 105 (1858); Hk. et Bak. Syn. Fil. 301 (1867); Bedd. Handb. Ferns, Brit. Ind. etc. 285 (1883).

Arthropteris ramosa Mett. Novora Exp. Bot. **1**: 214 (1876); Diels in Engl. u. Pratl: Nat. Pflanzenfam. **1**: 4. 208 (1899).

Rhizome slender, wide-creeping on the trunk of trees, to several m. tall, sparcely scaly; *frond* far apart, alternate, stipe 1-2 cm long, articulated to a prominent pseudopodium, lamina lanceolate, 20-40 cm long, 4-6 cm broad, simple pinnate under the distinct smaller end-pinna; *pinnae* numerous, alternate, 3-4 cm long, 1 cm broad, the lower ones smaller, deflexed, lanceolate-oblong, with deltoid auricle at the anterior base, obliquely cut away at the posterior base, apex bluntish or roundish, margin undulate-crenate towards apex, subsessile and articulated to the rachis, which is densely pustulately hirsute above with thick rod-like unicellular clear hairs, the same type of hairs also sparcely present on the midrib above; *texture* herbaceous; *veins* distinct, forked, ended some distance below the margin; sori round, terminating the anterior basal veinlet of each group, *indusium* reniform, reddish-brown, persistent.

Hainan: Nodoa, Sha Po Shan, *F. A. McClure 8157* (1921). Guangxi: Bako Shan, south of Peiseh, *R. C. Ching 7530* (1928).

Vietnam: Cao Bang, *Bourret 132*.

Also Malaysia-Polynesia, Sri Lanka, the Philippines, Australia and tropical Africa.

A very variable fern in respect to the shape and size of pinnae; very rare on the mainland, R. C. Ching No. 7530 and Bourret No. 132 are the only specimens seen by me.

The genus *Arthropteris* now comprising about 10 species chiefly in the tropical Asia, has been generally considered as closely related to *Nephrolepis* Schott; and this is true only in so far as the articulated pinnae, type of venation and soral conditions are concerned, but in point of anatomical features, and particularly, the type of stelar structure in rhizome, it shows a marked difference from that genus, nor is it comparable with *Oleandra* Cav. in this respect, to which it is also held as a close relative by workers to-day. Meanwhile, I consider it one of the fern *genera incertae sidis*.

Plate 106. Fig. 1. Habit sketch (natural size). 2. Pinna, showing venation and sori ($\times$ 2). 3. Sorus with indusium ($\times$ 16). 4. Scale from scandent rhizome ($\times$ 40).

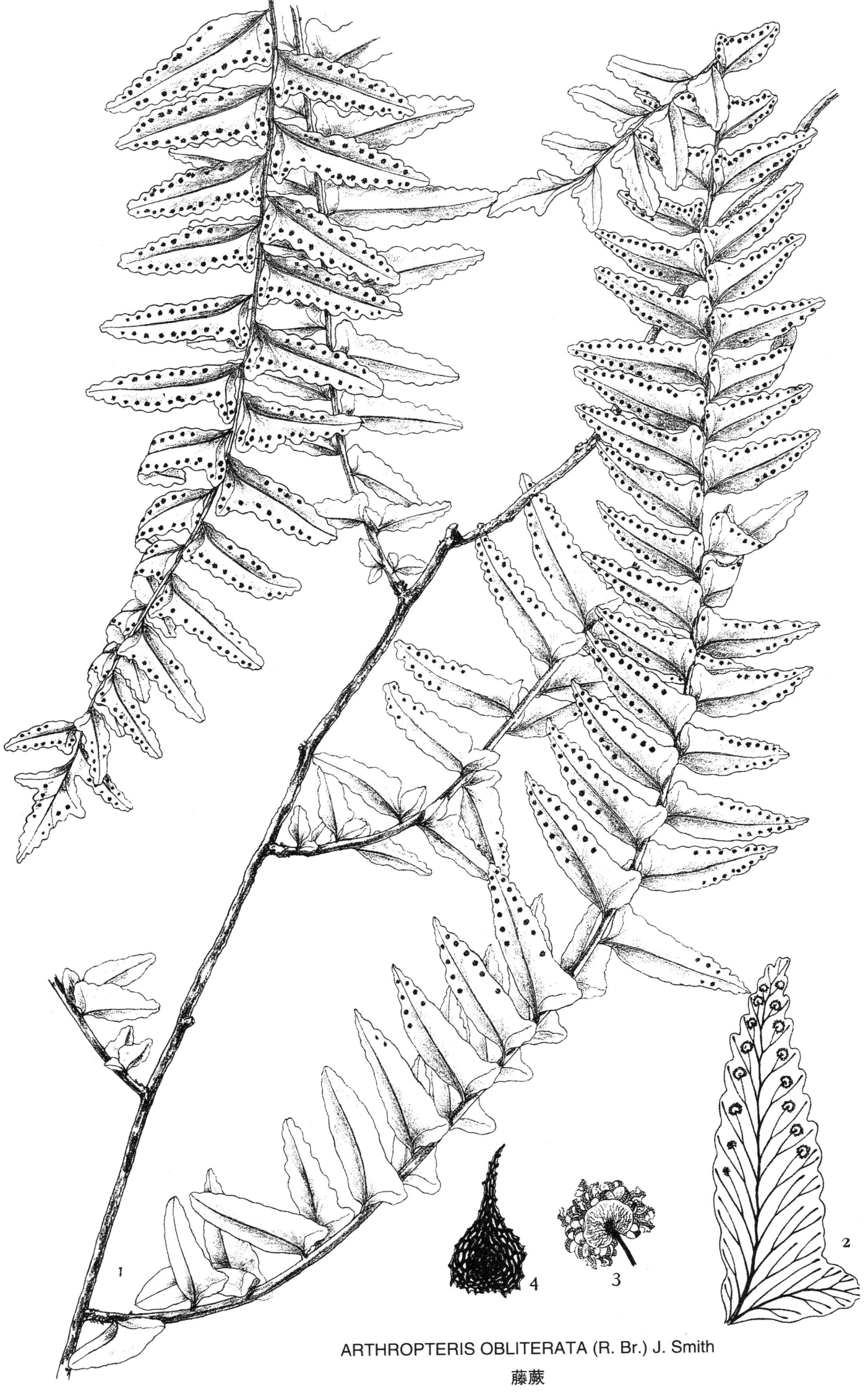

ARTHROPTERIS OBLITERATA (R. Br.) J. Smith
藤蕨

图版 107　过山蕨

根状茎短而直立,端具黝棕色之细长披针形鳞片,富于拆光性;叶簇生,二形,非子囊群叶体为卵形,钝头,长 1—4 cm,柄长 1—5 cm,纸质,绿色,全缘;叶脉网状,网眼二列,斜出,不具小脉;子囊群叶体为披针形,具较长之柄,叶端通常呈线状伸长而入土着根,基部不为心脏耳形,但为短楔形;叶脉亦为网状,两面光滑;子囊群形长,于中肋两侧成极不规则之 1—2 列,盖膜质,有向上方开者,有向下方开者,更有彼此相向开者。

分布:河北,山东,东三省,内蒙古;日本,朝鲜半岛,西伯利亚东部均产之。

本种为亚洲东北部之特产,其异于美洲东北部特产之 *Campt. rhizophyllus* (Linn.) 者,为其叶形较小,基部不为心脏耳形是也;本种特其细长着根之叶端,能由一地繁衍至他地,故特名之曰过山蕨。

图注:1. 本种全形(自然大);2. 着生子囊群叶体之一部,表明其叶脉及子囊群之位置(放大 6 倍);3. 根状茎先端之鳞片(放大 30 倍);4. 孢子(放大 100 倍)。

CAMPTOSORUS SIBIRICUS Ruprecht

CAMPTOSORUS SIBIRICUS Ruprecht, Distr. Crypt. Vasc. Ross. 45 (1845); Ledeb. Fl. Ross. 522 (1853); Milde, Fil. Europ. et Atlant. 95 (1867); Franch. Pl. David. **2**: 230 (1887); Komarov, Fl. Mansh, **1**: 177 (1901); C. Chr. Ind. Fil. 166 (1905); Kümmerle, Ann. Mus. Hungar. **24**: 90 (1926); Ogata, Ic. Fil. Jap. **2**: t. 59 (1929); Fomin, Fl. Sib. et Orient. Extr. **5**: 139 (1930).

Scolopendrium sibiricum Hk. 2nd. Cent. Ferns t. 35 (1861); Sp. Fil. **4**: 4 (1862); Syn. Fil. 248 (1867); Christ, Farnkr. d. Erde 218 (1897); in Warburg, Monsunia **1**: 73 (1900); Diels in Engl. u. Prantl: Nat. Pflanzenfam. **1**: 4. 231 (1899).

Antigramma sibirica J. Sm. Hist. Fil. 331 (1875).

Phyllitis sibirica O. Ktze, Rev. Gen. Pl. **2**: 818 (1891).

Asplenium rhizophyllum L. Sp. Pl. **2**: 1078 (1753), pro parte.

Camptosorus rhizophyllus var. *sibiricus* Christ ex Léveille, Bull. Acad. Géogr. Bot. (1910) 4.

Rhizome short, erect, densely radicose; *scales* at the apex dense, lanceolate, hair-pointed, fusco-brown, iridescent; *fronds* fasciculated, dimorphous, *sterile ones* ovate, rounded or acute, 1-4 cm long on stipes 1-5 cm long, herbaceous, green; *fertile ones* lanceolate, longer-stipitate, apex generally elongate and rooting, base not auricled but narrowed gradually or shortly cuneate; *venation* sub-biseriately reticulated along the prominent midrib, free towards margin, green and glabrous, except the underside being sparcely glandular; *sori* elongate, 1- or irregularly 2-seriate along midrib, *indusium* membranaceous, gray, some opening towards, and some against, the midrib, and still others, towards each other as in *Phyllitis*; *spores* bilateral, echinose.

Hebei (formerly Chihli): Peiping, *Bushell*, Oct. 1882; *Forbes* (1882); *Carles*, July 16, 1882; 1-yuan-kou, *Clemens 37036* (1913); Pinchow, Wang-mu Kou, *Chanet 451*; Peitaiho, *Cowdry 447*; Siawutai Shan, *C. L. Li*; ibid., *C. W. Wang 60589* (1933). Johoh: *David*. Shandong: Chefoo, *Faber* (1890); *Swinhoe* (1873); *Hancock 17* (1875); *Warburg*; Weihaiwei, *C. G. Matthew 410* (1904); Taishan, *C. Y. Chiao 21286* (1929); *Jacob 7*; *H. S. Hao 1810*, July 13, 1931; Tsingtau, *C. L. Tso* (1933). Inner Mongolia: Peitchely, E. *Licent* (1927). Northeast China: *Maximowicz 410* (1860); *James*, May-August, 1886; *Ross*, Oct. 1887; *Webster*, May, 1885; *Komarov 36*; Kirin, in the vicinity of Chingpohu, *H. W. Kung 2104*, August 14, 1931; Fengtien, *H. W. Kung 876*, August 1, 1930.

Korean peninsula: *Taquet 2479*.

Port Arthur: *Miss Möller* (1926).

Also Japan and Siberia.

The chief interest of this peculiar-looking little fern lies in its remarkable walking ability, that is, its frond generally prolongates into an elongate whiplike viviparous tip, by means of which it spreads itself all about. The second species of the genus is C. *rhizophyllus* (L.) Link, of North America, which differs from our fern in larger size, deeply auriculato-cordate base of frond and 2-3 rows of sori on each side of midrib.

Plate 107. Fig. 1. Habit sketch (natural size). 2. Portion of fertile frond, showing venation and position of sori ($\times$ 6). 3. Scale from rhizome ($\times$ 30). 4. Spores ($\times$ 100).

CAMPTOSORUS SIBIRICUS Ruprecht
过山蕨

图版 108 庐山蹄盖蕨

　　根状茎横行,黑色,先端及叶柄基部略具褐色之薄质披针形鳞片;叶散生,柄长 10—20 cm,稻秆色,光滑,叶体椭圆形或卵三角形,长 15—20 cm,宽 10—15 cm,一回羽状分裂,顶部渐尖头,深裂,纸质,绿色,光滑,唯中肋及主脉下面稍具短刚毛;小叶 3—8 对,镰形,下部数对无柄,上部数对合生,渐尖头,基部呈截形或短楔形,长 6—10 cm,宽 1.5—2 cm,叶缘具不规则之锐锯齿;叶脉明显,羽状分叉,伸入锯齿;子囊群小,椭圆形,着生于部一对叶脉上,盖肾形,具短刚毛,不久即脱落。

　　分布:江西,浙江,贵州,四川;日本及朝鲜半岛亦产之;最初发现于庐山。

　　本种通常散生于森林中,罕见有群生者,故采集时往往不能于一处得多数标本也。

　　图注:1.本种全形(自然大);2.基部小叶之一部,表示其叶脉及子囊群之位置(放大 5 倍);3.子囊群盖(高倍放大);4.根状茎上之鳞片(放大 20 倍)。

ATHYRIUM SHEARERI (Baker) Ching

ATHYRIUM SHEARERI (Baker) Ching in C. Chr. Ind. Fil. Suppl. Ⅲ. 44 (1934).
　　Nephrodium sheareri Baker, Journ. Bot. (1875) 200.
　　Dryopteris sheareri C. Chr. Ind. Fil. 292 (1905).
　　Nephrodium isolatum Baker, Gard. Chron. n. s. **14**: 494 (1880).
　　Aspidium polypodiforme Makino, Bot. Mag. Tokio **6**: 46 (1892, nom. nud.)
　　Nephrodium polypodiforme Makino, Bot. Mag. Tokio **13**: 58 (1899); Matsum. Ind. Pl. Jap. **1**: 323 (1904).
　　Dryopteris polypodiformis C. Chr. Ind. Fil. 285 (1905); Makino et Nemoto, Fl. Jap. 1621 (1925).
　　Athyrium polypodiforme Tagawa, Acta Phytotax. et Geobot. **1**: 158 (1932).
　　Aspidium otarioides Christ, Bull. Acad. Géogr. Bot. (1902) 247.
　　Dryopteris otarioides C. Chr. Ind. Fil. 282 (1905).
　　Dryopteris subsagenioides Christ, Bull. Acad. Géogr. Bot. (1910) 8; C. Chr. Ind. Fil. Suppl. Ⅰ. 40 (1912); Nakai, Fl. Kor. **2**: 396.

　　Rhizome wiry, wide-creeping, black, apex and base of stipe sparcely clothed in brown, thin, lanceolate *scales*; *frond* far apart, stipe 10-20 cm long, straminous, naked, lamina oblong or deltoid-ovate, 15-20 cm long, 10-15 cm broad, pinnate under pinnatifid, deltoid-lanceolate, acuminate apex; *texture* thin, herbaceous, green, glabrous except rachis and costa which are sparcely setaceous underneath; *pinnae* 4-8-jugate, subfalcate, acuminate, sharply serrate, the lower ones sessile, the upper ones adnate, truncate or cuneate at base, 6-10 cm long, 1.5-2 cm broad, lobato-in-cised; *lobes* oblong or rounded with aristate teeth; *veins* distinct, pinnate in lobes, 3-4-jugate, veinlets simple, curving up into the teeth; *sori* small, rounded, dorsal on the lower 2 pairs of veinlets, *indusium* reniform, setaceous, very fugaceous.

　　Jiangxi: Kiukiang, Lushan, *Dr. Shearer* (type) (1873); *Maires* (type of *Nephrodium isolatum*); *N. K. Up 1832*; Paradise Pools, *Charles E. DeVol 25*, July 27, 1933, edge of stream. Zhejiang: Hangchow, Fanchiao, *R. C. Ching 3715*. Hubei: Shin Shan Hsien, *S. S. Chien 8111* in herb. Univ. Nanking, July 29, 1922. Guizhou: Tenyen, *Y. Tsiang 5802*; Tinfan, *Cavalerie 2050*; Pinfa, *Cavalerie 46*, *1839*; Tsingay, *Bodinier 2050* (type of *Aspidium otarioides*). Sichuan: Nanchuan, *W. P. Fang 5794*.

　　Korean peninsula: Quelpaert, *Taquet 2382* (type of *Dryopteris subsagenioides*).

　　Also Japan, common.

　　This peculiarly distinct fern, known under many a name as cited above, has been subjected to much nomenclatural changes. Its position in *Athyrium* can, however, hardly be doubted, if the generic status of *Anisocampium* Presl is not recognized. Our species is closely related to *Athyrium Cumingianum* (Presl) Ching (*Dryopteris otaria* O. Ktze.), differs chiefly in the shortly setose rachis, costa and costules underneath, aristate serrature and free venation.

　　Plate 108. Fig. 1. Habit sketch (natural size). 2. Portion of basal pinna, showing venation and position of sori ($\times$ 5). 3. Indusium (greatly enlarged). 4. Scale from rhizome ($\times$ 20).

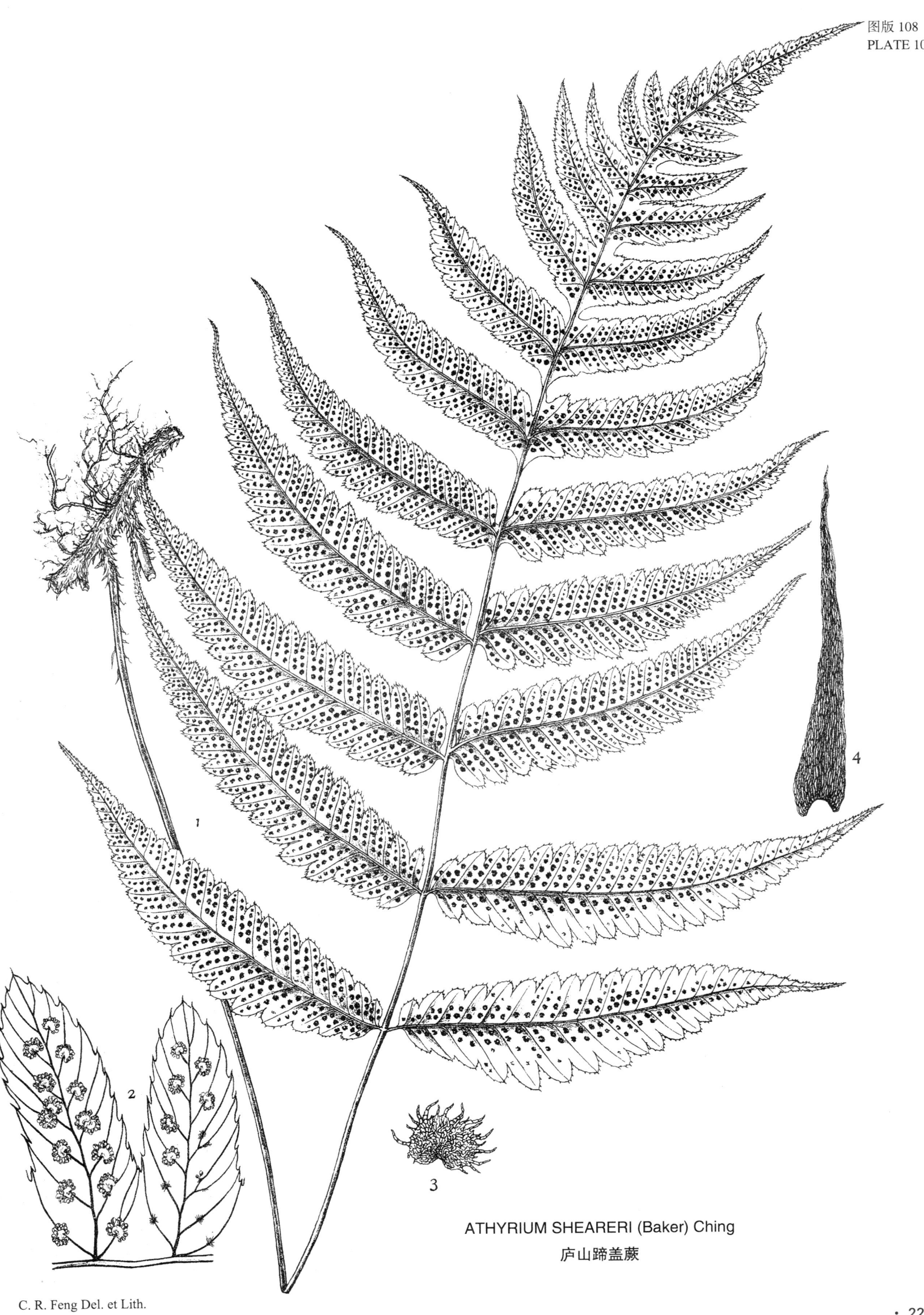

ATHYRIUM SHEARERI (Baker) Ching 庐山蹄盖蕨

图版 109　光蹄盖蕨

　　根状茎斜生，具黝暗色或深棕色之披针形鳞片；叶簇生，柄长 25—30 cm，宽 12—25 cm，呈淡紫色，基部具鳞片，叶体长 30—35 cm，二回羽状分裂，顶部尾状渐尖头，深裂，一回小叶约 10 对，具极短之柄，渐尖头，基部截形，披针形，长约 10 cm，宽约 2.5—4 cm，中肋基部下面呈淡紫色且略具腺毛，上面具肉刺；二回小叶为三角形，长 1.3—2.5 cm，短尖头，基部上方具耳形，下方斜形，叶缘具短锯齿；叶脉明显，三出或在上基部羽形分叉；厚纸质，两面光滑；子囊群长形，斜出，盖膜质，棕灰色。

　　分布：贵州，四川，湖北，安徽，广东；日本亦产之。

　　本种为一美丽之品，昔人误认为 Athyrium nigripes (Bl.) Moore，实则此种不产于中国及日本，乃系马来群岛种也。

　　图注：1. 本种全形（自然大）；2. 二回小叶（放大 7 倍）；3. 叶柄基部鳞片（放大 20 倍）。

ATHYRIUM OTOPHORUM (Miq.) Koidzumi

ATHYRIUM OTOPHORUM (Miq.) Koidzumi, Fl. Symb. Orient. Asiat. 40 (1930); C. Chr. Ind. Fil. Suppl. III. 43 (1934).

Asplenium otophorum Miq. Ann. Lugd. Bat. **3**: 175 (1867); Franch. et Sav. Enum. Pl. Jap 2: 229 (1877).

Diplazium otophorum C. Chr. Ind. Fil. 236 (1905).

Athyrium violascens Diels in Engl. Jahrb. **29**: 196 (1900); C. Chr. Acta Hort. Gothob. **1**: 76 (1924).

Athyrium wardii Christ (non Makino, 1857), Bull. Soc. Bot. France **52**; Mém. 1. 49 (1905).

Athyrium nigripes Christ (non Moore, 1857), Bull. Acad. Géogr. Bot. (1909) 174; C. Chr. Acta Hort. Gothob. **1**: 76 (1924).

　　Rhizome oblique, densely scaly; *scales* blackish or dark brown, linear-lanceolate, extending upwards some distance above the base of stipe; *fronds* caespitose, stipe 25-30 cm long, 3 mm thick above the base, stramineo-rufescent, naked, lamina 30-35 cm long, 20-25 cm broad, bipinnate; *lateral pinnae* 10-jugate below the cau-dato-acuminate pinnatifid apical part, very shortly petiolate with truncate base, elongate-lanceolate, long-acuminate, the basal ones only slightly shorter than those next above and decidedly narrowed towards the base, i. e. the basal pair of pinnules reduced, 10-12 cm long, 2.5-4 cm broad at the middle, rachilet straminous, glossy, shaded purplish, particularly towards the base, which is subglabrous with short, sparce glandular hairs and provided with appressed spines on the upper side; *pinnules* 14-17-jugate under the long-acuminate pinnatifid apex, 1.3-2.5 cm long, deltoid, acute or shortly acuminate, base strongly unequal, i. e. deltoidly auriculated above, obliquely cuneate below, with straight anterior side, sessile, rather regularly serrate with short, cuspidate teeth; *veins* distinct, forked, or pinnate in the auricle; *texture* rigidulously herbaceous, glabrous, except the rachis which is shortly glandular hairy above; *sori* elongate, oblique to costule, *indusium* membranaceous, brownish-gray.

　　Guizhou: Pinfa, *Cavalerie 2600*, *544*; Kweiyang, *Bodinier 2008*; Ganchow, *Cavalerie 3769*; Sanhoa, *Y. Tsiang 6414*, *6448*; Chengfeng, *Y. Tsiang 4568*; Tuyun, *Y. Tsiang 5810*. Hubei: *Henry 5163*. Sichuan: Mt. Omei, *W. P. Fang 2762*, *3119*; Nanchuan, *Rosthorn 1752* (type of A. *violascens*); *W. P. Fang 5709*; Kuan Hsien, *W. P. Fang 2161*; *Harry Smith 4888*; Tchenkoutin, *Farges 245*. Anhui: Hwang Shan, *S. S. Chien 1235*. Guangdong: North River, *C. L. Tso 20681*.

　　Japan: *C. G. Matthew 241* (1903), *forma typica*.

　　Type from Japan, now found to be a common fern throughout the Yangtze valley. Authors on Chinese ferns have generally considered this identical with *Athyrium nigripes* (Bl.) Moore of Malaysia-Polynesia, with which our fern has, in fact, nothing to do at all. The present species constitutes the type for the group of about ten closely related members in West China and Northwestern India, and holds an intermediate position between *A. Mackinnoni* (Hope) and *A. mengtzeense* Hieron. of the same region. The Japanese *A. rigescens* Makino appears hardly distinct from our fern (Cf. Ogata, Ic. Fil. Jap. 5: t. 216).

　　Plate 109. Fig. 1. Habit sketch (natural size). 2. Pinnule ($\times$ 7). 3. Scale from the base of stipe ($\times$ 20).

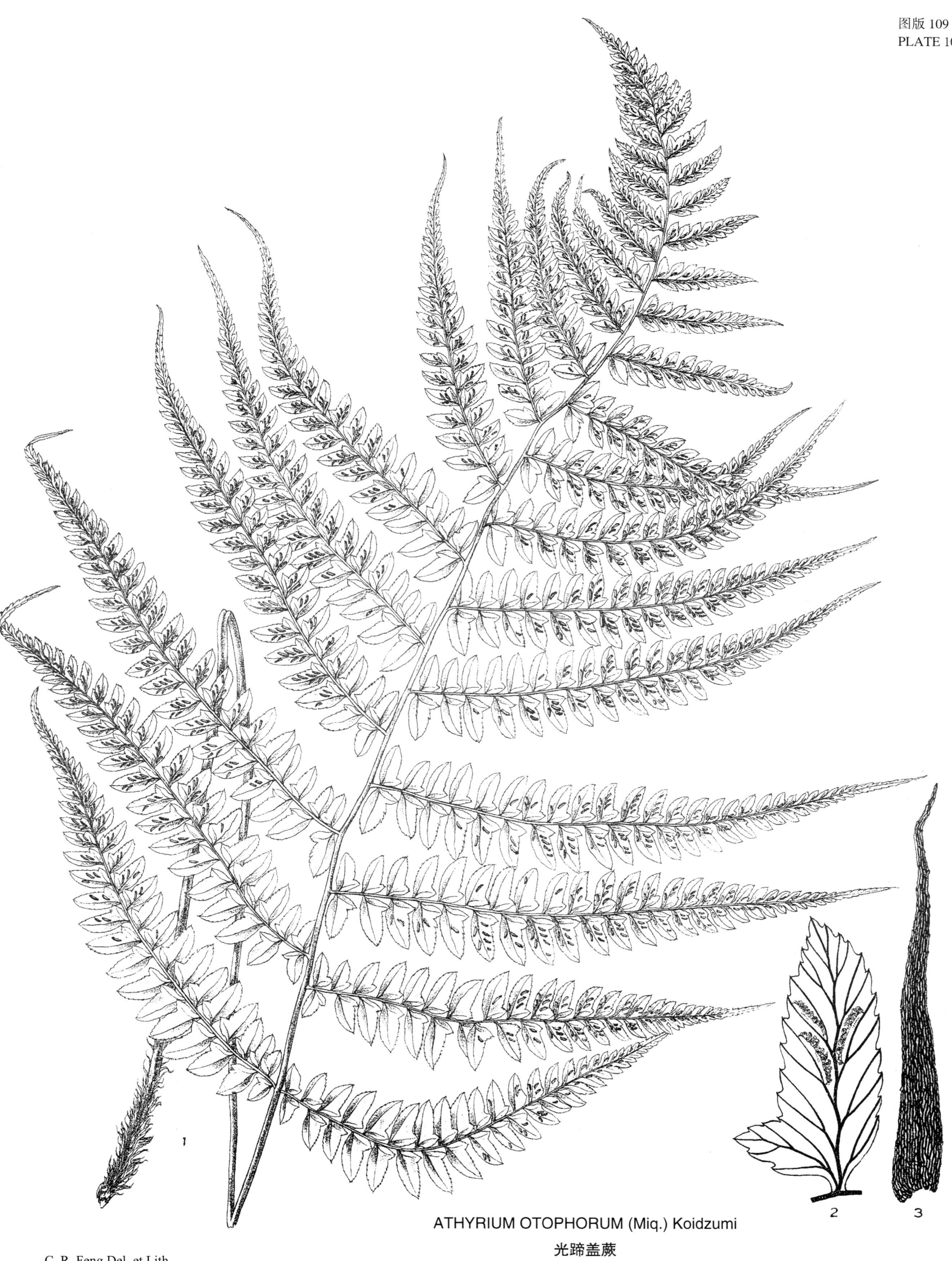

ATHYRIUM OTOPHORUM (Miq.) Koidzumi 光蹄盖蕨

根状茎斜出，鳞片为棕黄色，披针形；叶簇生，柄长 16—30 cm，淡稻秆色，光亮，叶体椭圆卵形，长 20—30 cm，宽 12—16 cm，渐尖头，三回羽状深裂；一回小叶 10—14 对，互生，具柄，斜出，基部一对稍短，长 7—10 cm，宽 3—4 cm，基部心脏形，渐尖头，二回小叶 8—11 对，具短柄，长 1.5—2 cm，椭圆形，短尖头或略为钝头，基部亚等边，羽状深裂达中肋，裂片 4—6 数，钝头，具锯齿，每一叶脉伸入一锯齿；薄质，淡绿色，一二各回中肋上面具肉刺；子囊群长形或马蹄形，盖膜质，形大。

分布：浙江，江苏，云南，四川，广东，福建；日本亦产之。

本种为本属复羽状分裂之一种，为我国东南各省常见之品，昔人误认为 *Athyrium nigripes* (Bl.) Moore，实则大谬不然也。

图注：1. 本种全形（自然大）；2. 二回小叶，表示其叶脉及子囊群（放大 8 倍）；3. 叶柄基部鳞片（放大 16 倍）。

ATHYRIUM GOERINGIANUM (Kunze) Moore

ATHYRIUM GOERINGIANUM (Kunze) Moore, Ind. Fil. 185 (1860); C. Chr. Ind. Fil. 143 (1905); Ogata, Ic. Fil. Jap. 2: t. 58 (1929).

Aspidium göringianum Kunze, Bot. Zeit. (1848) 557.

Lastrea göringianum Moore, Ind. Fil. 93 (1858).

Dryopteris göringiana Koidz. Bot. Mag. Tokio **43**: 382 (1929); Acta Phytotax. et Geobot. **1**: 233 f. 9-10 (1932).

Athyrium filix-foemina Christ (non Roth, 1799) in Warburg, Monsunia **1**: 75 (1900).

Athyrium iseanum Ros. in Fedde, Repert. Sp. Nov. **13**: 124 (1913); C. Chr. Ind. Fil. Suppl. Ⅱ. 7 (1916).

Rhizome oblique; *scales* brown, thin, lanceolate; *fronds* caespitose, stipe 16-30 cm long, pale straminous, glossy above the blackish scaly base, terete underneath, broadly flattened above, lamina oblong-ovate, 20-28 cm long, 12-16 cm broad, acuminate, tripinnatifid; *pinnae* 10-14-jugate, alternate, petiolate, obliquely patent, the basal ones somewhat shorter than those next above, 7-10 cm long, 3-4 cm broad, base cordate, apex acuminate; *pinnules* 8-11-jugate, shortly petiolulate, alternate, 1.5-2 cm long, oblong, acute or bluntish, base subunequal, pinnatifid nearly down to costule into 4-6 oblong serrato-incised blunt segments, with one simple veinlet to each tooth; texture soft herbaceous, light green, rachis and rachilets are provided with characteristic spines on the upper side; *sori* elongate or horse-shoe-shaped, *indusium* thin, grayish-brown, large; *spores* reniform, smooth.

Zhejiang: Ningpo, Tientai Shan, *Hancock 12* (1877); *C. Y. Chiao 14437*; *K. K. Tsoong 3713*. Sichuan: Mt. Omei, *W. P. Fang 2672, 3800* in herb. West China Acad. Sci. Yunnan: *H. T. Tsai 52696*. Fujian: Lin Fa Shan, *S. T. Dunn 3844*, June 6, 1904. Jiangxi: Kiukiang, Lushan, Incense Mill, *DeVol*, Aug. 5, 1933; Jiangsu: Wusih, Weishan, *R. C. Ching 3427*. Guangdong: Lohfau Shan, *C. O. Levine & McClure 6833, 6929*; *Levine 1518*; *Merrill 10344* (1916); *Gerlach* (without locality).

Japan: *Göring 115* (type) (1844); Ise, *Sakura 47* (type of *A. iseanum*).

A native of Japan and Eastern China and was generally considered in the past as identical with A. *nigripes* (Bl.) Moore, from which it differs in ampler frond as long as the stipe, incised or, at least, dentate-incised ultimate pinnules and not strictly costular sori. A still closer ally is A. *strigillosum*-Moore from Northwestern India and West China, which differs only in less divided frond and subcostular sori with uniform asplenioid indusium.

Plate 110. Fig. 1. Habit sketch (natural size). 2. Pinnule, showing venation and types of indusia ($\times$ 8). 3. Scale from the base of stipe ($\times$ 16).

ATHYRIUM GOERINGIANUM (Kunze) Moore
柯氏蹄盖蕨

图版 111　华中铁角蕨

根状茎短而直立，端具黝褐色狭披针形之密鳞片；叶簇生，柄长 5—10 cm，光滑或略具鳞片，叶体为三角卵椭圆形，长 5—13 cm，宽 2.5—5 cm，或较宽，三回分裂，一回小叶约 10 对，基部一对最大，卵形，长 1.5—3 cm，宽 1—2 cm，基部不等，末回小叶为披针形，2—3 裂，端具尖锯齿；子囊群 1—2 个，细长，盖膜质，全缘；叶为纸质，两面光滑，淡绿色。

分布：四川，湖北，江苏，江西，贵州；日本亦产之。

此为长江流域习见之种，形体变异甚大，与此种最相似者为 Asplenium pekinense Hance，然本种基部一回小叶较位于上部者较大，故得区别。

图注：1. 本种全形（自然大）；2. 同上，模式；3. 一回小叶（放大 6 倍）；4. 叶柄基部鳞片（放大 27 倍）。

ASPLENIUM SARELII Hooker

ASPLENIUM SARELII Hooker in Blakiston, Five Months on the Yangtze 363 (1862); C. Chr. Ind. Fil. 130 (1905).
  Asplenium Saulii Baker in Hk. Syn. Fil. ed. 2. 216 (1874); Christ, Farnkr. d. Erde 203 (1897); Diels in Engl. u. Prantl: Nat. Pflanzenfam. **1**: 4. 240 (1899); Ogata, Ic. Fil. Jap. **1**: t. 8 (1928).
  Asplenium Blakistoni Baker in Hk. Syn. Fil. 216 (1867); Hk. Ic. Pl. t. 1015, 1016 (1867).
  Asplenium pekinense Makino (non Hance, 1867), Bot. Mag. Tokio **9**: 245 (1895).

Rhizome short, erect, densely scaly at the apex; *scales* lanceolate-subulate, black; *fronds* caespitose, stipe 5-10 cm long, slender, naked or sparcely scaly, green, lamina deltoidly ovate-oblong, 5-13 cm long, 2.5-5 cm broad or broader at base, acuminate, tripinnate, the basal *pinnae* much the largest, ovate, with unequal base, 1.5-3 cm long, 1-2 cm broad, ultimate pinnules linear, 2-3-fid, with a fine tooth, with 1-2 elongate *sori*, *induium* thin, gray, entire; *texture* herbaceous, glabrous on both sides, color green even upon drying.

Sichuan: *Blakiston* (type), without exact locality; Mt. Omei, *E. Faber 1007*; ibid., *E. H. Wilson 5358, 5262* (ad f. typica). Hubei: Patung, *A Henry 3789*; Ichang, Nanto, *Henry 219*; Oupan Shan, *Silvestri 3249, 3248* (1905); Nang Hsien, *Silvestri 4029, 4030, 4032, 4033*. Jiangsu: Nanking, She Er Tung, *R. C. Ching 3499* (1927); *DeVol 63*, August 13, 1933; *L. F. Zee 11* in herb. Univ. Nanking; Ishing, *Y. L. Keng 2690*. Jiangxi: Fengcheng, *Y. Tsiang 10285*. Guizhou: Sanhoa, *Y. Tsiang 6243*; Pachai, *Y. Tsiang 6158*; Tsunyi, *Y. Tsiang 5298*.

Also Japan.

This distinct fern, common throughout the Yangtze valley, is closely related to A. *varium* Wall. from Northern India and West China, which differs in much narrower and bipinnate frond of a different outline with broadly oblong ultimate pinnules without so sharp teeth as in our fern. In the general shape of front, it appears very much like a small form of A. *tenuifolium* Don, differs in much thicker texture of front and entirely different shape of ultimate pinnules. A. *pekinense* Hance, common in North and Central China, seems to be specially distinct from our fern by its much reduced lower pinnae.

Plate 111. Fig. 1. Habit sketch (natural size). 2. The same, forma typica. 3. Pinnule of first order ($\times$ 6). 4. Scales from the base of stipe ($\times$ 27).

ASPLENIUM SARELII Hooker
华中铁角蕨

图版 112　岭南铁角蕨

　　根状茎肥短而直立,端具黑色披针形之鳞片;叶簇生,柄肉质,长 3—6 cm,略具黑色分叉鳞片,叶体为倒披针形,长 13—25 cm,宽 2—4 cm,渐尖头,向基部渐狭,二回羽状分裂,一回小叶 17—28 对,开张,几无柄,位于下部者为三角形,较上部者为小,中部小叶长 1.5—2.5 cm,宽约 1 cm,椭圆形,钝头,基部不等,羽状深裂,裂片 5—8 对,线形,具单脉,位于基部上方者通常二至三裂;纸质,绿色;子囊群狭长,每裂片一数,盖膜质,全缘,向上方开。

　　分布:广东,广西,贵州。

　　此为我国南部特有之种,其形体极类亚洲热亚产之 *Asplenium Belangeri* (Bory) Kze.,而叶体较小,且呈倒披针形,向下部渐狭,故易区别。

　　图注:1. 本种全形(自然大);2. 同上,大形;3. 中部一回小叶,表示其叶脉及子囊群(放大 16 倍);4. 根状茎上之鳞片(放大 50 倍);5. 叶柄上之鳞片(放大 76 倍);6. 中肋上之鳞片(放大 60 倍)。

ASPLENIUM SAMPSONI Hance

ASPLENIUM SAMPSONI Hance, Ann. Sci. Nat. V. **5**: 257 (1866); C. Chr. Ind. Fil. 130 (1905).

　　Asplenium belangeri Dunn & Tutch. (non Kze. 1848) Fl. Kwangt. & Hongk. 344 (1912).

　　Rhizome thick, short, erect, densely scaly; *scales* blackish, lanceolate, dense at apex; *fronds* caespitose, stipe fleshy, 3-6 cm long, sparcely clothed in blackish, irregularly armed scales, lamina lanceolate, 13-25 cm long, 2-4 cm broad, acuminate, gradually narrowed towards base; *pinnae* 17-28-jugate, patent, subsessile, the lower ones gradually reduced into a broadly deltoid outline below 1 cm long, the middle ones 1.5-2.5 cm long, 1 cm broad, oblong, obtuse, base unequal, being auricled at anterior side, deeply pinnatifid; *segments* 5-8-jugate, linear, pectinate, uni-nerved, the anterior basal one generally 2-3-forked; *texture* herbaceous, green; *sori* linear, one to each segment, *indusium* membranaceous, entire, open upward.

　　Guangdong: Kai Kun Shek, T. *Sampson 11165* (type), in herb. Hance, June, 1867; *K. K. Wang 436* (1928). Guangxi: Lungchow, *Morse 14*. Guizhou: *Esquirol 3731* (without exact locality).

　　This distinct endemic species is closely related to *A. Belangeri* (Bory) Kze. of tropical Asia, to which it was reduced by Baker (Cf. Syn. Fil. 223), from which it differs, however, in much smaller size with narrowly lanceolate frond, gradually narrowed towards the short fleshy scaly stipe, which is, as rachis, sparcely clothed in black, irregularly armed and distinctly reticulated scales. From *A. prolongatum* Hk., our fern differs in non-prolongated, and rooting apex of frond of a lanceolate outline, on much shorter stipe, and in very patent or subhorizontally spread, more copiously and regularly pinnatifid pinnae, of which the lower ones are gradually shortened.

　　Plate 112. Fig. 1. Habit sketch (natural size). 2. The same, a larger form. 3. A middle pinna, showing venation and sorus ($\times$ 16). 4. Scale from rhizome ($\times$ 50). 5. The same from stipe ($\times$ 76). 6. Scales from rachis ($\times$ 60).

ASPLENIUM SAMPSONI Hance
岭南铁角蕨

图版 113　长生铁角蕨

根状茎短而直立,端具黝褐黑色之狭披针形鳞片;叶簇生,柄长 8—15 cm,淡绿,光滑,叶体长 10—15 cm 或较长,宽达 3 cm,线披针形,端呈尾状,中肋延长至数厘米而生根,薄肉质,绿色,二回分裂,小叶多至十数,无柄,位于下部者略短,椭圆形,基部不等,裂片狭线形,略向上弯曲,钝头,具一数叶脉,其位于基部上方者,通常二至三裂;子囊群狭长,每裂片一个,盖膜质,向上开。

分布:香港,广东,广西,福建,浙江,湖北,贵州,四川,云南;印度北部,越南,日本及朝鲜半岛南部均产之。

本种为亚洲大陆温热带产,常生于湿润森林中之树干上或岩壁上,昔人误认为与非洲产之 Asplenium achilleifolium Lam. 相同,实则二者大相径庭。

图注:1. 本种全形(自然大);2. 一回小叶(放大 10 倍);3. 根状茎上之鳞片(放大 27 倍)。

ASPLENIUM PROLONGATUM Hooker

ASPLENIUM PROLONGATUM Hooker, 2nd. Cent. Ferns t. 42 (1860); Sp. Fil. **3**: 209 (1860); Christ, Bull. Soc. Bot. France **52**: Mém. 1. 54 (1905); C. Chr. Ind. Fil. Suppl. Ⅲ. 36 (1934); Blot, Bull. Soc. Hist. Nat. Toulouse (1932) 50.

Asplenium rutoefolium Hk. et Bak. Syn. Fil. 222 (1867); Christ, Farnkr. d. Erde 208 (1897); Diels in Engl. u. Prantl; Nat. Pflanzenfam. **1**: 4. 242 (1899), pro parte.

Asplenium rutoefolium Franch. et Sav. (non Kze. 1836) Enum. Pl. Jap. **2**: 222 (1876); Christ in Warburg, Monsunia **1**: 72 (1900); Makino, Phan, et Pterid. Jap. Ic. Illust. **1**: t. 55 (1899-1901).

Asplenium achilleifolium C. Chr. Ind. Fil. 99 (1905), pro parte

Asplenium bipinnatum var. *prolongatum* Bonap. Notes Pterid. Pt. **14**: 74 (1923).

Asplenium elongatum Christ, Bull. Acad. Géogr. Bot. (1910) 13.

Rhizome short, erect, scarcely scaly; *fronds* caespitose, stipe 8-15 cm long, glabrous, pale green, lamina 10-20 cm long or Inoger, to 3 cm broad, linear, suddenly terminated by a caudate, naked prolongation of rachis 2-5 cm long, and rooting at the apex, herbaceous, green, bipinnate, the lower pinnae only slightly shortened; pinnae subsessile, numerous, oblong with unequal base; *pinnules* narrow-linear, slightly curved, obtuse, uni-nerved, the anterior basal pinnule often bi-or tripartite; *sori* one to each pinnule, linear, *indusium* firm-membranaceous, pale green, opening upward.

Hong Kong: *Forbes 574*; *Ford* (1872); *Matthew 67* (1904), *368*; *Gibbs 1* (1928). Guangdong: Lohfau Shan, *C. O. Levine 515*; *N. K. Chun 41433*; Swatow, Thaiyong, *Dalziel*, July, 1901; Luntau Shan, Iu Village, *To & Tsang 12253*; Lokchong, *N. K. Chun 43684*, *42484*; Sunyi, *S. P. Ko 51350*; *Y. K. Wang 31146*; Yingtak, *H. Y. Liang 61364*; Yao Shan, Lokchong, *S. P. Ko 51949*; *N. K. Chun 43015*; *Gerlach* (without exact locality). Guangxi: Lin-yin Hsien, *R. C. Ching 7084*; San Fang, Luchen, *R. C. Ching 5751*, *5817*. Fujian: Kuliang, Foochow, *H. H. Chung 4284*, *4368*; *Hancock 6* (Kew No.). Zhejiang: Pinyang, *H. H. Hu 162*; Wenchow, *R. C. Ching 1855*. Hubei: Ichang, *A. Henry 3291*; *Wilson 2655*. Sichuan: Mt. Omei, *W. P. Fang 2553*; Tchenkoutin, *Farges 1262*. Guizhou: Tsingay, *Bodinier 2118*; Ganchow, *Michel 1026*; Petsen, *Cavalerie 1568*, *7002*; Chengfeng, *Y. Tsiang 4394*, *4610*, Tsunyi, *Y. Tsiang 5300*; Tuyun, *Y. Tsiang 5970*; Shihtsien, *Y. Tsiang 4116*; Vanching Shan, *Y. Tsiang 7783B*. Yunnan: Mengtze, *Hancock 37*; *Henry 9228*, *9228B*, *9228A*; south of Red River, *Henry 13659*; Shweli-Salwin, *Forrest 24492*.

Tsus-Sima: *Wilford 838*.

Also Northern India, Vietnam, Japan and South of Korean peninsula.

Type from Khasya (leg. *Hooker f. & Thomson*), or Mishmee (leg. *Griffith*). Wilford No. 838 from Tsus-Sima was also cited by Hooker. A well-marked species endemic in the temperate part of the mainland of Asia. It was unfortunate that Hooker himself later combined it in Syn. Fil. with *A. rutoefolium*, a synonym to *A. achilleifolium* Lam. from Africa, from which our fern differs in the characters as first emphasized by himself in Species Filicum **3**: **209**.

Plate 113. Fig. 1. Habit sketch (natural size). 2. Pinna ($\times$ 10). 3. Scale from rhizome ($\times$ 27).

ASPLENIUM PROLONGATUM Hooker
长生铁角蕨

根状茎短而直立,端具黑色披针形密鳞;叶簇生,柄长 7—10 cm,具深褐色细长鳞片,富于拆光性,叶体阔披针形,长 10—30 cm,宽 4—7 cm,一回羽状分裂,小叶 10—20 对,位于下部者渐短,中部者长 2—4 cm,卵斜方形,钝头或极尖头,基部不等,叶缘具不整齐之钝锯齿,厚纸质,干则呈黝褐色,中肋鳞片疏生,久则脱落,叶脉扇形分叉,上下面呈沟脊形;子囊狭长,斜出,盖为厚膜质,全缘,向上开,其位于下部者,常相对开。

分布:广东,香港,广西,云南,贵州;印度亦产之。

本种最初于广东发现,现则南部各省均产之,极似马来群岛产之 *Asplenium pellucidum* Lam. 或即为该种地理分布之一变形耳。

图注:1.本种全形(自然大);2.小叶,表明其叶脉及子囊群之位置(放大 2 倍);3.根状茎上之鳞片(放大 16 倍);4.叶柄上之鳞片(放大 10 倍);5.羽片(自然大)。

ASPLENIUM CRINICAULE Hance

ASPLENIUM CRINICAULE Hance, Ann. Sci. Nat. V. **5**: 254 (1866); Hk. et Bak. Syn. Fil. 208 (1867); Clarke, Tran. Linn. Soc. II. Bot, **1**: 479 (1880); Bedd. Handb. Ferns, Brit. Ind. 150 (1883); Christ, Bull. Herb. Boiss. **6**: 959 (1898); Diels in Engl. u. Prantl: Nat. Pflanzenfam. **1**: 4. 239 (1899); C. Chr. Ind. Fil. 106 (1905); Wu, Bull. Dept. Biol. Sun Yatsen Univ. No. **3**: 190 t. 87 (1932).

Asplenium hancei Baker in Hk. Syn. Fil. ed. 1. 208 (1867); Kuhn, Bot. Zeit. (1869) 130.

Asplenium beddomei Mett. ex Kuhn, Linnaea **36**: 93 (1869).

Asplenium falcatum Bedd. (non Lam., nec Thbg.) Ferns S. Ind. t. 141 (1864).

Asplenium polytrichum Christ, Bull. Acad. Géogr. Bot. (1909), Mém. XX. 172.

Asplenium adiantoides Wu, Bull. Dept. Biol. Sun Yatsen Univ. No. **3**: t. 88-90 (1932).

Rhizome short, erect, densely scaly; *fronds* caespitose, stipe 7-10 cm long, dark-brown, densely scaly throughout the rachis; *scales* dark rufo-brown, iridescent, clathrate, subulate from broad and often branched base, caducous, lamina lanceolate, caudate-acuminate, 10-30 cm long, 4-7 cm broad, simple pinnate; *pinnae* lanceolate, acuminate, 2-4 cm long, the lower ones gradually abbreviated, ovate-rhomboid with obtuse apex, base unequal, obliquely truncate above, cuneate below, margin irregularly erose-dentate; *texture* herbaceous, color opaque; *veins* flabellately forked, channelled; *sori* linear, oblique to costa, *indusium* firm, entire, lower ones often open towards each other.

Guangdong: Tingwu Shan, *Sampson 11203* (type) in herb. Hance, June 15, 1866; Whampoa, *Hance*; *Ford 228* in herb. Hong Kong: Lokchong, *C. L. Tso 21332*; Swatow, Thaiyong, *Dalziel*, July, 1899. Hong Kong: *Dr. Harland 91*; *Hance 7418* (1857); Taimo Shan, *Matthew*, Oct. 15, 1907. Fujian: Yengping, *H. H. Chung 2992*; Amoy, *De Grijs 1198* in herb. Hance; *T. S. Dunn 777*, May 15, 1905. Guangxi: Yao Shan, *S. S. Sin 331* (1928); Seh-feng Dar Shan, *R. C. Ching 8436*; Luchen, *R. C. Ching 5551*. Yunnan: Mengtze, *Hancock 182*; Szemeo, *Henry 11939*, 11939B, 11939C. Guizhou: Pin-fa, *Cavalerie 2844* (type of A. *polytrichum*); Tsin-gay, *J. Laborde*, Jan. 14, 1898; Tushan, *Y. Tsiang 6954*.

Also Northern India and Neilgheries.

A very variable fern as to the size and shape of pinnae, as has been well illustrated by Wu by half dozen plates under different names. In type, the pinnae are 4-5 cm long. A. *Hancei* Baker from Guangdong represents a small form, 15 cm long with lamina 2-3 cm broad, pinnae ovate-oblong with crenate margin. Habit exactly similar to A. *planicaule* Wall., which differs in glabrous stipe and rachis. I have strongly suspected (Cf. Blot, Asplen, d. Vietnam p. 36) that the present species from mainland may after all prove to be a geographic form of A. *pellucidum* Lam. of Malaysia-Polynesia and also frequent in Vietnam, which differs from our fern only in elongate lanceolate, acuminate pinnae, of which the lower ones are even more strongly reduced into an ovate outline.

Plate 114. Fig. 1. Habit sketch (natural size). 2. Pinna, showing venation and position of sori ($\times$ 2). 3. Scale from rhizome ($\times$ 16). 4. Scales from stipe and rachis ($\times$ 10). 5. Pinnae (natural size).

ASPLENIUM CRINICAULE Hance
毛铁角蕨

根状茎短而直立或斜生,端具红棕色狭披针形之鳞片;叶亚簇生,柄长 10—20 cm,深稻秆色,鳞片疏生,达于中肋,叶体长 12—28 cm,宽 7—11 cm,一回羽状分裂,小叶 5—12 对,顶上小叶形同侧叶,基部小叶较大,间具二三裂片,向上渐小,斜方形,渐尖头,基部延长;革质,两面光滑;叶脉扇形分叉,两面呈沟脊形;子囊群狭长,斜出,盖线形,淡褐色,质略厚,全缘,向上开,其位于下部者彼此相向开。

分布:广东,贵州,云南;越南亦产之。

本种为本区内特产,最初于广东发现,Hance 氏名之曰 *Asplenium comptum*,惜此名早为 Kunze 氏所用,不能存在耳;此种极类美洲热带产之 *Asplenium dimidiatum*,唯其小叶具较长之柄,叶缘深裂,故易分别。

图注:1.本种全形(自然大);2.小叶,表示其叶脉及子囊群(放大 2 倍);3.根状茎上之鳞片(放大 10 倍)。

ASPLENIUM SAXICOLA Rosenstock

ASPLENIUM SAXICOLA Rosenstock in Fedde, Repert. Sp. Nov. **13**: 122 (1913); C. Chr. Ind. Fil. Suppl. Ⅱ. 7. (1916).
  Asplenium comptum Hance, Ann. Sci. Nat. Ⅴ. **5**: 255 (1866); Kuhn, Bot. Zeit. (1869) 131; C. Chr. Ind. Fil. 105 (1905); Blot. Bull. Soc. Hist. Nat. Toulouse (1932) 39 (non Kze. 1852).
  Asplenium dimidiatum var. *comptum* Baker in Hk. Syn. Fil. ed. 2. 486 (1874).
  Asplenium affine var. *sinense* Christ, Bull. Acad. Géogr. Bot. (1905) 243.
  Asplenium dimidiatum Christ, (non Sw. 1788) Bull. Herb. Boiss. **7**: 9 (1899); Bull. Acad. Géogr. Bot. (1910) 13; Bonap. Notes, Pterid. Pt. **14**: 117 (1923).

Rhizome short, erect or oblique, densely scaly; *scales* dark rufo-brown, linear-lanceolate, subulate towards apex; *fronds* subcaespitose, stipe 10-20 cm long, dark straminous, sparcely scaly throughout rachis, lamina 12-28 cm long, 7-11 cm broad, simple pinnate with distinct end-pinna; *pinnae* 5-12-jugate, long-petiolate, the lower ones much the largest, gradually diminished towards apex, rhomboid with attenuate base and gradually acuminate towards apex, the lower ones often 1-2-lobed, margin erose-dentate; *texture* coriaceous, glabrous on both sides; *venation* distinctly canaliculated on both sides; *sori* linear, oblique, *indusium* of the same shape, brown, thick, entire, the lower ones open towards each other.

Guangdong: Kai Kun Sheh, West River, *Sampson 1190* in herb. Hance (type of *A. comptum*); North River, *C. Ford* (1879); 226 (1888); Lienchow, *Matthew*, Nov. 30, 1907; Yun Fou Hsien, West River, *K. K. Wang 434*; *K. K. Tsoong 4298*. Guangxi: Wangchin, Luchen, *R. C. Ching 5473*; Lanlon, *R. C. Ching 6398*; Lungchow, *Morse 40*; Lin Shan Hsien, *Sin & Wang 51* (1927). Guizhou: Tcehnfeng Hsien, *Cavalerie 3852* (type); Tiensen Kou, *Cavalerie 1853* (type of *A. affine* var. *sinense*); Tsehheng, Y. *Tsiang 9247*; Kouyhoua, *Cavalerie 1215*; Kweiyang, *Esquirol 901*; Ganchow, *Michel 992*, *1000*; Tchenlin, *Michel 1051*. Yunnan: Mengtze, *Henry 11542*.

Vietnam: Dong Dang, *Balansa 99*; Langnac, *Colani 3346*, frequent.

This is another of the distinct but little known Chinese species by Hance, but, unfortunately, his name, *A. comptum*, has to become invalidated by Kunze. Our fern is closely related to *A. dimidiatum* Sw. of tropical America, to which it was referred as a variety by Baker, as identical by Christ and others, but differs in long-petiolate pinnae with erose-dentate (not lacerate) margin. The type from Guizhou has lateral pinnae tend to lobato-pinnate; the same form is also seen from Guangdong.

Plate 115. Fig. 1. Habit sketch (natural size). 2. Pinna, showing venation and sori (× 2). 3. Scale from rhizome (× 10).

ASPLENIUM SAXICOLA Rosenstock
粤铁角蕨

图版 116　东方狗脊

　　根状茎木质肥大，直立，大形黄棕色之披针形鳞片密覆，长达 3—4 cm；叶簇生，叶柄粗厚，长 30 cm 或过之，下部鳞片密生，上部疏生，叶体长椭圆形，长 30—50 cm，宽约 20 cm，二回羽状深裂，顶部渐尖，深裂，小叶 6—8 对，亚对生，具短柄，长 10—13 cm，宽 4—5 cm，阔披针形，渐尖头，基部极不等，羽状深裂，裂片阔披针形，具锐强锯齿；叶脉不明显，网状，网眼二列，不具小脉；叶硬革质，两面光滑；子囊群长肾形，贴近裂片中肋，凹入叶质内，盖同形，厚膜质，隆起，宿存。

　　分布：福建，广东，香港，台湾；日本及菲律宾群岛均产之。

　　本种为东方特产，昔人误认为与欧美产之 *Woodwardia radicans* Smith 相同，余曾于昔年订正之，读者可参阅静生生物调查所汇报第二卷六至七页，即知两种之异同矣。

　　图注：1. 本种全形（自然大）；2. 裂片之一部，表示其叶脉及子囊群（放大 10 倍）；3. 叶柄基部鳞片（放大 60 倍）；4. 多子变种（自然大）；5. 自芽胞发生之幼植物（放大 16 倍）。

WOODWARDIA ORIENTALIS Swartz

WOODWARDIA ORIENTALIS Swartz in Schrad. Journ. Bot. 1800^2. 76 (1801); Syn. Fil. 117, 315 (1806); Willd. Sp. Pl. **5**: 418 (1810); Kze, Pterid. Jap. in Bot. Zeit. (1868) 323; Hk. Sp. Fil. **3**: 68 (1860); Syn. Fil. 188 (1867); Ching, Bull. Fan Mem. Inst. Biol. **2**: 3 t. 1 (1931); C. Chr. Ind. Fil. Suppl. Ⅲ. 196 (1934).

Blechnum radicans var. Houtt. Nat. Hist. **2**: t. 97 f. 1 (1783).

Woodwardia radicans var. *auriculata*, Kuhn, Journ. Bot. (1868) 268.

Woodwardia radicans var. *orientalis* C. Chr. Ind. Fil. 658 (1905).

Woodwardia angustiloba Hance, Journ. Bot. (1868) 176.

Woodwardia japonica Hk. Journ. Bot. (1857) 341, pro parte.

Woodwardia radicans Christ (non Sm. 1793) in Warburg, Monsunia **1**: 66 (1900).

Woodwardia prolifera Hk. et Arn. in Bot. Beech. Voy. 275 t. 56 (1836-1840); Nakai, Bot. Mag Tokio **39**: 105 (1925).

Woodwardia exaltata Nakai, Bot. Mag. Tokio 35: 149 (1925).

Rhizome thick, woody, erect, densely clothed at apex in very large, bright brown, lanceolate acuminate, membranaceous scales 2-4 cm long; stipe caespitose, thick, 30 cm long or longer, rufo-straminous, sparcely scaly throughout, lamina oblong-ovate, 30-50 cm long, 20 cm broad, bipinnatifid; pinnae 6-8-jugate under the deeply pinnatifid acuminate end-pinna with shortly decurrent base, subopposite, 10-13 cm long, 4-5 cm broad, broadly lanceolate, acuminate, base subequal or often strongly unequal, i. e. the lower basal 1-2 segments suppressed, leaving a wingless costa, the anterior basal segment more or less decurrent; segments oblong-lanceolate, shortly acuminate, sharply serrated; venation obscure, areolae imperfectly 2-rowed along costa; texture rigidly coriaceous; sori oblong-linear, immersed, uniseriate and close to costule, indusium brown, thick, vaulted.

　　Fujian: Foochow, *Gregory* (1857) in herb. Hance (type of *W. angustiloba*); *Warburg*; Santu Island, *Matthew*, Oct. 3, 1907; Diongloh Hsien, *H. H. Chung 1244*; Nanching, *Schindler 407*; Yengping, *H. H. Chung 2936*; Chuanchow, *H. H. Chung 3082*; Inghok, *H. H. Chung 2639*. Hong Kong: Victoria Peak, *Tutcher 670*. Zhejiang: Ningpo, *Hancock 5* (1877). Guangdong: Swatow, Thaiyong, *Dalziel*, July, 1899; Namyung, *S. P. Ko 50759*; Lokchong, *N. K. Chun 42727*, *42462*. Taiwan.

　　Also Japan and the Philippines.

　　Type from Japan (leg. Thunberg) and was first identified by Houttuyn under *Blechnum radicans* var. In the herbarium Thunberg, there are to be found four sheets of *Woodwardia*, of which only one marked β is the true *W. japonica*, while the other three are representing the present species. The type in herb. Swartz matches Houttuyn's figure exactly. Our fern has generally been considered as identical with *W. radicans* (L.) Sm. of Europe and America, but the differences between the two closely related species have already been contrasted by me (Cf. Bull. Fan Mem. Inst. Biol. **2**: 6-7, pls. Ⅰ-Ⅱ).

　　Var. **prolifera** Ching, Bull. Fan Mem. Inst. Biol. **2**: 6 (1931); C. Chr. Ind. Fil. Suppl. Ⅲ. 196 (1934).

Woodwardia prolifera Hk. et Arn. in Bot. Beech. Voy. 275 t. 56.

Woodwardia exaltata Nakia, Bot. Mag. Tokio **35**: 149 (1921).

　　Differs from the typical form only in the frond being copiously gemmiferous above.

　　Fujian: Foo-chow, *Fortune* (1853); *Carles* (1885); Min-chow, *H. H. Chung 1905*, *2086*; Amoy, *H. H. Chung 105*. Guangdong: Jao-ping, Lokchong, *N. K. Chun 42784*. *C. Wright*. Taiwan: Tamsui, *Oldham 20*, *245*.

　　Plate 116. Fig. 1. Habit sketch (natural size). 2. Portion of segment, showing venation and sori (× 10). 3. Scale from base of stipe (× 16). 4. Pinna of var. *prolifera* (natural size). 5. Young plant developed from a gemma (× 16).

WOODWARDIA ORIENTALIS Swartz
东方狗脊

根状茎短，斜生，具鳞片与长刚毛；鳞片披针形，红棕色，质硬，不透明，边缘具长刚毛；叶簇生，柄长 12—30 cm，稻秆色，基部具鳞片，遍体披长刚毛，叶体三角形或椭圆卵形，长 15—30 cm，宽 12—18 cm，羽状深裂或分裂，顶部渐尖头，深裂，基部心脏形；小叶 1—6 对或较多，无柄，阔披针形，长 8—10 cm，宽 2.5—4 cm，全缘或稍呈波形，渐尖头，基部圆形，下面密被长刚毛，上面短刚毛疏生，粗纸质，干则呈黝褐色；侧脉明显，小脉网状，网眼三列，六角形或四角形；子囊沿网脉散生，不具盖，圆球形，近顶部具二三刚毛。

分布：广东，广西，福建，贵州，台湾，浙江；日本，缅甸及印度北部均产之。

本属仅有一种，广布于亚洲大陆温暖带各地，昔人以之归入叉蕨属（*Tectaria*），实则其地位近于 *Bolbitis* 与 *Stegnogramma* 两属也。

图注：1. 本种全形（自然大）；2. 同上，叶体深裂；3. 小叶一部，表明其叶脉及子囊分布状况（放大 3 倍）；4. 叶柄下部鳞片（放大 20 倍）；5. 子囊（放大 140 倍）；6. 孢子（放大 140 倍）；7. 叶柄上之刚毛（放大 140 倍）；8. 中肋上之刚毛（放大 140 倍）。

DICTYOCLINE GRIFFITHII Moore

DICTYOCLINE GRIFFITHII Moore, Gard. Chron. (1855) 854; Ind. Fil. LIX (1857); J. Sm Hist. Fil. 149 (1875); Bedd. Ferns Brit. Ind. t. 155 (1866); Cop. Univ. Calif. Publ. Bot. **16**: 61 (1929); C. Chr. Ind. Fil. Suppl. III. 71 (1934).
Hemionitis griffithii Hk. f. et Thoms.; Hk. Sp. Fil. **5**: 193 (1864); Syn. Fil. 399 (1867); Bedd. Handb. Ferns, Brit Ind. etc. 415 (1883); Christ, Farnkr. d. Erde 62 (1897); Bull. Acad. Géogr. Bot. (1909) 177; Bull. Herb. Boiss. (1896) 674; in Warburg, Monsunia **1**: 58 (1900).
Dictyogramma griffithii Trev. Atti Ist. Veneto. V. **3**: 591 (1877).
Gymnogramme griffithii Hance, Journ. Bot. (1886) 14.
Aspidium griffithii Diels (non Bedd. 1876) in Engl. u. Prantl: Nat. Pflanzenfam. **1**: 4. 186 (1899); C. Chr. Ind. Fil. 76 (1905); Ogata, Ic. Fil. Jap. **1**: t. 6 (1928); Wu, Bull. Dept. Biol. Sun Yatsen Univ. No. **3**: 64 t. 23 (1932).
Hemionitis wilfordii Hk. Fil. Exot. t. 93 (1859).
Dictyocline wilfordii J. Sm. Hist. Fil. 149 (1875).
Hemionitis griffithii var. *Wilfordii* Hk. Syn. Fil. 399 (1867); Makino, Bot. Mag. Tokio **10**: 286.
Hemionitis griffithii var. *pinnata*, Makino, Bot. Mag. Tokio **10**: 286.
Dictyocline griffithii var. *tenuisissima* Ching, Bull. Fan Mem. Inst. Biol. **1**: 146 f. 1 (1930).

Rhizome short, oblique, densely scaly and setose hairy; *scales* lanceolate, rufo-brown, rigid, opaque, provided with gray, needle-like setose hairs on the margin; *fronds* caespitose, stipe 12-30 cm long, angular, straminous, sparcely scaly at base, setose hairy throughout the rachis, lamina deltoid to oblong-ovate, 15-30 cm long, 12-18 cm broad, deeply pinnatifid to pinnate under the deltoid acuminate pinnatifid apex, base cordate; *pinnae* free, 1-6-jugate, sessile lanceolate, 8-10 cm long, 2.5-4 cm broad, entire or slightly wavy, acuminate, base rounded, densely setose hairy underneath, sparcely and shortly so above; *texture* harsh-herbaceous, opaque upon drying; *lateral main veins* distinct to the edge, intervening veinlets anastomosing in 3 rows of hexagonal or quadri-angular and exappendiculate areolae; *sori* indefinite, reticulated, following the course of veinlets; *sporangia* globular, shortly stalked, provided with a few setose hairs near the top.

Guangdong: Taimo Shan (opposite Hong Kong), *Matthew*, Nov. 7, 1907; Lungtau Shan, Iu Village, *C. C. C. 12098*, *12225*; Swatow, *Dalziel*, July, 1901; Sam Koh Shan, Tsungfa-lungmon Hsiens, *W. T. Tsang 20484*, May 16, 1932; Yamfa Hsien, *S. P. Ko 50689*; Lohfau Shan, *C. L. Tso 20923*; Lokchong, *C. L. Tso 20923*; *N. K. Chun 42161*; Intak, *H. Y. Liang 61240*; *Gerlach*. Fujian: Yengping, *H. H. Chung 2948*, August 9, 1924; Yungchang, *Rankin*, *Dunn 3759* in herb. Hong Kong. Guangxi: Yao Mar Shan, Linyin Hsien, *R. C. Ching 7186* (1928); Yao Shan, *S. S. Sin 111*, May 25, 1928. Yunnan: Mengtze, *Henry 10422*; Hancock 214; *H. T. Tsai 52534* (without exact locality). Guizhou: Pinfa, *Cavalerie 2855*; Pana, Chengfeng, *Y. Tsiang 4283*, *4415*. Zhejiang: Pinyang, Yangtang Shan, *K. K. Tsoong 3750*, Oct. 19, 1920 (type of var. *tenuisissima*). Taiwan: Keelung, *Wilford 503* (type of *Hemionitis Wilfordii*).

Khasya: *Griffith* (type).

Upper Burma: Htawgaw, *Forrest 25414*.

Also Japan.

A peculiarly distinct genus of one single species with pinnatifid or pinnate fronds. *Hemionitis Wilfordii* represents a pinnatifid form of the fully pinnate Himalayan type; both forms are often seen from the same locality. The affinity of the genus to *Bolbitis* Schott and *Stegnogramma* Bl. seems to be most probable.

Plate 117. Fig. 1. Habit sketch (natural size). 2. The same, f. *Wilfordii* (Hk.). 3. Portion of pinna, showing venation and sori (× 3). 4. Scale from the base of stipe (× 20). 5. Sporangium (× 140). 6. Spores (× 140). 7. Hairs from stipe (× 140). 8. Hairs from rachis (× 140).

DICTYOCLINE GRIFFITHII Moore
圣蕨

图版 118　地耳蕨

根状茎细长，横行，密覆红棕色之披针形鳞片；叶散生，二形，非子囊群叶具长 3—5 cm 之柄，遍体披软毛，叶体舌形，长 6—9 cm，宽 2—3 cm，基部戟状心脏形，具一对卵形裂片，叶缘呈深波状，具睫毛；侧脉明显，小脉网状，网眼具分叉小脉，下面被软毛，上面短毛疏生；子囊群叶具长 10—18 cm 之细长柄，叶体三裂，叶片狭长，深波形或具二三小裂片，脉亦为网状；子囊群线形，一列，位于中肋与叶缘之间，不具盖，嗣向四面扩散。

分布：香港，广东，广西，海南，台湾；越南，印度南部及马来群岛等地均产之。

本属仅此一种，昔人以之隶于 *Leptochilus* 属，最近 Copeland 氏另辟一新属，殊不为过当，其与叉蕨属之关系最密。

图注：1—3. 本种全形（自然大）；4. 非子囊群叶之一部，表明其叶脉（放大 5 倍）；5. 子囊群叶之一部，表明其子囊幼时着生情形（放大 3 倍）；6. 根状茎上之鳞片（放大 24 倍）；7. 根状茎之横切面（放大 15 倍）；8. 叶柄基部横切面（放大 24 倍）；9. 孢子（放大 200 倍）；10. 叶柄上之毛（放大 80 倍）。

QUERCIFILIX ZEYLANICA (Houtt.) Copeland

QUERCIFILIX ZEYLANICA (Houtt.) Copeland, Phil. Journ. Sci. **37**: 408 (1928); Univ. Calif. Publ. Bot. **16**: 67 (1929); C. Chr. Ind. Fil. Suppl. Ⅲ. 169 (1934).

Ophioglossum zeylanicum Houtt. Nat. Hist. **14**: 43 (1783); Pfl. Syst. **13**: 47 t. 94 f. 1 (1786).

Leptochilus zeylanicus C. Chr. Ind. Fil. 388 (1905); Merr. Enum. Pl. Hainan in Lingnan Sci. Journ. **5**: 11 (1927); Ogata, Ic. Fil. Jap. **1**: t. 32 (1928).

Acrostichum quercifolium Retz. Obs. Bot. **6**: 39 (1791); Hk. Sp. Fil. **5**: 279 (1864); Ic. Pl. t. 905; Fil. Exot. t. 80 (1858); Syn. Fil. 418 (1867); Dunn & Tutch. Fl. Kwangt. & Hongk. 255 (1912).

Gymnopteris quercifolia Bernh. in Schard. Journ. Bot. **1806**[1]. 20 (1807); Presl, Tent. Pterid. 244 (1836); Bedd. Ferns S. Ind. t. 47 (1863); Handb. Ferns, Brit. Ind. etc. 403 (1883); J. Sm. Hist. Fil. 138 (1875); Christ, Farnkr. d. Erde 49 (1897); Diels in Engl. u. Prantl: Nat. Pflanzenfam **1**: 4. 200 (1899).

Leptochilus quercifolius Fée, Acrost. 88 (1845).

Dendroglossum quercifolium Fée, Gen. Fil. 80 t. 7B. f. 1 (1850-1852).

Rhizome creeping, slender, densely clothed in lanceolate, rufo-brown, rather thick *scales*; *frond* uniseriate, 1 cm apart, strongly dimorphous, *sterile ones* on stipe 3-5 cm long, angular, ferruginous-hairy, hairs soft, multi-cellular, spread, lamina ligulate with hastate base, 6-9 cm long, 2-3 cm broad, apex rounded, base cordate, generally with one free, deltoid blunt pinna on each side, the central part much the largest, inciso-crenate, lateral main veins distinct, intervening veinlets irregularly anastomosing with included forked veinlets, costa and veins underneath copiously provided with rufo-brown, spread long hairs, margin ciliate, subglabrous above; *fertile frond* on stipe 10-18 cm long, slender, subnaked, lamina trifoliolate, strongly contracted into linear segments, the lateral segment generally with one additional lobe below, the central segment much the longest, pinnatifid into a few shallow, roundish, remote lateral lobes; *texture* thin herbaceous; *veins* also reticulated but much simpler than in the sterile fronds; *sori* linear, continuous, uniseriate between costa and margin, confluent at last.

Hong Kong: *Matthew*, Nov, 1907. Guangdong: Lohfau Shan, *Pakwan 92* in herb. Hance (1871). Guangxi: Lungchow, *Morse 5*; Dar Wan, Tanlan, *R. C. Ching 6553* (1928). Yunnan: Szemeo, *Henry 13388*. Hainan: Nodoa, *F. A. McClure 7860*. Taiwan.

Also, Vietnam, Southern India, Sri Lanka and Malaysia-Polynesia.

The genus *Quercifilix*, of comparatively a recent date, contains only one terrestial little fern, fairly common in South China, on shaded moist banks of river. Its affinity to *Tectaria* is obvious according to Copeland.

Plate 118. Fig. 1-3. Habit sketch (natural size). 4. Portion of sterile lamina, showing venation ($\times$ 5). 5. Fertile frond, showing sori at young state ($\times$ 3). 6. Scale from rhizome ($\times$ 24). 7. Cross section of rhizome ($\times$ 15). 8. Cross section of the basal part of stipe ($\times$ 14). 9. Spores ($\times$ 200). 10. Hairs from stipe ($\times$ 80).

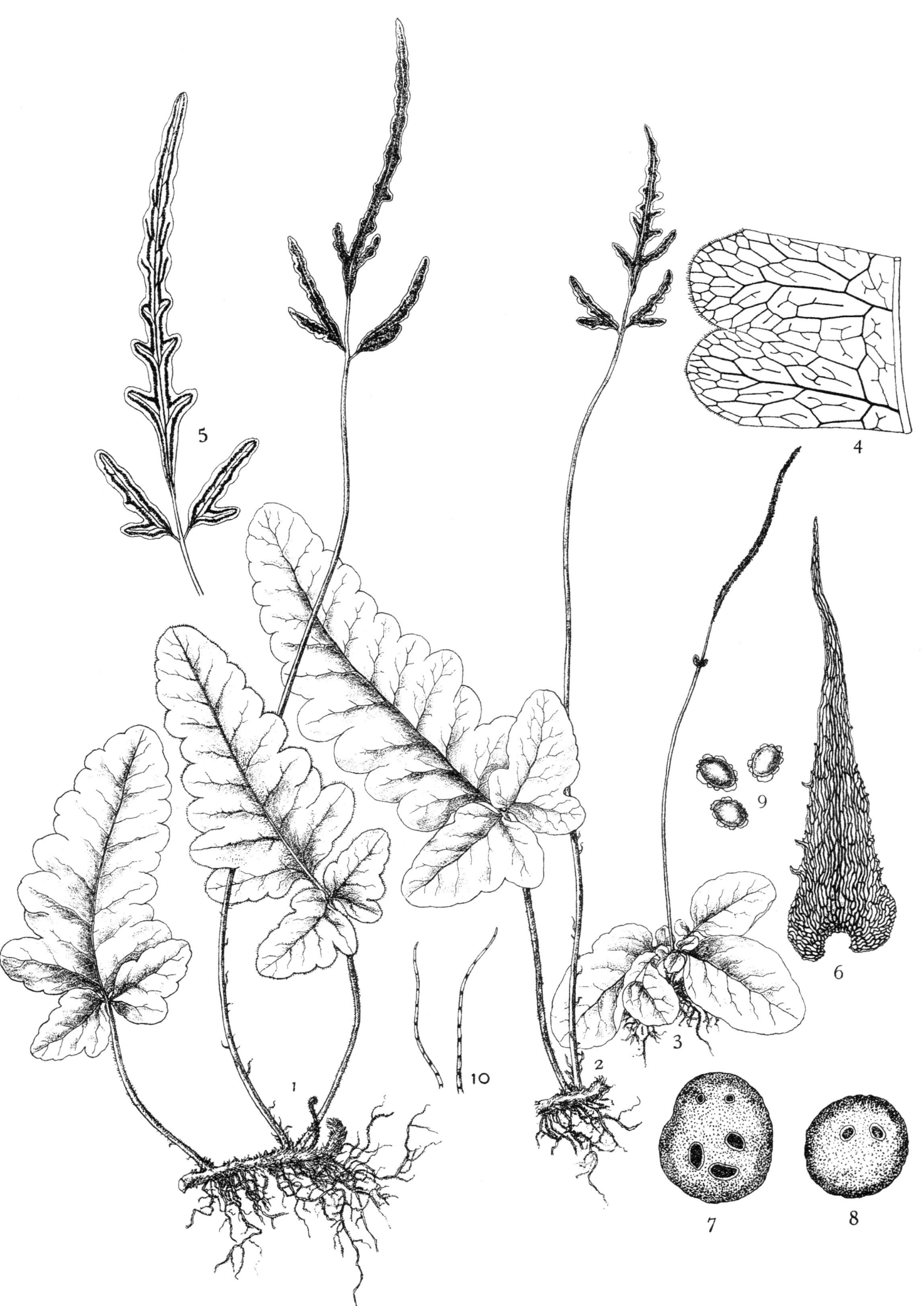

QUERCIFILIX ZEYLANICA (Houtt.) Copeland
地耳蕨

图版 119　长叶实蕨

根状茎肥厚，横行，具深棕色卵披针形之鳞片；叶亚散生，二形，非子囊群叶三裂或羽状分裂，或为披针形之单叶，小叶 1—5 对，顶端小叶与侧方小叶同形而较长，其顶常伸长成鞭形，入土着根，全缘或稍呈浅波状，侧脉明显，小脉网状，网眼三列，唯向叶缘则分散，薄纸质，干则呈浅黑色；子囊群叶与非子囊群叶同形而较小，柄较长，小叶较狭，子囊群初随网脉分布，继向四面扩散，不具盖。

分布：云南，广西，广东，香港，贵州，四川，海南及亚洲热带各地。

本种为本属特殊之一种，唯叶形变异甚大，由单叶至羽状分裂，干则常呈黑色。

图注：1. 本种全形（自然大）；2. 非子囊群叶之一部，表示其叶脉（放大 4 倍）；3. 子囊群叶之一部（放大 4 倍）；4. 根状茎上之鳞片（放大 20 倍）；5. 孢子（放大 300 倍）。

BOLBITIS HETEROCLITA (Presl) Ching

BOLBITIS HETEROCLITA (Presl) Ching in C. Chr. Ind. Fil. Suppl. Ⅲ. 48 (1934).

Acrostichum heteroclitum Presl, Rel. Haenk. **1**: 15 t. 2. f. 2 (1825).

Poecilopteris heteroclita Presl, Tent. Pterid. 242 (1836).

Heteroneuron heteroclitum Fée, Acrost. 92 (1845).

Chrysodium heteroclitum Kuhn, Ann. Lugd. Bat. **4**: 294 (1869).

Leptochilus heteroclitus C. Chr. Ind. Fil. 385 (1905).

Campium heteroclitum Cop. Phil. Journ. Sci. 37: 396 (1928).

Acrostichum flagelliferum Wall. ex Hk. et Grev. Ic. Fil. t. 23 (1827); Hk. Sp. Fil. **5**: 258 (1864); Hk. et Bak. Syn. Fil. 418 (1867); Clarke, Trans. Linn. Soc. Ⅱ. Bot. **1**: 579 (1880).

Bolbitis flagellifera Schott, Gen. Fil. ad t. 14 (1834).

Gymnopteris flagellifera Bedd. Ferns Brit. Ind. Suppl. 27 (1876); Christ, Farnkr. d. Erde 49 (1897); Diels in Engl. u. Prantl: Nat. Pflanzenfam. **1**: 4. 201 (1899).

Poecilopteris flagellifera Bedd. Ferns Brit. Ind. t. 112 (1865); Handb. Ferns, Brit. Ind. etc. 433 (1883); J. Sm. Hist. Fil. 137 (1875).

Rhizome thick, creeping, densely scaly; *scales* fusco-brown, ovate-lanceolate, peltately attached, subentire; *frond* 1 cm apart, dimorphous, stipe stramineous, bisulcate above, sparsely scaly, 15 cm long or longer, *sterile lamina* varies from simple lanceolate to pinnate with 1-5 pairs of pinnae under the much elongated lanceolate, free end-pinna, terminated in a prolonged rooting tip; *lateral pinnae* 10-15 cm long, 3-4 cm broad, subsessile, broadly lanceolate, acuminate, base rotundo-cuneate, margin undulate with a few remote, setiform teeth; *texture* thin herbaceous, blackish upon drying, glabrous on both sides; *lateral main veins* distinct, intervening veinlets angularly anastomosing in quadri-angular or hexagonal exappendiculate areolae, 3-rowed between main veins, veinlets free towards margin; *fertile frond* conform, on much longer stipe; *pinnae* greatly contracted; *sori* following the course of veinlets, confluent all over the under surface at last.

Yunnan: Mengtze, *Henry 10825*; Red River, *Hancock 176*; Szemeo, *Henry 12907*. Guangdong: North River, *Ford*, Sept. 1879. Sichuan: Mt. Omei, *E. Faber 1042*. Hainan: Hoichow, *Hancock 10*. Guizhou: Houangtsaopa, *Cavalerie 7018*, *7059*; Lofau, *Cavalerie 7645*. Guangxi: Tsin Lung Shan, Linyin Hsien, *R. C. Ching 6869* (1928).

Vietnam: Mt. Bana, *Balansa 1887* (1886).

Also India generally, Malaysia-Polynesia and the Philippines.

A very distinct but extremely variable fern with frond varying from simple to fully pinnate with terminal pinna always prolonged and rooting at apex. Leaves invariably turn blackish, when dried.

A form with leaf margin deeply crenate, var. **crenata** Ching, var. nov., is noted from Moulmein (leg. Parish, herb. Kew), Tenasserim (leg Geo. Gallatly **64**, herb. Kew) and Christmas Island (leg. C. W. Andrews **126**, herb. Mus. Brit.).

Plate 119. Fig. 1. Habit sketch (natural size). 2. Portion of sterile pinna, showing venation ($\times$ 4). 3. Portion of fertile pinna ($\times$ 4). 4. Scale from rhizome ($\times$ 20). 5. Spores ($\times$ 300).

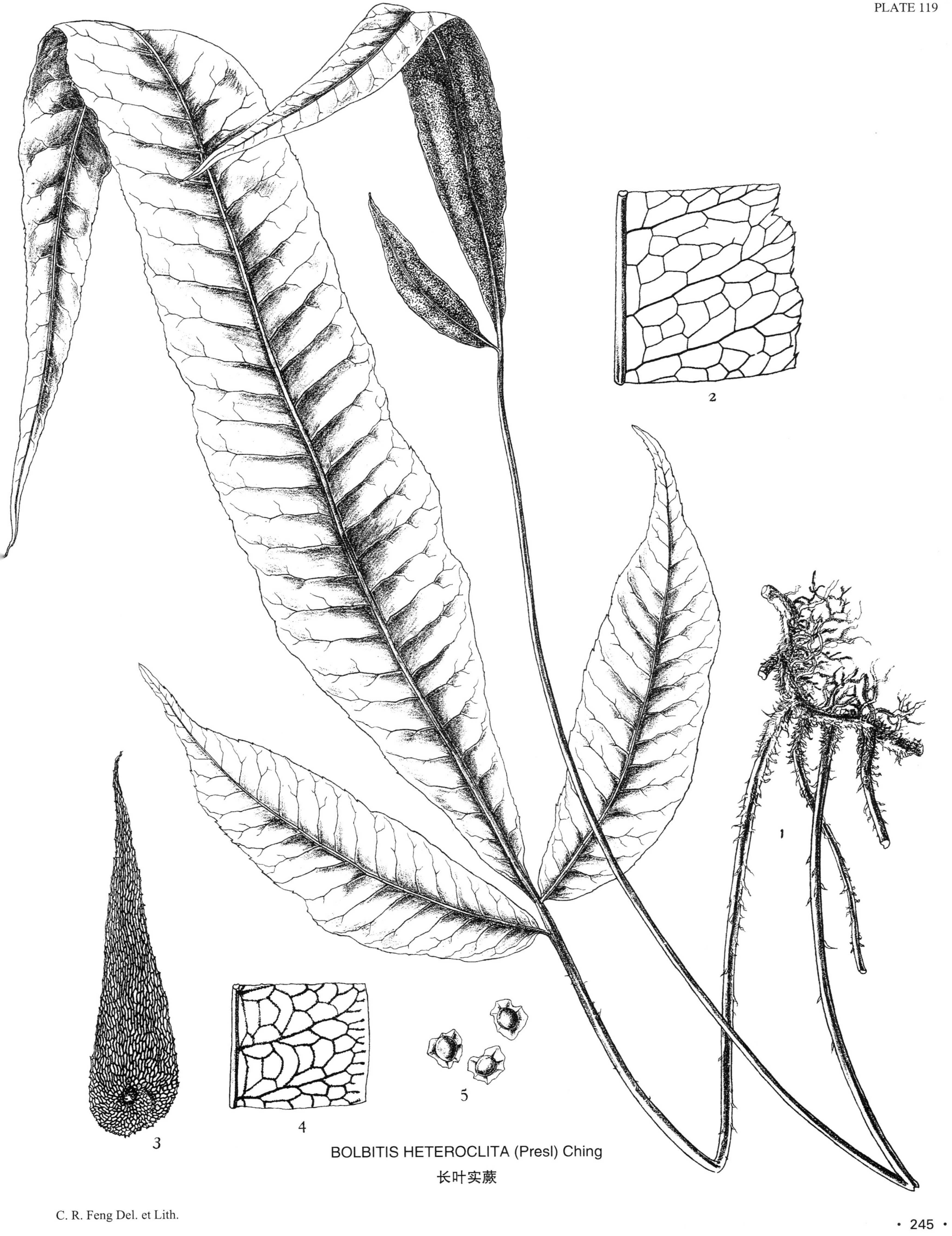

图版 119
PLATE 119

BOLBITIS HETEROCLITA (Presl) Ching
长叶实蕨

C. R. Feng Del. et Lith.

· 245 ·

根状茎肥厚，横行，具深棕色之卵披针形鳞片；叶簇生，二形，柄长 30—60 cm，略具鳞片，非子囊群叶长 20—50 cm，宽 15—28 cm，椭圆形，一回羽状分裂，小叶 4—10 对，披针形，长 9—20 cm，宽 2.5—5 cm，顶端小叶较大，且常伸长其端，入土着根，基部为圆截形，具短柄，渐尖头，叶缘呈深波形，具钝锯齿，而于缺口处具一强大之短肉刺，厚纸质，淡绿色，两面颇光滑；侧脉明显，小脉网状，网眼三列，内具小短脉一或否，向叶缘分开；子囊群叶与非子囊群叶同形，唯叶柄长，小叶较狭，长 6—8 cm，宽约 1 cm，子囊群初随网脉分布，继则满布全面；孢子卵圆形，具阔翅。

分布：海南，广东，广西，香港，台湾，福建；越南亦产之。

本种在香港及海南极为普通，生于林中深溪边，其叶端常延长回向土中着根。

图注：1—2. 本种全形（自然大）；3. 大形之非子囊群叶之一部（自然大）；4. 同上，表明其叶脉（放大 5 倍）；5. 子囊群叶之一部（放大 5 倍）；6. 孢子（放大 200 倍）；7. 根状茎上之鳞片（放大 16 倍）。

BOLBITIS SUBCORDATA（Cop.）Ching

BOLBITIS SUBCORDATA (Cop.) Ching in C. Chr. Ind. Fil. Suppl. III. 50 (1934).
    Campium subcordatum Cop. Phil. Journ. Sci. **37**: 369 f. 23, t. 16 (1928).
    Leptochilus subcordatum Wu, Bull. Dept. Biol. Sun Yatsen Univ. No. **3**: 92 t. 37 (1932).
    Acrostichum repandum Benth. (non Bl. 1828) Fl. Hongk. 443 (1961).
    Heteroneuron proliferum Hk. Florul. Hongk. in Journ. Bot. (1857) 339.
    Leptochilus cuspidatus var. *crenata* Ros. Hedwigia **56**: 348 (1915).
    Leptochilus sp. Merr. Enum. Pl. Hainan in Lingnan Sci. Journ. **5**: 10 (1927).

Rhizome thick, creeping, densely scaly; *scales* fusco-brown, ovate-lanceolate, acuminate, peltately, attached, clathrate, nearly entire; *frond* dimorphous, caespitose, stipe 30-60 cm long, terete underneath, bisulcate above, sparcely scary, *sterile lamina* 20-50 cm long, 15-28 cm broad, oblong, impari-pinnate; *pinnae* 4-10-jugate under the free trifurcate terminal one, which often prolongated and rooting at apex, 9-20 cm long, 2.5-5 cm broad, broadly lanceolate, opposite, the upper ones smaller and alternate, acuminate, base roundish or rotundo-truncate, shortly petiolate, margin crenate throughout, the rounded lobes bluntly serrate with a prominent seta at the sinus, rachis roundish underneath, broadly bisulcate above, glabrous on both sides; *lateral main veins* distinct, patent, 7 mm apart, intervening veinlets angularly anastomosing in 3 rows of areolae between main veins and 2 rows from costa outwards, areolae shortly appendiculate or not, veinlets free towards margin; *fertile frond* similar to the sterile one but much contracted, 7-10 cm broad; *pinnae* 6-8 cm long, about 1 cm broad; *sori* indefinite, following the course of veinlets, confluent all over the under surface at last; *spores* bilateral, ovoid, broadly winged all around.

Hainan: Five Finger Mt., *F. A. McClure 8725*, *9346*, *9370*, *9436* (type), Spring-May, 1922, shaded moist ravines; Tun Foo, *Eryl Smith 1641*, stream side, Jan. 14, 1923; Lingshui Hsien, *H. Fung 20146*, May 3-20, 1932. Hong Kong: *Hance 23*; *Dr. Naumann 763* (1869); *Schottmuller*; *Lamont 989*; *Dr. Harland*; *Bowring 22*; *Matthew 577*; *Robinson 8* (1925); *Katsumatta 6692*, June, 1909. Zhejiang: Pinyang, *H. H. Hu 156*. Guangdong: Swatow, Thaiyong, *Dalziel*, April, 1901; Lohfau Shan, *N. K. Chun 40626*; Yao Shan, Ku Koong, *Y. K. Wang 31559*; Shek Mang Tai Shan, *C. L. Tso 23591*, August 7, 1933; Hu Lang Hong, *W. Y. Chun 7117*, August 23, 1931. Fujian: Foochow, Kushan, *Metcalf 7380*, Dec. 5, 1927; *Alexander*. Guangxi: Yao Shan, *S. S. Sin 714*. Taiwan: *Faurie 281* (*L. cuspidatus* var. *crenata*); Tamsui, *Hancock 15*; Keelung, *C. W. Wilford 478*.

Vietnam: *Pételot*, Jan. 1922, a sterile frond.

Our fern, which is found to be abundant in shaded ravines in the Island Hong Kong, is closely related to *B. Quoyana* (Gaud.) Ching from Malaya, differs chiefly in its not coadunate but free apical pinna and rotundo-cuneate base of the lateral pinnae.

Plate 120. Fig. 1-2. Habit sketch (natural size). 3. Sterile frond of a larger form (natural size). 4. Portion of sterile pinna, showing venation (× 5). 5. Portion of fertile pinna (× 5). 6. Spores (× 200). 7. Scale from rhizome (× 16).

BOLBITIS SUBCORDATA (Cop.) Ching
海南实蕨

图版 121　肿足蕨

根状茎横行于岩隙中，密覆鲜明红棕色膜质披针形之鳞片，长逾 2—3 cm；叶近生，柄长 10—25 cm，稻秆色，其基部为瘤状膨大，为密鳞片所覆，不之见也。叶体五角形，10—25 cm 或过之，四回羽状深裂，纸质，两面被刚毛，一回小叶 5—10 对，具柄，基部一对最大，椭圆三角形，急尖头，末回裂片椭圆形，钝头，具大锯齿；叶脉明显，开离，羽状分叉；子囊群圆形，着于叶脉上，盖大，圆肾形，以下方缺口着生，具刚毛。

分布：本种广布于亚洲温热带各地，达于非洲之东南部及南洋群岛；在我国之云南，四川，贵州，湖北，广东，广西，福建，江苏，浙江，安徽均产之。

本种之特征为其好生于干燥之石岩隙中，其叶柄之瘤状基部为鲜红棕色之大鳞片所覆被，昔人以之隶于 Dryopteris 属，最近经余之研究，应另属以待之也。

图注：1. 本种全形（自然大）；2. 末回小叶，表示其叶脉及子囊群（放大 6 倍）；3. 子囊群盖（放大 20 倍）；4—5. 叶柄基部鳞片（放大 10 倍）；6. 叶下面之刚毛（放大 30 倍）；7. 孢子（高倍放大）。

HYPODEMATIUM CRENATUM (Forsk.) Kuhn

HYPODEMATIUM CRENATUM (Forsk.) Kuhn, v. Deck. Reis. Bot. 3^3: 37 (1879).

Polypodium crenatum Forsk. Fl. Aeg.-Arab. 185 (1775).

Aspidium crenatum Kuhn, Fl. Afr. 129 (1868); Christ, Farnkr. d. Erde 262 (1897); Bull. Herb. Boiss. 6: 195 (1898); in Warburg, Monsunia 1: 81 (1900); Bull. Soc. Bot. France 52: Mém. 1. 35 (1905).

Nephrodium crenatum Baker, Fl. Maur. 497 (1877); Clarke, Trans. Linn. Soc. II. Bot. 1: 524 (1880); Diels in Engl. u. Prantl: Nat. Pflanzenfam. 1: 4. 175 (1899); Dunn & Tutch. Fl. Kwangt. & Hongk. 348 (1912); Hand-Mzt. Symb. Sin. 6: 25 (1929).

Lastrea crenata Bedd. Ferns Brit. Ind. Suppl. 18 (1876); Handb. Ferns, Brit. Ind. etc. 258 (1883).

Dryopteris crenata O. Ktze, Rev. Gen. Pl. 2: 811 (1891); C. Chr. Ind. Fil. 258 (1905); Christ. Bot. Gaz. 51: 348 (1911); Sim, Ferns of S. Afr. 111 t. 22 (1915); Ogata, Ic. Fil. Jap. 3: t. 113 (1930); C. Chr. Acta Hort. Gothob. 1: 63 (1924); Pterid. of Madag. 56 (1933).

Hypodematium onustum Kze, Flora 1833^2: 690.

Aspidium eriocarpum Wall, List n. 342 (1828, nomen nudum); Mett. Farngatt. Pheg. u. Aspid. 60 (1857).

Nephrodium eriocarpum Dcne. Arch. Mus. 2: 185 (1841); Hk. Sp. Fil. 4: 141 (1862).

Lastrea eriocarpa Presl, Tent. Pterid. 77 (1836); Bedd. Ferns S. Ind. t. 95 (1862).

Nephrodium odoratum Baker in Hk. Syn. Fil. 280 (1867).

Dryopteris Fauriei Kpdama in Matsum. Ic. Pl. Koish. 2: 11 t. 90 (1914). For further synonymy see C. Chr. Ind. Fil. p. 258.

Rhizome 4-6 mm thick, wide-creeping, densely clothed in bright brown linear-lanceolate *scales* to 3 cm long; *frond* approximate, stipe 10-25 cm long, straminous, densely clothed at the swollen base in a cushion of golden brown scales similar to those on rhizome, glabrous upwards, deeply grooved and hirsute above, lamina deltoid-pentagonous, 10-25 cm long and broad, 4-pinnatifid; *pinnae* 5-10-jugate, the basal ones much the largest, petiolate; *pinnules* of 2nd. order oblong, bluntish, base rotundo-cuneate, equal, decurrent above the middle, deeply pinnatifid into oblong, rounded lobato-serrate *segments* with roundish apex; *texture* herbaceous, both sides densely setose hairy; *veins* distinct, free, pinnate in segment; *sori* large, rounded, dorsal on veinlets, *indusium* large, rotundo-reniform, attached by a deep sinus, densely setose hairy.

Widely distributed in Malaysia-Polynesia, Japan, the Philippines, French Indo-China, India, Arabia Abyssinia to South and East Africa, Cap Verde Islands and Maritius. In China, numerous specimens have been seen from Yunnan, Sichuan, Guizhou, Hubei, Guangdong, Guangxi, Fujian, Jiangsu, Zhejiang and Anhui.

Type from Arabia and now found non-existant in the herb. Forskal in the Botanical Museum, Copenhagen. Though variable in size, degree of pinnation and density of pubescense, etc., the different geographic forms agree well with each other in all essential characters.

Plate 121. Fig. 1. Habit sketch (natural size). 2. Ultimate pinnule, showing venation and sori ($\times$ 6). 3. Indusium ($\times$ 20). 4-5. Scales from the base of stipe ($\times$ 10). 6. Hairs from underside of lamina ($\times$ 30). 7. Spores (greatly enlarged).

HYPODEMATIUM CRENATUM (Forsk.) Kuhn 肿足蕨

本种形体酷似前种而瘦小，叶质较薄，柄长约二倍叶体，遍体不被长刚毛，仅球杆形之腺毛疏生，子囊群盖为膜质，形小，背及边缘亦具同样之腺毛，其末回小叶具较尖之锯齿。

分布：广东北部，他处则尚未之见也。

图注：1.本种全形（自然大）；2.叶柄之瘤状膨大基部（自然大）；3.末回小叶，表示其叶脉及子囊群（放大 7 倍）；4.子囊群盖（放大 30 倍）；5.着生于叶柄瘤状基部之鳞片（放大 10 倍）；6.7.8.根状茎之横切面（放大 8 倍）；9.叶柄瘤状基部之横切面（放大 10 倍）；10.叶柄上部横切面（放大）；11.茎下面之腺状毛（高倍放大）；12.孢子（高倍放大）。

HYPODEMATIUM FORDII（Baker）Ching

HYPODEMATIUM FORDII（Baker）Ching, comb. nov.

Nephrodium fordii Baker, Journ. Bot (1889) 177; Dunn & Tutch. Fl. Kwangt. & Hongk. 348 (1912).

Dryopteris crenata C. Chr. Ind. Fil. 258 (1905), pro parte.

Dryopteris fordii Ching, *Sinensia* **3**: 330 (1933).

Dryopteris crenata var. *Fordii* C. Chr. Ind. Fil. Suppl. Ⅲ. 86 (1934).

Rhizome slender, creeping, densely scaly; *scales* linear-lanceolate, golden brown, to 2 cm long; *frond* approximate, stipe 20-30 cm long, straminous, glossy, flexuose, the swollen base densely scaly, lamina deltoid-pentagonous, 4-pinnatifid; *pinnae* 6-9-jugate, petiolate, the basal ones much the largest, deltoid, short-acuminate; *pinnules* 6-jugate, petiolulate, deltoid-oblong, acute, base cordate; *pinnules* of 2nd. order oblong-ovate, the lower ones shortly petiolulate, the upper decurrent, deeply pinnatifid, with 3-6 pairs of oblong, rounded serrate or lobato-incised segments; *texture* thin herbaceous, both sides shortly glandular hairy; *veins* distinct, free, pinnate; *sori* rounded, dorsal on veinlets, *indusium* thin, membranaceous, gray, glandular hairy, rotundo-reniform, attached by a deep sinus, margin glandularly ciliate.

Guangdong: North River, *C. Ford 104* (type), Dec. 1888; Shiuhing, Tingwu Shan, *S. P. Ko 50597*, June 12, 1930.

This is a decidedly distinct fern, known so far only from Guangdong, differing from the preceding species in its gracil habit, thinner leaves, slender stipe almost twice as long as lamina, very membranaceous grayish, smaller indusium with glandular margin, and in its decidedly different type of hairs, which are not needle-like, but only shortly glandular with rounded or clavate, lemon-yellow enlarged apices. Dunn an Tutcher (*loc. cit.*) have distinguished this fern from the preceding by its "acute ultimate segments", neglecting altogether the characters as noted above.

Plate 122. Fig. 1. Habit sketch (natural size). 2. The swollen basal part of stipe (natural size). 3. Ultimate pinnule, showing venation and sori (× 7). 4. Indusium (× 30). 5. Scale from the swollen base of stipe (× 10). 6, 7, 8. Cross sections of rhizome (× 8). 9. Cross section of the swollen base of stipe (enlarged). 10. Cross section of the upper part of stipe (enlarged). 11. Glandular hairs from the underside of lamina (enlarged). 12. Spores (enlarged).

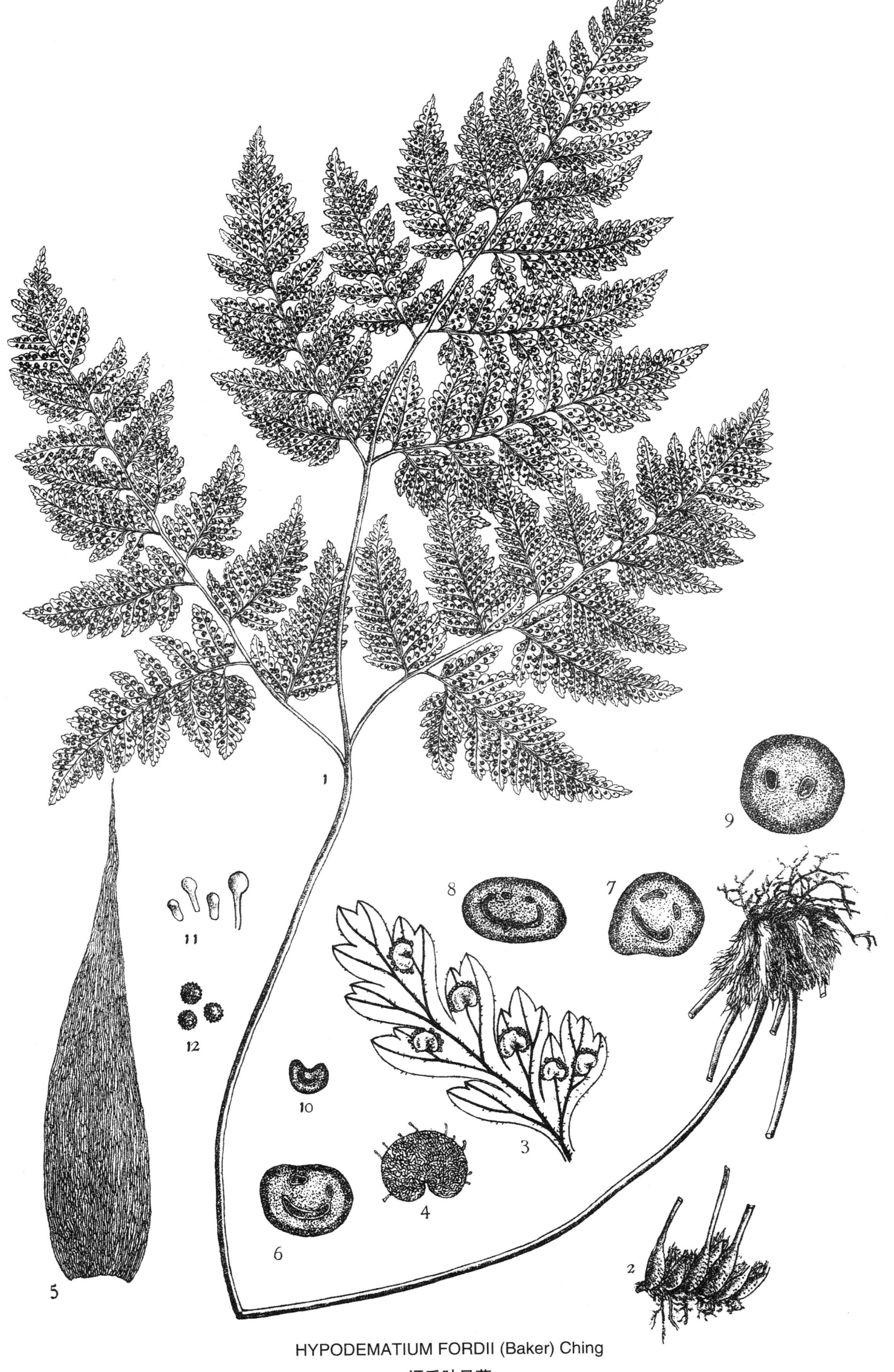

HYPODEMATIUM FORDII (Baker) Ching 福氏肿足蕨

图版 123　山东肿足蕨

　　本种形体酷类前种，唯其小叶及子囊群盖之形体大异，其小叶为狭披针形，或椭圆披针形，彼此不如前种之密接，顶为长渐尖头，基部呈强度之楔形，其子囊群盖不为圆肾形，但为三角状之披针形，或为卵形，或为卵状三角形，绝类 Cystopteris 属也。

　　分布：山东之济南，烟台，泰山，青岛一带及陕西中部，他处尚未之见也；此种为本属分布极北之一种，昔人误认为肿足蕨，实则大异；其模式标本藏于济南齐鲁大学生物系。

　　图注：1. 本种全形（自然大）；2. 末回小叶，表示叶脉及子囊群（放大 14 倍）；3. 着生于叶柄瘤状基部之鳞片（放大 10 倍）；4. 子囊群盖（放大 20 倍）；5. 孢子（高倍放大）；6. 腺状毛。

HYPODEMATIUM CYSTOPTEROIDES Ching

HYPODEMATIUM CYSTOPTEROIDES Ching, Sp. nov.

Aspidium crenatum Christ, Nuov. Giorn. Bot. Soc. Ital. n. s. (1897) 94; Bull. Bot. Soc. Ital. (1901) 297.

Status et configuratione cum H. Fordii (Baker) e China australis valde congruens, differt; indusiis tenuisissimis non rotundo-reniformibus sed valde variantibus, aut lanceolato-triangularibus, aut ovatis, aut ovato-triangularibus, basi vix cordatis, pinnulis anguste lanceolatis vel oblongo-lanceolatis, elongatis, sese multo separatis, longe acuminatis, basi magis cuneatis.

Rhizome slender, wide-creeping, densely scaly; *scales* rufo-brown, to 7 mm long, ovate-lanceolate, acuminate, entire, densely clothing the swollen base of stipe; *frond* approximate, long-stipitate, stipe 10-22 cm long, gracil, flexuose, palecolored, naked throught, lamina ovate-deltoid, 7-11 cm long and as broad, long-acuminate, tripinnate; *lateral pinnae* about 8-jugate under the pinnatifid acuminate apex, the basal ones much the largest, deltoid, opposite, to 10 cm long, 6 cm broad, basicopically produced, long-petiolate, the upper pinnae gradually abbreviated, oblong-lanceolate or lanceolate, all shortly petiolate with equal, subcuneate base, rather wide apart from each other, the second pair of pinnae from the base to 6 cm long, 2 cm broad near the base, pinnate; *pinnules* 7-jugate, rather far apart, oblong-lanceolate, to 1.2 cm long with subequal, cuneate, decurrent base and bluntly dentate apex, pinnatifid to the costule into 6-7 pairs of oblong or sparcely dentate *segments*; *veins* forked, ending in blunt teeth; both sides sparcely glandular hairy; *texture* thin herbaceous, color green; *sori* 5-7 pairs to each segment, medium-sized, rounded, *indusium* whitish, membranaceous, cystopteroid, triangular-lanceolate or ovate-deltoid with broad and hardly cordate base, cucullately attached to the receptacle.

Shandong: Tsinan, *A. P. Jacob 21*; ibid. Chien Fo Shan, specimens ex herb. Dept. Biol. Cheeloo Univ. (without collector's name); Taishan, *A. P. Jacob 1*; Chefoo, *E. Faber 1054* (herb. Kew, sub *Dryopteris crenata*). Shaanxi: Mt. Uansanpin prope "Pagoda", *Giraldi*, July 16, 1894; Seekintzuen, *Giraldi*, Dec. 28, 1895; Thaepaeshan, *Giraldi*, August, 1896.

Our fern is more closely related to H. Fordii (Baker) than to H. crenata (Forsk.) in gracil habit and type of hairs, but differs in its long-acuminate apices of frond and of lateral pinnae, which are elongate oblong-lanceolate with cuneate base in the middle pinnae, and in its cystopteroid indusium, a character which alone markes this species very unique in the genus.

Plate 123. Fig. 1. Habit sketch (natural size). 2. Ultimate pinnule, showing venation and sori (× 14). 3. Scale from the swollen base of stipe (× 10). 4. Indusium (× 20). 5. Spores (enlarged); 6. Glandular hairs.

HYPODEMATIUM CYSTOPTEROIDES Ching
山东肿足蕨

根状茎肥厚，横行，端具黝褐色之硬质披针形鳞针；叶近生，柄长 20—30 cm，深稻秆色，基部鳞片疏生，遍体具刚毛，叶体三角形或五角形，长宽各逾 25 cm，一回小叶 1—2 对，顶部小叶形亦相同，唯基部呈楔形，基部一对最大，具柄，三裂，中裂片最大，渐尖头，几为全缘，上部小叶或有或无，椭圆披针形，渐尖头，基部呈斜形，边缘呈深波状；纸质，干则呈黝褐色，下面具刚毛，上面唯各回中肋及主脉被深棕色之毛；叶脉网状，网眼或具小脉或否；子囊群圆形，颇大，着生于数小脉之交错点，一列，略近叶缘，盖圆肾形，质厚。

分布：广东，广西，香港，台湾，海南，福建。印度及越南亦产之。

本种为南方习见之品，而尤以香港为最多。

图注：1.本种全形（自然大）；2.叶体之一部，表示叶脉及子囊群之位置（放大 5 倍）；3.叶下面之刚毛（放大 120 倍）；4.叶柄基部之鳞片（放大 20 倍）；5.子囊群及盖（放大 50 倍）。

TECTARIA SUBTRIPHYLLA（Hk. et Arn.）Copeland

TECTARIA SUBTRIPHYLLA（Hk. et Arn.）Copeland, Phil. Journ. Sci. Bot. **2**：410（1907）; Ching, *Sinensia* **2**：33（1931）; C. Chr. Ind. Fil. Suppl. Ⅲ. 185（1934）.

Polypodium subtriphyllum Hk. et Arn. in Bot. Beech. Voy. 256 t. 50（1836-40）; Christ in Warburg, Monsunia **1**：75（1900）.

Aspidium subtriphyllum Hk. Sp. Fil. **4**：52（1862）; Bedd. Ferns Brit. Ind. Suppl. 14（1876）.

Nephrodium subtriphyllum Baker in Hk. Syn. Fil. 296（1874）.

Sagenia subtriphylla Bedd. Ferns S. Ind. t. 242（1863）.

Rhizome thick, creeping, clothed at the growing tip in blackish, rigid entire, lanceolate *scales*; *frond* approximate, stipe 20-30 cm long, dark straminous, scaly near the base, sparcely setose hairy throughout, lamina deltoid-pentagonous; *lateral pinnae* 1-2-jugate under the distinct, petiolate, pinnatifid large terminal pinna with cuneate base, the basal ones much the largest, petiolate, trifoliolate or trilobed with central lobe much the largest to 16 cm long, 12 cm broad, more or less lobed under the acuminate entire apex, base subrounded or cordate, the second pinnae (if there any) oblong-lanceolate, acuminate, lobato-sinuate, base oblique and slightly adnate; *texture* herbaceous, underside rather copiously setose hairy, upperside glabrous except costa and rachis clothed in short reddish hairs; *venation* distinct, anastomosing with or without included veinlets in areolae; *sori* large, rounded, compital, scattered or uniseriate in the lobes, leaving a broad sterile space along costa, *indusium* rotundo-reniform, thick, fallen off at last.

Besides those cited in my previous paper, additional specimens have since been noted as follows:

Hainan: Hung Mo Mt., *W. T. Tsang 18082*; Tai Hang, Lin Hwa Shan, *W. T. Tsang 15850*; Top of Dome, Five Finger Mt., *W. Y. Chun 6367* in herb. Univ. Nanking. Guangdong: Lokchong, *N. K. Chun 42789*; *C. L. Tso 21339*; Lohfau Shan, *N. K. Chun 41506*; *C. O. Levine 1405*; Honam Island, *C. O. Levine 1808*; Namhoi, *S. P. Ko 50051*; Tingwu Shan, *H. T. Ho 60081*. Hong Kong: *Y. Tsiang 701*. Fujian: Foochow, *Warburg*. Taiwan: *Drs. F. & C. Baker*, Dec. 20, 1914.

Vietnam: Lang Son, *Balansa 49*（1885）; *Colani 2745*; Thai Nguyen, *Colani 3401*; Bac Giang, *Colani 3406*, *2747*.

This is a very common fern in South China and particularly, the Island Hong Kong, and closely related to *T. variolosa* (Wall.) from which it differs by the characters as already noted in my previous paper.

Plate 124. Fig. 1. Habit sketch (natural size). 2. Portion of lamina, showing venation and position of sori (× 5). 3. Hairs from lamina (× 120). 4. Scale from the base of stipe (× 20). 5. Sorus with indusium (× 50).

TECTARIA SUBTRIPHYLLA (Hk. et Arn.) Copeland 三叉蕨

根状茎肥厚,横行,具红棕色之卵披针形鳞片;叶近生,柄长 25—35 cm,呈光亮深栗色,叶体三角形,二回羽状深裂,顶部羽状深裂;一回小叶 2—4 对,基部一对最大,具短柄,基部心脏形,三角披针形,长 10—20 cm,宽达 12 cm,羽状深裂,上部小叶概为合生,间为羽状深裂,裂片镰形,全缘,位于基部者为阔披针形,长达 6—7 cm,宽约 2 cm,波状深裂;薄纸质,上面被深棕色之短毛,下面有同样之毛疏生,叶缘具睫毛,各位中肋呈光亮之深栗色;叶脉网状,网眼具小脉或否;子囊群圆形,一列,位于裂片之中肋两侧,着生于网眼内之小脉端,盖椭圆形,棕色。

分布:云南,四川,贵州;印度北部,越南,泰国,斯里兰卡均产之。

本种形态稍似前种,唯其叶柄及各位中肋呈光亮之深栗色,叶体二回分裂,叶质较薄,上面及叶缘具深棕色之毛,子囊群着生于网眼内之小脉端,故易区别。

图注:1.本种全形(自然大);2.裂片,表示其叶脉及子囊群之位置(放大 10 倍);3.子囊群与盖(放大 20 倍);4.叶上面之毛(放大 80 倍);5.根状茎上之鳞片(放大 24 倍)。

TECTARIA MACRODONTA (Fée) C. Christensen

TECTARIA MACRODONTA (Fée) C. Christensen, Ind. Fil. Suppl. III. 181 (1934).

Sagenia macrodonta Fée, Gen. Fil. 313 (nomen) t. 24A f. 1 (1852).

Aspidium coadunatum Wall. (non Kaulf. 1824) Cat. 337 (1828, nom. nud.); Hk. et Grev. Ic. Pl. t. 202 (1831); Mett. Fil. Lips. 94 t. 22 f. 3-4 (1856); Christ, Bull. Acad. Géogr. Bot. (1910) 13.

Sagenia coadunata Bedd. Ferns S. Ind. t. 81 (1873); Christ, Bull. Soc. Bot. France **52**: Mém. 1. 34 (1905).

Tectaria coadunata C. Chr. Contr. U. S. Nat. Herb. **26**: 331 (1931); Ching, Sinensis **2**: 18 pls. 1-11 (1931).

Aspidium cicutarium var. *coadunatum* C. Chr. Ind. Fil. 65 (1905).

Nephrodium cicutarium Hk. Sp. Fil. **4**: 48 (1862); Baker in Hk. Syn. Fil. 299 (1867), pro parte.

Nephrodium cicutarium var. *coadunatum* Clarke, Trans. Linn. Soc. II. Bot. **1**: 540 (1880).

Aspidium cicutarium var. *tenerifrons* Christ, Bull. Acad. Géogr. Bot. (1902) 257.

Sagenia apiifolia Christ (non J. Sm. 1841), Bull. Acad. Géogr. Bot. (1906) 120.

Aspidium pin-faense Christ, Bull. Acad. Géogr. Bot. (1909) 169.

Aspidium kwanonense Hayata, Ic. Pl. Form. **8**: 137 f. 61 (1918).

Rhizome thick, creeping, densely scaly; *scales* rufo-brown, ovate-acuminate; *frond* approximate, stipe 25-35 cm long, glossy, dark castaneous, lamina deltoid, bipinnatifid under the coadunate pinnatifid terminal pinna; *lateral pinnae* 2-4-jugate, the basal ones much the largest, shortly petiolate, cordate, deltoid-lanceolate, 10-20 cm long, to 12 cm broad, pinnatifid to a narrow wing along costa, the upper pinnae more or less adnate, also deeply pinnatifid with entire, falcate segments, of which the lower ones broadly lanceolate, 5-7 cm long, about 2 cm broad, acuminate, lobato-incised; *lobes* falcate, rounded, entire; *texture* thin herbaceous, upperside densely pubescent with short, clear, reddish hairs, underside sparcely so, margin ciliate, rachis and costa shining castaneous underneath; *veins* anastomosing, areolae elongate, mostly without including veinlets; *sori* rounded, far apart, regularly uniseriate on each side of costule of segment, mostly apical on the including veinlets, *indusium* rotundo-reniform, large brown, fallen off at last.

As previously cited, the species is known from Yunnan, Sichuan and Guizhou in China, besides Northern India, Tailand and Peninsular India. Additional material recently examined are *K. K. Tsoong 2530* from Tali, Yunnan, and *Esquirol 1031* from Houangtsaopa, Guizhou.

Plate 125. Fig. 1. Habit sketch (natural size). 2. Segment, showing venation and position of sori ($\times$ 10). 3. Sorus with indusium ($\times$ 20). 4. Hairs from lamina ($\times$ 80). 5. Scale from rhizome ($\times$ 24).

TECTARIA MACRODONTA (Fée) C. Christensen 高山三叉蕨

根状茎肥短,斜出或直立,具光亮红棕色之卵形密鳞片,长逾 1 cm;叶簇生,柄长 15—30 cm,具密鳞片,并有披针形之狭鳞片混生,叶体椭圆形,长 15—45 cm,宽 10—17 cm,一回奇数羽状分裂,小叶 10—20 对,互生,具短柄,位于基部者与其上部者等长或略短,长可达 10 cm,宽 2 cm,镰形,长渐尖头,基部较宽,圆形或上方呈耳形,叶缘具细密锯齿,达于极顶,厚纸质,淡绿色,上面无光泽,中肋及叶柄具纤维状之鳞片;叶脉网状,网眼斜出,内具 1—2 数通直斜出之小脉;子囊群圆形,不规则排列,着生于网眼内之小脉上,盖圆形,质厚,灰色,盾状着生。

分布:长江流域各省均产之,北达陕西之南部,南及广东北部之乐昌县;日本及朝鲜半岛南部亦甚普通,最近在越南发现;欧美各国均盆栽以供玩赏。

本种为我温暖各省习见之品,且常群生一处,其形态变异甚大,小者高不及尺,小叶呈卵形,长仅 2—3 cm 耳。

图注:1.本种全形(自然大);2.阔叶变种(自然大);3.小叶变种(自然大)。

CYRTOMIUM FORTUNEI J. Smith

CYRTOMIUM FORTUNEI J. Smith, Ferns Brit. & Fore. 286 (1866); Hist. Fil. 205 (1875); L. H. Bailey, Manu. Cult. Pl. 71 (1924); C. Chr. Amer. Fern Journ. **20**: 49 (1930); Ind. Fil. Suppl. Ⅲ. 66 (1934); Tagawa, Acta Phytotax et Geobot. **3**: 61 f. 5-9 (1934).

Polystichum fortunei Nakai, Bot. Mag. Tokio **39**: 116 (1925).

Polystichum falcatum var. *Fortunei* Nichols; Matsum. Ind. Pl. Jap. **1**: 342 (1904); C. Chr. Ind. Fil. 581 (1905).

Cyrtomium falcatum (non Presl) C. Chr. Ind. Fil. Suppl. Ⅰ. 101 (1912); L. H. Bailey, Gentes Herb. (1920) 9.

Aspidium falcatum Hk. (non Sw.) in Blakiston, Five Months on the Yangtze 364 (1862); Baker in Hk. Syn. Fil. 257 (1867); Franch. et Sav. Enum. Pl. Jap. **2**: 234 (1879), pro parte.

Polystichum falcatum var. *polypterum* Diels in Engl. Jahrb. **29**: 195 (1900), pro parte.

Cyrtomium falcatum var. *polypterum* Christ, Bull. Soc. Bot. France **52**: Mém. 1. 33 (1905); Bull. Acad. Géogr. Bot. (1906) 250.

Aspidium falcatum var. *caryotideum* Christ, Nuov. Giorn. Bot. Soc. Ital. n. s. (1897) 93.

Cyrtomium lonchitoides Christ, Bull. Soc. Bot. France 52: Mém. 1. 33 (1905), pro parte.

Polystichum lonchitoides Nakai (non Diels, 1900), Bot. Mag. Tokio **23**: 77 (1914).

Cyrtomium vittatum Christ ex Léveille, Bull. Acad. Géogr. Bot. (1910, non 1905).

Rhizome short, oblique or erect, densely scaly; *scales* large, 1 cm or longer, rufo-brown or atro-castaneous, glossy, ovate, acuminate; *fronds* caespitose, stipe 15-30 cm long, strong, densely clothed in the similar scales, which are mixed with others of linear outline, lamina oblong, 15-45 cm long, 10-17 cm broad, pinnate under the free hastate end-pinna; *pinnae* 10-20-jugate, alternate, lower ones hardly shortened, in typical form, up to 10 cm long, 2 cm broad, falcate, long-acuminate, shortly petiolate, base broadened, rounded or auricled above, margin finely serrate throughout; *texture* chartaceous, light green, not glossy in living state, rachis fibrillose-scaly throughout; *veins* distinct, copiously reticulated, areolae hexagonal, oblique, each with 1-2 excurrent included soriferous veinlets; *sori* rounded, scattered over the underside, terminal or infra-apical on the veinlets, *indusium* large, coriaceous, grayish, rounded, peltately attached, depressed at centre, with wavy margin.

This is one of the most common ferns throughout the Yangtze valley and extends as far north as the southern part of Shaanxi and southwardly to the north of Canton (Lokchong, *W. T. Tsang 20706*), Also Japan and southern Korean peninsula; recently known from Vietnam.

A very distinct but very variable fern, generally regarded hitherto as a variety of *C. falcatum*, from which it can be easily told apart by its much thinner leaves of a light green color, not glossy, in living state, its generally more numerous, lanceolate-falcate or sometimes oblong-ovate (as in var. *polypterum*) pinnae always with minutely and regularly serrate margin from the base upwards. Generally grown as pot plant.

Plate 126. Fig. 1. Habit sketch (natural size). 2. Forma *latipinna* (natural size). 3. Forma *polypterum* (Diels).

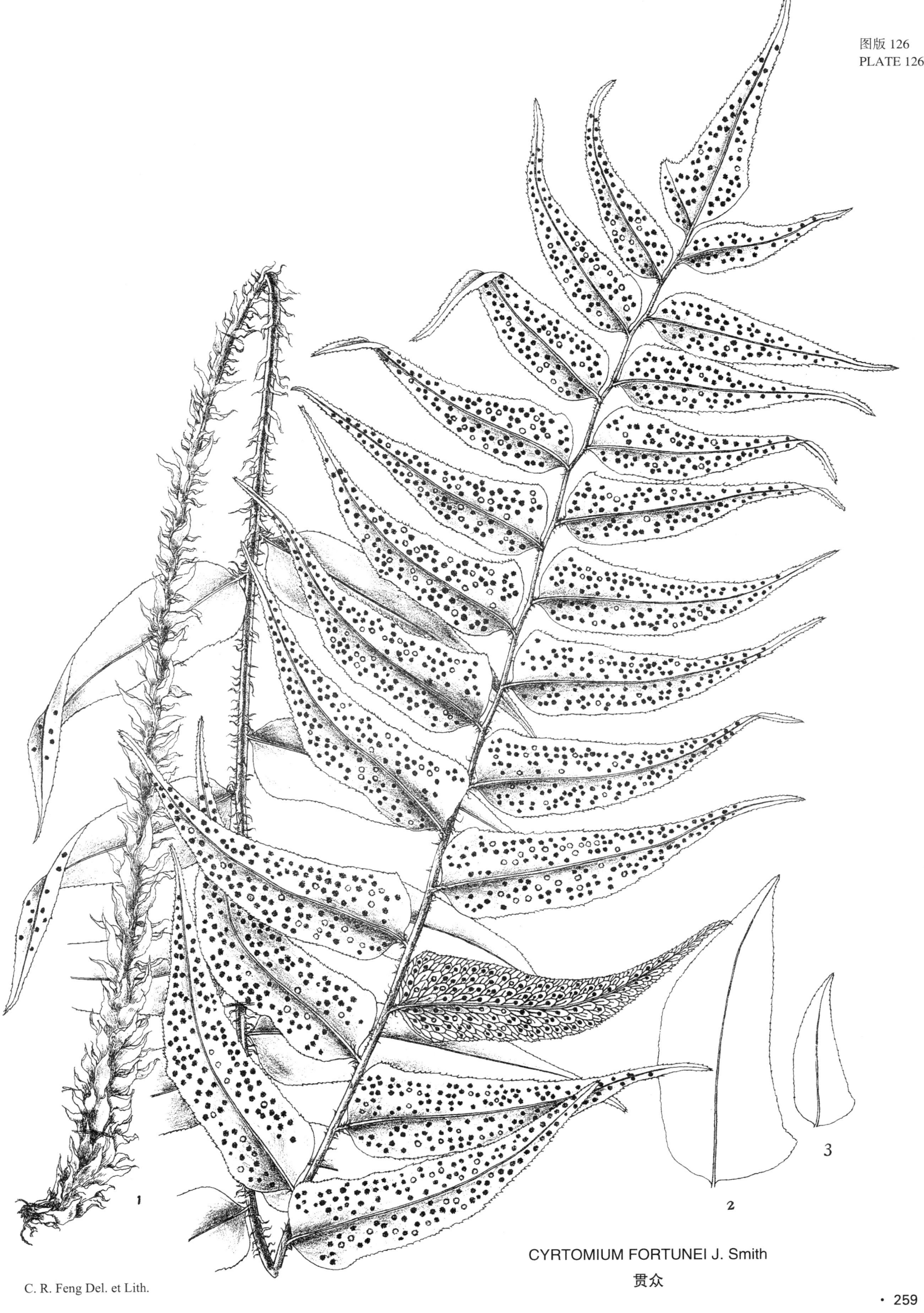

CYRTOMIUM FORTUNEI J. Smith 贯众

根状茎短肥，直立，具大卵形之红棕色密鳞片；叶簇生，柄长 20—30 cm，稻秆色如前种，具二形密鳞片；叶体椭圆形或椭圆披针形，长 15—40 cm，宽 6—13 cm，奇数羽状分裂，小叶 1—10 对，互生，卵状镰形，长渐尖头或尾状，基部圆形或上方呈圆耳形，具短柄，全缘，虽极顶无锯齿，基部小叶长 5—8 cm，宽 3—4 cm，上部者渐小，革质，上面呈光绿色，中肋及叶柄具纤维状之鳞片；叶脉网状，网眼及子囊群之着生一如前种，盖革质，圆形，盾状着生。

分布：江苏，浙江，福建，香港，广东，广西，贵州，台湾均产之，但不如前种之普通耳；日本，朝鲜半岛亦多习见；欧美各国均盆栽以供观赏。

此外有变种二种：一为长叶变种 Var. devexiscapulae Tagawa，产于日本及我国南部诸省，其小叶呈长披针形；一为裂片变种：Var. acutidens (Christ) C. Chr.，产于日本及朝鲜半岛，小叶形体如原种，唯边缘具大锯齿或披针形之裂片。

图注：1. 本种全形（自然大）；2. 下部小叶，表示叶脉及子囊群（自然大）；3. 子囊群盖（放大 24 倍）；4. 叶柄基部鳞片（放大 10 倍）；5. 中肋上之鳞片（放大 10 倍）。

CYRTOMIUM FALCATUM (L. fil.) Presl

CYRTOMIUM FALCATUM (L. fil.) Presl, Tent. Pterid. 86 (1836); Link, Sp. Fil. Hort. Berol. 164 (1841); Fée, Gen. Fil. 286 (1850-52); Moore, Ind. Fil, 277 (1861); Hk. Florul. Hongk. in Journ. Bot. (1857) 340; J. Sm. Ferns Brit. & Fore. 142 (1866); Hist. Fil. 204 (1875); C. Chr. Ind. Fil. Suppl. I. 101 (1912); Amer. Fern Journ. **20**: 48 (1930); Christ in Warburg, Monsunia **1**: 76 (1900); Bull. Acad. Géogr. Bot, (1910) 5; Tagawa, Acta Phytotax. et Geobot. **3**: 59 (1934).

Polypodium falcatum L. fil. Sp. Pl. Suppl. 446 (1781); Thunb. Fl. Jap. 336 (1784); Poir. Encyc. Bot. **5**: 527 (1804).

Aspidium falcatum Sw. in Schrad. Journ. Bot. (**1800**^{2}): 31 (1801); Syn. Fil. 43 (1806); Hk. et Arn. in Bot. Beech. Voy. 274 (1836); Hk. Lond. Journ. Bot. **1**: 494 (1842); Fil. Exot. t. 92 (1857); Sp. Fil. **4**: 40 (1862); Syn. Fil. 257 (1867); Benth. Fl. Hongk. 454 (1861); Lowe, Ferns **6**: t. 9 (1857); Franch. et Sav. Enum. Pl. Jap. **2**: 336 (1879); Christ, Frankr. d. Erde 232 (1897).

Dryopteris falcata O. Ktze, Rev. Gen Pl. **2**: 812 (1891).

Polystichum falcatum Diels in Engl. u. Prantl. Nat. Pflanzenfam. **1**: 4. 194 (1899); C. Chr. Ind. Fil. 581 (1905); Matsum. et Hayata, Enum. Pl. Form. 583 (1906); Yabe, Bot. Mag. Tokio **17**: 65 (1903); Nakai, Journ. Sci. Coll. Univ. Tokio **31**: 399 (1911); Brause ex Loess. Prod. Fl. Tsingtau. in Bot. Centralbl. **37**: 80 (1920).

Aspidium falcatum var. *genuinum* Makino, Bot. Mag. Tokio **10**: 212 (1896).

Polystichum falcatum var. *genuinum* Matsum. Ind. Pl. Jap. **1**: 342 (1904).

Polypodium japonicum Houtt. Nat. Hist. **14**: 167 t. 98 f. 3 (1783); Pfl. Syst. **13**: 187 t. 98 f. 3 (1786).

Rhizome short, erect, densely clothed in large, broadly ovate, suddenly acuminate, atro-brown, membranaceous *scales* with fimbriate margin; *fronds* caespitose, stipe 20-30 cm long, straminous, angular, densely scaly on the lower part, scales similar to those on rhizome and mixed with smaller, linear-lanceolate ones, lamina oblong-lanceolate, 15-40 cm long, 6-13 cm broad, impari-pinnate; *pinnae* 1-10-jugate, alternate, the lower ones much the largest, 5-8 cm long, 3-4 cm broad, broadly ovate-falcate, long-attenuated towards apex, base petiolate and rounded, the upper ones gradually diminished, the uppermost ones lanceolate-falcate, 5 cm long, 1.5 cm broad, subsessile, with the anterior base rounded or bluntly auricled, margin thickened, entire, or repando-undulate, never serrate (in f. typica) even at the apex; *texture* coriaceous, glossy green in living state, brownish when dried, rachis and particularly the base of petiole copiously fibrillose-scaly; *venation* rather distinct, copiously reticulated in the same manner as in *C. Fortunei*; *sori* rounded, scattered, *indusium* coriaceous, rounded, entire, deppressed at centre.

Jiangsu: Shanghai, on the way to Woosung, *Alexander*; Tseng Ming Hsien, *L. C. Tso 1542*, August 18, 1926; Haichow, Littoral Ledges, *Jacob 36*; Nanking, *10515* (pro parte), cultivated. Fujian: Foochow, *Dr. Grijs 56* in herb. Hance; Diongloh, *Chen Ping-en 2408*, July 24, 1925. Shandong: Tsingtau, Laushan, *Küntzel 45*; Chefoo, *Hancock 6*. Zhejiang: Shihpu, *C. Y. Chiao 14113*, July 11, 1927; *Mary Matthew 10218* in herb. Edinb. Univ. Guangdong: Swatow, *Gerlach*; *Dalziel*, Sept. 1899; Yingtak. Taishan, *C. C. C. 15074*. Taiwan.

Also Japan and Korean peninsula.

Type from Nagasaki, Japan, collected by Thunberg. In China, this fern is come across only occasionally. Its report as occuring in Africa. Madagascar and Southern India is a mistake for **C. micropterum** (Kunze) Ching, sp. nov. (*Aspidium anomophyllum* var. *micropteris* Kunze, Linnaea **24**: 278. 1851).

Var. **devexiscapulae** (Koidz. pro sp.) Tagawa, Acta Phytotax. et Geobot. **3**: 60 (1934).

Guangdong: Lienhsien, *S. P. Ko 51004*; ibid. Yang Shan, *S. P. Ko 51041*; ibid. Chen Hong, *S. P. Ko 50984*. Guangxi: Wangchin, Luchen, *R. C. Ching 5468*. Guizhou: Yao Ren Shan, Sanhoa, *Y. Tsiang 6308*.

Also Japan.

Differs from the type in much longer and narrower pinnae of a lanceolate outline.

The second variety, **acutidens** (Christ, pro sp.) C. Chr. Amer. Fern. Journ. **20**: 49 (1930), with laciniately serrate or lobed pinnae, is known only from Japan and Korean peninsula.

Plate 127. Fig. 1. Habit sketch (natural size). 2. Lower pinna, showing venation and sori (natural size). 3. Indusium ($\times$ 24). 4. Large scale from the base of stipe ($\times$ 10). 5. Small scale from rachis ($\times$ 10).

CYRTOMIUM FALCATUM (L. fil.) Presl
全缘贯众

图版 128 拟贯众

根状茎肥厚,斜出,具黄棕色之线状密鳞片,长达 2 cm,宽仅 1 mm,质软,边具小锯齿;叶簇生,柄长 30—40 cm,稻秆色或黄褐色,具黄色纤维状之鳞片,叶体椭圆形,奇数羽状分裂,小叶 10—15 对,互生,具短柄,节状着生于中肋,长 8—13 cm,宽达 3 cm,披针形,渐尖头,基部一对几与上部者等长,基部不等,上方呈圆耳形,下方斜形,叶缘向上部呈浅波状,纸质,绿色,中肋及主脉下面具纤维状密鳞片;叶脉分离,三至四叉,曲折;子囊群大圆形,着生于小脉上,三列,盖大,圆形,革质,黄褐色,楯状着生,宿存或久则脱落;孢子卵形,具阔翅。

分布:海南及越南特产。

本种形体颇似 C. Cumingiana(Fée) Ching,然后者之叶体具短柄,小叶密接,位于下部者之叶体具短柄,小叶密接,位于下部者渐短,其基部呈等形之圆耳形,故易与本种区别。

图注:1.本种全形(自然大);2.小叶之一部,表示其叶脉及子囊群之位置(放大 3 倍);3.叶柄基部鳞片(放大 10 倍);4.小叶在中肋上节状着生情形(放大 10 倍);5.孢子(高倍放大)。

CYCLOPELTIS CRENATA (Fée) C. Christensen

CYCLOPELTIS CEENATA (Fée) C. Christensen, Ind. Fil. Suppl. Ⅲ. 64 (1934).

Hemicardion crenatum Fée, Gen. Fil. 283 t. 22A f. 1 (1850-52).

Hemicardion cochin-chinense Fée, 1. c. (nom. nud.)

Cyclopeltis presliana C. Chr. Ind. Fil. 197 (1905), pro parte; Bonap. Notes Pterid. Pt. **7**: 99, 153 (1918); Merr. Enum. Hainan Pl. in Lingnan Sci. Journ. **5**: 9 (1927), non Berkeley, 1857.

Rhizome, short, thick, oblique, scaly; *scales* darkish brown, linear-subulate, 1.5-2 cm long, 1 mm broad, soft, with scarious margin; *fronds* caespitose, stipe 30-40 cm long, 5 mm thick above base, angular, straminous or brownish, fibrillose-scaly throughout the rachis, impari-pinnate; *lateral pinnae* 10-15-jugate, alternate, oblique, shortly petiolate, articulated to rachis, lower ones hardly shortened, about 5 cm apart from each other, the upper ones closer, 8-13 cm long, to 3 cm broad, lanceolate, acuminate, base unequal, *i. e.* the upperside oblique, the lower side cordate, with a rounded hamato-recurvate auricle against the rachis, margin sinuate, entire towards apex; *texture* chartaceous, green, costa fibrillose-scaly underneath; veins 4-forked, flexuose; *sori* large, rounded, dorsal on veinlets, 3-seriate on each side of costa, *indusium* large, coriaceous, dark brown, peltate, persistent; *spores* bilateral, broadly winged all around.

Hainan: Nodoa, Sha Po Ling, *F. A. McClure 8164*, *8168*, *8196*, moist shaded ravines of water edge; *Katsumatta*, in herb. Hong Kong; Ling Shui Hsien, *F. A. McClure 20068*, May 4-20, 1932; Yaichow, *F. C. How 71124*; *C. Wang 33487*, *34195*, *Eryl Smith 1620*, *1633*.

Vietnam: *Gaudichaud* (type); ibid., *J. & S. M. Clemens 4168*.

This decidedly distinct species, endemic in Vietnam and the Island Hainan, has hitherto been considered as identical with C. *Presliana* (J. Sm.) Berkeley, now should be known as **C. Cumingiana** (Fée) Ching, comb. nov., as the species was actually first properly described by Fée under his genus *Hemicardion* in Gen. Fil. 283 t. 22A. f. 2, from the Philippine Islands and Malaysia, which differs in more numerous lateral, narrower falcate and close pinnae, of which the lower ones are gradually shortened towards the very short stipe, with deeply cordate-auricled base at *both* sides and in much thicker leaves, turning blackish when dried.

Plate 128. Fig. 1. Habit sketch (natural size). 2. Portion of pinna, showing venation and position of sori (× 3). 3. Scale from the base of stipe (× 10). 4. Basal part of pinnae, showing articulation to rachis (× 10). 5. Spores (enlarged).

CYCLOPELTIS CRENATA (Fée) C. Christensen 拟贯众

图版 129　瓦鳞耳蕨

　　根状茎肥厚，直立，具光亮黄褐色之线形密鳞片，长逾 2 cm；叶簇生，叶长 15—18 cm，具大卵形之开张密鳞片；叶体为披针形，长达 30 cm，宽 5—6 cm，基部截形，一回羽状分裂，顶部为短渐尖头，羽状深裂；小叶 25—36 对，互生，密接或瓦覆，长 2.5—3 cm，宽 1.3 cm，镰椭圆形，钝头，基部具短柄，不等，上方呈尖耳形（有时分离），下方斜形，叶缘具长刺状之密锯齿，中肋及主脉下面具纤维状之密鳞片，革质，叶脉三叉或羽状分叉；子囊群圆形，一列，近于叶缘，着生于每群基部上方之一小脉之端，盖革质，圆形，黄褐色，盾状着生。

　　分布：贵州南部特产，他处尚未之见。

　　本种为本属特殊之一种，罕有与此相似之种，仅采得二次，盖为珍品也。

　　图注：1. 本种全形（自然大）；2. 小叶，表示其叶脉及子囊群着生情形（放大 2 倍）；3. 叶柄上之鳞片（放大 10 倍）；4. 中肋之鳞片（放大 12 倍）。

POLYSTICHUM FIMBRIATUM Christ

POLYSTICHUM FIMBRIATUM Christ, Bull. Acad. Géogr. Bot. (1906) 237; C. Chr. Ind. Fil. I. 64 (1912).

　　Rhizome thick, erect, clothed in a dense tuft of shining brown linear-subulate *scales* over 2 cm long; *fronds* caespitose, stipe 15-18 cm long, thick, densely clothed in large, ovate-acuminate, spread or deflexed, rufo-brown scales, lamina lanceolate, to 30 cm long, 5-6 cm broad, base truncate, apex short-acuminate, pinnate; *pinnae* 25-36-jugate, alternate, imbricate, very shortly petiolate, basal ones as long as those above, 2.5-3 cm long, 1.3 cm broad, oblong-falcate, blunt, base deltoidly auricled above (auricle sometimes free in the lower pinnae), cuneate below, with truncate upper inner edge parallel to, or imbricating on, the rachis, margin subentire with aristate teeth all around, rachis and underside of lamina copiously fibrillosely scaly; *texture* thick coriaceous; *veins* prominent underneath, pinnate in auricles and triforked upwards; *sori* rounded, uniseriate, nearer to the margin than to the costa, terminating the anterior basal veinlet of each group, *indusium* thick, brown, entire, rounded, peltately attached.

　　Guizhou: South of Tinfan, *Cavalerie and Fortunat 1842* (type), Oct. 1904; Pingchow, *Y. Tsiang 7134*, Sept. 14. 1930.

　　This species represented so far only by two collections from the southern part of Guizhou, proves to be so remarkably distinct that there is no known Chinese species to which it can be closely compared. It belongs to the group of *P. nepalense* (Spreng) from which it differs altogether in densely scaly stipe and rachis, in entire and aristately serrate pinnae with blunt apices and without cartilaginous teeth and the characteristically ovate-cuspidate appressed scales on the underside of pinnae and in supramedial sori with coriaceous and much smaller indusia. In the gross habit, our fern appears related to *Cyrtomium nephrole-pioides* Christ from the same region, but differs in its free venation, oblong-falcate pinnae with unequal auricled base and different scale characters.

　　Plate 129. Fig. 1. Habit sketch (natural size). 2. Pinna (× 2). 3a-b. Scales from stipe (× 10). 4. Scale from rachis (× 12).

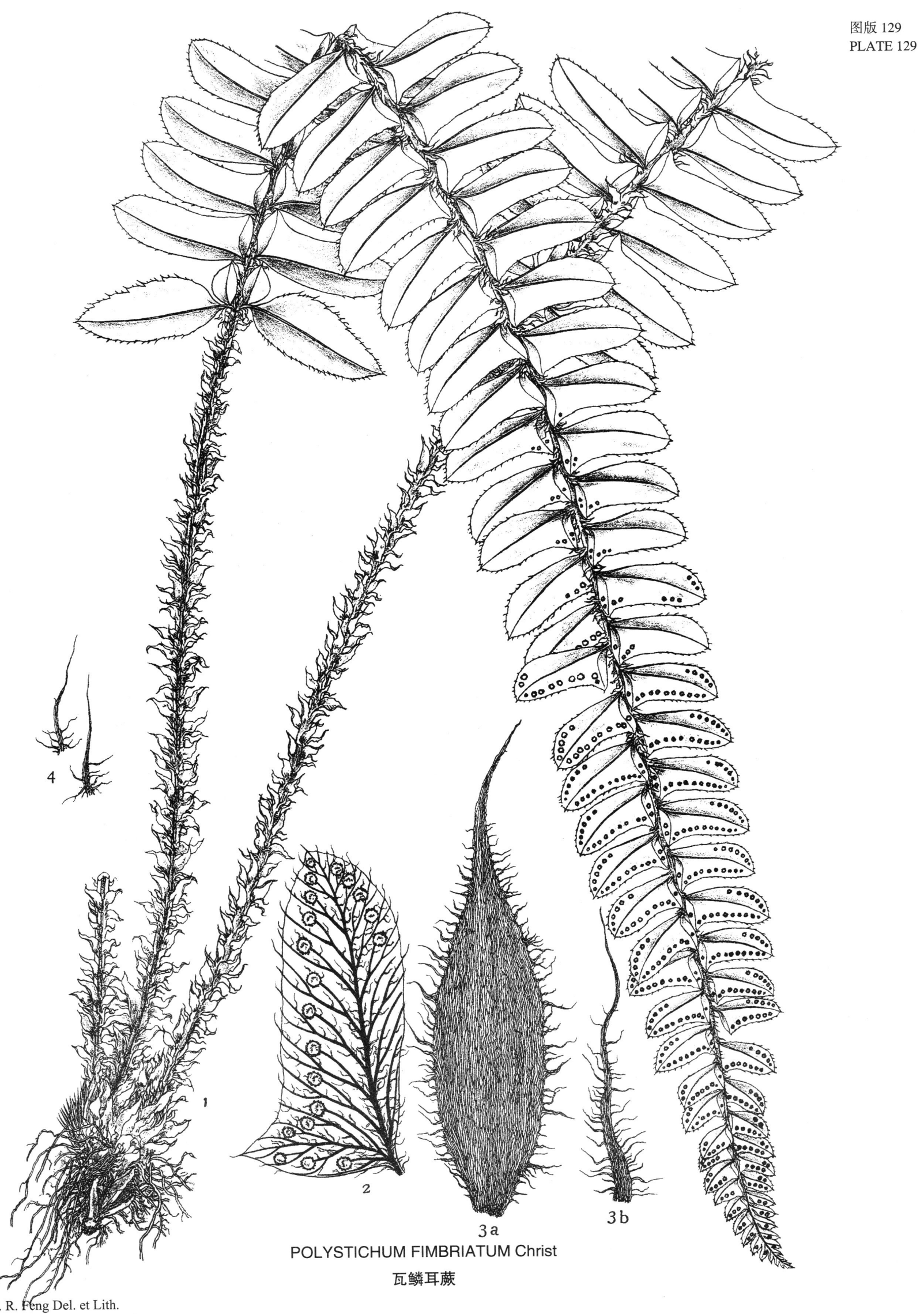

POLYSTICHUM FIMBRIATUM Christ
瓦鳞耳蕨

根状茎肥短，直立，具大卵形深棕色瓦覆状之密鳞片；叶簇生，柄长 5—10 cm，具二形密鳞片，叶体披针形，长 15—22 cm，宽 4—5 cm，一回羽状分裂，顶部渐尖头，一回深裂，小叶 20—30 对，密接，具短柄，开展，位于基部者下向，与上部者等长，长 2—2.5 cm，宽约 7 mm，披针形，渐尖头，基部不等，上方具尖耳形之深裂，合生或分离，下方斜形，叶缘具刺状刚锯齿，硬革质，中肋及主脉下面具纤维形之鳞片；叶脉三叉，在耳形小叶者羽状分叉；子囊群小圆形，近于叶缘，一列，着生于每群基部上方之一小脉之端，盖圆形，革质，灰色。

分布：四川西部高山特产，他处尚未之见。

本种亦为我国西部特种蕨类，略似欧洲北部产之 Polyst. lobatum（Linn），然各部则大异。

图注：1. 本种全形（自然大）；2. 大型小叶（自然大）；3. 模式标本之小叶（放大 3 倍）；4—5. 叶柄上之鳞片（放大 14 倍）。

POLYSTICHUM OTOPHORUM (Franch.) Beddome

POLYSTICHUM OTOPHORUM (Franch.) Beddome, Handb. Ferns, Brit. Ind. etc. Suppl. 42 (1892); Diels in Engl. u. Prantl: Nat. Pflanzenfam. **1**: 4. 190 (1899); C. Chr. Ind. Fil. 585 (1905); Acta Hort, Gothob. **1**: 68 (1924); Christ, Bot. Gaz. **51**: 345 (1911).

Aspidium otophorum Franch. Nouv. Arch. Mus. II. **10**: 116 (1887).

Rhizome thick, short, erect, densely scaly; *scales* large, dark brown, ovate-acuminate, fimbriate; *fronds* caespitose, stipe 5-10 cm long, clothed in dense spread, lanceolate blackish and fibrillose scales, lamina lanceolate, 15-22 cm long, 4-5 cm broad, pinnate under the short, pinnatifid, acuminate apex; *pinnae* 20-30-jugate, patent, shortly petiolate, the basal ones deflexed and as long as those above, 2-2.5 cm long, 7 mm broad above the base, which is provided above with a prominent auricle, either adnate or free, obliquely cuneate below, margin regularly serrate with long aristate teeth; *texture* rigidly coriaceous, rachis and underside of pinnae more or less fibrillose-scaly; *veins* pinnate in auricle, 3-forked upwards; *sori* small, nearer to the margin than to the costa, terminating the anterior basal veinlet of each group above the auricle, *indusium* coriaceous, rounded.

Sichuan: Moupin, *David* (type) (1890); Mt. Omei, *E. H. Wilson 2624*, *2596*, *2598*, *5363A*; *Brown 26*, *136*; Kuan Hsien, *F. T. Wang 20422* (f. typica); Tang Ho, *Legendre 1614*; Yuchikou, *Harry Smith 2354*; *T. T. Yü 4611*, *2742* (f. ad P. *xiphophyllum*).

A very pretty and distinct endemic species of the group of P. *lobatum* (L.) of Northern Europe. The nearest Chinese relative to this fern is P. *xiphophyllum* (Baker) from Mt. Omei, which differs chiefly in bipinnatifid frond of much larger size.

Plate 130. Fig. 1. Habit sketch (natural size). 2. A pinna from a larger form (natural size). 3. Pinna from the typical form (× 3). 4-5. Scales from stipe (× 14).

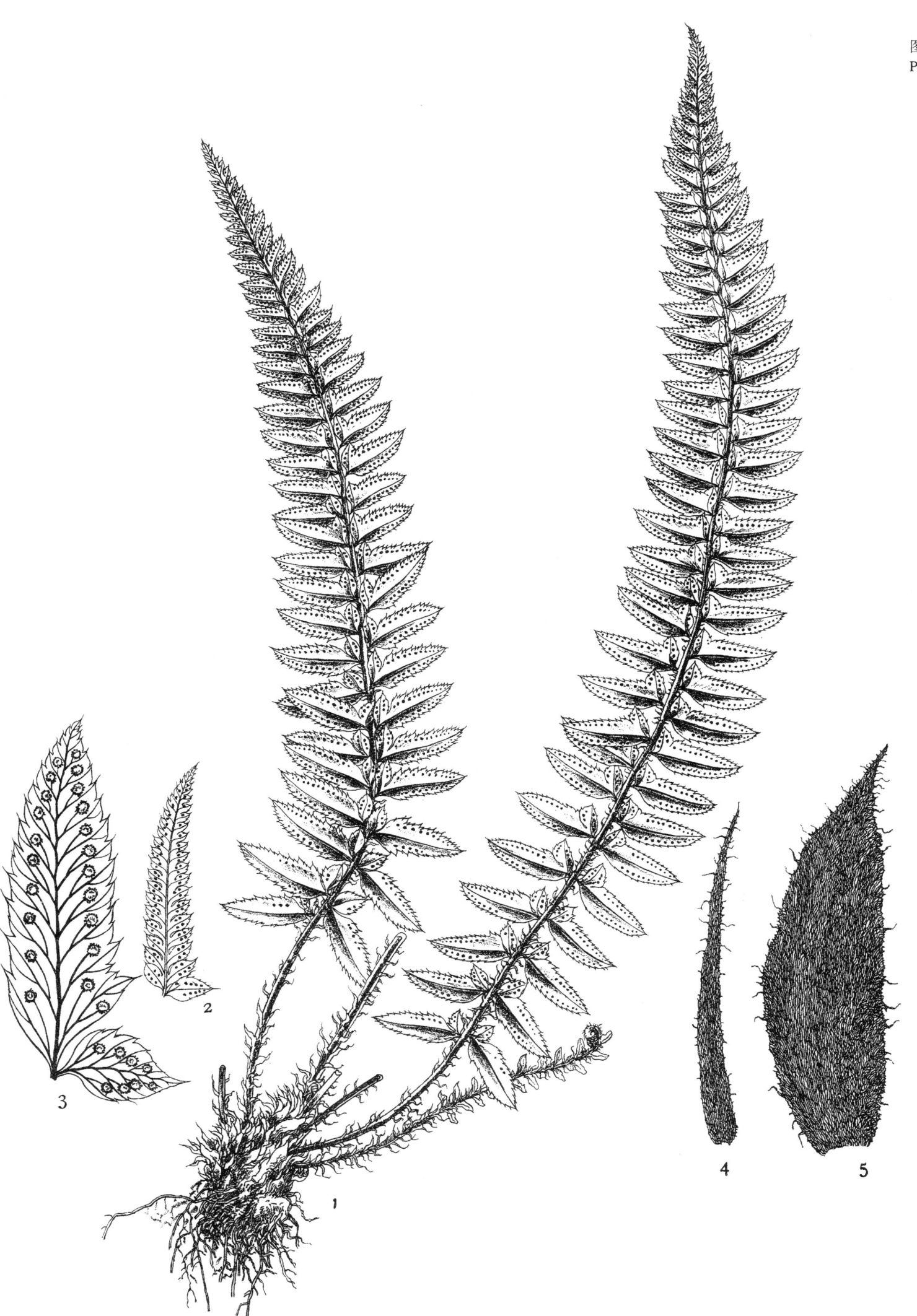

POLYSTICHUM OTOPHORUM (Franch.) Beddome
高山耳蕨

图版 131　峨眉耳蕨

根状茎短而直立或斜出，端具棕色卵形小鳞片；叶多数簇生，柄长达 13 cm，稻秆色，坚实，叶体倒披针形，长 15—25 cm，宽 4—7 cm，渐尖头，细密三回羽状分裂，一回小叶多数，密互生，几无柄，开展，披针形，长 2—3 cm，基部者稍短且下向；二回小叶 6—12 对，卵椭圆形，基部渐狭，循中肋延长，末回小叶 2—3 对，通常深二裂，裂片短，线形，宽不及 1 mm，锐尖头，全缘，具一小脉，不达于顶，纸质，绿色，下面略具小鳞片；子囊群小圆形，每裂片一个，着生于叶脉之端，盖圆形，厚膜质，与裂等宽或较宽，不久脱落。

分布：四川峨眉山，云南之蒙自及贵州南部。

本种为我国西部高山特产，其叶体细密分裂成线形尖头之小裂片，宽不及 1 mm，具一小脉，顶端着一个子囊群。

图注：1. 本种全形（自然大）；2. 二回小叶（放大 14 倍）；3. 子囊盖（放大 45 倍）；4. 根状茎上之鳞片（放大 40 倍）。

POLYSTICHUM OMEIENSE C. Christensen

POLYSTICHUM OMEIENSE C. Christensen (non Christ, 1909), Ind. Fil. 67 (1905), 585 (1906); Acta Hort. Gothob **1**: 72 (1924).

　　Aspidium carvifolium Baker, Journ. Bot. (1888) 228; Christ, Bull. Bot. Soc. Ital. (1901) 295; Bull. Acad. Géogr. Bot. (1906) 114 (non Kunze, 1851).

　　Polystichum carvifolium Christ, Bull. Herb. Boiss. **6**: 969 (1898); Diels in Engl. Jahrb. **29**: 194 (1900).

　　Polystichum faberi Christ in Lecomte, Not. Syst. **1**: 37 (1909).

Rhizome short, oblique, densely scaly at apex; *scales* small, brown, ovate-acuminate, subentire; *fronds* caespitose, 10-18 together, stipe to 13 cm long, straminous, terete, firm, sparcely clothed in small, ovate-acuminate, uniform scales, lamina slightly oblanceolate, 15-25 cm long, 4-7 cm broad, acuminate, finely tripinnate; *pinnae* numerous, close, alternate, subsessile, patent, lanceolate, 2-3 cm long, lower ones somewhat shortened and deflexed; *pinnules* 6-12-jugate, oblong-ovate, base long-attenuate and decurrent along deeply grooved and narrowly winged costa; *ultimate pinnules* 2-3-jugate, generally deeply bifid into linear-subulate uni-nerved *segments*, hardly over 1 mm broad, with sharp point; texture herbaceous, green, underside sparcely scaly; *sori* small, one to each segment, terminating the veinlet some distance below the apex of segment, *indusium* large, rounded, membranaceous, as broad as segment, soon falls off.

Sichuan: Mt. Omei, *E. Faber 1027* (type); *E. Faber 14* in herb. Hance; ibid *Scallan*; *Wilson 5267*; *W. P. Fang 2494*, *3155*; Nanchuan, *W. P. Fang 5836*, *6118*, *7446* (without exact locality). Yunnan: Mengtze, *Henry 9050*; *Hancock 20* (1893). Guizhou: Pinfa, *Cavalerie 2426*, *7094* (1907); Toushan, *Cavalerie 2536*.

This is one of the most distinct species of the group of *Polystichum* from West China and Himalaya, which is characterized by finely dissected fronds with linear-subulate uni-nerved entire and sharply pointed ultimate segments. *P. alcicorne* (Baker) from the same locality is a very close relative, which differs, however, in more coarsely dissected frond of a much stouter habit and in the presence on the stipe and rachis underneath of the large ovate, membranaceous brown persistent scales.

Plate 131. Fig. 1. Habit sketch (natural size). 2. Pinnule ($\times$ 14). 3. Indusium ($\times$ 45). 4. Scale from rhizome ($\times$ 40).

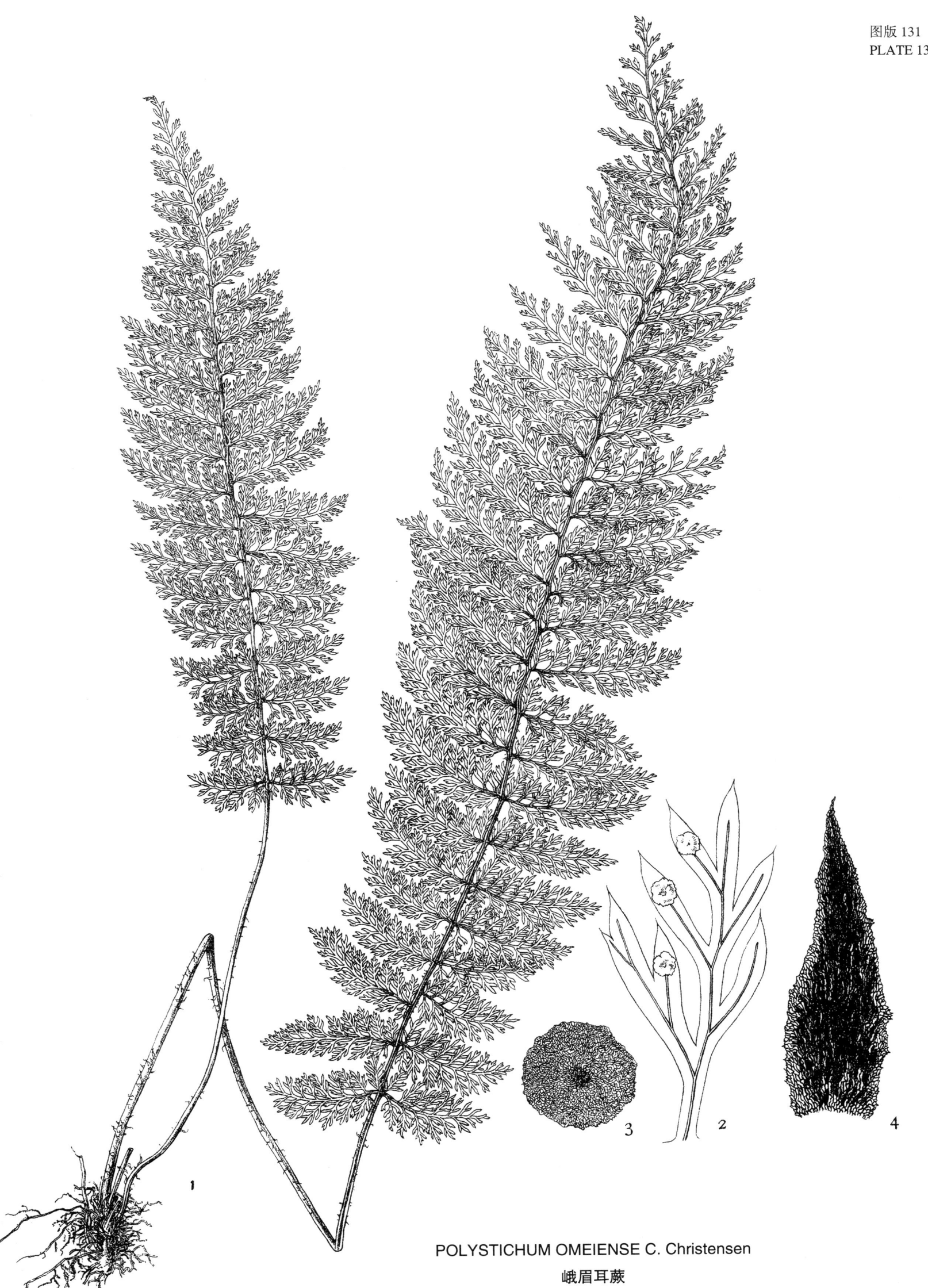

POLYSTICHUM OMEIENSE C. Christensen
峨眉耳蕨

根状茎短而直立,端具线形硬质棕色光亮小鳞片;叶多数簇生,柄长 6—10 cm,坚实,横切面圆形,略具鳞片,黄稻秆色,叶体卵椭圆形或三角形,长 4—6 cm,宽 4 cm,革质,二回羽状分裂,中肋上面具疏腺毛及一沟槽,一回小叶 3—4 对,对生或上部者亚对生,无柄,三角形,长宽达 2 cm,顶端小叶同形,羽状深裂,裂片 3—5 对,线形,急尖头,长 1—1.5 cm,宽 1—1.2 mm,全缘,基部沿中肋延长,主脉上面呈沟槽,下面突出,侧脉斜出,二叉;子囊群贴近叶缘,着生于小脉顶端,盖深褐色,膜质,线形,直达顶端,宽达于主脉,与其对面之盖相接。

分布:四川西部及云南西北部之高山,生于露出岩石上。

本种形体颇类 *Pellaea nitidula* Baker,然后者之叶柄为深栗色或乌木色,叶体分裂较密,中肋上面具较长较密之短刚毛,故易与本种区别。

图注:1.本种全形(自然大);2.裂片之一部,表示其叶脉及子囊群之位置及其盖(放大 6 倍);3.根状茎上之鳞片(放大 20 倍)。

PELLAEA SMITHII C. Christensen

PELLAEA SMITHII C. Christensen, Acta Hort. Gothob. **1**: 84 t, 18 (1924); Ind. Fil. Suppl. Ⅲ 135 (1934).

Rhizome short, erect, densely scaly at the apex and the base of stipe; *scales* small, linear, rigid, brown, shining, the outer ones are often shining black; *fronds* numerous, caespitose, stipe 6-10 cm long, firm, rounded, sparcely scaly throughout, brownish-straminous, lamina ovate-oblong or deltoid, 4-6 cm long, 4 cm broad, coriaceous, bipinnate, rachis glandular, grooved above; *lateral pinnae* 3-4-jugate under the pinnatifid deltoid apical part similar to the lateral ones, sessile, opposite or subopposite, deltoid, to 2 cm long and broad, the basal pair hardly larger than those above, pinnatifid to the costa into 3-5 pairs of linear entire, acute *segments* 1-1.5 cm long, 1-1.2 mm broad, the lower ones adnate, the upper ones decurrent, costa and costules deeply sulcate above, prominently raised underneath; *veins* in segments hidden, biforked, veinlets oblique; *sori* intra-marginal, distinct, terminating each veinlet, *indusium* linear, continuous till very apex, covering the entire under surface, gray at first, brown at last, persistent, with glandular-ciliate margin; *spores* small, minutely verrucose.

Sichuan bor-occid.: Hsuting, *Harry Smith 4799* (type), on sunny rocks; *E. H. Wilson 2664* (1908). Yunnan: east of Yungming, *Forrest 21238*, *10463*.

This peculiarly distinct little fern is closely related to *P. nitidula* (Wall.) Baker from Northern India and Southwestern China, differs in still simpler pinnation, much lighter-colored stipe and rachis, which are only sparcely glandular, and in much broader indusium. As first observed by its author, all the specimens cited above are found to be fertile in every segment.

Plate 132. Fig. 1. Habit sketch (natural size). 2. Portion of segment, showing venation, indusium and position of sori ($\times$ 6). 3. Scale from rhizome ($\times$ 20).

图版 132
PLATE 132

PELLAEA SMITHII C. Christensen
史氏旱蕨

C. R. Feng Del. et Lith.

图版 133　杜氏粉背蕨

　　根状茎短而直立，端具淡黑色而硬质不透明之披针形鳞片；叶簇生，柄长 14—25 cm，光亮无鳞毛，深栗色，横切面圆形，叶体卵三角形，半二回羽状分裂，长宽各 9—12 cm，淡黄色，革质，两面光滑，下面无白粉，中肋及主脉下面呈光亮之栗色，一回小叶 4—5 对，开展，顶部为线披针形全缘或羽状深裂，各小叶基部合生，唯基部一对分离，较大，下边羽状深裂，上边全缘，二回小叶线披针形，3—5 数，位于基部者最长，长约 4—5 cm，宽 5 mm，其上者渐短，有时下部小叶再羽状深裂，上部一回小叶为披针形，全缘或有时第二对亦为半羽状深裂，不具锯齿；侧脉细，数回对称分叉，不显明；子囊群贴近叶缘，着生于小脉之端，盖膜质，线形，达于裂片之极端。

　　分布：云南，四川峨眉山及贵州南部。

　　本种为我国西部特产，其形体酷似普通习见之粉背蕨（*Ch. argentea* Kze.），所不同者为其形体较大，叶体分裂较少及其开展之一回小叶，通常概为半羽状分裂是也，此外另有一变种：Var. sulphurea Ching，叶背面具硫黄色粉。

　　图注：1. 本种全形（自然大）；2. 同上，形体较单简（自然大）；3. 同上，幼植物（自然大）；4. 末回裂片之一部，表示其叶脉（放大 6 倍）；5. 叶柄基部之鳞片（放大 15 倍）。

CHEILANTHES DUCLOUXII (Christ) Ching

CHEILANTHES DUCLOUXII (Christ) Ching in C. Chr. Ind. Fil. Suppl. III. 54 (1934).

　　Doryopteris duclouxii Christ, Bull. Acad. Géogr. Bot. (1902) 231; Bull. Soc. Bot. France **52**: Mém. 1. 58 (1905); C. Chr. Ind. Fil. 244 (1906); Acta Hort. Gothob. **1**: 86 (1924).

　　Doryopteris muralis Christ, Bull. Acad. Géogr. Bot. (1904) 111; C. Chr. Ind. Fil. 244 (1905).

　　Doryopteris mairei Brause, Hedwigia **54**: 206 t. 4. f. J. (1914).

Rhizome short, erect; *scales* at the apex blackish, lanceolate, rigid and opaque; *fronds* caespitose, stipe 14-25 cm long, glabrous, glossy, dark castaneous, terete throughout, lamina deltoid, semi-bipinnate, 9-12 cm each way, falvous, coriaceous, glabrous on both sides, rachis, costa and costules underneath shining castaneous; *lateral pinnae* 4-5-jugate under the lanceolate, entire or pinnatifid apical part, all connected by a narrow wing along rachis, or in full-grown plants the basal pair separated from those above by wingless rachis, basal pinnae much the largest, sub- or semi-deltoid, 5-7 cm long, 4-5 cm broad, semi-pinnatifid, *i. e.* only the lower side of costa with 3-5 lanceolate *segments*, of which the basal ones much the longest, 4-5 cm long, 5 mm broad above base, entire throughout, or sometimes with a few additional short segments on the lower side, while the upperside of costa is entire or irregularly pinnatifid with short segments; the upper lateral pinnae generally simple, entire, falcate, or sometimes the second pair also semi-pinnatifid; *veins* in segments free, dichotomously forked, hidden; *sori* terminating the veinlets, *indusium* gray, membranaceous, entire, continuous to the very apex of segment; *spores* spherical, echinose.

　　Yunnan: *Ducloux* (type), Dec. 23, 1896; Mengtze, *Hancock 31*; *Tanant* (1897); *Maire 2746*, *2737*; *Forrest 298*; *Ducloux 1351* (1909); *K. K. Tsoong*; Makong; *E. E. Maire*, pro parte. Sichuan: Omei region, *H. D. Brown 80* (1928). Guizhou: Pinfa, *Cavalerie 4228*, *1212* (type of *Dory. muralis*).

　　A distinct species but closely related to *Ch. argentea* var. *obscura* Christ, differs in larger size and less divided frond with lanceolate, subhorizontally patent pinnae of falvous color, of which the lower ones are semi-pinnatifid.

　　Var. **sulphurea** Ching, var. nov.

　　Doryopteris duclouxii f. *argentea* Christ, Bull. Soc. Bot. France **52**: Mém. 1. 58 (1905).

　　Doryopteris mairei Brause, Hedwigia **54**: 206 t. 4. f. J (1914).

Differs from the type in frond being coated underneath with sulphur-yellow waxy powder.

　　Sichuan: Ninyuan Hsien, *Harry Smith 1830*; *Wilson 5297*; *Handel-Mazzetti 2790*. Yunnan: Tali, *Forrest 13499*; Ngaykio, *Ducloux 6343* (1909); Tchang Shan, *Ducloux 3363*; Nieou Ko Shan, *Ducloux 6965*; Pintchouan, *Ducloux 6930*; Mt. Mao Kou Tchang, near Tapintze, Tali, *Delavay 9* (1883).

　　This variety must not be confused with *Ch. Veitchii* (Christ) Ching from Sichuan, with which it is very closely related in habit, from which it differs above all in stipe being not glossy black and fronds not so finely dissected.

　　Plate 133. Fig. 1. Habit sketch (natural size). 2. The same, a simpler form (natural size). 3. The same, a young plant (natural size). 4. Portion of segment, showing venation ($\times$ 6). 5. Scale from the base of stipe ($\times$ 15).

CHEILANTHES DUCLOUXII (Christ) Ching
杜氏粉背蕨

图版 134　韩氏粉背蕨

根状茎短而直立，端具深褐色硬质狭披针形之鳞片；叶簇生，柄淡栗色，横切片面呈圆形，长 7—20 cm，基部略具鳞片，叶体五角形或三角形，长宽各 7—14 cm，三回羽状分裂或深裂，顶部渐尖头，羽状深裂，一回小叶 5—7 对，位于下部者具柄，基部一对最大，三角形，长 5—12 cm，二回小叶 5—7 对，基部下方一对最大，长 3—6 cm，具短柄，余皆无柄，位于中肋上边者均等大，合生；末回小叶椭圆形，钝头，边为深波状或深裂，裂片圆形，各位中肋上面具深沟槽，纸质；叶脉明显，2—3 叉，或单一；子囊群贴近叶缘，着生于小脉之端，盖膜质，椭圆形，呈波状。

分布：云南，四川。

本种为我国川滇特产，尤以在滇省为普通，其形体大小不一，故其名甚多。

图注：1. 本种小形全形（自然大）；2. 二回小叶，表明其叶脉及子囊群（放大 10 倍）；3. 本种模式标本全形（自然大）；4. 根状茎上之鳞片（放大 16 倍）。

CHEILANTHES HANCOCKII Baker

CHEILANTHES HANCOCKII Baker, Kew Bull. (1895) 54; Christ, Bull. Soc. Bot. France **52**: Mém. 1. 58 (1905); C. Chr. Ind. Fil. 175 (1905); Suppl. Ⅲ. 54 (1934).

Cheilanthes taliensis Christ, Bull. Soc. Bot. France **52**: Mém. 1. 58 (1905); C. Chr. Ind, Fil. 180 (1906); Acta Hort. Gothob. **1**: 89 (1924); Contr. U. S. Nat. Herb. **26**: 308 (1931).

Cheilanthes henryi Christ, Bull. Acad. Géogr. Bot. **16**: 133 (1906); C. Chr. Ind. Fil. Suppl. 1. 18 (1912).

Cheilanthes wilsoni Christ, I. c. 132; C. Chr. Ind. Fil. Suppl. Ⅰ. 18 (1912).

Cheilanthes bonatiana Brause, Hedwigia **54**; 203 t. 4 f. E (1914); C. Chr. Ind. Fil. Suppl. Ⅱ. 8 (1916).

Rhizome short, erect, densely radicose; *scales* at the apex dark brown, lanceolate-subulate, rigid, extending upward to some distance above the base of stipe; *fronds* caespitose, stipe castaneous, terete throughout, 7-20 cm long, lamina pentagonous or subdeltoid, 7-14 cm long and broad, tripinnate or tripinnatifid; *pinnae* 5-7-jugate, the lower ones petiolate, far apart, opposite, the basal ones much the largest, deltoid, 5-12 cm long; *pinnules* 5-7-jugate, the lower basal one much the largest and basicopically produced, 3-6 cm long, shortly petiolulate, the upper ones on the same side of costa sessile and adnate, gradually shortened, those on the upperside of costa are of about equal length, sessile, adnate, and much shorter than those below, alternate; *ultimate pinnules* oblong, blunt, lobato-incised with rounded *lobes*, or only crenate in small plants; rachis and rachilets deeply grooved above; *texture* herbaceous, *veins* distinct, 2-3-forked or simple towards apex in lobes; *sori* terminating the veinlets, distinct at first, finally subconfluent, *indusium* membranaceous, oblong-ovate, distinct or subcontinuous; *spores* bilateral, light brown.

Yunnan: Mengtze, *Hancock 63* (type) (1893); Szemeo, *Henry 12532* (type of *Ch. Henryi*; Shit Ping, *Henry 13223*; *Ducloux 1342*; Siaosulong, *E. E. Maire* (November); *2749, 1342*; *Ducloux 1343* (pro parte); environs de Yunnansen, *Ducloux & Bodinier 669* (1898); Tchangshan, *Ducloux 2361* (1906); *Cavalerie 4710* (1900-20); environs de Hay Tien, *Ducloux 2427* (1904); Laokong Shan, *Ducloux 5062*; Tcheou Kiatzetang, *Maire 1393* (type of *Ch. Bonatiana*); Tali, *Delavay 1187* (type of *Ch. taliensis*); Likiang Snow Range, *J. F. Rock 6000, 6047*. Sichuan occid: *Wilson 5290* (type of Ch. *Wilsoni*); Nuiyuan Hsien, at Lushan, *Harry Smith 1831*.

This distinct and endemic species belongs to the group of *Ch. tenuifolia* Sw. from which it differs altogether in the rigid, dark-brown lanceolate scales on rhizome, less divided lamina of a pentagonous outline, with ultimate segments of entirely different shape. All specimens cited above are very uniformly constant in all respects except size, in which the type represents a large form with pentagonous lamina and acuminate apex, on stipe twice as long; while *Ch. Wilsoni* and *Ch. taliensis* represent small, stunted form with less divided fronds.

Plate 134. Fig. 1. Habit sketch, smaller form (natural size). 2. segment, showing venation and sori (× 10). 3. Habit sketch from co-type (natural size). 4. Scale from rhizome (× 16).

CHEILANTHES HANCOCKII Baker
韩氏粉背蕨

图版 135　舟山粉背蕨

根状茎短而直立，端具红棕色狭披针形之厚质鳞片；叶簇生，柄长 2—5 cm，深栗色，光亮，上面具浅沟槽，遍体具小鳞片，叶体披针形，长 8—20 cm 或过之，宽 1.5—6 cm，短渐尖头，二回羽状深裂，一回小叶 10—15 对，几无柄，椭圆三角形钝头，长 1—2 cm，位于下部者渐短，边缘波形，纸质，两面光滑，无白粉，中肋及主脉下面均为栗色，叶脉 2—3 叉，子囊群圆球形，彼此不连接，位于每一小裂片之端，盖椭圆形，或半圆形，不连接。

分布：浙江，广东，福建，广西，江西，湖北，贵州，四川，江苏；日本，朝鲜半岛，菲律宾及越南亦产之。

本种为本区域内习见之种，形体大小不一，昔人误认为与印度产之 *Ch. mysuriensis* Wallich 相同，实则大相径庭。

图注：1. 本种全形（自然大）；2. 同上，大型（自然大）；3. 小叶（放大 3 倍）；4. 裂片，表示叶脉及子囊群之位置（放大 8 倍）；5. 叶柄基部之鳞片（放大 16 倍）；6. 叶轴上之鳞片（放大 13 倍）；7. 孢子（高倍放大）。

CHEILANTHES CHUSANA Hooker

CHEILANTHES CHUSANA Hooker, Sp. Fil. **2**: 95 t. 106B (1852); Ching, Bull. Dept. Biol. Sun Yatsen Univ. No. **6**: 26 (1933); C. Chr. Ind. Fil. Suppl. Ⅲ. 54 (1934).

Cheilanthes mysuriensis var. *chusana* Christ, Bull. Bot. Soc. Ital. (1901) 293; Bull. Acad. Géogr. Bot. (1907) 149; C. Chr. Acta Hort. Gothob. **1**: 91 (1924).

Cheilanthes mysuriensis Hk. et Bak. Syn. Fil. 135 (1867); C. Chr. Ind. Fil. 175 (1905), pro parte; Christ, Bull. Acad Géogr. Bot. (1906) 251; ibid. (1910) 13; Ogata, Ic. Fil. Jap. **2**: t. 60 (1929); Wu, Bull. Dept. Biol. Sun Yatsen Univ. No. **3**: t. 98 (1932), non Wall. 1828.

Cheilanthes fordii Baker, Journ. Bot. (1879) 304; Dunn & Tutch. Fl. Kwangt. & Hongk. 339 (1912).

Adiantopsis fordii C. Chr. Ind. Fil. 22 (1905).

Cheilanthes bockii Diels in Engl. Jahrb. **29**: 199 (1900); C. Chr. Ind. Fil. 172 (1905); Acta Hort. Gothob. **1**: 91 (1924).

Cheilanthes mysuriensis var. *Giraldii* Christ in Lecomte, Not. Syst. **1**: 51 (1909).

Cheilanthes tenuifolia Hk. (non Sw.) in Blakiston, Five Months on the Yangtze 362 (1862).

Cheilanthes boltoni Cop. in Perkins, Fragm. Fl. Phil. 186 (1905); C. Chr. Ind. Fil. 663 (1906).

Rhizome short, erect, densely scaly; *scales* rufo-brown, lanceolate-subulate; *fronds* tufted, stipe 2-5 cm long, dark castaneous, shining, terete with a channel above, densely scaly throughout, lamina lanceolate, 8-20 cm long or longer, 1.5-6 cm broad, shortly acuminate, sub-bipinnatifid; *pinnae* 10-15-jugate, subsessile, the lower ones more or less shortened, the middle ones 1.4 cm long, or longer, 0.5-2 cm broad, oblong-deltoid, acute or obtusish, pinnatifid nearly down to costa into 4-6 oblong, crenate or erosed *segments*, with 2-4 oblique and simple veinlets to each tooth, rachis more or less scaly, castaneous, costa of the same color underneath, both sides glabrous; *texture* herbaceous, greenish; *sori* globose, distinct on each crena with the reflexed margin of an oblong shape as *indusium*.

Numerous specimens have been seen from Zhejiang (including type, leg. Alexander in Chusan Island), Guangdong (including type of *Ch. Fordii*), Fujian, Guangxi, Jiangxi, Hubei, Guizhou, Sichuan (including type of *Ch. Bockii*, leg. Rosthorn 1766 in Nanchuan) and Jiangsu. A complete enumeration of the specimens from the above localities will be given later in my "Studies of Chinese Ferns."

Specimens from Japan (Nagasaki, leg. Maximowicz 25), Vietnam (Balansa 129), Luzon are also examined.

The present fern, now found to be very common in Eastern, Southern and Central China and Japan, has generally been considered as identical with *Ch. mysuriensis* Wall. of East India from which it differs in characters as already noted in my previous paper (Cf. Bull. Dept. Biol. Sun Yatsen Univ. No. **6**: p. 26).

Plate 135. Fig. 1. Habit sketch (natural size). 2. The same, a larger form (natural size). 3. Pinna (× 3). 4. Segment, showing venation and position of sori (× 8). 5. Scales from the base of stipe (× 16). 6. Scales from rachis (× 13). 7. Spores (enlarged).

CHEILANTHES CHUSANA Hooker
舟山粉背蕨

图版 136　毛粉背蕨

根状茎短而斜出,端具黑色厚质狭披针形之密鳞片;叶亚簇生,柄长 20—30 cm,乌木色或深栗色,上面具一浅沟槽,遍体被红棕色之短毛,叶体为长三角形,长 20—35 cm,宽 7—15 cm,三回羽状深裂,两面被同样之密毛,纸质,下面呈淡棕色,各位中肋呈强度曲折,一回小叶 6—10 对,互生,具柄,水平开展,或多少下向,顶部三角形,羽状深裂,基部一对最大,长 8—10 cm,宽 5 cm,长三角形,柄长达 1 cm,二回小叶 4—5 对,互生,长三角形,长达 3 cm,宽约 2 cm,基部截形,水平开展,羽状深裂,裂片二至四对,椭圆披针形,钝头,位于基部者长达 1 cm,宽约三类;叶脉扇状三叉,斜向上出;子囊群着生于叶边小脉之端,盖膜质,线形,通直,边呈波状,具睫毛。

分布:云南,四川及西藏东部。

本种为我国西部高山特产,极为奇特,与此相类者唯 Ch. Delavayi Baker,然其各位中肋不呈强度曲折,叶两面颇光滑,末回小叶之基部呈心脏形,故易与本种区别。

图注:1.本种全形(自然大);2.裂片,表示其叶脉及子囊群之位置(放大 10 倍);3.根状茎上之鳞片(放大 27 倍);4.叶面之红棕色毛(放大 40 倍)。

CHEILANTHES TRICHOPHYLLA Baker

CHEILANTHES TRICHOPHYLLA Baker, Ann. Bot. **5**: 211 (1891); Diels in Engl. u. Prantl: Nat. Pflanzenfam. **1**: 4. 276 (1899); C. Chr. Ind. Fil. 180 (1905); Acta Hort. Gothob. **1**: 91 (1924); Christ, Bull. Soc. Bot. France **52**: Mém. 1. 59 (1905); Bull. Acad. Géogr. Bot. (1906) 133; Hand-Mzt. Symb. Sin. **6**: 40 (1929).
Cheilanthes undulata Hope et C. H. Wright, Gard. Chron. Ⅲ. **34**: 397 (1903)

Rhizome short, oblique, densely radicose; *scales* at the apex dense, narrowly lanceolate, black, rigid; *fronds* subcaespitose, stipe 20-30 cm long, ebeneous or dark castaneous, terete, except a narrow groove above, densely clothed from the base upward and throughout the flexuose rachis in fine, reddish-brown, appressed articulated hairs, which also cover the both sides of leaves, lamina elongate-deltoid, 20-35 cm long, 7-15 cm broad at base, tripinnatifid; *pinnae* 6-10-jugate under the deltoid, pinnatifid apex, alternate, horizontally patent, or the lower ones more or less deflexed, all petiolate, the basal ones much the largest, 8-10 cm long, 5 cm broad, elongate-deltoid, on petiole 1 cm long; *pinnules* 4-5-jugate, alternate, broadly deltoid, shortly petiolulate, 2.5 cm long, 2 cm broad, base truncate, horizontally patent, pinnatifid to a short distance from costule into 2-4 pairs of oblong-lanceolate, blunt *segments*, of which the basal ones 1 cm long, 2-2.5 mm broad; *texture* herbaceous. brownish underneath, greenish above; *veins* flabellulately forked, very oblique to the flexuose costule of segment; *indusium* narrow, brown, continuous, crenate, ciliate.

Yunnan: Lan Kong, *Delavay* (type); Tapintze, near Tali, *Delavay 31, 1200, 1201*; Yaninshan, *Delaxay 1165, 1174*; Chonangcheteon, *Delavay 1165* (1883); Salwin, *G. Forrest 18331* (1917-19); *1034*; *Handel-Mazzetti 6303* (1914-18); Pinchow, *Ducloux 6986, 7042*; environs de Yunnan, *Ducloux 6338*; *A. Henry 13220* (type of *Ch. undulata*). Sichuan: *E. H. Wilson 5289* (1910). Tibet border: *Capt. Kingdon Ward 225*.

One of the most distinct endemic species of the genus, with which *Ch. Delavayi* Baker from the same locality is the only close relative, which differs in rachis and rachilets being not flexuose, glabrous surfaces of leaves and decidedly cordate pinnules. It appears to be a common fern in the north-western part of Yunnan, but not known elsewhere, besides Wilson's plant from western Sichuan.

Plate 136. Fig. 1. Habit sketch (natural size). 2. Segment, showing venation and position of sori ($\times$ 10). 3. Scale from rhizome ($\times$ 27). 4. Hairs from leaf surface ($\times$ 40).

CHEILANTHES TRICHOPHYLLA Baker
毛粉背蕨

C. R. Feng Del. et Lith.

根状茎短而斜出,端具红棕色之卵形渐尖头鳞片;叶簇生,柄长 10—25 cm,光亮,乌木色,光滑,叶体卵状三角形,长 20—30 cm,宽几如之,三回羽状分裂,一回小叶 5—8 对,互生,具柄,各位中肋呈强度曲折,基部一对最大,椭圆卵形,长 10—15 cm,宽 4—6 cm,二回小叶 6—7 对,具柄,互生,位于下部者,羽状分裂,上部者 2—3 裂,最上者不分裂;末回小叶长宽约各 1—1.5 cm,略呈扇形,顶端呈波形或 2—3 浅裂,具短柄,裂片无锯齿,向下凹,叶为薄纸质或厚膜质,光绿色,叶脉扇状分叉;子囊群长达 8 mm,每裂片一个,盖同形,向上弯曲,棕色,宿存。

分布:四川,贵州及云南。

本种为我国西部特产,极为美观,其异于普通习见之铁线蕨(Ad. capillus-veneris Linn.)者,为其簇生之叶,强度曲折之各位中肋,末回小叶仅具 2—3 个浅裂片与其较长之子群盖是也。

图注:1.本种全形(自然大);2.末回小叶(放大 4 倍);3—4.根状茎先端之鳞片(放大 14 倍)。

ADIANTUM REFRACTUM Christ

ADIANTUM REFRACTUM Christ, Bull. Acad. Géogr. Bot, (1902) 224; ibid. (1906) 136; C. Chr. Ind. Fil. 32 (1905); Acta Hort. Gothob. **1**: 93 (1924).

Adiantum capillus-veneris Diels (non. L.) in Engl. Jahrb. **29**: 202 (1900).

Adiantum capillus-veneris var. *sinuatum* Christ, Bull. Soc. Bot. France **52**: Mém. 1. 62 (1905); C. Chr. Acta Hort. Gothob. **1**: 93 (1924).

Rhizome short, oblique, densely clothed at apex in rufo-brown, ovate-acuminate, entire *scales*; *fronds* caespitose, stipe 10-25 cm long, glossy, ebeneous, naked, lamina deltoid-ovate, 20-30 cm long, nearly as broad, tripinnate; *pinnae* 5-8-jugate, alternate, with zigzag rachis, all long-petiolate, basal ones much the largest, oblong-ovate, 10-15 cm long, 4-6 cm broad; *pinnules* 6-7-jugate, petiolate, alternate, the lower ones pinnate, upper ones 2-3-foliolulate, the uppermost ones simple, rachilets also strongly flexuose; *pinnules* of ultimate order 1-1.5 cm each way, flabellate with sinuate or 2-3-incised outer margin, petiolulate; *lobes* entire with deeply incurvate outer edge; *texture* membranaceous, green; *veins* fine, flabellulately forked; *sori* one to each lobe, to 8 mm long, *indusium* more more or less curved, brown, persistent.

Sichuan occid: *Wilson 5259* (type); Nanchuan, *Rosthorn 954*; ibid. *W. P. Fang 5708*; between Haiting and Pingyipu, *Harry Smith 1753*, *1972*; Tchenkoutin, *Farges 730*; Kuan Hsien, *W. P. Fang 2357*. Guizhou: Ganpin, *Cavalerie 3725*; *Bodinier 1840*. Yunnan: environs de Yunnan, *Ducloux*.

This distinct Maiden-hair fern native of West China is closely related to A. *capillus-veneris* L. from which it differs in tufted leaves, zigzag rachis and rachilets, much less divided ultimate pinnules and much longer sori and indusium.

Plate 137. Fig. 1. Habit sketch (natural size). 2. Pinnule ($\times$ 4). 3-4. Scales from rhizome ($\times$ 14).

ADIANTUM REFRACTUM Christ 蜀铁线蕨

图版 138　白背铁线蕨

根状茎细长蔓生，横行，具深棕色光亮之卵披针形鳞片；叶散生，柄细长而坚实，横切面圆形，具光泽，呈栗色，长 10—20 cm，唯基部略具鳞片，余皆光滑，叶体卵三角形，长渐尖头，较柄短，三回羽状分裂，一回小叶 3—5 对，均具柄，基部一对最大，三角披针形，长达 7 cm，宽 3—4 cm，二回小叶椭圆形，具短柄，钝头，羽状分裂，末回小叶密接，扇形，长宽各 4—7 mm，上部叶缘具开展刺状锯齿，叶脉扇形分叉，各位中肋呈光亮之栗色，叶上面绿色，下面呈灰白或灰绿色，子囊群中大，通常每一小叶具一个，盖为圆肾形，黄色，厚膜质，长 2—3 mm，两侧叶缘具刺状锯齿。

分布：四川，云南，陕西南部及西藏东部；缅甸亦产之，唯极罕见耳。

本种为本属特殊之一种，其末回小叶彼此密迹或多小鳞覆，背现灰白色，至为美观。

图注：1. 本种全形（自然大）；2. 末回小叶（放大 8 倍）；3—4. 根状茎上之鳞片（放大 25 倍）。

ADIANTUM DAVIDI Franchet

ADIANTUM DAVIDI Franchet, Nouv. Arch. Mus. Ⅱ. **10**: 112 (1887); Diels in Engl. u. Prantl: Nat. Pflanzenfam. **1**: 4. 284 (1899); Christ, Bull. Soc. Bot. France **52**: Mém. 1. 62 (1905); Bull. Acad. Géogr. Bot. (1906) **136**; C. Chr. Ind. Fil. 25 (1905); Acta Hort. Gothob. **1**: 94 (1924); Contr. U. S. Nat. Herb. **26**: 310 (1931).

Adiantum aristatum Christ, Bot. Gaz. **51**: 345 (1911); C. Chr. Ind. Fil. Suppl. Ⅰ. 4 (1912); Bot. Gaz. **56**: 331 (1913).

Adiantum davidi var. *aristatum* C. Chr. Acta Hort. Gothob. **1**: 94 (1924).

Adiantum monochlamys var. *latedeltoideum* Christ, Nuov. Giorn. Bot. Soc. Ital. n. s. **4**: 88 (1897).

Adiantum latedeltoideum C. Chr. Acta Hort. Gothob. **1**: 94 (1924).

Adiantum venustum Christ, (non Don) Bull. Soc. Bot. France **52**: Mem. 1. 62 (1905).

Adiantum monochlamys C. Chr. Bot. Gaz. **56**: 331 (1913).

Rhizome wiry, wide-creeping, densely scaly at growing tip; *scales* nitide, atro-brown, ovate-lanceolate, entire; *frond* far apart, stipe slender, firm, terete, glossy, castaneous, 10-20 cm long, glabrous above the scaly base, lamina deltoid-ovate, much shorter than stipe, tripinnate; *pinnae* 3-5-jugate under elongate pin-natifid apex, all petiolate, basal ones much the largest, deltoid-lanceolate, to 7 cm long, 3-4 cm broad; *pinnules* oblong, petiolulate, obtuse, with 1-4 pairs of ultimate close pinnules which are of flabellate shape with rounded, aristately serrate outer margin, distinctly petiolulate, 4-7 mm each way, rachis, rachilets, costa and petiolules shining castaneous; *texture* crass herbaceous, surfaces glabrous, green above, more or less glaucous underneath; *veins* fine, flabellately forked, more or less projecting out from the sharp teeth; *sori* medium-sized, generally solitary to each ultimate pinnule, *indusium* brown, thick, rotundo-reniform, 2-3 mm long, with the sterile margin on both sides aristately serrate.

Sichuan: Moupin, *David* (type) (1870); Tchenkoutin, *Farges 689*; Yenguen, *Handel-Mazzetti 2789*; *Wilson 5254* (type of *A. aristatum*); Mo Tien Ling, *F. T. Wang 22453*; Tenghsiangying, *Harry Smith 1918*; Drogochi, *Harry Smith 4548*. Yunnan: The Red River, Mosoyu, *Delavay 1698*; Hokin, *Delavay 1198*; Lonkong, Yentzehay, *Delavay 1878*, *1687*; Teng Kou valley, *Forrest 12483*; Yungpeh Mt., *Forrest 17049*; *Ducloux 3368*. Shaanxi: Lunsan Huo, *Giraldi*, Oct. 1895; Miaonan Shan, *Giraldi*, Jan. 1899; Thaipei Shan, *Giraldi* August, 1893; August 10-20, 1894 & 1895; *Purdom 7978*; ibid., Hopingsze, *K. S. Hao 4272*, Sept. 12, 1932. Tibet Border: *Capt. Kingdon Ward 479* (1913).

Burma: Simila, Burhe 23 (1867).

This distinct endemic species belongs to the group of A. *venustum* Don of Northern India. The underside of leaves sometimes appears conspicuously glaucous, or sometimes only bluish as in A. *aristatum*. A. *latedeltoideum* C. Chr. from the southern part of Shaanxi is only a somewhat ampler form; and most specimens from the same region have ultimate pinnules rolled up, due evidently to the dry habitat.

Plate 138. Fig. 1. Habit sketch (natural size). 2. Pinnule (× 8). 3-4. Scales from rhizome (× 25).

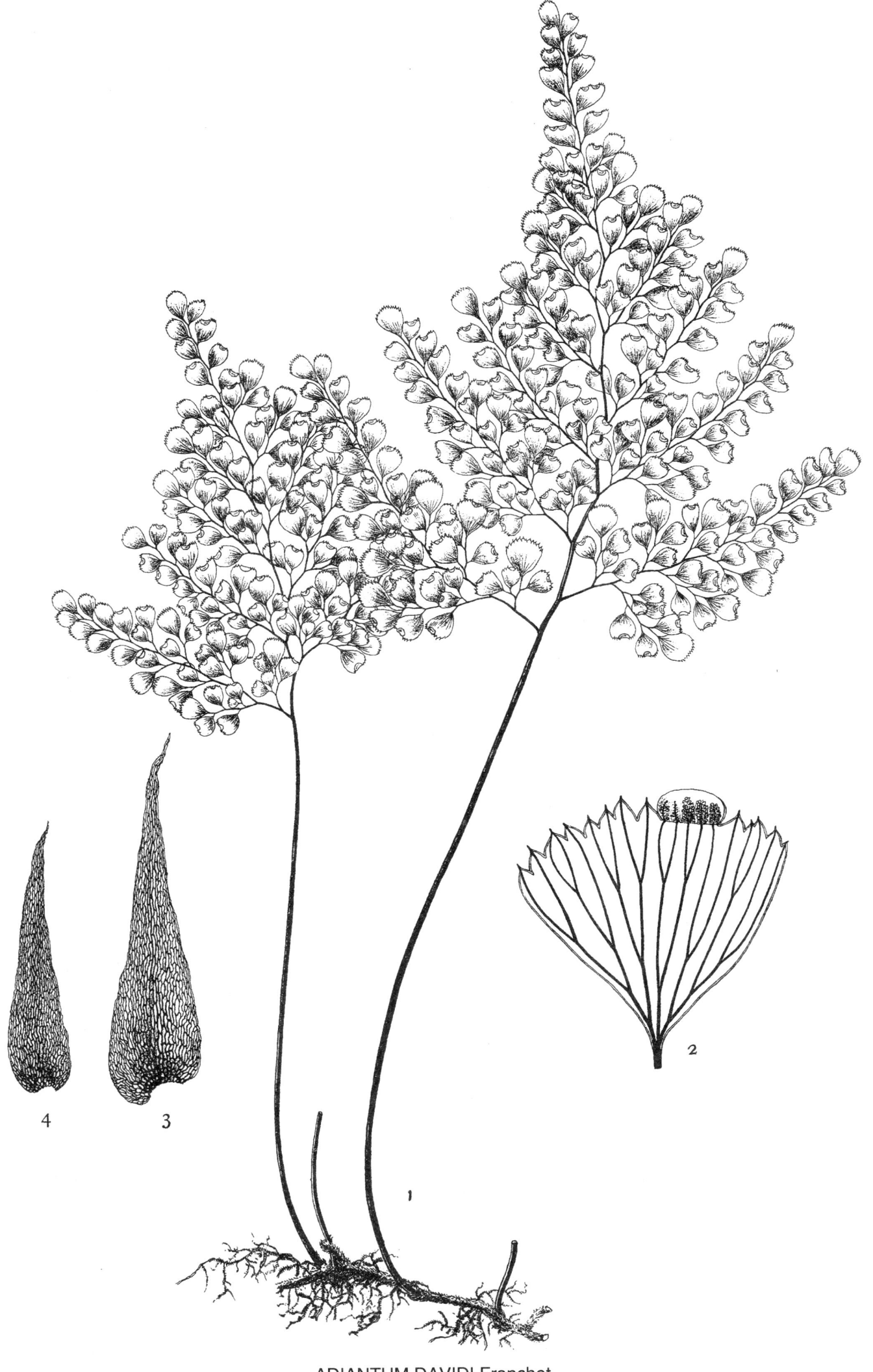

ADIANTUM DAVIDI Franchet
白背铁线蕨

图版 139　爱氏铁线蕨

根状茎短而直立，端部具黑色狭披针形之疏鳞片；叶簇生，柄细长，圆形，光亮，红栗色，长 10—18 cm，唯基部略具鳞片，叶体线披针形，长 10—23 cm，宽 2—2.5 cm，一回羽状分裂，其端常延长入土着根；小叶 10—40 对，开张，椭圆三角形，基部不等，长 1—2 cm，基部数对较小，下向，均具短柄，外缘呈多少浅裂，两面光滑，纸质；叶脉扇形分叉；子囊群每小叶数个，盖椭圆形，或亚肾形，黄色，光滑。

分布：云南，四川，贵州，山东，河北，台湾；越南，印度北部及菲律宾亦产之。

本种形体极类我国南部所产之 Ad. caudatum Linn. 所不同者，为其叶体不具刚毛，小叶外缘仅呈浅裂是也。

图注：1. 本种全形（自然大）；2. 同上，小型（自然大）；3. 小叶（放大 5 倍）；4. 根状茎上之鳞片（放大 24 倍）。

ADIANTUM EDGEWORTHII Hooker

ADIANTUM EDGEWORTHII Hooker, Sp. Fil. **2**: 14 t. 81B (1851); Syn. Fil. **472** (1867); Fée, Gen. Fil. 114 (1852); Moore, Ind. Fil. 25 (1857); Bedd. Ferns Brit. Ind. t, 17 (1865); J. Sm. Hist. Fil. 274 (1875); Christ, Bull. Soc. Bot. France **52**: Mém. 1. 61 (1905); Bull. Acad. Geogr. Bot. (1906) 136; ibid. (1910) 12; C Chr. Ind. Fil. **26** (1905); Suppl. Ⅲ. 19 (1934); Acta Hort. Gothob. **1**: 47 (1924); Contr. U. S. Nat. Herb. **26**: 310 (1931).

Adiantum caudatum var. *Edgeworthii* Bedd. Handb. Ferns, Brit. Ind. etc. 84 (1883).

Adiantum caudatum var. *rhizophyllum* (Wall) Clarke, Trans. Linn. Soc. Ⅱ. Bot. **1**: 453 (1880), non *Ad. rhizophyllum* Sw.

Adiantum guilelmi Hance, Journ. Bot. **5**: 261 (1867).

Adiantum spencerianum Cop. Phil. Journ. Sci. **1**: Suppl. Ⅱ. 154 t. 11 (1906); C. Chr. Ind. Fil. Suppl. Ⅰ. 4 (1912).

Rhizome short, erect, sparcely scaly at apex; *scales* black, lanceolate, subulate towards tip, rigid; *fronds* fasciculated, several together, stipe terete, glossy, castaneous, 10-18 cm long, wiry but firm, glabrous above scaly base, lamina linear-lanceolate, 10-23 cm long, 2-2.5 cm broad; *pinnae* 10-40-jugate, patent, basal ones somewhat smaller, deflexed, uppermost ones gradually diminished towards apex, which is sometimes elongate and rooting, all shortly petiolate, dimidiate, variable in size, apex rounded, anterior inner base truncate, margin more or less lobato-incised, in sterile ones cut nearly half way down into 3-5 oblong, roundish or bifid segments on the outer margin; *texture* thin herbaceous, glabrous in all parts; *veins* fine, flabellately forked; *sori* several to each pinna, *indusium* laterally oblong or subreniform, brown, glabrous.

Yunnan: Santchangkisu, Hokin, *Delavay 1714*; Chouanychetesu near Tapintze, *Delavay 7*; Maeulshan, *Delavay 1189*; Mekong, *Handel-Mazzetti 8508* (1914-18); *Maire 1390, 2736*; *Forrest 10787*; Mengtze, *Hancock*; Szemeo, *Henry 12821*; Tali, Mochetchin, *Delavay 1200*; *Schneider 2791*; *Ducloux 13, 171 1886*; *Cavalerie 1918, 1903, 1570, 4715, 7070*; environs de Yunnan, *Ducloux 5*; environes de Tengyueh, *J. F. Rock 7970*. Sichuan: *Wilson 526, 5258*; Mt. Ying Shanluan, *W. Fernusson* (1910); Yienpien Hsien, *T. T. Yü 1688*; Nihyuan Hsien, Lushan, *Harry Smith 1797*. Guizhou: Kianglong, *Michel 994*. Shandong: Taishan, *Clemens 1307* (1912); *Jacob 20*. Hebei: Peiping mountains, *Hancock*, Sept. 26, 1875; *S. W. Williams 13483* in herb. Hance; *Bushell*; *Bretschneider 77*; *W. R. Carles 235*; *C. L 206* in herb. Tsinghua Univ.; Changli, *Mrs. Clemens 6022, 6022B*, June 28, 1913; Mia Fan Shan. *C. T. Li 32178*, July 17, 1932; Hsinglung Shan, *T. N. Lion*, Sept. 15, 1930. Taiwan: *Hancock 113*.

Indo-China: Hermera, *F. Neuton 1080*.

N. W. India: Luni valley, alt. 5,000 ft., *Edgeworth* (type), common.

Also the Philippines.

This pretty distinct fern belongs to the group of *A. caudatum* L. with which it has not infrequently been confounded, and from which it differs in much less divided pinnae and naked leaves. Very variable in size and dimension of pinnae; Specimens from North China are generally of smaller size in every repect.

Plate 139. Fig. 1. Habit sketch (natural size). 2. The same, a smaller form (natural size). 3. Pinna ($\times$ 5). 4. Scales from rhizome ($\times$ 24).

ADIANTUM EDGEWORTHII Hooker
爱氏铁线蕨

图版 140　傅氏凤尾蕨

　　根状茎短而斜出，端具深棕色厚质披针形之密鳞片；叶亚簇生，柄长 30—40 cm，下部呈红稻秆色，上部为淡稻秆色，光滑，叶体阔卵三角形，长 30—45 cm，宽亦如之，二回奇数羽状分裂，小叶 4—7 对，几无柄，基部一对最大，中肋下面具 1—3 数小叶，中部一回小叶长 15—20 cm，宽 4.5—6 cm，中肋上面具肉刺，绿色，厚纸质，光滑，羽状深裂达于中肋，顶端呈尾形；裂片 20—30 对，长 2.5—4 cm，宽不及 1 cm，线状亚镰形，全缘，钝头，叶脉 13—17 对，显明，二叉，其基部下方之脉着生于一回小叶之中肋，3—4 叉；子囊群线形，贴近叶缘，几达叶顶，盖线形，宽约 1 mm，膜质，灰白色，全缘。

　　分布：广东，香港，海南，福建，台湾，浙江，湖北，云南，四川，广西；日本及越南亦产之。

　　本种为本区内习见之品，昔人误认为与斯里兰卡所产之 *Pt. quadriaurita* Retz. 同种，实则二者大相径庭。

　　图注：1. 本种全形（自然大）；2. 两裂片，表示及叶脉及子囊群之位置（放大 4 倍）；3. 小叶之一部，表示其中肋上面之肉刺（放大 2 倍）。

PTERIS FAURIEI Hieronymus

PTERIS FAURIEI Hieronymus, Hedwigia 55: 345 (1914); C. Chr. Ind Fil. Suppl. Ⅱ. 30 (1916).
　　Pteris quadriaurita Hk. Sp. Fil. 2: 179 (1852); Syn. Fil. 158 (1867); Diels in Engl. u. Prantl: Nat. Pflanzenfam. 1: 4. 292 (1899), pro parte.
　　Pteris quadriaurita Franch. et Sav. (non Retz.) Enum. Pl. Jap. 2: 214 (1877); Christ in Warburg, Monsunia 1: 69 (1900); Bull. Herb. Boiss. Ⅳ. 2: 612 (1904); Journ. d. Bot. France 19: 14 (1905); Bull. Acad. Géogr. Bot. (1910) 16; C. Chr. Acta Hort. Gothob. 1: 97 (1924).
　　Pteris biaurita C. Chr. Ind. Fil. 593 (1905). pro parte.
　　Pteris longipinnula Christ (non Wall. 1828), Bull. Acad. Géogr. Bot. (1906) 130; ibid., (1910) 16.

　　Rhizome short, oblique, densely scaly; *scales* dark brown, lanceolate, rigid, appressed; *fronds* subcaespitose, stipe 30-40 cm long, rufo-stramineous on the lower part, lighter-colored upwards, glabrous above the base, lamina broadly ovate-deltoid, 30-45 cm long, nearly as broad; *pinnae* 4-7-jugate under the free end-pinna, generally larger than those immediately below, the basal ones much the largest, deltoid, with 1-3 additional pinnules on the lower side of costa, all subsessile with shortly decurrent base, the middle ones 15-20 cm long, 4.5-6 cm broad, pinnatifid nearly down to costa into 20-30 pairs of linear-subfalcate, obtuse segments under the caudate, entire apex 2-4 cm long; *segments* 2.5-4 cm long, less than 1 cm broad at base with 13-17 pairs of biforked and very prominent *veins*, of which the posterior basal one springing from costa and 2-3-4-forked, the anterior basal one, from the costule of segment, biforked, with veinlets running to the margin some distance above the sharp, callous sinus; *texture* chartaceous, clear light green, glabrous on both sides, except a few brownish, thick, articulated hairs underneath, costa pale-colored, terete below, deeply sulcate and with one stout spine at the base of costule of segment above; *sori* continuous from near the sinus up to a short way from the sterile entire apex of segment, *indusium* membranaceous, gray, entire.

　　Numerous specimens have been seen from Guangdong, Islands Hong Kong and Hainan, Fujian, Zhejiang, Hubei, Yunnan, Sichuan, Guangxi and Taiwan, besides Japan, French Indo-China. A complete enumeration will appear later in my *Studies of Chinese Ferns*.

　　The present species has previously been generally considered the same as *Pteris quadriaurita* Retz. from Sri Lanka, the type of which, I had a chance to examine some years ago, is very different from our fern in general habit, but perfectly identical with *Pteris otaria* Bedd. Ferns S. Ind. t. 41, 219. Our plate is based upon a specimen from Guangdong (leg. Y. K. Wang No. 31655) and represents what Hieronymus has considered var. *rigida*, l. c. p. 346., a form, common in South China, which differs, according to its author, from the Taiwann type (leg. Faurie No. 628) in clear light green leaves and more prominently raised veinlets on both sides; but ample material from the region has convinced me that a varietal name is hardly justified. It is possible, however, that some specimens from Sichuan and Guizhou which have more pairs of lateral pinnae with much longer, linear caudate apices, may prove specifically distinct from the form from South China.

　　Plate 140. Fig. 1. Habit sketch (natural size). 2. Two segments, showing venation and sori (× 4). 3. A portion of pinna, showing the stout spines at the base of costules above (× 2).

PTERIS FAURIEI Hieronymus
傅氏凤尾蕨

图版 141　掌凤尾蕨

　　根状茎亚蔓生,交错,光滑,仅端具红棕色狭针形之疏鳞片;叶多数密生,柄长 15—30 cm,细长而质软,淡稻秆色,有光泽,全体光滑,基部呈深棕色,叶体圆卵形,掌状分裂,小叶 5—7 数,间为 3 数,线形,长 5—10 cm,宽 5—7 mm,几无柄,光绿色,中脉上面具沟槽,侧脉疏生,单一或二叉,达于叶缘锐锯齿之基部,非子囊群小叶及不生子囊群之部分之叶缘具锯齿;子囊群盖膜质,灰白色亚全缘,通直,宿存。

　　分布:云南,四川,西藏;印度北部之高山。

　　本种具秀雅之形体,至为美观,本属特殊品种之一也。

　　图注:1.本种全形(自然大);2.子囊群叶之一部(放大 8 倍);3.非子囊群叶之一部(放大 10 倍)。

PTERIS DACTYLINA Hooker

PTERIS DACTYLINA Hooker, Sp. Fil. **2**: 160 t. 130A (1858); Syn. Fil. 155 (1867); Baker, Journ. Bot. (1888) **226**; Diels in Engl. u. Prantl; Nat. Pflanzenfam. **1**: **4**. 292 (1899); Christ, Bull. Acad. Géogr. Bot. (1906) 130; C. Chr. Ind. Fil. 596 (1905); Acta Hort. Gothob. **1**: 96 (1924); Contr. U. S. Nat. Herb. **26**: 311 (1931); Hand-Mzt. Symb. Sin. **6**: 40 (1929); Clarke, Trans. Linn. Soc. Ⅱ. Bot. **1**: 463 (1880); Bedd. Ferns Brit. Ind. t. 23 (1865); Handb. Ferns, Brit. Ind. etc. 107 (1883).

Rhizome short-creeping, matted, glabrous except a few subulate, rufo-brown *scales* at growing tips; *fronds* numerous, tufted, stipe 15-30 cm long, slender, wiry, glossy, pale straminous, perfect glabrous throughout, shaded dark brown in the lower part, lamina digitate, consisting of 5-7, or rarely 3 linear elongate *pinnae* 5-10 cm long, 5-7 mm broad with deeply sulcate costa above, the barren ones sharply serrate towards apices; *veins* widely apart, distinct, simple or bifurcate, vein-lets rather patent, ended in a distinct clavate hydathode some distance below the base of tooth; *indusium* gray, broad, continuous, subentire, persistent.

　　Yunnan: Schilungba, *Handle-Mazzetti 275*; *Forrest 287*, *5983*, *12445*, *28569*, Likiang, *Schneider 2969*; *Delavay 1720*, August, 7, 1883; *J. F. Rock 4854*. Gansu: *F. N. Meyer 20106*. Sichuan: Mt. Omei, *Faber 1010*; *Wilson 5274*; Teng-hsiang-ying, *Harry Smith 1993*; between Hai-tang and Ping-yi-pu, *Harry Smith 1942*. Tibet: Muti, *Capt. Kingdon Ward 4773*.

　　Also Sikkim-Himalayas, common.

　　Type from Sikkim (leg. Drs. Hooker & Thomson). One of the most remarkably graceful slender fern of the group of *Pteris cretica* L. and is closely related to *P. stenophylla* Wall., differs in generally 5-7 much narrower pinnae on very wiry slender stipe and in its densely tufted leaves. As already pointed out by Hooker (Sp. Fil. **2**: 161), the frond of this fern is not strictly digitately dissected but is constructed on the same plane as is *P. cretica*, except the petiole of the central pinna is either totally suppressed or arrested from lengthening out as in the other species.

　　Plate 141. Fig. 1. Habit sketch (natural size). 2. Portion of fertile pinna (× 8). 3. Portion of sterile pinna (× 10).

PTERIS DACTYLINA Hooker
掌凤尾蕨

图版 142 拟凤尾蕨

根状茎亚蔓生,具棕色狭披针形之密鳞片;叶近生,柄长 6—20 cm,淡栗色,稍具光泽,光滑,叶体长 15—25 cm,椭圆形,奇数一回羽状分裂,间为线披针形之单叶;小叶 1—12 对,具短柄或几无柄,互生,长 6—13 cm,宽 1—2 cm 或较狭,线披针形,基部斜截形,两面光滑,纸质,全缘(非子囊群叶具锯齿);主脉明显,侧脉无,小脉网状,网眼 1—2 列,斜出,内不具小脉,向叶边分离;子囊群贴近叶缘,线形,盖两层,膜质,全缘,通直,向外开。

分布:广布于亚洲热带及澳洲,我国之香港,广东,海南亦产之。

本种酷似 Sch. heteropyllum (Dry.),所不同者,为其小叶成奇数,向叶端不渐缩小是也。

图注:1. 本种全形(自然大);2. 小叶之一部(放大 6 倍);3. 同上,具展开之双子囊群盖(放大 6 倍);4. 孢子(放大 500 倍);5. 根状茎上之鳞片(放大 10 倍)。

SCHIZOLOMA ENSIFOLIUM (Sw.) J. Smith

SCHIZOLOMA ENSIFOLIUM (Sw.) J. Smith. Journ. Bot. **3**: 414 (1841); Ferns Brit. & Fore. **231** (1846); Hist Fil. **271** (1875); Fée, Gen. Fil. 108 (1850-52); Bedd. Ferns S. Tnd. t. 25 (1863); Handb. Ferns, Brit. Ind. etc. 80 (1883); Diels in Engl. u. Prantl; Nat. Pflanzenfam. **1**: 4. 219 (1899); C. Chr. Ind. Fil. 618 (1905); v. A. v. R. Handb. Mal. Ferns 280 (1909).

Lindsaea ensifolia Sw. in Schrad. Journ. Bot. 1800^2: 77 (1801); Syn. Fil. 118 (1806); Willd. Sp. Pl. **5**: 420 (1810); Hk. et Grev. Ic. Fil. t. 3 (1827); Hk. Sp. Fil. **1**: 220 (1844); Gard. Ferns t. 62 (1840); Journ. Bot. (1857) 336; Syn. Fil. 112 (1867); Christ, Farnkr. d. Erde 297 (1897); Benth. Fl. Hongk. 446 (1861); Clarke, Trans. Linn. Soc. II, Bot. **1**: 452 (1880).

Lindsaya lanceolata Labill. Pl. Nov. Holl. 2: 98 t. 248. f. 1 (1806), *folia simplex*.

Lindsaya pentaphylla Hk. Sp. Fil. **1**: 219 t. 67 (1844).

Lindsaya griffithiana Hk. Sp. Fil. **1**: 219 t. 68B (1844); Bedd. Ferns Brit. Ind. t. 29 (1867), *folia simplex*.

Rhizome short-creeping, densely scaly; *scales* brown, lanceolate-subulate, entire; *frond* rather approximate, stipe 6-20 cm long, light castaneous, somewhat glossy, naked throughout, lamina 15-25 cm long, impari-pinnate or rarely simple (*Lindsaya Griffithiana*); *pinnae* 1-12-jugate, petiolate or subsessile, alternate, 6-13 cm long, from less than 1 cm to 2 cm broad, linear-lanceolate, base obliquely truncate, both sides glabrous; *texture* thin herbaceous, margin entire (sterile ones serrate); *midrib* prominent, lateral main veins none; *veins* anastomosing in 1-2 rows of oblique, angular exappendiculate areolae on each side of midrib, free towards margin; *sori* linear, continuous, uniting the apices of free veinlets, *indusium* double, membranaceous, gray, entire, continuous, opening outwardly; *spores* 4-angular, translucent.

Hong Kong; *Fortune 7, 27* (1848); *Faurie 15772* (1895); *C. Wright* (1853-56); *Bodinier 887*; *Lorain* (1856), pro parte; *Harland 63*; *Gower 2* (1896); *Matthew 304* (1904); Koolung, *C. T. Yong 291* in herb. Tsinghua Univ.; *Gardner, Dill.* Guangdong: Swatow, *Dalziel*, Sept. 1901; Lohfau Shan, *C. Ford* (1883); *N. K. Chun 40990*. Hainan: Hoichow, *Hancock 1* (1875); Dung Ka to Win Fa Shi, *C. L. Tso & N. K. Chun 43728*.

Widely distributed throughout tropical Asia, Australia and Polynesia. Very closely related to this fern is *Sch. heterophyllum* (Dry.), which differs in the lateral pinnae gradually decrescent towards the lobato-pinnatifid apex. Our fern is a very variable one as to the shape of frond, which varies from simple to fully pinnate with 10 pairs of pinnae of different width. Our plate is based upon a specimen from Guangdong (leg. N. K. Chun 40990), which agrees very well with type from Island Maritius, now to be found in the herb. Swartz.

Plate 142. Fig. 1. Habit sketch (natural size). 2. Portion of pinna ($\times$ 6). 3. The same, with open indusium ($\times$ 6). 4. Spores ($\times$ 500). 5. Scale from rhizome ($\times$ 10).

SCHIZOLOMA ENSIFOLIUM (Sw.) J. Smith
拟凤尾蕨

根状茎蔓生，横行，具棕色披针形之鳞片；叶散生，柄长 20—40 cm，光滑，淡稻秆色，上面具阔沟槽，叶体卵三角形，长 40—60 cm，宽 30—45 cm，奇数二回羽状分裂，一回小叶 3—10 对，互生，具柄，基部一对最大，对生，柄长达 2 cm，长逾20 cm，通常具小叶或为羽状分裂，二回小叶披针形，长 30—15 cm，宽 4—3 cm，顶部尾形或长渐尖头，上部一回小叶不分裂，或第二对为三裂，叶缘具密尖锯齿，两面光滑或下面被短疏毛，薄纸质，淡绿色；叶脉明显，斜出，自基部分二叉，平行，达于锯齿之基部或深入，子囊群淡黄色，线形，随叶脉分布，直达距叶边近处，无盖。

分布：云南，贵州，湖北，四川，广东，广西，东三省；日本，朝鲜半岛及印度东部亦产之。

本种为本区域内普通之种，因其叶体下面有无短毛可分为光叶(Var. glabra)与毛叶(Var. villosa)两变种。

图注：1. 本种全形（自然大）；2. 小叶之一部，表示其锯齿及叶脉（放大 12 倍）；3. 根状茎上之鳞片（放大 12 倍）。

CONIOGRAMME INTERMEDIA Hieronymus

CONIOGRAMME INTERMEDIA Hieronymus, Hedwigia **67**: 301 (1916); C. Chr. Ind. Fil. Suppl. II. 9 (1916); Contr. U. S. Nat. Herb. **26**: 307 (1931); Wu, Bull. Dept. Biol. Sun Yatsen Univ. No. **3**: 212 t. 97 (1932).

Coniogramme javanica Hk. Sp. Fil. **5**: 145 (1863); Syn. Fil. 381 (1867), pro parte.

Coniogramme fraxinea Diels in Engl. u. Prantl: Nat. Pflanzenfam. 1: 4. 262 (1899); C. Chr. Ind. Fil. 185 (1905), pro parte; Christ in Warburg, Monsunia **1**: 58 (1900); Fedde, Repert. Sp. Nov. **5**: 159 (1908); Bull. Acad. Géogr. Bot (1910) 5, 14; Komarov, Fl. Mansh. **1**: 140 (1900); Fomin, Fl. Sib. et Orient. Extr. **5**: 155 (1930); Ogata, Ic. Fil. Jap. **3**: t. 111 (1930), non Diels, pro parte.

Gymnogramme javanica Christ, Bull. Acad. Géogr. Bot. (1906) 130.

Rhizome wide-creeping, densely scaly; *scales* brown, lanceolate, acuminate, subentire; *frond* far apart, stipe 20-40 cm long, glabrous, pale straminous, terete below, broadly grooved above, lamina ovate-deltoid, 40-60 cm long, 30-45 cm broad, bipinnate at base, simple pinnate upwards under the free end-pinna, or rarely simple pinnate; *lateral pinnae* 3-10-jugate, alternate, the basal ones much the largest, opposite and long-petiolate (petiole to 2 cm long), over 20 cm long, generally trifoliolate or rarely pinnate, the upper ones simple or rarely the second pinnae trifoliolate or bilobed, petiolate or rarely the uppermost ones sessile or adnate, 30-15 cm long, 4-3 cm broad, broadly lanceolate, with caudate or attenuate apex, base rounded or rotundo-cuneate, margin sharply serrate; *texture* herbaceous, color light green, both sides naked or sparcely villose, rachis naked, glossy, pale straminous; *veins* free, distinct, oblique, generally biforked from near the base, veinlets fine, straight, parallel, extending till the base of teeth or well into the teeth and ended in an elongate clavate apex; *sori* pale brown, exindusiate, following veins and veinlets till a short distance from the margin, which remains sterile with the apex.

Var. **glabra** Ching, var. nov. Pinnis subtus glabris.

Yunnan: Mengtze, *Hancock 135*; Manmei, *Henry 9257B*; *Maire 2773*, *2808*, *2768*; *Delavay 4942*; Mekong, valley, *Forrest 12940*; San Tcha Ho, *Delavay* Oct. 13, 1887; Mao Kou Tchang, *Ducloux 13*, *1169*; Likiang, *J. F. Rock 7757*; Tchenfong Shan, *Delavay 3043* (1894). Guizhou: Pinfa, *Cavalerie 324*; Ganchow, *Michel 1044*; Chengfeng, *Y. Tsiang 4603*. Hubei: Ichang, Nanto, *Henry 2204*; Koi Hsien, *Silvestri 34*, *33*, *2839* (1909); Shin Shan Hsien, *S. S. Chien 8103* in herb. Univ. Nanking. Sichuan: Mt. Omei, *Wilson 5245*; *Brown 11*, *31*; *W. P. Fang 2707*, *2575*, *2576*, *2605*, *2548*; Nanchuan, *W. P. Fang 5738*. Hunan: Yun Shan, *Handel-Mazzetti 12250*. Guangdong: Lohfau Shan, *Levine & McClure 6831*; *Levine 1485*; Lokchong, *W. T. Tsang 20955*. Shaanxi: Tuikiosan, *Giraldi*, Sept. 1893; Sikutziusan, *Giraldi*, July, 1894; Weitzeping, *Licent 2545* (1916). Guangxi: Yao Shan, *S. S. Sin 468* (1928); Tsinlung Shan, Linyin Hsien, *R. C. Ching 6882*; Luchen, *R. C. Ching 5745*. Northeast China: Kirin, *Komarov 40* (1896); *James* (1886). Korean peninsula: *Taquet 2474*, *2475*. Also Japan, Vietnam and East India.

Var. **villosa** Ching, var. nov. Pinnis subtus plus minus pilis tenuisissimis, adpressis, articulatis onustis.

Yunnan: Mengtze, *Hancock 22*; Salwin valley, *Capt. Kingdon Ward*, Jan. 2, 1914; *Maire 2786*; Tianchong Shan, Tali, *K. K. Tsoong 2589*. Sichuan: *Henry 7107*; Mt. Omei, *W. P. Fang 2697*, *2064*; Marpien Hsien, *W. P. Fang 6331*; *T. F. Lou 342*; Kuan Hsien, *W. P. Fang 2064*. Guangdong: Lohfau Shan, *E. Faber*, Sept. 22, 1886, herb. Hance; Lokchong, *C. L. Tso 21067*; *W. T. Tsang 20892*. Guizhou: Pinfa, Kweiting, *Y. Tsiang 5350*.

Also Japan and India.

Our plate represents the var. *glabra* from Mt. Omei (leg. W. P. Fang No. 2548). Hieronymus is right in separating the Chinese and Japanese plants as distinct from true *C. fraxinea* (Don) from the Himalayas and Malaysia-Polynesia, which is characterized by entire pinna of still larger size and much thicker texture, but I can not agree with him in regarding Rosthorn's plant No. 1699 from Nanchuan and Henry's No. 2204 from Ichang (*C. Rosthornii*) as specifically distinct from var. *villosa* of the present species by merely having "shorter, stiff and easily broken hairs, consisting of 2-4 cells" on the underside of pinna, a character which seems to be too insignificent to justify a specific rank.

Plate 143. Fig. 1. Habit sketch (natural size). 2. Portion of pinna, showing serration and extension of veinlets ($\times$ 12). 3. Scale from rhizome ($\times$ 12).

CONIOGRAMME INTERMEDIA Hieronymus
华凤了蕨

根状茎颇肥，蔓生，具褐棕色且富曲光性之膜质披针形大鳞片；叶颇散生，长 30—35 cm，宽 1.5—2.5 cm，倒披针形，渐尖头，无柄，薄肉质，向基部渐狭，两面光滑，中肋上面明显，侧脉在光下可见，斜出，达叶缘近处彼此连接；子囊群线形，颇近叶边，隔丝密生，端具杯状之膨大细胞一个。

分布：云南；缅甸及越南。

本种为本区域内之特产，叶体宽广，为本属特异之种。

图注：1—2 本种全形（自然大）；3. 叶体之一部，表示叶脉及子囊群（放大 2 倍）；4. 根状茎上之鳞片（放大 12 倍）；5. 子囊群中之隔丝（放大 100 倍）。

VITTARIA FORRESTIANA Ching

VITTARIA FORRESTIANA Ching, *Sinensia* **2**: 191 f. **6** (1931); C. Chr. Ind. Fil, Suppl. Ⅲ. 194 (1934).

Vittaria doniana C. Chr. (non Hieron. 1915), Contr. U. S. Nat. Herb. **26**: 313 (1931).

Rhizome thick, creeping, densely scaly; *scales* fusco-brown, iridescent, lanceolate with hair-pointed tip, distinctly reticulated; *frond* rather far apart, 30-35 cm long, 1.5-2.5 cm broad at the broadest part, oblanceolate, acuminate, from one-third from the apex gradually narrowed and broadly decurrent along the stipe till base; *texture* fleshy in living state, coriaceous upon drying, midrib indistinct above, broad and prominent underneath; *venation* distinct against light, lateral veins fine, very oblique, regularly jointed each other towards margin; *sori* superficial, intra-marginal, leaving a broad, comparatively thin, plane sterile margin, *paraphyses*, filiform, with enlarged, cup-shaped apical cell.

Yunnan: Salwin divide, *G. Forrest 18347, 35106*, Sept. 1924; Northern Maikha-Salwin divide, *G. Forrest 27062*, July, 1925, on tree of dry rocks in mixed forest, 1,0100 ft. alt.

Burma: Between Sadon and the Yunnan border, *J. F. Rock 7423, 7496, 7398*.

Vietnam: Chapa, *Pételot 1598, 3901*.

In gross habit, our fern resembles *V. scolopendria* Thwaites from Malaysia-Polynesia, but differs above all in its superficial sori; it also closely related to *V. Doniana* Hieron. of East India, which differs in much narrower and linear elongate leaves twice as long with thick and strongly reflexed margin.

Plate 144. Fig. 1-2. Habit sketch (natural size). 3. Portion of frond, showing venation and position of sori ($\times$ 2). 4. Scale from rhizome ($\times$ 12). 5. Paraphyses from sorus ($\times$ 100).

VITTARIA FORRESTIANA Ching
宽书带蕨

图版 145　车前蕨

根状茎短而直立，端具褐棕色膜质且富曲光性之披针形鳞片；叶簇生，无柄，宽披针形，长 8—15 cm，宽 1.2—1.8 cm，渐尖头，向基部最狭，纸质，两面光滑，无中肋，小脉网状，网眼垂直，不具小脉；子囊群线形，随网脉分布而凹入叶质内，隔丝为宽带状。

分布：广东，广西及台湾。

本种为本区域内之特产，其异于本属其他之种者，为其宽带形之隔丝是也。

图注：1. 本种全形（自然大）；2. 叶体之一部，表示叶脉及子囊群（放大 3 倍）；3. 根状茎上之鳞片（放大 30 倍）；4. 子囊群中之隔丝（放大 80 倍）。

ANTROPHYUM FORMOSANUM Hieronymus

ANTROPHYUM FORMOSANUM Hieronymus, Hedwigia 57: 210 (1915); C. Chr. Ind. Fil. Suppl. Ⅱ. 4 (1916).
　　Antrophyum plantagineum Christ (non Kaulf.), Bull. Herb. Boiss. Ⅱ. 4: 610 (1914).

Rhizome short, erect, densely scaly at the base of stipe; *scales* lanceolate, fusco-brown, thin, iridescent, serrulate with hair-pointed tip; *fronds* fasciculated, sessile, lanceolate, 8-15 cm long, 1.2-1.8 cm broad, acuminate, gradually attenuate towards base; *texture* herbaceous, glabrous on both sides, no midrib; *venation* reticulated, fine but distinct against light; *sori* linear, following the course of grooved veins, *paraphyses* broad, ribbon-like.

Guangdong: Lohfau Shan, *N. K. Chun 41480*, July 2, 1930. Guangxi: Yao Mar Shan, Linyin Hsien, *R. C. Cling 7082* (1928). Taiwan: Kushaku, *Faurie 675* (type), June, 1903; *Hancock 2*; *Steere*; Takow, *A. Henry*; *Swinhoe* (1870); *Playfair* (1889).

From other species of the genus known in the region, our fern can be easily distinguished by its very broad and ribbon-like paraphyses in the sori and much thinner fronds. Our plate is based upon a specimen from Guangdong (leg. N. K. Chun 41480), which matches the type from Taiwan very well.

Plate 145. Fig. 1. Habit sketch (natural size). 2. Portion of frond, showing venation and sori ($\times$ 3). 3. Scale from rhizome ($\times$ 30). 4. Paraphyses from sorus ($\times$ 80).

ANTROPHYUM FORMOSANUM Hieronymus
车前蕨

根状茎细长，蔓生，缘直立之岩壁横行，具卵披针形之红棕色鳞片；叶散生，一列，长3—9 cm，宽1.5—3.5 mm，线形，无柄或具短柄，节状着生于茎上，硬革质，叶缘向下面强度反卷，上面具星芒状之疏毛，久则脱落，中肋上面呈沟状，下面凸出，密被黄色星芒状毛，小脉不见，成二列网状，网眼垂直，或具小脉或否；子囊群线形，位于中肋与叶边之间，幼时全为反卷之叶边所盖，嗣则由一裂缝露上。

分布：湖北，四川，陕西，广东，广西，浙江及台湾。

此为本属仅有之种，昔人以之隶于石苇属，据余最近之研究，觉其实与该属不类，故特辟一新属以位之。

图注：1. 本种全形（自然大）；2. 叶体一部，表示其叶脉及子囊群之位置；3. 湿润叶之横切面；4. 干叶之横切面；5. 根状茎之鳞片；6. 根状茎之横切面；7. 叶柄之横切面；8. 孢子；9. 叶下面之星芒毛；10. 叶上面之星芒状毛（2—10均放大）。

SAXIGLOSSUM TAENIODES (C. Chr.) Ching

SAXIGLOSSUM TAENIODES (C. Chr.) Ching, Contr. Inst. Bot. Nat. Acad. Peiping **2**: 2 t. 2 (1933); C. Chr. Ind. Fil. Suppl. Ⅲ. 170 (1934).

Cyclophorus taeniodes C. Chr. Ind. Fil. 201 (1905); Acta Hort. Gothob. **1**: 106 (1924).

Polypodium angustissimum Baker (non Fée, 1867), Ann. Bot. **5**: 472 (1891); Christ, Nuov. Giorn. Bot. Soc. Ital. n. s. **4**: 97 t. 3. f. 1. (1897).

Niphobolus angustissimus Diels in Engl. u. Prantl: Nat. Pflanzenfam. **1**: 4. 326 (1899); Engl. Jahrb. 29: 207 (1900); Giesen. Farngatt. Niph. 183 (1901); Christ, Bull. Acad. Géogr. Bot. (1902) 220 c. fig.

Niphobolus cavalerianus Christ, Bull. Acad. Géogr. Bot. (1904) 107.

Cyclophorus cavalerianus C. Chr. Ind. Fil. 198 (1905).

Cyclophorus sasakii Hayata, Ic. Pl. Form. **6**: 158 (1916); Ogata, Ic. Fil. Jap. **4**: t. 158 (1931).

Rhizome wiry, wide-creeping, densely scaly; *scales* at growing tips ovate-lanceolate, bright rufo-brown, with spread hair-pointed tip, on old parts, dark-brown, acuminate, peltately affixed; *frond* far apart, uniseriate 3-9 cm long, 1.5-3.5 mm broad, narrowly linear, sessile or shortly petiolate, articulated to rather a high pseudopodium, rigidly coriaceous, margin strongly revolved, upperside with a few stellate hairs at first, becoming naked at last, midrib deeply grooved, prominent underneath, underside densely clothed in brown wooly stellate hairs with fine, needle-like arms along the midrib; *venation* completely hidden, reticulated with 2-rows of elongated, exappendiculate areolae, veinlets free towards margin and ended in ovate hydathodes; *sori* linear, straight, one-rowed between margin and midrib, completely covered by the inwardly produced flap-like leaf-margin when young, which is more or less pushed open by the growing sori later; *spores* ovoid, smooth, transparent.

By the present distribution, this peculiarly distinct fern is known from Hubei, Shaanxi, Guizhou, Zhejiang, Guangdong, Guangxi, Sichuan and Taiwan. A complete enumeration of specimens was already given in my previous paper on the genus, to which readers are referred for more detailed information.

Plate 146. Fig. 1. Habit sketch (natural size). 2. Portion of front, showing venation and position of sori (after Giesenhagen). 3. Cross section of a soriferous frond, showing the indusium-like marginal lip and soral chamber (in moist state). 4. The same, in dry state. 5. Scale from rhizome. 6. Cross section of rhizome. 7. Cross section of the base of stipe. 8. Spores. 9. Stellate hair from the underside of frond. 10. The same from the upperside of frond (2-10 greatly enlarged).

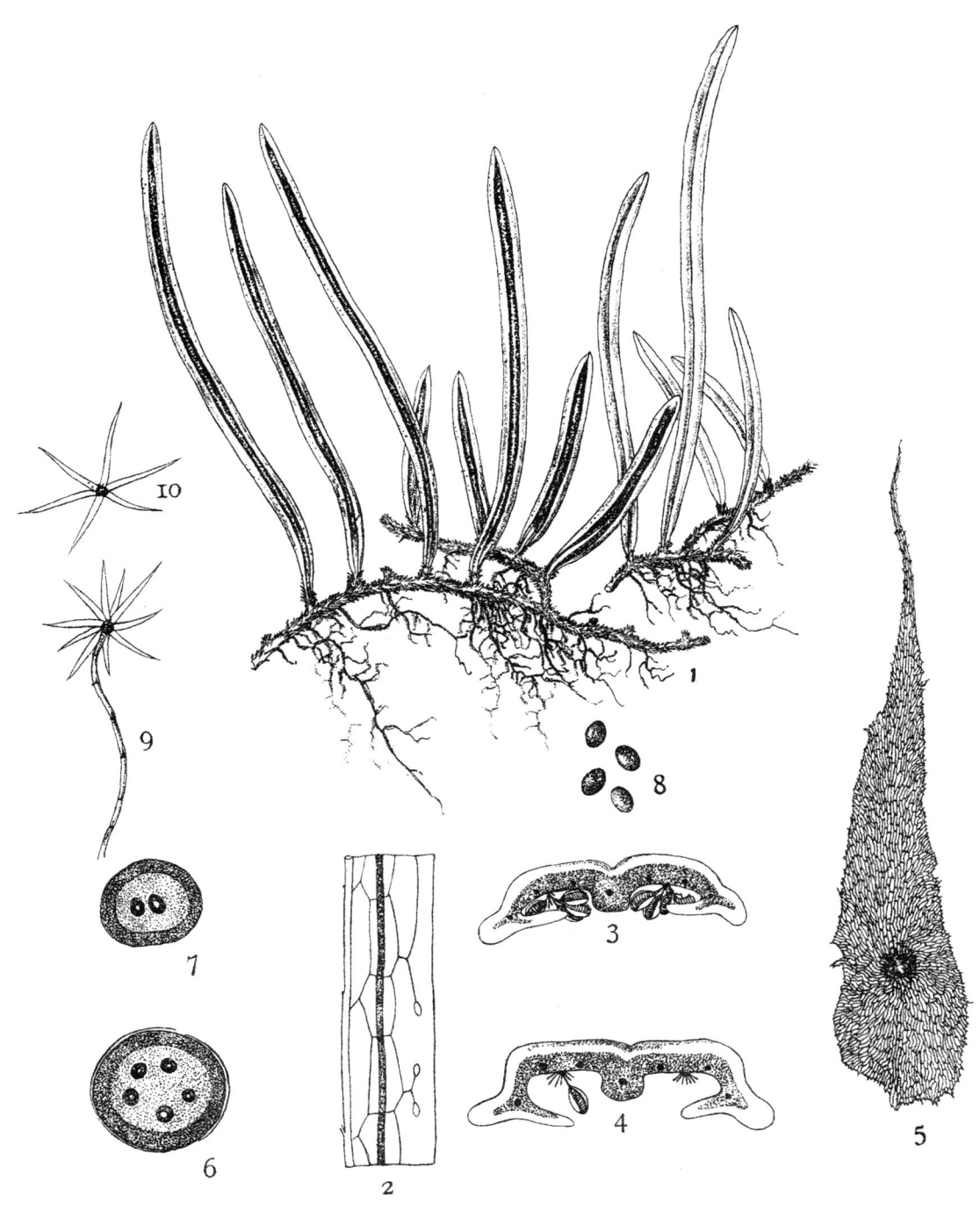

SAXIGLOSSUM TAENIODES (C. Chr.) Ching
拟石苇

根状茎肥厚，亚横行，具锈黄色之披针形鳞片；叶亚簇生，柄长 18—30 cm，坚强，基部节状着生，黝绿色，幼时具星芒状之长足刚毛，旋脱落，叶体为披针形，长 15—30 cm，宽 3—6 cm，渐尖头，向基部渐阔，呈不等耳形，厚革质，绿色，上面具小孔，初被星芒状毛，渐变光滑，下面具淡黄色宿存之厚星芒状毛，芒短而阔，在同一平面；侧脉上下可见，斜出，小脉网状，不见，网眼具小脉；子囊群圆形，着生于小脉之端或端下，初甚明显，旋呈亚合生。

分布：江西，湖南，浙江，贵州，湖北，四川，福建，台湾，云南，广东，广西；越南亦产之。

本种为本属特异之种，在庐山最为普通，其形体颇似后种，唯叶下面之毛较厚而松，色鲜黄，其芒细长如针，且不在同一平面，故易与本种区别。

图注：1. 本种全形（自然大）；2. 叶之一部，表示其叶脉（放大 4 倍）；3. 根状茎上之鳞片（放大 15 倍）；4. 叶下面之星芒状毛（放大 60 倍）；5. 柄上之星芒状毛（放大 60 倍）；6. 叶上面之星芒状毛（放大 60 倍）；7. 孢子（放大 300 倍）。

PYRROSIA SHEARERI (Baker) Ching

PYRROSIA SHEARERI (Baker) Ching, Bull. Bot. Soc. China **1**: 64 (1935).

Polypodium sheareri Baker, Journ. Bot. (1875) 201.

Niphobolus sheareri Diels in Engl. u. Prantl: Nat. Pflanzenfam. **1**: 4. 325 (1899).

Cyclophorus sheareri C. Chr. Ind. Fil. 201 (1905); Acta Hort. Gothob. **1**: 105 (1924); pro parte.

Polypodium drakeanum Christ, Bull. Herb. Boiss. **6**: 871 (1898), quoad pl. Faberi.

Cyclophorus drakeanus (non C. Chr. 1905). L. H. Bailey, Gentes Herb. (1920) 9.

Cyclophorus drakeanus f. *maxima* Wu, Bull. Dept. Biol. Sun Yatsen Univ. No. **3**: 338 t. 150 (1932).

Cyclophorus sheareri f. *maxima* C. Chr. Bull. Dept. Biol Sun Yatsen Univ. No. **6**: 18 (1933).

Rhizome thick, short-creeping, densely scaly; *scales* ferruginous, lanceolate, serrate; *fronds* subcaespitose, stipe 18-30 cm long, thick, stout, articulated at base, dark olive-green, when young, densely clothed in long-stipitate stellate hairs, lamina 15-30 cm long, 3-6 cm broad, broadly lanceolate, acuminate, somewhat broadened towards unequally subauriculated cordate or rounded base; *texture* rigidly coriaceous, green, densely pitted and also sparcely stellately hairy above when young, densely clothed underneath in thick, reddish-brown persistent indumentum consisting of appressed stellate uniform hairs with short broad arms on the same plan, midrib prominent on both sides, green underneath, dark oblive-green above; *lateral main veins* visible on both sides, oblique, *veinlets* completely hidden, copiously anastomosing with numerous included clavate veinlets; *sori* punctate, dorsal or subapical on included veinlets, distinct at first, finally subconfluent.

Jiangxi: Kiukiang, Lushan, *Dr. Shearer* (type) (1873); *Maires* (1880); *E. Faber* 10 (1897); *Miss Reid* (1899); ibid., Paradise Pool, *DeVol 49, 51*, July 15, 1933; *N. K. Up 1776*; *A. N. Steward 4649*; *R. C. Ching 10635* (1934); Fencheng, Talou Shan, *Y. Tsiang 10340* (1931). Hunan: Singning Hsien, *H. F. Chow 41*; Tienchaoping, *Ingvald Ofstad 7337* in herb. Univ. Nanking. Anhui: Chui Hua Shan, *R. C. Ching 7583* in herb. Univ. Nanking; Wu Yuan Hsien, *R. C. Ching 8842*; Hwang Shan, *R. C. Ching 8910*. Zhejiang: Ching Yuan Hsien, *R. C. Ching 2354* Tien Tan Shan, Ningpo, *C. Y. Chiao 14407*; *Sir E. Home* in herb. Hance; *Hancock 47* (1877); *C. W. Everard*. Guizhou: Pinfa, *Cavalerie 1809*; Tungtze, *Y. Tsiang 5049*; *Esquirol 635*; Tsingay, *Bodinier 1743*; Ganpin, *Perry* (1858); Kweiyang, *Cavalerie 7024*. Hubei: Patung, *Henry 5028, 1428, 2569, 2813*; *Wilson 309*. Sichuan: Nanchuan, *W. P. Fang 5840*; *Wilson 309*. Yunnan: Mengtze, *Henry 9116*; *Hancock 48*; Lan Kang, *Delavay 4954*; *Maire*; San Shan, *Ducloux 5102*; *H. T. Tsai 51491, 52338*. Fujian: Kuliang, *H. H. Chung 4294*. Guangdong: North River, Taitung, Lokchong, *S. P. Ko 51889*; ibid., Kau Tung, *W. T. Tsang 20908*; Yüyuan, *S. P. How 53747, 52811*. Guangxi: Miu Shan, Luchen, *R. C. Ching 6129*; Yao Shan, *S. S. Sin 438* (f. *maxima* Wu). Taiwan.

Also Vietnam.

One of the most distinct species of the genus, the following species is the only close Chinese ally which differs in generally shorter and broader lamina on much longer stipe and in very dense loose indumentum on the underside of leaves, consisting of long, needle-like arms not on the same plan.

Plate 147. Fig. 1. Habit sketch (natural size). 2. Portion of lamina, showing venation (× 4). 3. Scale from rhizome (× 15). 4. Stellate hairs from the underside of lamina (× 60). 5. The same from stipe (× 60). 6. The same from the upperside of lamina (× 60). 7. Spores (× 300).

PYRROSIA SHEARERI (Baker) Ching 庐山石韦

图版 148 毡毛石苇

本种形体颇类前种，所异者为叶体较宽而叶柄较叶体为长，间有二倍之者，为其叶下面之毡毛较厚而松，且其芒细长如针，不在同一平面，故得易于识别，再本种之分布概限于中国西南西北部之高山，与前者之限于中部及南部者显然有别。

分布：陕西，湖北，四川，西康，云南，西藏东部。

图注：1.本种全形（自然大）；2.叶体之一部，表示其叶脉（放大3倍）；3.叶体下面之星芒状毛（放大20倍）；4.叶柄上之星芒状毛（放大20倍）；5.根状茎上之鳞片（放大10倍）。

PYRROSIA DRAKEANA (Franch.) Ching

PYRROSIA DRAKEANA (Franch.) Ching, Bull. Bot. Soc. China **1**: 64 (1935).
 Polypodium drakeanum Franch. Nouv. Arch. Mus. Ⅱ. **7**: 165 (1883); Christ, Bull. Herb. Boiss. **6**: 871 (1898).
 Niphobolus drakeanus Diels in Engl. Jahrb. **29**: 206 (1900); Giesen. Farngatt. Niph. 117 (1901); Christ, Bull. Soc. Bot. France **52**: Mém. 1. 24 (1905).
 Cyclophorus drakeanus C. Chr. Ind. Fil. 198 (1905); Contr. U. S. Nat. Herb. **26**: 326 (1931).
 Polypodium sheareri Christ (non Baker), Nuov. Giorn. Bot. Soc. Ital. n. s. **4**: 95 (1897).
 Cyclophorus sheareri C. Chr. Acta Hort. Gothob. **1**: 105 (1924, non 1905), pro parte.
 Niphobolus sheareri Christ (non Diels), Bull. Soc. Bot. France **52**: Mém. 1. 26 (1905).
 Niphoholus inaequalis Christ, Bull. Soc. Bot. France **52**: Mém. 1. 25 (1905); Bull. Acad. Géogr. Bot. (1906) 109.
 Cyclophorus inaequalis C. Chr. Ind. Fil. 199 (1905).

Rhizome thick, short-creeping, densely scaly; *fronds* subcaespitose, stipe thick, stout, 13-17 cm long or much longer, ferruginously stellate hairy, lamina oblong-ovate or oblong-lanceolate, 13-16 cm long, or longer, to 6 cm broad, short acuminate, base obliquely truncate or rotundo-cuneate, slightly decurrent; *texture* rigidly coriaceous, upperside green and sparcely hairy when young, underside densely clothed in thick wooly brown persistent and loose indumentum consisting of long needle-like arms not on the same plan, midrib prominent on both sides, hairy underneath; *lateral main veins* scarcely visible; *sori* punctate, subconfluent at last.

Shaanxi merid: *David* (type), March, 1873; Chuikiosuen, *Giraldi*, Sept. 29, 1897; *Purdom 97*; Thaipei Shan, *Giraldi*. Gansu: Kuatsa, *Meyer 1818*, Nov. 5, 1914; *Licent 4934* (1919). Hubei: Patung, *Henry 3684*; *Wilson 1909, 1727, 2629*; *Henry 1428* (type of *Niph. inaequalis*); Shin Shan Hsien, *S. S. Chien 8324* in herb. Univ. Nanking; Ichang, *Henry 2569*. Guizhou: Ganpin, *Bodinier 1743*; Pinfa, *Cavalerie 512*; Esquirol *635, 699*; *Cavalerie 1807, 1809* (without exact locality). Sichuan: Mt. Omei, *Wilson 5323*; *Faber 1076*; *Brown*; *W. P. Fang 2710*, on tree; Nanchuan: *Bock & Rosthorn 1713, 1059, 3123*; Tachienlu, *Harry Smith 4930*; *Prince Henri d'Orleans*; *Y. Chen*; Pinyipu, *Harry Smith 1984, 4930, 4825*; Moupin, *David* (1870); *T. Tang 787* (1932). Yunnan: Tsekou, *Monberg 275* (1908); *Souliè* (1898); Mengtze, *Henry 9114, 9116*; *Hancock 51* (Kew No); Salwin, *Forrest 16195, 28940*; Mekong, *Forrest 13119* (1914), *19842, 19841*; Kiang Yu Shan near Lanpou, *Delavay 1734*; Tapintze, *Delavay 3250, 4518*; Peyentsin, *Simon Ten*; *Ducloux 4984*; Inon, *Maire*. Tibet orient. : *Souliè 588*.

A distinct pretty fern known only from mountains of comparatively high altitude in the western and north-western part of China; its distinction from *Py. Sheareri* (Baker) is already noted above. *Niph. inaequalis* Christ represents a slender form with lanceolate lamina on stipe nearly twice as long.

Plate 148. 1. Habit sketch (natural size). 2. Portion of lamina, showing venation ($\times$ 3). 3. Stellate hairs from the underside of lamina ($\times$ 20). 4. The same from stipe ($\times$ 20). 4. Scale from rhizome ($\times$ 10).

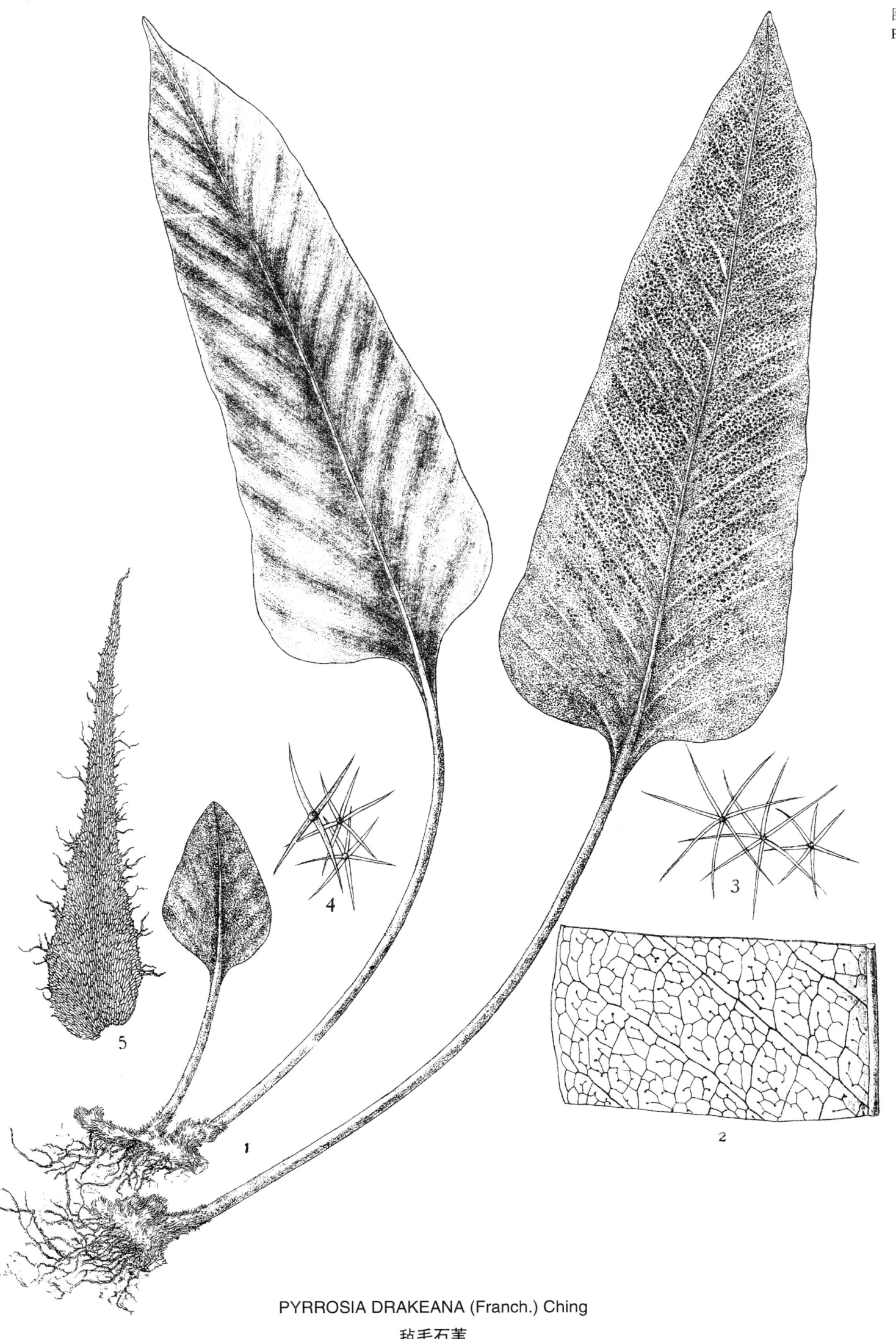

PYRROSIA DRAKEANA (Franch.) Ching
毡毛石苇

根状茎肉质,肥厚,蔓生,灰白色;鳞片疏生,线形,具卵圆形之基部,黄色;叶疏生,柄长达 15 cm,坚实,深稻秆色,有光泽,光滑,叶体长 20—45 cm,宽 15—20 cm,奇数羽状分裂,小叶 1—5 对,节状着生于中肋,对生,无柄,长 12—16 cm,宽 3—4 cm,阔披针形,尾状,基部圆形或亚心脏形,全缘,具一极宽之薄膜质,两面绒毛密生,侧脉明显,斜出,下面具长刚毛,小脉网状,厚纸质;子囊群圆形,呈不规则之二列,无盖。

分布:云南;缅甸及印度北部。

本种为本属特殊之种,在我国唯云南仅见之;此外尚有一种变种,产于锡金,其小叶四缘深裂。

图注:1.本种全形(自然大);2.同上;3.小叶之一部,表示其叶脉及子囊群(放大 5 倍);4.根状茎上之鳞片(放大 20 倍);5.侧脉下面之刚毛(放大 80 倍);6.叶下面之绒毛(放大 80 倍);7.羽片基部之关节(放大 10 倍)。

ARTHROMERIS HIMALAYENSIS (Hk.) Ching

ARTHROMERIS HIMALAYENSIS (Hk.) Ching, Contr. Inst. Bot. Nat. Acad. Peiping **2**: 99 (1933).

Polypodium himalayense Hk. Sp. Fil. **5**: 91 (1863); Syn. Fil. 369 (1867); Diels in Engl. u. Prantl: Nat. Pflanzenfam. **1**: 4. 321 (1899); Christ, Farnkr. d. Erde 114 (1897); Bull. Soc. Bot. France **52**: Mém. 1. 19 (1905), pro parte; C. Chr. Ind. Fil. 533 (1905).

Pleopeltis himalayensis Bedd. Ferns Brit. Ind. t. 318 (1869); Handb. Ferns, Brit. Ind. etc. 372 (1883).

Polypodium venustum Wall. Cat. n. 305 (1828, nom. nud.); Clarke, Trans. Linn. Soc. Ⅱ. Bot. **1**: 566 (1880), non Desv. 1811.

Pleuridium venustum J. Sm. Cat. Cult. Ferns 10 (1857).

Arthromeris venusta J. Sm. Hist. Fil. 111 (1875).

Polypodium venustum var. *niphoboloides* Clarke, I. c. 567.

Rhizome thick, woody, wide-creeping, glaucous, sparcely scaly; *scales* linear from orbicular base, bright brown; *frond* far apart, stipe firm, glossy, dark straminous, 1-15 cm long, glabrous, lamina 20-45 cm long, 15-20 cm broad, imparipinnate; *pinnae* 1-5-jugate, articulated to rachis, opposite. 5-7 cm apart, sessile, 12-16 cm long, 3-4 cm broad, oblong-lanceolate, caudate, base roundish or sub-cordate, margin entire but wavy and provided with a very broad whitish membrane; *texture* subcoriaceous, both sides densely glandular hairy, costa underneath with long spread hairs; *lateral main veins* fine but distinct, areolae fine, with copious included veinlets; *sori* large, rounded, irregularly 2-seriate, of 3-4 each between main veins, exindusiate.

This distinct and pretty fern can easily be distinguished from its relatives by oblong-lanceolate and caudate, densely glandular hairy pinnae with a very broad whitish membranaceous margin.

Yunnan: South of Red River, Manmei, *Henry 9747*; Tsetchoupa, *Maire*; Song-san, *Delavay 2612*.

Sikkim-Himalayas, from 6,000 to 10,000 ft. elevation, common.

Var. **furcans** Ching, I. c. p. 100.

Habit of *Cyrtomium caryotideum* Presl, with irregularly lacerato-forked pinnae. Known only from Sikkim (leg. Duthie 5177).

Plate 149. Fig. 1. Habit sketch (natural size). 2. The same (natural size). 3. Portion of pinna, showing venation and sori (× 5). 4. Scale from rhizome (× 20). 5. Hairs from costa (× 80). 6. Glandular hairs from underside of lamina (× 80). 7. Rachis with articulated base of the lateral pinna (× 10).

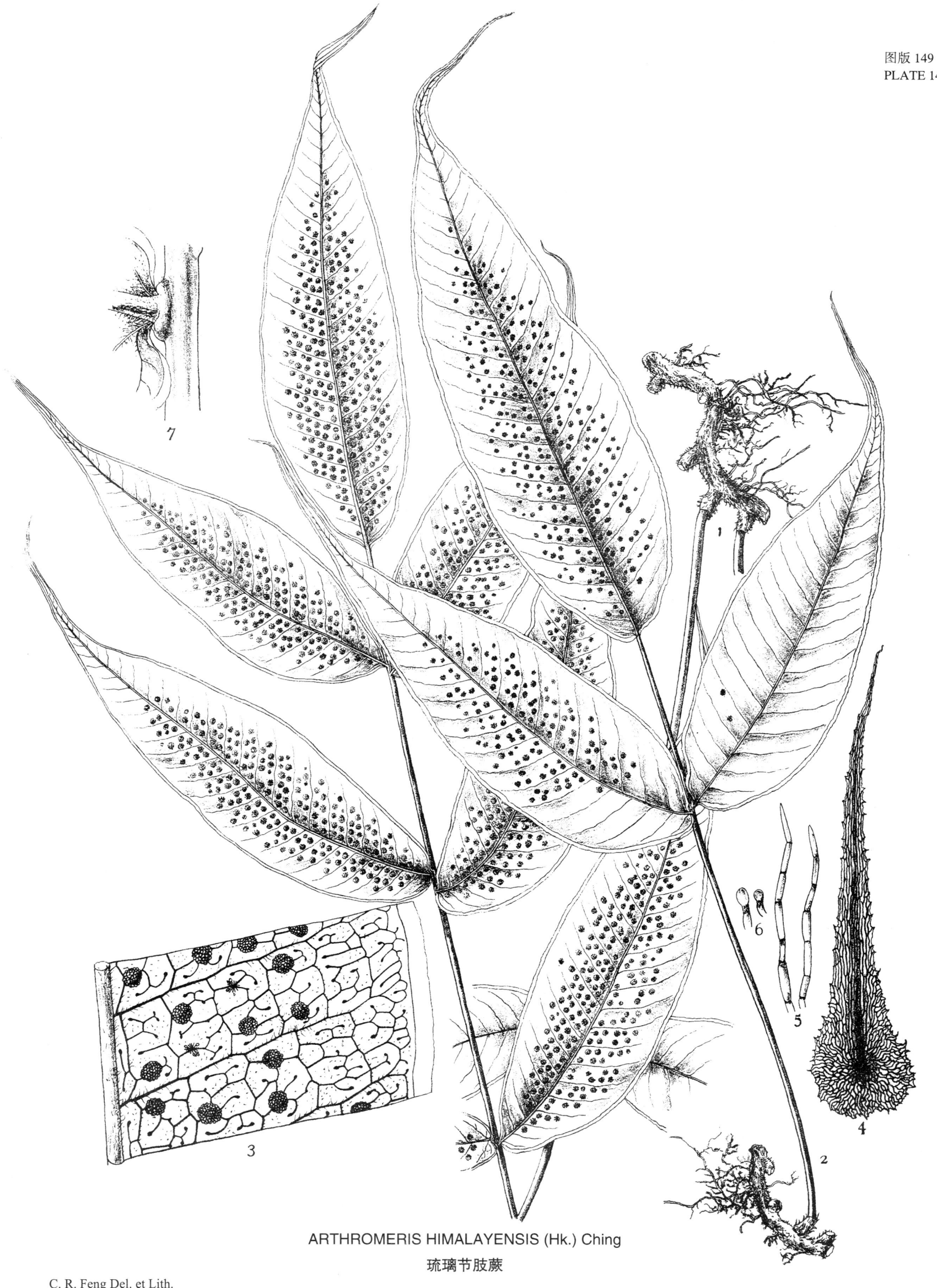

ARTHROMERIS HIMALAYENSIS (Hk.) Ching
琉璃节肢蕨

本种形体颇类前种，唯较小，小叶仅具线状骨质叶缘，两面仅刚毛疏生，故得易于识别。

分布：广州北部山中。

图注：1. 本种全形（自然大）；2—3. 同上；4. 小叶之一部分，表示叶脉及子囊群（放大 5 倍）；5. 根状茎上之鳞片（放大 20 倍）；6. 叶体上之刚毛（放大 60 倍）。

ARTHROMERIS LUNGTAUENSIS Ching

ARTHROMERIS LUNGTAUENSIS Ching, Contr. Inst. Bot. Nat. Acad. Peiping **2**: 98 (1933).

Polypodium lungtauense Ching in C. Chr. Ind. Fil. Suppl. Ⅲ. 152 (1934).

Rhizome wide-creeping, 4-5 mm thick, terete, atratous; *scales* dense, ovate, acuminate, grayish-brown with scarious margin; *frond* far apart, stipe 3-10 cm long, dark straminous, terete, slender, naked, lamina oblong or deltoid, 7-30 cm long, impari-pinnate; *pinnae* 1-3-8-jugate, the terminal pinnae similar to the lateral ones, which are patent, sessile, 6-10-16 cm long, 2-3 cm broad in the middle, gradually acuminate, base equal, cordate, subamplexicaulous, margin entire with a cartilaginous line all around; *lateral veins* prominent, 4 mm apart, obliquely patent, transversal veinlets 4-6 between costa and margin, intervening veinlets anastomosing with included veinlets; *texture* chartaceous, brownish when dried, both sides moderately villose; *sori* rather small, rounded, dark brown, bi-seriate, of 3-4 each between main veins, glabrous, with a few protruding hairs from vein-lets underneath.

Guangdong: Lungtau Shan, Iu village, *To Kang Ping 12430* type ex herb. Lingnan Univ., June 7, 1924; Tsungfa-Lungmoon Districts, *Tsang Wei Tak 20508*, May 18, 1932; Yüyuan, *S. P. Ko 53074*, July 13, 1933; Lohfau Shan, *C. Ford 14* (1883); *C. O. Levine & McClure 6828*; *Levine 1545*; *B. C. Henry 22210* in herb. Hance, May, 1863.

A critical species, which differs from *A. himalayensis* (Hk.) in not densely pubescent surfaces of pinnae of a different outline with a narrowly cartilaginous margin; from *A. Lehmanni* (Mett.) it differs in sparcely villose leaf surfaces and much smaller size.

Plate 150. Fig. 1. Habit sketch (natural size). 2-3. The same (natural size). 4. Portion of pinna, showing venation and sori ($\times$ 5). 5. Scale from rhizome ($\times$ 24). 6. Hair from lamina ($\times$ 60).

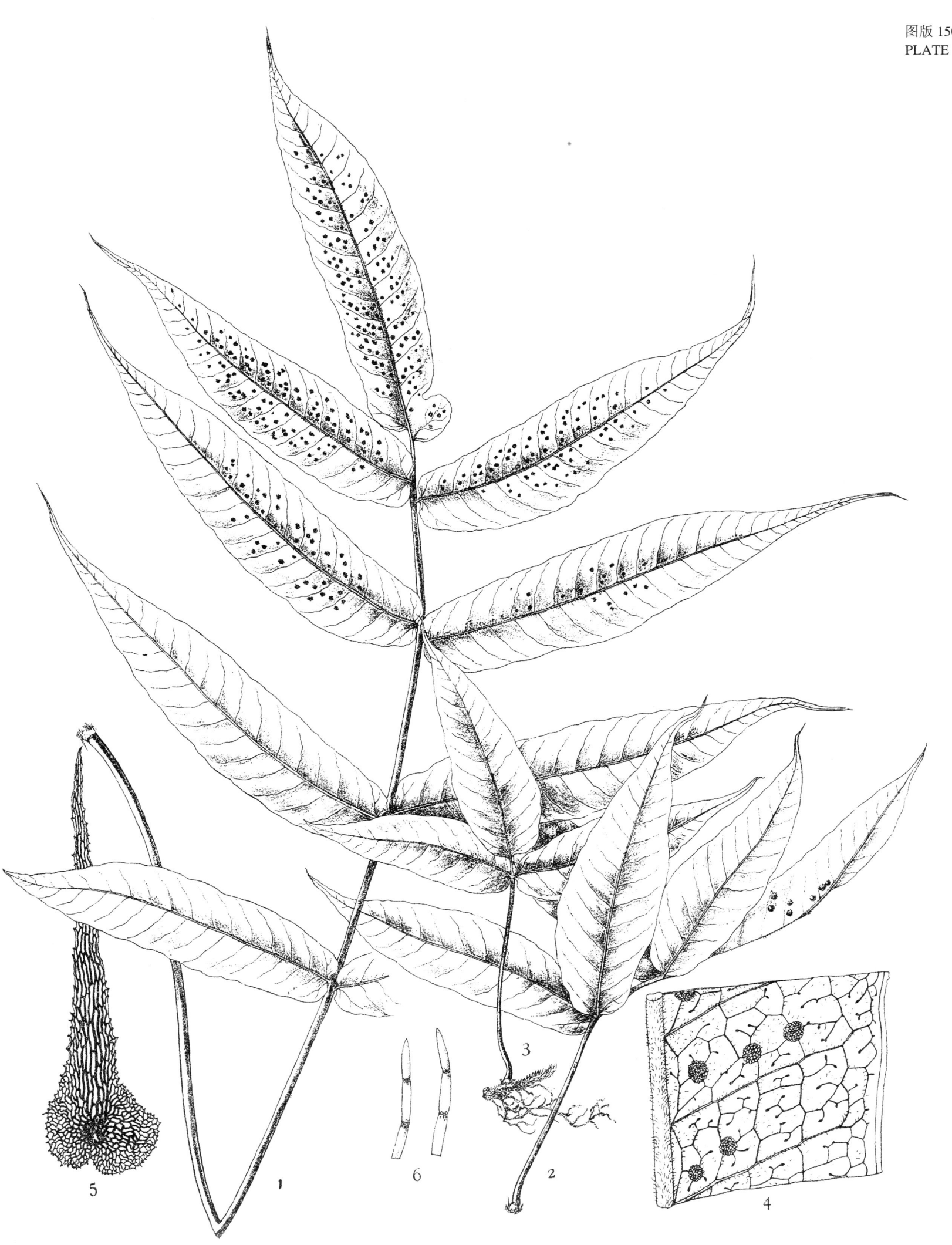

ARTHROMERIS LUNGTAUENSIS Ching
粤节肢蕨

中国蕨类植物图谱

ICONES FILICUM SINICARUM

BY

REN-CHANG CHING, B. S.

KEEPER

LU-SHAN ARBORETUM AND BOTANICAL GARDEN

FASCICLE 4, PLATES 151-200

第四卷

TO

PROF. WOON-YOUNG CHUN, M. S.

DIRECTOR OF BOTANICAL INSTITUTE

NATIONAL SUN YATSEN UNIVERSITY, CANTON

AND

MY FORMER TEACHER

IN RECOGNITION OF

HIS EXEMPLARY WORK IN BOTANY OF CHINA

AND

HIS UPLIFTING INFLUENCE UPON THE

YOUNGER GENERATION OF CHINESE BOTANISTS

THIS FOURTH FASCICLE OF ICONES FILICUM SINICARUM

IS RESPECTFULLY DEDICATED

根状茎横行；叶柄高达 2 m，酒红色，光滑无毛，叶分叉；一回羽叶对生，长达 80 cm，宽 40 cm 或过之，腋间具一被栗黑色鳞片之大芽；二回羽叶多数，互生，开展，具短柄，长达 30 cm，宽 4 cm，线状披针形，基部为截楔形，等宽，向顶渐尖，羽状深裂至中肋；裂片 60—70 对，线形，亚斜出，长达 2 cm，宽 2 mm，全缘，强度反卷，钝头，叶脉约 20 对，均由基部分叉，上面光滑，下面淡粉白色，且被星状毛，中轴亦呈酒红色，稍具阔披针形之深栗色鳞片；子囊群中生，位于上方小脉，由 2—4 个子囊组成之。

分布：广东信宜县及黄埔产之。

此种为广东特产，其形态极类里白(G. glauca)，唯形体较大，叶柄及中轴呈酒红色，二回小叶长达 30 cm，宽 4 cm，且具柄，裂片亚斜出，边缘强度反卷，叶腋间之芽所被之鳞片之长仅半之耳。

图注：1. 本种全形(自然大)；2. 裂片，表示叶脉及子囊群之位置(放大 5 倍)；3. 子囊(放大 50 倍)；4. 叶腋间之芽之鳞片(放大 50 倍)；5. 叶下面所被之星状毛(放大 50 倍)。

GLEICHENIA CANTONENSIS Ching（GLEICHENIACEAE）

GLEICHENIA CANTONENSIS Ching, Lingnan Sci. Journ. 15：391 (1936).

A large straggling fern to several meters tall; *rhizome* creeping, *stipe* up to 2 meters long, stout, thick as a finger near base, wine-colored, subnitid, glabrous, forked; *primary pinna* opposite, to 80 cm long, 40 cm or broader, axillary bud large, densely clothed in imbricate, lanceolate, atro-castaneous *scales* about 4 mm long, with densely fimbriate margin; *secondary pinna* numerous, alternate, 3-3.5 cm apart, patent, shortly petiolate, up to 30 cm long, 4 cm broad, linear-lanceolate, base equally truncato-cuneate, gradually narrowed towards acuminate apex, pinnatifid nearly down to rachilet; *segments* 60-70-jugate, linear, suboblique, to 2 cm long (sometimes longer), 2 mm broad, margin entire but strongly revolute, apex obtuse, often subemarginate, separated from each other by somewhat broader sinuses; *veinlets* 20-jugate, regularly forked above base, rachis subnitid, wine-colored, with a few broad-lanceolate, deciduous, appressed, castaneous scales; *texture* herbaceous, glabrous and light green above, glaucescent and sparingly stellate hairy beneath; *sori* medial on anterior veinlets, consisting of 2-4 (generally 3) large, globular, pale lemon-yellow sporangia, *receptacle* elonagte, naked.

Guangdong: Suni, *Y. K. Wang 30967* (type), roadside, July 17, 1931; Whampoa, one specimen without collector's name in Herb. Rigsmuseum at Stockholm.

This distinct and pretty fern is closely related to *G. glauca* Hk., from which it differs in enormously larger size, wine-red stipe and rachis, the petiolate secondary pinna up to 30 cm long, 4 cm broad, and the longer and narrower oblique segments with strongly revoluted margin and broader sinuses. The scales covering the axillary bud are only half as long as those in its relative.

Plate 151. Fig. 1. Habit sketch (natural size). 2. Segments, showing venation and position of sori ($\times$ 5). 3. Sporangium ($\times$ 50). 4. Scale from axillary bud ($\times$ 50). 5. Stellate hairs on the underside of leaf ($\times$ 50).

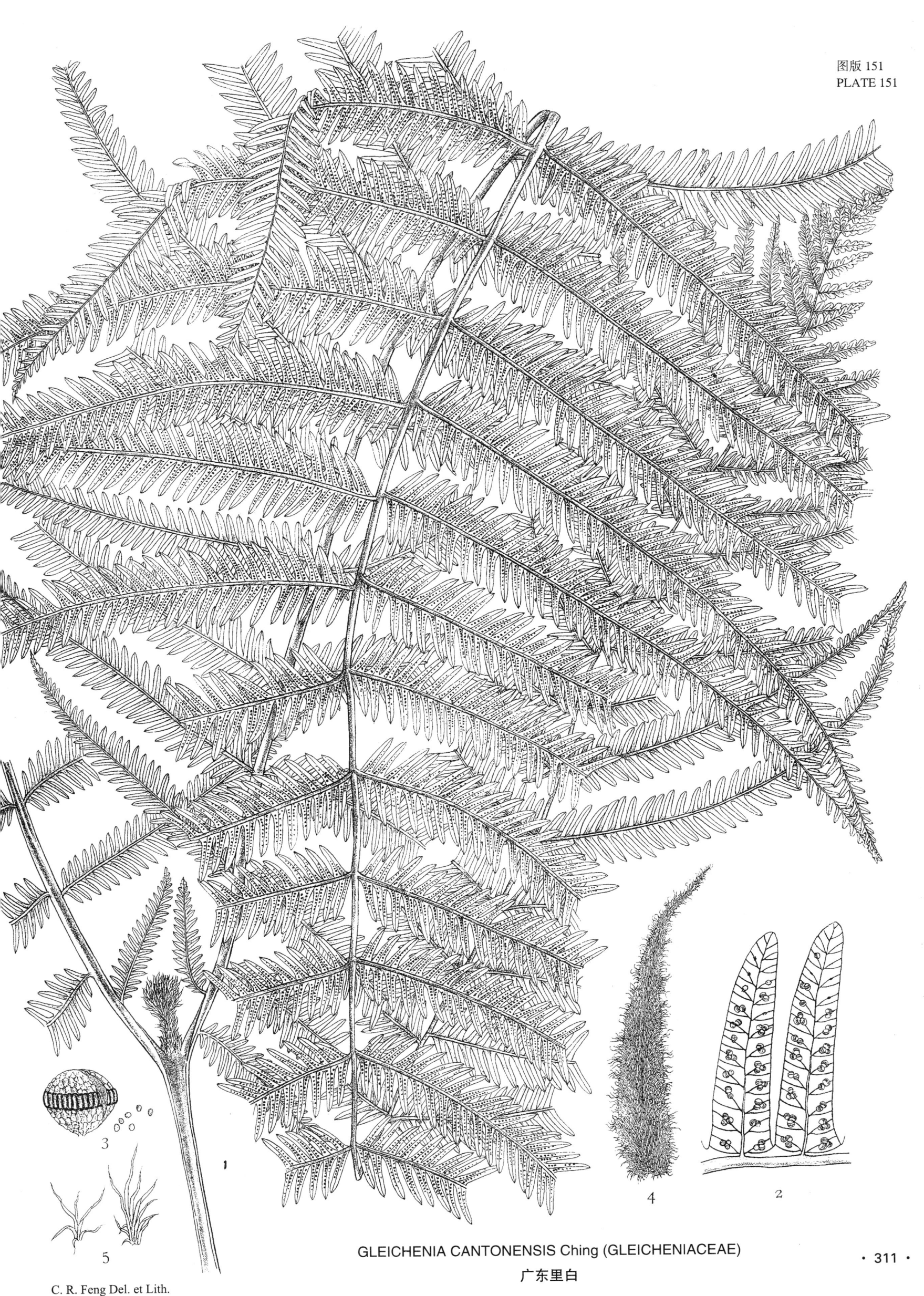

GLEICHENIA CANTONENSIS Ching (GLEICHENIACEAE)
广东里白

根状茎横行,密被红棕色之细长鳞片;叶散生,柄长达 50 cm,淡绿色,光亮,叶体卵形,二裂,小叶长达 40 cm,宽16 cm,椭圆形,渐尖头,二回羽状分裂;叶腋间具一大芽,被淡黄色之密鳞片;二回小叶线状披针形,渐尖头,具短柄,基部外侧数对最长,长达 12 cm,宽 2 cm,内侧一二对最短,渐尖头,羽状深裂,裂片栉篦排列,长 1—1.5 cm,宽 2 mm,尖头,斜出,边缘强度反卷,厚纸质,两面光滑,绿色,叶脉分叉;子囊群小,圆形,生于上方小脉,由 3—4 个淡黄色之子囊组成之。

分布:贵州,江西,浙江,广东,广西,海南;日本亦产之。

此为本属特殊之种,全体光绿,甚易识别。

图注:1.本种全形(自然大);2.两裂片,表示叶脉及子囊群之位置(放大 3 倍);3.子囊群(放大 50 倍);4.根状茎上之鳞片(放大 16 倍);5.腋芽之鳞片(放大 16 倍);6.根状茎之横切面,表示维管束之布置(放大 4 倍)。

GLEICHENIA LAEVISSIMA Christ

GLEICHENIA LAEVISSIMA Christ, Bull. Acad. Géogr. Bot. (1902) 268; C. Chr. Ind. Fil. 322 (1905).

Mertensia laevissima Nakai, Bot. Mag. Tokio **39**: 182 (1925).

Gleichenia kiusiana Makino, Bot. Mag. Tokio **18**: 139 (1904); C. Chr. Ind. Fil. Suppl. Ⅰ. 44 (1906-12); Ogata, Ic. Fil. Jap. **4**: pl. 179 (1931).

Rhizome wide-creeping, densely scaly; *scales* lanceolate, long-acuminate, rufo-brown, nitid, thick, entire; *frond* 2-4 cm apart, stipe 30-50 cm long, 3 mm thick near base, green or pale straminous, smooth, glabrous above base, rounded beneath, lower part flattened with sharp edge above and upper part deeply grooved, lamina ovate, bifurcate at the tip of stipe into two similar pinna of oblong outline, 30-40 cm long, 13-16 cm broad, acuminate, axillary bud large, densely scaly, scales ovate, entire, with long subulate apices; *pinnules* linear-lanceolate, acuminate, short-petiolate, the lower ones on the exterior side of rachilet much the longest, to 12 cm long, 2 cm broad, gradually shortened upward, the interior basal one or two pairs generally much smaller, pinnatifid down to costa; *segments* numerous, pectinate, oblique, linear with sharp apices, 1-1.5 cm long, 2 mm broad, lower base decurrent, margin revolute, with rounded and as broad sinus; *texture* rigidly herbaceous. pleasing green, naked throughout, bluish beneath; *veins* in segments 10-15-jugate, suboblique, all forked above base, *sori* small, medial, borne on the anterior veinlets above forking, consisting of 3-4 globular brownish sporangia.

Guizhou: Kweiyang, *Bodinier 2095* (type), March 18, 1898; Pinfa, *Cavalerie 528*; Vanchingshan, Yinkiang, *Y. Tsiang 7677*; Kweiting, *Y. Tsiang 5481*; Tuyun, *Y. Tsiang 5889B*, *5889A*. Sichuan: without locality, *T. Tang 22775* (1930). Yunnan: Without locality, *Delavay 46*. Jiangxi: Kiukiang, Lushan, *Dr. Shearer* (1887); Three Falls, *C. E. DeVol 47*, August 7; 1933, steep damp hill side. Zhejiang: Tientai Shan, *R. C. Ching 1427* (1923), under forest. Guangdong: Lokchong, *N. K. Chun 42513*. Guangxi: Luchen Hsien, Miu Shan, Dar Siar Ping, *R. C. Ching 6189*; Ling Yen Hsien, Loh Hoh Tsuen, *A. N. Steward & H. C. Cheo 134*. Hainan Island: without locality, mountain summit, *C. Wang 35525*.

Also Japan: Kinsin, Oosumi, *G. Koidzumi*, April 14, 1923.

One of the most distinct species of the genus, differing from the previous one in much smaller size, narrower and more oblique segments with revolute margin and sharply pointed apices and in the scales on the axillary bud being entire, light-brown, with hair-pointed apices. By the present distribution, this species is now known from central, southern and eastern Chinese provinces.

Plate 152. Fig. 1. Habit sketch (natural size). 2. Two segments, showing venation and sori (× 3). 3. Sorus (× 50). 4. Scale from rhizome (× 16). 5. Scale from axillary bud (× 16). 6. Cross section of rhizome, showing solenostele (× 4).

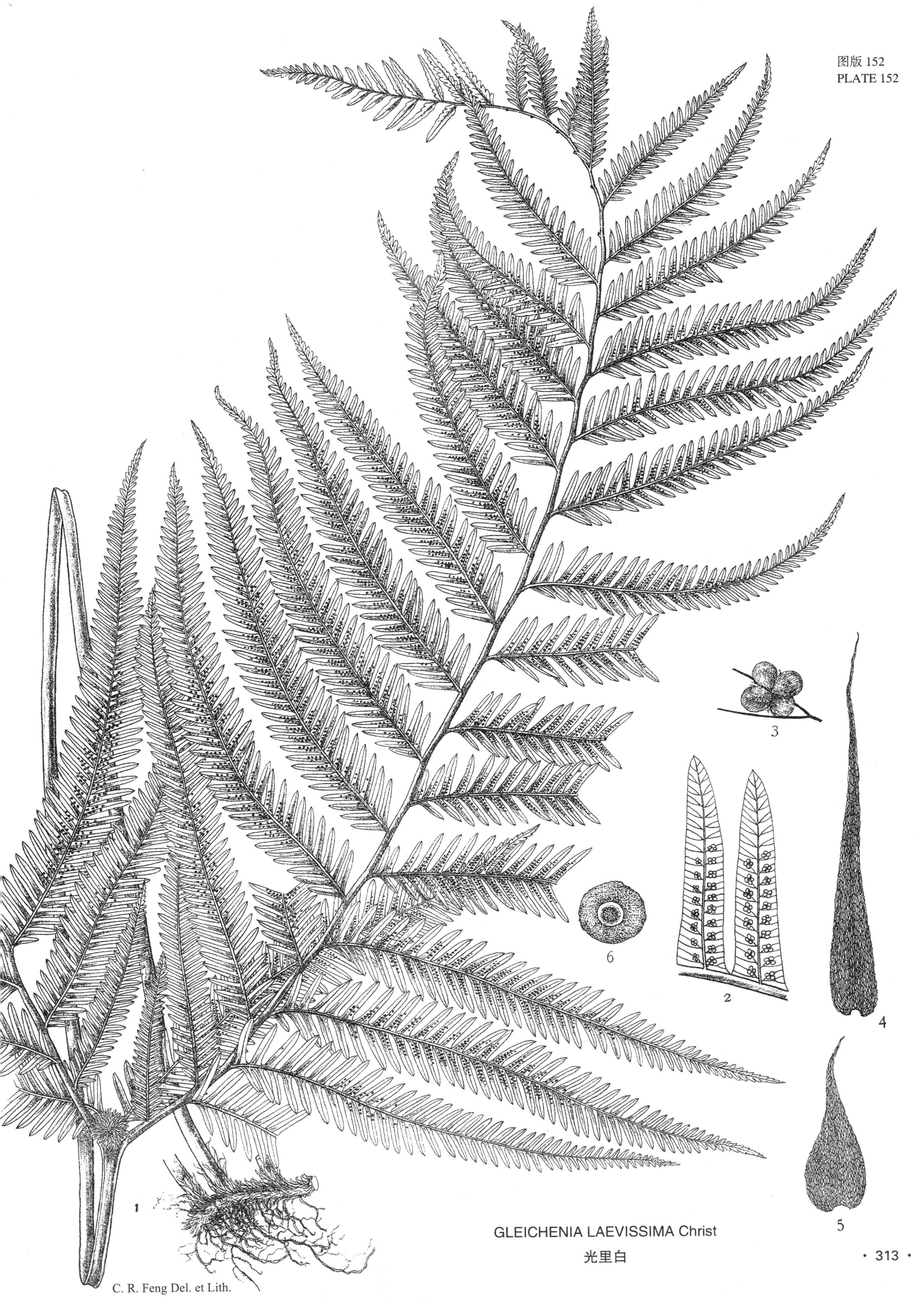

GLEICHENIA LAEVISSIMA Christ
光里白

根状茎横行,光亮,粗糙;鳞片深棕色,线形,簇生;叶长一至数米,直立或为蔓性,叶柄深稻秆色,光亮,无毛,高自 30 cm 至数米,叶体数回分叉,每分叉处之基部下方具一对小叶片;一回小叶长 30—40 cm,宽 10—16 cm,顶渐尖,基部渐狭,无柄,枥状深裂;裂片长 5—10 cm,基部之宽达 1 cm,线状披针形,端略钝,边缘呈浅波形,亚革质,下面稍呈白色,叶脉每组具 5—6 数平行小脉;子囊群为不规则之二列,中生,由 10—20 个子囊组成之。

分布:云南,广东,香港,广西;缅甸,越南均产之。

本种形态极类普通之枥里白(*G. linearis*),唯其各部形体特大,裂片长达 10 cm,宽 1 cm,边缘呈波形,故易识别。

图注:1. 本种全形(自然大);2. 裂片之一部,表示叶脉及边缘(放大 4 倍)。

GLEICHENIA SPLENDIDA Hand-Mzt

GLEICHENIA SPLENDIDA Hand-Mzt. Akad. Anz. Akad. Wien (1924) 81; Symb. Sin. **6**: 16 (1929); C. Chr. Ind. Fil. Suppl. Ⅲ. 106 (1934).

Gleichenia linearis C. Chr. (non Clarke, 1880), Contr. U. S. Nat. Herb. **26**: 271 (1931), pro parte.

Rhizome wide-creeping, dull brown, subnitid, muricate, densely scaly; *scales* atro-brown, linear-subulate, rigid, spreading, tufted, deciduous; *frond* 1 to several meters tall, erect or trailing, stipe dark straminous, shining, glabrous, 4-5 mm across, 30 cm to several meters long, lamina repeatedly di- or trichotomously forked, with a distinct pair of pinnae arising from the base of the primary forking branches and two smaller foliaceous and coriaceous ones at the axil covering the scaly terminal bud; *pinnae* 30-40 cm long, 10-16 cm broad at the middle, gradually acuminate towards apex, base decrescent either on both sides or the upper side only, sessile, pectinately pinnatifid; *segments* 5-10 cm long, up to 1 cm broad above the broadened base, linear-lanceolate, with obtusish apex, and repando-undulate margin; *texture* subcoriaceous, glaucescent beneath; *veinlets* 5-6 in each group, parallel; *sori* irregularly 1-2-rowed, medial on veinlets, consisting of 10-20 sporangia.

Yunnan: Tibet-Burmese border, *H. Handel-Mazzetti 9351* (type); Tengyueh, *J. F. Rock 7172*; Mengtze, *Hancock 71*. Guangdong: Sunyi, *Y. K. Wang 30920*. Guangxi: Lin Yen Hsien, Tsinlung Shan, *R. C. Ching 7048*; ibid., Yeo Mar Shan, *R. C. Ching 7256*. Hong Kong: Victoria Peak, *Lamont 975*.

Burma: *Lace 4748* ex Herb. Bedd.

Khasia: *Hooker f. et Thomson*.

Vietnam: Laos, *M. Poilane*, November 6, 1921.

A gigantic species of the group of G. *linearis* (Burm.) Clarke and has previously been considered as identical with that common fern, from which it differs in enormously larger size in all parts, the segments being up to 10 cm long and nearly 1 cm broad, with repando-undulate margin, and more sporangia in each sorus, which is, however, rarely found present.

Plate 153. Fig. 1. Habit sketch (natural size). 2. Portion of segment, showing venation and undulate margin (× 4).

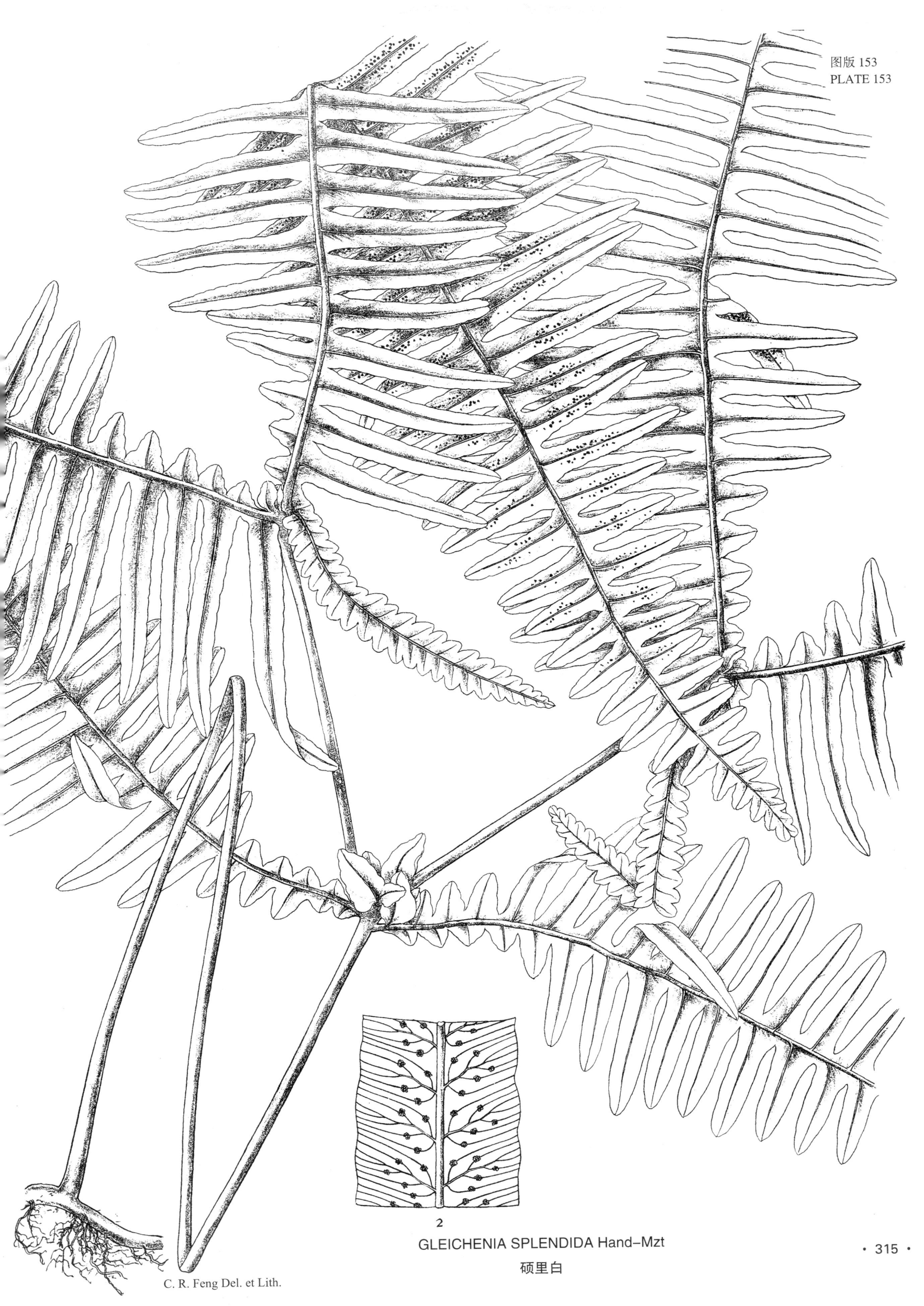

2. GLEICHENIA SPLENDIDA Hand-Mzt
硕里白

图版 154　双扇蕨

　　根状茎横行,粗健,木质,被深棕色之紧覆细长厚鳞片;叶散生,柄长 30—60 cm,木质,叶体二裂,成等大之两扇形,各扇复四至五深裂,渐尖头,各裂片一次深裂,具粗锯齿,干纸质,上面光滑,下面叶脉略具深棕色之短毛,主脉数回扇状分叉,连以显明之横脉,网脉颇明显,各网眼具分叉或单小脉;子囊群形圆而小,散生,不具盖,线状体密生,端呈膨大之伞形,深棕色。

　　分布：贵州,广西,云南,藏边产之。

　　此为本属特殊之一种,为中国特产,其形体略似亚洲热带产之 *D. conjugata*,唯较小,叶之分裂较少,下面不呈粉白色,子囊群中之线状体不呈球杆形,故易识别。

　　图注：1.本种全形(自然大);2.叶片之一部,表示叶脉及子囊群(放大 5 倍);3.子囊(放大 150 倍);4.孢子(放大 150 倍);5.根状茎上之鳞片(放大 40 倍);6.根状茎之横切面,表示维管束之布置(放大 5 倍);7.叶下面脉上之短毛(放大 30 倍);8.子囊群中之隔丝(放大 150 倍)。

DIPTERIS CHINENSIS Christ

DIPTERIS CHINENSIS Christ, Bull. Acad. Géogr. Bot. (1904) 104 cum fig. et tab.; C. Chr. Ind. Fil. 242 (1905); Ching, Bull. Dept. Biol. Coll. Sci. Sun Yatsen Univ. No. **6**: 23 (1933). *Dipteris Horsfieldii* Christ (non Bedd. 1869), Bull. Herb. Boiss. **6**: 880 (1898).

Dipteris conjugata Hand.-Mzt. (non Reinw. 1924), Sym. Sin. **6**: 28 (1929); Wu, Polyp. Yaoshan, in Bot. Dept. Biol. Coll. Sci. Sun Yatsen Univ. No. **3**: 94 t. 38 (1932).

Rhizome wide-creeping, woody, densely scaly; *scales* atro-brown, rigid, setaceous, linear-subulate, nitid, appressed; *frond* far apart, stipe 30-60 cm long, dark straminous, glabrous above base, woody, rounded below, broadly grooved upward on the upper side, lamina 20-30 cm high, 30-60 cm broad, bipartite into nearly equal fan-shaped halves, each again cleft into 4-5 broad acuminate lobes, 5-8 cm broad, the exterior one divided to three-quarters of the way down, the middle one or two, less deep, and the interior one, only one-third way down, each lobe may be once forked with grossly serrated margin; *texture* crass herbaceous or subcoriaceous, turning brownish when dried; underside green or slightly bluish and with some dark brown articulated deciduous hairs along veins; main *veins* prominent, dichotomously branched, connected by finer transversed ones, with intermediate veinlets anastomosing copiously into hexagonal areolae with simple or forked included veinlets; *sori* small, punctiform, superficial, campital, scattered, exindusiate, *sporangia* intermixed with atro-brown cup-shaped stalked paraphyses.

　　Guizhou: Pinfa *Gavalerie 7641*, *341* (type); Cheugfeng, *Y. Tsiang 4706*; without locality, *Esquirol 3139*, *667*. Guangxi: Yao shan, *S. S. Sin 647*, June 29, 1928; Tseung Hsien, *C. Wang 39494*, June 22, 1936; Luchen Hsien, Miu Shan, *R. C. Ching 6140*. Yunnan: Mengtze, *Hancock 213*; A, *Henry 9041*, *9041A*; Tzitzoti, *Forrest 27693* (1925); Pingpien Hsien, *H. S. Tsai 55420*, *60250*, in thickets. Tibet-Burmese border; *H. Handel-Mazzetti 9347*.

　　Upper Burma: Htawgaw, *Forrest 24613*.

　　This distinct endemic species can be easily distinguished from the tropical Asiatic *D. conjugata* Reinw. by much less divided leaves with fewer and broader ultimate segments, green or slightly bluish under surface and not clavate but cup-shaped paraphyses in sorus.

　　Plate 154. Fig. 1. Habit sketch (natural size). 2. Portion of lamina, showing venation and sori ($\times$ 5). 3. Sporangium ($\times$ 150). 4. Spores ($\times$ 150). 5. Scale from rhizome ($\times$ 40). 6. Cross section of rhizome, showing solenostele ($\times$ 5). 7. Hairs from veins on the under side of leaf ($\times$ 30). 8. Paraphyses in sorus ($\times$ 150).

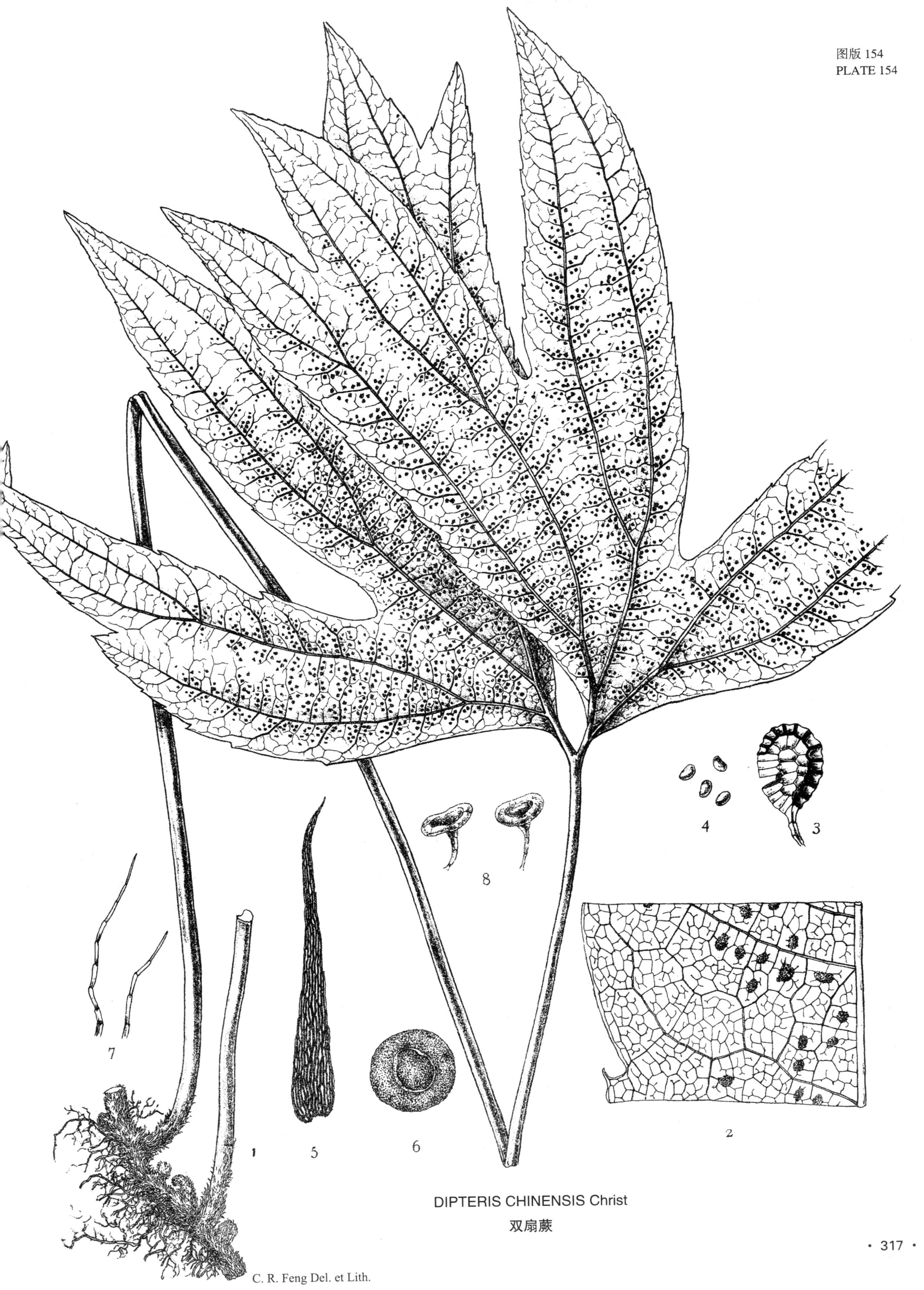

DIPTERIS CHINENSIS Christ
双扇蕨

图版 155　峨眉瘤足蕨

根状茎木质，粗厚，不具鳞片；叶簇生，二形，不生子囊群叶之柄长达 20 cm，坚硬，深稻秆色，上面具阔沟槽，基部扁形，背部具脊，两侧各具三个瘤状凸起，叶体椭圆披针形，长达 30 cm，宽约 10 cm，基部较狭，一回羽状分裂，顶部渐尖，羽状深裂；小叶 30—35 对，线状披针形，基部向上延长合生，基部数对强度下向，且较短，中部各对长约 8 cm，宽 8 mm，水平开展，上部各对渐短，斜出，渐尖头，全缘，唯向顶部略具疏小锯齿，全体光滑，上面绿色，下面呈灰白色，（幼时为绿色），叶脉多数分叉，斜出；生子囊群之叶具较长之柄，叶体较狭，小叶强度紧缩，成细长线形，长达 3 cm，宽约 2 mm，基部膨大合生，下面中肋两侧子囊群满布，不具盖。

分布：四川西南部高山特产。

图注：1. 本种全形（自然大）；2. 小叶，表示叶脉及锯齿（放大 2 倍）；3. 生子囊群小叶之一部，表示子囊着生情形（放大 16 倍）；4—5. 子囊及孢子（放大 150 倍）。

PLAGIOGYRIA ASSURGENS Christ

PLAGIOGYRIA ASSURGENS Christ, Bull. Soc. Bot. Ital. (1901) 293; C. Chr. Ind. Fil. 495 (1905).; Cop. Phil. Journ. Sci. **38**: 398 (1929).

Lomaria deflexa Baker (non Col. 1844, nec Liebm. 1849), Journ. Bot. (1888) 226.

Blechnum faberi C. Chr. Ind. Fil. 153 (1905).

Rhizome thick, woody; *fronds* caespitose, dimorphous, *sterile* one with stipe to 20 cm long, rigid, dark straminous, terete beneath, broadly grooved above, base flattened, carinate, with 3 aerophores on each of exterior sides, lamina oblong-lanceolate, quite narrowed at base, to 30 cm long, 10 cm broad at middle, simple pinnate with pinnatifid acuminate apical part; *pinnae* 30-35-jugate, linear-lanceolate with upper side of base running upward, the lower several pairs gradually shortened and strongly deflexed, the middle ones to 8 cm long, 8 mm broad, horizontally patent, entire except the acuminate apex being with a few obsure teeth, the upper pinnae gradually shortened and oblique; *texture* subcoriaceous, glabrous in all parts, green above, glaucous or bluish-white (or green when young) underneath; *veins* distinct on both sides, oblique, mostly forked; *fertile* frond strongly contracted, on longer stipe (to 40 cm long), lamina linear-lanceolate, 5 cm broad; *pinna* contracted, linear, to 3 cm long, 2 mm broad, base broadly adnate; *sori* indefinite, completely covering the under surface, except the midrib.

Sichuan: Tientosan, *Scallan* (type); Mt. Omei, *Wilson 5284*; *E. Faber 1023* (type of *Lomaria deflexa*); Tah Liang Shan, *T. T. Yü 4045*, Sept. 22, 1934.

This distinct endemic fern, so far collected only thrice in mountains in the southwestern part of Sichuan, is closely related to P. *adnata* of the same region in general outline, differs in shortened and strongly deflexed lower pinnae and the glaucous underside of lamina, less serrated apice of sterile pinnae and broadly adnate base of fertile pinnae. *Lomaria deflexa* Baker represents only a young state of this species, having thinner leaves, less prominent veins and green color beneath.

Plate. 155. Fig. 1. Habit sketch (natural size). 2. pinnae, showing venation and serrature ($\times$ 2). 3. Portion of fertile pinna, showing position of sori ($\times$ 16). 4—5. Sporangium with spores ($\times$ 150).

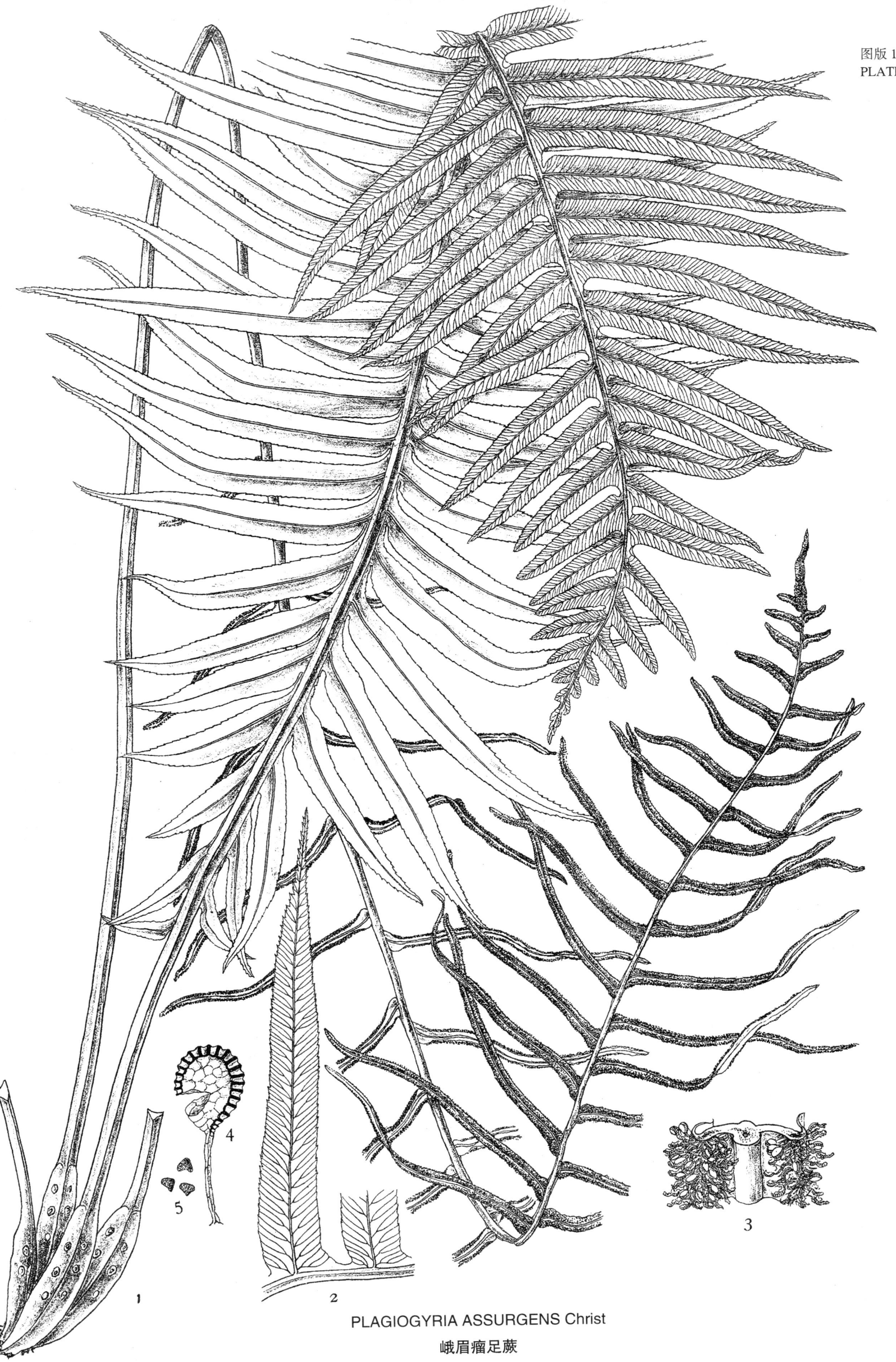

PLAGIOGYRIA ASSURGENS Christ
峨眉瘤足蕨

根状茎短而横行，略被组织简单之深棕色之小鳞片；叶近生或亚簇生，柄长 5—10 cm，淡绿色，光滑，具四棱，叶体线状披针形，长 20—30 cm，宽 2—2.5 cm，渐尖头，向基部稍狭，一回羽状分裂；小叶多数，开展，互生，长约 1 cm，宽约 6 mm，呈半卵形，向基部延长，具短柄，下边平直，上边向顶略呈弧形，且具浅裂片，纸质，光滑；叶脉分离，扇形分叉；子囊群一个，生于裂片之顶，线形或略呈弧形，盖膜质，同形，全缘，向外开，比叶边为狭。

分布：海南；越南，泰国，缅甸及南洋群岛均产之。

图注：1. 本种全形（自然大）；2. 小叶，表示叶脉及子囊群之位置（放大 5 倍）；3. 根状茎上之鳞片（放大 40 倍）；4. 叶柄之横切面（放大 10 倍）。

LINDSAYA LOBBIANA Hooker

LINDSAYA LOBBIANA Hooker, Sp. Fil. **1**: 205 t. 62C. (1846); C. Chr. Gardens Bull. Str. Settl. **4**: 396 (1929); Ind. Fil. Suppl. III. 122 (1934), c. syn.

Lindsaya cultrata Hk. et Bak. Syn. Fil. 105 (1868), pro parte; Christ (non Sw.), Journ. Bot. d. France **19**: 9 (1905); Merr. Enum. Hainan Pl. in Lingnan Sci. Journ. **5**: 13 (1927).

Lindsaya gracilis C. Chr. Ind. Fil. 393 (1905), pro parte.

Rhizome creeping, slender, sparsely scaly; *scales* brown, small, oblong-lanceolate, consisting 3-4 rows of elongate luminae entire; *fronds* aggregate or approximate, stipe 5-10 cm long, pale green or light straminous, naked, prominently 4-angular, lamina linear-lanceolate, 20-30 cm long, 2-2.5 cm broad, acuminate, slightly narrowed towards base, pinnate; *pinnae* numerous, close, horizontally patent, alternate, about 1 cm long, 6 mm broad, broadly half-ovate, short-petiolate, base attenuate, upper side truncate, apex rounded, lower edge straght, upper edge convex and lobato-incised, gradually decrescent towards acuminate apex, lower ones rather far apart and smaller; *texture* herbaceous, light green even when dried, glabrous on both sides, rachis also 4-augular; *veins* fine, distinct green, flabellulately forked; *sori* one to each lobe, transversally linear or slightly curved, *indusium* linear, entire, greenish, narrower than the leaf-edge.

Hainan: Tun Fao, Kachek, *Eryl Smith 1466*, on stream banks at low altitude; Chim Fung Ling, *S. K. Lau 3596*; *C. Wang 33509, 34571*, on rocks along stream side.

Vietnam: Tahl Nguyen, *Pételot* s. n.; *Eberhardt 2028. Cadier 63*; Tourane, *Gaudichaud* (1837).

Thailand: Koh Chang, *Johs Schmidt 779*.

Assam: *Griffith*. Also South India, Malaysia and Java (type locality).

A fairly common fern in the localities noted. In general habit, it is closely related to *L. cultrata* (Willd.), but differs always by pale green and prominently 4-angular stipe and rachis throughout.

Plate 156. Fig. 1. Habit sketch (natural size). 2. Pinna, showing venation and sori (× 5). 3. Scales from rhizome (× 40). 4. Cross section of stipe (× 10).

LINDSAYA LOBBIANA Hooker
洛氏鳞蕨

图版 157　网脉林蕨

根状茎颇细短而横行或卷曲生,略被组织简单之红褐色小鳞片;叶近生,多数,柄挺直,淡绿色,四棱形,上面具沟槽,长 10—20 cm,叶体变异甚大,或为线状披针形之一回羽状分裂,或为掌状分叉,具 2—6 对一回羽状分裂线状披针形之小叶,末回小叶多数,长 1—2 cm,阔半之,半卵形,位于下部者,常强度下向,下方全缘,上方向端常浅裂,纸质,淡绿色,两面光滑无毛,叶脉网状,具 1—2 列斜出网眼,内不具小脉;子囊群边生,线形,条直或呈弧形,每小裂片一个,盖为淡绿色,全缘,较叶缘为狭。

分布:亚洲热带各地均产之,最近在中国之海南发现。

图注:1.本种全形(自然大);2a—2b.末回小叶,表示叶脉及子囊群之位置(放大 5 倍);3.根状茎上之鳞片(放大 30 倍)。

LINDSAYA DECOMPOSITA Willdenow

LINDSAYA DECOMPOSITA Willdenow, Sp. Pl. 5: 425 (1810); C. Chr. Ind. Fil. 393 (1905), pro parte; Gardens Bull. Str. Settl. 7: 236 (1934); v. A. v. R. Handb. Mal. Ferns 274 (1909); Merr. Enum. Hainan Pl. in Lingnan Sci. Journ. 5: 12 (1927); Holttum. Gardens Bull. Str. Settl. 5: 66 (1930).

For synonymy see C. Chr. Ind. 392.

Rhizome rather wiry, short-creeping, matted, sparingly scaly; *scales* small, brown, of very simple structure; *fronds* approximate, numerous, stipe firm, erect, naked, pale green, 4-angular, deeply grooved above, 10-25 cm long, lamina varies from linear-lanceolate and simple pinnate up to 30 cm long, 2.5 cm broad to palmately divided with 1-3 pairs of lateral linear, pinnate branches; ultimate *pinnae* numerous, 1-2 cm long, half as broad, half-ovate, the lower ones often strongly decurved, the lower margin entire, the outer and upper shallowly lobato-incised enough to interrupt the sori; *texture* thin, herbaceous, light green and glabrous on both sides; *veins* anastomosing in 1-2 rows of angularly elongate oblique exappendiculate areolae; *sori* marginal, one to each shallow lobe, linear, straight or curved, *indusium* greenish, entire, narrower than the outer leaf-margin.

Hainan: Five Finger Mt., *F. A. McClure 9482*; *Eryl Smith 1469*, beside a stream; *Hancock 119*; *C. Wang 35711*, Dec. 20, 1933.

Widely dispersed throughout tropical Asia to Polynesia.

A very variable fern as to habit, leaves sometimes simple, sometimes 2-6-forked; the incision of pinna are also very variable, being nearly entire in type in the herbarium Willdenow.

Plate 157. Fig. 1. Habit sketch (natural size). 2a—2b. Ultimate pinna, showing venation and sori ($\times$ 5). 3. Scale from rhizome ($\times$ 30).

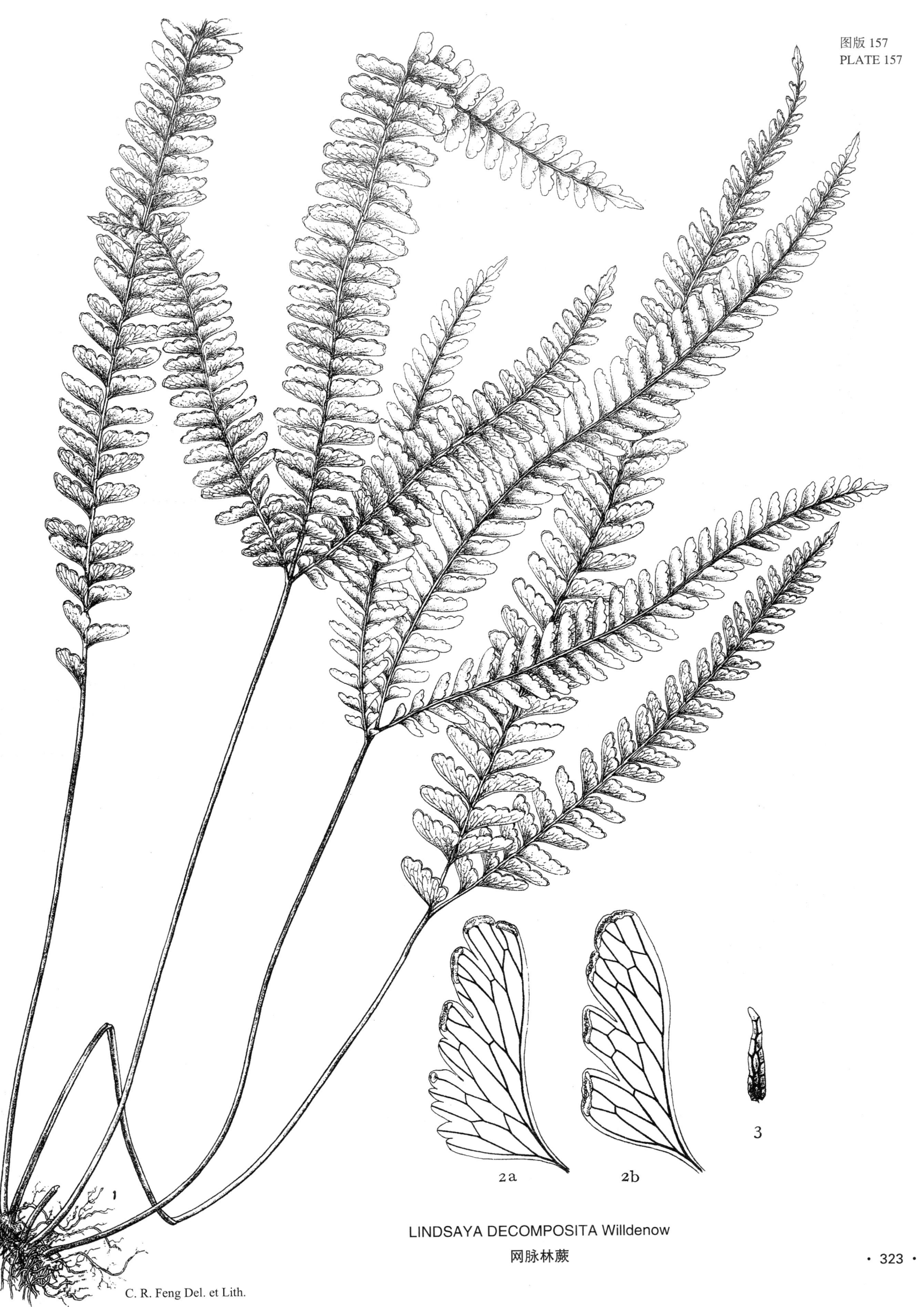

LINDSAYA DECOMPOSITA Willdenow
网脉鳞始蕨

根状茎短而直立,顶部被黑色之细长鳞片;叶簇生,柄长 2—4 cm,细如铜丝,光亮,栗黑色,叶体长 2—5 cm,宽约 1.5 cm,一回奇数羽状分裂;小叶 3—6 对,长达 6 mm,宽 3—5 mm,倒卵形,互生,具柄,全缘,与叶柄着生处有肢节,纸质,两面光滑无毛,下面稍呈粉白色;叶脉扇形分叉,达于角质之边缘;子囊群一个,生于小叶之顶,盖长椭圆形,着生于小叶顶部之缺刻。

分布:仅产于广东之北部。

此为特殊之种,本属其他之种鲜有类此者。

图注:1. 本种全形(自然大);2. 小叶,表示叶脉及子囊群之位置(放大 10 倍);3. 叶柄基部之鳞片(放大 40 倍)。

ADIANTUM GRAVESII Hance

ADIANTUM GRAVESII Hance, Journ. Bot. (1875) 197; Christ, Farnkr. d. Erde 140 (1897); Diels in Engl. u. Prantl: Nat. Pflanzenfam. **1**: 4. 284 (1899); C. Chr. Ind. Fil. 27 (1905); Dunn & Tutcher, Fl. Kwangt. & Hongk. 338 (1912)

Adiantum monochlamys Christ (non Eaton, 1858) in Warburg, Monsunia **1**: 67 (1900).

Rhizome short, erect, densely radicose, copiously scaly at apex; *scales* small, linear-subulate, almost black, scarious along the upper margin; *fronds* fasciculated, many together, stipe wiry, atro-brown or almost black, terete, 2-4 cm long, glabrous, lamina 2-5 cm long, about 1.5 cm broad with wiry blackish rachis, impari-pinnate; *pinnae* 3-6-jugate, to 6 mm long, 3-5 mm broad, obovate or broadly obovate, alternate, patent, entire, petiolate, petiole 2-3 mm long, capillaceous (the terminal pinna with longer stalk), articulated at the base of pinnae; *texture* papyraceous, glabrous on both sides, glaucescent beneath; *veins* fine but distinct against light, flabellulately forked, extending to the narrowly cartilaginous margin; *indusium* large, reniform or transversal oblong, blackish, coriaceous, one to each pinna, attached to the deeply notched apex, persistent.

Guangdong: North River, 175 miles from Canton, *R. H. Grave 18831* (type); Lienchow, *B. C. Henry* (1881); *Rev. J. Lamont*, Oct. 1876; *Matthew* (1907), on limestone rocks; *Gerlach*; Yüyuen Hsien, *S. P. Ko 53788*, Oct. 21, 1933.

A very distinct and one of the most slender species of the genus, to which might safely be referred a series of Chinese forms known as *A. Mariesii* Baker from Ichang, *A. Leveillei* Christ from Guizhou, *A. Greenii* Ching and *A. nanum* Ching both from Guangxi, which all differ from type only in the general outline and relative size of pinna.

Plate 159. Fig. 1. Habit sketch (natural size). 2. Pinna, showing venation and sorus (× 10). 3. Scales from apex of rhizome (× 40).

粤铁线蕨

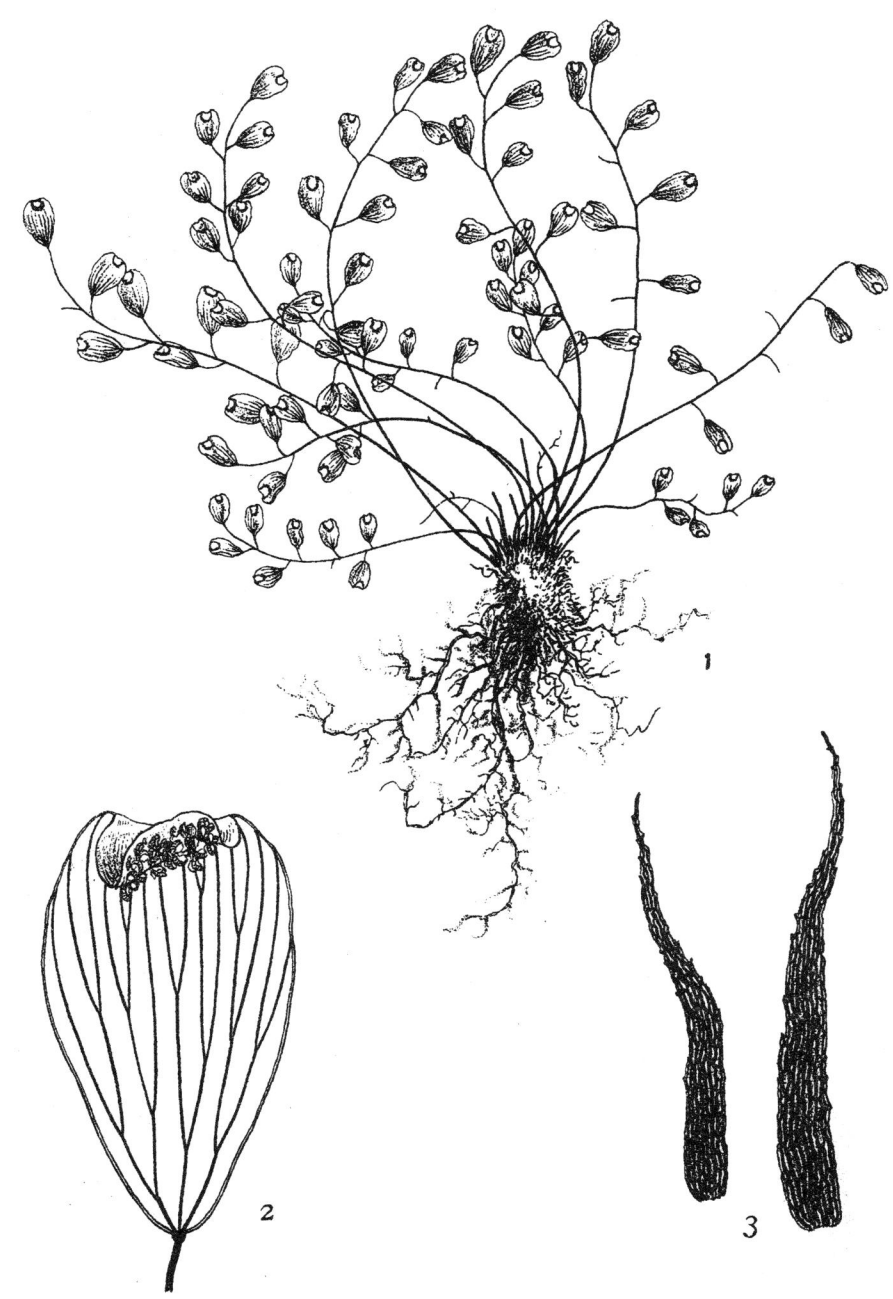

ADIANTUM GRAVESII Hance
粤铁线蕨

图版 159　钱氏铁线蕨

根状茎短,直立,端被深棕色之线状鳞片;叶多数簇生,柄长 5—7 cm,栗黑色,光亮,无毛,叶体线状披针形,长 13—18 cm,宽 3—4 cm,奇数羽状分裂;小叶 7—9 对,开展,亚对生,倒卵状三角形,长 1.2—1.6 cm,宽亦如之,全缘,具柄,亚革质,上面光亮,下面呈淡粉白色,全体光滑,叶脉扇形分叉,多数,直达骨质之边缘;子囊群一个,长 5—8 mm,位于小叶之截形顶部,盖革质,黑色,全缘,宿存。

分布：仅产于广东之北江。

本种形体,甚似莱氏铁线蕨(A. Leveillei),然各部均较大,故易识别。

图注：1. 本种全形(自然大);2. 同上,而较小(自然大);3. 小叶,表示叶脉及子囊群(放大 5 倍);4. 叶柄基部之鳞片(放大 30 倍)。

ADIANTUM CHIENII Ching

ADIANTUM CHIENII Ching, Sinensia 1: 50 (1930); C. Chr. Ind. Fil. Suppl. Ⅲ. 18 (1934).

Rhizome short, erect, densely radicose and scaly; *scales* linear-subulate, atro-brown, rigid, entire; *fronds* caespitose, numerous together, stipe 5-7 cm long, ebeneous, shining, glabrous, breaking off at the middle, lamina linear-lanceolate, impari-pinnate, 13-18 cm long, 3-4 cm broad, base not any broader than above, truncate, slightly narrowed towards apex; *pinnae* 7-9 on each side under the smaller terminal one, patent, petiolate (petiole 2 mm long), subopposite, 2-3 cm apart, the lower ones broadly deltoid-obovate, 1.2-1.6 cm each way, with rounded base, entire margin, the upper ones narrower, obovate with cuneate base, all with truncate apex; *texture* subcoriaceous, lustrous green above, pale or bluish beneath, glabrous in all parts; *veins* distinct, flabellulately forked, veinlets numerous, fine, reaching somewhat thickened margin; *sori* large, 5-8 mm long, one to each pinna, transversally linear, terminating the truncate apex, *indusium black*, coriaceous, entire, persistent.

Guangdong: North West River, **Lo-aqwai**, February, 3, 1890.

This endemic fern, only collected once, resembles A. Leveillei Christ in general habit, but differs in enormously larger size, much more stoutly built, with more numerous and much larger pinna.

Plate 159. Fig. 1. Habit sketch (natural size). 2. the same but young form (natural size). 3. Pinna, showing venation and sorus (× 5). 4. Scales from base of stipe (× 30).

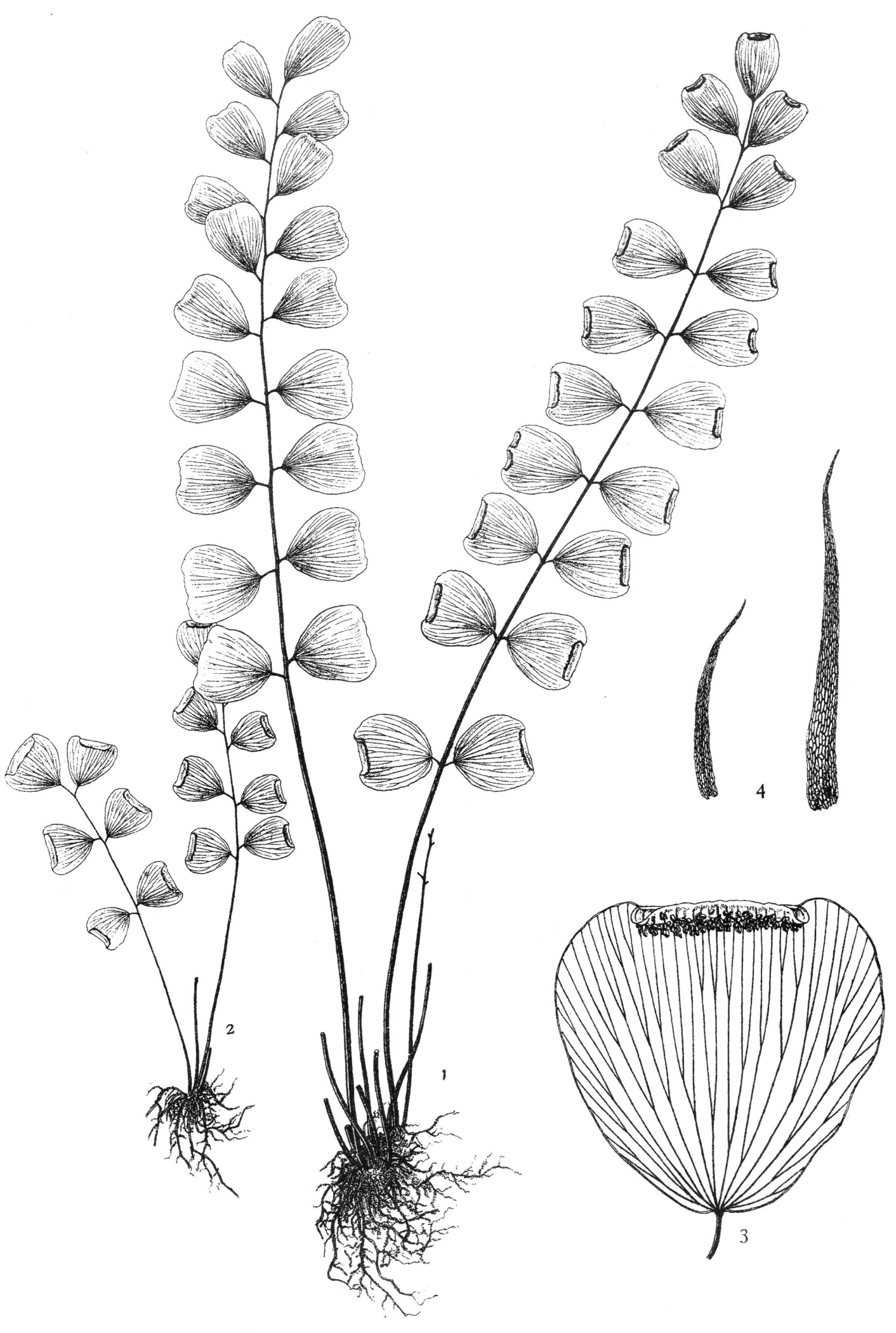

ADIANTUM CHIENII Cling
钱氏铁线蕨

图版 160　陇铁线蕨

根状茎短而直立，被深棕色之细长鳞片；叶簇生，柄长 6—10 cm，细而圆，栗黑色，光滑，叶体与柄等长，宽 2—3 cm，披针形，基部二回或亚三回分裂，顶部一回分裂，各回小叶均具柄；一回小叶 5—7 对，长 2—2.5 cm，宽 1.5—2 cm，三角形；二回小叶 3—4 对，其基部一对通常分裂，末回小叶甚小，三角形或卵形，基部楔形，绿色，纸质，光滑，叶脉扇形分叉，不达于叶边；子囊群通常每小叶两个（有时仅一个），盖圆形，深棕色，革质，着生于小叶顶部之深缺刻内。

分布：仅甘肃西部产之。

本种形体极似峨眉铁线蕨（A. Faberi），唯形体纤长，叶体不为卵状三角形，末回小叶较小，故易分别。

图注：1. 本种全形（自然大）；2. 同上，幼形（自然大）；3. 末回小叶，表示叶脉及子囊群之位置（放大 10 倍）；4. 叶柄基部之鳞片（放大 30 倍）。

ADIANTUM ROBOROWSKII Maximowicz

ADIANTUM ROBOROWSKII Maxomowicz, Mél. Biol. **11**：867 (1883); C. Chr. Ind. Fil. 33 (1905); Journ. Wash. Acad. Sci. **17**：498 (1927).

Rhizome short, erect, densely scaly; *scales* rufo-brown, narrowly lanceolate, entire; *fronds* caespitose, stipe 6-10 cm long, slender, terete, castaneous, shining, rigid, lamina as long as stipe, 2-3 cm broad, lanceolate, bipinnate or subtripinnate at base; *pinnae* 5-7-jugate, under simple pinnate apical part, alternate, oblique, petiolate, the basal ones larger, 2-2.5 cm long, 1.5-2 cm broad, deltoid, rachilet castaneous, flexuose, *pinnules* 4-3-jugate, the basal pair generally forked, or very rarely pinnate, the upper ones simple, *ultimate pinnule* triangular or broadly ovate, entire, with cuneate base, and capillaceous castaneous petiole, the middle pinnae generally simple pinnate with 1-2 pairs of pinnules; *texture* herbaceous, green, glabrous in all parts; *veins* visible against light, flabellulately forked, not reaching leaf-margin; *sori* generally 2 to each segment (not infrequently one), *indusium* orbicular, rufo-brown, coriaceous, attached to a deep notch at apex.

Gansu：Tangut, *N. M. Przewalski* (1880), type; without locality, *Purdom 78* (1910); Jarganar, south of Old Taochow, *R. C. Ching 902* (1923); Möping, Manyueszai, *Dr. D. Hummel 2292, 2310*.

A distinct endemic species of the group of *A. venustum* Don and especially closely related to *A. Faberi* Baker, differing chiefly in smaller size, narrower leaves, smaller segments of generally triangular shape.

Plate. 160. Fig. 1. Habit sketch (natural size). 2. The same but young form (natural size). 3. Ultimate pinnule. showing venation and sori ($\times$ 10). 4. Scale from base of stipe ($\times$ 30).

ADIANTUM ROBOROWSKII Maximowicz
陇铁线蕨

图版 161　高山乌蕨

根状茎短,直立或卧生,顶部被淡栗色之披针形鳞片;叶亚簇生,柄长 25—35 cm,淡稻秆色,而基部呈黑色,光滑无毛,叶体广卵形,渐尖头,长 15—25 cm,宽亦如之,五回羽状分裂,一回小叶 8—14 对,基部一对最大,三角形,渐尖头,具柄,各回小叶彼此密接,末回小叶为线状披针形,具短尖头,全缘,长 3—5 mm,纸质,淡绿色,两面光滑,叶脉通直明显;子囊群短线形,通常由 4—6 个子囊组成,盖大,膜质,灰白色,全缘,达于叶脉,宿存。

分布:云南,西藏;泰国,尼泊尔及印度东北二部之高山均产之。

此种昔日学者多认为与普通之乌蕨(O. japonicum)相同,实则其叶体之分裂度更细密,叶柄基部常为黑色,子囊群较短,盖短阔,故易分别。

图注:1. 本种全形(自然大);2. 不生子囊群小叶之一部(放大 10 倍);3. 生子囊群小叶之一部(放大 10 倍);4. 同上(放大 16 倍);5. 叶柄之横切面(放大 6 倍);6. 根状茎上之鳞片(放大 27 倍);7. 根状茎之横切面(放大 6 倍)。

ONYCHIUM CONTIGUUM (Wall.) Hope.

ONYCHIUM CONTIGUUM (Wall.) Hope, Journ. Bomb. Nat. Hist. Soc. **13**: 444 (1901); Ching Lingnan Sci. Journ. **13**: 498 (1934).

Cheilanthes contigua Wall. List. no. 72 (1828, nom. nud.).

Onychium japonicum var. *intermedia* Clarke, Trans. Linn. Soc. Ⅱ. Bot. **1**: 457 (1880); Kümmerle. Amer. Fern Journ. **29-30**: 135 (1929-30).

Onychium lucidum Bedd. (non Spr. 1827) Ferns Brit. Ind. t. 21 (1865); C. Chr. Ind. Fil. Suppl. Ⅲ. 133 (1934).

Onychium japonicum Hk. et Bak. Syn. Fil, 143 (1867); Bedd. Handb. Ferns Brit. Ind. (1883); C. Chr. Ind. Fil. 469 (1905) pro parte.

Onychium japonicum var. *lucidum* Kümmerle, l. c., pro parte.

Onychium cryptogrammoides Christ in Lecomte, Not. Syst. **1**: 52 (1911).

Rhizome short, erect or procumbent, densely redicose, apex clothed in lanceolate, light castaneous *scales*; *fronds* subcaespitose, stipe 25-35 cm long, pale straminous, always black near the base, glabrous, lamina 15-25 cm each way, broadly ovate, pentagonous, acuminate, very finely 5-pinnate; *pinnae* 8-14-jugate, the basal pair much the largest, triangular, acuminate, all long-petiolate, very oblique, *pinnules* of second and third orders all petiolate, confert; ultimate *segments* linear-lanceolate, apiculate, entire, 3-5 mm long; *texture* herbaceous, pale green, naked on both sides, *veins* fine, one to each segment; *sori* linear, short, consisting of 4-6, or rarely 9 sporangia on each side, *indusium* large, broad, membranaceous, pale gray, entire, reaching costule from both sides and persistent.

Yunnan: Kiao-kia, *Duclonx 6971*, *5049* (type of O. *cryptogrammoides* Christ), August, 1911; Tongchow, *E. E. Maire 1379*, *2096*, *2774*, *1484* (1913); Without locality, *G. Forrest 285*, *6068*. Sichuan: Hueili Hsien, *T. T. Yu 1479*, Sept. 10, 1932, under woods; *W. P. Fang 6869*; Fenghsiangying, *Narry Smith 1880*. Tibet: Yatung, *Hobson* (1897).

Thailand: Without locality, *H. B. J. Garrett 453*.

North-eastern India and Himalayas generally.

Type from Nepal.

In my recent monograph, I have treated at some length of the nomenclatural confusion for this very distinct fern, which was generally considered as identical with the widely dispersed O. *japonicum* Kze., from which our fern can always be distinguished by more finely divided lamina of a pentagonous outline, on proportionally longer pale-colored stipe always with nearly black basal part and by shorter sori with larger, broader, nearly bullate persistent indusium reaching the costule from both sides.

Plate 161. Fig. 1. Habit sketch (natural size). 2. Portion of sterile frond, showing venation ($\times$ 10). 3. Portion of fertile frond ($\times$ 10). 4. Soriferous segment, with one indusium open ($\times$ 16). 5. Cross section of stipe ($\times$ 6). 6. Scale from rhizome ($\times$ 27). 7. Cross section of rhizome ($\times$ 6).

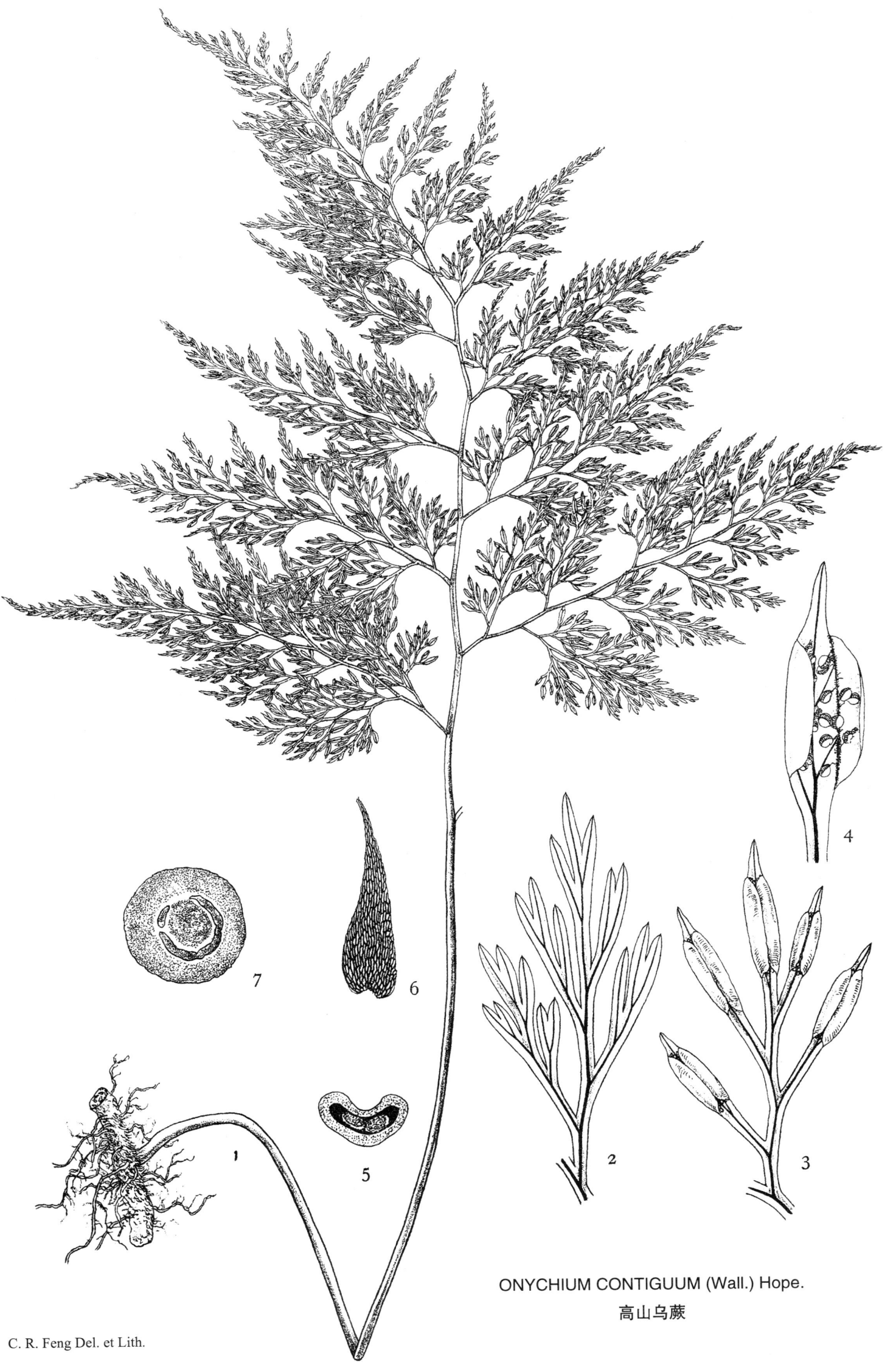

ONYCHIUM CONTIGUUM (Wall.) Hope.
高山乌蕨

图版 162　木坪乌蕨

根状茎细长,横行,被红棕色之细长厚质鳞片;叶散生,亚二形,不生子囊群叶之柄长达 10 cm,细长,淡稻秆色,叶体为狭线形,长 10—15 cm,宽 2—3 cm,向基部较阔,二回羽状分裂;一回小叶 10—15 对,斜方形,具柄,基部一对长达 3.5 cm,基部亚楔形,二回小叶 2—3 对,形小,具 2—3 个锯齿,质颇坚厚,绿色;生子囊群之叶较长而宽,柄长 15—20 cm,稻秆色,叶体长亦如之,阔披针形,中部以上呈尾状狭长,基部阔达 6 cm,三回羽状分裂;一回小叶 10—15 对,具长柄,基部为不等楔形,下部一对长达 10 cm,宽 3 cm,长形渐尖头,顶部一回羽状分裂;末回裂片少数,扁圆形,长约 1 cm,宽 1.5 mm,渐尖头;子囊群线形,长 5—7 mm,盖膜质,全缘。

分布:四川宝庆县高山特产。

本种异于普通乌蕨（O. japonicum）者,为其狭长亚二形之叶,其分裂度较少,顶部呈尾形细长是也。

图注:1.本种全形(自然大);2.不生子囊叶之一部,表示叶脉(放大 8 倍);3.生子囊群叶之一部(放大 8 倍);4.生子囊群之末回小叶,表示子囊群之位置及盖(放大 16 倍);5.根状茎上之鳞片(放大 27 倍)。

ONYCHIUM MOUPINENSE Ching

ONYCHIUM MOUPINENSE Ching, Lingnan Sci. Journ. **13**: 500 (1934)

Rhizome wide-creeping, 1.5 mm thick, densely scaly; *scales* rufo-brown, subulate, firm; *frond* distant, dimorphous, *sterile* one narrowly linear elongate, stipe to 10 cm long, slender, pale straminous, lamina 10-15 cm long, 2-3 cm broad, slightly broader towards base, bipinnate; *pinna* 10-15-jugate, rhombic, petiolate, basal pair to 3.5 cm long, upper ones 2 cm long, subunequally cuneate at base; *pinnules* 2-3-jugate, subrhombic, base slightly decurrent, apex obtuse; *segments* 2-3-jugate, small, 2-3-dentate; *texture* rather rigid, color green; *fertile frond* much longer and broader, stipe 15-20 cm long, straminous, lamina as long as stipe, broadly lanceolate, long-attenuate towards above middle, base to 6 cm broad, tripinnate; *pinnae* 10-15-jugate, long-petiolate, base unequally cuneate, basal pair to 10 cm long, 3 cm broad, long-acuminate towards simple pinnate apex; *segments* few, sili-queform, 10 mm long, 1.5 mm broad, acuminate; *sori* linear, 5-7 mm long, *indusium* conforms, broad, gray, entire, reaching costule from both sides.

Sichuan: Moupin, *David* (type).

This species, confined to the highland of western Sichuan, differs from O. japonicum Kze. in much narrower and less divided frond and particularly the sterile one, which is of linear-elongate outline to 3 cm broad from somewhat broader base, and the fertile frond with extremely long-attenuate upper part.

Plate 162. Fig. 1. Habit sketch (natural size). 2. Portion of sterile frond, showing venation (× 8), 3. Portion of fertile frond (× 8). 4. Ultimate segment of fertile frond, showing position of sori and indusia (× 16). 5. Scale from rhizome (× 27).

ONYCHIUM MOUPINENSE Ching
木坪乌蕨

根状茎短而横行，颇粗肥，被线形淡黄色之狭鳞片；叶亚簇生，亚二形，柄长 6—15 cm，光滑，叶体椭圆披针形，长 15—25 cm，基部三回羽状分裂，上部二回羽状分裂；一回小叶 5—8 对，具柄，斜出，长 5—10 cm，披针形；二回小叶斜方形，羽状分裂；裂片 3—5 对，线形，长 5—7 mm，宽仅 1.5 mm，渐尖头，边缘具不明显之锯齿，坚革质，颇明亮，两面光滑，叶脉两面显特，羽状分裂；子囊群长 3—5 mm，盖膜质，灰白色，边缘凹凸不齐；不生子囊群叶具较短之柄及较宽之裂片。

分布：云南特产。

本种略似普通之乌蕨(O. japonicum)，然其叶簇生，叶体细长，分裂度较少，茎上鳞片呈淡黄色，末回裂片之边缘具不甚显明之锯齿及子囊群盖，具凹凸不齐之边缘，故易分别。

图注：1. 本种全形（自然大）；2. 小叶之一部表示叶脉（放大 10 倍）；3. 着生子囊群小叶之一部放大，表示叶脉及子囊群之着生情形，并刮去一部分之孢盖（其边缘有锐锯齿）（放大 16 倍）；4. 根状茎上之鳞片（放大 27 倍）。

ONYCHIUM TENUIFRONS Ching

ONYCHIUM TENUIFRONS Ching, Lingnan Sci. Journ. **13**：500 (1934).
Onychium japonicum var. *Delavayi* Christ, Bull. Soc. Bot. France **52**：Mém. Ⅰ. 60 (1905).
Onychium lucidum Kümmerle, Amer. Fern Journ. **20**：135 (1930); C. Chr. Ind. Fil. Suppl. Ⅲ. 133 (1934) proparte.

Rhizome short-creeping, rather thick, densely clothed in pale brown, linear-subulate scales; *fronds* subcaespitose, subdimorphous, stipe firm, erect, straminous, 6-15 cm long, flexuose, naked, lamina oblong-lanceolate, 15-25 cm long, tripinnate at base, bipinnate towards acuminate apex; *pinnae* 5-8-jugate, petiolate, oblique, 5-10 cm long, lanceolate; *pinnules* rhombic, pinnate; *segments* 3-5-jugate, linear, 5-7 mm long, 15 mm broad, acuminate, margin obscurely denticulate; *texture* rigidly coriaceous, subnitente, glabrous on both sides; *veins* prominently raised above, one to each segment, pinnate; *sori* 3-5 mm long, *indusium* gray, reaching costule from both sides, margin deeply erosed; *sterile* leaves on much shorter stipe, with broader and confert pinnules and segments.

Yunnan: Shweli-Salween divide, *G. Forrest 24175* (type), July. 1924; Loko Chan, *Delavay 1715*; Tapintze, *Delavay 32*; Tyly, *Ducloux 5821*; Taitsienteen, *E. E. Maire*, Sept. 1913; Tchongsan, *Ducloux 3372*, Nov. 2, 1909; Nieou Ko Chan region, Pinchow, *Ducloux 6973*. Sichuan: On the Yunnan border, *W. P. Fang 9191*, Oct. 20, 1930.

This endemic species is closely related to O. japonicum Kze. differs from that or other related species in oblong-lanceolate and once less pinnate fronds, subcaespitose leaves, pale brown scales on rhizome, more or less denticulated ultimate segments and the deeply erosed margin of indusium, which last character has so far been known only in O. melanolepis (Decsn), a species from Abyssine.

Plate 163. Fig. 1. Habit sketch (natural size). 2. Portion of sterile frond, showing venation (× 10). 3. Soriferous segment, showing attachment of sori and indusium with erose-dentate margin (× 16). 4. Scale from rhizome (× 27).

ONYCHIUM TENUIFRONS Ching
狭叶乌蕨

图版 164　叶氏乌蕨

本种略似木坪乌蕨(图版162),唯其形体更为细长,叶不为二形,末回裂片及子囊群均较短,故易识别。

分布:湖北古城县之乌龙山特产。

图注:1.本种全形(自然大);2.生子囊群叶之一部(放大8倍);3.不生子囊群叶之一部(放大8倍);4.生子囊群叶之末回小叶,表示子囊群之位置及盖之着生情形(放大16倍);5.根状茎上之鳞片(放大27倍)。

ONYCHIUM IPII Ching

ONYCHIUM IPII Ching, Lingnan Sci. Journ. **15**: 282 (1936).

Rhizome wide-creeping, 2 mm thick; *scales* imbricate, lanceolate, atro-brown, nitente; *fronds* approximate, stipe slender but firm, erect, pale green, naked, sulcate above, 10-20 cm long, lamina narrowly lanceolate with deltoid base, attenuate towards apex, 10-20 cm long, 3-7 cm broad, tripinnate at base, simple pinnate towards apex; *pinnae* 10-13-jugate, basal ones much the largest, 4-7 cm long, 3-4 cm broad at base, deltoid, long-acuminate, oblique, petiole to 1 cm long; *pinnules* subrhombic, acute, 1.5 cm long, 1 cm broad, petiolulate, pinnatisect; *segments* rhombic, below 5 mm long, anterior basal one much the largest, lobato-incised with 2-4 acute soriferous teeth, the middle pinnae lanceolate, to 3 cm long, about 1 cm broad, very oblique, bipinnatifid; *texture* rigidly herbaceous, light green and glabrous on both sides; *veins* prominently raised, pinnate, one to each soriferous tooth; *sori* short, 2 mm long, *indusium* conforms, gray, membranaceous, entire, completely covering sorus.

Hubei: Koo Chen Hsien, Wu Leng Shan, *K. C. Chow 3982*, Oct. 14, 1935, in shade.

This endemic species is closely related to *O. moupinense* Ching, differs in its decidedly slender habit with uniform fronds, shorter ultimate segments with very short sori.

Plate 164. Fig. 1a-b. Habit sketch (natural size). 2. Portion of fertile frond, (× 8). 3. Portion of sterile frond (× 8). 4. Ultimate segment of fertile frond, showing position of sori and indusium (× 16). 5. Scale from rhizome (×27).

ONYCHIUM IPII Ching 叶氏乌蕨

地上茎细长,横行,分叉,被红棕色之细长鳞毛;叶散生,柄长 1.5—5 cm,细柔,稻秆色,具红棕色之密毛,叶体椭圆形或椭圆卵形,急尖头,或钝头,基部等宽,高 1.5—6 cm,宽 1—2 cm,二回羽状分裂;小叶 4—7 对,具柄,长 5—7 mm,三角形,钝头,基部楔形,羽状深裂;裂片 1—2 对,椭圆舌形,全缘或顶端稍呈缺刻,具一数小脉,纸质,中轴及叶之两面均被红棕色之密睫毛;子囊群线形,循小脉及主脉生,无盖。

分布:四川,陕西;日本,朝鲜半岛及西伯利亚东部。

本属仅此一种,附生于树干或林中岩石上之苔藓中,在中国仅四川及陕西产之。

图注:1.本种全形(自然大);2.根状茎之横切面(放大 30 倍);3.叶柄之横切面(放大 30 倍);4.一回小叶,表示叶脉,睫毛及子囊群之位置(放大 20 倍);5.叶体上之睫毛(放大 16 倍);6.根状茎上之毛(放大 16 倍);7.子囊及孢子(放大 150 倍)。

PLEUROSORIOPSIS MAKINOI (Maxim.) Fomin

PLEUROSORIOPSIS MAKINOI (Maxim.) Fomin, Bull. Jard. Bot. Kieff **11**: 8 (1929); Fl. Sib. et Orient. Extr. **5**: 215 (1930); C. Chr. Ind. Fil. Suppl. Ⅲ, 142 (1934).

Gymnogramme makinoi Maxim.; Makino, Bot. Mag. Tokio **8**: 481 c. tab. (1894); Phan. et Pterid. Jap. Icon. Illustr. **1**: pl. 47 (1899-1901); Christ, Bull. Acad. Géogr. Bot. (1906) 129.

Anogramma makinoi Christ in C. Chr. Ind. Fil. 58 (1905); Ogata, Ic. Fil. Jap. **1**: pl. 2 (1928).

Rhizome epigaeous, wiry, branched, wide-creeping, densely clothed in rufo-brown, hair-like, unicellular, softly shaggy *hairs*; *fronds* far apart, erect, stipe 1.5-5 cm long, slender, straminous, densely clothed throughout in similar hairs, lamina oblong or oblong-ovate, acute or bluntish, base not narrowed. 1.5-6 cm long, 1-2 cm broad, bipinnate; *pinna* 4-7-jugate under pinnatifid apex, petiolate, 5-7 mm long, deltoid, obtuse, base cuneate, pinnate with 1-2 pairs of oblong-ligulate entire or slightly notched uninerved decurrent segments under the trilobed terminal part; *texture* thin herbaceous, rachis, petiole and both sides copiously clothed in reddish-brown, spreading, septate, transparent hairs; *veins* visible against light, one to each lobe, falling far short below apex; *sori* linear, along veins and costa of pinnae, exindusiate; sporangium broadly subglobular, shortly stalked; *spores* ovate-reniform, discolored, bilateral, and smooth (with both perispore and exospore).

Sichuan: Mt. Omei, *E. H. Wilson 5274*. Shaanxi: Mt. Huan Ton Shan, *Giraldi*. Also Japan and southern part of Korean peninsula (Quelpaert, *Taquet 3946*).

This is a singularly interesting little fern, epiphytic on tree trunks or growing in mosses on rocks under forest. The genus comprising only one species, differs from *Anogramma* in wide-creeping and branched rhizome, with distant leaves, dense reddish-brown articulated hairs in all parts and the bilateral reniform-ovate spores.

Plate 165. Fig. 1. Habit sketch (natural size). 2. Cross section of rhizome ($\times$ 30). 3. Cross section of stipe ($\times$ 30). 4. pinnae, showing venation, hairs and position of sori ($\times$ 20). 5. Hairs on lamina ($\times$ 16). 6. The same from rhizome ($\times$ 16). 7. Sporangium with spores ($\times$ 150).

PLEUROSORIOPSIS MAKINOI (Maxim.) Fomin
睫毛蕨

根状茎横行，粗肥如指，被线状深棕色之鳞片；叶大散生，柄长达 1 m，基部粗达 1 cm，稻秆色，叶体长达 1 m，宽约 50 cm，卵形或卵状椭圆形，一回奇数羽状分裂；小叶 6—10 对，对生或亚互生，斜出，具柄，长达 30 cm，宽 6—8 cm，椭圆披针形，端呈尾状，基部楔形，亚等形，上部数对无柄，较小，全缘而质薄，不为骨质，纸质，两面光滑无毛，绿色，叶脉下面明显，多数分叉，平行，端直而膨大，不达于叶边；子囊群线形，分叉，自中肋几达于叶边。

分布：台湾，云南；印度，爪哇，菲律宾群岛均产之。

此为本属极大之一种，在中国仅产于云南西部高山及台湾，其产于其他各省昔人认为此种者，实为华凤了蕨（C. intermedia）也。

图注：1. 本种全形（自然大）；2. 小叶之一部，表示叶脉及边缘（放大 4 倍）；3. 根状茎上之鳞片（放大 16 倍）。

CONIOGRAMME FRAXINEA (Don) Diels

CONIOGRAMME FRAXINEA (Don) Diels in Engl. u. Prantl: Nat. Pflanzenfam. **1**: 4. 262 (1899); C. Chr. Ind. Fil. 185 (1905), pro parte; Suppl. Ⅱ, 9 (1913-17); Contr. U. S. Nat. Herb. **26**: 307 (1931); Hieron. Hedwigia **57**: 286 (1916).

Diplazium fraxineum Don. Prod. Fl. Nepal. 12 (1825).

Gymnogramme fraxinea Bedd. Ferns Brit. Ind. Suppl. 24 (1876), excl. Ferns Brit. Ind. t. 232, and Ferns S. Ind. t. 57.

Syngramme fraxinea Bedd. Handb. Ferns Brit. Ind. 386 (1883), pro parte.

Neurogramme fraxinea Christ, Farnkr. d. Erde 63 (1897), pro parte.

Gymnogramme javanica Bl. Enum. Pl. Jav. 112 (1828); Fl. Jav. 95 t. 41 (1828).

Coniogramme javanica Fée, Gen. Fil. 167 t. 14 B, f. 1 (1850-52).

Rhizome creeping, thick as a finger, densely scaly; *scales* linear-subulate, atro-brown, thick, entire; *frond* ample, 2-3 cm apart, stipe up to 1 meter long, 1 cm thick and scaly near base, prominently bisulcate on the upper side and terete beneath, straminous or dark straminous; lamina over 1 meter long, 50 cm broad, ovate or oblong-ovate, simple imparipinnate, *pinnae* 6-10-jugate, opposite or subalternate, 10-15 cm apart, oblique, lower ones long-petiolate (petiole 1.5 cm long), uppermost ones nearly sessile, basal ones generally simple, or very rarely bifid, to 30 cm long (sometimes longer), 6-8 cm broad, oblong-lanceolate, base cuneate (unequally so in lower ones), long-caudate at apex, margin entire to the very tip, generally repand, thin, and not cartilaginous; *texture* chartaceous, glabrous and green on both sides; *veins* distinct beneath, mostly forked above base, vein-lets parallel, ended in large clavate straight hydathodes some distance from the thin leaf-margin; *sori* linear, forked, extending from costa to near the margin.

Yunnan: Tengyueh, *G. Forrest 9496*, *26688*; between Muang Hun and Muang Hai, *J. F. Rock 2401*; without locality, *H. T. Tsai 56934*. Taiwan.

Also Sikkim-Himalayas, S. India, Java and Philippines.

One of the largest species of the genus, characterized by generally simple pinnate leaves and large subopposite pinnae with very entire, thin margin and long-caudate apex. From the available herbarium material, this distinct fern seems by no means abundant in the localities noted and has hitherto generally been utterly misunderstood by authors in the past. *Coniogramme fraxinea* of authors on Indian and Chinese ferns generally represents a mixture of a number of species, while its previous report from different parts of China has mostly been a mistake for *C. intermedia* Hieron. (cf. pl. 143 of this Icones), the specimens cited above from Yunnan constituting the first and only authentic record of the species from China. The other and the only species, which is similar to our fern in size, general habit and entire leaf-margin, is *C. macrophylla* (Bl.) Hieron. var. *Copelandii* (Christ) Hieron. (l. c. 292) of the Philippine Islands and recently collected in the Island Hainan (F. A. McClure 2147), which differs, however, in its veins ended in similarly prominent but somewhat arcuate hydathodes connected with the broadly cartilaginous margin by sclerenchymatous cells.

Plate 166. Fig. 1. Habit sketch (natural size). 2. Portion of pinna, showing veins with prominent clavate hydathodes and entire thin leaf-margin ($\times$ 4). 3. Scale from rhizome ($\times$ 16).

CONIOGRAMME FRAXINEA (Don) Diels
全缘凤了蕨

图版 167　毛叶凤了蕨

　　根状茎横行,粗肥如指。被深棕色披针形之密鳞片;叶亚散生,柄长30—60 cm,光滑无毛,叶体卵状三角形,长30—50 cm,宽如之,一回奇数羽状分裂,基部二回分裂,小叶通常2—3对,长14—20 cm,宽3—4 cm,阔披针形,具柄,基部呈楔形,顶为尾形,顶部一小叶几等大,同形,基部一对通常2—3裂,柄长达2 cm,厚纸质,边缘具刺状骨质之密锯齿,上面光滑,下面被密毛,叶脉明显,一回或二回分叉,直达于锯齿之端;子囊群线形,自中肋外出,达于叶边。

　　分布:云南,西藏及喜马拉雅山区产之。

　　此为本属特殊之种,叶之下面被密毛,其叶脉直达于刺状骨质之密锯齿之顶,小叶2—3对,端为长尾形,故易识别。

　　图注:1.本种全形(自然大);2.小叶之一部,表示叶脉,子囊,锯齿及被毛之下面(放大4倍);3.叶下面之毛(放大40倍);4.根状茎上之鳞片(放大20倍);5.子囊(放大40倍)。

CONIOGRAMME CAUDATA (Wall.) Ching

CONIOGRAMME CAUDATA (Wall.) Ching in C. Chr. Ind. Fil. Suppl. III. 56 (1934).

　　Grammitis caudata Wall. List no. 4 (1828, nom. nud.).

　　Gymnopteris caudata Presl, Tent. Pterid. 218 (1836, nom. nud.); Ettingash, Farnkr 57 t. 37 f. 7, t. 38 f. 13 (1865).

　　Gymnogramme javanica var. *spinulosa* Christ, Bull. Soc. Bot. France **52**: Mém. I. 55 (1905).

　　Coniogramme spinulosa Hieron. Hedwigia **57**: 311 (1916); C. Chr. Ind. Fil. Suppl. II. 10 (1913-17); Contr. U. S. Nat. Herb. **26**: 307 (1931).

　　Coniogramme pubescens Hieron., l. c. p. 314; C. Chr., l. c.

　　Gymnogramme serrulata Wall, (non Bl.), List no. 134 (1828, nom. nud.).

　　Gymnogramme javanica Bedd. Ferns S. Ind. 77 t. 232 (1864).

　　Gymnogramme fraxinea var. *pilosa* Clarke (non Brack.), Trans. Linn. Soc. II. Bot. **1**: 569 (1880).

　　Rhizome wide-creeping, thick as a small finger, densely scaly; *scales* narrowly lanceolate, entire, dark-brown, thick; *frond* 1-3 cm apart, stipe 30-60 cm long, dark straminous, naked, deeply grooved above, lamina ovate-deltoid, 30-50 cm each way, simple pinnate or bipinnate at base; *pinnae* generally 1-3-jugate under the terminal one similar to the lateral, 14-20 cm long, 3-4 cm broad, broadly lanceolate, petiolate, attenuate-cuneate at base, caudate at apex, the basal pair long-petiolate (petiole to 2 cm long), generally 2-3-foliolate with the lower one or basal pair of pinnules somewhat smaller than the upper or central one; *texture* thickly chartaceous, margin regularly and prickly serrate with deltoid cartilaginous teeth, glabrous above, densely pubescent beneath; *veins* fine, distinct, once or twice forked, veinlets parallel, extending into the serrature, each provided with a large prominent brown hydathode at the tip; *sori* linear, extending from costa to near the margin.

　　Yunnan: Tsanschan, *Delavay 4212* (type of *C. spinulosa*), 5043, August, 1894; February 22, 1889; Maikha-Salween divide, *G. Forrest 18332*; Shweli-Salween divide, *G. Forrest 24628*, 27987; Tchen Fong Shan, *Delavay 5043*; Shangpa, *H. T. Tsai 58819*, 58780, in forest. Southeastern Tibet: forest of Doyan Longba, *J. F. Rock 11627*.

　　Nepal: *Wallich 4* (type). Also Sikkim-Himalayas, common.

　　This distinct fern, now found to be common in Sikkim-Himalayas and the western part of Yunnan, is characterized by simple pinnate frond often with bipinnate base, only 1-3 pairs of pinnae with long-caudate apex, regularly and prickly serrated margin and densely pubescent under surface.

　　Plate 167. Fig. 1. Habit sketch (natural size). 2. Portion of pinna, showing venation, sori, serrature and pubescent under surface (× 4). 3. Hairs from under surface (× 40). 4. Scale from rhizome (× 20). 5. Sporangium (× 40).

CONIOGRAMME CAUDATA (Wall.) Ching
毛叶凤了蕨

图版 168　高山凤了蕨

根状茎横行，粗肥如小指，略被淡黄色之披针形鳞片；叶散生，柄长达 60 cm，光滑，淡稻秆色，叶体甚大，长逾 60 cm，宽约 50 cm，二回奇数羽状分裂；小叶约 10 对，开展，具柄，基部一对最大，长达 30 cm，宽 9 cm，奇数羽状分裂，二回小叶 10—13 对，对生，长约 5 cm，宽 2 cm，无柄，或多少合生，端呈尾形，基部为阔圆截形，边缘锯齿整齐，上部小叶渐短，二回小叶渐少，其位于叶顶下部数对为单叶，薄纸质，绿色，两面光滑，叶脉明显，仅达于锯齿之基部；子囊群线形，分叉，自中肋达于叶之半阔。

分布：云南及台湾高山；尼泊尔产之。

此为强度二回羽状分裂之种，其基部小叶具十数对以上之二回对生小叶，叶为薄纸质，绿色，两面光滑无毛，最易识别。

图注：1. 本种全形（自然大）；2. 二回小叶之一部，表示锯齿及子囊群（放大 4 倍）；3. 根状茎上之鳞片（放大 30 倍）。

CONIOGRAMME PROCERA (Wall.) Fée

CONIOGRAMME PROCERA (Wall.) Fée, 10 Mém. 22 (1865); Hieron. Hedwigia **57**: 317 (1916); C. Chr. Ind. Fil. Suppl. Ⅱ. 10 (1913-17); Contr. U. S. Nat. Herb. **26**: 307 (1931).

Grammitis procera Wall. List no. 3 (1828, nom. nud.).

Gymnogramme javanica Hook. Sp. Fil. **4**: 145 (1862), pro parte.

Coniogramme parvipinnula Hayata, Ic. Pl. Form. **4**: 237 f. 166 (1914); C. Chr. Ind. Fil. Suppl. Ⅱ. 10 (1913-17).

Rhizome creeping, thick as a small finger, sparsely scaly; *scales* light brown, lanceolate, entire; *frond* distant, stipe to 60 cm long, over 0.5 cm thick near base, light straminous, deeply bisulcate above, lamina ample, over 60 cm long, 50 cm broad at base, ovate-deltoid, fully bipinnate under the simple pinnate apical part; *pinnae* about 10-jugate, patent, petiolate, the basal ones much the longest, to 30 cm long, 9 cm broad, impari-pinnate with 10-13 pairs of opposite (or subopposite) pinnules to 5 cm long, 2 cm broad, with sessile or slightly adnate, broadened rotundo-truncate base, caudate apex and crenate-serrated margin, the upper several pairs of pinnae gradually shortened with 9-7-5 pairs of pinnules respectively under the much longer terminal one, the upper middle pinnae with only 3-2-1 pairs of pinnules under still longer terminal one, the uppermost 3-5 pairs generally simple, under the similar terminal one; *texture* thin herbaceous, green and glabrous on both sides; *veins* fine, forked above base, veinlets parallel, with slightly enlarged tip, extending to the base of serrature; *sori* linear, forked; extending to little over half way to the margin.

Yunnan: Mengtze, *Hancock 16*; Schweli-Salween divide, *G. Forrest 25233*; *Handel-Mazzetti 7057*; without locality, *Delavay* (1886); Salween, *G. Forrest 26234* (1925); *H. T. Tsai 52486*, *51765*, Feb. 1932; Tchen Fong Shan, *E. E. Maire*; between Tengyueh and Burmese border, *J. F. Rock 7328*. Taiwan: Arisan, *B. Hayata et S. Sasaki*, Jan. 1912.

Nepal: *Wallich 3* (type).

One of the most distinct species of the genus, being characterized by fully bipinnate lower half of lamina, with the basal pinna having 10-13 pairs of small, opposite, sessile or slightly adnate pinnules with caudate apex and broadened rotundo-truncate base, and by thin herbaceous light green glabrous leaves. Tsai's no. 52486 represents an unusual large form with pinnules in lower pinna to 15 cm long, 3.5 cm broad.

Plate 168. Fig. 1., Habit sketch (natural size). 2. Portion of pinnule, showing serrature and sori (× 4). 3. Scale from rhizome (× 30).

CONIOGRAMME PROCERA (Wall.) Fée
高山凤了蕨

图版 169　华南蒹蕨

根状茎横行，缘石而生，被瓦覆状之金黄色披针形之鳞片；叶亚散生，柄深稻秆色，长 2—4 cm，基部上面 1—2 cm 处有显特之肢节，叶体线状披针形，长 15—30 cm，宽 2—3 cm，渐尖头，下部渐狭，边缘为骨质而呈浅波状，纸质，两面具多少之细毛，叶脉细长，显明，自基部分叉，达于叶边；子囊群圆形，为不规则之一列，盖为肾形，棕色，具短毛。

分布：广东，香港；菲律宾群岛及南洋群岛均产之。

本种形体极似高山蒹蕨(*O. Wallichii*)，唯其茎上之鳞片，彼此瓦覆，叶柄较长，叶缘通常光滑无毛，故易分别。

图注：1. 本种全形(自然大)；2. 叶体之一部，表示叶脉，子囊群之位置与被毛之下面(放大 4 倍)；3. 根状茎上之鳞片(放大 16 倍)；4. 叶体下面之细毛(放大 50 倍)。

OLEANDRA CUMINGII J. Smith

OLEANDRA CUMINGII J. Smith, Journ. Bot. **3**: 413 (1841, nom. nud.); Presl, Epim. Bot. 41 (1849); Hk. Sp. Fil. **4**: 158 (1860); Hk. et Bak. Syn. Fil. 303 (1874); Diels in Engl. u. Prantl: Nat. Pflanzenfam. **1**: 4. 204 (1899); C. Chr. Ind. Fil. 466 (1905); Copel. Polyp. Phil. Isl. 49 (1905); Dunn & Tutcher, Fl. Kwangt. & Hongk. 349 (1912). *Oleandra chinensis* Hance, Ann. Sci. Nat. IV. **18**: 238 (1861); C. Chr. Ind. Fil. Suppl. III. 132 (1934).

Rhizome wide-creeping, densely scaly, *scales* lanceolate, long-acuminate, imbricate, ferruginously brown, margin long-fimbriate; *fronds* 0.5-1 cm apart, stipe dark straminous, hairy, articulated at 1-2 cm above base, lamina linear-lanceolate, 15-30 cm long, 2-3 cm broad, acuminate, gradually narrowed downward, with narrowly cartilaginous, wary and generally naked margin; *texture* papyraceous, rather sparcely and shortly pubescent beneath and with a few very fine appressed hairs above; *veins* fine, distinct, forked above base; *sori* irregularly 1-rowed, subcostal, *indusium* reniform, brown, hirsute.

Guangdong: Sichu Shan, *Sampson 1998* in Herb. Hance (type of *O. chinensis*); Lofau Shan, *Ford* (1883); *N. K. Chun 40930*; North River, Feiloy Hap, *Matthew*, Nov. 25, 1907; Canton, Peiyun Shan, *H. Y. Liang 60252*, July 3, 1930; White Cloud Hill, *W. Hillebrand*. Hong Kong: Mt. Parker, *Matthew*, Oct. 12, 1907.

Philippine Islands: Luzon, *Cuming 60* (type). Also Malaysia-Polynesia.

Closely related to *O. undulata* (Willd.) Ching, from which it differs in characters as will be noted under that species. Upon a close comparison made in the herbarium at Kew in winter, 1930, I could see no tangible differences between the Philippine type and the southern Chinese plant, known as *O. chinensis* Hance.

Plate 169. Fig. 1. Habit sketch (natural size). 2. Portion of lamina, showing venation, sori and hairy under side (× 4). 3. Scales from rhizome (× 16). 4. Hairs from the under side of lamina (× 50).

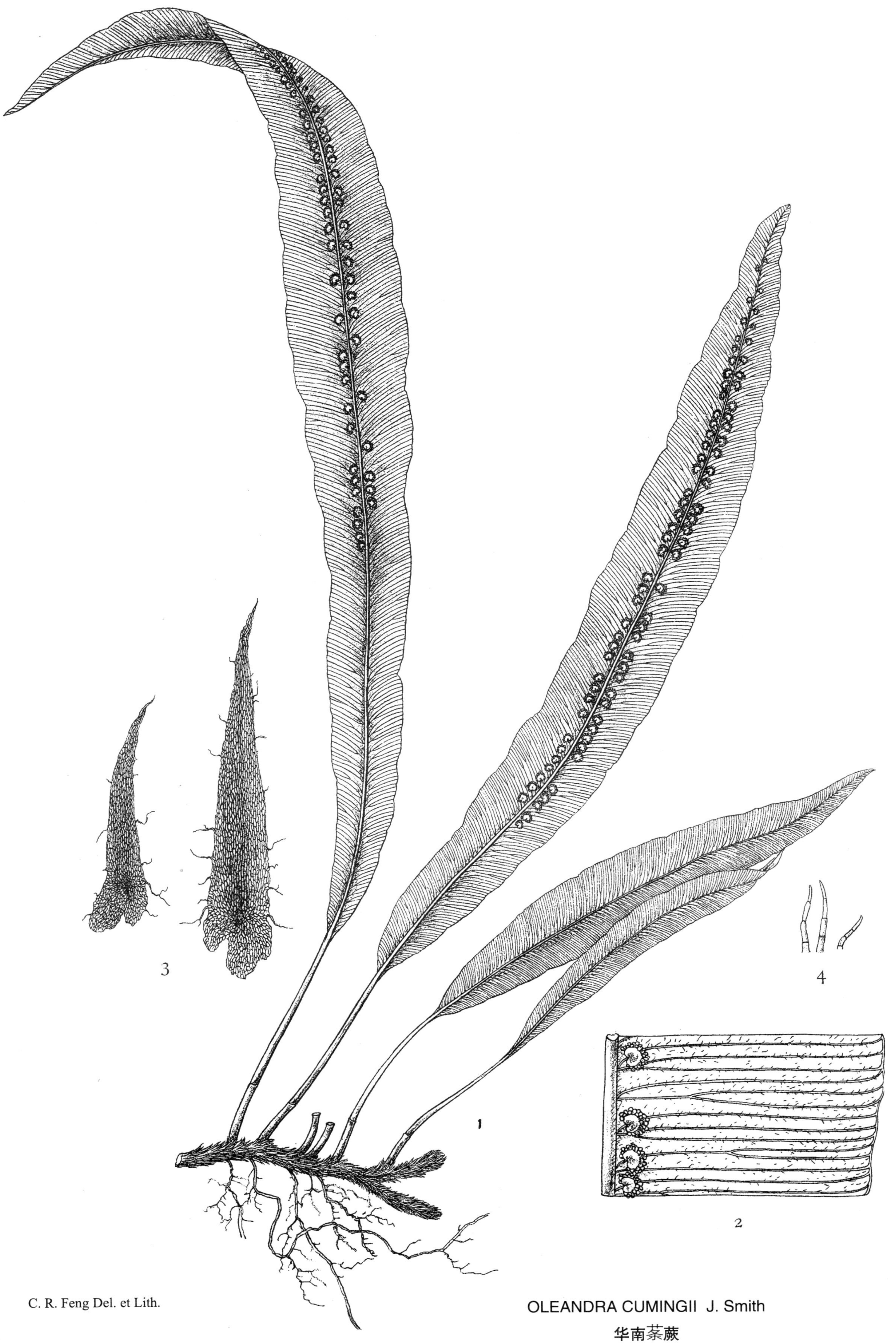

图版 169
PLATE 169

C. R. Feng Del. et Lith.

OLEANDRA CUMINGII J. Smith
华南条蕨

· 347 ·

图版 170　瑶山篠蕨

本种形体颇似图版 169，唯其叶为簇生，两面光滑无毛（或中肋下面略具一二细毛），叶边几呈平行，子囊群较大，距中肋较远，盖无毛，故易分别。

分布：仅产于广西平南县之瑶山。

图注：1.本种全形（自然大）；2.叶体之一部，表示叶脉及子囊群之位置（放大 4 倍）；3.根状茎上之鳞片（放大 16 倍）；4.中肋下面之鳞片（放大 27 倍）；5.中肋下面之细毛（放大 50 倍）。

OLEANDRA WHANGII Ching

OLEANDRA WHANGII Ching, Bull. Dept. Biol. Coll. Sci. Sun Yatsen Univ. No. **6**: 23 (1933); C. Chr. Ind. Fil. Suppl. III. 133 (1934).

Oleandra musifolia Wu (non Bl. 1828), l. c. No. **3**: t. 37 (1932).

Rhizome wide-creeping along the rock surface, densely scaly; *scales* rusty brown, lanceolate, thick, dorsally affixed, margin subscarious, densely imbricate; *fronds* subfasciculated, 3-4 together, stipe 1-4 cm long, pale stramnious, nitid, naked, articulated above base, lamina broadly linear-lanceolate, 25-35 cm long, 3-5 cm broad with almost parallel margin except towards both ends being slowly narrowed, margin subundulate; *texture* chartacaous, green, glabrous on both sides except the costa beneath being sparcely provided with brown scales and a few short articulated hairs; *veins* fine, but distinct, mostly forked from base, parallel, extending to the cartilaginous margin; *sori* irregularly 1-rowed some distance from costa, *indusium* reniform, large, brown, membranaceous, glabrous.

Guangxi: Pin Nam, Yao Shan, *S. S. Sin & Whang 300* (type), June 2, 1928, ibid. *C. Wang 39287*, July 3, 1936, 3000 ft. alt, on rocks.

This distinct endemic species, known so far only from one single locality, differs from its all Chinese relatives in subfaciculated fronds on very short stipes, naked surfaces of lamina (or with very few short hairs along costa beneath) of broadly linear-lanceolate outline with nearly parallel edges, and large naked indusium at some distance from the costa.

Plate 170. Fig. 1. Habit sketch (natural size). 2. Portion of lamina, showing venation and position of sori ($\times$ 4). 3. Scale from rhizome ($\times$ 16). 4. Scales from under side of costa ($\times$ 27). 5. Hairs from the under side of costa ($\times$ 50).

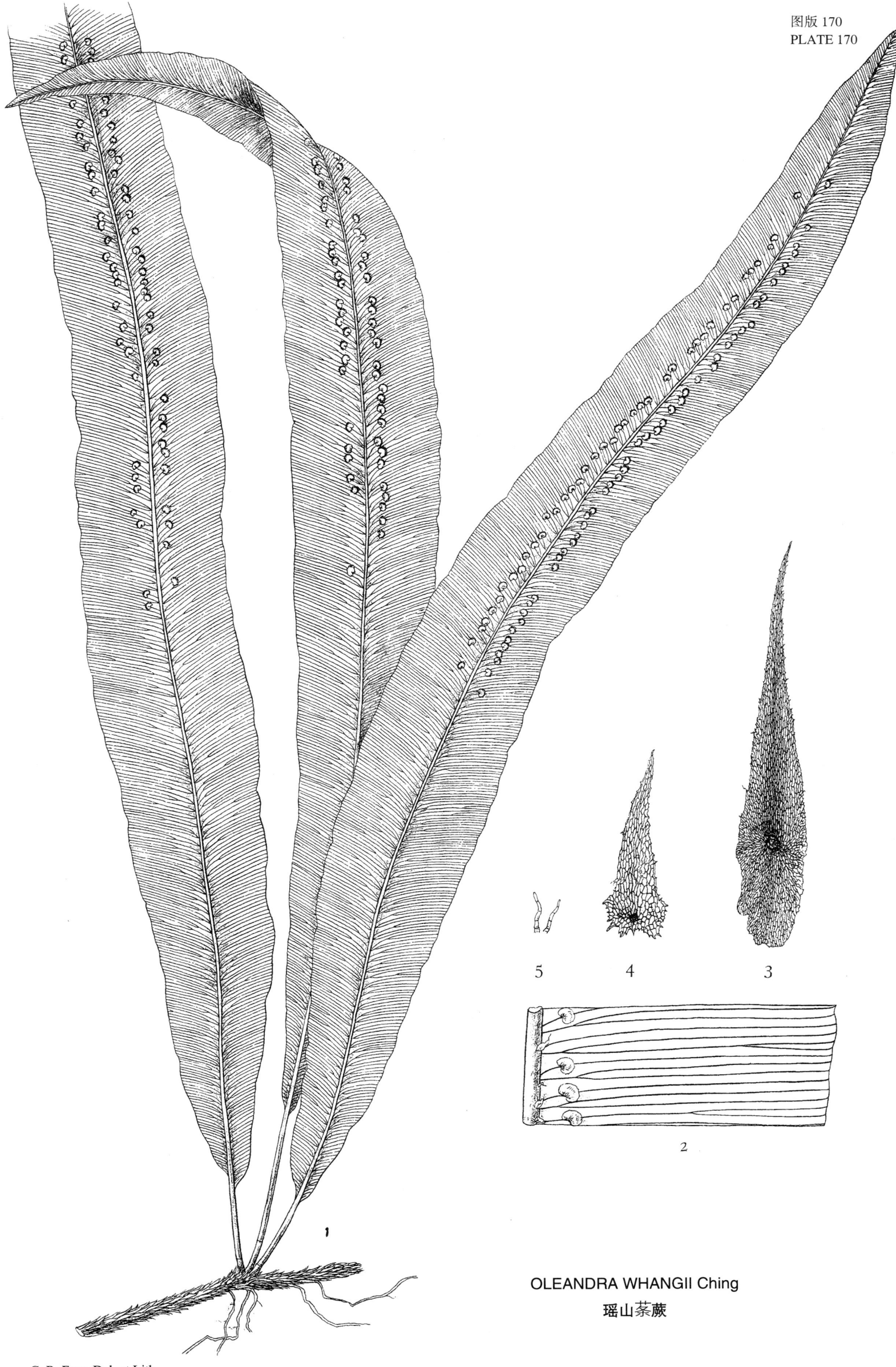

OLEANDRA WHANGII Ching
瑶山条蕨

图版 171　长柄荙蕨

　　本种形体,亦如图版169,唯叶柄甚长,其肢节位于基部3—6 cm处,叶体较阔,其最阔处在中下部,基部呈楔形或稍下延,边缘呈波状反卷,叶质亦较厚,故易识别。

　　分布:广东之海南,台湾;泰国,缅甸及印度东部。

　　图注:1.本种全形(自然大);2.叶体之一部,表示叶脉,子囊群之位置与被刚毛之下面(放大4倍);3.根状茎上之鳞片(放大16倍);4.叶体下面之刚毛(放大50倍);5.根状茎之横切面,表示维管束之布置(放大10倍)。

OLEANDRA UNDULATA (Willd.) Ching

OLEANDRA UNDULATA (Willd.) Ching, Lingnan Sci. Journ. **12**: 565 (1933); C. Chr. Ind. Fil. Suppl. Ⅲ. 132 (1934).

　　Polypodium undulatum Willd. Sp. **5**: 155 (1810).

　　Oleandra cumingii var. *longipes* Hk. Sp. Fil. **4**: 158 (1860); Bedd, Ferns Brit. Ind. t. 135. (1866); C. Chr. Contr. U. S. Nat. Herb. **26**: 290 (1931).

　　Oleandra cumingii (non J. Sm.) Hk. et Bak. Syn. Fil. 303 (1868); Clarke, Trans. Linn. Ⅱ. Bot. **1**: 542 (1880); Bedd. Handb. Ferns Brit. Ind. 288 (1883); C. Chr. Ind. Fil. 466 (1905), pro parte.

　　Oleandra pubescens Cop, Univ. Calif. Publ. Bot. **12**: 397 pl. 52a (1931).

Rhizome thick, wide-creeping, densely scaly; *scales* linear-subulate, ferruginous-brown, imbricate, margin sparcely villose-fimbriate; *frond* 1-3 cm apart, stipe 13-20 cm long, dark straminous, naked, articulated at 4-6 cm above the base, lamina broadly lanceolate, 20-26 cm long, 3-4.5 cm broad at the lower middle, being the broadest part, acuminate, base cuneate, shortly decurrent, margin naked, repando-undulate; *texture* thick chartaceous, densely pubescent on the under side (hairs on costa spreading), glabrous and subnitid above; *veinlets* fine, distinct, forked from base; *sori* irregularly 1-rowed some distance from costa, *indusium* reniform, large, dark brown, hirsute.

　　Hainan: Chim Shan, Fan Maan Tsuen, *F. A. McClure 20061*, May 4-20, 1932; Ue Lung Shan, Changkiang Hsien, *S. K. Lau 3108*, Jan. 9, 1934; Ka Chik Shan, ibid., *S. K. Lau 1490*, April 8, 1933. Taiwan: Mt. Arisan, *Faurie 483*, May, 1914.

　　Thailand: Doi Chang, *Eryl Smith 1072* (type of *O. pubescens*); Kao Sabap, *Eryl Smith 531*, June 29, 1931.

　　Burma: Keng Teng Territory, *J. F. Rock 2026*, *2026A*. Assam: *Griffith*.

　　India orientalis: Tranquebar, *Klein 887* (1800, type); *Helfer, Meebold, Wight*. Also French Indo-China.

　　A close relative of O. Cumingii J. Sm., from which it can easily be distinguished by its decidedly longer stipe with articulation at 3-6 cm above base, broadly lanceolate (broadest at the lower middle) lamina of thicker texture, with repando-undulate margin and cuneate, or shortly decurrent base, more densely pubescent under and glabrous upper, surfaces.

　　Plate 171. Fig. 1. Habit sketch (natural size). 2. Portion of lamina, showing venation, sori and strigose hairy under side ($\times$ 4). 3. Scale from rhizome ($\times$ 16). 4. Hairs from under side of lamina ($\times$ 50). 5. Cross section of rhizome, showing the arrangement of steles ($\times$ 10).

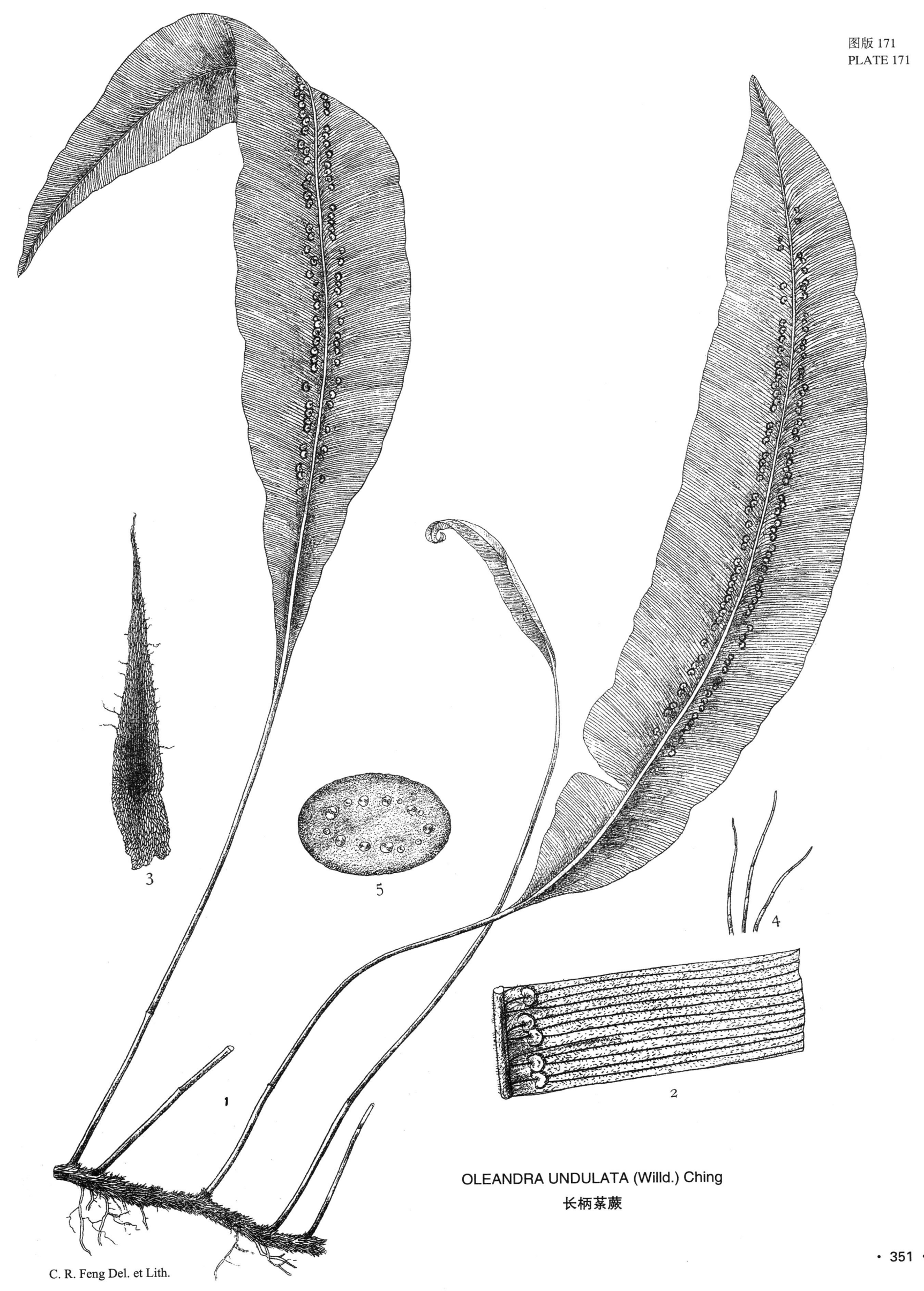

OLEANDRA UNDULATA (Willd.) Ching
长柄蒫蕨

根状茎细长,横行,仅端部被卵状披针形淡黄色之膜质鳞片;叶远生,柄长15—26 cm,细长,光滑,淡稻秆色,基部被稀疏鳞片,余皆光滑,叶体三角形,长10—18 cm,宽亦如之,三回羽状深裂,顶部渐尖头,羽状深裂;一回小叶5—8对,对生,基部以肢节着生于中轴,下部二对具柄,余皆无柄,长7—13 cm,宽3—5 cm,卵状椭圆形;二回小叶5—8对,对生,无柄,或基部上方一片具短柄,亚全缘或羽状深裂;一回小叶之第二对较小,阔披针形,长达8 cm,宽2 cm,一回小叶椭圆形,无柄,羽状深裂或具缺刻,亚纸质,淡绿色,两面光滑无毛,唯中轴与小叶着生处略具淡黄色之球形腺,叶脉匀细而明显,分叉或羽状分裂;子囊群圆或椭圆形,黄色,无盖,贴近叶缘。

分布:中国西北二部各省,东三省,台湾;日本,朝鲜半岛及印度北部均产之。

图注:1.本种全形(自然大);2.叶之一部,表示叶脉及子囊群(放大8倍);3.小叶与叶柄节状着生之情形(放大6倍);4.根状茎上之鳞片(放大10倍)。

GYMNOCARPIUM REMOTI-PINNATUM (Hayata) Ching

GYMNOCARPIUM REMOTI-PINNATUM (Hayata) Ching, Bull. Chin. Bot. Soc. **1**: No. 2, xiv (1935).

Dryopteris remoti-pinnata Hayata, Gen. Ind. Ic. Pl. Form. 108 (1917); C. Chr. Ind. Fil. Suppl. Ⅲ. 96 (1934).

Dryopteris remota Hayata, Mater. Fl. Form. in Journ. Coll. Sci. Imp. Univ. Tokio **30**: 421 (1911); Ic. Pl. Form. **4**: 177 (1914); C. Chr. Ind. Fil. Suppl. Ⅱ. 16 (1913-16).

Gymnocarpium remotum Ching, Contr. Biol. Lab. Sci. Soc. Chin. Bot. **9**: 41 (1933).

Aspidium dryopteris var. *longulum* Chrisr, Bull. Herb. Poiss. Ⅱ, **2**: 830 (1902); Bull. Soc. Bot. France **52**: Mem. Ⅰ. 35 (1915).

Dryopteris linnaeana C. Chr. Ind. Fil. 275 (1905), pro parte; Acta Hort. Gethob. **1**: 42, 55 (1924).

Dryopteris robertiana C. Chr. (non Index, 1905) Acta Hort. Gothob. **1**: 55 (1924).

Dryopteris continentalis Petrov, Fl. Jakutiae 15 c. ic. 1930.

Rhizome slender, wide-creeping, densely scaly on new shoots; *scales* ovate-lanceolate, light brown, membranaceous, fimbriate, long-acuminate; *frond* far apart, or sometimes approximate, stipe 15-26 cm long, slender, firm, pale straminous, sparcely scaly in the lower part; lamina deltoid, 10-18 cm each way, tripinnatifid at base; *pinna* 5-8-jugate under the deeply pinnatifid acuminate apical part, opposite, subpatent, all prominently articulated to rachis, the lowest two pairs generally petiolate, the upper ones sessile, the basal pair much the largest, 7-13 cm long, 3-5 cm broad, ovate-oblong, on petiole to 2 cm long, bipinnatifid under the deeply pinnatifid acuminate apical part; *pinnules* 5-8-jugate, opposite, sessile or adnate or petiolulate in the anterior basal one, which is the longest, patent, deltoid-lanceolate, 1-2-3 cm long, 1-1.5 cm broad at base, acuminate, pinnatifid down nearly to costa into 5-8 pairs of oblong, rounded, entire or inciso-crenate *segments* under the pinnatifid apex; the second pair of pinnae much smaller, broadly lanceolate, to 8 cm long, 2 cm broad at base, pinnules oblong, sessile, pinnatifid or incised, the third pair and further upper pinnae linear-lanceolate, sessile, pinnatifid or pinnate at base with oblong crenate or entire pinnules with rounded apex; *texture* submembranaceous, light green, glabrous on both sides, except rachis near the insertion of pinnae being sparingly and shortly glandular; *venation* fine, distinct, forked or pinnate; *sori* roundish, brown, exindusiate, much nearer to the margin.

The species is now found to be fairly common in North, Northeastern and Northwestern China and Taiwan province; Sibiria, Sahalin, Korean peninsula, Japan and Northwestern Himalayas. In China, it has been reported from provinces Hebei, Shaanxi, Shanxi, Gansu, Sichuan, Xinjiang, Taiwan and Northeast China (cf. my Monograph of Gymnocarpium p. 41).

In general habit, this fern resembles G. Robertianaum (Hoffm.) Newman of northern Europe and North America, differs, above all, in leaves being eglandular, or sometimes only rachis near the articulated inseration of pinnae being sparingly and shortly glandular.

Plate 172. Fig. 1. Habit sketch (natural size). 2. Ultimate pinnule, showing venation and position of sori ($\times$ 8). 3. Portion of rachis, showing the articulation of the base of pinnae ($\times$ 6). 4. Scale from rhizome ($\times$ 10).

GYMNOCARPIUM REMOTI-PINNATUM (Hayata) Ching
肢节蕨

根状茎短而直立,须根丛生;端被深棕色之卵状披针形鳞片;叶簇生,柄长2—4 cm,细如铜丝,绿色光滑,叶体长4—6 cm,线状披针形,宽约1 cm,二回羽状深裂,顶部常延长具一芽,着地生根;小叶8—11对,对生或亚对生,几不具柄,阔卵形,长约5 mm,宽亦如之,向上渐小,深裂,裂片1—2对,下者2—3裂,上者不分裂,顶部2—3裂,全缘,薄纸质,光滑,淡绿色,叶脉简单,每裂片一脉,不达于顶;子囊群短线形,生于上部之裂片者向中肋开,生于基部之裂片者向下开。

分布:贵州,云南,四川特产。

本种生于阴湿之石灰岩洞中,颇类后种,唯形体较为细长,叶之分裂较少,故易分别。

图注:1.本种全形(自然大);2a—2e.自基部至顶部之各小叶,表示其形态,分裂度,叶脉及子囊群(放大10倍);3.根状茎上之鳞片(放大27倍)。

ASPLENIUM FUGAX Christ

ASPLENIUM FUGAX Christ, Bull. Soc. Bot. France **52**. Mém. 1. 53 (1905); Bull. Acad. Géogr. Bot. (1910) 13; C. Chr. Ind. Fil. 112 (1905); Acta Hort. Gotheb. **1**: 82 (1924).

Rhizome short, erect, densely radicose; *scales* fusco-brown, ovate-lanceolate, entire; *fronds* caespitose, several together, stipe wiry, slender, green, naked, 2-4 cm long, lamina linear-lanceolate, 4-6 cm long, about 1 cm broad, bipinnatifid, often with prolongated and viviparous nodding apex; *pinnae* 8-11-jugate, opposite or subopposite, subsessile, broadly ovate, 5 mm each way, gradually smaller upwards, deeply pinnatifid into 1-2 pairs of oblong-ovate, entire or bifid segments under the 3-2-fid or entire terminal segment; *texture* thin herbaceous, green, glabrous; *veins* simple and uninerved to each segment, not reaching the acute apex of segments; *sori* short-linear, one to each segment, *indusium* linear, membranaceous, entire, those on the upper segments opening towards costa of pinnae, while those on the lower segments generally opening downward.

Guizhou: Ouanly, *Esquirol 3213* (type). Sichuan: Tchenkouting, *Farges 657*. Yunnan: without locality, *Henry*.

A distinct endemic fern, inhabiting dripping calcareous rocks and only closely related to *A. exiguum* Bedd. from which it differs in much slender habit and less divided fronds.

Plate 173. Fig. 1. Habit sketch (natural size). 2a-2e. Lateral pinnae from base upwards, showing shape, pinnation, venation and sori (× 10). 3. Scale from rhizome (× 27).

ASPLENIUM FUGAX Christ
阴地铁角蕨

根状茎短而直立，具黑色小线状披针形之鳞片；叶簇生，叶柄长 1.5—3 cm，深栗色，密被狭鳞片，叶体为线状披针形，长 10—30 cm，宽 1—3 cm，向两端渐狭，二回羽状深裂，中轴光滑，其顶部常延长而着地生根；一回小叶 15—28 对，具短柄，卵状椭圆形或椭圆披针形，开展，长 5 mm—2.5 cm，深裂成 3—6 对椭圆形之裂片，其基部上方一裂片较大，锯齿尖锐，薄纸质，两面光滑，叶脉分叉，直达锯齿；子囊群形长，盖为膜质，灰白色，全缘，向中肋开，唯生基部上方之二回小叶者，则向其主脉开。

分布：云南，四川，贵州，西藏；喜马拉雅山区，印度南部及菲律宾群岛均产之。

本种分布甚广，形体大小变异极大，异名迭出，实皆一种也。

图注：1. 本种全形（自然大）；2. 一回小叶（放大 6 倍）；3. 叶柄基部鳞片（放大 24 倍）。

ASPLENIUM EXIGUUM Beddome

ASPLENIUM EXIGUUM Beddome, Ferns S. Ind. t. 145 (1863); Hope, Journ. Bomb. Nat. Hist. Soc. **13**: 663 (1900-1); C. Chr. Ind. Fil. 110 (1905); Suppl. Ⅲ. 32 (1934).

Asplenium fontanum var. *exiguum* Bedd. Handb. Ferns Brit. Ind. 158 (1883).

Asplenium yunnanense Franch. Bull. Soc. Bot. France **32**: 28 (1885); Diels in Engl. u. Prantl: Nat. Pflanzenfam. **1**: 4. 241 (1899); C. Chr. Ind. Fil. 138 (1905); Acta Hort. Gothob. **1**: 80. 1924; Blot, Aspl. du Vietnam 42 t. 4 f. 1-4 (1932).

Asplenium fontanum var. *yunnanense* Bedd. Handb. Ferns Brit. Ind. Suppl. 31 (1892).

Asplenium Loherianum Christ, Bull. Herb. Boiss. **6**: 152 (1898); C. Chr. Ind. Fil. 118 (1905).

Asplenium woodsioides Christ, Bull. Soc. Bot. Ital. (1900) 261; C. Chr. Ind. Fil. 138 (1905).

Asplenium lushanense C. Chr. Acta Hort. Gothob. **1**: 80 t. 16 f. e-g (1924).

Asplenium fontanum Clarke, Trans. Linn. Soc. Ⅱ. Bot. **1**: 484 (1880), pro parte.

Rhizome short, erect, densely radicose and scaly; *scales* linear-lanceolate, dark brown, thin, iridescent; *fronds* tufted, stipe 1.5 -3 cm long, atro-castaneous, densely fibrillose-scaly throughout, lamina linear-lanceolate, 1-3 cm broad, varying from 10-30 cm long, narrowed towards both ends, bipinnatifid or rarely subbipinnate, rachis quite glabrous, castaneous below, green towards apex which often prolongated and rooting at tip; *pinnae* 15-28-jugate, shortly petiolate, ovate-oblong to oblong-lanceolate, patent, 0.5-2.5 cm long, deeply incised into 3-6 pairs of oblong, dentate segments with the anterior basal segment not infrequently being the largest and subpinnatifid; *texture* herbaceous, both sides glabrous; *veins* obscure, each tooth with one veinlet; *sori* elongate, *indusium* membranaceous, gray, entire, opening mostly towards the costa of pinnae, but those on the anterior basal pinnules often towards costules.

Yunnan: Lankong, *Delavay*, April 7, 1883 (type of *A. yunnanense*); Mengtze, *A. Henry 10106*, *13603*; *Hancock 56* (1893); Puseh Cliff, *Henry 13392*; Chungtien Plateau. *Forrest 13043*; Mekong, *Forrest 15279A*; Kintchong Chow, *E. E. Maire 2805*; Yunnanfu near Laka Tiang, *Schneider 458* (1914); Yungling Mt., *Forrest 15244*, Hockiang, *Schneider 2789*; TcheouKiaTzeTang, *Maire 1412A*. Sichuan: Moupin, *David*; *Wilson 2658*, *5350*, *5349*; NinYuanFu, *Harry Smith 1801* (type of *A. lushanense*). Guizhou: Majo, *Cavalerie* (1908); Pinfa, *Cavalerie 660* (pro parte); Kianglong, *Michel 992*; *Cavalerie*, Jan. 1910; without locality, *Esquirol 799*. Tibet: Muti, *Capt. Kingdom Ward 4327*.

Also Himalayas, South India, Vietnam and the Philippine Islands (leg. *Loher*, type of *A. Loherianum*).

A distinct but very variable fern, now known rather extensively in Asia. The frond varies from scarcely 5 mm to over 3 cm in width and leaf-apex sometimes prolongated and rooting at tip. The nearest relative is evidently *A. fontanum* (L.) Bernh. from which it differs in less pinnatifid fronds of dark green color, in costal sori and sometimes prolongated and rooting leaf-apex. The type based upon a specimen from Mt. Nilgari, South India, represents a small and simple form with some fronds having prolongated and rooting apex and agrees well with *A. lushanense* C. Chr. The Philippine plant described under *A. Loherianum* differs from the mainland form in no respect. The Mexican *A. Glenniei* Baker has been found not specifically different from the typical form of our fern, as already pointed out by Hope (l. c.) long ago.

Plate 174. Fig. 1. Habit sketch (natural size). 2. Lateral pinnae ($\times$ 6). 3. Scale from the base of stipe ($\times$ 24).

ASPLENIUM EXIGUUM Beddome
低头铁角蕨

图版 175　南海铁角蕨

　　根状茎颇粗肥，短而直立，被披针形之棕色疏鳞片；叶簇生，柄长 20 cm 或以上，淡稻秆色，略具细长鳞片，扁形，叶体卵状椭圆形，长 17—24 cm，宽 7—14 cm，一回奇数羽状分裂（间为披针形之单叶）；小叶 2—4 对，长 10—15 cm，宽 1.6—2.5 cm，披针形，长渐尖头，基部亚等形，稍下延，具短柄，边缘具疏缺刻形之锯齿，唯基部及端为全缘，软纸质，淡绿色，下面鳞片疏生，侧脉明显，分叉，斜出，不达于叶边；子囊群直线形，长约 6 mm，生于上方小脉，斜出，达于小叶宽 2/3，盖膜质，全缘，宿存。

　　分布：广东之信宜县及海南，台湾；越南亦产之。

　　此种在其分布区域内甚为普通，其形体极似印度南部产之 A. Wightianum Wall.，唯小叶数较少较狭，基部为亚等边，顶部全缘，缺刻形之锯齿疏生，故易分别。

　　图注：1. 本种全形（自然大）；2. 小叶之一部，表示叶脉及锯齿（放大 4 倍）；3. 叶柄上之鳞片（放大 16 倍）；4. 叶下面之鳞片（放大 20 倍）。

ASPLENIUM LORICEUM Christ

ASPLENIUM LORICEUM Christ in C. Chr. Ind. Fil. 119 (1905).

　　Asplenium formosae Christ (non *A. formosanum* Baker, 1891), Bull. Herb. Boiss. Ⅱ. **4**: 613 (1904).

　　Diplazium makinoi Yabe in Matsum. et Hayata, Fnum. Pl. Form. in Journ. Coll. Sic. Imp. Univ. Tokio **22**: 600 (1906).

　　Asplenium makinoi Hayata, Ic. Pl. Form. **4**: 224 f. 154 (1914); C. Chr. Ind. Fil. Suppl. Ⅰ. 6 (1913-17).

　　Asplenium wightianum Merr. (non Wall. 1828), Emum. Hainan Pl. in Lingnan Sci. Journ. **5**: 15 (1927).

　　Rhizome rather thick, short, erect, densely radicose, sparcely scaly; *scales* lanceolate, brown, fimbriate; *fronds* caespitose, stipe 20-24 cm long, pale straminous, herbaceous, sparcely scaly, with rachis compressed upon drying, lamina ovate-oblong, 17-24 cm long, 7-14 cm broad, impari-pinnate (sometimes simple and lanceolate); *pinnae* 2-4-jugate, 10-15 cm long, 1.6-2.5 cm broad, lanceolate, long-acuminate, base subequal, short-attenuate on petiole about 5 mm long, margin remotely incise-serrate above base and below long-acuminate entire apex; *texture* soft herbaceous, pale green, with a few small scales on the under side; *costa* prominent on both sides; *veins* quite distinct, mostly forked, veinlets oblique, parallel, extending to some way below leaf-margin; *sori* straight, oblique, about 6 mm long, borne on the anterior veinlet of each group, extending from costa to one-third way from margin, *indusium* gray, linear, entire, persistent.

　　Hainan: Ng Chi Leng, *F. A. McClure 8406*, *8554*; *Katsumada 6687* ex Herb. Hong Kong; *W. Y. Chun 6624* in Herb. Univ. Nanking; *Eryl Smith 1429*; Sha Po Leng, *W. T. Tsang 16185*. Guangdong: Sunyi, Sick Toun, *C. Wang 32032* in moist place in revine. Taiwan: Urai, *Faurie 669* (type), *159*; Taihoku, *S. Sasaki 21525*; *Y. Shemada 114* (1915).

　　Vietnam: Thue Lui, *Cadier 100*, *161*; *Chevalier 38*, *718*. *Billet 7727*.

　　This distinct species is closely related to *A. Wightianum* Wall. from S. India and Sri Lanka, differs in fewer and broader pinna with subequal base and only a few remote incisions on the margin below the entire long-acuminate apex.

　　Plate 175. Fig. 1. Habit sketch, (natural size). 2. Portion of pinnae, showing venation and serrature ($\times$ 4). 3. Scale from base of stipe ($\times$ 16). 4. Scale from under side of pinna ($\times$ 20).

根状茎短而直立，被深栗黑色之细长鳞片；叶簇生，柄长 12—15 cm，栗色或绿色，光滑，叶体三角卵形，渐尖头，长 10—15 cm，宽亦如之，下部三回羽状分裂；小叶 5—7 对，具柄，开展，基部一对最大，长 9 cm，宽 6 cm，三角形，基部不等边，柄长 1 cm，羽状分裂，二回小叶 3—4 对，具柄，基部下方一片长 2—3 cm，宽如之，卵状三角形，钝头，羽状深裂；裂片 2—3 数，卵形，有锯齿，薄纸质，光绿无毛，叶脉扇状分叉，每锯齿一脉，唯不达于尖端；子囊群直线形，或稍呈弧形，长 5 mm，盖膜质，全缘。

分布：贵州原产；近发现于越南。

此种形体稍似欧洲产之 A. adiantum nigrum L.，唯叶质较薄，分裂较少耳。

图注：1. 本种全形（自然大）；2. 同上，幼形（自然大）；3. 上部小叶，表示叶脉及子囊群（放大 3 倍）；4. 叶柄基部之鳞片（放大 27 倍）。

ASPLENIUM INTERJECTUM Christ

ASPLENIUM INTERJECTUM Christ, Bull. Acad. Géogr. Bot. (1902) 241; (1907) 149; C. Chr. Ind. Fil. 116 (1905); Blot, Aspl. d. Vietnam 41 t. 2. f. 3 (1932).

Asplenium cuneifolium var. *vegetius* Christ, loc. cit. p, 240.

Asplenium interjectum var. *elatum* Christ, Bull. Acad. Géogr. Bot. (1907) 149.

Asplenium cuneifolium Christ (non Viv. 1806), Bull. Acad. Géogr. Bot. (1910) 13.

Asplenium longkaense Rosenst. in Fedde, Repert. Sp. Nov. **13**: 123 (1913); C. Chr. Ind. Fil. Suppl. Ⅱ. 6 (1913-17).

Rhizome short, erect, densely radicose, sparcely scaly; *scales* linear-subulate, blackish, thin, clathrate, iridescent; *fronds* caespitose, stipe 15-12 cm long, atro-castaneous throughout, or green, naked, herbaceous, lamina deltoid, 10-15 cm long, and nearly as broad, tripinnate at base; *pinnae* 5-7-jugate under simple pinnate and acuminate apex, petiolate, patent, the basal pair much the largest, to 9 cm long, 6 cm broad, deltoid, with unequal base, petiolate (petiole 1 cm long), bipinnate; *pinnules* 3-4-jugate, anadromously arranged (i. e. the pinnule nearest to the rachis is borne on the upper side of rachilet), petiolulate, the lower ones 2-3 cm each way, deltoid-ovate, obtuse, pinnatifid or deeply lobed into 2-3 ovate, rounded, dentate lobes, the upper ones rhombic, incised and eroso-dentate; *texture* thin herbaceous, light green, glabrous on both sides; *veins* flabellately forked, with veinlet running into each tooth but falling short of the tip; *sori* linear, straight or slightly curved, to 5 mm long, *indusium* narrow, gray, entire, opening towards costule of ultimate lobes, or towards costa on the apical portion of pinnae, or of pinnule.

Guizhou: Tsingay, *Bodinier 2094* (type); Tchenfau, *Esquirol 801*, *1017*, *707*; Hoangkochou, *Bodinier 2557*; Longka, *Cavalerie 3773* (type of A. *longkaense*).

Also Vietnam.

A distinct endemic fern, closely related in habit to *A. adiantum nigrum* L. of Europe, differing in much less divided green leaves of thin herbaceous texture.

Plate 176. Fig. 1. Habit sketch (natural size). 2. The same but young form (natural size). 3. Upper pinnae, showing venation, and sori ($\times$ 3). 4. Scale from base of stipe ($\times$ 27).

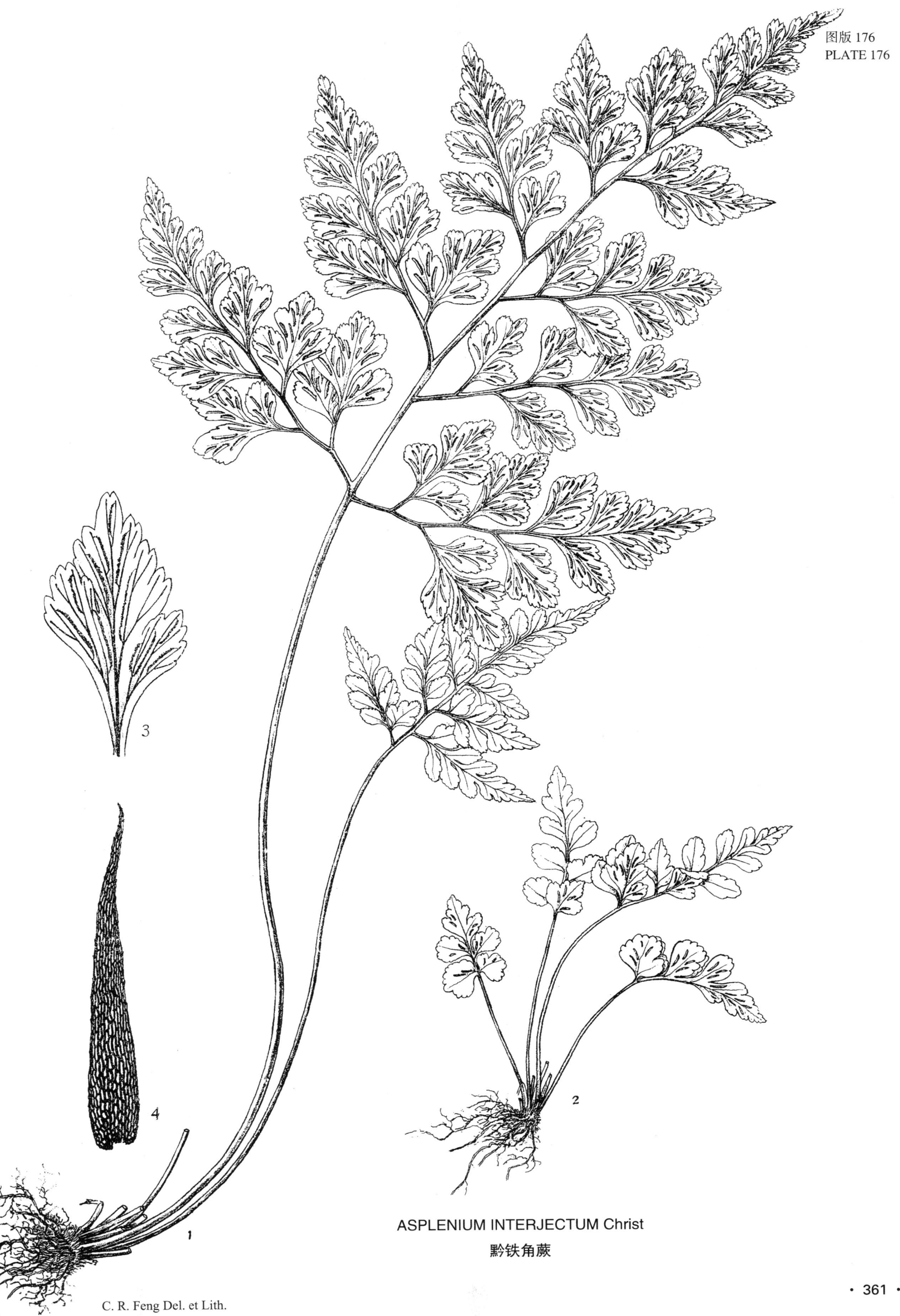

ASPLENIUM INTERJECTUM Christ
黔铁角蕨

根状茎木质，粗肥，斜出或卧生，被亮棕色之卵状大鳞片，长达 1 cm；叶簇生，柄长 30—45 cm，基部粗达 1 cm，下部粗糙并具鳞片，稻秆色，光亮，叶体大卵形，长达 70 cm，宽亦如之，渐尖头，四回羽状分裂；一回小叶约 10 对，对生，开展，无柄，下部数对相距约 12 cm，基部一对最大，长达 40 cm，宽约 30 cm，卵形，渐尖头，中轴向上弯曲；二回小叶约 10 对，无柄，开展，基部一对对生，且甚短，紧靠二回小叶之中轴，第二对长达 3—5 cm，宽约 1.2 cm；末回小叶 6—9 对，椭圆卵形，顶圆，基部楔形，羽状深裂，裂片 1—3 对，圆形，具棱角，具小脉一数，纸质，光滑，唯各回之基部着生处具一二心脏形之膜质大鳞片；子囊群小，圆形，每裂片一个，生于小脉之顶，盖小，圆卵形，膜质，仅基部着生。

分布：云南，四川，贵州，广东，广西，台湾；越南及印度均产之。

图注：1. 本种全形（自然大）；2. 二回小叶，表示叶脉，子囊群位置及基部下面着生之鳞片（放大 3 倍）；3. 末回小叶，表示子囊群及盖着生情形（放大 10 倍）；4. 着生于各回小叶基部下面之鳞片（放大 16 倍）；5. 中肋上面之毛（放大 76 倍）；6. 叶上面之刺状毛及其着生情形（放大 76 倍）；7. 叶柄基部之鳞片（放大 10 倍）。

ACROPHORUS STIPELLATUS (Wallich) Moore

ACROPHORUS STIPELLATUS (Wallich) Moore, Gard. Chron. (1854) 135; C. Chr. Ind. Fil. 4 (1905) pro parte; Contr. U. S. Nat. Herb. **26**: 273 (1931); Wu Polyp. Yaoshan. in Bull. Dept. Biol. Coll. Sci. Sun Yatsen Univ. No. **3**: 20 t. 1 (1932); Ogata, Ic. Fil. Jap. **5**: t. 201 (1933).

Davallia stipellata Wallich List no. 260 (1828, nom. nud.).

Acrophorus nodosus J. Sm. Hist. Fil. 222 (1875); Christ Farnkr. d Erde 285 (1897); Diels in Engl. u. Prantl: Nat. Pflanzenfam. **1**: 4, 164 (1899) pro parte; Bedd. Ferns Brit. Ind. t. 93 (1865).

Davallia nodosa Hk. sp. Fil. **1**: 157 (1846); Hk. Journ. Bot. (1857) 9 t. 10; Syn. Fil. 92 (1867), pro parte.

Leucostegia nodosa Bedd. Ferns Brit. Ind. Suppl. 4 (1876); Handb. Ferns Brit. Ind. 56 (1883).

Rhizome subterraneous, thick, woody, oblique or short-creeping, densely scaly; *scales* large, bright brown, over 1 cm long, ovate-acuminate, thin, entire, extending upward to some distance above base of stipe; *fronds* caespitose, stipe 30-45 cm long, nearly 1 cm thick at base, straminous, subnitid, lower part densely scaly and muricated by transversed scars from the persistent base of fallen scales, lamina immense, ovate, to 70 cm long, nearly as broad, acuminate, 4-pinnate; *pinnae* about 10-jugate, opposite, horizontally patent, sessile, the lower pairs 12 cm apart, the basal pair much the largest, to 40 cm long, 30 cm broad, ovate, acuminate, rachilet curved upward, *pinnules of first order* about 10-jugate, sessile, patent, basal pair opposite and much shortened, 4 cm long, the second one on posterior side much the largest, to 15 cm long, those on the anterior side all much smaller than those on the other side, oblong-lanceolate, acuminate, far apart; *pinnules of second order* about 10-jugate, perdendicular to rachilet, oblong-lanceolate, sessile, basal pair opposite, shortened, against rachilet of first order, the second pair 3-5 cm long, 1.2 cm broad; *ultimate pinnules* 6-9-jugate, oblong-ovate, cuneate, rounded, lobato-incised with 1-2-3 pairs of rounded angular uninerved soriferous *lobes*; *texture* herbaceous, pale green or brownish, glabrous beneath except base of pinnae and pinnules of different order being provided with a few large broadly ovate acuminate deeply cordate scales at the point of insertion, upper side of rachis and rachilets pustulately hairy, of ultimate segments with a few short, appressed, rufo-red, articulated hairs, of costa somewhat spinulose; *veins* in ultimate pinnule distinct, pinnate, one to each lobe, falling far short from margin; *sori* small, rounded, one to each lobe, terminating the veinlet some distance below margin, *indusium* small, gray, membranaceous erosed at top, free on all sides except being cucculate at base; *spores* broadly winded.

Yunnan: Between Tengyueh and Burmese border, *J. F. Rock 7353*. Guizhou, Pinfa, *Cavalerie 2857*; Tuhshan, *Y. Tsiang 6959, 6720*; Vanchin Shan, *Steward et Chiao 858*. Guangdong: Lokchong, *N. K. Chun 42404*. Guangxi: Yao Shan, *S. S. Sin 459A*. Sichuan: Without locality, *W. P. Fang 8257*. Taiwan: Mt. Arisan, *Drs. F. et C. Baker*, Nov. 1914, 2300 ft. alt., (pro parte); *U. Faurie 662*.

Also Himalayas, Khasia and Vietnam.

The genus *Acrophorus* comprises to-day two species, the other being *A. Blumei* Ching (*Aspidium nodosum* Bl.) from Malaysia-Polynesia, which differs from the continental one in smaller size, more finely dissected leaves of an opaque color when dried, more copiously scaly and hairy lamina and much larger indusium of brown color and rigider consistancy.

Fern students have been much divided in their opinions as to the systematic position of this rather isolated genus, but in the light of anatomical and morphological evidences, I am convinced that its natural position falls with Davallioid ferns, and especially *Leucostegia* (*L. perdurans* Christ, for instance).

Plate 177. Fig. 1. Habit sketch (natural size). 2. Pinnule of 2nd. order, showing venation, position of sori and the large orbicular scale at its base beneath (× 3). 3. Ultimate pinnule, showing venation, position of sori and manner of indusial attachment (× 10). 4. Scale detached from the base of pinnule of each order (× 16). 5. Hairs from the opper side of rachilet (× 76). 6. Hairs from the upper side of pinna (× 76). 7. Scale from base of stipe (× 10).

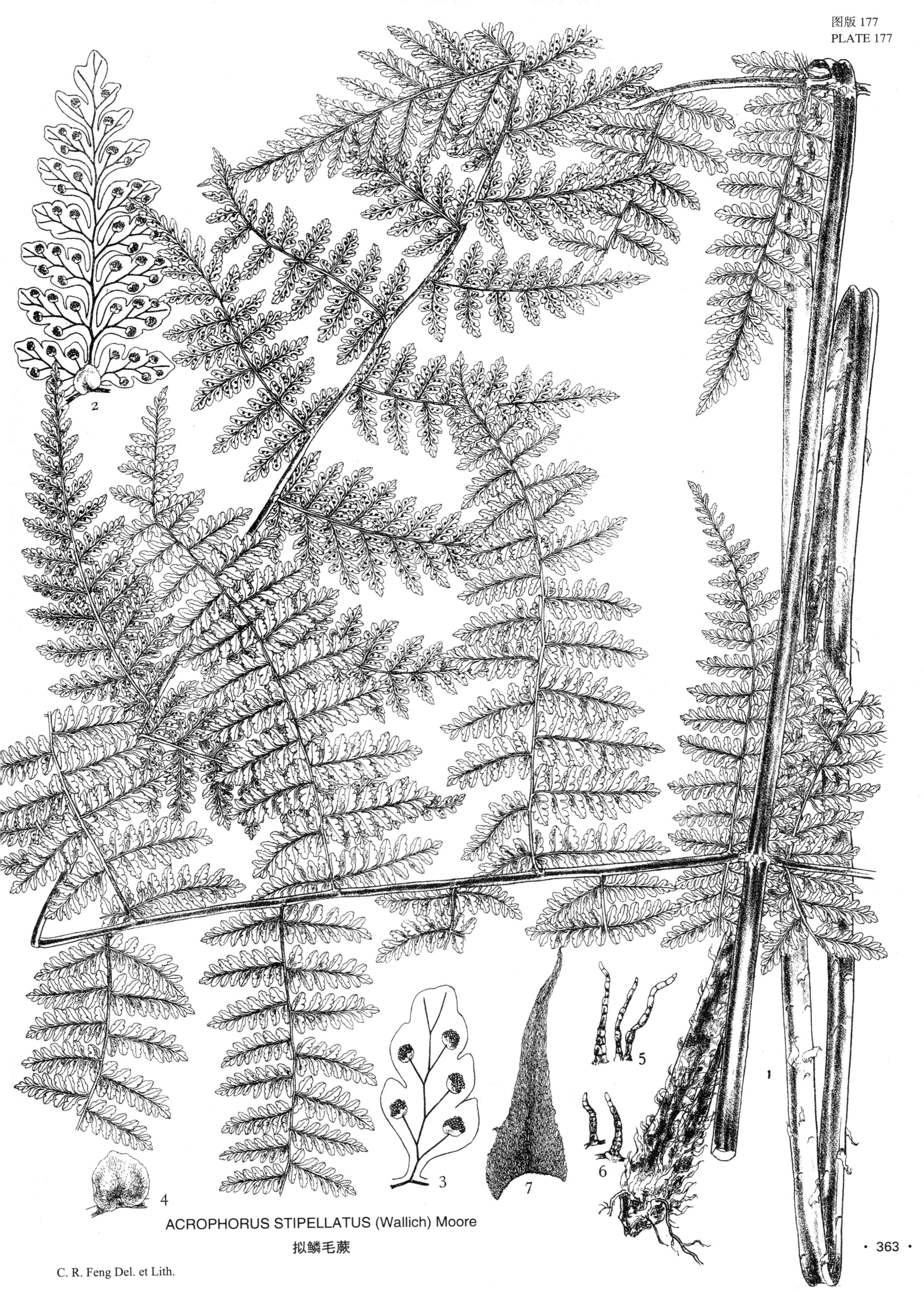

ACROPHORUS STIPELLATUS (Wallich) Moore
拟鳞毛蕨

图版 178　高山鳞毛蕨

根状茎短而直立或斜生,被阔披针形之深棕色或栗褐色鳞片;叶簇生,柄长达 14 cm,深稻秆色,略具鳞片,叶体长达 20 cm,宽 8 cm,椭圆披针形,渐尖头,基部截形,不甚狭缩,亚二回羽状分裂;一回小叶 6—10 对,开展,具短柄,椭圆披针形,渐尖头,长达 4 cm,宽 2 cm,基部呈截形,略为心脏形,羽状深裂;二回小叶达十对,无柄或稍合生,开展,卵状椭圆形,圆顶,基部截形,长约 1 cm,宽 6 mm,羽状深裂,裂片 3—4 对,圆形,具开展之锐锯齿,纸质,光滑,唯中轴及中肋略具细长鳞片,叶脉明显,羽状分叉;子囊群小,圆形,生于二回小叶,一列,大盖,圆肾形,膜质,边缘具刺状之锯齿。

分布:四川,湖北,西藏,台湾高山;喜马拉雅山区产之。

图注:1.本种全形(自然大);2.小叶,表示叶脉及子囊群之位置(放大 5 倍);3.子囊群盖(放大 20 倍);4.叶柄基部之鳞片(放大 16 倍)。

DRYOPTERIS SERRATO-DENTATA（Beddome）Hayata

DRYOPTERIS SERRATO-DENTATA (Beddome) Hayata, Ic. Pl. Form. **4**: 179 f. 116 (1914); C. Chr. Ind. Fil. Suppl. Ⅲ. 98 (1934).

Lastrea filix-mas var. *serrato-dentata* Bedd. Handb. Ferns Brit. Ind. Suppl. 55 (1892).

Nephrodium serrato-dentatum Hope. Journ. Bomb. Nat. Hist. Soc. **12**: 622 t. 10 (1899).

Nephrodium filix-mas var. *odontoloma* Baker (non *Lastrda odontoloma* Moore, 1858), Syn. Fil. ed. 2, 498 (1874), pro parte; Bedd. Ferns Brit. Ind. Suppl. t. 373 (1876).

Nephrodium odontoloma Clarke (non *Lastrea*, Moore, 1858), Trans. Linn. Soc. Ⅱ. Bot. **1**: 521 (1880).

Lastrea odontoloma Bedd. (non Moore, 1858) Handb. Ferns Brit. Ind. 248 f. 128 (1883).

Woodsia veitchii Christ, Bull. Acad. Géogr. Bot. (1906) 121; C. Chr. Ind. Fil. Suppl. Ⅰ. 74 (1912).

Rhizome short, erect or oblique, densely scaly; *scales* broadly lanceolate, finely acuminate, denticulate, atro-brown or castaneous; *fronds* caespitose, stipe to 14 cm long, soft; dark-straminous below with lax lanceolate smaller scales, lamina to 20 cm long, 8 cm broad at base, oblong-lanceolate, acuminate, base hardly narrowed, truncate, subbipinnate; *pinnae* 6-10-jugate, patent, short-petiolate, oblong-lanceolate, acuminate, to 4 cm long. 2 cm broad at truncate-cordate base, pinnate; *pinnules* to 10-jugate, sessile or more or less adnate, patent, ovate-oblong; rounded, base truncate, to 1 cm long, 6 mm broad, pinnatifid half-way down with 3-4-jugate, rounded, sharply dentate teeth; *texture* thin, soft, becoming hyaline towards margin, rachis and costa of pinnae fibrillose-scaly, otherwise glabrous; *veins* very distinct, pinnate in ultimate segments, one to each tooth, but not reaching the tip; *sori* rather small, rounded, one-rowed and nearer to costa of pinnule than margin, borne on the lower middle of anterior basal veinlet of each ultimate segment, *indusium* fairly large, rounded, deeply notched, gray, membranaceous, with fimbriate margin.

Sichuan: Without locality, *E. H. Wilson 5400* (type of *Woodsia Veitchii*). Yunnan: Tibet border, *Capt. Kingdom Ward 730* (1913). Hubei: Ichang, *A Henry* (1889). Taiwan.

Himalayas, Sikkim and Bothan; very common in Sikkim.

A distinct rather little fern of the genus, characterized by, above all, spinulose-dentate serrature and gray membranaceous indusium with rather long-fimbriate margin.

Plate 178. Fig. 1. Habit sketch (natural size). 2. pinnae, showing venation and position of sori (× 5). 3. Indusium (× 20). 4. Scale from base of stipe (× 16).

DRYOPTERIS SERRATO–DENTATA (Beddome) Hayata
高山鳞毛蕨

图版 179　史氏鳞毛蕨

根状茎颇粗，短而直立，被大披针形之细长黑色稠密鳞片；叶簇生，柄长 25—35 cm，稻秆色，基部被披针形之黑色密鳞片，向上部细长鳞片疏生，叶体椭圆形，长 25—35 cm，宽 15—20 cm，渐尖头，基部等宽，一回羽状分裂，顶部三角形，渐尖头，羽状深裂；小叶 6—10 对，披针形，渐尖头，长 10—15 cm，宽 1.5—2 cm，或稍宽，几无柄，基部圆截形，开展，互生，边缘具整齐之钝锯齿，纸质，上面光滑无毛，中轴及肋下面略具细长鳞片，叶脉明显，侧脉曲折，羽状分裂，小脉 3—4 对，急斜出，除基部一对外余均达叶边；子囊群圆形，为不规则之 2—3 列，生于小脉上，无盖。

分布：云南，贵州，广东，广西，台湾；越南及印度西北部均产之。

此种在以上各地极为普通，其形体极类 D. hirtipes，唯子囊群无盖，小叶数较少，距离较远，具钝锯齿，叶质较薄，中轴略具细长鳞片，故易分别；且 D. hirtipes 在中国，越南及喜马拉雅山区均未见之。

图注：1. 本种全形（自然大）；2. 小叶之一部，表示叶脉及子囊群之位置（放大 5 倍）；3. 叶柄基部之鳞片（放大 16 倍）；4. 中轴上部之鳞片（放大 16 倍）；5. 根状茎之横切面，表示维管束之布置（放大 4 倍）。

DRYOPTERIS SCOTTII (Bedd.) Ching

DRYOPTERIS SCOTTII (Bedd.) Ching, Bull. Dept. Biol. Coll. Sci. Sun Yatsen Univ. No. **6**: 3 (1933); C. Chr. Ind. Fil. Suppl. Ⅲ. 97 (1934).

Polypodium Scottii Bedd. Ferns Brit. Ind. t. 345 (1870).

Phegopteris Scottii Bedd. Ferns Brit. Ind. Suppl. 19 (1876).

Phegopteris grossa Christ, Bull. Herb. Boiss. **7**: 13 (1899).

Dryopteris grossa C. Chr. Ind. Fil. 269 (1905).

Dryopteris hirtipes C. Chr. Ind. Fil. 270 (1905), pro parte; Wu (non O. Ktze.) Polyp. Yaoshan in Bull. Dept. Biol. Coll. Sci. Sun Yatsen Univ. No. **3**: 26 pl. 4 (1932).

Dryopteris subdecipiens Hayata, Ic. Pl. Form. **3**: 181 f. 119 (1914); C. Chr. Ind. Fil. Suppl. Ⅱ. 17 (1913-16).

Rhizome short, thick, erect, densely scaly; *scales* black, large, lanceolate, hair-pointed, entire; *fronds* caespitose, stipe 25-35 cm long, straminous, basal part densely clothed in black lanceolate scales, sparingly fibrillose-scaly upwards, lamina oblong, 25-35 cm long, 15-20 cm broad, acuminate, base not narrowed, simple pinnate under the large, deltoid acuminate adnate apical pinna with lower part pinnatifid; *pinnae* 6-10-jugate, lanceolate, acuminate, 10-15 cm long, 1.5-2 cm broad or broader, subsessile, base rotundo-truncate, patent, alternate, margin regularly serrated with crenate-cuspidate teeth; *texture* herbaceous, glabrous above, sparingly fibrillose-scaly on rachis as well as costa beneath; *veins* distinct, lateral main vein flexuose, veinlets 3-4-jugate, ascending-oblique, all reaching margin except the basal pair, which stop somewhere midway; *sori* rounded, 2-3 irregularly seriate, dorsal on veinlets, exindusiate.

Yunnan: Mengtze, *A. Henry 10266*, *11558* (type of *Phegopteris grossa* Christ); *Hancock*, Oct. 1893; Souan-tsai-owen, *Maire*, alt. 600 m. Guizhou: Pinfa, *Cavalerie 2874*; *Esquirol 918*; Kenngfeng, *Y. Tsiang 4249*; Sihfeng, *Y. Tsiang 8727*; Chenfeng, *Y. Tsiang 4249*; Vanchin Shan, *Y. Tsiang 7874, 7864, 7682, 7768*; Tuhshan, *Y. Tsiang 7022*; Siaotchangonglan, *Cavalerie 4221*. Guangdong: Lohfau Shan, *N. K. Chun 42435*; North River, *C. L. Tso 20692*; Sunyi, *Y. K. Wang 31217*; Yao Shan, *S. P. Ko 51967*, *N. K. Chun 42823, 42435*. Guangxi: Ping-nam, Yao-shan, *S. S. Sin 442B*. Taiwan: *Faurie 401*.

Vietnam: Chapa, *A. Petélot 3309* (1929); *Colani 2825, 3309*; Lang-Bian, *Chevalier 30886*.

N. W. India: Kashima, *C. B. Clarke 2882*, Nov. 7, 1885.

The species, now found to be common in the localites cited, resembles *D. hirtipes* (Bl.) O. Ktze. in general habit, to which it has generally been referred, from which, however, it differs in exindusiate sori, fewer and more distant lateral pinnae with crenato-cuspidate serrature, thinner texture and sparingly fibrillose-scaly rachis and upper part of stipe. *D. hirtipes* (Bl.) has never been found in China, Himalayas and Vietnam and its report therefrom by authors in the past has chiefly been a mistake for the present fern.

Plate. 179. Fig. 1. Habit sketch (natural size). 2. Portion of pinna, showing venation and position of sori (×5). 3. Scale from base of stipe (×16). 4. The same from upper part of stipe (×16). 5. Cross section of rhizome, showing arrangment of steles (×4).

DRYOPTERIS SCOTTII (Bedd.) Ching
史氏鳞毛蕨

根状茎粗肥，木质，直立或斜生；叶簇生，柄长45—60 cm，基部厚达1 cm，被红棕色亚二形之披针形薄质鳞片，长逾1 cm，叶体长椭圆形，长80—100 cm，宽达30 cm，一回羽状分裂，顶部为短三角形，羽状分裂；小叶约20对或较多，长20 cm，宽2.5 cm，位于基部者等长，向顶部者长约8 cm，宽达1.2 cm，阔披针形，渐尖头，基部为斜截形，几无柄，开展，互生，彼此相距3—5 cm，边缘具钝锯齿，纸质，绿色，中轴及下面被细长之密鳞片，上面光滑，叶脉分离，明显，曲折，羽状分裂，5对，斜出，其基部一对仅达于小叶之中部，子囊群较小，圆形，为不规则之2—3列，无盖，沿中肋两侧分布。

分布：广西罗城县三防镇西二十里之九万山及广东信宜县产之。

本新种为本属特殊之种，其异于前种者，为其形体特大，小叶多至20余对，全体密被细长薄鳞片是也。

图注：1.本种全形（自然大）；2.小叶之一部，表示叶脉及子囊群（放大3倍）；3.叶柄上之鳞片（放大10倍）；4.小叶中肋下面之鳞片（放大10倍）。

DRYOPTERIS LIANKWANGENSIS Ching

DRYOPTERIS LIANKWANGENSIS Ching, sp. nov.

Species *D. Scottii* (Bedd.) Ching proxime affinis, differt multo majore, pinnis lateralis numerosis, longioribus, paleis stipitis rachisque rufobrunneis creberris, majoribus, persistentibusque.

Rhizome thick, woody, erect or procumbent; *fronds* caespitose, stipe 45-60 cm long, over 1 cm thick near base, broadly grooved above, densely clothed throughout in sub-dimorphous rufo-brown, lanceolate, hair-pointed, membranaceous, spreading scales to 1.2 cm long, lamina oblong-elongate, 80-100 cm long, 30 cm broad, simple pinnate under the rather short, caudate, deltoid, pinnatifid and acuminate apical part; *pinnae* 20-jugate or more, basal ones not shortened, 20 cm long, 2.5 cm broad, the uppermost ones to 8 cm long, 1.2 cm broad, broadly linear, acuminate, base truncate, slightly oblique, subsessile, patent, alternate, 3-4-5 cm apart, incisely crenato-serrate with large roundish teeth; *texture* herbaceous, green, rachis densely scaly with similar but smaller scales as those on stipe, glabrous above, fibrillosely scaly beneath, and especially on the lower part of costa; *venation* free, distinct, lateral main veins flexuose, pinnate with 5 pairs of obliquely ascending veinlets, of which the basal pair stop short midway, the rest extending towards margin; *sori* rather small, rounded, irregularly 2-3-seriate, exindusiate, leaving rather a broad sterile margin.

Guangxi: San Fan, Chu Fen Shan, north of Luchen Hsien, *R. C. Ching 5832* (type), in deep wooded ravine, very common. Guangdong: Suni, *Y. K. Wang 31042*, July 22, 1931. in moist ravine.

This distinct endemic species of the group of *D. hirtipes* (Bl.) resembles none but perhaps *D. Scottii* (Bedd.), with which it has in common exindusiate sori, from which, however, it differs in enormously larger size with numerous close, longer pinnae and very dense reddish-brown linear-lanceolate large thin scales not only on stipe but also on rachis.

Plate 180. Fig. 1. Habit sketch (natural size). 2. Portion of pinna, showing venation and sori ($\times$ 3). 3. Scales from stipe ($\times$ 10). 4. The same from costa beneath ($\times$ 10).

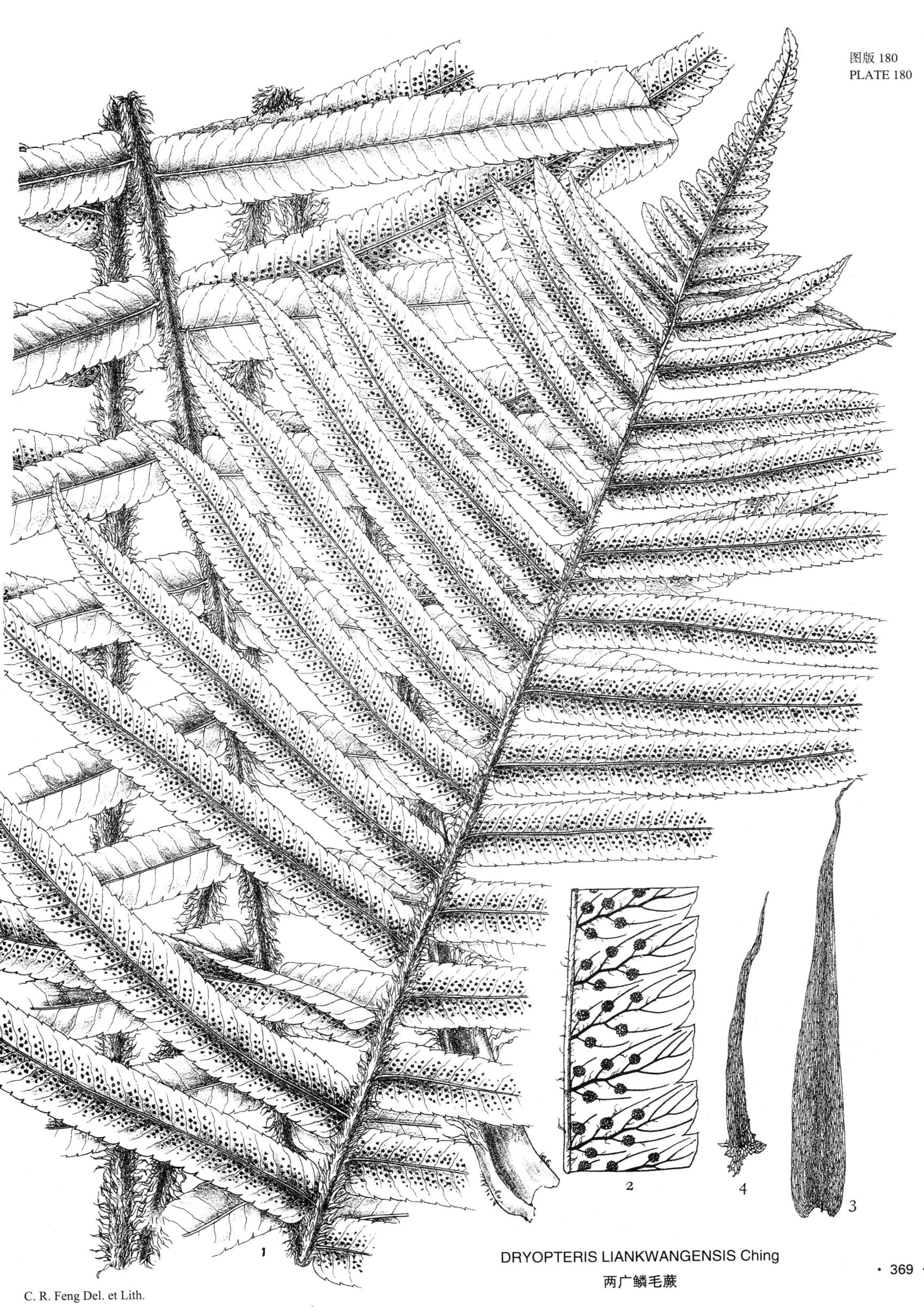

DRYOPTERIS LIANKWANGENSIS Ching
两广鳞毛蕨

图版 181　张氏鳞毛蕨

根状茎木质,粗肥,直立或斜生,具红黄色之阔披针形膜质大鳞片,长达 1 cm;叶簇生,柄长 20—35 cm,粗强,棕稻秆色,红黄色之大鳞片密生,质粗糙,叶体椭圆卵形,长 35—60 cm,宽 17—25 cm 或过之,渐尖头,基部圆形,二回羽状分裂;小叶 8—14 对,长 10—20 cm,宽 3—4 cm,披针形,渐尖头,亚斜出,基部一对对生,上部各对互生,彼此相距 5—6 cm,具短柄,基部截形,略呈心脏形,羽状分裂,二回小叶 14—18 对,长 1.5—3.5 cm,宽 5—10 mm,椭圆卵形,亚对生,无柄,钝头,基部两侧耳形膨大。几等长,亚全缘或具钝锯齿或为羽状深裂,亚革质,上面光滑,下面略具小鳞片,叶脉不甚明显,羽状分叉;子囊群圆形,一列,常贴近叶边生,盖肾圆形,其一深缺刻,革质,最后脱落。

分布：江苏,浙江,福建,广东,香港,江西,安徽,湖北,台湾等省均甚普通；日本亦产之。

此为我国温暖各省习见之蕨种,昔人常误为 D. erythrosora,实则以其叶柄及其他各部所具红黄色之大鳞片与二回小叶之基部两侧呈耳形凸起,颇易识别。

图注：1. 本种全形(自然大)；2. 小叶,表示叶脉及子囊群与锯齿(放大 8 倍)；3. 子囊群盖(放大 40 倍)；4. 叶柄上之鳞片(放大 16 倍)；5. 中肋上之鳞片(放大 16 倍)。

DRYOPTERIS CHAMPIONI (Benth.) C. Christensen

DRYOPTERIS CHAMPIONI (Benth.) C. Christensen apud Ching, *Sinensia* **3**: 327 (1933), C. Chr. Ind. Fil. Suppl. Ⅲ. 83 (1935).
　Aspidium Championi Benth. Fl. Hongk. 456 (1861).
　Polypodium rheosorum Baker (non 1884), Ann. Bot. **5**: 457 (1891).
　Nephrodium rheosorum Hand-Mzt. Symb. Sin. **6**: 24 (1929).
　Dryopteris lepidorachis C. Chr. Ind. Fil. 274 (1905).
　Aspidium erythrosorum var. *amoyense* Christ in Warburg, Monsunia **1**: 80 (1900).
　Dryopteris erythrosora var. *Cavaleriei* Rosenst. in Fedde, Repert. Sp. Nov. **13**: 131 (1914).
　Dryopteris mingetsuensis Hayata, Ic. pl. Form. **5**: 281 f. 109 (1915).
　Dryopteris erythrosora auctt. plur. quoad plant. chin.

Rhizome thick, woody, oblique or erect, densely scaly; *scales* bright ferruginous-brown, to 1 cm long, broadly lanceolate, long-acuminate, membranceous, fimbriate, mixed with smaller linear ones; *fronds* caespitose, stipe 20-35 cm long, stout, rufo-stramineous, densely clothed throughout in similar but somewhat smaller spreading curled scales, which extend upward over entire rachis and base of pinna beneath, lamina oblong-ovate, 35-60 cm long, 17-25 cm broad, acuminate, base not narrowed, bipinnate; *pinnae* 8-14-jugate, 10-20 cm long, 3-4 cm broad, linear-lanceolate, acuminate, subpatent, basal ones opposite, upper ones subopposite, 5-6 cm apart, short-petiolate, base truncate, slightly cordate, fully pinnate under deeply pinnatifid apical part; *pinnules* 14-18-jugate, 1.5-3.5 cm long, 5-10 mm broad, deltoid-oblong, subopposite, sessile, obtuse with auriculately broadened base on both sides, basal ones in the basal pinnae rarely any longer than neighbouring ones, subentire, or crenate-serrate or, in large forms, lobato-pinnatifid half-way down into 4-6 pairs of oblong truncate lobes under serrate acuminate apex; *texture* coriaceous, light green, glabrous above, rather copiously fibrillose-scaly on costa of pinnae beneath, stipe and rachis moderately muricate by the persistent base of fallen scales; *venation* obscure, veins in pinnules or lobes pinnate; *sori* rounded, medium-sized, brown, one-rowed midway between costa and margin or often much nearer to the margin, dorsal on the anterior basal veinlet of each group; *indusium* rotundo-reniform, notched, brown, subcoriaceous and fallen off at last.

Jiangsu: Shanghai, Fen Wang Shan, *Forbes 531*; *C. G. Matthew*, June 1, 1904; Tai Ho, *Schindler 254*; Chang Cho, Yü Shan, *J. R. Chu 8024*; Ishing, Lung Chi Shan, *R. C. Ching & C. L. Tso 497*; ibid., *Y. L. Kong 2389*; Nanking, Tsehsia Shan, *Nos. 32, 37 69 ex Herb. Metrop. Mus. Nat. Hist.* Zhejiang: Ningpo, *Forbes 531*; Staunnton; *C. G. Matthew 142*; *Everard* (1874); Sia Kan, Fen Chiao. *R. C. Ching 3696*; Hangchow, Ling-yin Tze; *R. C. Ching 3801*; *T. F. Yü 9630*; Yan-tan Shan, *C. Y. Chiao 14757*; Tien-mo Shan, *K. K. Tsoong 455*. Fujian: Amoy, *Gerlach 5509* ex Herb. Warburg; Sam Sa Inlet, *Matthew*, Oct. 5, 1907; *Grijis 10150* in Herb. Hance; Chuan Chow, *H. H. Chung 3088*; Yengping, *H. H. Chung 3522*; *T. S. Dunn 3874*; Foochow, Kushan, *T. S. Ging 5865, 5335*. Jiangxi: Kiukiang, Lushan, *Forbes 1078* (1874); *Schindler 377* (1908); *Miss Reid 6*; *Maires*; *Staunton*, *Dr. Shearer*; *C. E. DeVol 124, 119, 120*; *R. C. Ching*; Lienchu Shan, Singping, Lingchuan, *Y. Tsiang 9944*. Hubei: Ichang, Nanto, *A. Henry 257* (1881). Guizhou: Pingchow, *Esquirol 3607*; *Cavalerie 3771, 7307*; Pinfa, *Cavalerie 1060*; Tsingay, *Cavalerie 1238, 452*; Kweiyang, *Bodinier 1184*; Gan-pin, *Martin 2029*; Vanchin Shan, *Y. Tsiang 7807*; Tuhshan, *Y. Tsiang 6619*; Yuyun, *Y. Tsiang 5596*. Sichuan: Mt. Oemi, *W. P. Fang 3080*. Anhui: Yüting, Lantien Hsien, *K. K. Tsoong 4520, 4820*; Chu Hwa Shan, *R. C. Ching 8411*; Hwang Shan, *A. N. Steward 7148*, Guangdong: Canton, Honam Island, *E. D. Merrill 10085*; *Y. Tsiang 2029*; Lohfau Shan, *C. O. Levine 1495*, North River, *Tutcher 10767, 5112*; *C. L. Tso 20420*; Lokcong, *N. K. Chun 42362*; *Y. K. Wang 31649*; *Hance 8275*; *Staunton*; Macao, *Gaudichaud*. Hong Kong: *Champion* (type); *Urquahart* in Herb. Hk. Taiwan.

Also Japan.

This distinct fern, now found to be very common in the eastern and south-eastern parts of China, resembles in general habit *D. erythrosora* (Eaton) with which it has hitherto been considered as identical and from which, however, it can always be distinguished by, above all, the characteristically golden brown, spreading, broadly lanceolate, fimbriate, curled, dense large scales from base of stipe upwards throughout the entire length of rachis and the underside of costa, by somewhat oblong-deltoid falcate pinnules with auriculately broadened base and thicker texture.

Plate 181. Fig. 1. Habit sketch (natural size). 2. Sorus with indusium (× 40). 3. Scale from stipe (× 16). 4. Scale from rachilet (× 16).

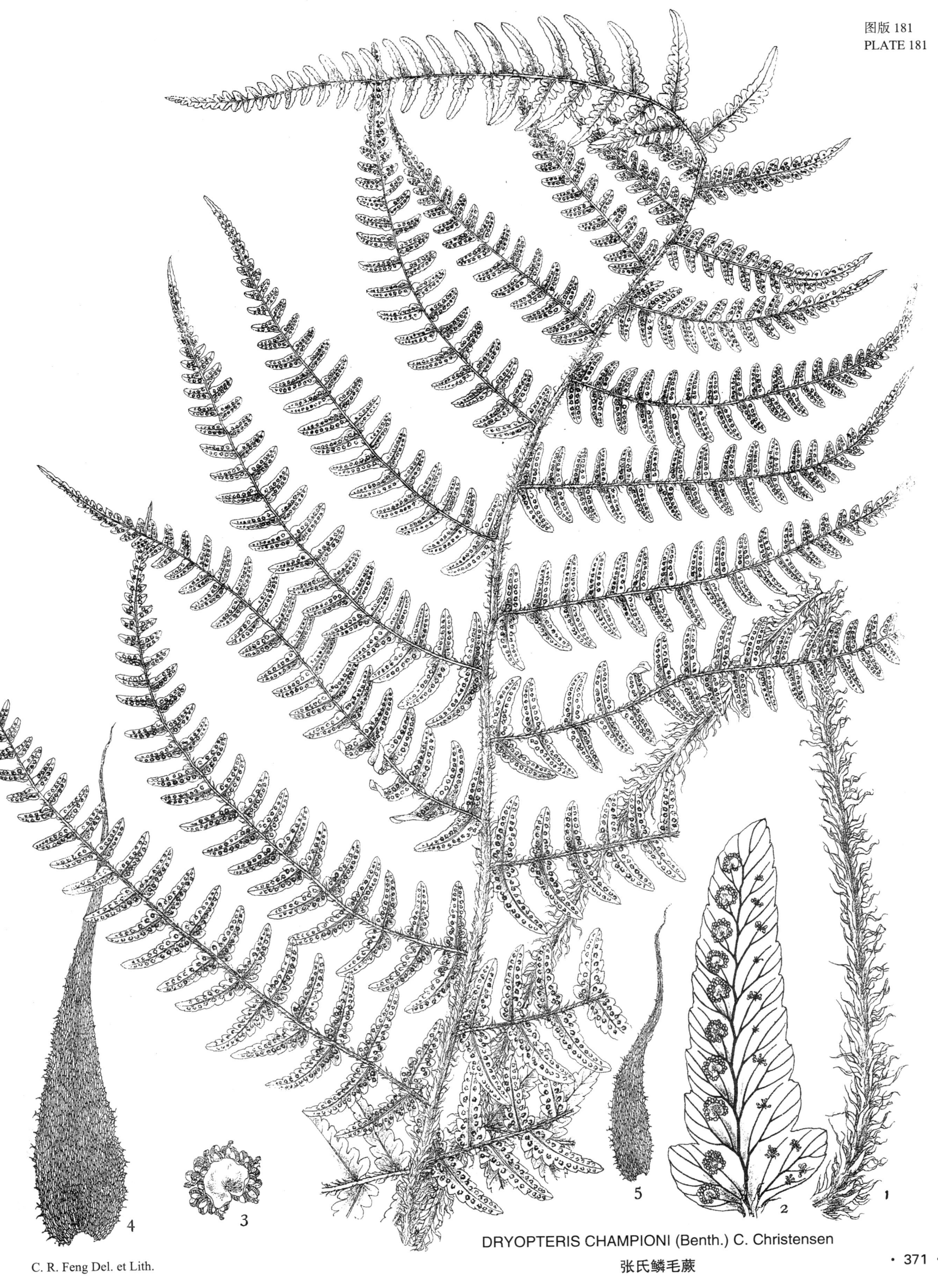

DRYOPTERIS CHAMPIONI (Benth.) C. Christensen
张氏鳞毛蕨

图版 182　滇耳蕨

　　根状茎短肥，直立，被线状披针形之黄褐色密鳞片；叶簇生，柄长 16—30 cm，稻秆色，或下部呈红棕色，光亮，被大卵形之亮栗棕色鳞片，叶体椭圆形，长 25—35 cm，宽 10—14 cm，基部等大，一回羽状分裂，顶部短渐尖头，三角形，羽状深裂；小叶 13—17 对，水平开展，彼此接近，长 6—8 cm，宽 1.5 cm，镰状披针形，长渐尖头，基部数对等长，下向，向顶部渐短，几无柄，基部上方呈尖锐三角形凸出，下方为楔形，边缘具疏短锐锯齿，革质，绿色，上面光滑无毛，中轴及叶下面被细长鳞片，叶脉不显明，侧脉曲折，羽状分裂，小脉 4—6 对，其基部上方一脉仅达小叶之中部，余达于叶边；子囊群小，圆形，不规则二列，生于小脉上，盖小，圆形，早落。

　　分布：此为云南特产。

　　本新种为本属特殊之种，其形体极似贯众属之 *Cyrtomium Balansae*，然其叶脉不为网状，故易区别。

　　图注：1. 本种全形（自然大）；2. 小叶，表示叶脉及子囊群（放大 2 倍）；3—4. 叶柄上之鳞片（放大 10 倍）；5. 小叶下面之鳞片（放大 10 倍）；6. 子囊群盖（放大 20 倍）。

POLYSTICHUM CHINGAE Ching

POLYSTICHUM CHINGAE Ching, sp. nov.

　　Species *P. xiphophylli* Baker proxime affinis, differt paleis stipitis rachisque castaneo-brunneis, ovatis, pinnis majoribus, falcatis, basi anteriore auricula deltoidea magna acutissima instructa, soris utraque costae latere biseriatis.

　　Rhizome short, erect, densely scaly; *scales* linear-lanceolate, ferruginous brown; *fronds* caespitose, stipe 16-30 cm long, straminous, or rufo-brown on the lower part, nitente, sparcely clothed in large ovate, castaneous-brown shining scales, lamina oblong, 25-35 cm long, 10-14 cm broad, base not attenuate, simple pinnate under rather short acuminate, coadunate apex; *pinnae* 13-17-jugate, horizontally patent, close, 6-8 cm long, 1.5 cm broad at middle, falcate-lanceolate, long-acuminate, the basal ones as long as next above, more or less deflexed, the uppermost ones shortened, subsessile, anterior side provided with a large deltoid, sharply pointed auricle, the posterior side cuneate, margin remotely serrate with low, sharply pointed and appressed teeth; *texture* coriaceous, color green, glabrous above, under side and rachis copiously clothed in light brown, lanceolate, fimbriate, appressed scales; *venation* not distinct on both sides, lateral veins flexuose, pinnate, 4-6 in each group, the anterior basal one stops midway, the rest extend to margin; *sori* small, rounded, irregularly 2-rowed, dorsal on veinlets, *indusium* small, rounded, fugaceous.

　　Yunnan: Without locality, *H. T. Tsai 51800*, 51643 (type).

　　This remarkably distinct species, resembling *Cyrtomium Balansae* Christ in general habit but with free venation, finds no close relative in the genus but *P. xiphophyllum* Baker, which differs in much smaller size, coal-black subulate scales on stipe and rachis, in smaller pinnae with at most bluntly auriculated anterior base and always uniseriate sori on each side of costa.

　　The species is named after my wife, in recognition of her untired assistance in ably typewriting my manuscripts for this Icones and many other papers so far published.

　　Plate 182. Fig. 1. Habit sketch (natural size). 2. Pinna, showing venation and sori ($\times$ 2). 3-4. Scales from stipe ($\times$ 16). 5. The same from costa beneath ($\times$ 16). 6. Indusium ($\times$ 20).

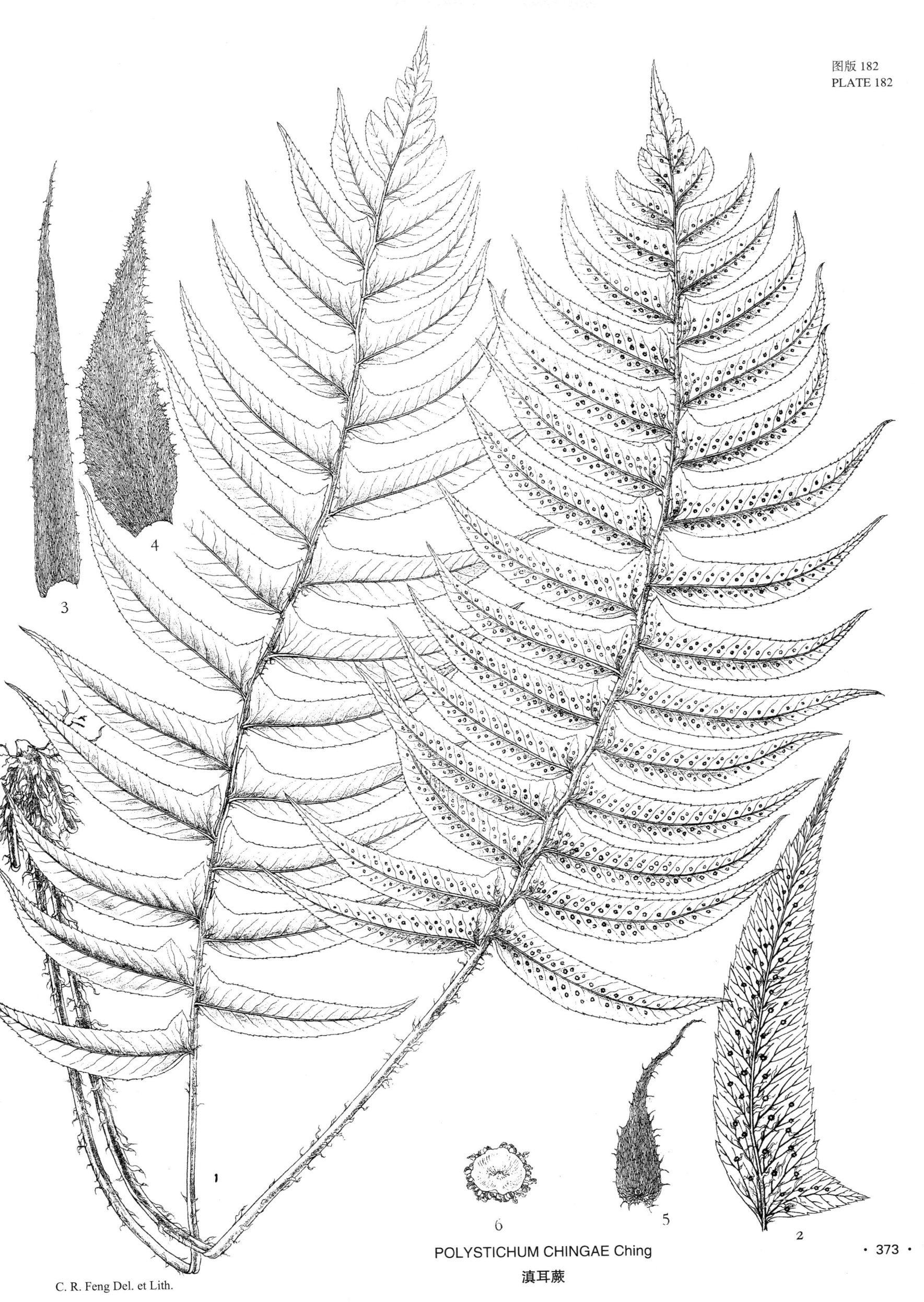

PLATE 182

POLYSTICHUM CHINGAE Ching 滇耳蕨

图版 183　滇贯众

　　根状茎短粗，斜生，遍被卵形光亮栗色大鳞片，而间以细长之小鳞片；叶簇生，柄长 15—25 cm，土褐色，下部被密鳞，上部较疏，叶体长椭圆形，长 20—40 cm，宽达 15 cm，奇数羽状分裂；小叶 4—6 对，长 10—15 cm，宽 2—3 cm，亚对生或互生，阔披针形，具短柄，向顶部尖长，基部为等楔形，或亚圆形，叶缘具小尖锯齿，顶部一小叶较大，具长柄，通常为三裂，纸质，侧脉隐约可见，屈折，小脉网状，网眼为多角形，斜出，中有二数外出之着生子囊群之单脉；子囊群圆形，散生，盖圆形，革质。

　　分布：云南特产。

　　本种形体极类刺叶贯众(C. Caryotideum)，唯小叶为披针形，基部上方不为耳形凸出，故易识别。

　　图注：1. 本种全形（自然大）；2. 子囊群及盖（放大 16 倍）；3. 孢子（放大 150 倍）；4. 叶柄基部之鳞片（放大 8 倍）。

CYRTOMIUM AEQUIBASIS (C. Chr.) Ching

CYRTOMIUM AEQUIBASIS (C. Chr.) Ching, Bull. Chin. Bot. Soc. **2**: 99 (1936).

Cyrtomium caryotideum var. *aequibasis* C. Chr. Amer. Fern Jour. **20**: 51 (1930).

Aspidium falcatum Christ (non Sw.), Bull. Herb. Boiss. **6**: 969 (1898).

Rhizome short, thick, oblique, densely scaly; *scales* large, ovate-oblong, acuminate, intermixed with linear-lanceolate ones, black or atro-brown, sparingly fimbriate, extending over half way up the stipe; *fronds* caespitose, stipe 15-25 cm long, sordid brown, densely scaly on the lower part, sparcely so upwards, lamina oblong, 20-40 cm long, to 15 cm broad, impari-pinnate; *pinnae* 4-6-jugate, 10-15 cm long, 2-3 cm broad, subopposite or alternate, lanceolate, shortly petiolate, long-attenuate towards apex, base equal, cuneate or subrounded, margin regularly minutely cuspidate-serrate above base, the terminal pinna much the largest, long-petiolate, hastately tri-lobed, base cuneate; *texture* thin chartaceous, glabrous, sparcely fibrillose-scaly on rachis, petiole and costa beneath, color brownish when dried; lateral *veins* subconspicuous, flexuose, areolae generally with 2 included soriferous veinlets; *sori* rounded, scattered, *indusium* rounded, brown, coriaceous.

Yunnan: Mengtze, *Hancock 8* (type), *25*, *130*, in a deep dark glen, very rare; Szemeo, *Henry 9123*, *9123A*, *9123B*; Without locality, *Handel-Mazzetti 6852*; *H. T. Tsai 56333*; Longky, *E. E. Maire* (pro parte).

An endemic species, closely related to *C. caryotideum* (Wall.) Presl, differing chiefly in lanceolate lateral pinnae with equal and exauriculate base, which varies from cuneate to subrounded.

Plate 183. Fig. 1. Habit sketch (natural size). 2. Sorus with indusium (× 16). 3. Spores (× 150). 4. Scales from base of stipe.

CYRTOMIUM AEQUIBASIS (C. Chr.) Ching
滇贯众

根状茎短粗,直立,遍被深栗褐色之卵形大鳞片;叶簇生,柄长 16—20 cm,下部被密鳞,上部较稀,叶体椭圆形,长 20—35 cm,宽 13—16 cm,奇数羽状分裂,小叶 2—5 对,亚对生或互生,卵状椭圆形,略呈镰形,基部一对最大,长达10 cm,宽 5—7 cm,渐尖头,基部圆形或圆截形,两边几等大,或上边稍呈圆耳形,具短柄,上部数对小叶渐小,顶部一小叶甚大,三裂,边缘为亚全缘或小锯齿疏生,直达叶顶,厚纸质,中肋及叶柄具细长鳞片,侧脉可见,小脉网状,网眼具 2—3 数外出之生子囊群之单脉;子囊群圆形,散生,盖大,圆形,革质,全缘。

分布:云南,四川,贵州,湖北山林中均产之,在日本亦甚普通。

本种形体极类刺叶贯众(*C. Caryotideum*),唯较大,小叶基部不具三角形之尖耳形凸起,边缘不具刺状之密齿,故易分别。

图注:1—2. 本种全形(自然大);3. 小叶之一部,表示叶脉及子囊群(放大 1.5 倍);4. 子囊群及盖(放大 16 倍);5. 叶柄基部鳞片(放大 8 倍);6. 叶柄基部横切面(放大 8 倍)。

CYRTOMIUM MUTICUM（Christ）Ching

CYRTOMIUM MUTICUM (Christ) Ching in C. Chr. Ind. Fil. Suppl. III. 66 (1933).
    Cyrtomium falcatum var. *muticum* Christ in Lecomte, Not. Syst. **1**: 37 (1909).
    Polystichum falcatum var. *macropterum* Diels in Engl. Bot. Jahrb. **29**: 195 (1900); C. Chr. Acta Hort. Gotheb. **1**: 72 (1924).
    Cyrtomium falcatum var. *macropterum* Christ, Bull. Soc. Bot. France **52**: Mém. I. 32 (1905); Bull. Acad. Géogr. Bot (1906) 115.
    Aspidium falcatum var. *macrophyllum* Makino, Bot. Mag. Tokio **16**: 90 (1902).
    Polystichum falcatum var. *macrophyllum* Matsum. Ind. Pl. Jap. **1**: 342 (1904).
    Polystichum caryotideum var. *macropterum* Nakai, Bot. Mag. Tokie **29**: 115 (1925).
    Polystichum macrophyllum Tagawa, Acta Phytotax. et Geobot. **2**: 194 (1933).
    Cyrtomium macrophyllum Tagawa, ibid. **3**: 63 t. 3 f. 5-7 (1924).

Rhizome short, thick, erect, densely radicose and scaly; *scales* large, fusco-brown or nearly black, shining, ovate, acuminate, densely fimbriate, extending, when young, over the whole length of stipe; *fronds* caespitose, stipe 16-20 cm long, dark straminous, densely scaly near the base, lamina oblong, 20-35 cm long, 13-16 cm broad, pinnate with a large, hastate, cuneate terminal pinna; pinnae 2-5-jugate, subopposite or alternate upwards, falcate, basal ones broadly ovate, the upper ones oblong-ovate, 10 cm long, 5-7 cm broad, acuminate, base rounded or rotundo-cuneate, nearly equal or much broadened above, shortly petiolate, the uppermost ones under the 3-lobed end-pinna only slightly smaller, margin obscurely or minutely cuspidate-serrate above the middle; *texture* chartaceous, green, glabrous above, moderately fibrillose-scaly on rachis, petiole and costa beneath; lateral *veins* distinct, oblique, flexuose, intervening veinlets anastomosing in 1-rowed angular areoale each with 2-3 (only 1 in the costal areolae) excurrent soriferous included veinlets after the goniophlebioid type; *sori* rounded, scattered, dorsal or subapical on included veinlets, *indusium* large, gray, coriaceous, subentire.

Yunnan: Sanshan près Tchenhiong, *Ducloux 5098* (type); *E. E. Maire*; Shweli-Salwin divide, North of Hotou, *Forrest 26341*; Guon-Kay, *Delavay 1724*, *Sept. 1885*; Taton près Tapintze, *Delavay 2311*, Lanping Hsien, *H. T. Tsai 54021*; Weise Hsien, *H. T. Tsai 57825*; without locality, *H. T. Tsai 51234, 52758, 52773, 52288, 51044, 50872*; Guizhou: Sihfeng; *Y. Tsiang 8723*; Tuyun; *Y. Tsiang 5816*; Kiangkow, foot of Vanching Shan, *Y. Tsiang 7649*; ibid., Huang Chia Wan, *Steward, Chiao & Cheo 444*; Loumongtouan, *Cavalerie 1565*; *Perny* (1858). Sichuan: *Farges 656A, 4937*; Haitang, *Harry Smith 1983*; Nanchuan, *Rosthorn 45* (var. *macropterum* Diels); Mt. Omei, *E. Faber 1058*; *W. P. Fang 2488*; *Wilson 5339*; Lepo Hsien, *T. T. Tu 3559*. Hubei: Patung, *Henry 3687*; *Wilson 195, 2628, 2634*.

Hiamalayas: Simla District of Bashahr State, *R. N. Parker 3018*.

Japan, common.

A large fern of the habit of *C. caryotideum* (Wall.), from which it differs in generally larger, oblong-ovate pinnae with rounded or rotundo-cuneate and almost equal-sided base without deltoid auricle, subentire margin from base upward and a few remote small teeth towards apex, and in indusium with subentire margin. A very common fern in West China from where I have seen numerous specimens.

Plate 184. Fig. 1-2. Habit sketch (natural size). 3. Portion of pinna, showing venation, and sori ($\times$ 1.5). 4. Sorus with nearly entire indusium ($\times$ 16). 5. Scale from base of stipe ($\times$ 8). 6. Cross section of basal part of stipe ($\times$ 8).

CYRTOMIUM MUTICUM (Christ) Ching
大叶贯众

根状茎粗厚，木质，斜出或卧生，被线状深棕色厚质鳞片，长达 1 cm；叶簇生，二形，不生子囊群叶之柄为红棕色，光亮，长 10—25 cm，基部具鳞片，叶体卵形，长 16—35 cm，宽几如之，基部由柄下延，奇数羽状深裂或分裂（间为单叶）；小叶 1—2 对，对生，阔披针形，基部下延或否，而具短柄，长达 20 cm，宽 5 cm，渐尖头，向基部渐狭，全缘，顶部一小叶较长，亚革质，干则变为棕绿色，两面光滑，侧脉明显，小脉网状，网眼内具分叉或简单小脉；生子囊群叶之柄较长，叶体羽状分裂（或为单叶），小叶收缩，长仅 10 cm，宽 2 cm，下面子囊满布，仅中肋可见。

分布：香港，广东，海南，台湾；越南亦产之。

本属在亚洲大陆仅此一种，其形体颇似叉蕨，唯叶为二形，子囊不成群，唯散布于叶之下面，且无盖。

图注：1. 本种全形（自然大）；2. 不生子囊群叶之一部，表示叶脉（放大 4 倍）；3. 生子囊群叶之一部，表示叶脉及子囊着生情形（放大 4 倍）；4. 叶柄基部之鳞片（放大 8 倍）；5. 根状茎之横切面，表示维管束之布置（放大 10 倍）。

HEMIGRAMMA DECURRENS (Hooker) Copeland

HEMIGRAMMA DECURRENS (Hooker) Copeland, Phil. Journ. Sci. **37**: 404 (1928); C. Chr. Ind. Fil. Suppl. III. 109 (1934), cum. syn.

Gymnopteris decurrens Hk. Journ. Bot. **9**: 359 (1857); Fil. Exot. t. 94 (1859); Benth. Fl. Hongk. 443 (1861) (non Hk. Gard. Ferns t. 6, 1862).

Acrostichum decurrens Hk. Sp. Fil. **5**: 274 (1864); Syn. Fil. ed. 2, 118, (1874); Dunn & Tutcher, Fl. Kwangt. & Hongk. 355 (1912).

Leptochilus harlandii C. Chr. Ind. Fil. 385 (1905), cum. syn.

Polypodium dimorphum Baker (non Link, 1833), Ann. Bot. **5**: 477 (1891).

Polypodium hainanenae C. Chr. Ind. Fil. 531 (1905).

Gymnopteris bonii Christ, Bull. Herb. Boiss. II. **4**: 610 (1904); Copel. 1. c. 405.

Leptochilus kanashiroi Hayata, Ic. Pl. Form. **5**: 298 f. 120 (1915).

Tectaria dictyosora Cop. Phil. Journ. Sci. **38**: 187 (1929).

Hemigramma distinctipetiolata Ching, Bull. Fan Mem. Inst. Biol. **1**: 156 (1930).

Rhizome thick, woody, oblique or short-creeping, densely scaly at extremity and base of stipe; *scales* linear-subulate, over 1 cm long, atro-brown or castaneous, shining, rather thick and firm; *fronds* caespitose; strongly dimorphous, *sterile* ones with rufo-brown or castaneous shining stipe, 10-25 cm long, broadly winded over two-thirds way towards scaly base, lamina ovate, 16-35 cm long, less broad; pinnatifid or pinnate at base (sometimes simple) with 1-2 or 3 pairs of opposite, broadly lanceolate, acuminate, entire pinnae to 20 cm long, 5 cm broad, narrowed towards decurrent or sessile base, the terminal pinna much the largest, narrowed towards both ends and connected with the lower lateral pair by broad decurrent wing on each side; *texture* subcoriaceous, firm, rich brown-green, glabrous on both sides, costa and rachis shining brown or light castaneous beneath; *venation* distinct, lateral main veins oblique, connected by finer transverse ones, intervening veinlets copiously anastomosing in several rows of large angular areolae with divaricate clavate veinlets; *fertile fronds* with stipe to 40 cm long, lamina conform but smaller with contracted decurrent pinnae to 10 cm long, 2 cm broad, areolae mostly without included veinlets; *sori* indefinite, appear at first in a medial band along veinlets between main veins, finally confluent over the entire under surface.

Hong Kong: *Harland* (type); *Wilford 316*; *G. G. Matthew*, March 25, 1907; *Hance 94*; *Forbes 581*. Guangdong: North River, *C. Ford*, May 26, 1888 (f. simplex); Lungtau Mt, *C. O. Levine 1949* (type of *Tectaris dictyosora* Cop.); Kochow, *Y. Tsiang 2752* (type of *H. distinctipetiolata* Ching); Swatow, Thaiyong, *Dalziel*; Teiloy, *Matthew*, Nov. 25, 1907; Ting Wu Shan, *S. P. Ko 50544*; *S. Y. Lau 20246*; Namhoi, Sai Chiu Shan, *S. P. Ko 51536*. Hainan: *H. Y. Liang 64687, 63457*; Rev. *A. G. Henry 86B* (type of *Polypodium dimorphum* Baker). Taiwan: Tamsui, *Hancock 47*; *Jutsugetsutan 370*, Oct. 2. 1929.

Vietnam: *Cadier 62* (type of *Gymnopteris Bonii* Christ).

The species represents type of the genus *Hemigramma* Copeland now comprising 4 or 5 species in the warm parts of Asia and Polynesia. As a genus, which is closely related to *Bolbitis* Schott, *Hemigramma* is characterized by thick short woody rhizome, tectarid type of scale and venation, dimorphous leaves with strongly contracted fertile ones and indefinite sori.

A variable fern as to the degree of pinnation. *Gymnopteris Bonii* Christ differs in frond having 2-3 pairs of lateral pinnae, of which the lower 1-2 pairs not decurrent along stipe and rachis, but sessile or short-petiolate in basal pairs. *H. distinctipetiolata* Ching, based upon Y. Tsiang's No. 2782 from southwestern part of Guangdong, proves to be the same, as represented by our plate.

Plate 185. Fig. 1. Habit sketch (natural size). 2. Portion of sterile frond, showing venation ($\times$ 4). 3. Portion of fertile frond, showing venation and position of sori ($\times$ 4). 4. Scales from base of stipe ($\times$ 8). 5. Cross section of rhizome, showing arrangement of steles ($\times$ 10).

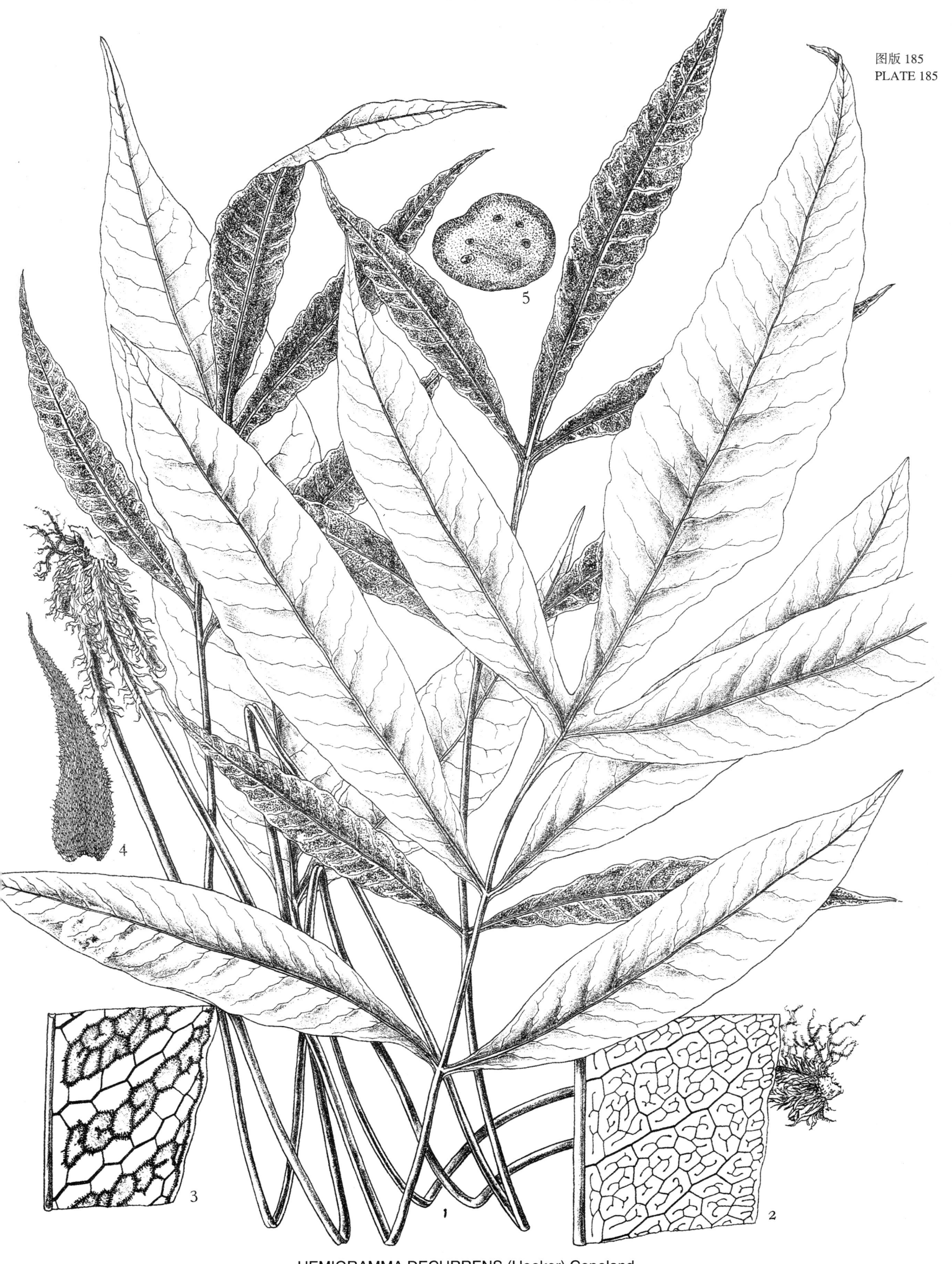

HEMIGRAMMA DECURRENS (Hooker) Copeland
拟叉蕨

根状茎木质,横行,略被线状披针形之膜质黄色鳞片;叶远生,柄长 20—35 cm,稻秆色,光滑,无毛,叶体卵状三角形,渐尖头,长 25—35 cm,宽亦如之,三回羽状分裂;小叶约 10 对,斜出,具柄,互生,基部一对最大,长三角形,长 12—20 cm,宽 6—10 cm;二回小叶约 10 对,基部下方一片最大,均具柄;三回小叶长 1—1.5 cm,宽达 1 cm,亚斜方卵形,无柄,钝头,基部楔形,深裂,裂片具钝锯齿,纸质,淡绿色,两面光滑无毛,叶脉羽状分叉,小脉每锯齿一数,不达于顶;子囊群形大,每三回小叶具 1—2 个,生于小脉之顶,盖大,为半圆形,膜质,全缘,宿存,仅基部着生于叶质。

分布:云南,台湾;印度北部,越南,泰国及菲律宾群岛均产之。

图注:1.本种全形(自然大);2.末回小叶,表示叶脉及子囊群之位置(放大 10 倍);3.根状茎上之鳞片(放大 27 倍)。

LEUCOSTEGIA IMMERSA (Wallich) Presl

LEUCOSTEGIA IMMERSA (Wallich) Presl, Tent. Pterid. 95 t. 4, f. 11 (1836); Hk. Gen. Fil. t. 52A (1840); J. Sm. Hist. Fil. 84 (1875); Bedd. Handb. Ferns Brit. Ind. 51 (1883); Cop. Phil. Journ. Sci. **34**: 240, 252 (1927); C. Chr. Contr. U. S. Nat. Herb. **26**: 293, 331 (1931); Ind. Fil. Suppl. Ⅲ. 120 (1934).

Davallia immersa Wallich, List no. 256 (1828, nom. nud.); Hk. Sp. Fil. **1**: 156 (1846); Fil. Exot. t. 79 (1858); Hk. et Bak. Syn. Fil. 91 (1865); Clarke, Trans. Linn. Soc. Ⅱ. Bot. **1**: 443 (1880); Christ, Farnkr. d. Erde 302 (1897); C. Chr. Ind. Fil. 211 (1905).

Acrophorus immersus Moore, Proc. Linn. Soc. **2**: 286 (1854); Ind. Fil. 2 (1857); Bedd. Ferns S. Ind. t. 11 (1863).

Humata immersa Mett. Fil. Hort. Lips. 102 (1856); Diels in Engl. u. Prantl; Nat. Pflanzenfam. **1**: 4, 209 (1899).

Rhizome thick, woody, wide-creeping, hypogaeous (subterranean); *scales* linear-lanceolate, thin, rusty brown, sparce or rather copious at growing tip or base of stipe; *frond* far apart, stipe 20-35 cm long, firm, erect, dark-straminous or pale colored, smooth, nitente, glabrous from base upwards, lamina deltoid-ovate, acuminate, 25-35 cm long, and nearly as broad at base, tripinnate, *pinnae* about 10-jugate, oblique, long-petiolate, alternate, the basal pair much the largest, elongate-deltoid, 12-20 cm long, 6-10 cm broad, bipinnate; *pinnules* about 10-jugate under pinnate acuminate apex, the posterior basal one much the largest and produced, all petiolulate; *ultimate pinnules* 1-1.5 cm long, to 1 cm broad, subrhombic-ovate, sessile, cuneate, apex roundish, lobato-incised with 2-3 ovate lobes with bluntly dentate teeth; the upper pinnae narrowly oblong-lanceolate and gradually shortened; *texture* herbaceous, pale green, glabrous in all parts; *veins* in ultimate pinnules fine, repeatedly branched, one to each tooth, but not reaches tip and ended in a clavate hydathode; *sori* large, 1-2 to each ultimate pinnule, terminating veinlet near the margin; *indusium* large, semi-orbicular, gray, membranaceous, entire, persistent, free on all sides except the lower side.

Yunnan: Shweli-Salween divide, *G. Forrest 25329*, Szemeo, *Henry 10083A*; Mengtze, *Hancock 63* (Kew No.); Yunnan, *Ducloux 1326, 6329*, between Tengyueh and Lungling, *Rock 7240*; east of Tengyueh, *Rock 7618*. Taiwan.

Also North India generally, Indo-China, Thailand, the Philipping Islands and Malaysia-Polynesia.

This distinct fern resembles none of the genus, to which it belongs, by subterranean habit, pale green leaves with broad segments and large membranaceous indusium.

Plate 186. Fig. 1. Habit sketch (natural size). 2. Portion of frond, showing venation and position of sori (× 10). 3. Scale from rhizome (× 27).

LEUCOSTEGIA IMMERSA (Wallich) Presl
膜盖蕨

图版 187　霍氏膜盖蕨

根状茎木质，粗肥，横生，被金黄色之卵状披针形之密鳞片；叶亚散生，柄长 5—10 cm，红棕色，光亮，虽干枯而仍宿存，光滑或具疏鳞片，叶体椭圆三角形，长 7—15 cm，宽达 7 cm，四回羽状分裂或深裂；一回小叶 10 对，开展，无柄，长达 5 cm；末回小叶羽状深裂；裂片 3—4 数，线状披针形，尖头，具小脉一数，薄纸质，光滑；子囊群小，圆形，生于裂片之基部，盖小，膜质，宿存，宽过于长。

分布：云南，西藏；印度北部。

本种形体极似 L. Delavayi (Bedd)，唯其茎上鳞片不为卵形而为卵状披针形，具长渐尖头，故易分别。

图注：1. 本种全形（自然大）；2. 小叶之一部，表示叶脉及子囊群之着生情形（放大 10 倍）；3. 根状茎上之鳞片（放大 16 倍）。

LEUCOSTEGIA HOOKERI (Moore) Beddome

LEUCOSTEGIA HOOKERI (Moore) Bedd. Hendb. Ferns Brit. Ind. 52 (1883).
  Acrophorus hookeri Moore, Ind. Fil. 2 (1857, nom. nud.); Bedd. Ferns Brit. Ind. t. 95 (1865).
  Davallia clarkei Baker in Hk. et Bak. Syn. Fil. ed. 2, 91 (1874); C. Chr. Ind. Fil. 208 (1905), pro parte.
  Leucostegia Clarkei (Baker) C. Chr. Contr. U. S. Nat. Herb. **26**: 294 (1931); Ind. Fil. Suppl. III. 120 (1934).
  Araiostegia clarkei Cop. Phil. Journ. Sci. **34**: 241 (1927).
  Davallia dareaeformis Levinge ex Clarke, Trans. Linn. Soc. II. Bot. **1**: 443 (1880), pro parte.
  Leucostegia dareaeformis Bedd. Ferns Brit. Ind. Suppl. 4 (1876), pro parte.
  Araiostegia parva Cop. Univ. Calif. Publ. Bot. **12**: 399 pl. 53A (1931).
  Leucostegia parva C. Chr. Ind. Fil. Suppl. III. 121 (1934).

Rhizome thick, woody, wide-creeping, epigaeous, densely scaly; *scales* dense, golden brown, ovate-lanceolate, finely acuminate, spreading; *frond* approximate, stipe 5-10 cm long, reddish-brown, persistent, nitente, glabrous or with a few large deciduous scales, lamina deltoid-oblong, 7-15 cm long, to 7 cm broad, 4-pinnate or pinnatifid; *pinna* 10-jugate, patent, sessile, to 5 cm long, *ultimate pinnules* pinnatifid with 3-4 small ligulate acute uninerved *segments*, 1-2 mm long, 0.5 mm broad; *texture* thin herbaceous, pale green, glabrous; *sori* small at the base or forking of ultimate lobes, *indusium* small, membranaceous, gray, persistent, broader than long.

Yunnan: Hokin, *Delavay*, July 24, 1883; *G. Forrest 15220*; Leilung Shan, *Forrest. 15228* (1917); Muli, west of Yalung River, *Rock 17850*. Tibet: Yatung, *Hobson* (1897); Yunnan-Tibetan border, *Capt. Kingdom Ward 780*.

North India generally: Sikkim, *Hooker fil et Thomson 315* (type).

Rather a small fern characterized by the dense, large, broadly lanceolate scales with spreading long-acuminate tips and the dead persistent, reddish-brown soft stipes, which often break at 2-3 cm above base. In scale the species is very closely related to L. *perdurans* (Christ) Hieron. which differs by much larger size, without so characteristically persistent dead stipes of previous years. In habit and size, it resembles L. *Delavayi* (Bedd.) Ching, but differs in rhizomatic scales being not ovate and imbriate. From L. *dareaeformis* (Hk.) Bedd., our fern differs in sessile pinnae, indusiate sori and shape and color of scales.

The nomenclature of this fern has been very much confused. By priority, *Acrophorus Hookeri* Moore is found the legitimate name, because Moore's *nomen nudum* was subsequently effectively described and illustrated by Beddome in 1865, and is much older than *Davallia Clarkei* Baker. It was, however, unfortunate that Beddome himself later (Handb. p. 316) withdrew the figure under *Acrophorus Hookeri* in his Ferns Brit. Ind. t. 95 as being a mistake for *Polypodium dareaeforme* Hk., an exindusiate species, but his plate represents, in fact, a fern with fairly large indusia and, in this respect alone, agrees well with Moore's species based upon a specimen collected in Sikkim by Hooker and Thomson.

Plate 187. Fig. 1. Habit sketch (natural size). 2. Portion of frond, showing venation and position of sori ($\times$ 10). 3. Scale from rhizome ($\times$ 16).

LEUCOSTEGIA HOOKERI (Moore) Beddome
霍氏膜盖蕨

图版 188　毛膜盖蕨

根状茎木质,粗肥,横行,被卵状椭圆形之红黄色大鳞片;叶远生,柄长 15—25 cm,深稻秆色,光亮,略具鳞片,叶体卵状三角形,渐尖头,长 35—90 cm,宽 17—30 cm,三回羽状分裂;一回小叶 10—15 对,开展,相距甚远,互生,具柄,基部一对较大,长 15—25 cm,宽 7—10 cm,椭圆三角形,渐尖头,基部亚等形,二回羽状分裂;二回小叶约 10 对,互生,具柄,基部一对最大,长 5—7 cm,宽 3—4 cm,基部不等形,三角形,渐尖头,羽状深裂;末回小叶十对,密接,基部上方一片最大,深裂成 4—7 对椭圆形急尖头之裂片,薄纸质,各回小叶基部具一二卵形膜质大鳞片,叶体上面被密短毛,叶脉明显,每裂片具一数小脉;子囊群小,位于锯齿之基部,盖小,马蹄形,膜质,早落,唯基部着生。

分布:云南;尼泊尔及印度东北二部产之。

此种异于本属其他各种者,为其被密短毛之叶体是也。

图注:1.本种全形(自然大);2.小叶之一部,表示叶脉及子囊群之位置(放大 10 倍);3.根状茎上之鳞片(放大 20 倍);4.中肋上面之毛(放大 150 倍);5.中肋下面之鳞片(放大 27 倍)。

LEUCOSTEGIA MULTIDENTATA (Wallich) Beddome

LEUCOSTEGIA MULTIDENTATA (Wallich) Beddome, Ferns Brit. Ind. Suppl. 4 (1876); Handb. Ferns Brit. Ind. 51 (1883); C. Chr. Ind. Fil. Suppl. Ⅲ. 121 (1934).

Aspidium multidentatum Wallich, List no 346 (1828, nom. nud.).

Davallia multidentata Hk. Syn. Fil. 91 (1867); Clarks, Trans. Linn. Soc. Ⅱ. Bot. **1**: 443 (1880); C. Chr. Ind. Fil. 212 (1905).

Humata multidentata Diels in Engl. u. Prantl: Nat. Pflanzenfam. **1**: 209 (1899).

Araiostegia multidentata Cop. Phil. Journ. Sci. **34**: 241 (1927).

Acrophorus thomsoni Moore, Ind. Fil. 4 (1857, nom nud.).

Microlepia pteropus Bedd. Ferns Brit. Ind. t. 313 (1869).

Rhizome thick, wide-creeping, epigaeous, densely scaly; *scales* large, ovate-oblong, acuminate, bright brown, thin; *frond* distant, stipe 15-25 cm long, dark-straminous, nitente, densely at base and sparingly upwards clothed in large broadly ovate, acuminate, thin, brown scales, lamina ovate-deltoid, acuminate, 35-90 cm long, 17-30 cm broad near base, tripinnate; *pinnae* 10-15-jugate, patent, far apart, subalternate, petiolate (petiole about 1 cm long), the basal pair larger, 15-25 cm long, 7-10 cm broad, deltoid-oblong, acuminate, base subequal, bipinnate; *pinnules* to 10-jugate, alternate, petiolulate, the basal pair much the largest, 5-7 cm long, 3-4 cm broad at unequal base, deltoid-acuminate, pinnate to a narrow wing along costa; *ultimate pinnules* about 10-jugate, close, anterior basal one much larger, to 2 cm long, 1 cm broad, ovate-oblong, acute, deeply pinnatifid into 4-7-pairs of oblong acute segments, the lower ones again inciso-serrate; *texture* thin herbaceous, rachis and rachilets glabrous except the base of pinnae, pinnules and costa being clothed with a few large, ovate membranaceous, brown scales, glabrous or glandular beneath, rachis, rachilets and costa above densely hirsute; *veins* fine, distinct, in segments pinnate, one to each sharp tooth, but never reaches tip; *sori* small, 2-12 to each ultimate pinnule, placed at the base of its teeth on the upper side of veinlets; *indusium* small, horse-shoe-shaped, brown, membranaceous, fugaceous, free on all sides except the base.

Yunnan: Shweli-Salween divide, *G. Forrest 24701*; Tengyueh, *G. Forrest 27182* (1925); Htawgaw, *G. Forrest 27010*; Mengtze, *Hancock 17* (1893)

Himalayas: From Nepal to Bothan. Also Khasia, common.

Another distinct species of the genus, characterized by the presence of large broadly ovate thin brown scales on stipe, rachis, rachilets, and costa beneath and densely hirsute above. In hairiness, the species is closely related to another Himalayan species, *L. membranulosa* Wall., which differs by much smaller size, pinnate or bipinnatifid lanceolate leaves and pale brown lanceolate scales on rhizome.

Plate 188. Fig. 1. Habit sketch (natural size). Portion of frond, showing venation and position of sori ($\times$ 10). 3. Scale from rhizome ($\times$ 20). 4. Hairs from the costa of pinnule above ($\times$ 150). 5. Scale from the costa of pinna beneath ($\times$ 27).

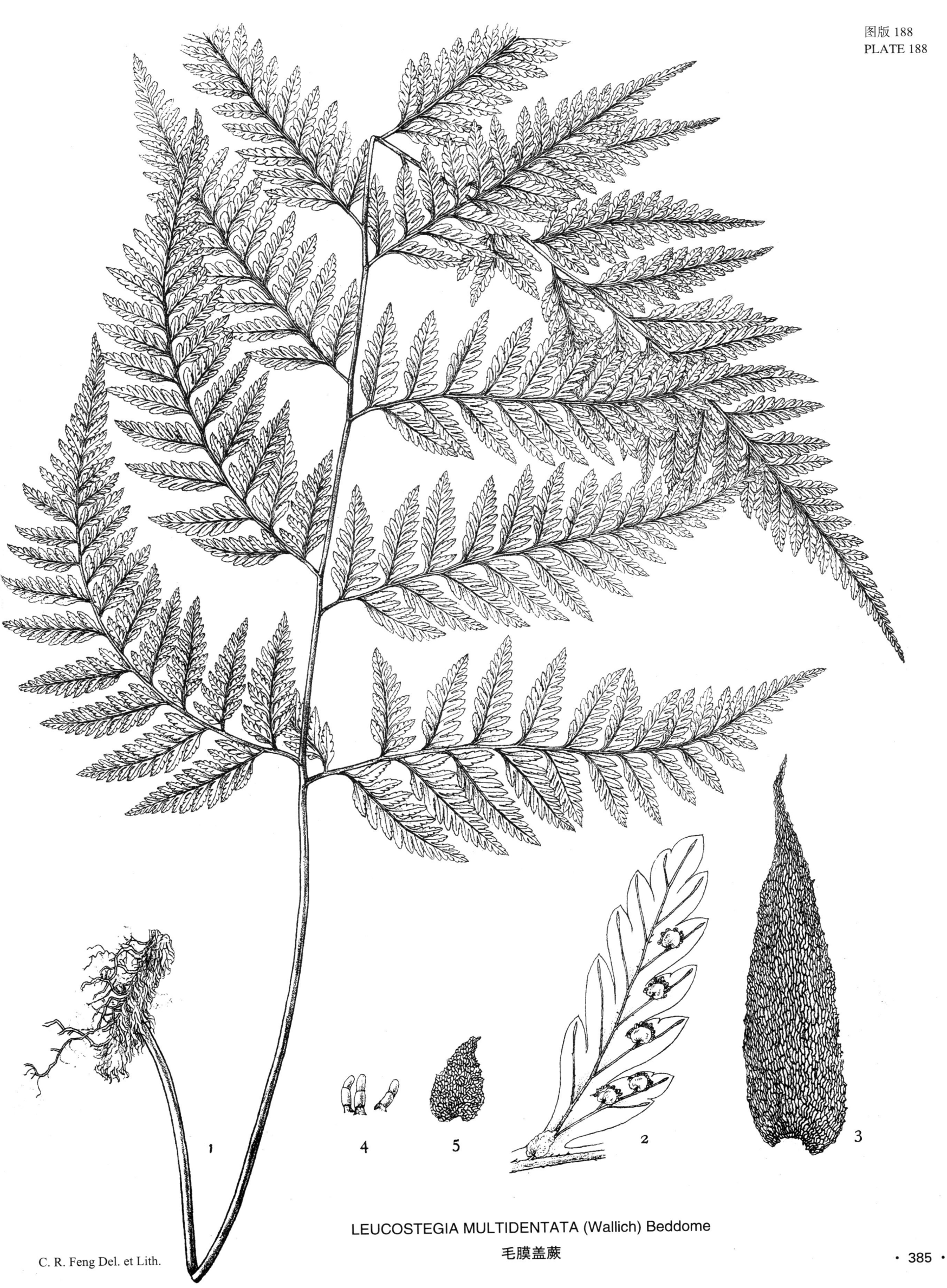

LEUCOSTEGIA MULTIDENTATA (Wallich) Beddome
毛膜盖蕨

图版 189 小叶剑蕨

根状茎细长,横行,被深褐色之披针形膜质鳞片;叶远生,长 3—10 cm,宽 5—10 mm,倒披针形,急尖头,顶部最宽,向下渐狭,延长至基部,全缘,肉质,中肋上面显凸,下面稍隆起,叶脉不见,网眼 2 列;子囊群线形,2—4 对,急斜出,位于叶顶部之最宽处,不达于叶边。

分布:湖北,四川,江西,云南;日本亦产之。

此为本属极小之种之一,具倒披针形之叶,以顶部为最阔,具 2—4 对急斜出之子囊群,最易识别。

图注:1.本种全形(自然大);2.叶体之一部,表示叶脉及子囊群之位置(放大 5 倍);3.根状茎上之鳞片(放大 28 倍)。

LOXOGRAMME GRAMMITOIDES (Baker) C. Christensen

LOXOGRAMME GRAMMITOIDES (Baker) C. Christensen Ind. Fil. Suppl. II, 21 (1916); III. 125 (1934).

Gymnogramme grammitoides Baker, Journ. Bot. (1889) 178.

Polypodium grammitoides Diels in Engl. Bot. Jahrb. **29**: 209 (1900); C. Chr. Ind. Fil. 530 (1905).

Selliguea grammitoides Christ, Bull. Herb. Boiss. II. **3**: 510 (1903).

Gymnogramme lanceolata var. *minor* Baker; Makino, Bot. Mag. Tokio **10**: 178 (1896).

Loxogramme minor Mak. Bot. Mag. Tokio **19**: 139. (1905)

Polypodium yakushimae Christ, Bull. Herb. Boiss. II. **1**: 1014 (1901); C. Chr. Ind. Fil. 575 (1905); Kodama in Matsum. Ic. Pl. Koisik. **1**: no. 3, pl. 42 (1912).

Loxogramme yakushimae C. Chr. Ind. Fil. Suppl. II. 22 (1916).

Loxogramme spatulata Cop. Phil. Journ. Sci. **30**: 331 (1926).

Rhizome epigaeous, slender, wide-creeping, densely clothed in fusco-brown, lanceolate, acuminate, thin, clathrate *scales*; *fronds* distant, 3-10 cm long, 0.5-1 cm broad at the broadest part in the uppermost part, oblanceolate, much broadened below acute or short-acuminate apex, gradually long-attenuate until base, margin entire, thin; *texture* subcarnose, greenish-brown when dried; midrib prominently raised above, only slightly keeled or not visible beneath, *veins* not seen, forming two rows of elongate oblique areolae on each side of midrib; *sori* linear-oblong, very oblique, 2-4 pairs, confined to the uppermost broadest part, subcostal, not reaching margin.

Hubei: Ichang, *A. Henry 5451* (type), *5451A*; *Wilson 620* (type of *L. spatulata*). Guizhou: Kianghow, foot of Vanchin Shan, *Y. Tsiang 7556*, *7899*. Sichuan: Hungya Hsien, *W. P. Fang 8496*, August, 1930. Jiangxi: Wang Lung Tze, *R. C. Ching*, Oct. 1935, on wet mose-clad rock cliff under woods. Yunnan: Long-ki, *Delavay*, August, 1899; *E. E. Maire*.

Japan: Yokohama, *Maximowicz 11* (1862), and other localities.

This fern represents one of the smallest species of the genus, being characterized by small sessile oblanceolate or spathulate leaves, much broadened in the upper one-fifth part, thence gradually narrowed and attenuate until base, by a few pairs of short and very oblique subcostal sori, confined to the uppermost broadest part. The Japanese *L. Yakushimae* (Christ) appears not specifically different from Chinese type.

Plate 189. Fig. 1. Habit sketch (natural size). 2. Portion of frond showing venation, and sori ($\times$ 5). 3. Scales from rhizome ($\times$ 28).

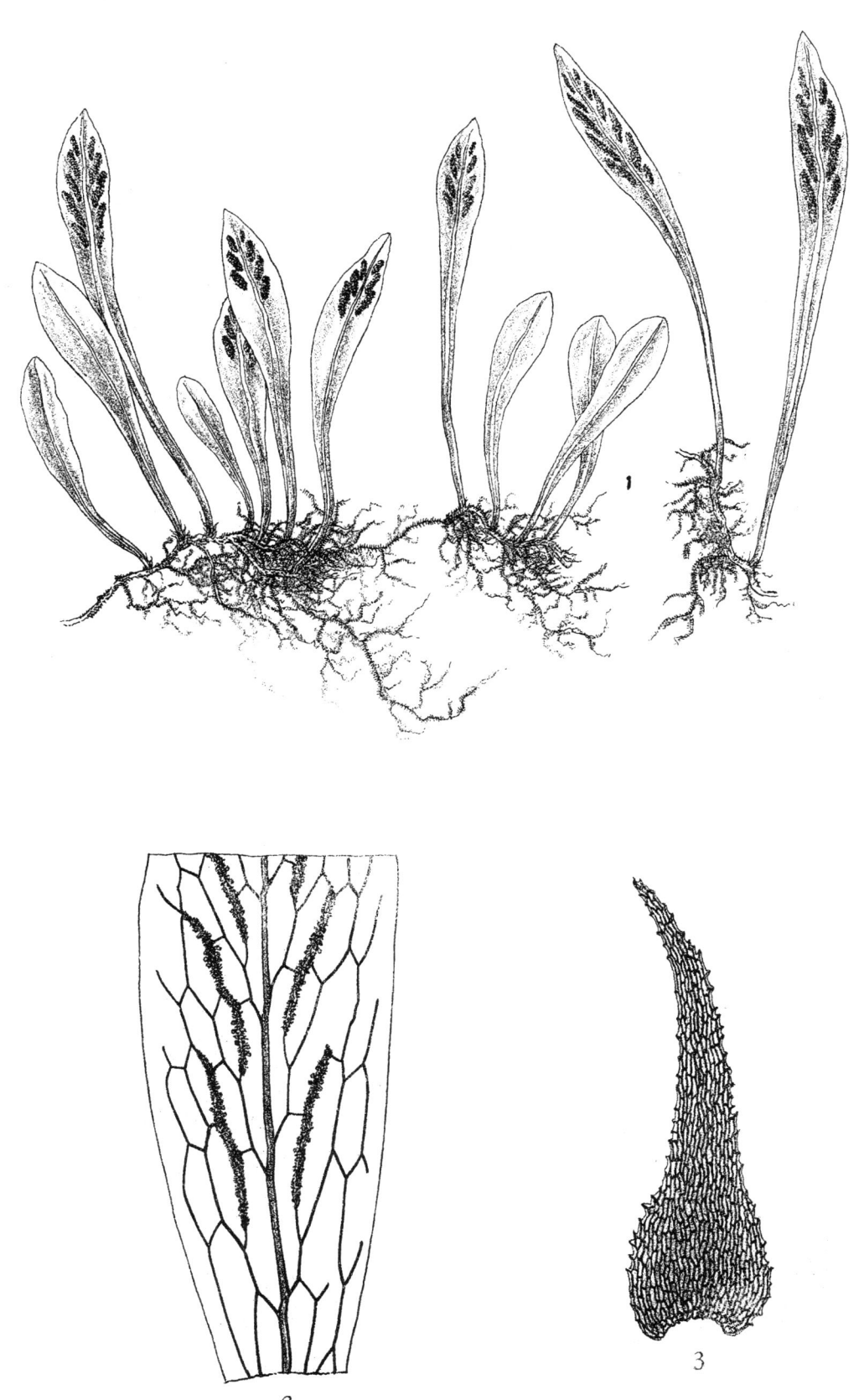

LOXOGRAMME GRAMMITOIDES (Baker) C. Christensen
小叶剑蕨

根状茎细长，横行，略被卵状披针形之深棕色鳞片；叶散生，长 15—35 cm，宽 1.2—2.5 cm，长披针形，上部 1/3 处最阔，向下渐狭，沿叶柄下延，达于离基部 2—6 cm 处，全缘，亚革质，光滑，呈淡黄色，中肋上面凸起，下面隆起，网脉不见；子囊群线形，细长，斜出，自中肋达叶边。

分布：云南，四川，湖北，贵州，广东，广西，福建，江西，香港，台湾；朝鲜半岛，日本，越南及喜马拉雅山区亦产之。

此为我国温暖各省习见之种，形体大小不一，异名甚多，皆同种也。

图注：1. 本种全形（自然大）；2. 叶体之一部，表示叶脉及子囊群之位置（放大 8 倍）；3. 根状茎上之鳞片（放大 16 倍）；4. 根状茎之横切面，表示维管束之布置（放大 16 倍）。

LOXOGRAMME SALICIFOLIA Makino

LOXOGRAMME SALICIFOLIA Makino, Bot. Mag. Tokio **19**: 138 (1905); Ching Bull. Dept. Biol. Coll. Sci. Sun Yatsen Univ. no. **6**: 31 (1933); C. Chr. Contr. U. S. Nat. Herb. **26**: 324 (1931); Ind. Fil. Suppl. Ⅲ. 125 (1934).

Gymnogramme salicifolia Makino, Phan, Plerid. Jap. Ic. Pl. 34 (1899).

Polypodium makinoi C. Chr. Ind. Fil. 339 (1905); 543 (1906).

Loxogramme makinoi C. Chr. Ind. Fil. Suppl. Ⅱ. 22 (1913-17).

Loxogramme duclouxii Christ, Bull. Acad. Géogr. Bot. (1907) 140; C. Chr. Ind. Fil. Suppl. Ⅲ. 125 (1934).

Polypodium succulentum C. Chr. Ind. Fil, Suppl. 1. 63 (1907-12).

Loxogramme fauriei Copel. Phil. Journ. Sci. **9**: 232 (1914); C. Chr. Ind. Fil. Suppl. Ⅱ. 21 (1913-17).

   Gymnogramme involuta Bak. (non Hk. 1864) Journ. Bot. (1888) 231; Franch. Pl. David. In Nouv. Arch. Mus. Ⅱ. **10**: 123 (1887).

Selliguea involuta Christ (non Kze. 1858) Bull. Soc. Bot. France **52**: Mém, Ⅰ. 21 (1905); Bull. Acad. Géogr. Bot. (1906) 108.

Loxogramme involuta C. Chr. (non Presl. 1836), Acta Hort. Gothob. **1**: 104 (1924).

Polypodium scolopendrinum Wu (non C. Chr. Index) Polyp. Yaoshan. in Bull. Dept. Biol. Coll. Sci. Sun Yatsen Univ. No. **3**: t. 155 (1932).

Rhizome slender, wide-creeping, densely radicose. sparcely scaly; *scales* brown or fusco-brown; ovate-lanceolate, acuminate, entire; *frond* 1-3 cm apart, uniseriate along the rhizome, 15-35 cm long, 1.2-2.5 cm broad or rarely broader, lanceolate, broadest at the upper third, gradually narrowed downward on each side of the costa until 2-6 cm above the base of stipe, apex caudate-acuminate, margin entire, thin and slightly revolute; *texture* coriaceous, naked on both sides; *midrib* keeled beneath and prominently raised above, *veins* hidden, areolae elongate, oblique, rarely with one short included veinlet; *sori* linear, rather slender, very obliquely extending from midrib to near the margin.

Yunnan: Mengtze, *Hancock 111*; A. *Henry 9059*, *9059A*; Szemeo, *Henry 10343*; Hay-Y près Loa Lan, *Ducloux 133* (type of *L. Duclouxii*); Tchen Fong Chan, *Delavay*, August, 1894; Maeulchan, *Delavay*, *3880*; Ami, Y. *Tsiang 13096*, *13180*; Weise Hsien, *H. T. Tsai 59885*; Tsekou, *Souliè 1665*; Maokou Tchang, *Delavay 17*, *1199*; Without locality, *S. Ten (1915)*; Salween, *Capt. Kingdom Ward*, Jan. 2, 1914; between Tengyueh and Lungling, *J. F. Rock 7295*; between Kambaiti and Tengyueh, *J. F. Rock 7543*. Sichuan: Tchenkoutin, *Farges 179*; Moupin, *David*; Mt. Omei, *Faber 1019* (pro parte); *Wilson 5348* (pro parte); *W. P. Fang 7453*; *T. Tang 23594*; Hungya Hsien, *W. P. Fang 8061*; Nanchuan Hsien, *W. P. Fang 5807*; ibid., Nos. *3151*, *4995*, *4851*, ex Herb. of West China Acad. Sci. Guizhou: Kao po, *Laborde et Bodinier 1978*; Ganchow, *Cavalerie 877*; Vanchin Shan, *Y. Tsiang 7561*; *7904*; Tuhshan Y. *Tsiang 6925*; Sinwen, *Y. Tsiang 8688*; Pinfa; *Cavalerie 877*. Hubei: Hsing Shan Hsien, *Wilson 2661*; Wushan Hsien *Wilson 615*; without locality, *Silvestri 59*. Jiangxi: Lushan, Whang Lung Tze, *R. C. Ching 11591*. Fujian: Inghok, *F. P. Metcalf 820*, May 1, 1925; ibid., *H. H. Chung 2646*, April 24, 1924; Samsa Inlet, *Matthew*, Oct. 5, 1907. Guangdong: Lafau Shan, *N. K. Chun 40907*; *Ford 33*; Sam Kok Shan, Tsungfa Hsien, *W. T. Tsang 20578*; Swatow, Thaiyong, *Dalziel*, July, 1901. Guangxi: Lin Yen Hsien, Yeo Mar Shan, *R. C. Ching 7242*. Hong Kong: Lantao Island, one specimen without collectors name in Herb. Taiwan: *Hancock 20*; Arisan, *Faurie 464*, *465*.

Korean peninsula: *Faurie 74*; Tsus-sima, *Wilford 775*; Quelpaert, *Taquet 3690*.

Japan: Kyoto, *Kiyabe 16134*; Oosumi, Kyushu, *Koidzumi*, Sept., 1921.

Vietnam: Chapa, *Eberhardt 5145*.

Assam: Manipur, *G. Watt 6133* (1882). Bothan: *Griffith*. Khasia: *Hooker f*.

A quite variable species as to size, specimens from Southeast China and Japan being generally smaller than those from West and Southwest China, but all agree in essential characters, Some specimens (Henry 9095) from Yunnan almost approaches *L. involuata* (Don) Presl in size, but differs in wide-creeping rhizome, distant leaves, thicker texture and prominently raised midrib above.

Plate 190. Fig. 1. Habit sketch (natural size). 2. Portion of lamina, showing venation, and sori (× 8). 3. Scale from rhizome (× 16). 4. Cross section of rhizome, showing the arrangement of steles (× 16).

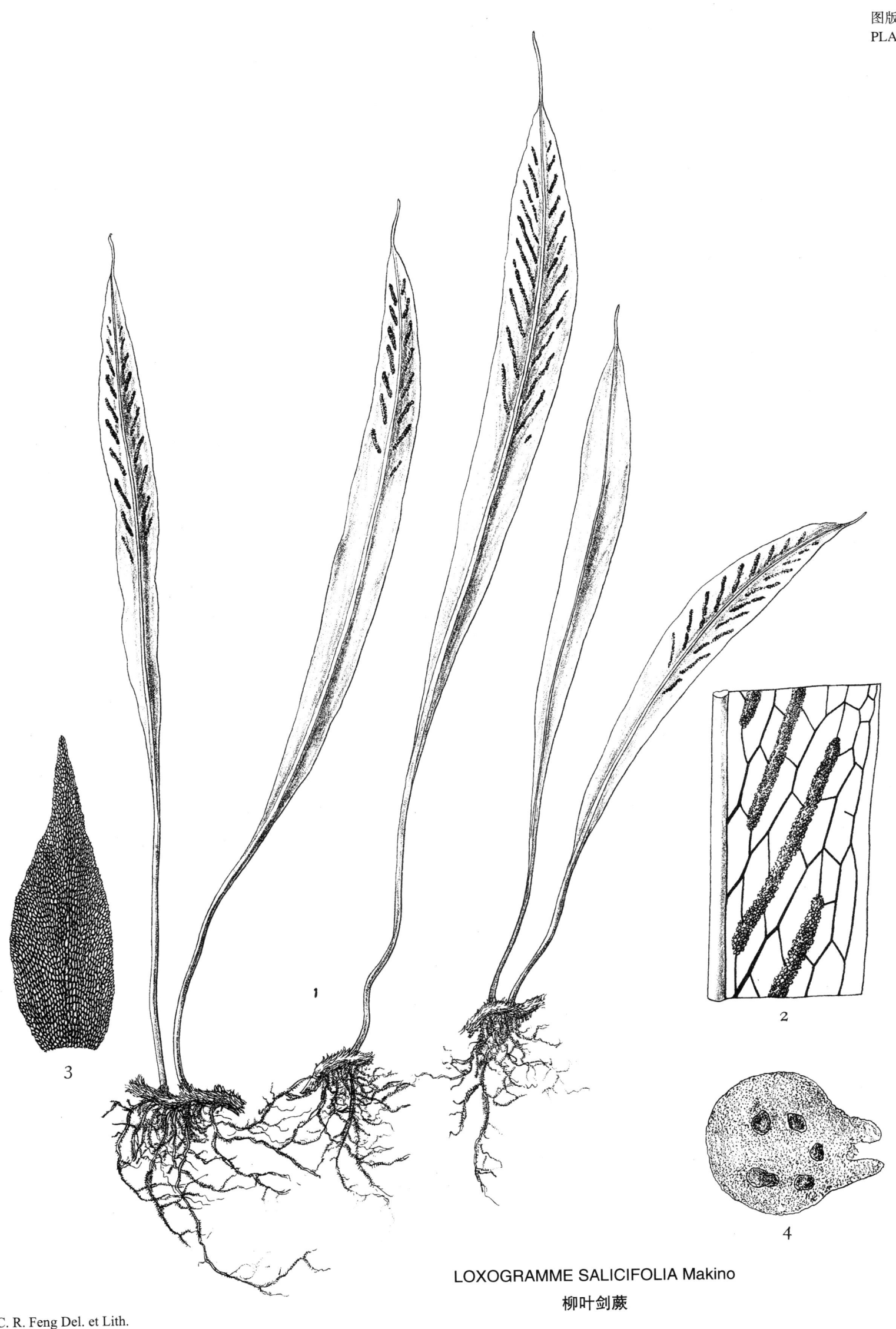

LOXOGRAMME SALICIFOLIA Makino 柳叶剑蕨

根状茎短而直立；鳞片为阔卵形，渐尖头，淡黄色；叶簇生，柄长仅 2 cm，扁形，叶体倒披针形，长尖头，下部渐狭，沿叶柄下延，长 20—25 cm，宽 3—3.5 cm，以上部 2/3 处为最阔，亚革质，两面光滑，中肋上面凸出，下面隆起，叶脉网状，不见，网眼长形，间具一数单脉；子囊群线形，斜出，行距约 5 mm，位于叶之上部，自中肋达叶阔 2/3。

分布：四川及贵州产之。

本种极似印度北部产之 L. involuta，然其茎上鳞片较小，网眼亦异，子囊群彼此相离甚远，不达于叶边，故可分别。

图注：1. 本种全形（自然大）；2. 叶体之一部，表示叶脉及子囊群（放大 3 倍）；3. 根状茎上之鳞片（放大 10 倍）。

LOXOGRAMME ENSIFORMIS Ching

LOXOGRAMME ENSIFORMIS Ching, sp. nov.

Species magnitudine et configuratione *L. involutae* (Don) himalayensae magis similis differt paleis rhizomatis duplo parvioribus, late ovatis, apice breve acuminatis (luminae parvioribus rotundatis, creberrimis); soris multo brevioribus, crassioribus, a se magis distantibus, nec costae nec marginem frondis attingentibus; costa centralis supera prominenti.

Rhizome short, erect, densely scaly, *scales* light brown, broadly ovate, short acuminate, entire, 5 mm long, consisting of numerous small roundish brown luminae; *fronds* caespitose, stipe short, thick, 2 cm long, compressed, lamina oblanceolate, long acuminate, 20-25 cm long, 3-3.5 cm broad at the upper two-thirds part, gradually attenuate along a long narrowly winged stipe; *texture* chartaceous, greenish, naked, mibrib distinct on both sides, slightly keeled below and raised above; *veins* not seen but distinct against light, the lateral veins oblique, areolae elongate, only occasionally with one short included veinlet; *sori* oblique, thick, brown, raised, about 5 mm apart, confined to the upper half of leaf, extending from near the costa to one-third way from the margin.

Sichuan orientalis: Without locality, *T. F. Lou 231* (type); without locality *Wilson 5348* (pro parte). Guizhou: Lintchang, *Cavalerie 3389*, April, 1909; Pinfa to ganchow, *Cavalerie 1303*, 478.

The present fern appears so alike the Himalayan *L. involuta* (Don) that it might well be passed for that species, from which, however, it can easily be distinguished by its broadly ovate and half as long scales with short acuminate apex and much smaller, round and more numerous brown luminae, by leaves of green color, gradually attenuate downward from the broad upper half, and by much thicker, shorter, fewer and more widely separate sori extending from near the costa to only one-third way from the margin. The costa is slightly raised on the upper side. *L. involuta* (Don) has broadly lanceolate rhizomatic scales to 1 cm long, consisting of large, clear elongate luminae, much longer, narrower and very oblique sori extending from costa to near the margin and not raised midrib above.

Plate 191. Fig. 1. Habit sketch (natural size). 2. Portion of lamina, showing venation, and sori($\times$ 3). 3. scale from rhizome ($\times$ 10).

PLATE 191

LOXOGRAMME ENSIFORMIS Ching
阔叶剑蕨

图版 192 槲蕨

一种附生于树干或石壁上之蕨类，根状茎横行，肉质，肥厚如指，被金黄色之卷曲狭长鳞片；叶二形，其不生子囊群之叶为圆卵形，无柄，彼此瓦覆，长约 5—7 cm，宽 3—6 cm，灰褐色，干厚革质，边缘浅裂，叶脉显凸，网状，其通常生子囊群之叶为绿色，长 25—40 cm，宽 14—18 cm，长椭圆形，具有翅之短柄，叶体向基部渐狭，厚纸质，两面光滑，羽状深裂，裂片 7—13 对，长 7—9 cm，宽 2—3 cm，渐尖头，基部二三对缩为耳形，边缘具浅疏缺刻，叶脉网状，显凸；子囊群大，略呈圆形，数列，无盖。

分布：浙江，江西，湖北，云南，四川，广东，广西，福建，台湾；越南亦产之。

此为中国温暖各省普通之蕨种，常附生于树干或干燥之石壁上，以其肉质之茎及槲树形之枯叶吸收雨水与落叶以供其养料，至饶兴味。

图注：1. 本种全形（自然大）；2. 叶之一部，表示叶脉及子囊群之位置（放大 6 倍）；3. 根状茎之鳞片（放大 16 倍）；4. 不生子囊群叶下面之毛（放大 76 倍）。

DRYNARIA FORTUNEI (Kze.) J. Smith

DRYNARIA FORTUNEI (Kze.) J. Smith in Bot. Voy. Herald. 425 (1857); Diels in Engl. u. Prantl. Nat. Pflanzenfam. **1**: 4. 330 (1899); C. Chr. Ind. Fil. 247 (1905); Acta Hort. Gotheb, **1**: 106 (1924).

Polypodium Fortunei Kze. apud Mett. Farngatt. Polyp. 121 t. 3 f. 42-45 (1857); Hk. Sp. Fil **5**: 95 (1864); Hk. et Bak. Syn. Fil. 367 (1868); Christ, Farnkr. d. Erde 119 (1897); Baker, Journ. Bot (1888) 230; Franch. Pl. David. in Nouv. Arch. Mus. Ⅱ. **10**: 121 (1887); Christ, in Warburg, Monsunia **1**: 63 (1900); Bull. Soc. Bot. Ital. (1901) 297.

Drynaria quercifolia Hk. (non J. Sm.) Journ. Bot. (1857) 357.

Polypodinm biforme Lour. Fl. Cochinch. 827 (1790); Sw. Syn. Fil. 62 (1806).

Polypodium quercifolia Hk. (non L. 1753) in Blakiston, Five Months on the Yangtze 366 (1682).

An epiphytic fern on tree trunks or rocks. *Rhizome* wide-creeping, fleshy, thick as a finger, densely clothed in bright ferruginous, frizzy, linear-subulate *scales* with long-fimbriate margin; *fronds* dimorphous, the sterile ones sessile, reddish-brown (without chlorophyll), dry, coriaceous, nitente, imbricate, with coarse venation, 5-7 cm long, 3-6 cm broad, broadly ovate, cordate at base, acute at apex, margin crenate below, lobato-pinnatid in the upper half with 4-6 pairs of deltoid, very acute, entire lobes 1-1.5 cm long, shining glabrous above, shortly pubescent on costa and veins beneath; the *fertile* ones 25-40 cm long, including winged stipe 5-8 cm long, 14-18 cm broad at middle, oblong, acute, pinnatifid nearly down to rachis; *segments* patent, 7-13-jugate under the caudate apex, 7-9 cm long, 2-3 cm broad above the broadened base, lanceolate, acute or obtusish, margin with remote incisions, the lowest ones somewhat shorter, followed by a few auricles, the upper ones gradually shortened, sinuses broad, roundish or acute at bottom; *texture* crass chartaceous, green and glabrous on both sides; *venation* prominent on both sides, lateral veins distinct, oblique, connected by transverse veins, forming 4-5 quadri-angular soriferous areolae, filled with free or netted included veinlets; *sori* large, roundish, copious, regularly seriate, 2-4 between costa and margin, one in each 4-angular areola, exindusiate.

Zhejiang: Chusan, *Robert Fortune*; Ningpo, *Hancock 25*; *Cooper* (1884); Taichow, *R. C. Ching 1580*; Pingyang Hsien, *H. H. Hu 96*; Wenchow, *K. Ling 7407* (1924); Siachw Hsien, *R. . C. Ching 1580*. Jiangxi: Kiukiang, Lushan, *A. N. Steward 2661*; Kwaiin Chiao (Goddess of Mercy Bridge), *R. C. Ching*, numerous specimens; *C. E. DeVol 22*, August 1, 1933; Tsoongjen, *Y. Tsiang 10228*. Guizhou: Ganchow, *Cavalerie 3711*, *7797*. Hubei: Ichang, *Maires* (1880); Patung Hsien, *Henry 3704*; *Wilson 2646*. Sichuan: Mt. Omei, *E. Faber 1072*; *Brown 73*; Chungchow, *Limprichte* (1913); *Col. Sarel*; *R. Francis* (1870); *Blakiston* in Herb, Hk.; Hochuan Hsien, *Hopkinson 108*, May 2, 1930. Yunnan: Mengtze, *Hancock 112*; Mile, *Henry 10177A*, *10177B*. Guangdong: North River, *Tutcher 10626*; Lofau Shan, *C. Ford* (1833); Swatow, Thaiyoung, *Dalziel*, Sept. 1898; *Gerlach*; Lienchow, *Matthew*, Dec. 1907; Lokchong, *N. K. Chun 42423*; *C. L. Tso 21538*; Yingtak, Wantong Shan, *H. Y. Liang 60590*; Lungtau Shan, *Y. K. Wang 31707*. Guangxi: Lungchow, *Morse 2*; Wuchow, *S. S. Sin & K. K. Wang 6* (1926); Luchen Hsien, Tze Poo, *R. C. Ching 5554*. Fujian: Foochow, *R. Fortune 34* (type); *La Touche*, *Forbes 2492*; *L. Y. Tai 11170*; *T. S. Ging 7104*; Kushan, *T. S. Ging 6848*; Lunglau, *Alexander*; Amoy, *Medhurst* in Herb. *Hance 1409*; Yuenfu, *Warburg*; Changchow, *H. H. Chung 903*; Hinghwa, *H. H. Chung 971*; Samsa Inlet, *Matthew*, Oct. 6, 1907. Taiwan.

Also Vietnam.

A common epiphytic fern in the warm parts of China and differs from *D. quercifolia* (L.) J. Sm. of Tropic Asia in much smaller size in all parts and the large uniseriate sori between lateral veins.

Plate 192. Fig. 1. Habit sketch (natural size). 2. Portion of segment, showing venation and position of sori (× 6). 3. Scale from rhizome (× 16). 4. Hairs on underside of sterile leaf (× 76).

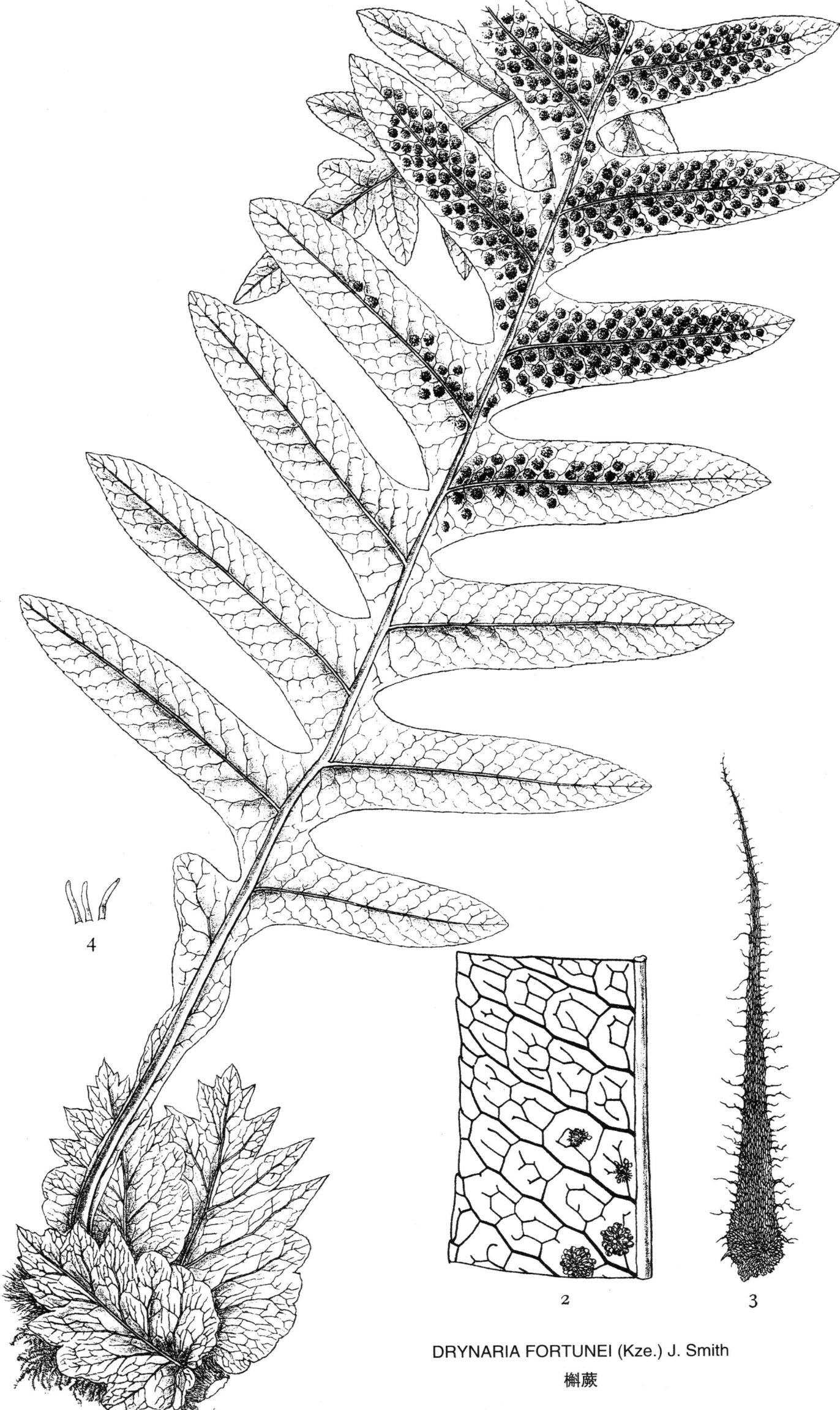

DRYNARIA FORTUNEI (Kze.) J. Smith
槲蕨

本种形体颇类前种,唯其不生子囊群之叶疏生或往往罕见,且其形体与生子囊群之叶无大异,唯较小,为椭圆披针形,呈黄绿色或淡黄色,其生子囊群之叶之两面被甚多之短毛(尤以中肋上面为甚),子囊群一列,位于中肋两侧,故易分别;又此种之生态为半附生或土生,因其茎直接与土壤相接,可以利用土中养料,故其不生子囊群之叶往往变为生子囊群者。

分布:四川,云南,陕西,甘肃。

此为本属分布极北而且极耐寒之种,甚类喜马拉雅山区所产之毛槲蕨(D. mollis Bedd.)唯其叶柄下部不为死稻秆色,叶面之毛较疏,故尚易鉴别。

图注:1. 本种全形(自然大);2. 生子囊叶之一部,表示叶脉,子囊群之位置及锯齿(放大 6 倍);3. 根状茎上之鳞片(放大 16 倍);4. 叶之中肋上面之短毛(放大 76 倍);5. 根状茎之横切面,表示维管束之布置(放大 4 倍)。

DRYNARIA SINICA Diels

DRYNARIA SINICA Diels in Engl. Jahrb. **29**: 208 (1900); C. Chr. Ind. Fil. 249 (1905); Acta Hort. Gotheb. **1**: 106 (1924); Journ. Wash. Acad. Sci. **17**: 498 (1927).

Polypodium baronii Christ (non Baker 1886), Nuov. Giorn. Bot. Itat. n. s. **4**: 100 t. 2. (1897) Farnkr. d. Erde 120 (1897).

Drynaria baronii Diels in Engl. u. Prantl. Nat. Pflanzenfam. **1**: 4. 330 (1899); Christ, Bull. Soc. Bot. France **52**: Mém. Ⅰ. 23 (1905).

Drynaria reducta Christ in C. Chr. Ind. Fil. 247 (1905), C. Chr. Bot. Gaz. **56**: 332. 1913.

Rhizome wide-creeping, fleshy, thick as a small finger, densely scaly; *scales* bright ferruginous, frizzy, lanceolate-subulate, densely fimbriate; *fronds* dimorphous, the *sterile* ones rather scarce, pale green or light brown, chartaceous, or subcoriaceous, sessile, to 10 cm long, 4-5 cm broad at middle, oblong-lanceolate, acuminate, pinnatifid down nearly to rachis with deltoid-lanceolate, acute segments 2-3 cm long, with the lowest ones much reduced, glabrous beneath, pubescent above; *fertile* fronds distant, stipe 8-15 cm long, dark straminous, with narrow wing on each side running down nearly to the base, lamina 17-40 cm long, 7-11 cm broad, oblong-elongate, pinnatifid down nearly to rachis, 14-20-jugate, patent, broadly linear-lanceolate, acute or bluntish, or rounded, the middle ones 4-6 cm long, 1-1.5 cm broad above the dilated base, the basal 1-2 pairs shortened or reduced into more pair of auricles, margin finely and closely serrate with low arcuate sharp teeth; *texture* crass chartaceous, green, more or less pubescent especially on rachis and costa above; *venation* distinct on both sides, lateral veins erecto-patent, intervening veinlets anastomosing in 3-4 rows of angular areola; occasionally with one short included veinlets; *sori* large, roundish, costal, uniseriate on each side, near the upper base of lateral veins.

Sichuan: Nanchuan, *Rosthorn 3121* (type); Mt. Omei, *Scallan*: Tachienlu, *Soulie 512* (1893); without locality, *Wilson 5335*; Maochow, *F. T. Fang 21818*; Tungnan Hsien, *W. P. Fang 1452*; Kangtien Hsien, *W. P. Fang 3687*; Hung-yuen Hsien, *W. P. Fang 9081*, *9090*; Drogochi, *Harry Smith 4502*; *C. S. Liu 705*, July 13, 1934; Mapien Hsien, *T. T. Yü 2509* (pro parte). Yunnan: Tongchow, *E. E. Maire 1373*, *1383* (1913); Tsekou, *Monbeig 277*. Shaanxi: Ki Shan, *Giraldi* (type of *Polypodium Baronii*); Mt. Zulu, *Giraldi*, August, 1894; *Purdom 87*. Gansu: Pingfan Hsien, *R. C. Ching 481*, forming dense carpet on moist foothill.

This endemic species is closely related to the Himalayan *D. Mollis* Bedd., differs chiefly in less pubescent fertile leaves of thicker texture and the basal part of stipe being not of dead straw-colored appearance.

Plate 192. Fig. 1. Habit sketch, (natural size). 2. Portion of segment showing venation. position of sori and serrature (× 6) 3. Scale from rhizome (× 16). 4. Hairs from the upper side of rachis (× 76). 5. Cross section of rhizome, showing arrangement of steles (× 4).

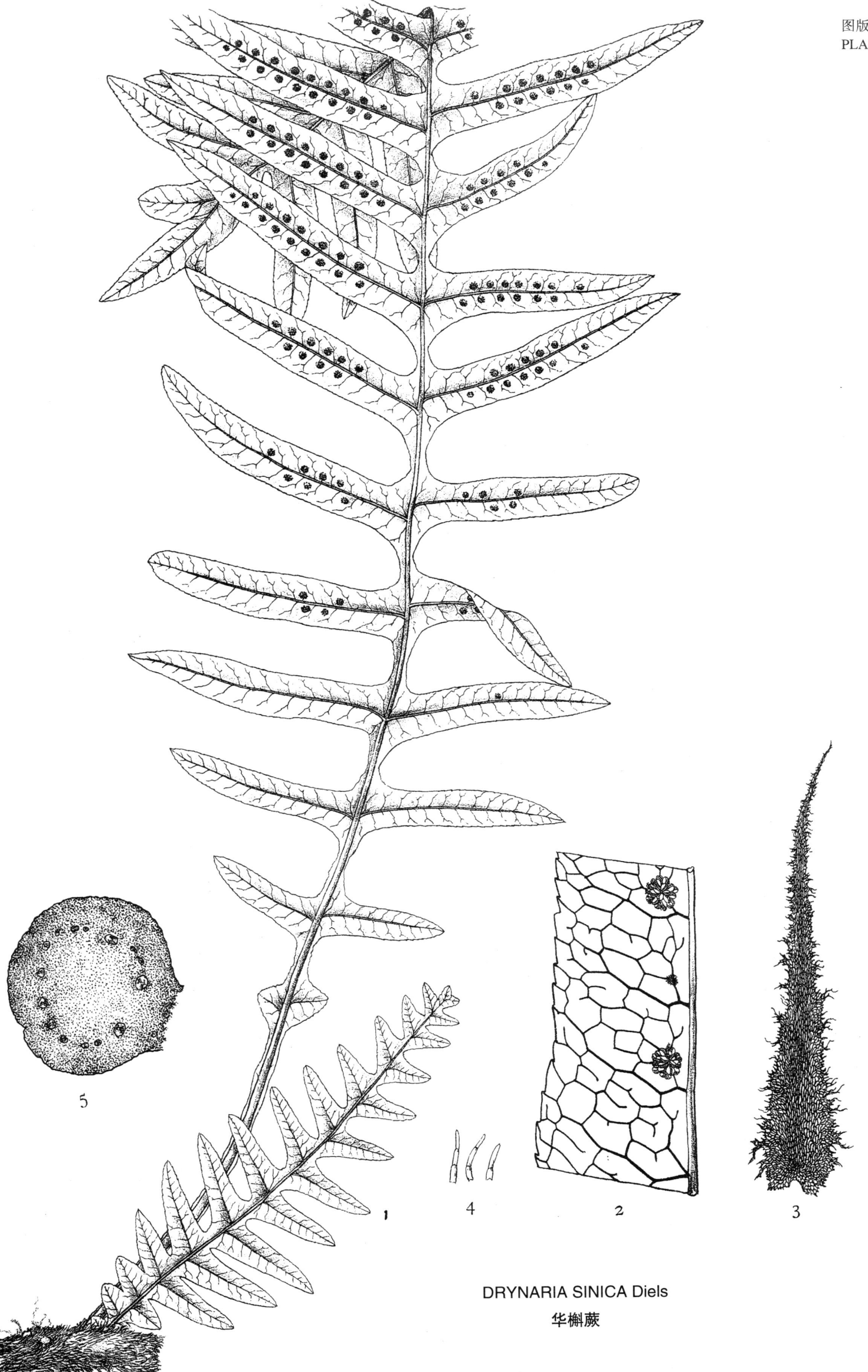

DRYNARIA SINICA Diels
华槲蕨

根状茎横行,被深棕色之线状披针形鳞片;叶散生亚二形,柄长 30—35 cm,稻秆色,光滑,叶体长 15—20 cm,基部宽 5—8 cm,卵状椭圆形,渐尖头,基部为圆截形,向下稍延长,叶边呈浅波状,纸质,两面光滑,侧脉明显,屈折,网脉可见;子囊群线形,自中肋达叶边;不生子囊群之叶体较宽,具较短之柄。

分布:广东及广西之瑶山产之。

此为稀见之种,其卵状椭圆形之叶体具二倍长之叶柄,易与本属各种区分。

图注:1.本种全形(自然大);2.叶体之一部,表示叶脉及子囊群(放大 1.5 倍);3.根状茎上之鳞片(放大 30 倍)。

COLYSIS WUI (C. Chr.) Ching

COLYSIS WUI (C. Chr.) Ching, Bull. Fan Mem. Inst. Biol. **4**: 322 (1933).

Polypodium sp. nov. Wu, Polyp. Yaoshan, in Bull. Dept. Biol. Coll. Sci. Sun Yatsen Univ. No. **3**: 318 t. 150 (1932).

Polypodium wui C. Chr., l. c. No. **6**: 17 (1933); Ind. Fil. Suppl. III. 161 (1934)

Rhizome wide-creeping, densely scaly; *scales* fusco-brown, linear-lanceolate from rounded base, thin, clathrate; *frond* 1-2 cm apart, scarcely subdimorphous, the *fertile* one with stipe 30-35 cm long, straminous, glabrous, lamina 15-20 cm long, 5-8 cm broad at base, oblong-ovate, gradually acuminate, base rotundo-truncate, decurrent a short way along stipe, margin narrowly cartilaginous, repando-undulate; *texture* herbaceous, green, glabrous on both sides; *lateral veins* distinct, erecto-patent, flexuose, veinlets anastomosing on each side in a row of elongate areolae with divaricate included veinlet; *sori* linear, oblique, extending regularly from costa to margin; *sterile* fronds conform, but with somewhat broader lamina on shorter stipe.

Guangxi: Yao Shan, Shengtang Ling, *S. S. Sin et K. K. Wang 613* (type), June 15, 1928; Szeloh Hsien, *Guangxi Natural History Museum No. 27*. Guangdong: Tung Shin Hsien, Nanlien, *K. K. Tsoong 1967*; Tailung Tung, Eu Wai Shan, *C. L. Tso 22410* (1929).

An endemic species, closely related to *C. pedunculata* (Hk. et Grev.) from Sikkim-Himalayas, differing in broadly ovate-oblong leaves being scarcely dimorphous, with rotundo-truncate base shortly decurrent along stipe and narrower sori being more wide apart from each other.

Plate 194. Fig. 1. Habit sketch (natural size). 2. Portion of lamina, showing venation and sori ($\times$ 1.5). 3. Scale from rhizome ($\times$ 30).

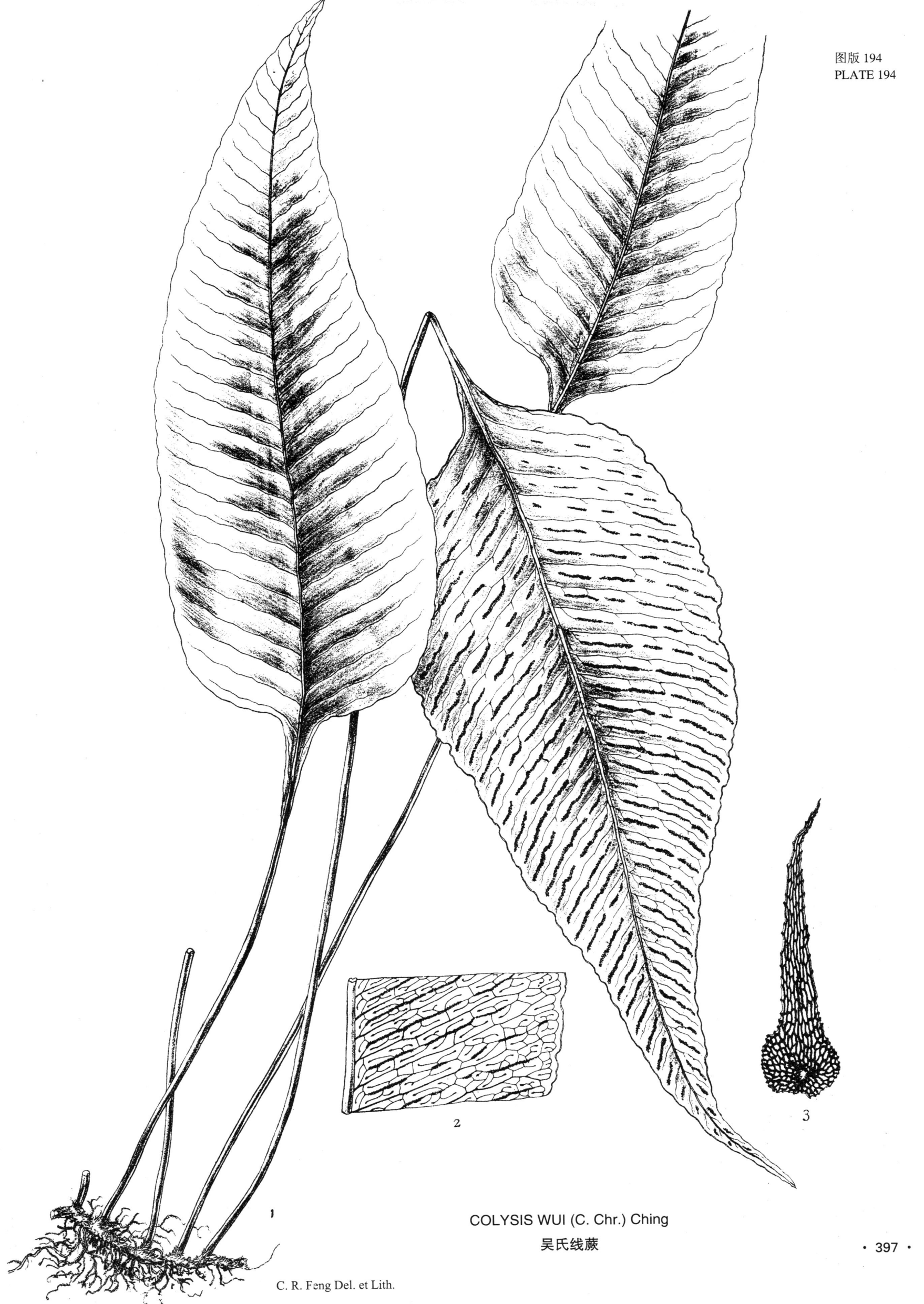

COLYSIS WUI (C. Chr.) Ching
吴氏线蕨

图版 195 断线蕨

根状茎横行,被深棕色之卵状披针形疏鳞片;叶疏生,长 40—60 cm,宽 5—7 cm,阔披针形,渐尖头,全缘,下部渐狭,沿叶柄下延几达基部,纸质,光滑,干则呈淡黄色,叶脉明显,侧脉亚斜出,曲折,间距 8 mm,细脉网状,眼具分叉小脉;子囊群大,椭圆形,短线形或卵圆形,一行排列,不具盖。

分布:广东,广西,海南,香港,云南,贵州,台湾;越南,印度及菲律宾群岛均产之。

本种异于本属其他各种者,为其子囊群不成通直线形而常断为椭圆形或卵圆形是也。

图注:1.本种全形(自然大);2.叶体之一部,表示叶脉及子囊群之情形(放大 2 倍);3.根状茎上之鳞片(放大 16 倍);4.叶柄基部之鳞片(放大 16 倍)。

COLYSIS HEMIONITIDEA (Wallich) Presl

COLYSIS HEMIONITIDEA (Wallich) Presl, Epim. Bot. 147 (1849); Ching, Bull. Fan Mem. Inst. Biol. 4: 320 (1933).

Polypodium hemionitideum Wallich, List no. 284 (1828, nom. nud.); Mett. Farngatt. Polyp. 122 (1857); Hk. Sp. Fil. 5: 73 (1863); Syn. Fil. 360 (1864); Clarke, Trans. Linn. Soc. II. Bot. 1: 651 (1880); Christ, Farnkr. d. Erde 105 (1807); Diels in Engl. u. Prantl, Nat Pflnzenfam. 1: 4. 315 (1899); C. Chr. Ind. Fil. 532 (1905); Christ, Journ. Bot. de France 19: 21 (1905); ibid. II. 1: 9 (1908); Takeda, Notes, R. Bot. Gard. Edinb. 8: 308 (1915); Wu, Polyp. Yao-shan. in Bull. Dept. Biol. Coll. Sci. Sun Yatsen Univ. No. 3: 282 pl. 132 (1932).

Selliguea hemionitidea Presl, Tent. Pterid. 216 t. 9 f. 17 (1836).

Pleopeltis hemionitidea Moore, Ind. Fil. 436 (1862); Bedd. Ferns S. Ind. t. 182 (1866); Handb. Ferns Brit. Ind. 359 (1883).

Gymnopteris feei f. *anomala* Bedd. Ferns Brit. Ind. t. 274 (1868).

Rhizome wide-creeping, sparcely scaly; *scales* rufo-brown, ovate-lanceolate, acuminate, thin, clathrate; *frond* distant, 40-60 cm long, 5-7 cm broad, broadly lanceolate, acuminate, entire, gradually narrowed and long-decurrent down to near the base of sparcely scaly stipe; *texture* herbaceous, brownish-green upon drying; *venation* distinct, lateral main veins subpatent, parallel, flexuose, about 8 mm apart, the intervening veinlets anastomosing in 3-rowed rectangular areolae with divaricate included veinlets; *sori* large, oblong, short linear or roundish, one-rowed between each pair of lateral veins, exindusiate.

Guangdong: Lohfau Shan, *N. K. Chun 41586*, *41293*, *40466*; *C. Ford*; *C. O. Levine 506*. Swatow, Thaiyong, *Dr. Dalziel*, July, 1901; Taimo Shan, *C. G. Matthew*, Oct. 15, 1907; Yingtak, Tai Chun, *C. L. Tso 22026*; ibid., Jewhan, *H. Y. Liang 61296*; Sunyi, *S. P. Ko 51262*; Tsingtan Hsien, *K. K. Tsoong 1286*. Guangxi: Tzepoo, Luchen Hsien, *R. C. Ching 5574*; Lin Yen Hsien, Tsinglung Shan, *R. C. Ching 6907*; Yao Shan, Pingnam Hsien, *S. S. Sin 104A*, *104B*. Hainan: Ng Chi Leng, *F. A. McClure 9342*. Hong Kong: Lantao Island, *C. Ford* (1874); *Tutcher 642* (1909), Yunnan: Mengtze, *Henry 10342*, *11488A*; *Hancock 50* (1893). Guizhou: Without locality, *Cavalerie 3396*. Taiwan.

Also Vietnam, East India and the Philippine Islands.

A distinct and also perhaps a linking species between the genera *Colysis* and *Microsorium*, as indicated by its unstable soral conditions, which are generally interrupted into oblong or roundish shape.

Plate 195. Fig. 1. Habit sketch (natural size). 2. Portion of frond, showing venation and soral conditions ($\times$ 2). 3. Scale from rhizome ($\times$ 16). 4. Scale from base of stipe ($\times$ 16).

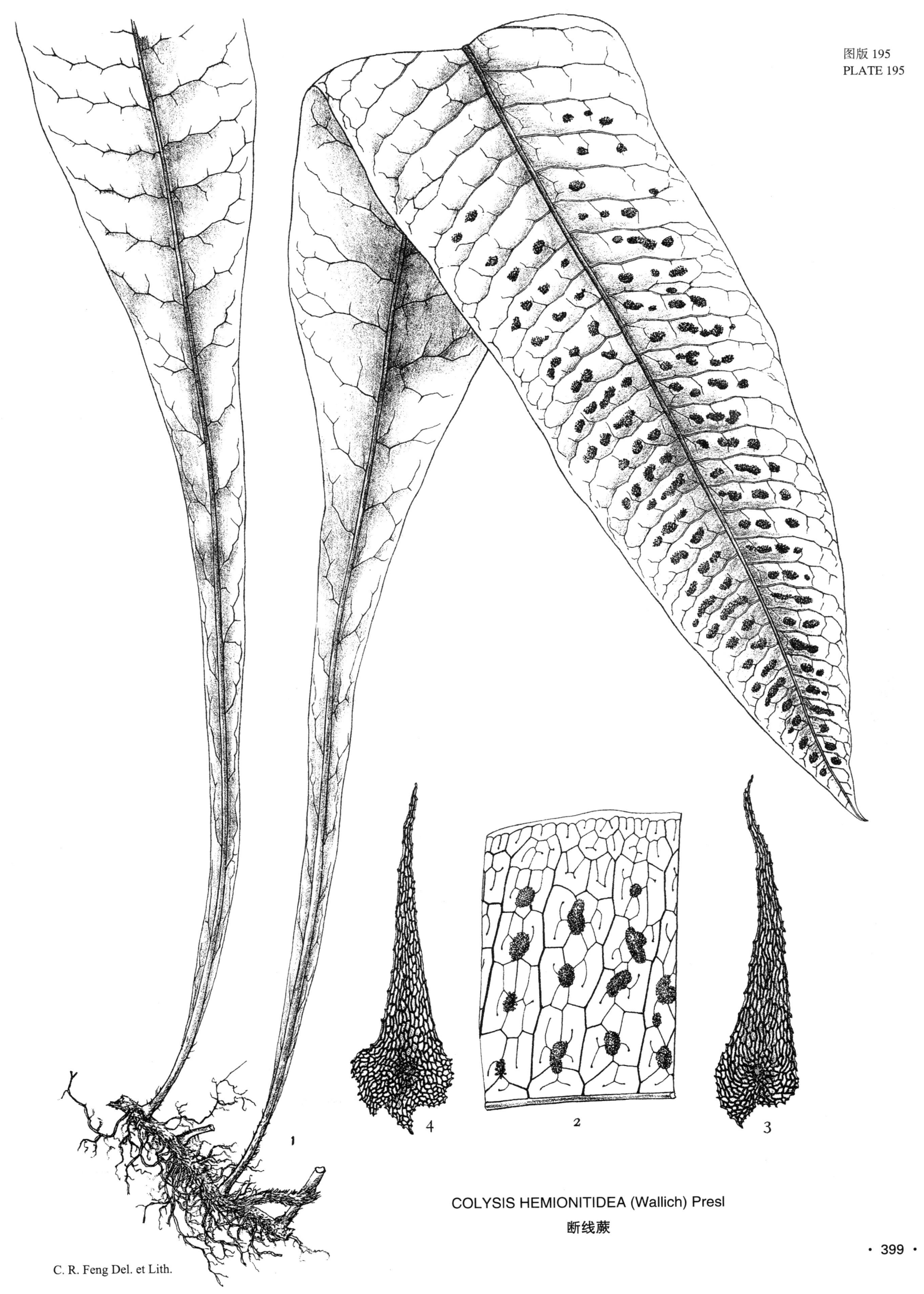

COLYSIS HEMIONITIDEA (Wallich) Presl
断线蕨

图版 196　莱氏线蕨

根状茎横行；鳞片为披针形，黑褐色，质薄，网眼明显；叶散生，长 25—35 cm，宽 3—4 cm，披针形，渐尖头，基部自叶柄两侧下延，仅具 2—5 cm 长无翅之柄，边缘呈波形，侧脉屈折，网脉显明，薄纸质，干则呈黑褐色；子囊群线形，自中肋达叶边。

分布：广东，香港，台湾；越南亦产之。

此为本属特殊之蕨种，其异于前种者，已详于该种，兹不赘述。

图注：1. 本种全形（自然大）；2. 叶体之一部，表示叶脉及子囊群（放大 2 倍）；3. 根状茎上之鳞片（放大 30 倍）。

COLYSIS WRIGHTII (Hooker) Ching

COLYSIS WRIGHTII (Hooker) Ching, Bull. Fan Mem. Inst. Biol. **4**：324 (1933).

Gymnogramme wrightii Hooker, Sp. Fil **5**：160 t 303 (1864); Syn. Fil. 388 (1867).

Polypodium wrightii Mett. ex Diels in Ehgl. u. Prantl: Nat. Pflanzenfam. **1**：4. 316 (1899); C. Chr. Ind. Fil. 575 (1905).

Selliguea wrightii J. Sm. Hist. Fil. 102 (1875).

Polypodium kusukusense Hayata, Ic. Pl. Form. **5**：320 f. 131 (1915); C. Chr. Ind. Fil. Suppl. II. 26 (1912-16).

Rhizome wide-creeping, densely scaly; *scales* fusco-brown, lanceolate from rounded base, thin, clathrate; *frond* distant, 25-35 cm long including wingless stipe 2-5 cm long (sterile leaves almost without wingless stipe), lanceolate, 4-3 cm broad, acuminate, rather gradually narrowed and decurrent along stipe in a broad wing on each side, margin repando-undulate; *texture* thin herbaceous or submembranceous, glabrous on both sides, turning blackish upon drying; *venation* distinct, lateral main veins subpatent, flexuose, intermediate veinlets anastomosing in 2-rowed elongate areolae with simple or divaricate included veinlets; *sori* linear, extending from costa to near the margin.

Hong Kong: *Bodinier 145*, May 2, *1898*. Guangdong: Tung Shing Hsien, *K. K. Tsoong 4843*, *4905*, *1950*, *1153*; Yao Shan, *C. L. Tso*; North River, *C. G. Matthew*, Nov. 26, 1907; Swatow, Thaiyong, *Dr. Dalziel*, August, 1897; July, 1901; Sept. 1899. Zhejiang: Pinyang Hsien, *H. H. Hu 1898*. Taiwan.

Also Vietnam.

A very distinct species which, by its present distribution, seems to be more common in the Islands Taiwan and Loochoo than on the mainland. It is most closely related to *C. Leveillei* (Christ) Ching but differs in its submembranaceous leaves with repando-undulate margin, always turning blackish upon drying, of which the fertile ones are generally provided with short wingless stipes.

Plate 196. Fig. 1. Habit sketch (natural size). 2. Portion of lamina, showing venation and sori ($\times$ 2). 3. Scale from rhizome ($\times$ 30).

COLYSIS WRIGHTII (Hooker) Ching
莱氏线蕨

根状茎横行,被深棕色之披针形鳞片;叶散生,柄长 20—25 cm(不生子囊群之叶之柄长仅半之),略被细长鳞片,叶体卵形三角形,长 10—16 cm,基部宽达 6—10 cm,渐尖头,边缘下部浅裂成 2—7 对线状披针形之裂片,基部极狭而下延,薄纸质,干则常呈黑褐色,侧脉屈折,网脉可见;子囊群线形,自中肋达叶边,唯常断续。

  分布:广东北部,福建中部,广西东部产之。

  图注:1. 本种全形(自然大);2. 同上,唯少分裂(自然大);3. 叶体之一部,表示叶脉及子囊群(放大 1.5 倍);4. 根状茎上鳞片(放大 40 倍);5. 柄上鳞片(放大 40 倍);6. 叶体下面中肋上之鳞片(放大 40 倍);7. 叶体下面小脉上子囊群中之鳞片(放大 40 倍)。

COLYSIS HEMITOMA (Hance) Ching

COLYSIS HEMITOMA (Hance) Ching, Bull. Fan Mem. Inst. Biol. **4**: 326 (1933).

  Polypodium hemitomum Hance, Journ. Bot. (1883) 269; C. Chr. Ind. Fil. 532 (1905).

  Polypodium macrophyllum var. *fokiense* Cop. Phil. Journ. Sci. Bot. **3**: 283 (1908).

  Polypodium sp. nov. Wu. Polyp. Yaoshan. in Bull. Dept. Biol. Coll. Sci. Sun Yatsen Univ. No. **3**: 316 t. 149 (1932).

  Rhizome wide-creeping, densely scaly: *scales* fusco-brown, lanceolate from rounded dentate base, thin, clathrate; *frond* far apart, stipe 20-25 cm long, (half as long and broadly winged throughout in sterile leaves), straminous, sparcely scaly, winged half way down, lamina broadly lanceolate, acuminate, generally with hastate base, thence broadly decurrent downward, entire or more frequently with 1-2 pairs of lanceolate, horizontally patent lobes, or sometimes (as in type) regularly lobato-laciniate with 5-6 linear-lanceolate lobes on each side, margin entire, but undulate; *texture* herbaceous, glabrous above, more or less sparcely scaly on veins and costa beneath when young; *venation* distinct, lateral veins oblique, flexuose, veinlets anastomosing along main vein in one row of elongate areolae with divaricate included veinlet; *sori* linear, flexuose, extending from costa to margin, often interrupted.

  Guangdong: Lienchow, B. C. Henry (1881), 22104 in Herb. Hance (type); ibid., Fak Shan, *C. L. Tso 22626* (typical), Oct. 5, 1930; Lungtau Shan, Iu village, *To & Tsang 12159*, *12231*, May 27, 28, 1924; ibid. Ku Koong, *Y. K. Wang 31688*; North River, Lanfang Kan, *N. K. Chun 5824*; Yintak, *C. L. Tso 22626*; Lokchong, Kook Kiang, *S. P. Ko 50217*; *Y. K. Wang 31481*; *N. K. Chun 43036*. Fujian: Central kart, *S. T. Dunn 3894*. Guangxi: Yao Shan, *S. S. Sin et K. K. Wang 102*; ibid., *S. S. Sin 3761* (f. integra); May 26, 1928, ibid., *Y. J. Wang 5209*, *5309*, *5342*, *113*.

  A peculiarly distinct endemic fern, only related to C. *Wrightii* (Hk.) Ching, differing in hastate or lobato-laciniate lamina on a long and broadly winged stipe, and green color of leaves with less undulate margin. There is an entire-leaved form (f. *integra*) which resembles C. *Wrightii* so closely that it can be distinguished from that species only by proportionally short broad lamina rather suddenly narrowed towards base and much longer stipe, which is generally winged only half way down.

  Plate 197. Fig. 1. Habit sketch, representing typical form (natural size). 2. The same but with only 1-2 lobes on each side (natural size). 3. Portion of lamina, showing venation and sori ($\times$ 1.5). 4. Scale from costa beneath ($\times$ 40). 5. The same, from stipe ($\times$ 40). 6. The same, from costa beneath ($\times$ 40). 7. The same, from veinlet beneath, intermixed with sorus ($\times$ 40).

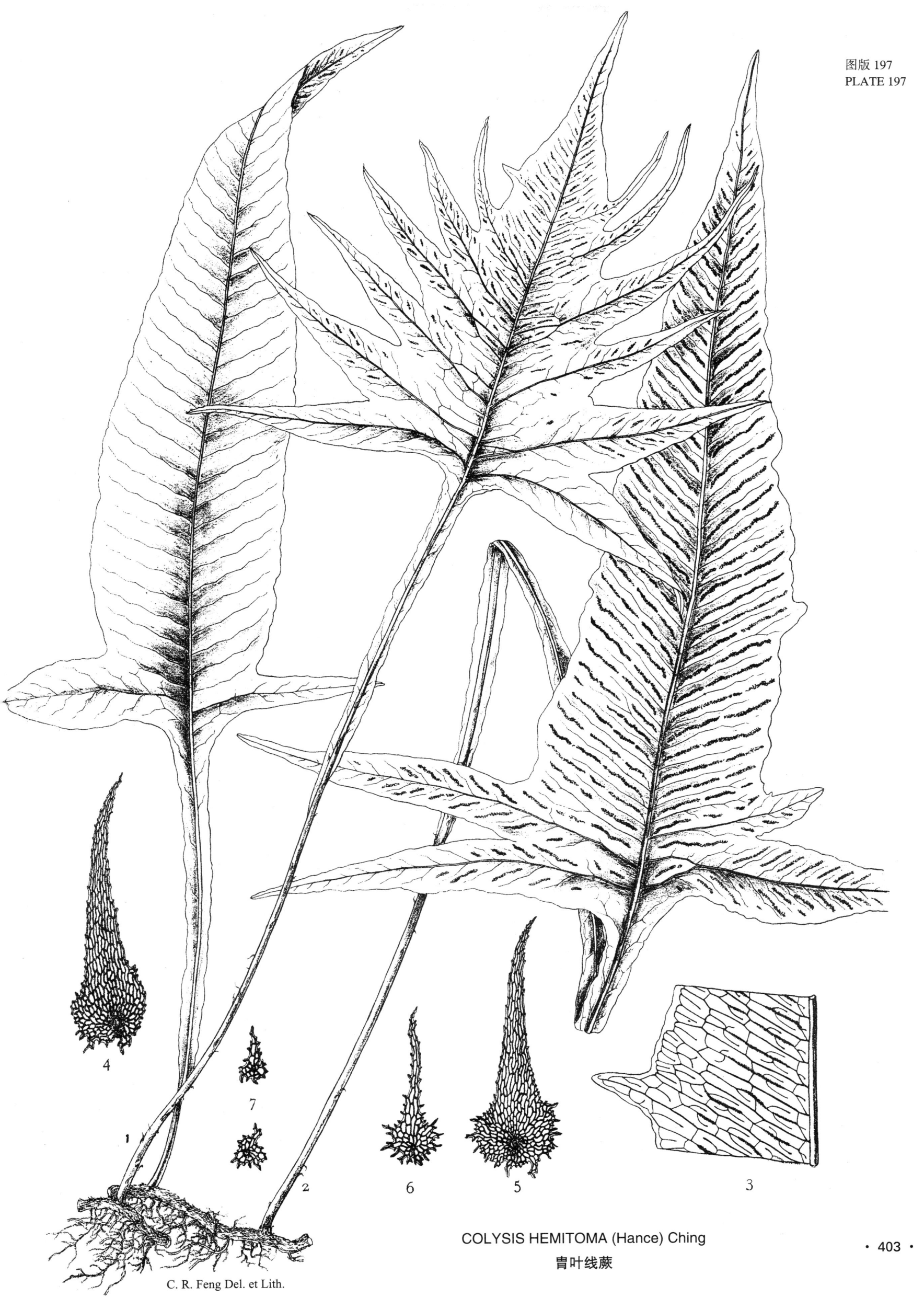

COLYSIS HEMITOMA (Hance) Ching
胄叶线蕨

根状茎横行,具黑褐色之披针形鳞片;叶散生,亚二形,叶柄长 20—30 cm,淡稻秆色,光滑,叶体长 10—18 cm,宽亦如之,掌状分裂(间为 2—3 裂,或单叶),基部略延长,裂片 3—5 数,披针形,渐尖头,长 10—16 cm. 宽 1.5—3 cm,基部稍狭,全缘唯呈浅波状,纸质,两面光滑,侧脉斜出,曲屈可见,网脉二列,内具小脉;子囊群线形,斜出,自中肋达叶边;不生子囊群之叶具较短之柄与较阔之裂片。

分布:广西,海南;越南亦产之。

本种通常因其掌状分裂之叶,故易于区别,然间有 2—3 裂者,或甚有不分裂者,是当注意耳。

图注:1. 本种全形(自然大);2. 裂片之一部,表示叶脉及子囊群之位置(放大 3 倍);3. 子囊(放大 108 倍);4. 根状茎上之鳞片(放大 36 倍);5. 根状茎之横切面,表示维管束之布置(放大 20 倍)。

COLYSIS DIGITATA (Baker) Ching

COLYSIS DIGITATA (Baker) Ching, Bull. Fan Mem. Inst. Biol. **4**: 328 (1933).
  Gymnogramme digitata Baker, Journ. Bot. (1890) 267.
  Polypodium digitatum C. Chr. Ind. Fil. 522 (1905).
  Grammitis finlaysoniana Wall. List 248, No. 776 (1829, nom. nud.).
  Selliguea finlaysoniana Moore, Ind. Fil. LXVI (1857); Christ, Journ. Bot. d. France 2^0. sér. **1**: 11 (1908)
  Gymnogramme finlaysoniana Baker, Ann. Bot. **5**: 486 (1891).
  Colysis tridactylis Fée, Gen. Fil. 176 (1850-52, nom. nud.).
  Polypodium Vietnamense Christ, Journ. Bot. d. France **19**: 77 (1905); C. Chr. Ind. Fil. 508 (1906); Merr. Enum.
    Hainan Pl. in Lingnan Sci. Journ. **5**: 18 (1927).
  Polypodium ampelidium Christ, I. c. p. 78; C. Chr. I. c.; Merr. I. c. p. 17.
  Polypodium podopterum Christ, I. c. p. 125; C. Chr. I. c. 555.
  Polypodium cadieri Christ, I. c. p. 75; C. Chr. I. c. p. 515.

Rhizome wide-creeping, copiously clothed in lanceolate hair-pointed atro-brown and clathrate thin *scales*; *fronds* subdimorphous, 1-3 cm apart, stipe 20-30 cm long, pale straminous, naked, base articulated, lamina 10-18 cm each way, generally palmately divided (sometimes 2-3-lobed or simple), base rarely shortly decurrent, *segments* 3-5, lanceolate, acuminate, 10-16 cm long, 1.5-3 cm broad, base somewhat narrowed, margin thickened, entire but repandulous; *texture* chartaceous, glabrous on both sides, pale green; lateral *veins* oblique, flexuose, visible; *sori* linear, oblique, between lateral main veins, extending from costa to margin; *sterile frond* conform, but on much shorter and often winged stipe and with broader segments.

Guangxi: Lungchow, *Morse 45*; Luchen Hsien, *R, C. Ching 5620*, on rocks along stream under forest. Hainan Island: Lea Mui, *Eryl Smith 1513*, Jan. 4, 1923, on stream side; South of Fan Ta, *Tsang Wai-tak 17797* (1929); Five Finger Mt., *F. A. McClure 8080*, *8552*; *W. Y. Chun 6804* (in Herb. Univ. Nanking); *Eryl Smith 2542* (f. simplex), on tree; Hoichow, *Hancock 20*; Linfa Shan, *F. A. McClure 8070*; *Tsang Wai-tak 17024*, *15839*; *W. Y. Chun 6605*; Pat Ka Shan, *F. A, McClure 8552* (f. simplex); Tun Kao, *Eryl Smith 1510*, *1518*, *1511*; Huploha, *W. Y. Chun 1316*; Lohoe, *Miss Moninger 225*.

Vietnam: *Balansa 102* (type); Lang Biang, *Eberhardt 106*; *Gaudichaud*; *Mrs. Clemens 4365*. *Cadier 45* (type of *Polypodium Vietnamense*); *Cadier 103* (type of *P. Cadieri*).

A very distinct and pretty fern, now found to be fairly common in the localities noted, but still unknown elsewhere. It differs from the other species of the genus in its palmatifid leaves with 3-4-5 lanceolate entire segments and wingless stipe, but forms with 2-3-lobed or even simple leaves have also been found not uncommon even in the same collection or on the same rhizome. *Polypodium Vietnamense* Christ differs only in the slightly winged upper stipe, while *P. Cadieri* Christ has simple or 2-3-lobed leaves with much contracted linear fertile segments.

Plate 198. Fig. 1. Habit sketch (natural size). 2. Portion of fertile segment, showing venation and position of sori ($\times$ 3). 3. Sporangium ($\times$ 108). 4. Scale from rhizome ($\times$ 36). 5. Cross section of rhizome, showing the arrangement of steles ($\times$ 20).

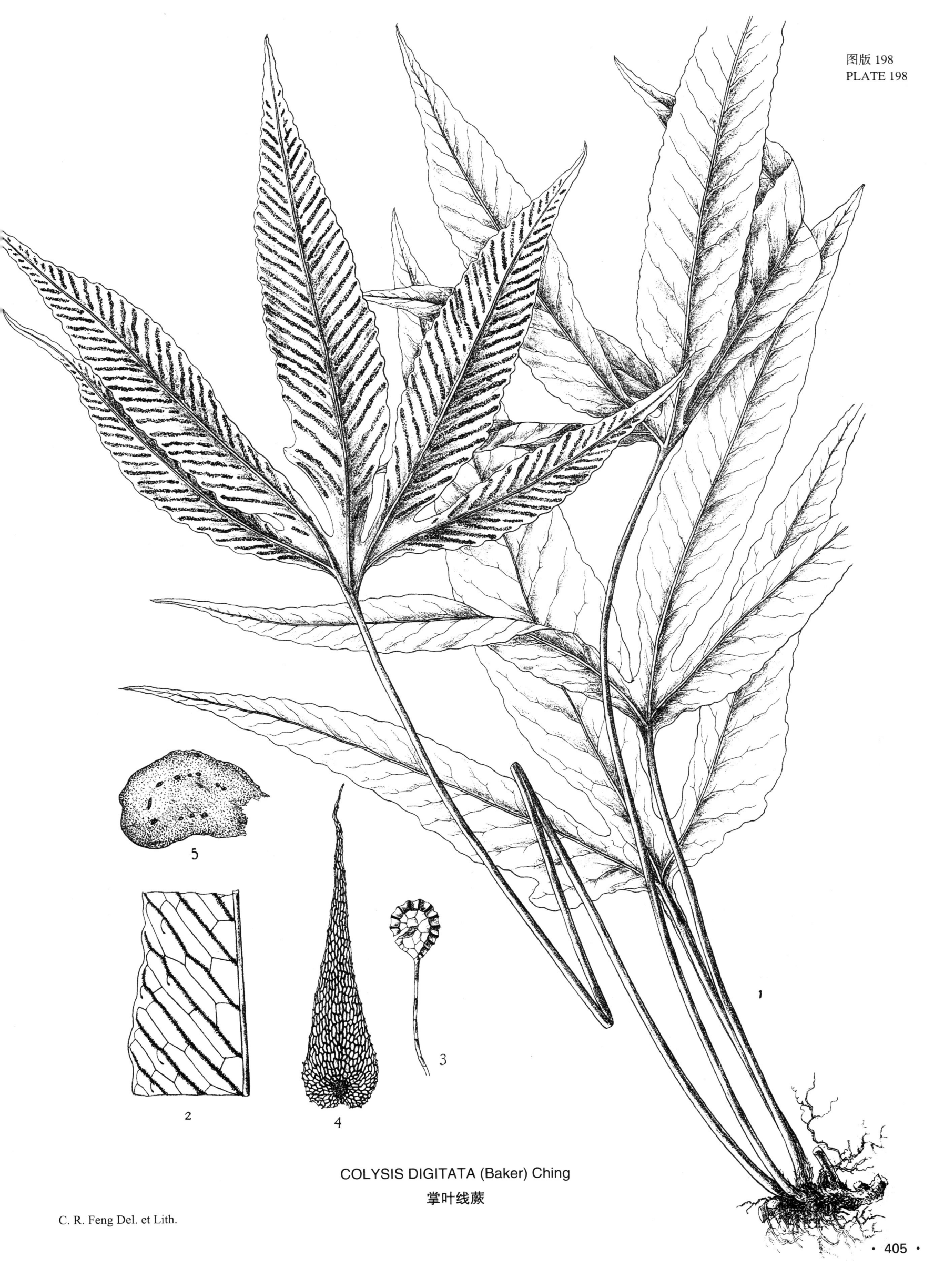

COLYSIS DIGITATA (Baker) Ching
掌叶线蕨

根状茎横行，被卵形渐尖头淡棕色之鳞片；叶散生，柄长 35—55 cm，稻秆色，光滑，叶体卵形，长 15—20 cm，宽约如之，羽状分裂；小叶 1—3 对，对生，长 14 cm，宽达 3 cm，阔披针形，渐尖头，基部渐狭，沿中肋延长，叶边全缘而稍反卷，纸质，光滑，叶脉明显，网脉可见；子囊群线形，斜出，自主脉外出，达叶宽 2/3。

分布：云南特产。

本种形体极类椭圆线蕨（*C. elliptica*）之阔叶变种（var. pothifolia），唯其叶柄特长，叶体较短，仅具 1—3 对之小叶，茎上鳞片为卵状披针形，呈淡黄色，故易识别。

图注：1. 本种全形（自然大）；2. 小叶之一部，表示叶脉及子囊群之位置（放大 2.5 倍）；3. 根状茎上之鳞片（放大 20 倍）。

COLYSIS PENTAPHYLLA (Baker) Ching

COLYSIS PENTAPHYLLA (Baker) Ching, Bull. Fan Mem. Inst. Biol. **4**: 332 (1933).

Gymnogramme pentaphylla Baker, Kew Bull. (1898) 233

Polypodium pentaphyllum Christ (non Baker, 1891), Bull. Acad. Céogr. Bot. (1906) 248.

Polypodium ellipticum var. *pentaphyllum* C. Chr. Ind. Fil. 524 (1905).

Polypodium mediosorum Ching, Bull. Fan Mem. Inst. Biol. **2**: 19 t. 4 (1931); C. Chr.. Ind. Fil. Suppl. III. 153 (1934).

Rhizome wide-creeping, densely radicose and scaly; *scales* ovate-acuminate, light brown, iridescent, thin, clathrate, dorsally affixed; *frond* distant, stipe 35-55 cm long, straminous, glabrous above base; lamina ovate, 15-20 cm long, nearly as broad, pinnate; *pinnae* 1-3-jugate, or rarely more, 14 cm long, to 3 cm broad, broadly lanceolate; opposite, acuminate, base attenuate and decurrent along rachis, equal-sized, margin entire, slightly repand; *texture* herbaceous, green, glabrous; *costa* prominent on both sides, *lateral main veins* visible above, veinlets anastomosing in 4 rows of elongate, oblique areolae with included recurrent simple veinlets; *sori* linear, oblique, extending over two-thirds way to the margin.

Yunnan: Mengtze, *A. Henry 9033A*, *9033* (*type*), *9295*; Wen Shan Hsien, Louchin Shan, *H. T. Tsai 51373*; ibid., Da Tsin, *H. T. Tsai 51647*.

A distinct endemic species, closely related to *C. elliptica* (Thbg.) var. *pothifolia* (Don) Ching in habit, differing in very long-stipitate leaves and proportionally short ovate lamina with only 1-3 pairs of broadly lanceolate pinna, thick and rather short sori and the broadly ovate-acuminate, light brown scales on rhizome.

Plate 199. Fig. 1. Habit sketch (natural size). 2. Portion of pinna, showing venation and sori ($\times$ 2.5). 3. Scale from rhizome ($\times$ 20).

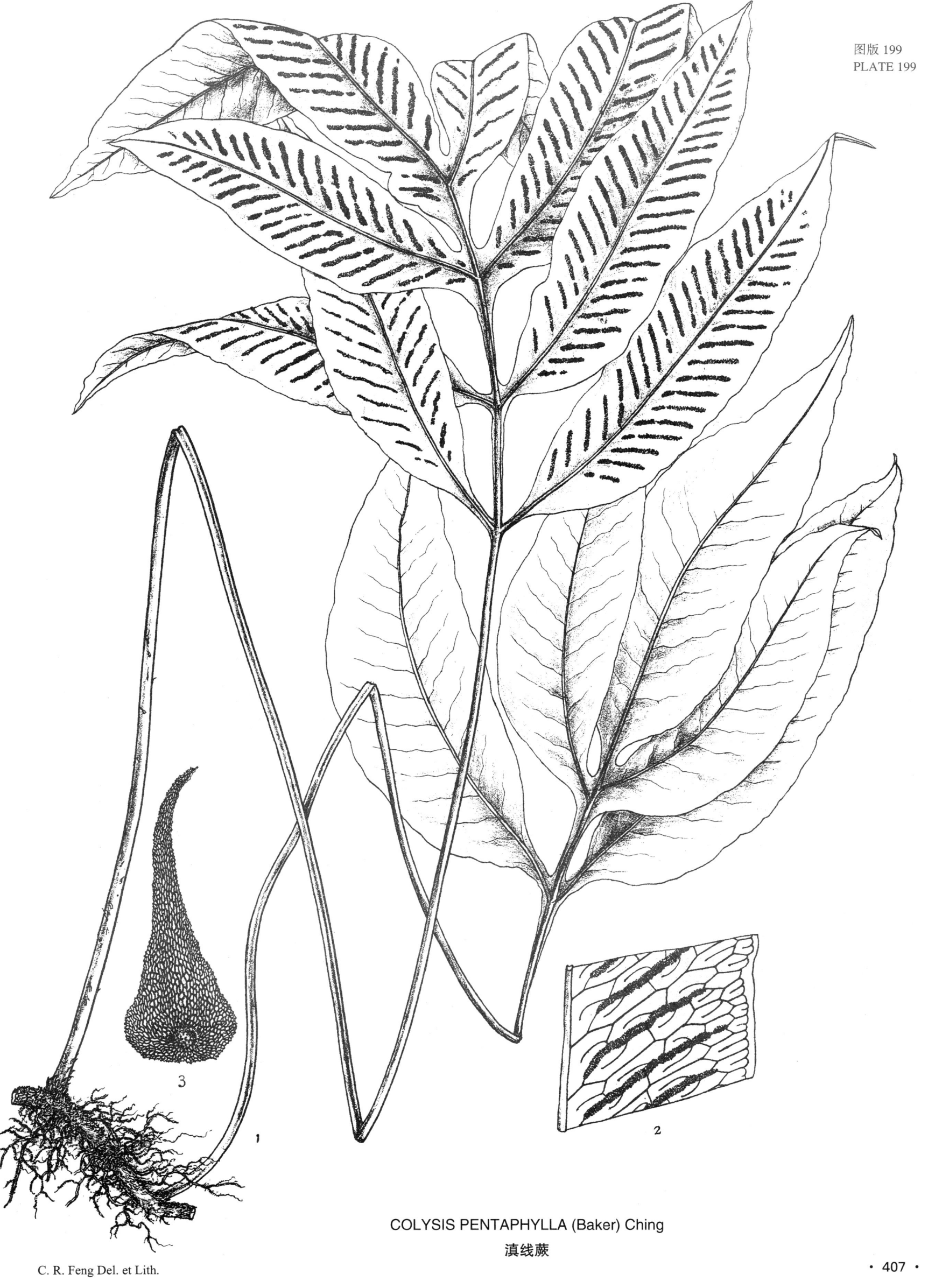

COLYSIS PENTAPHYLLA (Baker) Ching
滇线蕨

图版 200　马氏线蕨

根状茎横行,被深棕色线状披针形之鳞片;叶散生,亚二形,柄长达 20 cm,光滑,叶体长 20—24 cm,宽达 8 cm,长椭圆形,端呈尾状,羽状分裂,小叶达 16 对,亚对生,狭披针形,长达 5 cm,宽 7 mm,基部数对等长,向叶端渐短,渐尖头,全缘,基部较狭而下延,彼此分离,唯上部数对间各有狭翅连接,纸质,光滑,侧脉缺如,小脉成二列斜出之网眼;子囊群线形,斜出;不生子囊群之叶之柄长达 10 cm,叶体较宽,小叶宽达 1 cm,中轴具狭翅。

分布:广西及越南产之。

本种形体极似椭圆线蕨,唯其叶为显著之二形,小叶数较多而较狭,故易分别。

图注:1.本种全形(自然大);2.生子囊群之小叶,表示叶脉及子囊群(放大 2 倍);3.根状茎上之鳞片(放大 50 倍)。

COLYSIS MORSEI Ching

COLYSIS MORSEI Ching, Bull. Fan Mem. Inst. Biol. **4**: 330 (1933).

Polypodium Morsei Ching, Bull. Fan Mem. Inst. Biol. **2**: 17 t. 1 (1931); C. Chr. Ind. Fil. Suppl. Ⅲ. 154 (1934).

Rhizome wide-creeping, densely scaly; *scales* fusco-brown, linear-lanceolate from ovate base, clathrate, entire; *frond* far apart, subdimorhpous, the fertile one with stipe to 20 cm long, straminous, glabrous, lamina 20-24 cm long, to 8 cm broad, oblong-elongate, with caudate apex, simple pinnate; *pinnae* to 16 pairs, subopposite, obliquely patent, to 5 cm long, 7 mm broad, basal ones not shortened, gradually abbreviated towards the caudate apex, narrowly lanceolate, acuminate, entire, considerably constricted above decurrent base, the upper ones are connected by a narrow wing along rachis; *texture* herbaceous, green, glabrous on both sides; *veins* anastomosing only in two rows of oblique areolae along costa; *sori* linear, oblique, extending from near the costa to margin; *sterile fronds* conform but much shorter, on stipe 10 cm long, lamina 15 cm long, to 7 cm broad; *pinnae* 3.5 cm long, 1 cm broad, lanceolate, narrowed above decurrent base.

Guangxi: Lungchow, Ah Chin, n. w. hills, *H. B. Morse 22, 64* (type); Lin Yen Hsien, *R. C. Ching 6633*.

Vietnam: Than Moi, *Balansa 36, 100*, sur les roches calcaires, Jan. 19, 1886; *Pételot 4107*.

A pretty and gracil fern, closely related to the typical form of *C. elliptica* (Thbg.), differing chiefly in more pronounced dimorphism of leaves, more numerous and much narrower pinnae with anastomosed venation of simpler type.

Plate 200. Fig. 1. Habit sketch (natural size). 2. Fertile pinna, showing venation and sori ($\times$ 2). 3. Scale from rhizome ($\times$ 50).

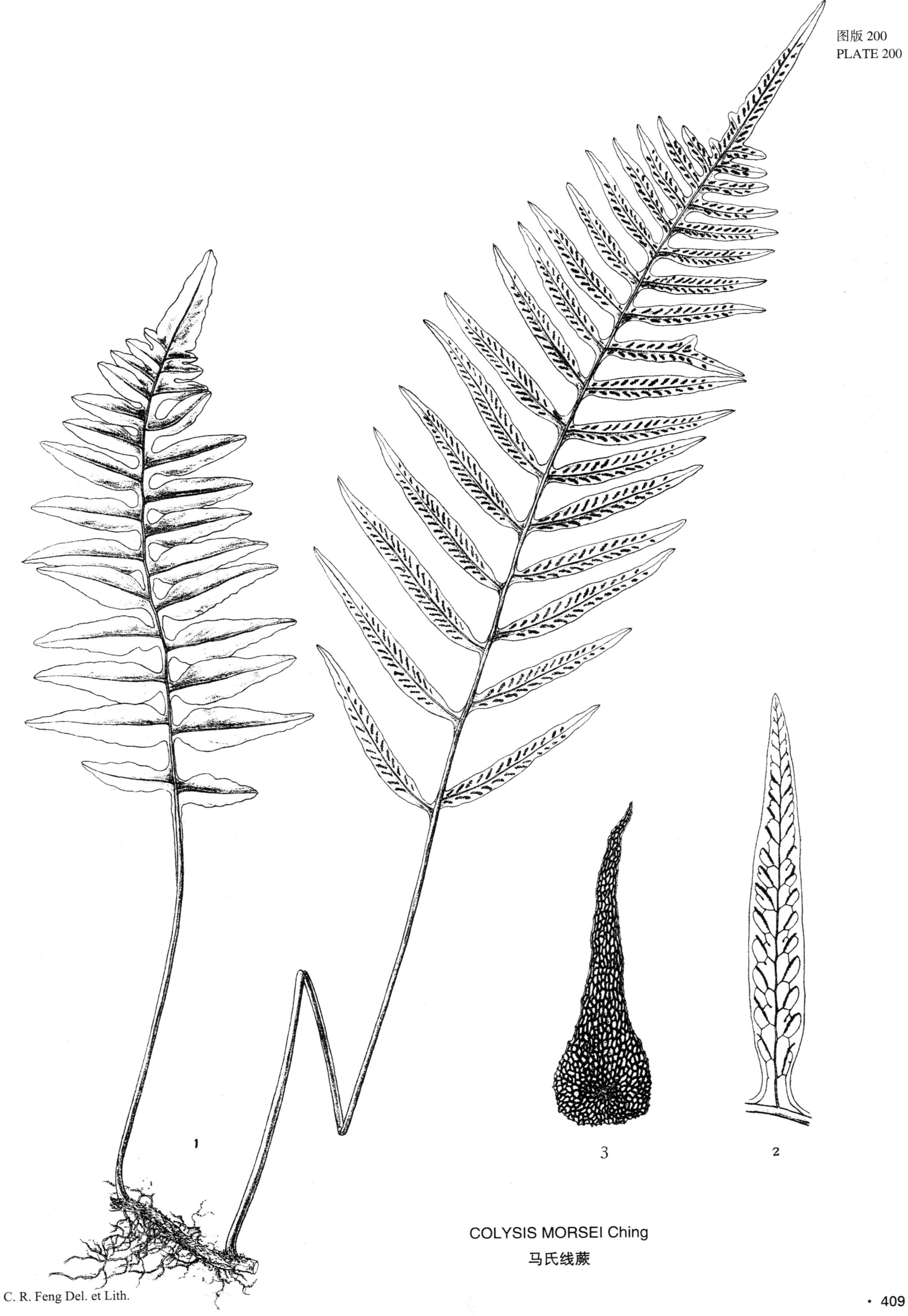

COLYSIS MORSEI Ching
马氏线蕨

中国蕨类植物图谱

ICONES FILICUM SINICARUM

BY

REN-CHANG CHING

PROFESSOR

DIVISION OF TAXONOMY & PHYTOGEOGRAPHY

INSTITUTE OF BOTANY

ACADEMIA SINICA

第五卷

FASCICLE 5, PLATES 201-250

高大的附生植物,根状茎为粗大匍匐状,厚肉质,密生蓬松深锈色的长鳞片,为钻状长线形,边缘有睫毛,另有毛茸的线状根混生于鳞片间,弯曲的根状茎盘结成为大块的垫状物,由此生出一丛无柄,坚硬而略张开的革质叶,形成一个圆而中空的高冠,形体极似鸟巢蕨 Neottopteris Nidus (L.) J. Sm.;叶一型,长 80—120 cm 或过之,中部宽 20—30 cm,长圆状倒披针形,向顶端渐尖,向下渐变狭,至 2/3 处又逐渐扩张成为膨大圆心脏形的基部,宽约 15—25 cm,有宽缺刻或浅裂的边缘,基部以上叶体为深羽裂,再向上几乎深裂到叶轴;裂片多数,被圆缺刻分开,中部的裂片长达 15—22 cm,宽 2—3.5 cm,向上斜出,披针形,基部较宽,向顶端渐尖,全缘,有加厚的叶缘,下部裂片较中部为短,分裂较浅,为急尖头或圆头,为阔圆形的缺刻分开;叶为硬革质,两面光滑无毛,干后硬而有光泽,裂片往往由深褐色光亮的叶轴以关节脱落;叶脉粗而很明显,一回侧脉斜出,开展,突出,通直,相隔 4—5 mm,向外达于加厚的叶缘,二回侧脉与一回侧脉直角相交,成一回网眼,再分割一次成 3 个长方形小网眼,内有棍棒状顶端的分叉小脉;子囊群位于小叶脉交接处,限于叶体上半部,其下半部为不育,4—6 个生于一回侧脉之间,但并不位于正中央,而是略偏近下脉,每一个网眼里有一个子囊群,在中肋与叶缘间排成一长列,亚圆球形或长圆形,分开,但成熟后常多少汇合成一条连贯的囊群线,孢子二面形,肾状长圆形,光滑透明。

分布:云南,广东,广西,海南,台湾,香港;越南,缅甸,马六甲。

本种主要产于亚洲大陆热带及亚热带的雨林或季雨林中,在华南地区相当普遍,向北分布到贵州南部,南达南太平洋的马六甲。植株高大,挺拔,附生于树干或溪边岩石上,为极美丽的观赏植物,适于温室栽培,其粗大肉质根状茎可供药用。

图注:1. 本种全形(自然大);2. 能育裂片的一部,表示叶脉及子囊群(放大 4 倍);3. 同上,放大。4. 根状茎上的鳞片(放大 10 倍);5—6. 孢子囊及孢子(放大 100 倍)。

PSEUDODRYNARIA CORONANS (Wallich) Ching

PSEUDODRYNARIA CORONANS (Wallich) Ching in Sunyatsenia **5**: 357. 1940; **6**: 10. 1941; Copel. Gen. Fil. 201. 1947.
  Polypodium coronans Wall. List n. 288. 1828 (nom. nud.); Hook. Exotic Ferns t. 95. 1857; Mett. Farngatt, Polyp. 121 n. 242 t. 3, f. 40, 41. 1857; Benth. Fl. Hongk. 459. 1861; Christ, Farnkr. d. Erde 117. 1898; Diels in Engl. u. Prante, Nat. Pflanzenfam. **1**: iv. 319. 1899; C. Chr. Ind. Fil. 518. 1905; Merr. Enum. Hainan Pl. in Lingnan Sci. Journ. **5**: 17. 1927; Wu, Wong et Pong, Polyp. Yaoshan. in Bull. Dept. Biol. Sunyatshan Univ. No. **3**: 304 t. 141. 1932.
  Phymatodes coronans Presl, Tent. Pterid. 198. 1836 (nom. nud.).
  Drynaria coronans J. Sm. in Hook. Journ. Bot. **4**: 61. 1841 (nom. nud.).
  Aglaomorpha coronans Copel. in Univ. Calif. Publ. Bot. **16**: 117. 1929.
  Polypodium conjugatum Bak. in Hook. & Bak. Syn. Fil. 366. 1868 (non Poir. 1804); Dunn & Tutcher, Fl. Kwangt. and Hongk. 352. 1912.
  Drynaria conjugata Bedd. Ferns Brit. Ind. Correction, 1870.
  Drynaria esquirolii C. Chr. in Bull. Acad. Géogr. Bot. Mans **13**: 139. 1913.

A large and strong epiphyte. *Rhizome* creeping, thick, fleshy, stout, and interwoven, covered by dense, shaggy, dark ferruginous, long linear-subulate scales with ciliate margin and intermixed with wiry and hairy roots, holding tortuous branches of the rhizome together and thus forming a large cushion-like mass, from which arises, in a perfect circle, a tall crown of sessile, stiffy coriaceous fronds after the manner of *Neottopteris Nidus* (L.) J. Sm. *Fronds* monomorphous, 80-120 cm or even longer, 20-30 cm broad at the middle, oblong-oblanceolate, acuminate, gradually narrowed downwards to about 2/3 way down, thence again expanding gradually towards the dilated broad, rounded and cordate base, 15-25 cm across with broadly sinuate or shallowly lobed margin, above which the whole leaves are deeply pinnatifid and further up almost down to the rachis with numerous *segments*, of which the middle ones 15-22 cm long, 2-3.5 cm broad, oblique, separated by round sinuses, lanceolate from the broadened base, gradually acuminate with entire thickened margin throughout, the lower ones gradually shortened, cut down less and less deep with acute or rounded apex and separated by round broad sinuses; *texture* stiffy coriaceous, glabrous, glossy and firm when dried and all segments become detached from the strong, brown and shining rachis; *venation* distinct and conspicuous beneath, the *primary veins* in the segments patent, oblique, quite straight, prominent, reaching the thickened margin, the *secondary veins* transverse with the primary veins and form the *primary areolae*, which are 4-angled and are again divided at right angles and their areolae are occupied by the ultimate divaricate veinlets with clavate apex; *sori* compital, confined to the upper half of the frond, its lower half being sterile, 4-6 between the primary veins, not exactly medial (but nearer to the lower vein), one in each primary areole forming a single row between the costa and margin, sub-globose or oblong, and upon maturity often more or less confluent in a continuous line; *spores* bilateral, reniform-oblong, smooth, hyaline.

In China numerous specimens have been collected from the south and southwestern Provinces including Yunnan, Guangxi, Guizhou, Guangdong, Hainan, Taiwan and also Hong Kong. Also common in Northern India (subtropical regions), Burma, Thailand, Indo-China and Malacca.

The present fern is the sole species of the genus *Pseudodrynaria* Ching and a majestic and beautiful epiphyte, endemic in the subtropics of the Asiatic mainland extending southwards to Malacca. It inhabits monsoon forests, epiphytic on tree trunks or large rocks by stream side. Highly recommended for greenhouse culture and conservatories.

Plate 201. 1. Habit sketch (natural size). 2. Portion of fertile segment, showing venation and sori ($\times$ 4). 3. The same, enlarged. 4. Scale from the rhizome ($\times$ 10). 5-6. Sporangium with spores ($\times$ 100).

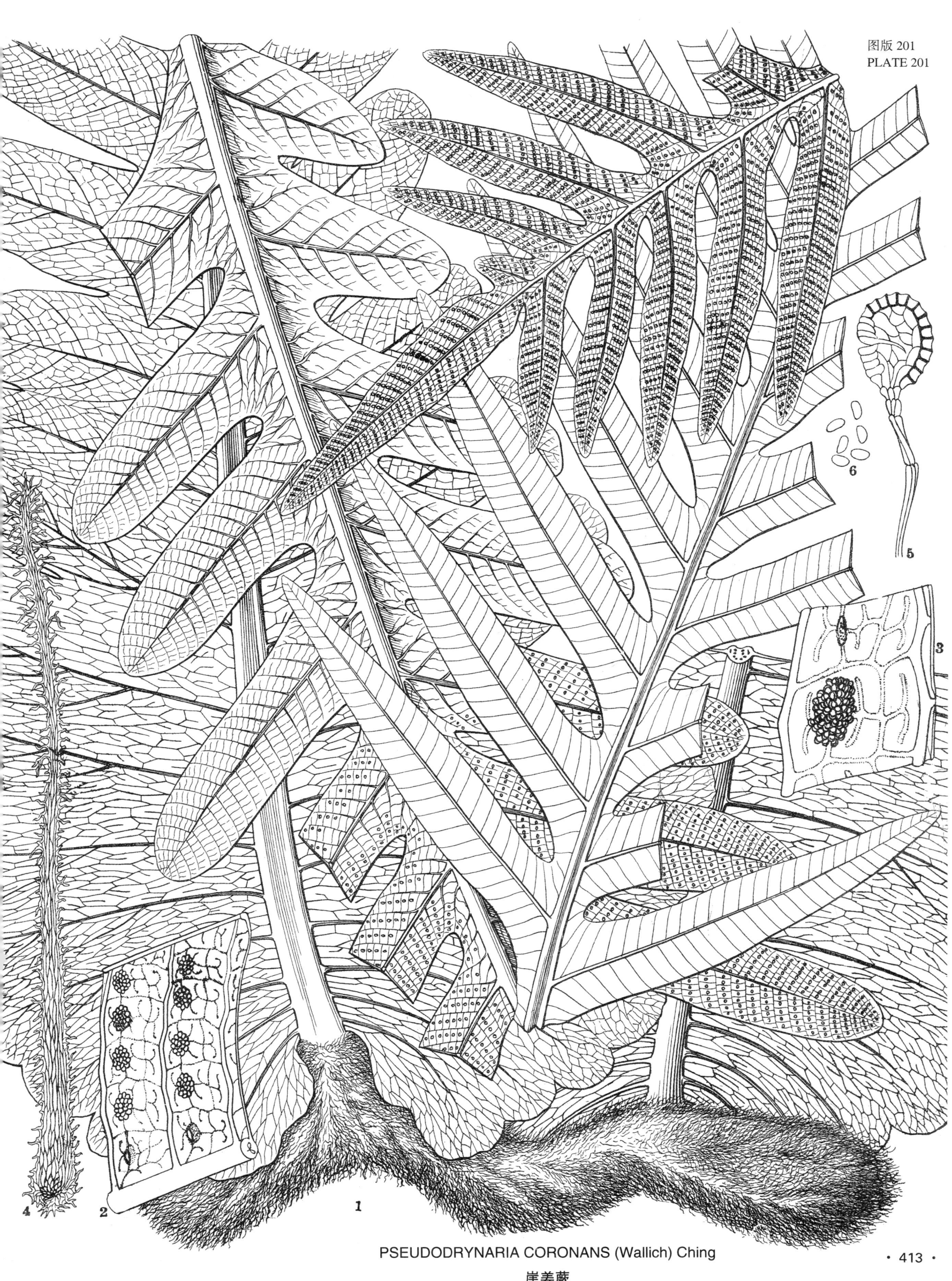

PSEUDODRYNARIA CORONANS (Wallich) Ching
崖姜蕨

高大植物,达 1 m 以上;叶柄长约 50 cm,直径 2 cm 以上,叶体宽广,为宽卵形,长与阔各 60 cm 以上,羽片 5—7 对,长 50—60 cm,宽 14—18 cm,有小叶柄(长约 3—4 cm),奇数羽状;小羽片多数,颇开展,具短柄,为宽达小羽片之半的缺刻分开,长 7—9 cm,宽 1—1.5 cm,披针形,渐尖头,基部亚截形或圆形,顶部向上微弯,下部小羽片较短,近基部的小羽片长仅 3 cm 或过之,顶端小羽片一个分离,有小叶柄,和下面的小羽片同形;叶轴向顶部有狭翅,叶缘具有规则的浅三角形锯齿;叶为草质,绿色,两面光滑无毛;叶脉开展,分叉或罕有单脉,明显,无倒行小脉;子囊群为边内生,与叶缘相隔一短距离,长圆形,由 8—10 个或少有较多的孢子囊组成,宿存,无夹毛。

分布:福建,广东,广西,贵州及香港,生林下溪沟边,甚为普通。

图注:1.叶柄的一部(自然大);2.羽片(自然大);3.小羽片的一部,表示叶脉及子囊群(放大 4 倍);4.子囊群的纵切面,表示子囊群托(放大 20 倍);5.子囊群横切面(放大 20 倍);6.孢子(放大 100 倍)。

ANGIOPTERIS FOKIENSIS Hieronymus

ANGIOPTERIS FOKIENSIS Hieronymus in Hedwigia **61**: 275. 1919; C. Chr. Ind. Fil. Suppl. III. 25. 1933.
Angiopteris evecta Benth. Fl. Hongk. 440. 1861 (non Hoffm. 1796).

Tall plants, over a meter high. *Stipe* over 2 cm thick, about 50 cm high, lamina ample, broadly ovate, over 60 cm each way, bearing 5-7 *pinnae* 50-60 cm long, 14-18 cm broad, petiolate (petiole about 3-4 cm long), imparipinnate; *pinnules* numerous, very patent, subsessile, separated by sinuses half as broad as the pinnules, which are 7-9 cm long, 1-1.5 cm broad, lanceolate, acuminate, slightly up-curved towards apex, base subtruncate or rounded, gradually shortened downwards with the lowest pinnules about 3 cm long or longer, the terminal pinnule free, petiolulate, similar to those below, rachis narrowly winged towards apex, margin regularly serrate with low deltoid teeth; *texture* herbaceous, green, both sides glabrous; *veins* patent, forked or rarely simple, distinct beneath, spurious recurrent veinlets none; *sori* intramarginal, leaving a narrow sterile margin, oblong, consisting of 8-10 sporangia or rarely more, not paraphysate.

Fujian: Foochow, *De Grijs 20* ex herb. Hance (type); *C. B. Rickett*; *S. T. Dunn 3769*; Yenping, *H. H. Chung 2826, 3592 (1925)*. Guangdong: Lok Chong Hsien, Kau Fung, *W. T. Tsang 20833*; *N. K. Chun 42667*; ibid., Tai Tung, *N. K. Chun 42873*; Lung Tau Shan, Iu Village, *To & Tsang 12114*; Yai Hsien, Youngling Shan, *S. K. Lau 6247*; Watsa Hsien, Tou Ngok Shan, *W. T. Tsang 23226*; Sunyi Hsien, Foun Dar Lee, *C. Wang 32180*; *Y. K. Wang 30952*; Ku Koong Hsien, Lung Tau Shan, *Y. K. Wang 31674*; Oonyuan Hsien, Wan Shi Shan, *S. K. Lau 958*; Polo Hsien, Lofou Shan, *T. M. Tsui 75*; Kweiyang Hsien, Lin Fa Shan, *W. T. Tsang 25704*; Shek Mang, Tai Shan, *C. L. Tso 23361*, alt. 450 m.; Wantong Shan, *H. Y. Liang 60495*; Sinto Hsien, Po Yin Shan, *W. T. Tsang 23094*; Tapu Hsien, Tung Koo Shan, *W. T. Tsang 21647*; Wang Hang Hsien, Tai Mu Shan, *S. P. Ko 51242*; Jaoping, Taiping, *N. K. Chun 42667*; Ingtak Hsien, *Hsui Hsiang-hou 8175*; Youngling Shan, *S. K. Lau 6247*. Guangxi: Kweiping Hsien, Univ. Guangxi Herb. *No. 121, 40230*; Sanhoa, Yao Ren Shan, *R. C. Ching 6255*, alt. 400 m.; Luchen Hsien, Mung Tung Kou, *R. C. Ching 5501*, alt. 350 m. in densely shaded ravine; Lin Yin Hsien, *R. C. Ching 7288*, alt. 900 m.; Hu Hsien, *Chen Lang Hsiang 500195*; Lohsiang, Yao Shan, *S. S. Sin 187A*; *C. Wang 40633*. Guizhou: Shanho Hsien, *S. C. Hou 1701*; River of Tathay, *Esquirol 2706*; Lypo, *Cavalerie*; Tsehen, *Esquirol 2653, 2766*. Also Hong Kong: *Hance (1856)*; *Forbes 566*.

Plate 202. 1. Portion of stipe (natural size). 2. Pinna (natural size). 3. Portion of pinnule, showing venation and sori (× 4). 4. Longitudinal section of sori, showing receptacle (× 20). 5. Cross section of sori (× 20). 6. Spores (× 100).

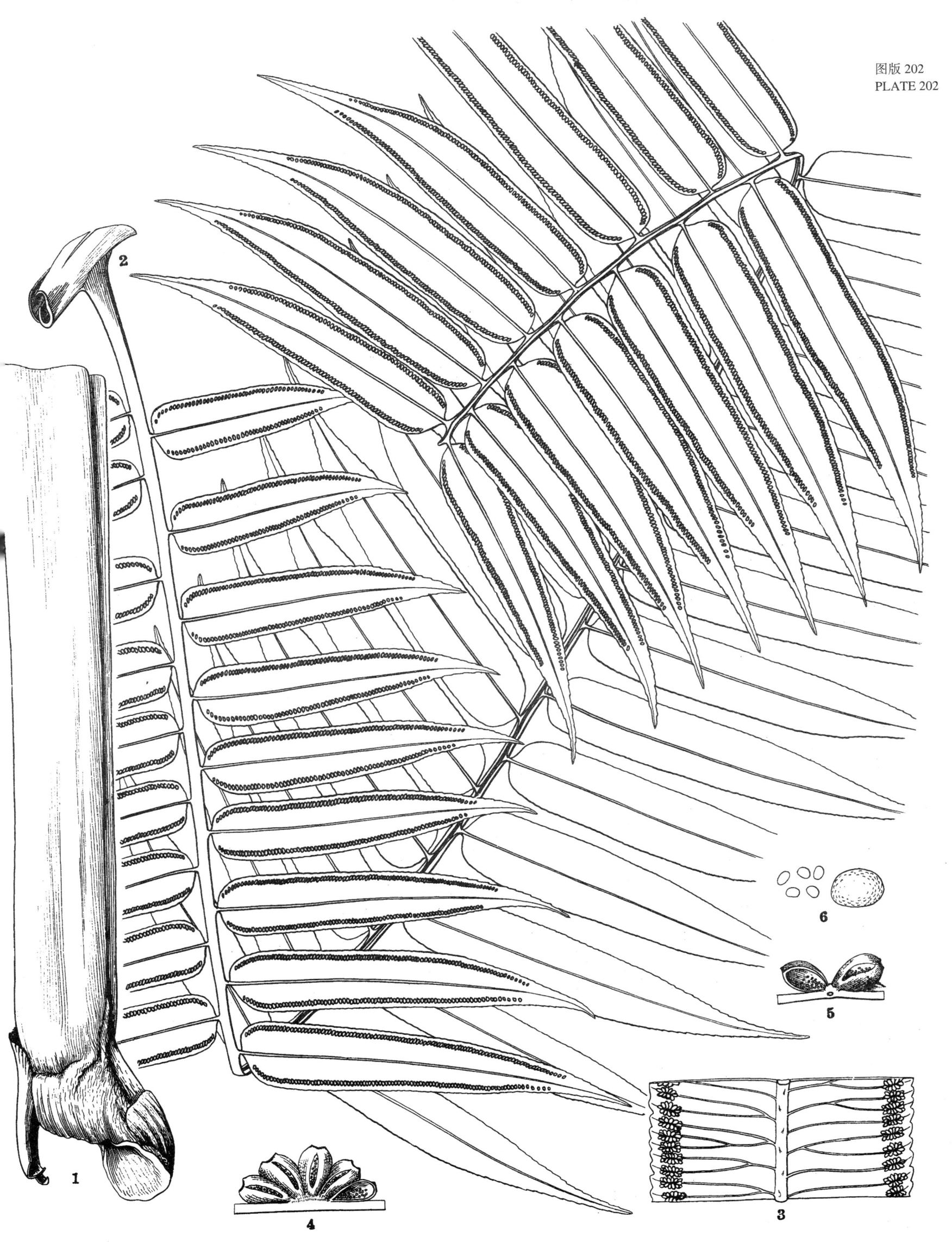

ANGIOPTERIS FOKIENSIS Hieronymus
福建观音座莲

图版 203　二回原始观音座莲

叶柄长 60—70 cm，直径约 4 mm，腹面有深沟，淡绿色，草质，下部略有紧贴的暗棕色披针形长尖头的鳞片，基部以上约 20—34 cm 处有一个膨大的节，叶体三角状卵圆形，长 40—50 cm，中部宽约 22 cm，基部为二回羽状，向上为一回奇数羽状；羽片 10—12 对，基部一对或二对羽片特大，长 16—19 cm，宽 6—7 cm，有 2.5—3 cm 长的小叶柄，羽裂为 2—7 对侧生小羽片，宽披针形，渐尖头，几无柄，开展，并有粗齿牙，长 2—3 cm，基部以上宽约 1 cm，圆楔形，顶端小羽片大形，长 7—10 cm，宽 2.8 cm；上面的一回羽片有叶柄（长 4—6 mm），线状披针形，向顶端渐狭为渐尖头，长 12—17 cm，近中部宽 2—2.8 cm，基部圆楔形，叶缘全部具有规则的粗齿牙；叶轴干后压扁，向上端两边有狭翅；顶生羽片与相邻的同形；叶为草质，干后仍为绿色，除叶轴，中肋下面及叶柄膨大处有一些棕色小鳞片外，全为光滑；叶脉上下两面明显，脉间距离 2 mm，一般为单脉或分叉，几乎成直角从中肋伸出，直行，达于叶缘的每一齿牙；子囊群线形，生于上部一回羽片上的，长约 5 mm，由近中肋向外伸展到距叶缘 4 mm 处，沿生单脉上或分叉脉上，由 20—40 个孢子囊组成，在孢子囊下面有许多密生分枝的夹毛，长等于或过于孢子囊；生于小羽片上的子囊群较短，由 10—12 个孢子囊组成，从小中肋出发，几达叶缘；孢子暗色，半透明，圆而有棱角，表面有粗疣状突起。

分布：云南东南部，马关县、金口（老君山），冯国楣号码 13679，生杂木林下，少见，海拔 1100—1300 m。

本种为本属中最特殊的一种，一般形体酷似观音座莲；在原始观音座莲属中发现二回羽叶的种还是第一次，这是有很大意义的，因为由这一种的发现，在一定程度上改变了过去对属的界说，打破了过去总是认为一回羽叶为本属的特征之一的传统看法。

图注：1. 本种全形（自然大）；2. 一回羽叶的一部，表示叶脉及子囊群（放大 4 倍）；3. 叶柄下部的鳞片（放大 40 倍）；4. 两个孢子囊着生于有夹毛的托上（放大 40 倍）；5. 两个分离的孢子囊，表示腹面的纵裂缝（放大 10 倍）；6. 子囊群下的夹毛（放大 100 倍）；7. 孢子（放大 100 倍）。

ARCHANGIOPTERIS BIPINNATA Ching

ARCHANGIOPTERIS BIPINNATA Ching, sp. nov.

Species distinctissima ab omnibus affinibus adhuc cognitis differt parte basali frondis bipinnata superiore simpliciter pinnata, pinnulis pinnarum inferiorum 3-8 jugis, parvis ad 3 cm longis ultra 7 mm medio latis, fere sessilibus, pinnis lateralibus terminalique conformibus lineari-lanceolatis ad 15 cm longis 2.5 cm latis longe acuminatis, marginibus pinnarum etiam pinnularum e basi usque ad apicem grosse dentatis, soris brevibus vix ultra 5 mm longis e sporangiis 20-40 vel in lamina pinnularum tantum 10-20 compositis.

Stipe 60-70 cm long, about 4 mm across, deeply grooved above, pale green, herbaceous, sparsely clothed in the lower part in dark brown lanceolate long-acuminate appressed scales and provided with one nodose swelling at about 20-34 cm above the base; lamina deltoid-ovate, 40-50 cm long, about 22 cm broad at the middle, bipinnate at base, simply impari-pinnate upwards; *pinnae* 10-12 pairs, of which the basal 3-4 pairs much the largest, about 5 cm apart, oblique, alternate, petiolate, the basal pair or two pairs 16-19 cm long, 6-7 cm broad, on petioles 2.5-3 cm long, pinnate with 2-7 pairs of small, broadly lanceolate, acuminate, patent, coarsely dentate and subsessile lateral *pinnules* which are 2-3 cm long, about 1 cm broad above the round-cuneate base, terminal pinnule large, 7-10 cm long, to 2.8 cm broad; the upper simple pinnae on petioles 4-6 mm long, linear-lanceolate, gradually long-acuminate, 17-12 cm long, 2-2.8 cm broad near the middle, base round-cuneate, margin from base up to apex regularly and prominently dentate; rachis compressed when dry, narrowly winged on each side towards the apex, terminal pinna similar to the lateral ones immediately below; *texture* herbaceous, green, when dry glabrous except for a few small brown scales on the rachis and costa beneath, also on the inflated petiole; *veins* distinct on both sides, about 2 mm apart, generally simple or forked from above the base, diverging almost at a right angle to the costa and running straight towards the margin into each tooth; *sori* linear, straight, those on the upper simple pinnae to about 5 mm long, extending from near the costa to about 4 mm within the leaf-margin, simple or forked along the forked veins, consisting of about 20-40 sporangia with dense brown branched paraphyses underneath and as long as the sporangia, those on the pinnules of the lower pinnae shorter, of about 10-12 sporangia, starting from the costule but fall short of the leaf-margin; *spores* pale, hyaline, round but angular with verrucose surface.

Southeastern Yunnan: Markuan Hsien, Chingkou (Lao Ching Shan), *K. M. Feng* 13679, November 7, 1947, in mixed forest, rare, alt. 1100-1300 m.

One of the most distinct species in the genus, resembling *Marratia* in general habit. The discovery of a bipinnate-leaved species in the genus *Archangiopteris* is very important in that it goes to change to a certain extent our previous generic concept as to the degree of pinnation, which is simply pinnate in all previously known species.

Plate 203. 1. Habit sketch (natural size). 2. Portion of simple pinna, showing venation and sori ($\times$ 4). 3. Scale from the lower part of stipe ($\times$ 40). 4. Two sporangia attached to the receptacle with paraphyses ($\times$ 40). 5. Two sporangia detached, showing longitudinal slit on the ventral side ($\times$ 10). 6. Paraphyses detached from the sori ($\times$ 100). 7. Spores ($\times$ 100).

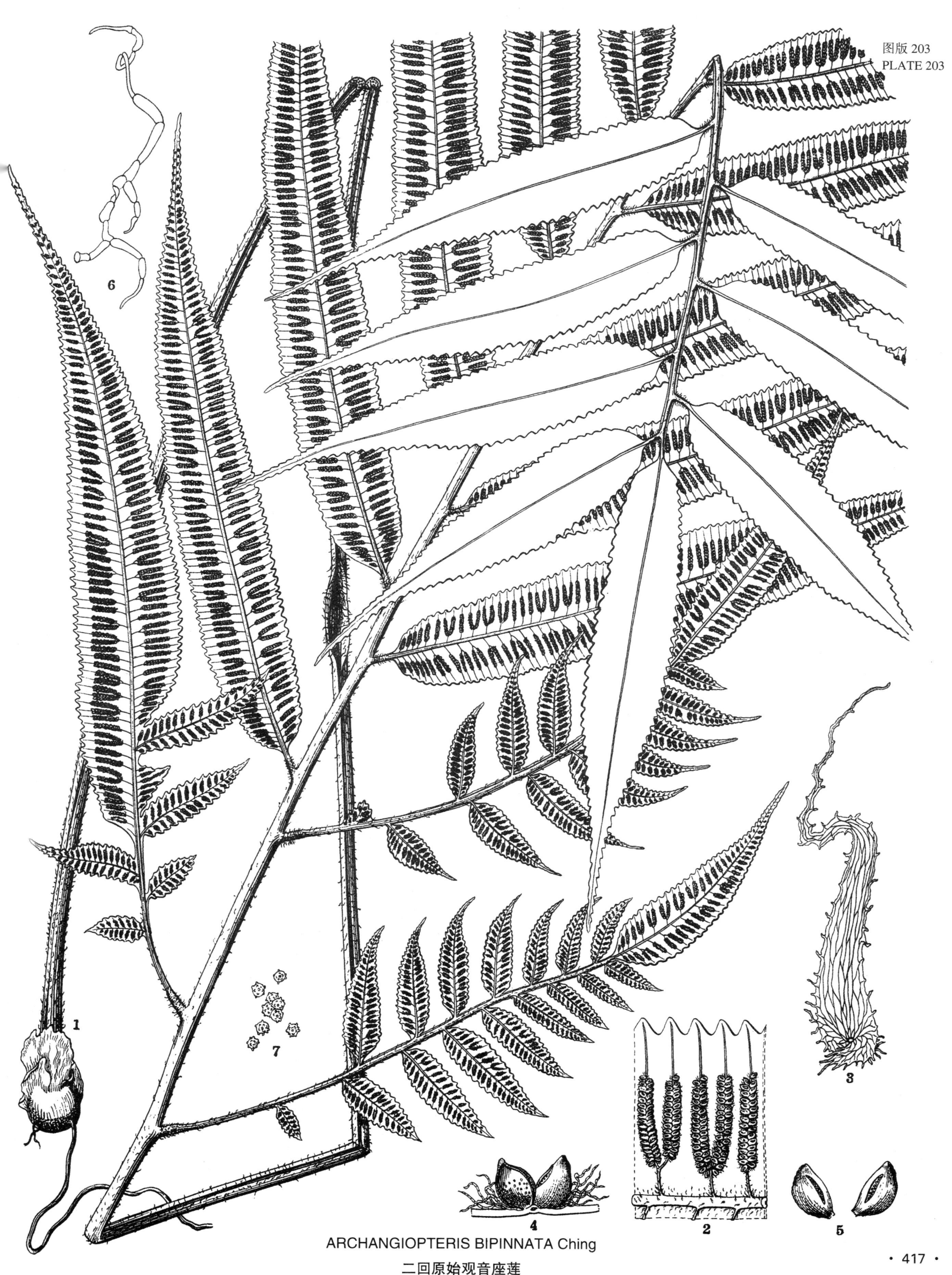

ARCHANGIOPTERIS BIPINNATA Ching
二回原始观音座莲

图版 204　河口原始观音座莲

根状茎粗大，肉质，亚直立，直径 3—4 cm，下面具铁丝状厚肉质的黑色不分枝的粗根，叶簇生，柄长达 50 cm，厚达 5 mm，肉质，绿色，有 4—5 个膨大具沟槽而干后为黑色的节状膨大，各间的距离大致相等，另外被有一些卵状披针形而基部为圆心脏形的深棕色鳞片，边缘有长锯齿；叶体为宽卵形，长达 30 cm，宽约 38 cm，一回奇数羽状，顶端小羽片较大，长 20—22 cm，宽 7—9 cm，侧生小羽片 2—3 对，同形，对生或亚对生，间距约 5—6 cm，长 15—20 cm，中部宽 5—7 cm，阔椭圆披针形，有小柄，长约 1.5 cm，膨大，淡黑色，略具鳞片，顶部为短尾状渐尖头，并有粗锯齿，向基部渐狭，成楔形，边缘有波状浅齿或波状齿牙；叶为纸质，上面深绿，下面淡绿，并有相当多的节状细毛覆盖；叶脉细长，颇开展，明显，大都分叉，间为单一，近叶边向上弯弓，并伸入锯齿；子囊群线形，长 3—3.5 cm 或较长，彼此颇接近，由 160—240 个子囊成二列组成，不育边缘和中肋两侧宽达 5 mm，夹毛线形，稠密，节状分枝，由 10—15 个细胞组成，长过于子囊；孢子矩圆形，透明，具密集的刺状突起。

分布：云南东南部，河口，南溪镇，朱维明号码 1726，生于潮湿浓阴的林下沟中，海拔 150 米，普通。

本种为一个独特的种，其叶柄在基部以上具有 4—5 个节状膨大，小叶很宽，椭圆披针形，边缘有波状钝锯齿，下面密生节状毛；子囊群极长，并具很长的密毛茸状的夹毛，几乎完全覆盖着初生的子囊群。

图注：1. 本种全形（自然大）；2. 羽片的一部，表示叶脉，锯齿，子囊群和下面的节状毛（放大 3 倍）；3. 叶柄下部的鳞片（放大 40 倍）；4. 两个着生于托上的子囊和夹毛（放大 20 倍）；5. 两根夹毛（放大 60 倍）；6. 孢子（放大 100 倍），另一个（放大 400 倍）。

ARCHANGIOPTERIS HOKOUENSIS Ching

ARCHANGIOPTERIS HOKOUENSIS Ching, sp. nov.

Rhizomate crasso, suberecto, 3-4 cm diamtero, radicanti, radicibus incrassatis, teretibus, nigrescentibus, simplicibus e facie inferiore rhizomatis abundanter oriundis obtecto; *frondibus* fasciculatis, stipite ad 50 mm longo, 5 mm crasso, carnoso-herbaceo, virescenti, per totam longitudinem nodis 4-5 tumidis geniculatis in sicco nigris inter se plus minus fere aequaliter remotis instructo, paleis ovato-lanceolatis basi cordatis atrobrunneis copiose onusto praeditis, lamina ambito late ovata, ad 30 cm longa, 38 cm lata, impari-pinnata cum pinna terminali majore, 20-22 cm longa, 7-9 cm lata; *pinnis* lateralibus 2-3 jugatis, omnibus conformibus, oppositis vel. subalternis, ca. 5-6 cm inter se separatis, 15-20 cm longis, 5-7 cm medio latis, late elliptico-lanceolatis, petiolatis (petiolo ca. 1.5 cm longo, tumido, nigricante et paleaceo), apice breviter caudato-acuminatis et grosse serratis, basin versus gradatim cuneatis, marginibus grosse arcuato-serratis aut crenato-dentatis; *textura* chartacea, folia supra atro-viridi, subtus pallidiora, et pilis minutis articulatis modeste conspersa; *venis* lateralibus subrecto-patentibus, conspicuis, plerisque furcatis, rarius simplicibus, prope marginem antrorsim curvatis et fere in dentes protensis; soris linearibus, 3-3.5 cm longis, vel longioribus, subcontiguis, e 160-240 sporangiis 2-seriatim compositis, parte sterili secus marginem costamque ca. 5 mm lata, paraphysibus filiformibus ramosis, articulatis, 10-15-cellularibus, magis quam sporangiis longioribus, densissime obtectis; *sporis* rotundo oblongis, minutissime denseque echinatis, hyalinis.

Southeastern Yunnan: Hokou, Nanchi, *Chu Ve-ming 1726*, July 1955, in densely wooded humid ravine, alt. 150 m. common.

A remarkable species, differing from all other known species of the genus in having 4 or 5 nodose swellings above the base of the fleshy stipe, the unusually broad elliptico-lanceolate serrate pinnae densely glandular hairy on the under surface, the very long sori which extend from near the costa outwards to a short distance from the margin, and particularly in the very long dense branched shaggy paraphysate hairs almost completely covering the young developing sori.

Plate 204. 1. Habit sketch (natural size). 2. Portion of pinna, showing venation, serrature and sori, also hairy under surface (× 3). 3. Scale from the lower part of stipe (× 40). 4. Two sporangia attached to the receptacle with paraphyses (× 20). 5. Two paraphyses removed from the sori (× 60). 6. Spores (× 100) with one enlarged (× 400).

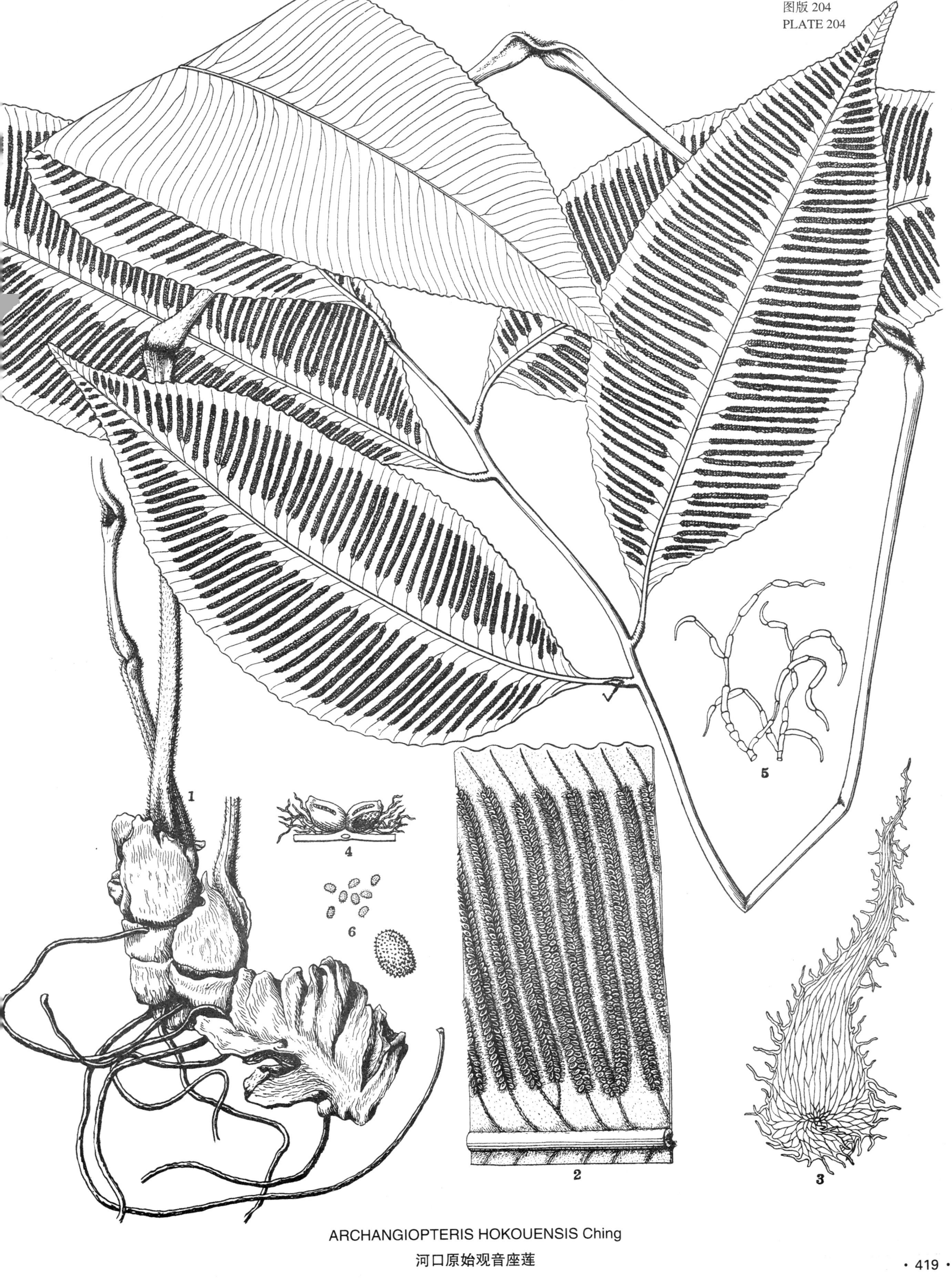

ARCHANGIOPTERIS HOKOUENSIS Ching
河口原始观音座莲

叶柄长 50—60 cm，有疏生鳞片，尤以下部为多，基部以上约 25 cm 处有膨大的节，叶体长卵圆形，奇数羽状，羽片 2—3 对，顶生羽片与下部的同形，长 20 cm，宽 4 cm，为微倒披针形，中部以下渐狭，基部略为宽楔形，顶端渐尖，尾状，长 3—4 cm，互生，有小叶柄（柄长约 1 cm），叶缘浅波状，基部以上有不规则的圆齿状齿牙；叶为厚草质或坚纸质，干后棕色，几无鳞毛；叶脉颇开展，单脉与分叉脉相间；子囊群线形，长 1.5 cm，位于中部，中肋两边及叶缘下边为不育空间，夹毛短于孢子囊，隐匿，从表面不易见到；孢子圆形或长圆形，颇微小，具疣状突起，无色透明。

分布：台湾特产。

本种为台湾南部，中部及北部高山森林中一种很普通的蕨类，并与云南东南部的亨利原始观音座莲（A. Henryi Christ et Gies.）颇相似，但因羽片为厚草质，长倒披针形，有圆齿状齿牙和叶缘为浅波状，故有不同。

图注：1. 本种全形（自然大）；2. 羽片的一部，表示叶脉，锯齿及子囊群（放大 4 倍）；3. 叶柄下部的鳞片（放大 30 倍）；4. 叶柄下部的横切面（放大 4 倍）；5. 叶柄中部的横切面（放大 4 倍）；6. 叶柄上部的横切面（放大 4 倍）；7. 两个孢子囊着生于有夹毛的托上（放大 20 倍）；8. 子囊群下的夹毛（放大 100 倍）；9. 孢子（放大 100 倍），另一个（放大 400 倍）。

ARCHANGIOPTERIS SOMAI Hayata

ARCHANGIOPTERIS SOMAI Hayata, Ic. Pl. Form. **5**: 256. 1915; **6**: 154, fig. 60, t. 19. 1916; C. Chr. Ind. Fil. Suppl. I. 4. 1916; III. 26. 1933; Nakai in Bot. Mag. Tokio **41**: 78. 1927; M. Genkai, Short Fl. Form. 1. 1936.

Angiopteris somai Hayata apud Makino et Nemota, Pl. Jap. 1563. 1925.

Protangiopteris somai Hayata in Bot. Mag. Tokio **42**: 308. 1928.

Stipe 50-60 cm long, sparsely scaly especially in the lower part, with one nodose swelling at about 25 cm above the base; lamina oblong-ovate, impari-pinnate with 2-3 pinnae on each side under the similar terminal one; *pinnae* 20 cm long, 4 cm broad, slightly oblanceolate, very gradually narrowed from the middle downwards with a rather broadly cuneate base, apex caudate-acuminate with a linear acumen 3-4 cm long, alternate, petiolate (petiole about 1 cm long), margin repand, rather irregularly crenate-dentate above the base; *texture* thickly herbaceous or chartaceous, dry brown, nearly glabrous; *veins* very patent, alternately forked and simple; *sori* linear, to 1.5 cm long, medial, leaving a broad sterile space on each side of costa and along the margin, paraphyses shorter than sporangia and concealed when viewing from above; *spores* round or round-oblong, very minutely verrucose, hyaline.

Taiwan: Bahao, *T. Somai* (type), December 1910; Rengechi, Taichu, *S. Sasaki*, March 3, 1927; *Yamamota and Mori*, 2-3 November, 1932; Taihoku-syu, Bunzongun, Kanica, *Y. Yamamota*, July 24, 1938.

A fairly common fern in the mountain forests in the southern, middle and northern parts of Taiwan Island and is quite near *A. Henryi* Christ et Gies. from Southeast Yunnan and seems to differ in elongate oblanceolate pinnae of a thicker texture with crenate-dentate and often repand margin.

Plate 205. 1. Habit sketch (natural size). 2. Portion of pinna, showing venation, serrature and sori ($\times$ 4). 3. Scale from the lower part of stipe ($\times$ 30). 4. Cross section from the lower part of stipe ($\times$ 4). 5. Cross section from the middle of the stipe ($\times$ 4). 6. Cross section from the upper part of the stipe ($\times$ 4). 7. Two sporangia attached to the receptacle with paraphyses ($\times$ 20). 8. Paraphyses removed from the sori ($\times$ 100). 9. Spores ($\times$ 100), one enlarged ($\times$ 400).

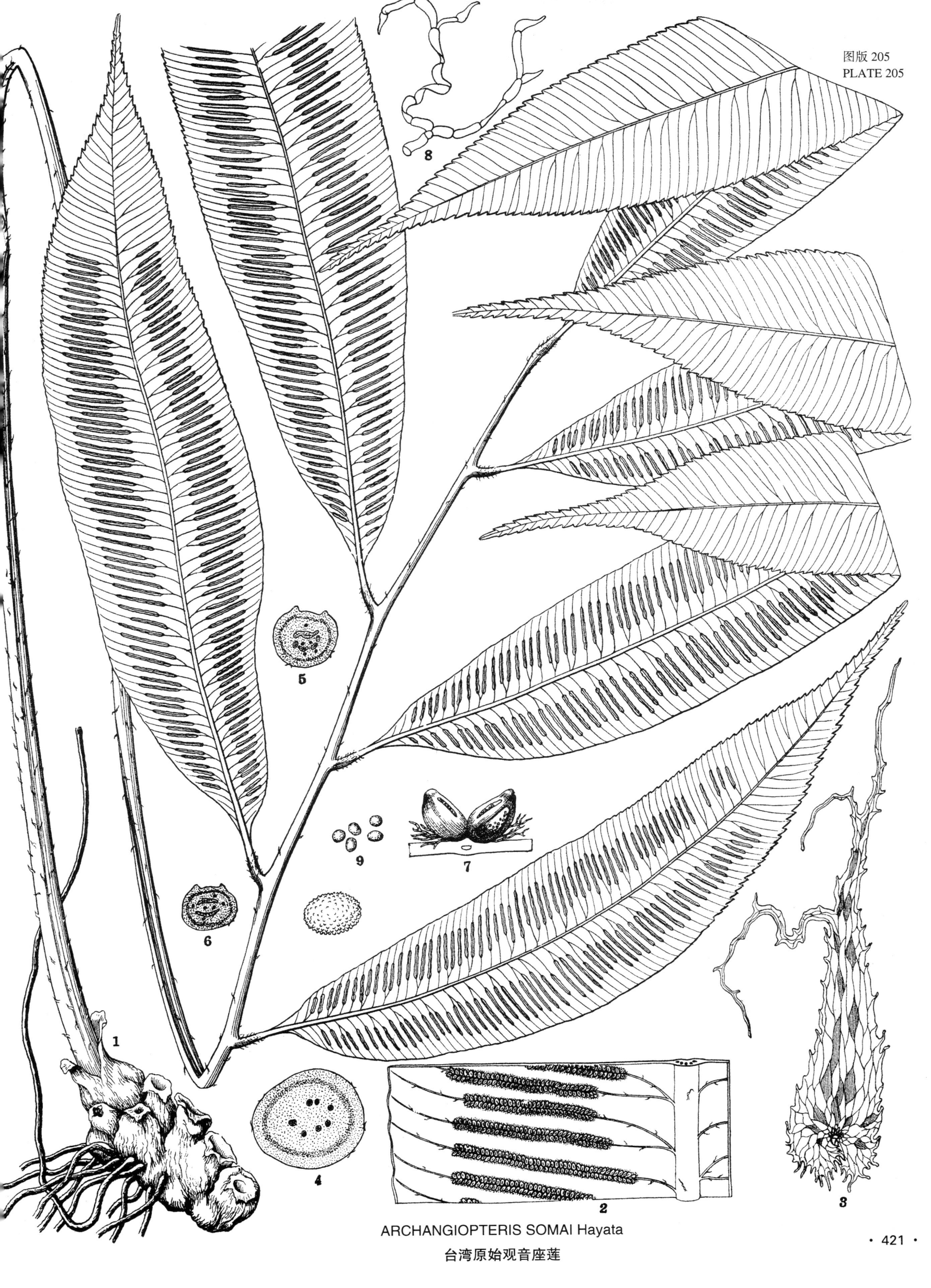

ARCHANGIOPTERIS SOMAI Hayata
台湾原始观音座莲

叶柄长 36—60 cm,淡绿色,腹面有深沟,有淡红棕色的线形鳞片,尤以下部为多,边缘有睫毛,基部以上 20—30 cm 处有膨大的节,叶体长与宽各约 40 cm,宽卵形,奇数羽状,羽片 4—6 对,顶生羽片与侧生的同形,侧生羽片为互生,相距约 4 cm,有小叶柄(柄长 7 mm,膨大),基部一对甚短(长 14 cm)镰刀状,向上弯弓,与上部的同形,基部圆形,上部羽片长 22—25 cm,中部宽 5 cm,平行阔披针形,渐尖头,基部几不变狭,圆形或亚圆形,不为楔形,叶全缘或至多略呈波状,顶端有锯齿;叶为薄草质,绿色,除中肋有鳞片疏生外,均为光滑;叶脉开展,明显,相距 4 mm,单脉或分叉,近叶缘显然向上弯弓;子囊群线形,一般长 8—12 mm,相距甚远,位于边缘与中肋之间,两端有等宽的不育空间,有密生分枝的节状夹毛,长过于孢子囊;孢子长圆肾形,有细密刺头突起,透明。

分布:云南东南部,西畴县,冯国楣号码 12019,生杂木林下,海拔 1500—1600 m。

本种形体颇似亨利原始观音座莲(A. Henryi Christ et Gies.),但羽片不为披针形而为平行阔披针形,其下部的为镰刀状,基部羽片远较上部的为短,边缘为全缘或在顶端以下有时为波状,基部圆形或亚圆形,不为明显的变狭。

图注:1. 本种全形(自然大);2. 羽片的一部,表示叶脉,叶缘及子囊群(放大 4 倍);3. 叶柄下部的鳞片(放大 30 倍);4. 两个孢子囊着生在有夹毛的托上(放大 20 倍);5. 二条夹毛(放大 100 倍);6. 孢子(放大 100 倍),另一个(放大 400 倍)。

ARCHANGIOPTERIS SUBROTUNDATA Ching

ARCHANGIOPTERIS SUBROTUNDATA Ching, sp. nov.

Species configuratione gregis *A. Henryi* Christ et Gies., differt pinnis nec lanceolatis sed late lineari-lanceolatis, inferioribus falcatis, infimis quam superioribus multo brevioribus, marginibus integris vel inter-dum undulatis, basin versus paulo dilatatis et rotundatis vel subrotundatis, vix conspicue angustatis.

Stipe 36-60 cm long, greenish, deeply grooved on the upper side, with one nodose swelling at 20-30 cm above the base, copiously clothed especially in the lower part in reddish-brown linear ciliate *scales*; lamina about 40 cm long and as broad, broadly ovate, impari-pinnate; *pinnae* 4-6-jugate below the similar terminal pinna, alternate, about 4 cm apart, petiolate (petiole ca. 7 mm long, inflated), the basal pair much shortened (to 14 cm long) and like those next above rather strongly falcate, base rounded, the upper ones 22-25 cm long, to 5 cm broad at the middle, broadly linear-lanceolate, gradually acuminate, hardly narrowed towards the rounded or subrounded, not cuneate base, margin entire, or at most slightly wavy below the dentate-serrate apical part; *texture* thin herbaceous, green, glabrous except the sparsely scaly costa beneath; *veins* patent, distinct beneath, 4 mm apart, forked or simple, decidedly curved upwards near the margin; *sori* linear, generally 8-12 mm long, far apart, situated midway between the costa and margin, leaving an equally broad sterile space on each side, densely paraphysate with freely branched articulate hairs longer than sporangia; *spores* oblong-reniform, finely echinate, hyaline.

Southeastern Yunnan: Sichour Hsien, Faadocn, *K. M. Feng 12019*, Sept. 27, 1947, in mixed forest, alt. 1500-1600 m.

Plate 206. 1. Habit sketch (natural size). 2. Portion of pinna, showing venation, leaf-margin and sori ($\times$ 4). 3. Scale from the lower part of stipe ($\times$ 30). 4. Two sporangia attached to the receptacle with paraphyses ($\times$ 20). 5. Two paraphyses removed from sori ($\times$ 100). 6. Spores ($\times$ 100), one enlarged ($\times$ 400).

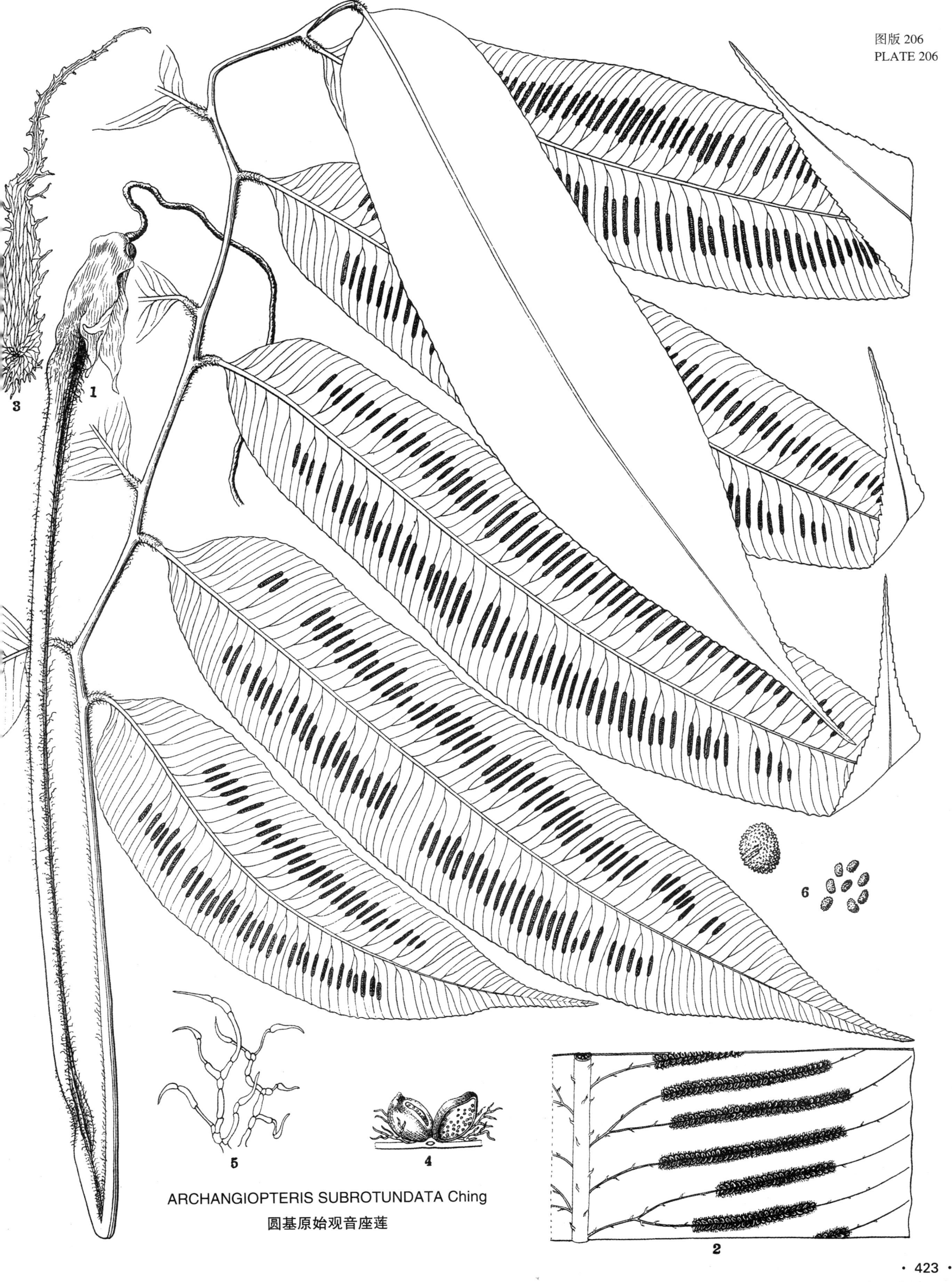

ARCHANGIOPTERIS SUBROTUNDATA Ching
圆基原始观音座莲

叶柄长 47—55 cm，直径 5 mm，腹面有深沟，下部鳞片颇多，基部以上约 20 cm 处有膨大的节，叶体卵状长圆形，长 35—45 cm，宽 26—30 cm，奇数羽状；羽片 2—3 对，互生，相距 4—5 cm，斜出，基部一对较上部的为短，镰刀状，向上弯弓，上部羽片长 20 cm，宽 5—5.5 cm，中部最宽，阔卵状披针形，向下渐狭为楔形基部，顶端为短渐尖或渐尖头，并有锯齿，叶全缘或多少为微波状；小叶柄长 1 cm，略为膨大，几无鳞毛；叶为厚纸质，上下两面光滑，唯沿叶轴和中肋下面略有鳞片；叶脉疏生，开展，单脉或分叉，近叶缘向上弯弓，明显；子囊群长 1—1.5 cm，位于叶缘与中肋之间，两边相距各约 7 mm，列间有较宽间隙，夹毛密生，略短于孢子囊；孢子亚圆形，有密生的微小疣状突起。

分布：云南东南部，屏边县，蔡希陶号码 60299，生林下，海拔 1200 m。

本种表面上颇似亨利原始观音座莲（A. Henryi Christ et Gies.），并曾被认为同种，但叶为厚纸质，羽片远较宽，边缘呈微波状，基部一对显然为镰刀状和子囊群的间距较宽。

图注：1. 本种全形（自然大）；2. 羽片的一部，表示叶脉，锯齿及子囊群（放大 4 倍）；3. 叶柄下部的鳞片（放大 30 倍）；4. 两个孢子囊着生于有夹毛的托上（放大 20 倍）；5. 子囊群下的夹毛（放大 100 倍）；6. 孢子（放大 100 倍），另一个（放大 400 倍）。

ARCHANGIOPTERIS LATIPINNA Ching, sp. nov.

ARCHANGIOPTERIS LATIPINNA Ching, sp. nov.

Species gregis *A. Henryi* Christ et Gies., a qua differt frondis textura duriore, pinnis late ovato-lanceolatis, inferioribus falcatim recurvatis, ad apicem potius breviter acuminatis et insuper soris longioribus.

Stipe 47-55 cm long, 5 mm across, deeply sulcate above, moderately scaly in the lower part, with one nodose swelling at about 20 cm above the base; lamina oblong-ovate, 35-45 cm long, 26-30 cm broad, impari-pinnate; *pinnae* 2-3-jugate, alternate, 4-5 cm apart, oblique, the basal pair only slightly shorter than those next above, the lower ones falcately curved, the upper ones to 20 cm long, 5-5.5 cm broad, the broadest part being near the middle, broadly ovate-lanceolate, gradually narrowed towards the cuneate base, apices rather short-acuminate, or acuminate and serrate, margin entire and more or less repandulate, petiole about 1 cm long, scarcely inflated and subglabrous; *texture* thickly chartaceous, both sides glabrous except a few small scales along the rachis and costa underneath; *veins* lax, patent, forked or simple, antrorsely curved towards the margin, distinct on both sides; *sori* 1-1.5 cm long, midway between the margin and costa, leaving a sterile space about 7 mm broad equally as wide along the margin as along the costa, and separated from each other by broader interstitial sterile spaces, paraphyses quite dense, somewhat longer than the sporangia; *spores* roundish rather than oblong, minutely and densely verrucose.

Southeastern Yunnan: Pingpien Hsien, *H. T. Tsai 60299*, June 19, 1934 in woods, alt. 1200 m.

This new species resembles, in a superficial way, so closely *A. Henryi* Christ et Gies. that the type number on which it is based has actually been filed for sometime under that species. *A. latipinna* differs from the generic type described in 1899 in several constant and correlated characters, namely, thickly chartaceous texture of the fronds, repandulate margin of the much broader pinnae, of which the lower ones are decidedly falcate and the more widely separated sori due to the more lax venation with broader interstitial spaces.

Plate 207. 1. Habit sketch (natural size). 2. Portion of pinna, showing venation, serrature and sori ($\times$ 4). 3. Scale from the lower part of stipe ($\times$ 30). 4. Two sporangia attached to the receptacle with paraphyses ($\times$ 20). 5. Paraphyses removed from the sori ($\times$ 100). 6. Spores ($\times$ 100), one enlarged ($\times$ 400).

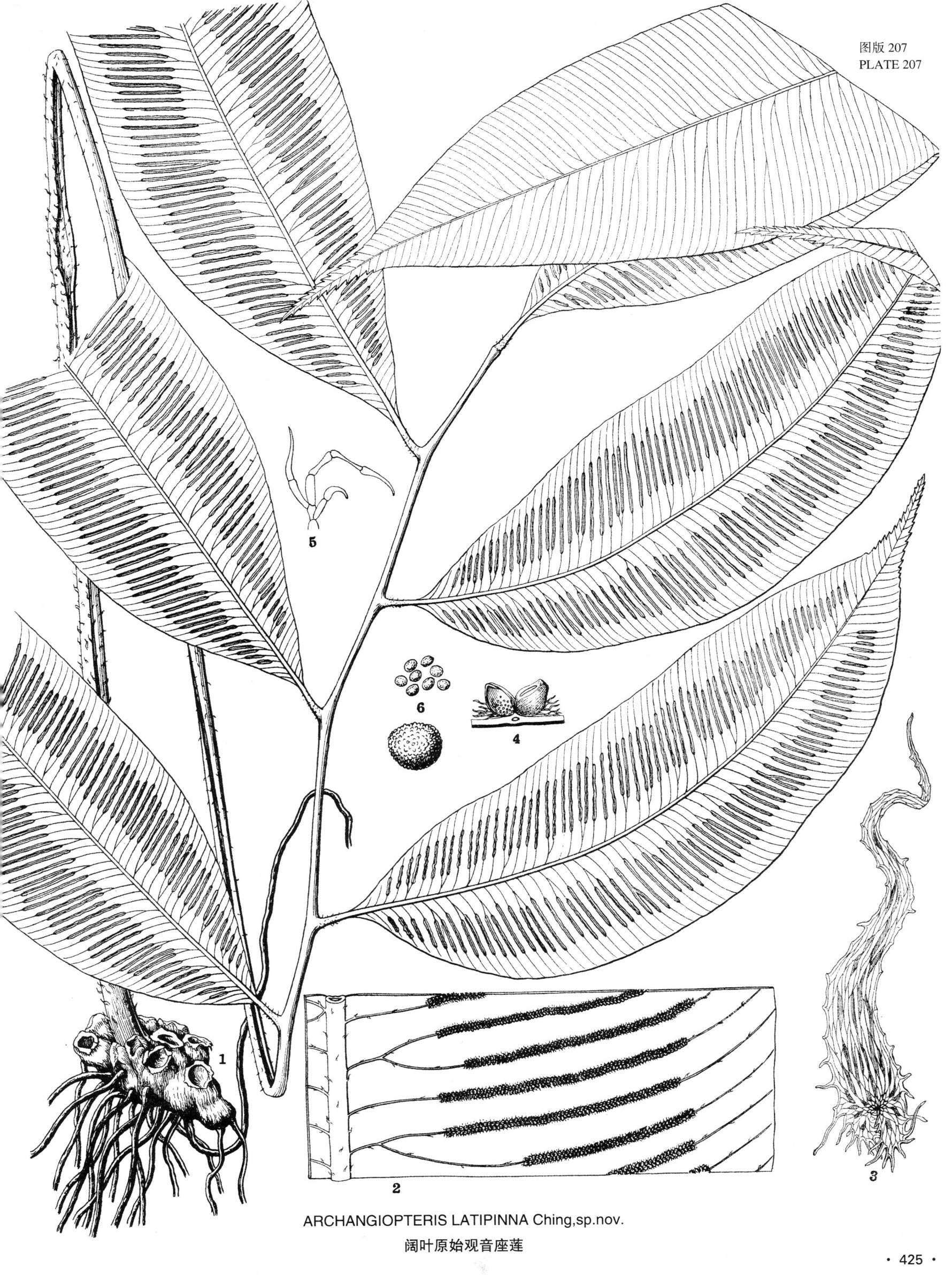

ARCHANGIOPTERIS LATIPINNA Ching, sp. nov.
阔叶原始观音座莲

图版 208　尾叶原始观音座莲

根状茎亚直立,粗达 2 cm,有粗根,叶柄长 30—45 cm,直径 3—4 mm,绿色,几无鳞片,在基部以上 30 cm 处有膨大的节,叶体为卵形,长 30 cm,宽约 25 cm,奇数羽状;侧生羽片在所见标本为 2 对,亚对生,相距 5—6 cm,顶生羽片稍较大,长 16—18 cm,中部宽 4.5—5 cm,侧生羽片皆为同形或基部一对微较短,阔倒披针形,向楔形的基部渐变狭,顶部突变为尾状(尾头长约 1.5 cm,线形,有锯齿),叶边自基部以上有粗齿牙;叶为草纸质,绿色,上下两面光滑,但沿中肋及叶脉有披针形小鳞片疏生;叶脉亚开展,与中肋成 60°的上角,大多数分叉,或单脉,近叶缘向上弯弓而达于锯齿;子囊群长 8—10 mm,离中肋及叶边各约 6 mm,亚近生,由 40—70 个子囊组成,罕有更多的,托稍隆起,夹毛远较子囊为短,红棕色,有节,由基部分枝很多;孢子球圆形,具细密刺头,透明。

分布:广西南部,明江,乾牛山,前广西大学生物系采,生半山以上的林内沟中。

本种是一明显的地方种,其不同于亨利原始观音座莲者为倒披针形的羽片突变成尾状顶端,叶边具粗齿牙和子囊群的夹毛非常短而由基部分枝繁密,不易从上方见到。

图注:1.本种全形(自然大);2.羽片的一部,表示叶脉,锯齿及子囊群(放大 4 倍);3.叶柄下部的鳞片(放大 20 倍);4.子囊群横断面,表示生于托上的二个子囊和短夹毛(放大 20 倍);5.夹毛(放大 100 倍);6.孢子(放大 100 倍),另一个(放大 400 倍)。

ARCHANGIOPTERIS CAUDATA Ching

ARCHANGIOPTERIS CAUDATA Ching, sp. nov.

Rhizomate suberecto, ca. 2 cm crasso, radicoso; stipite 30-45 cm longo, 3-4 mm diametro, virescenti, subglabro, circ. 16 cm supra basin tumide nodoso et geniculato; lamina ambitu ovata, 30 cm longa, ca. 25 cm lata, impari-pinnata; *pinnis* uti videtur 2-jugis, suboppositis, 5-6 cm inter se remotis, pinna terminali paulo majore, 16-18 cm longa, 4.5-5 cm medio lata, lateralibus omnibus similibus, vel infimis vix brevioribus, late oblanceolatis, basin versus gradatim angustatis, basi cuneatis, apice subito caudatis (cauda 1.5 cm longa, lineari, serrata), margine e basi sursum grosse dentatis; *textura* herbaceo-chartacea, virescenti, utraque facie subglabra, secus costam et venulam squamis minutis lanceolatis sparsis conspersis; *venis* subrecte patentibus, sub angulo ca. 60° excurrentibus, plerisque furcatis, nonnullis simplicibus, intra marginem antrorsim curvatis et in dentes intrantibus; *soris* 8-10 mm longis, ca. 6 mm a margine et costa remotis, subcontiguis; sporangiis 40-70 pro soro vel rarius pluribus, receptaculo paulo elevato, pilis quam sporangiis multo brevioribus, rufis, articulatis, e basi dense ramosis praeditis; *sporis* globulo-tetraedribus, minutissime denseque echinatis.

Southern Guangxi: Min River, Kang Nu Shan, specimens ex Herb. Guangxi University, in ravine under forest, about halfway up the mountain, Oct. 2, 1935.

A very distinct local species, differing from *A. Henryi* Christ et Gies. in oblanceolate pinnae with abruptly caudate apices, coarsely dentate leaf-margin and in the very short and profusely branched paraphyses so completely hidden under the sori that their presence can hardly be detected from above, or without the sporangia being carefully removed off the veins.

Plate 208. 1. Habit sketch (natural size). 2. Portion of pinna, showing venation, serrature and sori ($\times$ 4). 3. Scale from the lower part of stipe ($\times$ 30). 4. Two sporangia attached to the receptacle with very short paraphyses ($\times$ 20). 5. Paraphyses removed from the sori ($\times$ 100). 6. Spores ($\times$ 100), one enlarged ($\times$ 400).

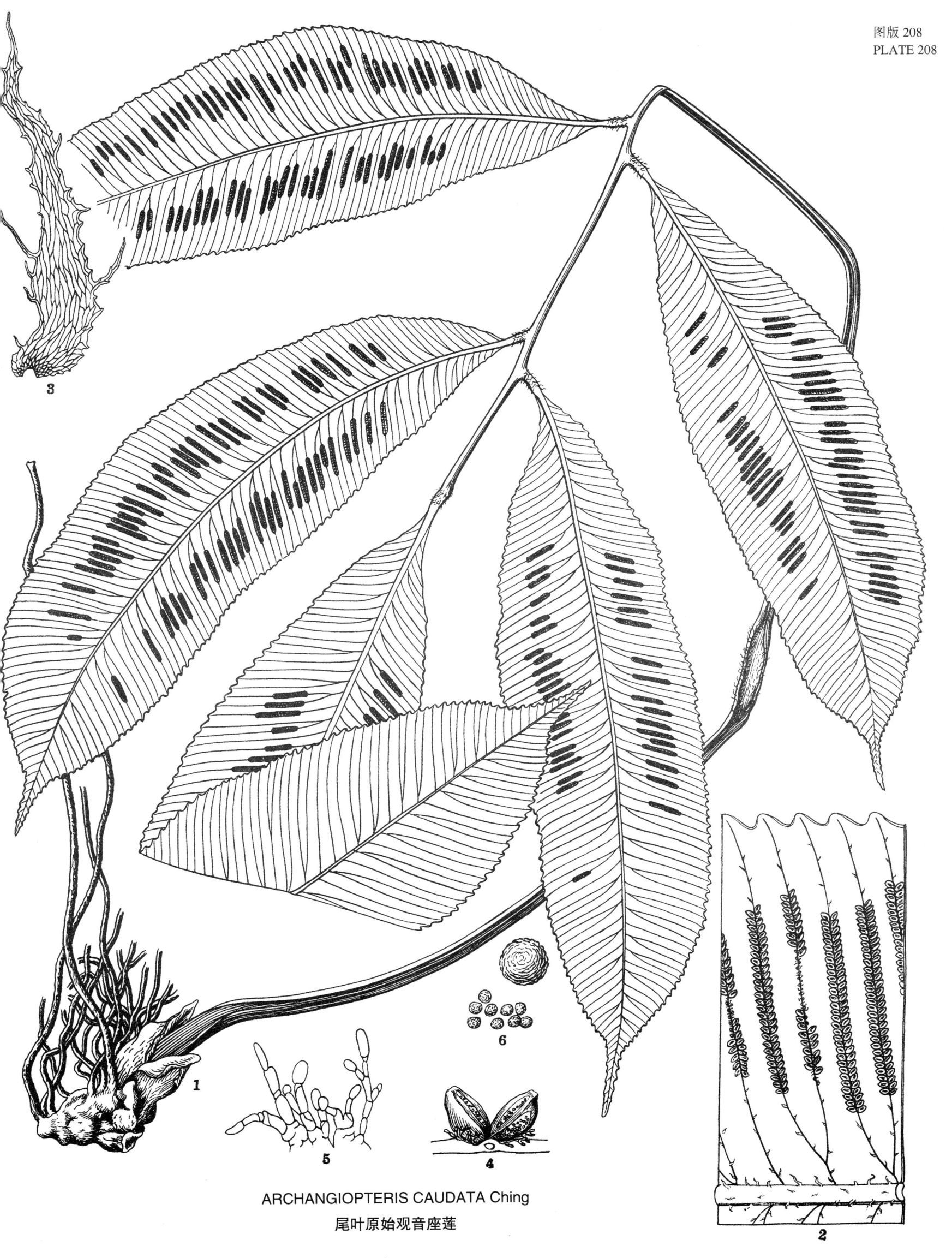

ARCHANGIOPTERIS CAUDATA Ching
尾叶原始观音座莲

叶柄长 40—45 cm,淡绿色,腹面有宽沟;有疏生棕色披针形而有齿牙的鳞片,下部较多,基部以上 20—30 cm 处有膨大的节,叶体阔卵圆形,远短于叶柄,奇数羽状,顶生羽片有长柄,与侧生的同形,侧生羽片 2—4 对,对生,斜出或斜开展,卵状披针形,长 20—25 cm,宽 4—5 cm,有小叶柄(柄长约 5 mm,膨大,干后近黑色),中部最宽,向两端渐狭,顶端为长渐尖头,基部短楔形,叶缘有弧曲形尖锯齿;叶多少为厚坚纸质,腹面光滑无鳞毛,背面沿中肋有鳞片疏生;叶脉甚开展,很明显,分叉或单脉,顶部向上弯弓,直达锯齿;子囊群长 7—10 cm,线形,位于叶缘与中肋之间,其细密而分枝的节状夹毛短于孢子囊,易脱落;孢子长圆形,有微小密刺头突起,透明。

分布:海南岛及越南。

本种颇似亨利原始观音座莲(*A. Henryi* Christ et Gies.),但叶柄远较叶体为长(约 2 倍),羽片卵状披针形,对生,边缘具有规则的弧曲锯齿,夹毛远短于孢子囊,故易区别。

图注:1.本种全形(自然大);2.羽片的一部,表示叶脉,锯齿及子囊群(放大 4 倍);3.叶柄基部的鳞片(放大 30 倍);4.两个孢子囊着生于有夹毛的托上(放大 20 倍);5.子囊群下的夹毛(放大 100 倍);6.孢子(放大 100 倍),另一个(放大 400 倍)。

ARCHANGIOPTERIS TONKINENSIS (Hayata) Ching

ARCHANGIOPTERIS TONKINENSIS (Hayata) Ching, comb. nov.

Protomarratia Vietnamensis Hayata in Bot. Gaz. **67**: 88, fig. 1. 1919.

Archangiopteris tamdaoensis Hayata in Bot. Gaz. **67**: 90, fig 2. 1919; C. Chr. Ind. Fil. Suppl. Ⅲ. 26. 1933; C. Chr. et. Tard. in Lecomte, Not. Syst. **5**: 5. 1935; Tard. et C. Chr. in Fl. Ind. **7**: 17. 1939.

Protangiopteris tamdaoensis Hayata in Bot. Mag. Tokio **42**: 309. 1928.

Stipe 40-45 cm long, pale green, broadly grooved in its whole length above, sparsely clothed in brown lanceolate dentate *scales* especially in the lower part, with a nodose swelling at 20-30 cm above the base, lamina broadly ovate in outline, much shorter than stipe, impari-pinnate with one large and long-petiolate terminal pinna similar to those immediately below; *pinnae* 2-4 pairs, opposite, oblique or obliquely patent, ovate-lanceolate, 20-25 cm long, 4-5 cm broad, petiolate (petiole about 5 mm long, inflated and blackish), broadest at the middle, gradually narrowed towards both ends, apex long-acuminate, base shortly cuneate, margin serrate with rather sharp and arcuate teeth; *texture* more or less thickly chartaceous, glabrous above, very sparsely scaly along the rachis and on the under side; *veins* very patent, quite distinct below, forked or simple, curved upward, one into each tooth; *sori* 7-10 mm long, linear, midway between the costa and margin, finely paraphysate with branched articulate hairs much shorter than sporangia, finally becoming abraded; *spores* roundish-oblong, minutely and densely echinose, hyaline.

Hainan: Five Finger Mt., Shan Ah Ping, *C. L. Tso 44184*, Oct. 24, 1932, in woods, alt. 3000 ft.; southern slope of the same locality, *F. A. McClure 9470*, May 19, 1922, in wooded ravines; *C. Wang 35723*, Dec. 20, 1933, by stream side.

Vietnam: Tam Dao, *B. Hayata* (type of *A. tamdaoensis* Hayata), août, 1917; *Pételot 3956*, Mai. 1913; environs de Chapa on the border of Yunnan, *Pételot*, août, 1913, vers alt. 1300 m. Cua Tung, *Cadière*, Mai, 1934.

The present species closely resembles *A. Henryi* Christ et Gies., differing in the much longer stipe (about twice as long as the lamina), ovate-lanceolate opposite pinnae with regularly and rather arcuately serrate margin and in having paraphyses much shorter than the sporangia.

Plate 209. 1. Habit sketch (natural size). 2. Portion of pinna, showing venation, serrature and sori (× 4). 3. Scale from the basal part of stipe (× 30). 4. Two sporangia attached to the receptacle with paraphyses (× 20). 5. Paraphyses removed from the sori (× 100). 6. Spores (× 100), one enlarged (× 400).

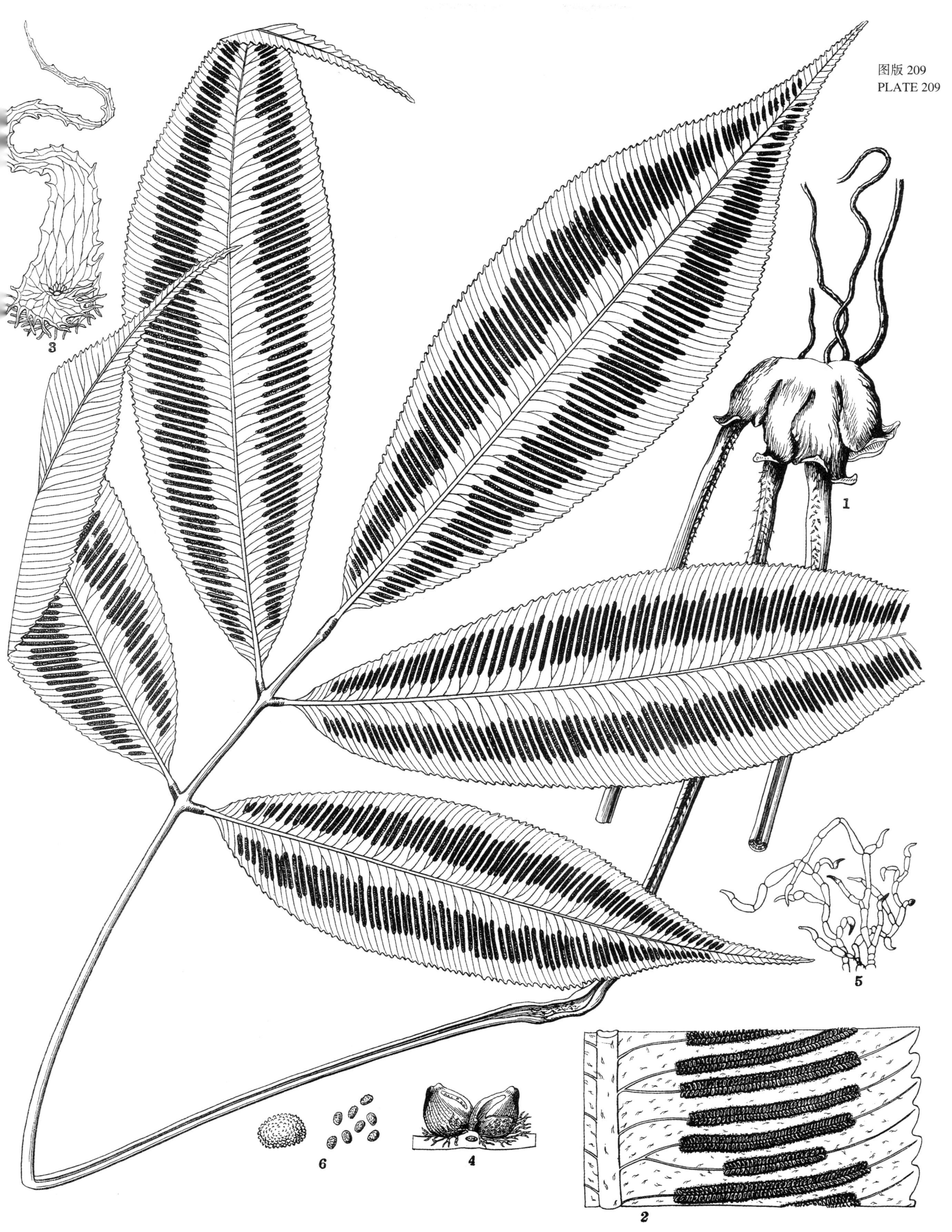

ARCHANGIOPTERIS TONKINENSIS (Hayata) Ching
尖叶原始观音座莲

根状茎短而直立,有密的根丛,叶簇生,直立,敧斜或伏地;叶柄长 5—12 cm,似铁线形,开展,暗栗色,下部有鳞片和粗长毛密生,鳞片深棕色,披针形,全缘;粗长毛棕色有节,叶体线状披针形,长 15—25 cm,基部不变狭,或几不变狭,顶端渐尖,或通常中轴延长成鞭状匍匐茎,向下着地生根,一回羽状;羽片多数,篦齿状,开展,基部的相距较远,长 1.2—1.5 cm,宽 7 mm,几无小柄,对开式,下边切成水平直线,上边为圆形,并为或深或浅的分裂,经常多次深锐裂成为许多有钝圆顶的接近裂片;叶为粗糙硬纸质,中轴和叶的上下面有粗长毛,叶脉下面明显,多次二叉分枝;子囊群略为圆形或为横生于各裂片顶端的圆肾形,假囊群盖上也有短粗毛。

分布:福建,贵州,云南,四川,浙江,湖北,江西,湖南的南部,台湾,香港;印度,缅甸,越南,泰国,斯里兰卡,马来西亚和菲律宾群岛也有分布,但不产于日本;尤常见于华南和云南南部,喜钙质土壤。

图注:1.本种全形(自然大);2.羽片,表示叶脉及子囊群(放大 4 倍);3.叶柄下部的鳞片(放大 30 倍);4.叶柄上的毛(放大 50 倍);5—6.孢子囊及孢子(放大 100 倍)。

ADIANTUM CAUDATUM Linné

ADIANTUM CAUDATUM Linné Mant. 308. 1771; Sw. Syn. Fil. 122. 1806; Willd. Sp. Pl. 5: 431. 1810; Hook. Exot. Pl. 2: t. 104. 1828; Sp. Fil. 2: 13. 1851; in Blakiston on the Yangtze 362. 1862; in Journ. Bot. **1857**: 336; Hook. & Bak. Syn. Fil. 115. 1867 (pro parte); Benth. Fl. Hongk. 447. 1861; Bedd. Ferns South. Ind. t. 2. 1863; Handb. Ferns Brit. Ind. 83. 1883; Milde, Fil. Europ. Atl. 29. 1867; Henry, List Pl. Form. 110. 1896; Christ in Bull. Herb. Boiss. **6**: 958. 1898; Diels in Engl. u. Prantl. Nat. Pflanzenfam. **1**: iv. 283. 1899; Christ in Bull. Soc. Bot. France **52**: Mém. I. 61. 1905; C. Chr. Ind. Fil. 24. 1905; in Acta Hort. Gotheb. **1**: 92. 1924; in Contr. U. S. Nat. Herb. **26**: 310. 1931; in Danske Bot. Ark. **7**: 125. 1932; Dunn & Tutcher, Fl. Kwangt. and Hongk. 338. 1912; Matsum. et Hayata, Enum. Pl. Form. 415. 1906; v. A. v. R. Handb. Mal. Ferns 324. 1908; Hand-Mzt. Symb. Sin. **6**: 38. 1929; Ogata, Ic. Fil. Jap. **2**: t. 51. 1929; Tagawa in Journ. Jap. Bot. **14**: 309. 1938 (pro parte); Tard. et C. Chr. in Fl. Gen. Ind. **7**: 180. 1940 (pro parte); Dickson in Ohio Journ. Sci. **46**: 135. 1946; Ching in Acta Phytotax. Sinica **6**: 313. 1957.

Adiantum lyratum Blanco, Fl. Phil. 832. 1837.

Adiantum balansae Christ in Bull. Soc. Bot. France **52**: Mém. I. 61. 1905 (non Baker).

Rhizome short, erect, densely radicose; *fronds* tufted, erect or spreading, stipe 5-12 cm long, wiry, spreading, dark castaneous-brown, densely scaly and strigose in the lower part, *scales* dark brown, lanceolate, entire, *hairs* long, brown, septate, lamina linear-lanceolate, 15-25 cm long, base not narrowed, or hardly so, apex acuminate but more often prolonged into a whip-like stolon and rooting at the apex, simply pinnate; *pinnae* numerous, pectinate, patent, the lower ones rather widely separated, 1.2-1.5 cm long and 7 mm broad, nearly sessile, dimidiate, the lower line cut straight and horizontal, the upper roundish, more or less cut, often deeply and repeatedly incised with numerous contiguous segments having blunt apex; *texture* rigidly chartaceous, the rachis and both surfaces of the frond villose; *veins* prominent below, repeatedly dichotomous; *sori* roundish or transversely oblong at the tip of segments, one to each segment, indusium also shortly strigose.

A pan-tropical and subtropical fern. In China numerous specimens have been seen from Guangdong, Hainan, Taiwan, Guangxi, Fujian, Guizhou, Southern part of Yunnan, Sichuan, Hubei, Zhejiang, Jiangxi and Hunan, Hong Kong, being most abundant in South China and Southern Yunnan, prefering calcareous soil.

Also India, Burma, Indo-China, Thailand, Sri Lanka, Malaya and the Philippines, but not yet known from Japan.

Plate 210. 1. Habit sketch (natural size). 2. Pinna, showing venation and sori ($\times$ 4). 3. Scale from the lower part of stipe ($\times$ 30). 4. Hairs from the stipe ($\times$ 50). 5-6. Sporangium with spores ($\times$ 100).

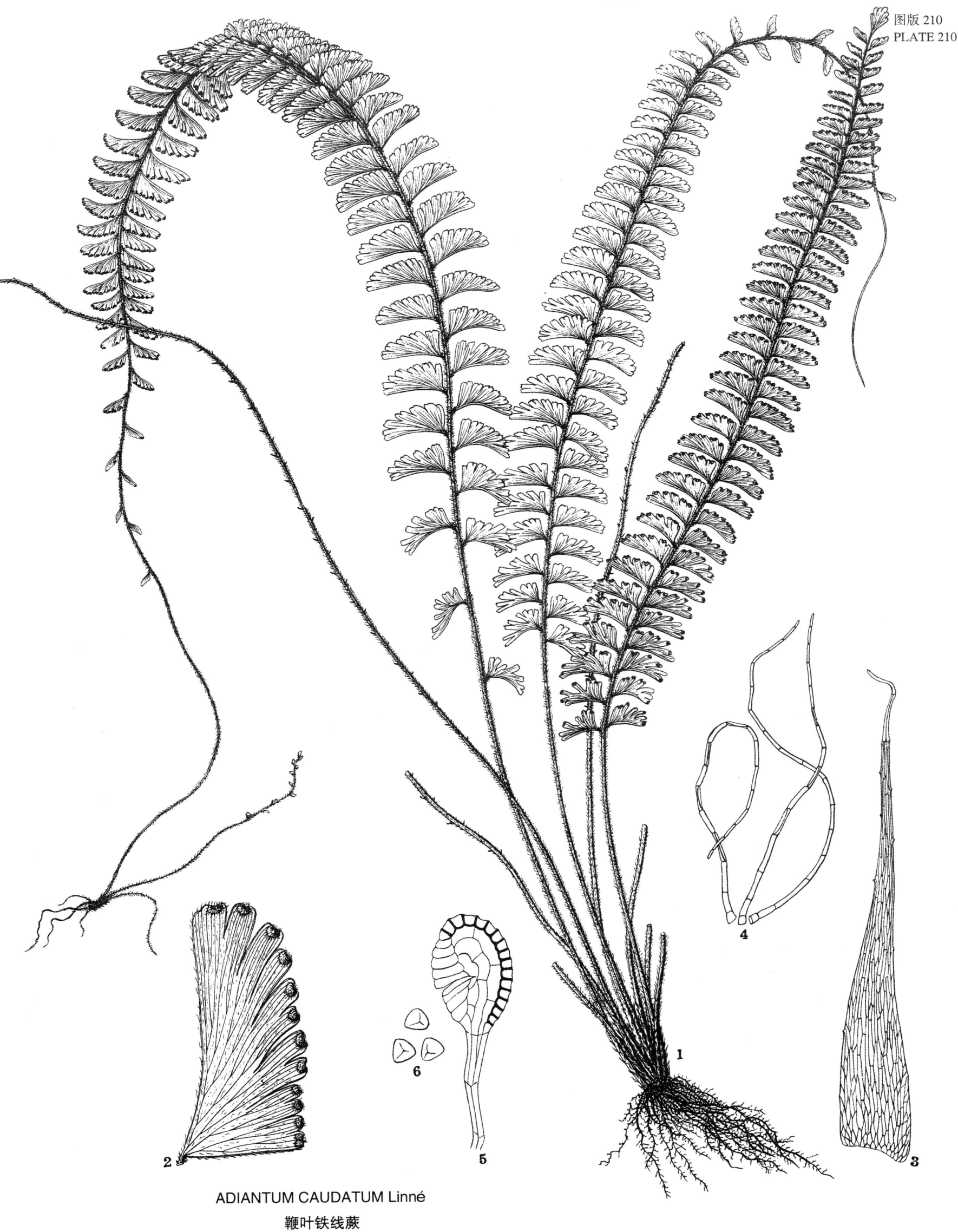

ADIANTUM CAUDATUM Linné
鞭叶铁线蕨

根状茎短而直立,有密的根丛,鳞片钻状线形,微亮黑色,有狭棕色的边缘,覆瓦状密生;叶簇生,硬而直立,叶柄长 5—12 cm,粗 1.5 mm,栗黑色,光亮,同叶轴一样,全部密生有节状淡红色刚毛状的张开长毛,基部混生有与根状茎上同样的鳞片,叶体长 12—27 cm,宽 1.7—2.5 cm,线状披针形,向顶端渐狭,稍呈钝头,基部几不变狭,一回羽状,羽片 24—32 对,开展,近生或亚篦齿状,无柄,基部一对几不变小,多少下向,中部羽片长 1—1.5 cm,基部宽 5—6 mm,对开式,下边切成直线,基部上方截形,顶端略为圆形,外边微弯弓,并浅裂为 4—5 个宽裂片;顶生羽片为扇形,浅三裂,较大于下面侧生羽片;叶为粗草质,干后淡绿色,上下两面有长而硬的刚毛状的浅色毛;叶脉细而明显,二叉分枝,每小裂片约有 8 条小脉,子囊群在每羽片有 3—4 个,或每小裂片有 1 个,罕为 2 个,圆形或横生长圆形,顶端为不育,假囊群盖黑棕色,坚硬,生有与叶面上同样的长硬毛。

分布:云南西北部,邓川县,苍山,海拔 1500 米,生石上。

本种为云南大理苍山西北部特有种,一般形体酷似爱氏铁线蕨(*Adiantum Edgeworthi* Hook.),但在毛茸上又极似鞭叶铁线蕨(*Adiantum caudatum* L.),唯叶体顶端有一扇形羽片,侧生羽片仅浅裂成三四个小裂片,叶柄基部的栗黑色光亮的鳞片有浅棕色的边缘及较长而粗硬开展的淡红色刚毛,故易与鞭叶铁线蕨区别。

图注:1.本种全形(自然大);2.羽片,有子囊群(放大 4 倍);3.不育羽片(放大 4 倍);4.叶柄基部的鳞片(放大 30 倍);5.叶柄上的刚毛(放大 100 倍);6.羽片上的刚毛(放大 100 倍);7—8.孢子囊及孢子(放大 100 倍)。

ADIANTUM SINICUM Ching

ADIANTUM SINICUM Ching in Bull. Fan Mem. Inst. Biol. New Series **1**: 267. 1949; in Acta Phytotax. Sinica **6**: 315. 1957.

Rhizome short, erect, densely radicose, *scales* linear-subulate, shining blackish with narrow brown margin, densely imbricate; *fronds* caespitose, erect, rigid, stipe 5-12 cm long, 1.5 mm across, dark castaneous, shining, densely clothed in reddish articulate setaceous spreading long hairs throughout the rachis, mixed in the basal part with similar scales as those on rhizome, lamina 12-27 cm long, 1.7-2.5 cm broad, linear-lanceolate, gradually attenuate towards the bluntish apex, base scarcely narrowed, simply pinnate; *pinnae* 24-32 pairs, patent, close or subpectinate, sessile, the lowest ones scarcely reduced, more or less deflexed, the middle pinna 1-1.5 cm long, 5-6 mm broad at base, dimidiate, cut straight below, anterior base truncate, apex roundish, outer edge slightly curved, shallowly lobato-incised into 4-5 broad lobules, the terminal pinna flabellate, shallowly trilobed, larger than those immediately below; *texture* thickly herbaceous, dry greenish, both sides rather copiously hirsute with setaceous stout hairs of paler color; *veins* fine but distinct on both sides, flabellately forked with about 8 veinlets to each lobule; sori 3-4 to each pinna, or 1 (rarely 2) to each lobule, orbicular or transversely oblong, leaving apical part of pinna sterile, indusium dark-brown, firm, similarly hirsute as the leaf surfaces.

Northwestern Yunnan: Whangchiaping, Tangchuan Hsien, north of Tali, *R. C. Ching 24897*, September 12, 1940, on the edge of rocks, alt. 1500 m.

A remarkable species, sharing the general habit with *Adiantum Edgeworthi* Hook., but in the type of hairs it closely resembles A. *caudatum* L. from which it can be easily distinguished by the fronds terminated by a flabellate pinna, the much less divided pinna, the blackish shining scales with light-colored margin and by the longer, coarser and spreading reddish setose hairs along the stipe and rachis.

Plate 211. 1. Habit sketch (natural size). 2. Pinna, soriferous ($\times$ 4). 3. Pinna, sterile ($\times$ 4). 4. Scale from the base of stipe ($\times$ 30). 5. Hairs from the stipe ($\times$ 100). 6. Hairs from the under side of pinna ($\times$ 100). 7-8. Sporangium with spores ($\times$ 100).

ADIANTUM SINICUM Ching
苍山铁线蕨

图版 212　仙霞铁线蕨

根状茎短而直立，有密的根丛和疏生的黑棕色披针形厚质鳞片；叶簇生，直立，多数，长 15—18 cm（包括长 5—7 cm，栗色，光亮圆形的叶柄），叶体长 10—12 cm，宽约 3 cm，一回奇数羽状；羽片 7—9 对，对生，或向上为亚对生，斜出开展，节间相距 2—3 cm，有小叶柄（柄长 3 mm），长 1 cm，宽 1.2—1.5 cm，为阔亚圆三角形，全缘或外边缘略为微波状，上部羽片较小，基部都以关节着生；叶为亚革质，上面绿色，下面灰绿色；叶脉两边颇明显，数次二叉分枝；每羽片有 4—6 个假囊群盖，亚圆形或横生肾形，革质，黑色，宿存。

分布：福建北部与浙江南部交界处的仙霞岭，少见。

本种为一明显的蕨种，颇近似广东省的钱氏铁线蕨（*Adiantum Chienii* Ching），但羽叶为宽扇形，宽过于长，每羽片上有 4—6 个子囊群和亚圆形或横生肾形的假囊群盖，故易区别；在一般形体上本种更近于团叶铁线蕨（*Adiantum capillus-junonis* Rupr.），唯叶直立，有较粗健的叶柄，叶为亚革质，有短假囊群盖（圆形或横生肾形），中轴不延长成为鞭状匍匐茎，故大不相同。

图注：1. 本种全形（自然大）；2. 羽片，表示叶脉及子囊群（放大 3 倍）；3. 根状茎上的鳞片（放大 40 倍）；4—5. 孢子囊及孢子（放大 100 倍）。

ADIANTUM JUXTAPOSITUM Ching

ADIANTUM JUXTAPOSITUM Ching in Acta Phytotax. Sinica **6**: 312. 1957.

Rhizome short, erect, densely radicose, sparsely clothed in dark-brown lanceolate firm *scales*; *fronds* caespitose, upright, numerous together, 15-18 cm long including stipe, which is 5-7 cm long, stout, castaneous, shining and glabrous, lamina 10-12 cm long, about 3 cm broad, simply impari-pinnate; *pinna* 7-9 pairs, opposite or subopposite upward, oblique-patent, 2-3 cm apart, petiolate (petiole 3 mm long), 1 cm long, 1.2-1.5 cm broad, smaller towards apex, orbicular-triangular with round, entire or slightly wavy outer edge, the terminal pinna smaller, all articulate at the base; *texture* subcoriaceous, green above, bluish below; *veins* flabellately forked, rather distinct on both sides; *indusia* 4-6 to each pinna, suborbicular or transversely reniform, blackish, coriaceous, persistent.

Northern Fujian: SihYaLing, on the border of Zhejiang, growing in dry rock crevices, rare.

A distinct endemic species, proximately related to *Adiantum Chienii* Ching (see Ching, Ic. Fil. Sin. **4**: t. 159) from Guangdong Province, but differing in the broader fan-shaped pinnae, which are broader than long and in having more sori to each pinnae with suborbicular or transversely reniform indusia. In general habit, this species resembles *Adiantum capillus-junonis* Rupr. more than any others, yet differing in upright fronds of much stouter build with much thicker stipe and subcoriaceous leaf texture, shorter (orbicular or transversely reniform) indusia and in not prolongated and rooting apex of fronds.

Plate 212. 1. Habit sketch (natural size). 2. Pinna, showing venation and sori ($\times$ 3). 3. Scale from the rhizome ($\times$ 40). 4-5. Sporangium with spores ($\times$ 100).

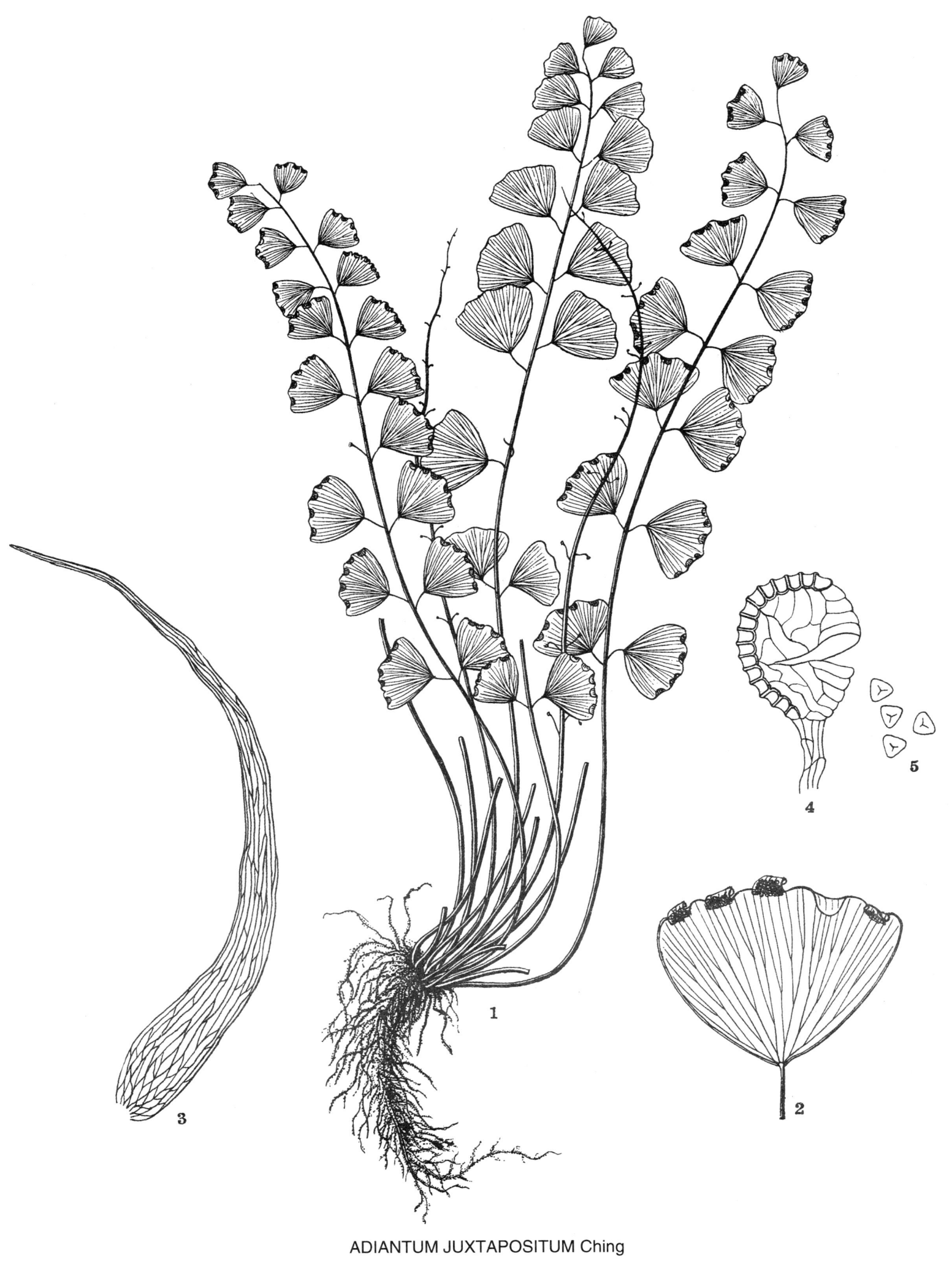

ADIANTUM JUXTAPOSITUM Ching
仙霞铁线蕨

根状茎短而直立,有密的根丛,顶覆深棕色披针形的鳞片;叶多数簇生,向四面展开,并且大都为匍匐状;叶柄长 2—6 cm,细铁丝形,光亮,浅黑色,平滑,叶体长 8—15 cm,宽 2—3 cm,阔披针形,一回奇数羽状,或者中轴通常延长变成鞭状匍匐茎,顶端向下,着地生根,发生幼株;羽片 3—7 对,对生或通常互生,远分开,有小柄(柄长约 3 mm),长宽各约 1—1.5 cm 或宽过于长,亚圆形或为阔扇形,基部圆形或短楔形,上边为阔圆形,全缘或通常多少缺刻成宽浅裂片,具微小而均匀的齿牙,上部的羽片渐变小,扇形,在鞭状匍匐茎上也有少数同样而疏生的羽片;叶为薄草质,光滑,上面嫩绿色,下面淡灰绿色;叶脉为细丝状,多次二叉,有许多明显小脉,分别达于小锯齿;子囊群通常 2—5 个,生于羽片上边缘,略为圆形或横长圆形或短线形,被缺刻分开,假囊群盖为深棕色,质坚实。

分布:本种主要为中国特有种,最初在北京西山发现,现知广布于河北,山东,河南,甘肃,四川,贵州,云南,广西,广东和台湾。日本及朝鲜半岛也有。喜生于湿润的石灰岩脚阴凹处或古墙脚下,为一美丽的蕨种。

图注:1.本种全形(自然大);2.能育羽片(放大 3 倍);3.不育羽片(放大 3 倍);4.叶柄基部的鳞片(放大 40 倍);5—6.孢子囊及孢子(放大 100 倍)。

ADIANTUM CAPILLUS-JUNONIS Ruprecht

ADIANTUM CAPILLUS-JUNONIS Ruprecht, Distr. Crypt. Vasc. Ross. 49. 1845; Milde in Bot. Zeit. **1867**: 148; Fil. Europ. Alt. 29. 1867; Hook & Bak. Syn. Fil. 114. 1874; Franch. Plant. David. **1**: 348. 1884; Diels in Engl. u. Prantl, Nat. Pflanzenfam. **1**: iv. 283. 1899; in Bot. Jahrb. **29**: 200. 1900; Christ in Warb. Monsunia 67. 1900; in Bull. Soc. Bot. France **52**: Mém. I. 61. 1905; C. Chr. Ind. Fil. 24. 1905; in Acta Hort. Gotheb. **1**: 93. 1924; Matsum. et Hayata, Enum. Pl. Form. 612. 1906; Dunn & Tutcher, Fl. Kwangt. and Hongk. 338. 1912; Hand-Mzt. Symb. Sin. **6**: 38. 1929; Tagawa in Acta Phytotax. et Geobot. **1**: 101. 311. 1932; in Journ. Jap. **14**: 312. 1938; Kitagawa in Rep. 1st. Sci. Exped. Manch. **4**: 87. 1935; Kitagawa, Lineamenta Fl. Mansh. 26. 1939; Ching in Acta Phytotax. Sinica **6**: 317. 1957.

Adiantum cantoniense Hance in Ann. Sci. Nat. Ser. 4. **15**: 129. 1861; Hook. & Bak. Syn. Fil. 114. 1867.

Rhizome short, erect, densely radicose below, crowned with dark brown lanceolate *scales*; *fronds* fasciculate, numerous together, spreading and mostly creeping, stipe 2-6 cm long, wiry, polished, blackish, glabrous, lamina 8-15 cm long, 2-3 cm broad, broadly lanceolate, simply pinnate with a terminal pinna, or more often prolonged into a whip-like stolon and rooting at the apex; *pinna* 3-7 pairs, distant, opposite or generally alternate, petiolate (petiole 3 mm long), 1-1.5 cm long, and as broad or broader, suborbicular, or flabellate, base cuneate or subrounded, the outer edge broadly rounded, entire or slightly and broadly lobed, minutely denticulate, the upper pinnae smaller, fan-shaped, the stolons bear a few distant similar pinnae below; *texture* pellucido-herbaceous, glabrous, light tender green above, bluish-green below; *veins* fine, repeatedly forked with distinct veinlets, one to each denticule; *sori* usually 2-5 round the outer edge, roundish, transversely oblong, or linear, separated by a notch, indusium dark brown, firm.

A very pretty and shade-loving fern in dense patches on shaded moist limestone or calcareous soil. One of the common ferns in China, ranging from the Provinces of Hebei, Shandong, Henan, Gansu, Sichuan, Hubei, Guizhou down to Yunnan, Guangxi, Guangdong and Taiwan.

Also Japan and Korea peninsula. Type from the mountains around Beijing.

Plate 213. 1. Habit sketch (natural size). 2. Pinna, fertile ($\times$ 3). 3. Pinna, sterile ($\times$ 3). 4. Scale from the base of stipe ($\times$ 40). 5-6. Sporangium with spores ($\times$ 100).

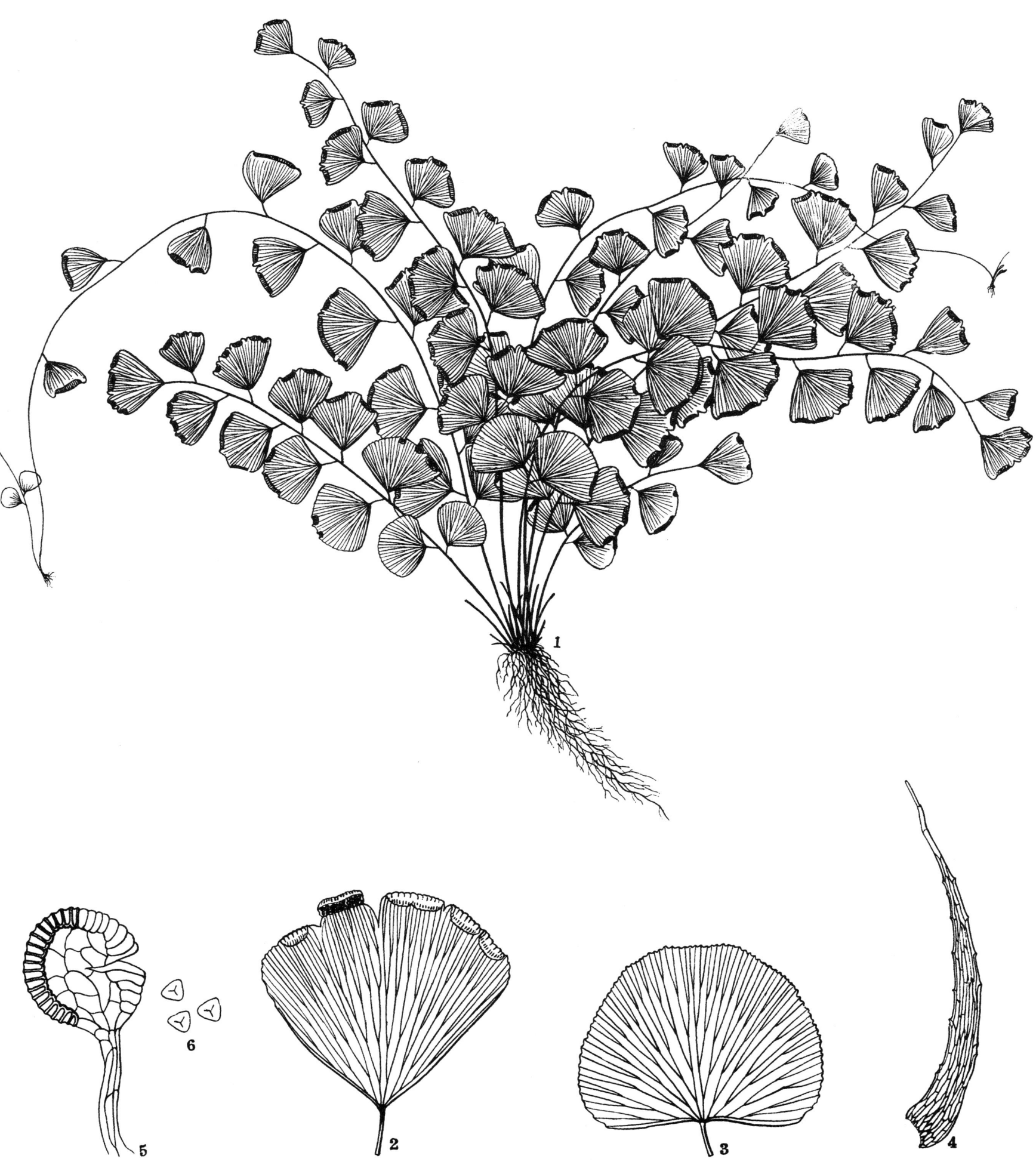

ADIANTUM CAPILLUS-JUNONIS Ruprecht
团叶铁线蕨

图版 214 半月形铁线蕨

根状茎短而直立,有密的根丛,顶部有深棕色狭披针形的鳞片疏生;叶簇生,向四面开展或向下着地;叶柄长 10—15 cm,光亮,深栗棕色,基部以上光滑,叶体长 12—30 cm 或更长,宽 4—8 cm,披针形到长圆披针形,一回奇数羽状或者中轴往往延长变成无叶的鞭状匍匐茎,顶端向下着地生根,发生幼株;羽片 7—12 对,开展,接近,长 1.5—4 cm,宽 1—2 cm,略为半圆形或新月形,对开式,下边缘几切成直线或与小叶柄成倾斜角,上边缘圆形与钝形顶端有几个或深或浅的缺刻,但无锯齿;小叶柄长达 1—1.5 cm,细铁丝形,几与中轴成直角,以关节着生于小羽叶下边缘的中央偏内或自内向外 1/3 处;顶生羽片(如有)为扇形,大小几等于侧生羽片;鞭状中轴下部通常有少数退缩的小羽片疏生;叶为纸草质,中轴及上下面均光滑无毛;叶脉下面明显,数次二叉分枝;子囊群成长线形沿上边缘着生,仅被少数的缺刻割断。

分布:本种广泛分布于热带及亚热带的亚洲,澳洲及非洲;在中国产海南,台湾,广东,广西,云南及贵州南部,四川西南部(峨眉山麓)和香港。喜生于湿润酸性红壤,常成块状群生。

本种为一突出的经典种,由于它的半圆形或新月形的羽片,易与其他之种识别,但要指出,本种植株的高矮和羽片的大小形态变异极大,鉴定时应加以注意。

图注:1.本种全形(自然大);2.能育羽片(放大 2 倍);3.不育羽片(放大 2 倍);4.叶柄基部的鳞片(放大 40 倍);5—6.孢子囊及孢子(放大 100 倍)。

ADIANTUM PHILIPPENSE Linné

ADIANTUM PHILIPPENSE Linné Sp. Pl. **2**: 1094. 1753; Sw. Syn. Fil. 120 1806; Hook. Sp. Fil. **2**: 3. 1861; Copel. Polyp. Phil. 93. 1905; C. Chr. in Contr. U. S. Nat. Herb. **26**: 310. 1931; Ind. Fil. Suppl. Ⅲ. 19. 1933; Tagawa in Journ. Jap. Bot. **14**; 310. 1938; Tard. et C. Chr. in Fl. Gen. Ind. **7**: 182. 1940; Ching in Acta Phytotax. Sinica **6**: 318. 1957.

Adiantum lunulatum Burm. Fl. Ind. 235. 1768; Sw. Syn. Fil. 120. 1806; Hook, and Grev. Ic. Fil. t. 104. 1831; Benth. Fl. Hongk. 446. 1861; Hook. & Bak. Syn. Fil. 114. 1867; Milde, Fil. Europ. Alt. 28. 1867; Bedd. Ferns South. Ind. t. 1. 1863; Handb. Ferns Brit. Ind. 82. 1883; Diels in Engl. u. Prantl, Nat. Pflanzenfam. **1**; iv. 283. 1899. C. Chr. Ind. Fil. 29. 1905; v. A. v. R. Handb. Mal. Ferns 325. 1908; Dunn & Tutcher, Fl. Kwangt. and Hongk. 338. 1912; Hand-Mzt. Symb. Sin. **6**: 38. 1929; Ogata, Ic. Fil. Jap. **5**: t. 204. 1933.

Pteris lunulata Retz. Obs. **2**: 28, t. 4. 1761.

Rhizome short, erect, sparsely crowned at the apex with dark brown narrowly lanceolate *scales*; *fronds* fasciculate, spreading, or nodding, stipe 10-15 cm long, polished, dark chestnut-brown, glabrous above the base, lamina 12-30 cm or longer, 4-8 cm broad, lanceolate to oblong-lanceolate, impari-pinnate, or often elongated and rooting at the tip of a wiry stolon; *pinnae* 7-12 pairs, spread, close, 1.5-4 cm long, 1-2 cm broad, semi-orbicular, or lunate, subdimidiate, the lower edge cut nearly in a straight line or slightly oblique with the petiole, the upper edge rounded and, like the bluntly-rouned outer edge, usually with a few remote incisions, or more or less lobed, not serrate, petioles of the lower pinnae 1-1.5 cm long, wiry, almost form a right angle with the rachis, attached by an articulation to the pinnae at about one-third way from the inner edge; the terminal pinna (if present) flabellate, cuneate, as large as, or larger than the lower ones; the whip-like stolon usually bears a few distant small pinna below; *texture* papyraceo-herbaceous, the rachis and surfaces glabrous; *veins* distinct below; *sori* in a continuous line along the upper edge, only interrupted by a few narrow notches.

Widely distributed in tropical and subtropical regions of Asia, Australia and Africa. In China, this fern has been known from Yunnan, Sichuan (Mt. Omei), Guizhou, Guangxi, Guangdong, Hainan, Taiwan and Hong Kong, preferring a shaded position and forming large patches in acid soil.

A very distinct species, not easily to be confounded with other related species by dint of its peculiarly shaped pinnae, which, however, vary a great deal in size and depth of incision along the upper margin.

Plate 214. 1. Habit sketch (natural size). 2. Fertile pinna ($\times$ 2). 3. Sterile pinna ($\times$ 2). 4. Scale from the base of stipe ($\times$ 40). 5-6. Sporangium with spores ($\times$ 100).

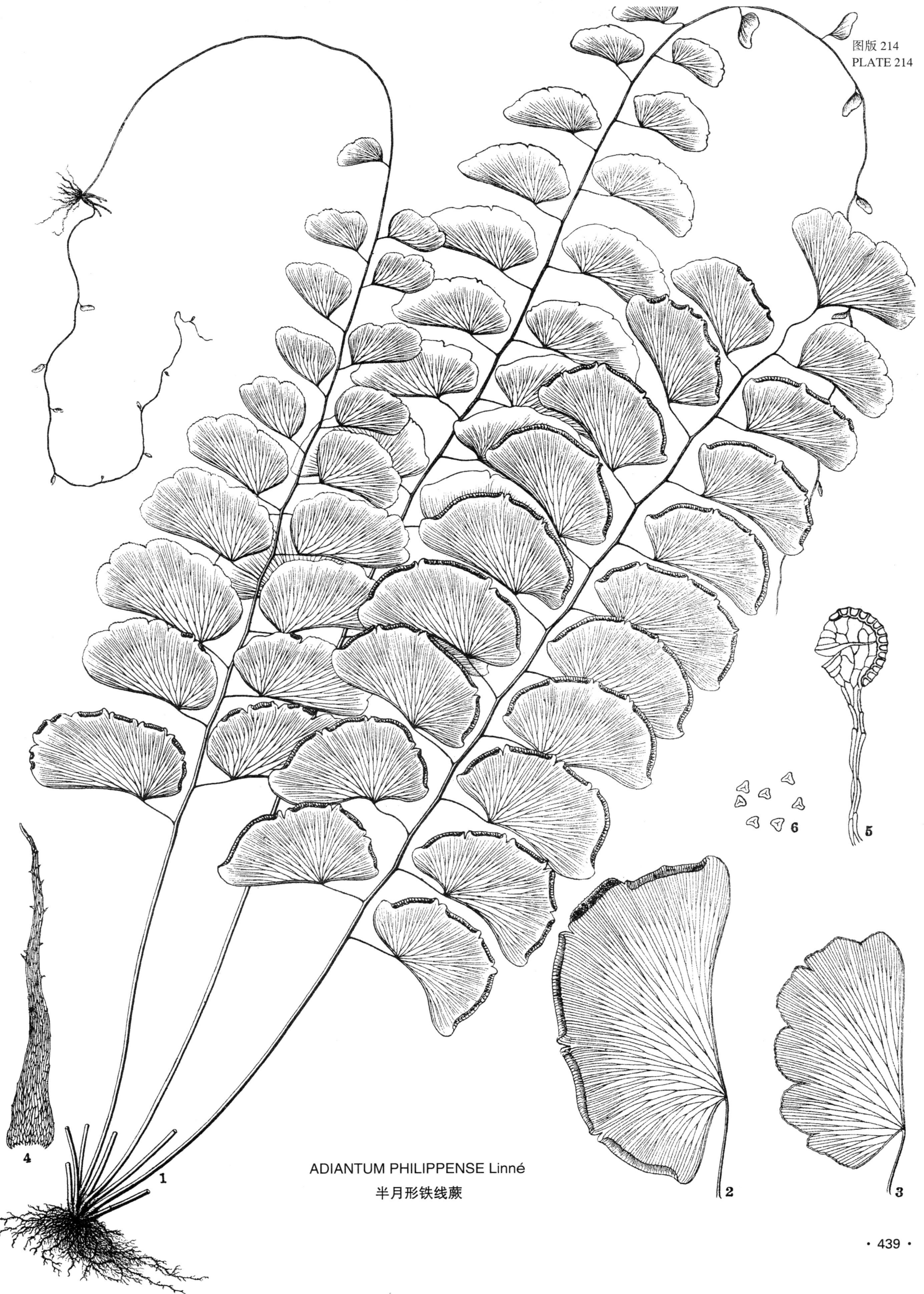

ADIANTUM PHILIPPENSE Linné
半月形铁线蕨

图版 215　翅柄铁线蕨

　　根状茎短而直立,有密的根丛,顶部被深棕色狭披针形的硬鳞片;叶簇生,直立,罕下垂;叶柄长 12—16 cm,暗栗棕色,粗强,两侧有淡棕色的狭翅,鳞片疏生或几光滑,叶体长等于柄,阔披针形,向基部不变狭,一回奇数羽状,或中轴有时延长变成鞭状的匍匐茎,向下着地生根;羽片 7—9 对,分开,开展,或者下部的多少向下倾斜,长 2—3 cm,宽 1—1.3 cm,亚长方长圆形,对开式,下边几割成直线,圆头的上基部为切形,上边几与下边平行,但多少缺刻为 3—5 个圆形无锯齿的裂片;小叶柄长 2—3 mm,两侧也有淡棕色波形狭翅;顶生小羽片(如有)为扇形,大小不一;鞭状延长的中轴下部常具几个疏生小羽片;叶为厚纸质,两面光滑,干后呈淡棕绿色;叶脉下面颇明显;子囊群为横生长圆形或肾形。

　　分布:本种也是广泛分布于东亚热带及亚热带和非洲的蕨种。在中国分布于海南,台湾和云南及广西南部,但甚稀见。越南,印度,缅甸,爪哇及菲律宾群岛等地也有。原种模式标本最初发现于印度北部。从引证的文献可见本种过去有过许多异名,十分混乱。

　　图注:1. 本种全形(自然大);2. 能育羽片(放大 2 倍);3. 不育羽片,表示全缘裂片(放大 2 倍);4. 叶柄横断面,表示两侧的翅(放大);5. 叶柄基部的鳞片(放大 50 倍);6—7. 孢子囊及孢子(放大 100 倍)。

ADIANTUM SOBOLIFERUM Wallich

ADIANTUM SOBOLIFERUM Wallich, List n. 74. 1828 (nom. nud.); Hook. Sp. Fil. **2**: 13, t. 74 A. 1851; Bedd. Ferns Brit. Ind. t. 16. 1867; C. Chr. Ind. Fil. 33. 1905; Suppl. Ⅲ. 197. 1933; Tagawa in Journ. Jap. Bot. **14**: 309. 1938; Tard. et C. Chr. in Fl. Gen. Ind, **7**: 181. 1940; Ching in Acta Phytotax. Sinica **6**: 320. 1957.

Adiantum caudatum Hook. & Bak. Syn. Fil. 115. 1874 (pro parte).

Adiantum caudatum var. *soboliferum* (Wall.) Bedd. Handb. Ferns Brit. Ind. 84. 1883.

Adiantum mettenii Kuhn, Fil. Afr. 65. 1865 (nom. nud.); Hook. & Bak. Syn. Fil. ed. 2. 472. 1874; v. A. v. R. Handb. Mal. Ferns 326. 1908; C. Chr. Ind. Fil. 30. 1905; in Dansk Bot. Ark. **7**: 124, t. 48, fig. 4. 1932; Ogata, Ic. Fil. Iap. **2**: t. 53. 1929.

Adiantum lunulatum Burm. var. *Mettenii* (Kuhn) Bedd, Ferns Brit. Ind. Suppl. 6, t. 354. 1876; Handb. Ferns Brit. Ind. 83. 1883.

Adiantum balancae Bak. in Journ. Bot. **28**: 263. 1890; Christ in Bull. Herb. Boiss. **6**: 597. 1898.

Adiantum alatum Copel. in Perkins Fragm. 192. 1905; Polyp. Phil. 93. 1905; A. Peter in Fedde, Rep. Sp. Nov. **40**: 43. 1929.

Adiantum lunulatum Ogata, Ic. Fil. Jap. **2**: t. 53. 1929. (non Burm. 1768).

Rhizome short, erect, densely radicose, clothed at the growing apex with some small, dark brown, firm, linear-lanceolate *scales*; *fronds* fasciculate, erect, or rarely nodding, stipe 12-16 cm long, dark chestnut-brown, stout, narrowly winged on each side, sparsely scaly or subglabrous, lamina as long as stipe, broadly lanceolate, not narrowed towards the base, simply impari-pinnate, or sometimes the rachis prolongated into a whip-like stolon and rooting at the tip; *pinnae* 7-9 pairs, distant, spreading, or the lower ones somewhat deflexed, 2-3 cm long, 1-1.3 cm broad, subrectangular-oblong, subdimidiate, cut quite straight below, upper side of the base oblique-truncate, roundish, the upper edge also quite in a line but more or less deeply lobed especially towards the blunt apex; *lobes* 3-5, round, entire, petiole 2-3 mm long, stout and, like the entire rachis, is provided on each side with a brown, narrow often wavy wing, the terminal pinna (if present) flabellate, cuneate, the stolon (when present) often with a few smaller distant pinna below; *texture* rather thickly chartaceous, both sides glabrous, dry brownish green; *veins* quite distinct below; *sori* transversely oblong or reniform, placed in sinuses, one to each lobe, about 3-5 to each pinna.

Another widely distributed fern in Tropical and Subtropical Asia and Africa. In China, specimens have been seen from Southern Yunnan, Guangxi, Hainan and Taiwan. Also Indo-China, India, Burma, Java and the Philippines. Type from N. India.

The present species, though quite similar to A. *philippense* L. in general habit, can be easily distinguished by the presence on each side of stipe, rachis and petiole of a narrow brown wing, which is more or less wavy on the petiole, much thicker leaf texture and shorter sori.

Plate 215. 1. Habit sketch (natural size). 2. Fertile pinna (× 2). 3. Sterile pinna, showing entire margin of lobes (× 2). 4. Cross section of stipe with a wing on each side. 5. Scale from the base of stipe (× 50). 6-7. Sporangium with spores (× 100).

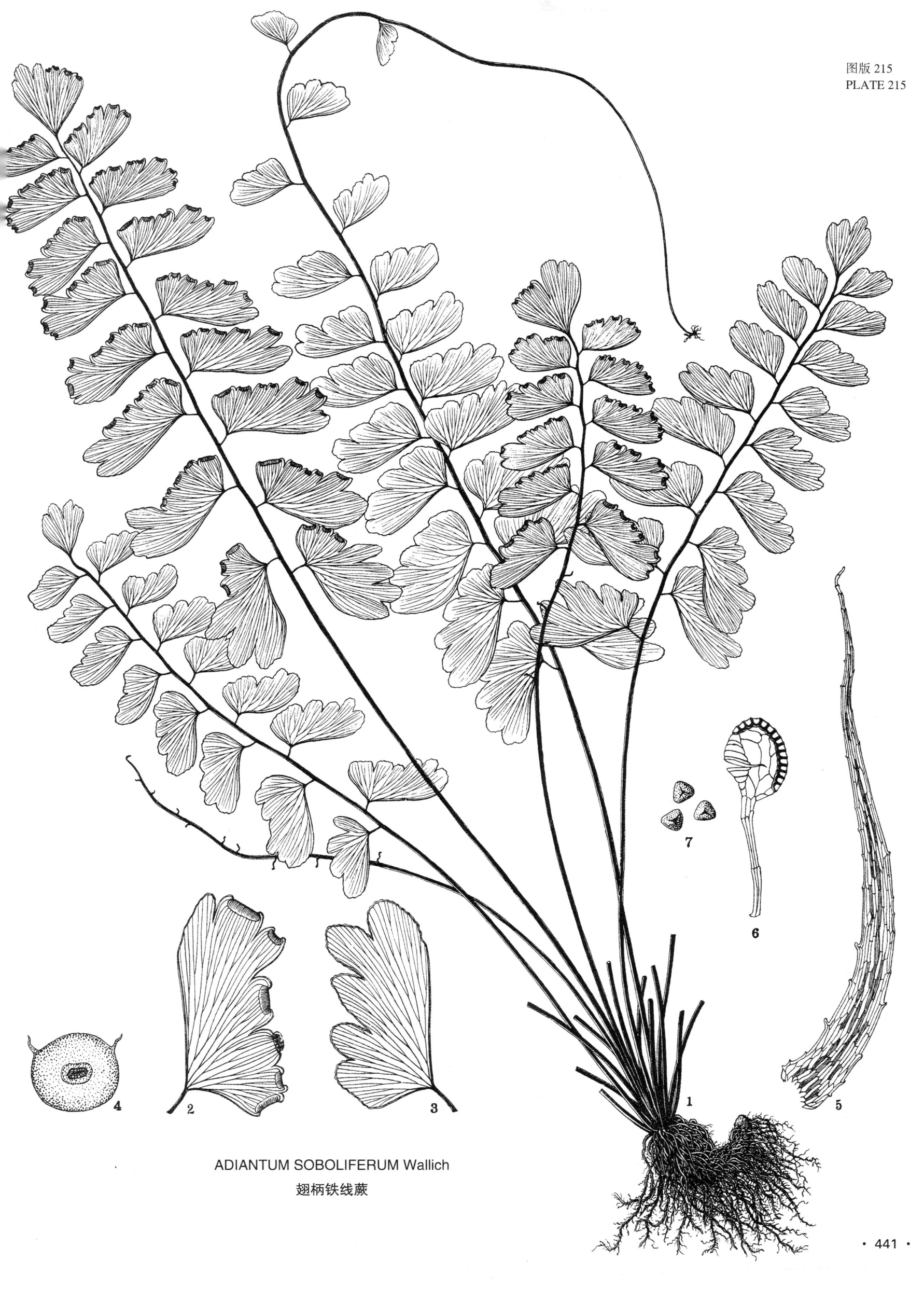

ADIANTUM SOBOLIFERUM Wallich
翅柄铁线蕨

根状茎匍匐，颇细长，有红棕色的狭披针形的鳞片密生；叶近生或较远生，直立；叶柄细长，长 10—16 cm，光亮，淡黑色，基部以上光滑，叶体为三角状卵形，大小不一，通常长 13—25 cm，宽 6—15 cm，二或三回羽状，顶部一回羽状，有时为一或二回羽叶；羽片 8—13 对，互生，有柄，斜出向上，接近，基部一对最大，二回羽状，或在小型植株为一回羽状，有小柄；末回小羽片宽 1—2 cm，大都为扇形或为亚长方形的对开式，基部楔形，有小柄，上边缘常为不规则的深裂；裂片钝头，不育裂片有小齿牙，顶生小羽片多少为扇形，基部为长楔形；叶为薄草质，嫩淡绿色，两面光滑；叶脉明显，对称二叉分裂为细丝状的小脉，在不育小叶达于锯齿；子囊群亚圆形或横生长圆形，生于宽浅的凹陷边缘，每裂一个，而为缺刻所分开。

分布：本种为世界种，而形体的变异性，分裂度及小羽片的锐裂深度，往往变异极大。在中国由北向南有大量标本采自河北，陕西，山西，河南，甘肃的南部和长江以南各省直达广东，海南，台湾，由东向西，自东海边的江浙起经长江流域直达四川，云南，更向西至缅甸，喜马拉雅山区。日本，鲜朝半岛都有分布。喜钙质土或石灰岩，为美丽的蕨种，并为世界著名的室内盆栽观赏植物之一。

图注：1. 本种全形（自然大）；2. 能育小羽片，表示假囊群盖（放大 3 倍）；3. 不育小羽片，表示边缘小齿牙（放大 3 倍）；4. 根状茎上的鳞片（放大 30 倍）；5—6. 孢子囊及孢子（放大 100 倍）；7. 团叶变型（自然大）；8. 细裂叶变型（自然大）。

ADIANTUM CAPILLUS-VENERIS Linné

ADIANTUM CAPILLUS-VENERIS Linné, Sp. Pl. **2**；1096. 1753；Sw. Syn. Fil. 124. 1806；Hook Gen. Fil. t. 60 B. 1836；Sp. Fil. **2**：36. 1851；Hook & Bak. Syn. Fil. 123. 1867；Milde, Fil. Europ Alt, 30. 1867；Bedd. Ferns South Ind. t. 4. 1863；Handb. Ferns Brit. Ind. 84. 1883；Christ, Farnkr. d. Erde 138. 1897；Diels in Engl u. Prantl, Nat. Pflanzenfam. **1**：iv. 284. 1899；C. Chr. Ind. Fil. 24. 1905；Ogata, Ic. Fil. Jap. **1**：t. 1. 1928；Ching in *Sinensia* 3：340. 1933；Tagawa in Journ. Jap. Bot. **14**：313. 1936；Tard. et C. Chr. in Fl. Gen. Ind. **7**：185. 1940；DeVol, Ferns East. China 115. 1945；Ching in Acta Phytotax. Sinica **6**：341. 1957.
Adiantum submarginatum Christ in Bull. Herb. Boiss. Ⅱ. **3**：511. 1903 (pro parte).
Adiantum michelii Christ in Bull. Acad. Géogr. Bot. Mans. **1910**：10.

Rhizome creeping, rather slender, densely clothed in rufo-brown linear-lanceolate *scales*; *fronds* proximate, or quite distant, stipe slender, 10-16 cm long, polished, blackish, naked above the base, lamina deltoid-ovate, quite variable in size, generally 13-25 cm long, 6-15 cm broad, 2-3 pinnate under the simply pinnate apical part, sometimes quite simply pinnate; *pinnae* 8-13 pairs, alternate, petiolate, oblique, close, the basal pair much the largest, fully bipinnate, or pinnate in small plants, petiolulate; the *ultimate pinnules* 1-2 cm broad, mostly fan-shaped, or subrectangular and dimidiate, cuneate, petiolulate, the outer margin often irregularly and deeply lobed from the circumference in the direction of the centre, the *lobes* obtuse and dentate in the sterile part, the terminal pinnule more or less fan-shaped and long-cuneate; *texture* thin herbaceous, light green, glabrous; *veins* distinct, with fine dichotomous veinlets, one to each tooth in the sterile pinnule; *sori* roundish or transversely oblong, placed in roundish sinuses, one to each lobe.

A cosmopolitan species, very variable in size, degree of pinnation and depth of segmentation of pinnules. In China numerous specimens have been seen from Hebei, Shaanxi, Henan, Gansu and all the provinces south of the Yangtze southwards down to Guangdong, Taiwan, and from the sea coast westwards to Sichuan and Yunnan. Also Japan, Korea peninsula and the Himalayas.

A most graceful shade-loving calciphilous (lime-loving) plant, often forming large patches in dripping limestone nitches or crevices. Extensively grown as favourite pot plant for house decoration and conservatories.

Plate 216. 1. Habit sketch (natural size). 2. Fertile pinna, showing the shape of indusia (× 3). 3. Sterile pinna, showing serrature (× 3). 4. Scale from the rhizome (× 30). 5-6. Sporangium with spores (× 100). 7. Forma *dissecta* (Mart. et Galeot.) Ching (natural size). 8. Forma *fissa* (Christ) Ching (natural size).

ADIANTUM CAPILLUS-VENERIS Linné
铁线蕨

根状茎短匍匐或斜行，较粗健，有红棕色或深棕色阔披针形渐尖的鳞片覆盖；叶亚簇生，多数成丛；叶柄长 12—20 cm，粗过于 1 mm，暗栗色，光亮，干后多少呈压扁形，基部以上光滑，叶体长 12—18 cm，宽 5—12 cm，三角卵形或为狭长圆形，顶端为钝头，2—3 回羽状；羽片 4—6 对，生于一回羽状的由 2—4 对小羽片组成的顶部之下，互生、向上斜出，接近，有小叶柄（基部小叶柄长达 1.5 cm），基部一对较大，长 5—10 cm，一至二回羽状，长圆形（如为一回羽状）或为三角形（如为二回羽状）；小羽片 3—6 对，不分裂，有小柄，长 5 mm；末回小羽片几乎都为同形，阔 1—1.2 cm，高 8—10 mm，扇形，瓦覆于小轴，基部楔形，下部的有略斜的基部，有短小柄，上边缘圆形，经常两裂，不具锯齿；叶为草质或坚草质，干后为绿色，上下面同色而且光滑，不育的裂片为全缘或至多为波状；叶脉下面颇明显；子囊群一般每小羽片有两个，为长圆形或线形，横生于上边的宽深凹处。

分布：本种为我国西南部高山针叶林下的特有种，产云南西北部，四川西南部（峨眉山）及贵州中部，海拔自 3000—3400 m 的石灰岩地区。

本种为一明显的地区性种，在形体上极似普通的铁线蕨（*Adiantum capillus-veneris* L.），但形体较小，末回羽片一般具两个长形的子囊群横生于阔深的缺口内，而且特别是不育小羽片为全缘，不具锯齿，至多为波状，故易于识别。

图注：1. 本种全形（自然大）；2. 能育小羽片，表示子囊群在发育初期不甚下陷于缺口中（放大 4 倍）；3. 同上，表示子囊群成熟后深陷于缺口情况（放大 3 倍）；4. 不育小羽片，表示全缘或略成波状边缘（放大 3 倍）；5. 根状茎上的鳞片（放大 30 倍）；6—7. 孢子囊及孢子（放大 100 倍）。

ADIANTUM EDENTULUM Christ

ADIANTUM EDENTULUM Christ in Bull. Soc. Bot. France **52**: Mém. Ⅰ. 63. 1905; C. Chr. Ind. Fil. 661. 1905; in Contr. U. S. Nat. Herb. **26**: 310. 1931; Ching in Acta Phytotax. Sinica **6**: 338. 1957.

Adiantum delavayi Christ, l. c; C. Chr. l. c.

Adiantum capillus-veneris L. var. *sinuatum* Christ in Bull. Soc. Bot. France **52**: Mém. Ⅰ. 61. 1905; in Bull. Acad. Géogr. Bot. Mans. **1906**: 62; C. Chr. in Acta Hort. Gotheb. **1**: 93. 1924.

Rhizome short-creeping or oblique, rather thick, clothed in rufo-brown or fusco-brown broadly lanceolate acuminate *scales*; *fronds* subcaespitose, several together, stipe 12-20 cm long, a little over 1 mm thick, dark castaneous, glossy, dry somewhat compressed, glabrous above the base, lamina 12-18 cm long, 5-12 cm broad, deltoid-ovate to narrowly oblong, apex obtuse, 2-3-pinnate; *pinnae* 4-6 pairs (below the simply pinnate apical part consisting of 2-4 pairs of simple pinnae), alternate, very oblique, contiguous, petiolate (petioles on the basal pinnae to 1.5 cm long), the basal pinna larger, 5-10 cm long, oblong (when simply pinnate), or deltoid (when bipinnate), varying from 1-2-pinnate; *pinnules* 3-6 pairs, simple, on petiolules 3 mm long, or the lower 1-2 pairs again pinnate, on petiolules 5 mm long; the *ultimate pinnules* nearly all similar, 1-1.2 cm broad, 8-10 mm high, fan-shaped, cuneate, imbricate on the costules, the lower ones with slightly oblique base, shortly petiolulate, the upper margin rounded, entire or more or less 2-lobed; *texture* herbaceous or firmly so, dry green, concolorous and glabrous on both sides, the lobes of sterile pinnule rounded, entire or slightly undulate; *veins* quite distinct below; *sori* usually 2 to each pinnule, transversely oblong to linear, placed in deep sinuses especially at maturity.

Numerous specimens have been seen from Northwest Yunnan, Southern Sichuan (Mt. Omei) and Guizhou (Vanchin Shan). Type from Tapintze, West Yunnan, *Delavay 1523*. A common fern in the region, growing on moss-clad rocks under forests at elevations of 2500 to 3400 m.

A very distinct local species, but was much confused by Christ by giving different names to different growing stages of the same species. It is obviously a close relative to *A. capillus-veneris* L., differing in smaller ultimate bifid pinnules generally with 2 elongate sori transversely placed in deep sinuses and, above all, in the entire or at most wavy margin of the sterile pinnules.

Plate 217. 1. Habit sketch (natural size). 2. Fertile pinna, showing the less deeply sunk indusia in younger state ($\times$ 4). 3. The same, showing more deeply sunk indusia in mature state ($\times$ 3). 4. Sterile pinna, showing entire or slightly wavy margin of the lobes ($\times$ 3). 5. Scale from the rhizome ($\times$ 30). 6-7. Sporangium with spores ($\times$ 100).

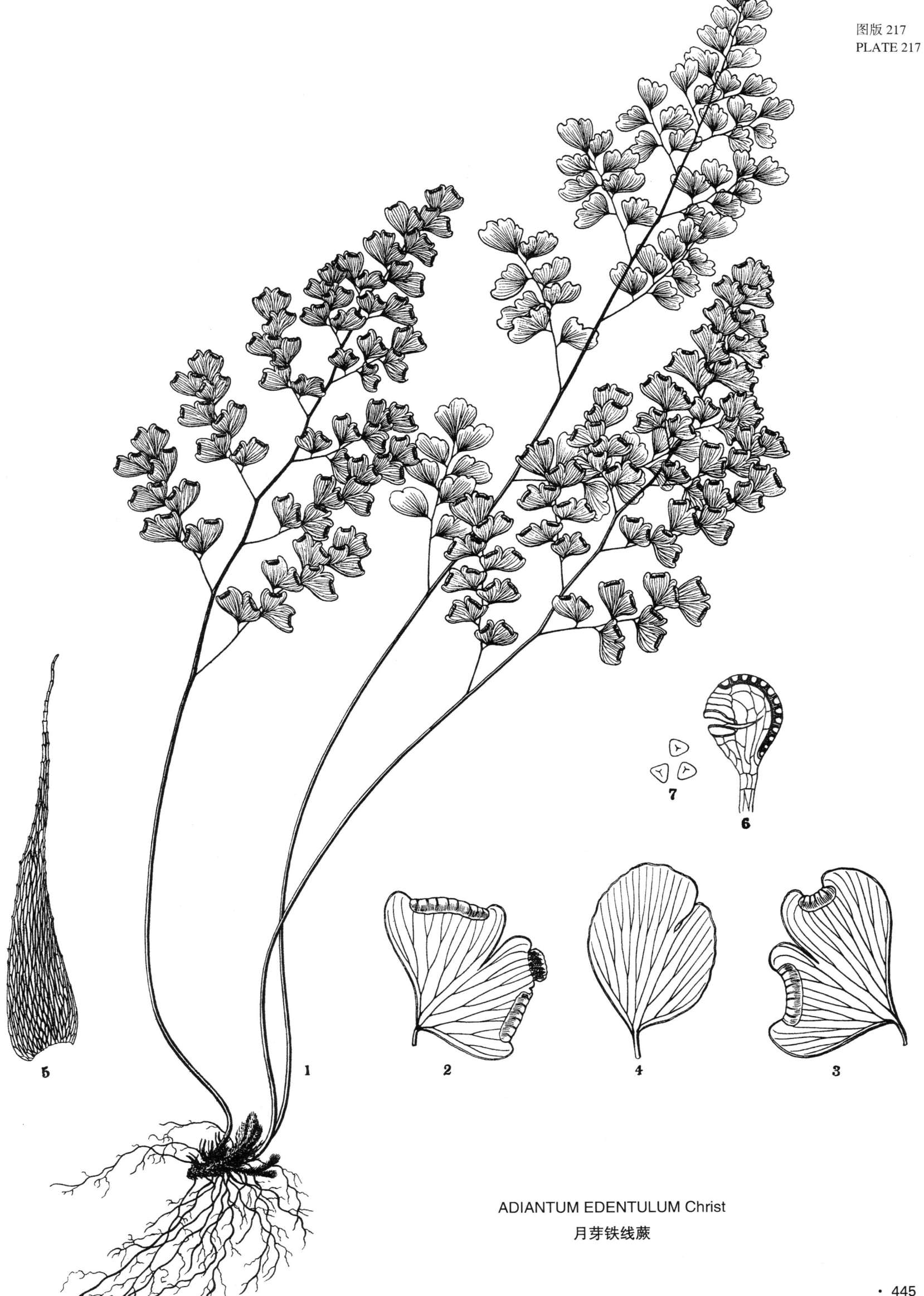

ADIANTUM EDENTULUM Christ
月芽铁线蕨

根状茎细长而匍匐，斜出，有黑褐色钻状线形小鳞片；叶近生，多数，叶柄为细弱毛发状，亮栗色，基部以上光滑，长 4—6 cm，叶体几与叶柄等长，宽 1—2 cm，钝头，狭长圆形，二回奇数羽状；羽片 5—6 对，长 12—20 mm，宽 7—10 mm，有小叶柄，斜出，亚覆瓦状，羽片在一回羽状顶端下面略为羽状，或仅由 2—3 片小羽片组成；末回小羽片都同样，长 3—5 mm，宽 2—4 mm，扇状，基部楔形，上边截形或亚圆形，有 7—10 个略为钝头的长三角形齿牙状的深锯齿；叶为薄草质，干后绿色，上下两面光滑；叶脉明显，二叉分枝，每一齿牙有一条小脉；每一小羽片有一个横生长圆形笔直的子囊群，长几乎等于小羽片外边缘，两端仅留 1—2 齿牙为不育部分，假囊群盖全缘，但有微波状边缘，质薄，灰白色，宿存。

分布：云南西北部，丽江东北部的持古雪山，海拔 3400 m，生林下湿润岩石上。

本种为细叶铁线蕨（*Adiantum venustum* Don）群的最小巧的也是最特出的一种，不易和其他相近的种混淆。特别引人注意的是孢子囊有 40 个以上的加厚细胞所组成的环带和很短的柄。

图注：1. 本种全形（自然大）；2. 两片末回小羽片，表示子囊群及假囊群盖（放大 10 倍）；3. 不育小羽片，表示叶脉及有锯齿的边缘（放大 10 倍）；4—5. 根状茎上的鳞片（放大 30 倍）；6—7. 孢子囊及孢子（放大 100 倍）。

ADIANTUM FENGIANUM Ching

ADIANTUM FENGIANUM Ching in Bull. Fan Mem. Inst. Biol. New Series **1**: 267. 1949; in Acta Phytotax Sinica **6**: 328. 1957.

Adiantum fimbriatum Christ in Bull. Soc. Bot. France **52**: Mém. I. 62. 1905; C. Chr. Ind. Fil. 661 1905 (pro parte).

Rhizome slender, creeping, oblique, clothed in dark brown, linear-subulate small *scales*; *fronds* proximate, numerous, stipe capillary, weak, light castaneous, glabrous above the base, 4-6 cm long, lamina nearly as long as the stipe, 1-2 cm broad, narrowly oblong, obtuse, bipinnate; *pinnae* 5-6 pairs, 15-20 mm long, 7-10 mm broad, petiolate, very oblique, subimbricate, pauci-pinnate or 2-3-foliolate under the simply pinnate and short apical part; *ultimate pinnules* all similar, 3-5 mm long, 2-4 mm broad, fan-shaped, cuneate, truncate or subrounded along the upper margin and deeply dentate with 7-10 elongate, deltoid teeth having rather bluntish apex; *texture* pellucido-herbaceous, dry green, glabrous on both sides; *veins* distinct, fine, dichotomously forked with one veinlet to each tooth; *sori* one to each pinnule, transversely oblong, straight, nearly as long as the width of the outer margin except for the 1-2-toothed sterile edge on each side, indusium thin, whitish, entire, with straight, slightly wavy margin, persistent.

Northwestern Yunnan: Chichungloo, Chukoo Snow Mountain, N. E. of Likiang city, upon wet rocks under forest, *K. M. Feng* 9206, August 29, 1942, alt. 3400 m.; Bois de Santcha Ho, *Delavay* 1, June 17, 1887 (*A. fimbriatum* Christ, pro parte).

This distinct endemic species is one of the smallest and most elegant of the group of *A. venustum* Don, not easily to be confounded with any other species by dint of its delicate and very small stature. Attention may be called to the striking feature of the sporangium which has a very broad annulus consisting of over 40 incrassate cells and the short stalk, while other species under observation generally has long-stalked sporangia with narrower annulus of about 20 thickened cells.

Plate 218. 1. Habit sketch (natural size). 2. Two ultimate pinnules, showing sori and indusia ($\times$ 10). 3. Sterile pinna, showing venation and dentate margin ($\times$ 10). 4-5. Scales from the rhizome ($\times$ 30). 6-7. Sporangium with spores ($\times$ 100).

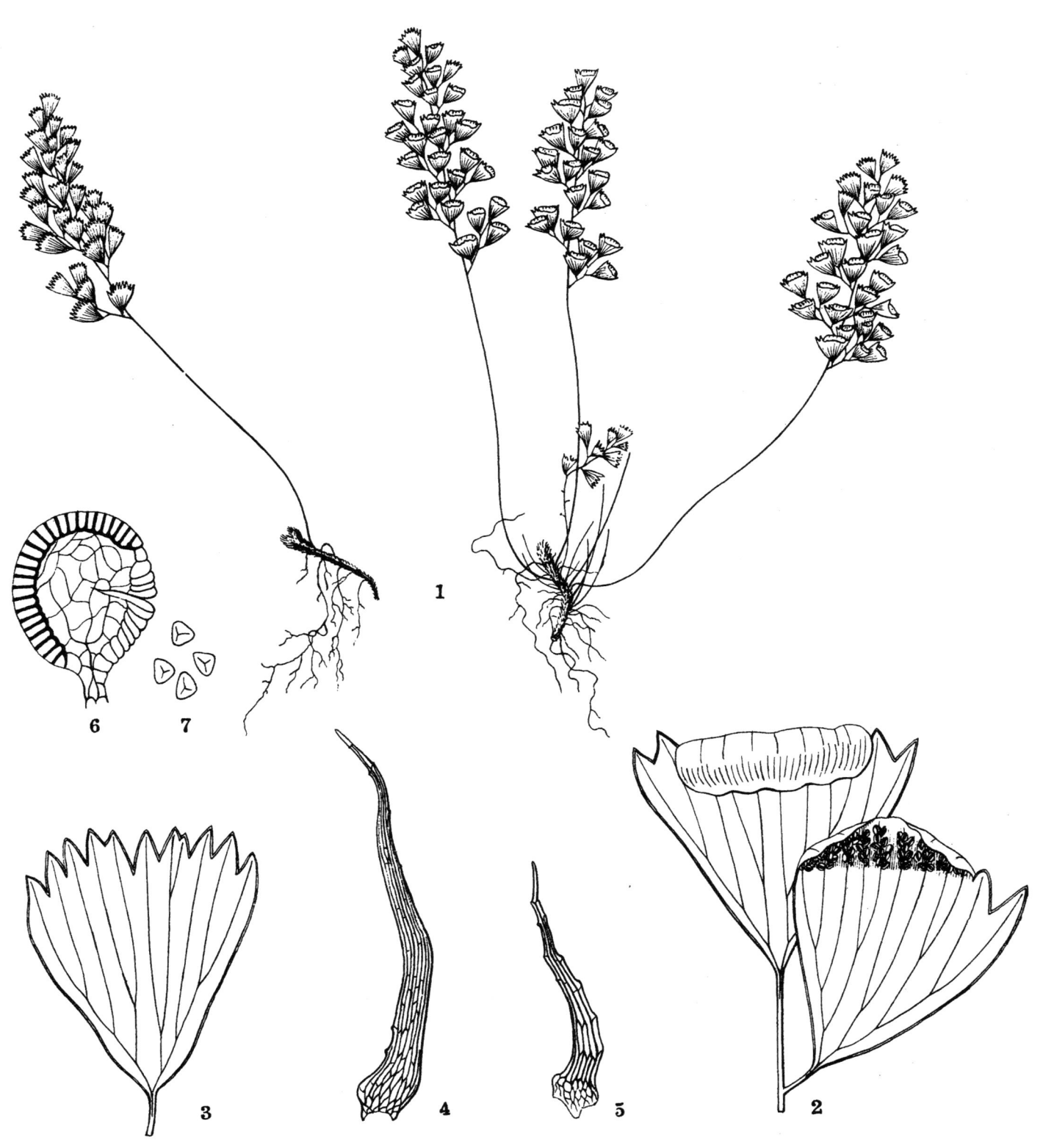

ADIANTUM FENGIANUM Ching
冯氏铁线蕨

图版 219　长盖铁线蕨

　　根状茎细长,横生,有红棕色卵状披针形的鳞片,渐尖头,全缘;叶散生,二列,叶柄直立,圆柱状,初为栗红色,渐变为栗黑色,有光泽,全体光滑无毛,长 15—20 cm,粗约 1.5 mm,叶体为阔卵状三角形,基部亚圆形,顶端急尖或钝头,长与宽各 16—22 cm,三回羽状;羽片 5—7 对,互生,间距 2—3 cm,卵状椭圆形,有小叶柄,斜出,基部一对较大,长 8—11 cm,宽 4—5 cm,上部的较小;一回小羽片有小叶柄,顶部为奇数羽状,下有 3—4 对一回小羽片,上先出排列,长 3—4 cm,宽约 1.5 cm,基部上方一片直立,多少盖在中轴上面,有 3—4 个小羽片;末回小羽片为宽倒卵形或扇形,基部楔形,上边为半圆形,并有深而密的锯齿,长 1—1.3 cm,宽 7—9 mm 或过之,基部相等,有小叶柄;锯齿为长三角形,有短芒,直立,多而密,在不育叶约有 30—35 个;叶为薄草质,干后为绿色;末回小羽片上有假囊群盖 1—3 个或更多,不是肾形,而是半圆形,或常为长圆形,上边笔直,下边半圆形,膜质,生于较深的缺刻内。

　　分布：四川北部,云南西北部(丽江),西康西南部,甘肃南部,生亚高山针叶林下,海拔 2500—3000 m。

　　本种为青藏高原的普通蕨种,过去曾误认为喜马拉雅山区的细叶铁线蕨(Adiantum uenustum Don),其实后者在中国少见。本种向北分布直达陕西秦岭太白山及山西大五台山,但形体较模式远小而简单,可名为陕西变种(var. Shensiense Ching)。

　　图注：1.本种全形(自然大);2.末回小羽片,有长圆形的假囊群盖(放大 3 倍);3.末回小羽片,有半圆形假囊群盖(放大 3 倍);4.末回小羽片,表示叶脉及锯齿(放大 3 倍);5.根状茎上的鳞片(放大 30 倍);6—7.孢子囊及孢子(放大 100 倍)。

ADIANTUM SMITHIANUM (C. Chr.) Ching

ADIANTUM SMITHIANUM (C. Chr.) Ching in Acta Phytotax. Sinica **6**: 335. 1957.

Rhizome slender, wide-creeping, clothed in ovate-lanceolate acuminate, rufo-brown entire *scales*; *fronds* far apart, distichous, stipe strong, erect, terete, atro-castaneous, glossy and glabrous above the base, 15-20 cm long, about 1.5 mm across, lamina broadly deltoid-ovate, base broader, cuneate-rounded, apex rather obtuse, 16-22 cm long and as broad, tripinnate; *pinna* 5-7-jugate under a short, simply pinnate apex, alternate, contiguous, petiolate, ovate-oblong, 2-3 cm apart, oblique, the basal pair much the largest, 841 cm long, 4-5 cm broad, becoming smaller upwards; *pinnules of the 1st. order* petiolulate, 3-4-jugate under the simply impari-pinnate apex, anadromously disposed, 3-4 cm long, about 1.5 cm broad, the anterior basal one erect and imbricating on the rachis, pinnate; *pinnules of ultimate order* broadly cuneate-obovate or flabellate, semi-circular round the upper margin, 1-1.3 cm long, 7-9 mm broad or even broader, base equal, petiolulate, the outer margin deeply and densely serrate; *teeth* elongate-deltoid, aristate, erect, about 30-35 to each sterile pinnule; *texture* thin herbaceous, dry green; indusia 1-2-3 (or rarely more) to each ultimate pinnule, not reniform but semi-orbicular or quite often oblong, and straight above, semi-orbicular below, membranaceous, placed in rather deep sinuses.

Sichuan: Northern part, near Kuanyinmiao, *Harry Smith 3671*, alt. 3900 m. in stony and shrubby meadows; northwestern part, Drogochi, *Harry Smith 4546*, *4712* (type), alt. 2500 m. in open mixed forest by the road leading to Tsaga. Gansu: Takachang, *Dr. D. Hummels 1336*. Yunnan: Likiang Snow Range, northwestern flank, *S. K. Chiao 30217*; *K. M. Feng*, alt. 2900 m.; on the border of Sikang, *T. T. Yü 9490*; Atuntze, *T. T. Yü 5711*, *9490*, *7936*; *C. W. Wang 69868*, alt. 2700 m. under forest; Weisih Hsien, *C. W. Wang 63993*, alt. 3500 m.; Dokerla, Mekong-Salwin Divide, *G. Forrest 19974* (pro parte). Sikang: Tsawarung, *C. W. Wang 65241*, *66276*, alt. 3400 m. under forest.

This is a very common fern in the mountains in the Eastern Tibetan Plateau and was previously always confounded with the Himalayan *A. venustum* Don, which proves to be rare in W. China. The species ranges northwards as far as Tai-pei Shan in Shaanxi Province and Tawutai Shan in Shanxi Province in North China, where it has become a much smaller and simpler form and is known as var. *Shaanxiense* Ching (l. c.).

Plate 219. 1. Habit sketch (natural size). 2. Ultimate pinnule with elongate oblong indusia ($\times$ 3). 3. The same with semi-orbular indusia ($\times$ 3). 4. Ultimate sterile pinnule, showing venation and serrature ($\times$ 3). 5. Scale from the rhizome ($\times$ 30). 6-7. Sporangium with spores ($\times$ 100).

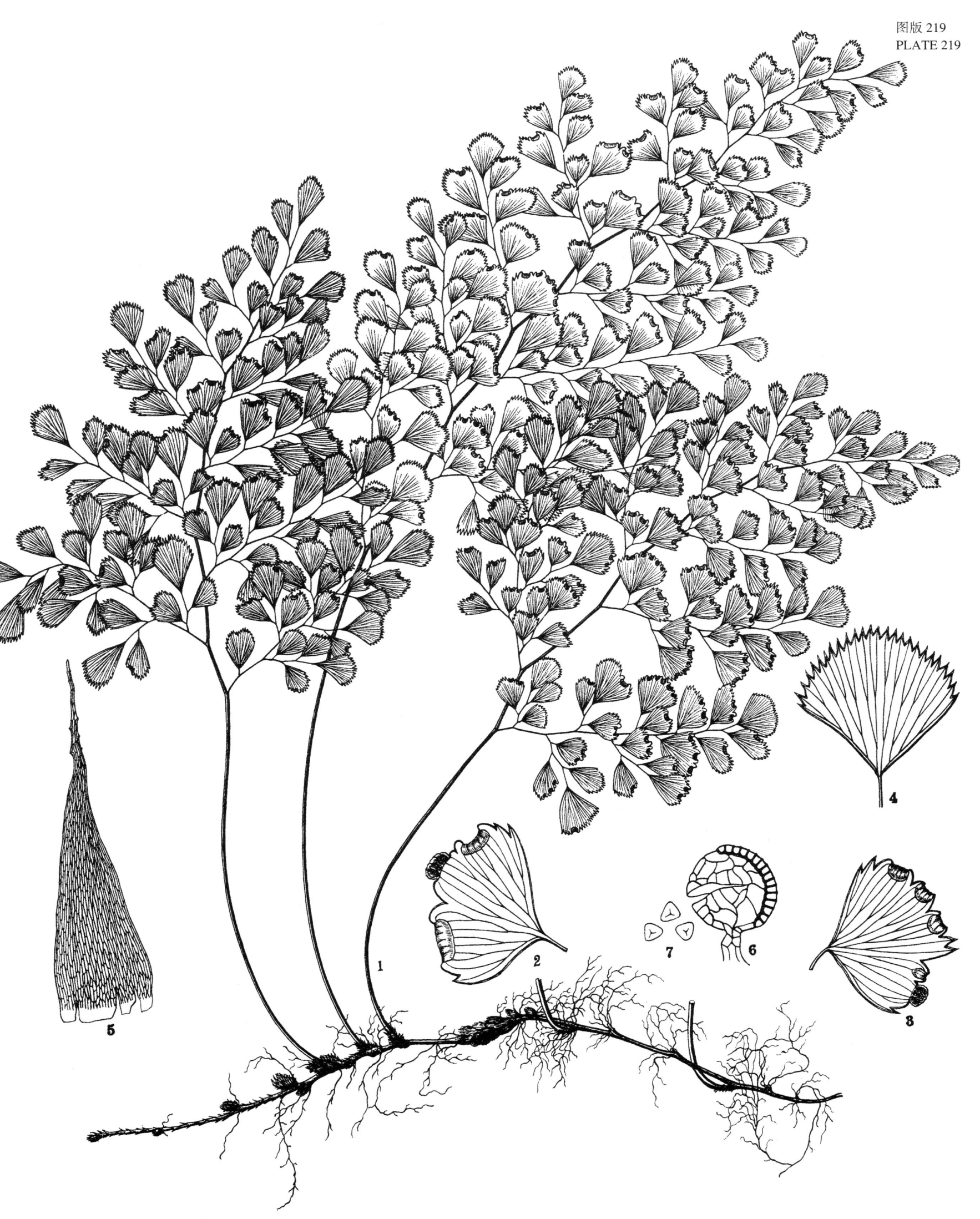

ADIANTUM SMITHIANUM (C. Chr.) Ching
长盖铁线蕨

根状茎细长而广匍匐状,有红棕色阔披针形的鳞片,尤以生长顶端为多;叶远生或近生,直立,叶柄长 15—20 cm,光亮,红棕色,光滑无毛,上面有深沟,叶体长圆状卵形或三角卵圆形,长 15—25 cm,宽与长几相等,三回羽状,在羽状的短顶端以下有羽片 5—6 对,互生,斜出,有小叶柄,基部一对最大,三角状卵圆形,在羽状顶端下为二回羽状;一回小羽片 4—5 对,下部的长 5—6 cm,宽约 3 cm,有小叶柄,长圆形,近基部羽裂为 2—3 片;末回小羽片倒卵状,基部楔形,长宽各为 6—8 mm,或长过于宽,有小叶柄(柄长 1—2 mm,丝状),上边圆形或不规则,有许多小三角形急尖齿牙,顶端无芒;叶为薄草质,干后为淡绿色,但常常在成熟后呈灰绿色或淡灰白色;叶脉细而明显,每一齿牙里有一条小脉;子囊群位于每一小羽片的深缺刻里,常有 1—3 个,假囊群盖圆形或圆肾形,亚膜质,周围有淡灰色的边缘。

分布:四川西北部,西康,云南西北部,甘肃南部,青海东部;喜马拉雅山区的东北部及阿富汗也产。

本种在中国西部少见,颇近似长盖铁线蕨(*Adiantum smithianum* Ching),但末回小羽片较狭小,有急尖头三角形齿牙,通常有 1—3 个子囊群,假囊群盖为亚圆形,生于末回小羽片上边缘的深凹里,叶下面为灰绿色或淡灰白色,故得以区别。

图注:1. 本种全形(自然大);2. 能育羽片(放大 4 倍);3. 不育羽片(放大 4 倍);4. 根状茎上的鳞片(放大 30 倍);5—6. 孢子囊及孢子(放大 100 倍)。

ADIANTUM VENUSTUM Don

ADIANTUM VENUSTUM Don, Prod. Fl. Nepal. 17. 1825; Hook. Sp. Fil. **2**: 40 t. 96 B. 1851; Hook. & Bak. Syn. Fil. 125. 1867; Clarke in Trans. Linn. Soc. II. Bot. **1**: 453. 1880; Bedd. Ferns Brit. Ind. t. 20. 1865; Handb. Ferns Brit. Ind. 86. 1883; Christ, Farnkr. d. Erde 139. 1897; Diels in Engl. u. Prantl, Nat. Pflanzenfam. **1**: iv. 284. 1899; C. Chr. Ind. Fil. 35. 1905; in Journ. Wash. Acad. Sci. **17**: 498. 1927; Ching in Acta Phytotax. Sinica **6**: 334. 1957. *Adiantum microphyllum* Roxb. in Cal. Journ. **4**: 513. 1844.

Rhizome slender, wide-creeping, clothed especially at the growing apex in reddish-brown broadly lanceolate *scales*; *fronds* distant or proximate, erect, stipe 15-20 cm long, glossy, chestnut-red, glabrous, deeply grooved above, lamina oblong-ovate or deltoid-ovate, 15-25 cm long, and nearly as broad, tripinnate; *pinnae* 5-6 pairs under the simply pinnate short apex, alternate, oblique, petiolate, the basal pair larger than those above, deltoid-ovate, bipinnate under the simply pinnate apex; *pinnules of the 1st. order* 4-5 pairs, the lower ones 5-6 cm long, about 3 cm broad, petiolulate, oblong, pinnate or 2-3-foliolate below; *ultimate pinnules* obovate-cuneate, 6-8 mm each way or longer than broad, petiolulate (petiolule 1-2 mm long, filiform), the upper edge round or irregularly so, provided with numerous small, deltoid, acutely dentate teeth, never aristate at apices; *texture* thin herbaceous, dry pale green, but often bluish or glaucescent beneath at maturity; *veins* fine, distinct, with one veinlet to each tooth; *sori* usually 1-2 or 3 to each pinnule placed in quite deep sinuses, indusium orbicular or round-reniform, submembranaceous, brown with thinner gray margin all round.

Northwestern Sichuan: Soongpan Hsien, *K. S. Fu 1699*, alt. 3200 m.; *Harry Smith 4546*. Sikang: Tsawarung, *C. W. Wang 65723*, alt. 3200 m.; *Tsai Yu-wen 5217*, August 11, 1951. Yunnan: Tibetan border, *Kingdon-ward 662* (1913); *Ducloux 1322*; Santcha Ho, *Delavay 2*, June 12, 1887 (*A. fimbriatum* Christ, pro parte); Mekong-Salwin Divide, Dokola, *K. M. Feng 5885*, alt. 3100 m.; *T. T. Yü 57729*; Likiang, Snow Range, east flank, Blackwhite Water River, *K. M. Feng 9151*; Atuntze, Beima Shan, *C. W. Wang 69721*, alt. 3000 m.; Weisih Hsien, Yehchih, *C. W. Wang 68721*, alt. 3600 m. in rock crevices. Gansu: Sining, Lanzecheon Kou, south of the city, *R. C. Ching 587*, alt. 2650-3100 m. July 23, 1923, in moist forest.

Also Northeast Himalayas and Afghanistan.

This species, which is rather rare in West China, is closely related to *A. Smithianum* Ching, differing in the much smaller and narrower ultimate pinnules with acute deltoid teeth, and usually 1-2 or 3 sori with roundish indusia sunk in deep notches, and in bluish or glaucescent under side of the mature leaves.

Plate 220. 1. Habit sketch (natural size). 2. Fertile pinna ($\times$ 4). 3. Sterile pinna ($\times$ 4). 4. Scale from rhizome ($\times$ 30). 5-6. Sporangium with spores ($\times$ 100).

ADIANTUM VENUSTUM Don
细叶铁线蕨

根状茎细长而广匍匐,有密生暗褐色钻状线形鳞片,尤以生长顶端为多,且有细毛混生;叶为二列,直立,相距 1—3 cm,叶柄长 25—30 cm,直径 2 mm,红栗色,极光亮,圆形,基部密生深棕色的节状长刚毛,开展,但易擦落,向上光滑,叶体宽大,有散开广阔的羽片,长与宽各为 25—35 cm,三角状卵形,钝头,四回羽状;羽片 6—8 对;互生,斜出,下部一对相距 6—10 cm,长 13—19 cm,宽 6—9 cm,长圆卵形,顶端钝形,有长柄,三回羽状;所有羽片的第一回小羽片的排列都是显著的上先出分枝,即上方基部的一回小羽片约距中轴 1 cm,下方基部的距中轴约 3—4 cm;一回小羽片在羽状的短而钝形顶端下有小羽片 6—7 对,长 4—5 cm,宽 3 cm,长圆卵形,有小柄,再次二回羽状,二回小羽片斜出,而上方基部一片显著地和中轴及小轴平行并重叠;二回小羽片有 4—5 对,下部的 2—3 对再羽裂为 3—5 数末回小羽片,彼此接近,长与宽为 7—10 mm,扇形,基部楔形,也有小叶柄(柄长 2—4 mm),上边缘圆形,并有许多宽三角形齿牙,其末端有张开的软骨质长尖刺头,这些刺头有的向左偏倾,有的向右,或有时两个合在一起;叶为薄草质,柔软,干后为绿色,下面呈微灰绿色,上下两面光滑;叶脉细而明显,每一齿牙有一条小脉;子囊群小形,每一末回小羽片上边有 1—2 个(有时为 3 个),假囊群盖圆肾形,膜质,灰色,宿存,并保持平坦。

分布:特产中国的西南部,包括四川南部,西康和云南西北部,向东分布到昆明的西山和筇竹寺,生常绿阔叶林缘,冬天枯死。

本种与细叶铁线蕨(*Adiantum venustum* Don)的区别是:叶柄基部有很多淡红色的节状长刚毛,叶体远较宽大和本种特有的长刺头锯齿向左方或右方敧斜或二齿基部合生。

图注:1. 本种全形(自然大);2—3. 能育末回小羽片(放大 4 倍);4. 不育末回小羽片,表示特有的锯齿型(放大 4 倍);5. 根状茎上的鳞片(放大 30 倍);6. 叶柄基部的长毛(放大 30 倍);7—8. 孢子囊及孢子(放大 100 倍)。

ADIANTUM BONATIANUM Brause

ADIANTUM BONATIANUM Brause in Hedwigia **54**: 206 t. 4, k. 1914; C. Chr. Ind. Fil. Suppl. Ⅱ. 1. 1916; Hand-Mzt. Symb. Sin. **6**: 39. 1929; Ching in Sunyatsenia **6**: 13. 1941; in Acta Phytotax. Sinica **6**: 337. 1957.

Adiantum venustum (non Don) Christ in Bull. Acad. Géogr. Bot. Mans. **1906**: 131; C. Chr. Hort. Gotheb. **1**: 92. 1924.

Rhizome slender, wide-creeping, clothed especially at the growing tip in dense, dark brown, linear-subulate *scales* intermixed with long fine hairs; *fronds* distichous, 1-3 cm apart, erect, stipe 25-30 cm long, 2 mm thick, chestnut-red, highly polished, terete, copiously clothed near the base with easily abraded, spreading atro-brown, fine, setose, articulate, long *hairs*, becoming glabrous upwards, lamina ample, with widely separated pinna, 25-35 cm long and as broad, deltoid-ovate with blunt apex, quadri-pinnate; *pinnae* 6-8 pairs, alternate, oblique, the lower ones 6-10 cm apart, 13-19 cm long, 6-9 cm broad, oblong-ovate with bluntish apex, long-petiolate, tripinnate; the pinnules from the first order onwards in all pinnae are strongly anadromously branched, i. e. the anterior basal pinnule about 1 cm away from the rachis, while the posterior basal one about 34 cm away from the rachis; *pinnules of the 1st. order* in the basal pinna 6-7 pairs under the simply pinnate short blunt apex, 4-5 cm long, 3 cm broad, oblong-ovate, petiolulate, again bi-pinnate under the simply pinnate and blunt apex, oblique, but the anterior basal one strongly deflexed and overlapping the rachis; *pinnules of the 2nd. order* 4-5-jugate, the lower 2-3 pairs again pinnate with 3-5 *ultimate pinnules*, which are contiguous, 7-10 mm each way, cuneate-flabellate, all petiolulate (petiolule 2-4 mm long), rounded on the outer margin and provided with numerous broadly deltoid teeth ended in spreading, cartilaginous long aristate spines, of which some often in a curious manner tip to the right, while others to the left, or sometimes two together; *texture* thin herbaceous, dry green, faintly bluish-green beneath, glabrous; *veins* fine, distinct, one to each tooth; *sori* small, as a rule 1-2 (occasionally 3) to each ultimate pinnule, indusium orbicular-reniform, or reniform, membranaceous, gray, persistent and remains flat.

Type from Tchong Shan, Tali. In distribution this distinct but heretofore much confused species is confined to Southwest China: Southern Sichuan, Sikang and N. W. Yunnan, eastwards to Kunming, where it is a common fern under evergreen forest in Sihshan, west of the city. From *A. venustum* Don this species can be easily distinguished by the presence at the basal part of stipe of the copious spreading long reddish setaceous hairs to be easily abraded, the much more ample fronds and the characteristically aristate serrature.

Plate 221. 1. Habit sketch (natural size). 2-3. Fertile pinna, showing sori and characteristic teeth ($\times$ 4). 4. Sterile pinna, showing peculiar type of serrature ($\times$ 4). 5. Scale from the rhizome ($\times$ 30). 6. Hairs from the base of stipe ($\times$ 30). 7-8. Sporangium with spores ($\times$ 100).

ADIANTUM BONATIANUM Brause
毛足铁线蕨

图版 222　鹤庆铁线蕨

根状茎短匍匐状,斜出,有适量的鳞片,红棕色,狭披针形,渐尖头,略坚硬;叶亚簇生,柄长 13—15 cm,直径 1.5 mm,深栗色,光亮,基部以上光滑,叶体长 16—20 cm,宽 10—13 cm,三角状卵形,短渐尖,基部二回羽状或亚三回羽状,在一回羽状顶部以下有羽片 6—7 对,互生,斜出,有长小柄(基部的羽片小柄长 1.5 cm),亚近生,基部一对稍长于上方的,三角状长圆形,长 9—11 cm,宽 3—4 cm,二回羽状或有时为一回羽状;顶部以下有小羽片 5—7 对,互生,斜出,有小叶柄(柄长 4 mm,丝状),一回羽状或下部的 1—2 对小羽片常由 2—4 个小羽片组成,向上渐变小;末回小羽片宽 1—1.5 cm,长 1 cm,宽扇形,基部楔形,顶端的较小,有小叶柄(柄长 4 mm,丝状),深二裂超过一半,每裂片又浅裂一次,小裂片或再为浅二裂或为不裂,每小裂片有一个亚圆形,肾形或圆肾形的假囊群盖联结 3—5 条小脉,每个末回小羽片有 4—7 个子囊群,位于深弯缺刻内;叶为薄草质,干后亮绿色,上下两面光滑;不育小羽片为全缘或波状,不具锯齿。

分布:云南西北部,鹤庆县,生林下沟内,海拔 3000 m。

本种在一般形体上略似月芽铁线蕨(A. edentulum Christ),但更似铁线蕨(A. capillus-veneris L.),而因其不育小羽片为全缘或波状边缘,子囊群较短和末回小羽片多次分裂,故与铁线蕨有所不同,其与月芽铁线蕨的不同点,则为其叶体远较大,宽三角状卵形,末回小羽片多次分裂,每片有许多圆形子囊群。

图注:1. 本种全形(自然大);2. 能育小羽片(放大 3 倍);3. 同上,分裂较少(放大 3 倍);4. 不育小羽片,表示全缘和波状边缘(放大 3 倍);5. 根状茎上的鳞片(放大 20 倍);6—7. 孢子囊及孢子(放大 100 倍)。

ADIANTUM MUTICUM Ching

ADIANTUM MUTICUM Ching in Bull. Fan Mem. Inst. Biol. New Series **1**: 268. 1949; in Acta Phytotax. Sinica **6**: 339 1957.

Rhizome short-creeping, oblique, moderately scaly, *scales* reddish-brown, narrowly lanceolate, acuminate, rather firm; *fronds* subcaespitose, stipe 13-15 cm long, 1.5 mm across, dark castaneous, glossy, glabrous above the base, lamina 16-20 cm long, 10-13 cm broad, deltoid-ovate, short-acuminate, bipinnate or subtripinnate below; *pinnae* 6-7 pairs under the simply pinnate apical part, alternate, very oblique, long-petiolate, (petiole in the basal pinnae 1.5 cm long), subcontiguous, the basal pair only slightly larger than those next above, deltoid-oblong, 9-11 cm long, 3-4 cm broad, bipinnate or sometimes simply pinnate; *pinnules* 5-7 pairs under the smaller terminal one, alternate, oblique, petiolulate (petiolule 4 mm long, filiform), simple or the lower 1-2 pairs often 2-4-foliolate, gradually decrescent upwards; *ultimate pinnules* 1-1.5 cm broad, 1 cm high, broadly flabellate, cuneate, much smaller towards apex, petiolulate (petiolule 4 mm long), deeply bifid over halfway down with the each half again slightly bifid, the lobules may be again bifid or entire, each bearing one roundish, reniform or round-reniform *sori* connecting 3-5 veinlets, 4-7 sori to each ultimate pinnule, and placed in deep sinuses; *texture* pellucido-herbaceous, dry light green, both sides glabrous; the sterile pinnules with entire or wavy margin, never serrate.

Northwestern Yunnan: Hokin, Shanchang, Chupeiho, in wooded ravine, *R. C. Ching 23383*, July 27, 1940, common, alt. 3000 m.

This interesting fern more resembles A. *capillus-veneris* L. than A. *edentulum* Christ in general habit, from the former it differs in the sterile pinnules with entire or wavy margin, the shorter sori and the repeatedly lobed pinnules, from the latter in much ample fronds, with broadly fan-shaped and repeatedly lobed ultimate pinnules and the numerous roundish sori to each pinnule.

Plate 222. 1. Habit sketch (natural size). 2. Fertile pinnule (× 3). 3. The same, less divided (× 3). 4. Sterile pinnule, showing entire and wavy margin (× 3). 5. Scale from the rhizome (× 20). 6-7. Sporangium with spores (× 100).

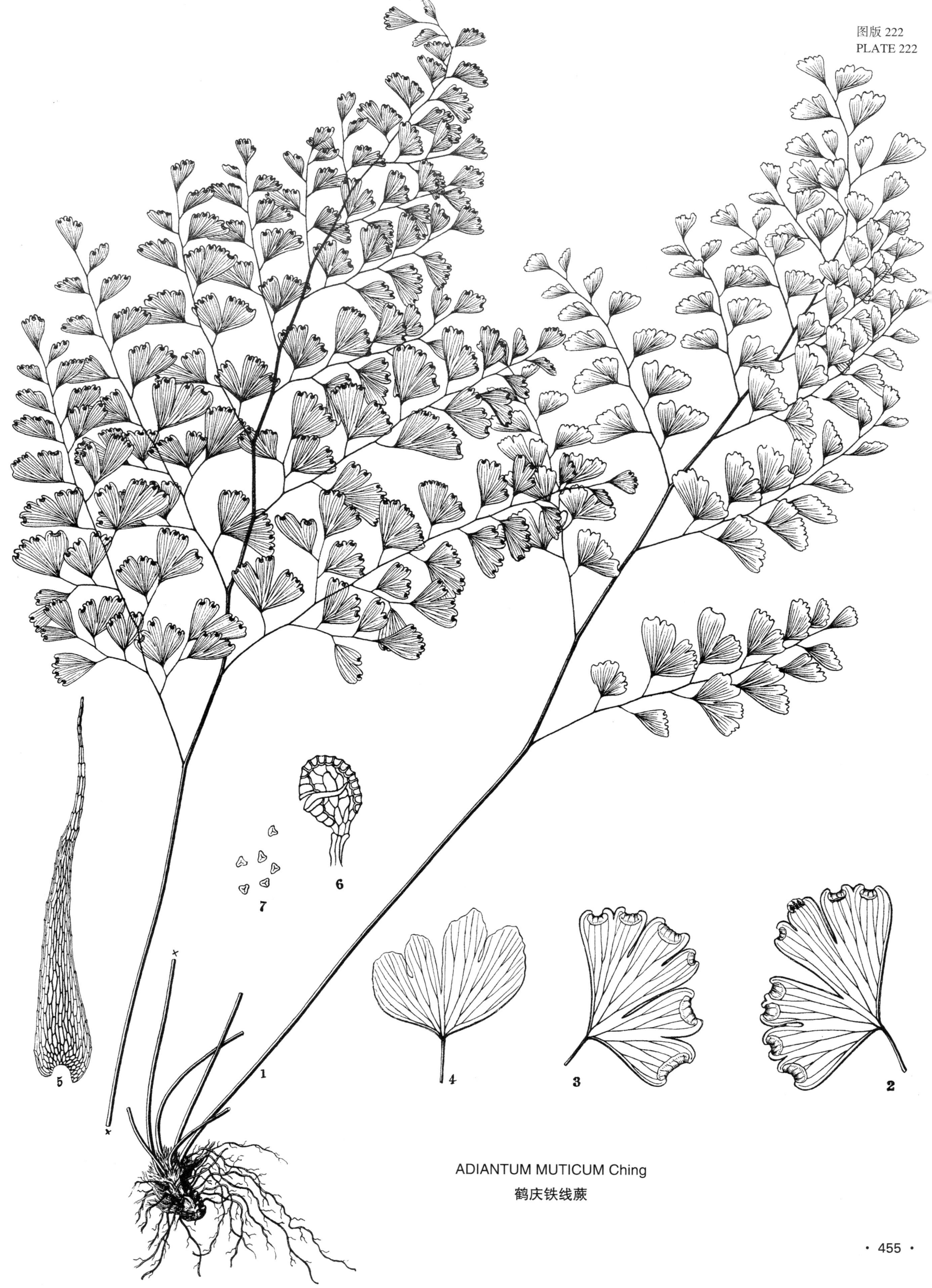

ADIANTUM MUTICUM Ching
鹤庆铁线蕨

根状茎短而直立或斜出，顶部密生锈色的狭钻状披针形鳞片；叶簇生，直立，不育叶柄长约 10 cm，能育叶柄长 25 cm，近黑色、坚硬、光亮、圆柱状，但腹面有宽沟，向上部尤为明显，基部以上光滑无毛，但上部腹面沟里有暗锈色短硬毛，叶体为二至三回二叉分枝或有时为亚二叉分枝，中间的一对羽片较长于侧生的，长 10—16 cm，宽 1.5 cm，线状披针形，一回羽状，下部侧生羽片渐短，同样为一回羽状，末回小羽片为篦齿状，有小叶柄，在中间的长羽片有许多小羽片，长 8—10 mm，宽 5—7 mm，对开式或亚菱形，基部短楔形，下边切成直线形，外边与上边同为圆形，不育小羽片有微尖锯齿；叶为革质，干后为棕色，上下两面光滑无毛，上面略为光亮，下面微灰白色（新鲜标本为灰绿色），但叶轴，小叶轴和叶柄上面沿沟槽有暗锈色短毛密生；叶脉扇形分叉，上下两面显著；子囊群圆肾形或横生长圆形，每小羽片有 2—5 个，假囊群盖黑色，革质，为浅缺刻分开。

分布：本种为亚洲热带和亚热带的广布种，在中国分布于台湾，香港，海南，广东，广西，贵州，福建和江西，湖南，四川，云南，浙江等省的南部。为一种喜光植物，生在干旱丘陵地区的酸性土壤的杂木林缘。

图注：1.本种全形（自然大）；2.能育羽片（放大 4 倍）；3.不育羽片，表示有微尖锯齿的边缘（放大 4 倍）；4.叶柄基部的鳞片（放大 50 倍）；5—6.孢子囊及孢子（放大 100 倍）。

ADIANTUM FLABELLULATUM Linné

ADIANTUM FLABELLULATUM Linné Sp. Pl. **2**：1095，1753；Sw. Syn. Fil. 121. 1806；Hook. Sp. Fil. **2**：30. 1851；in Journ. Bot. **1857**：330；Hook. & Bak. Syn. Fil. 120. 1867；Benth. Fl. Hongk. 447. 1861；Bedd. Ferns South. Ind. t. 218. 1864；Handb. Ferns Brit. Ind. 88. 1883；Fr. et Sav. Enum. Pl. Jap. **2**：454. 1880；Diels in Engl. u. Prantl, Nat. Pflanzenfam. **1**：iv. 284. 1899；C. Chr. Ind. Fil. 26. 1905；Matsum. et Hayata, Enum. Pl. Form. 617 1906；Dunn & Tutcher, Fl. Kwangt. and Hongk. 338. 1912；C. Chr. in Acta Hort. Gotheb. **1**：95. 1924；Hand-Mzt. Symb. Sin. **6**：39. 1929；Ogata, Ic. Fil. Jap. **2**：t. 52. 1929；Wu, Pong & Wong, Polyp. Yaoshan. in Bull. Dept. Biol. Sunyatshan Univ. No. **3**：226, t. 104. 1932；Tard. et C. Chr. in Fl. Gen. Ind. **7**：189. 1940；Ching in Acta Phytotax. Sinica **6**：326. 1957.

Adiantum fuscum Retz, Obs. **2**：28, t. 5. 1781.

Adiantum amoenum Wall. List n. 78. 1828 (nom. nud.); Hook. & Grev. Ic. Fil. t. 103. 1829; Hook, in Lond. Journ. Bot. **1**：294. 1840.

Rhizome short, erect or oblique, densely clothed in ferruginous-brown, narrowly lanceolate, subulate *scales* attached by base; *fronds* tufted, erect, stipe of the sterile about 10 cm long, of the fertile to 25 cm long, blackish, firm, glossy, terete beneath, but broadly grooved on the upper side especially towards the upper part, glabrous above the basal part, but rusty-hirsute in the groove upwards, lamina 2-3 dicho-tomously forked, or nearly so, with the *central pinnae* longer than the lateral ones, 10-16 cm long, 1.5 cm broad, linear-lanceolate, simply pinnate, the other lower pinnae progressively shortened downwards and similarly pinnate; *ultimate pinnules* pectinate, petiolulate, numerous in the central pinna, 8-10 mm long, 5-7 mm broad, dimidiate, subrhomboidal, base short-cuneate, the lower edge straight, the inner oblique, the outer roundish and, like the upper edge, rather sharply dentate in the sterile part; *texture* subcoriaceous, dry brown, both sides glabrous, slightly glossy above, faintly glaucescent below (bluish in living state), upper part of rachis, rachilets and petiole always densely and shortly rusty-hirsute on the upper side; *veins* flabellately forked, prominent on both sides; *sori* roundish, reniform or transversely oblong, 2-5 to each pinnule, indusium dark-colored, coriaceous, separated by shallow sinuses.

A widely dispersed species in tropical and subtropical regions of Asia. In China numerous specimens have been seen from Hainan, Guangdong, Guangxi, Guizhou, southern part of Fujian, Jiangxi, Hunan, Sichuan, Yunnan, Zhejiang, Taiwan and Hong Kong. A light-demanding fern growing in exposed hillsides with acid soil.

Plate 223. 1. Habit sketch (natural size). 2. Fertile pinna ($\times$ 4). 3. Sterile pinna, showing dentate margin ($\times$ 4). 4. Scale from the base of stipe ($\times$ 50). 5-6. Sporangium with spores ($\times$ 100).

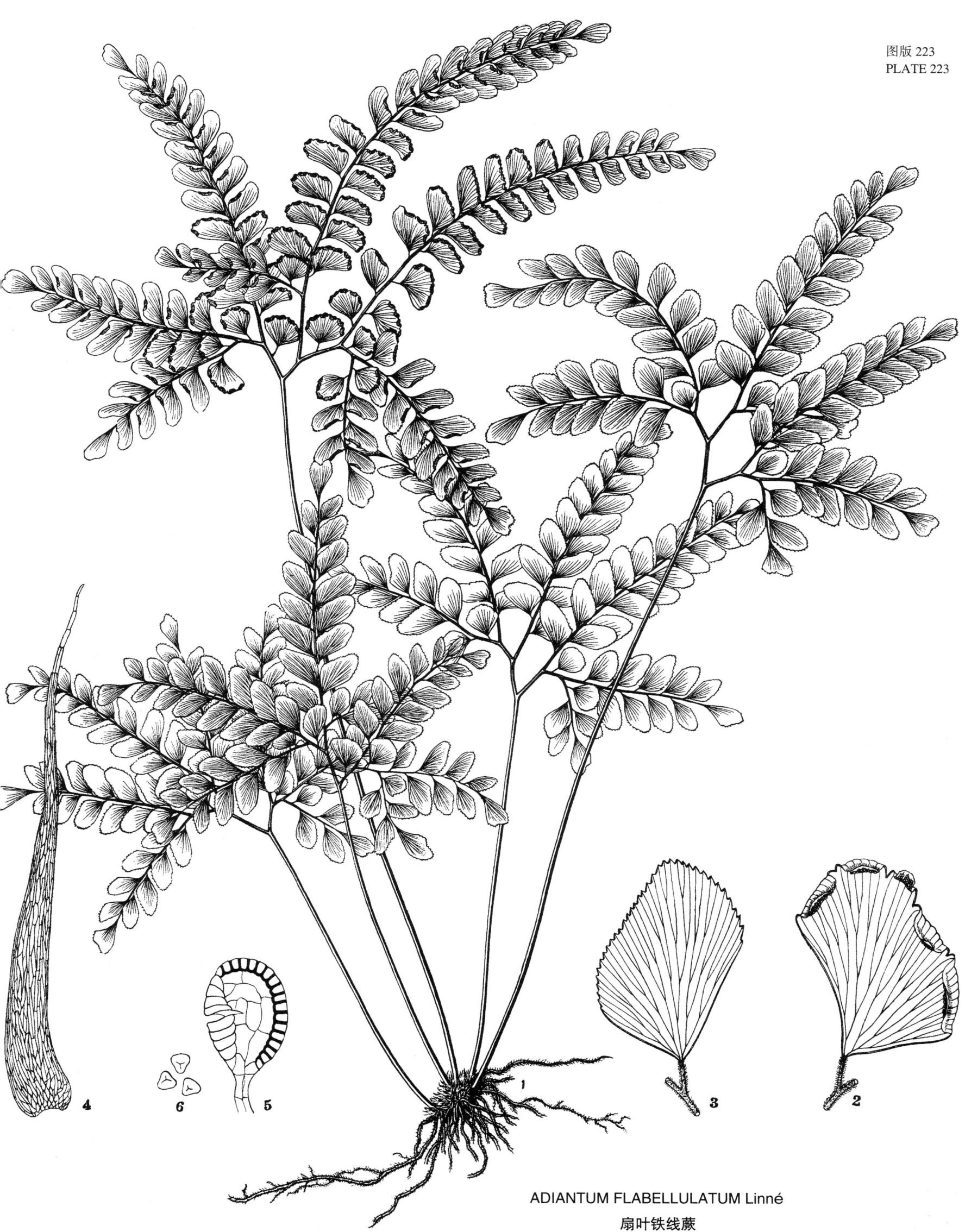

ADIANTUM FLABELLULATUM Linné
扇叶铁线蕨

体形一般与扇叶铁线蕨（*Adiantum flabellulatum* Linné）极近似，但因叶体为羽状分枝而不是二叉分枝，叶柄，叶轴，及小叶轴为圆柱形（上面不具纵沟槽）和末回小羽片几为圆形或亚圆形，不育叶的小羽片上有微钝齿牙或圆齿状齿牙，故有不同。

分布：海南，广西南部和越南北部。

本种在我国为稀见的蕨种，只发现于海南岛及广西南部，这可以进一步证明海南岛植物区系的许多组成种类与越南北部有相当密切关系。

图注：1.本种全形（自然大）；2.能育羽片（放大4倍）；3.不育羽片，表示有圆齿状齿牙的叶缘（放大4倍）；4.叶柄基部的鳞片（放大40倍）；5—6.孢子囊及孢子（放大100倍）。

ADIANTUM INDURATUM Christ

ADIANTUM INDURATUM Christ in Journ. Bot. France **1**：233，265. 1908；C. Chr. Ind. Fil. Suppl. Ⅰ. 4. 1912：Tard. et C. Chr. in Fl. Gen. Ind. **7**：187. 1940；Ching in Acta Phytotax. Sinica **6**：327. 1957.

In the gross habit, the present species very closely resembles *A. flabellulatum* L., but differs chiefly and specifically in the fronds being not dichotomously but rather pinnately branched, in the terete stipe, rachis and rachilets and in the almost orbicular ultimate pinnules even in the sterile fronds with sterile pinnules not sharply but rather bluntly dentate or even crenate-dentate.

Hainan：Changkiang Hsien, Kachik Shan, *S. K. Lau 2925*, December 22, 1933, in the thickets on dry gentle hillside of clayey soil, rare；*Liang-Chian-ta 66493*, under brush wood, December 20, 1933. Guangxi, Peiseh Hsien, Ecological Survey Party, Bot. Inst. Academia Sinica, without number.

Vietnam：*Eberhardt* (type)；*Chevalier 3062*.

A rather rare fern in the region. Its discovery in Hainan may serve as further evidence that the flora of the island is closely related to that of the northeastern part of Indo-China.

Plate 224. 1. Habit sketch (natural size). 2. Fertile pinna ($\times$ 4). 3. Sterile pinna, showing crenate-dentate margin ($\times$ 4). 4. Scale from the base of stipe ($\times$ 40). 5-6. Sporangium with spores ($\times$ 100).

ADIANTUM INDURATUM Christ
海南铁线蕨

图版 225　长尾铁线蕨

　　根状茎短而直立，有密的根丛，顶端有颇多的红棕色小鳞片，披针形，渐尖头，全缘；叶簇生，柄长 10—18 cm，直立光滑，铁丝状，几为黑色，基部有疏生鳞片，而向上部光滑，腹面有宽纵沟，叶体与叶柄等长，分裂度的变化很大，有时为线状披针形，一回羽状，有时为不完全的二回羽状，基部发生 1—3 条有柄的一回羽状的较短羽片，宽 2—3 cm，顶端钝渐尖头；在中央的主枝有许多羽片，近生，斜出，有短小叶柄，长 1—2 cm，宽 7—10 cm，对开式、亚菱形，下边切成直线，里边为截形或微内弯，外边为圆形或斜形，如同上边一样，有圆齿（在能育羽片上），或有齿牙（在不育羽片上）；叶质薄而近乎半透明，深绿色，干后暗绿色，叶轴光滑无毛；但在小羽片下面的小脉间有许多红棕色伏贴的单细胞的细刚毛，叶脉明显，光滑无毛，假囊群盖圆肾形，暗色，革质，着生在深而狭的弯缺里，每小羽片的外边和上边共有 5—8 个，彼此分开。

　　分布：本种广布于印度尼西亚，马来西亚，澳大利亚，新西兰及热带和亚热带亚洲，在中国仅产海南、广东、福建及台湾；越南也有。

　　本种形体特殊，不易与本属中的其他中国蕨种混淆，虽然与毛叶铁线蕨（*Adiantum pubescens* Schkuhr）有许多相似之点，但后者的叶柄和叶体上有密生的毛和叶体为二叉分枝，故易区别；本种与其他邻近的种也有不同，即叶柄腹面有宽沟（不是圆柱形）并且孢子囊不仅着生在小脉上而且也着生于小脉间的叶面上。

　　var. *affine* v. A. v. R. (l. c.)变种。本变种不同于原种之点仅为其叶体下面光滑无毛，干后为橄榄绿。分布于我国海南和福建；印度尼西亚和马来西亚也有。

　　图注：1. 本种全形（自然大）；2. 能育羽片（放大 3 倍）；3. 不育羽片，表示有齿牙的叶缘（放大 3 倍）；4. 叶柄基部的横切面（放大）；5. 叶柄基部的鳞片（放大 60 倍）；6. 叶体下面的细刚毛（放大 100 倍）；7. 假囊群盖的下面，表示孢子囊及细刚毛的分布（放大 100 倍）；8—9. 孢子囊及孢子（放大 100 倍）。

ADIANTUM DIAPHANUM Blume

ADIANTUM DIAPHANUM Blume, Enum. Pl. Javae 215. 1828; Hook. Sp. Fil. **1**: 10 t. 80 C. 1851; Hook. & Bak. Syn. Fil. 117. 1867; Benth. Fl. Austr. **7**: 725. 1878; Diels in Engl u. Prantl, Nat. Pflanzenfam. **1**: iv. 286. 1899; C. Chr. Ind. Fil. 25. 1905; Copel. Polyp. Phil. 93. 1905; Matsum. et Hayata, Enum. Pl. Form. 617. 1906; v. A. v. R. Handb. Mal. Ferns 323. 1908; Ogata, Ic. Fil. Jap. **6**: t. 251. 1935; Tagawa in Journ. Jap. Bot. **14**: 393. 1938; Tard. et C. Chr. in Fl. Gen. Ind. **7**: 183. 1940; Ching in Acta Phytotax. Sinica **6**: 321. 1957.
Adiantum setulosum J. Sm. in Bot. Mag. **72**: Comp. 22. 1846.
Adiantum affine Hook. Sp. Fil. **1**: 32. 1851 (non Willd.).

Rhizome short, erect, densely radicose, moderately clothed at the apex in reddish-brown, small, lanceolate-acuminate, entire *scales*; *fronds* fasciculate, stipe 10-18 cm long, erect, polished, wiry, nearly black, sparsely scaly near the base but glabrous upwards, with a broad groove on the upper side, lamina as long as the stipe, varies from linear-lanceolate and simply pinnate to bipinnate at the base with 1-3 additional stalked simply pinnate shorter pinna, 2-3 cm broad, apex bluntly acuminate; *pinnae* in the main central branch numerous, contiguous, oblique, shortly petiolate, 1-2 cm long, 7-10 mm broad, dimidiate, subrhom-boidal, the lower edge quite straight, the inner truncate or slightly recurved, the outer roundish or oblique and crenate like the upper edge with round entire teeth (in the soriferous pinnae), or dentate (in the sterile pinnae); *texture* thin, nearly pellucid, dark green, dry opaque green, rachis glabrous, with moderate reddish-brown appressed bristle-like unicellar hairs between the veinlets on the under side of pinnules; *veins* distinct, glabrous; *sori* small, round-reniform, indusium dark-colored, coriaceous, attached to the deep narrow sinuses, 5-8 round the outer and upper edge, distant.

Hainan: Five Finger Mt. *Mrs. Eryl Smith 1471*, . *1464*; *Tsang and Tung 17829* ex Herb. Lingnan Univ.; Mei Hsien, Yam Na Shan, *W. T. Tsang 21556*. Fujian: Foochow, *Forbes 2487*; *H. H. Chung 3798*, August 18, 1925; Minchow, *H. H. Chung 2292*; Amoy, *Swinhoi* in Herb. Hance (1857); *Hance 1411*; Inghock, *H. H. Chung 1340*, *2634*, June 14, 1923; Yungchung, *Rankin* (1913); Hinghwa, *Lin Pi 6538*; Lienkiang Hsien, *D. C. Liu 84*; Nantsin Hsien, *H. C. Chao F22*, April 20, 1954. Taiwan: *Hancock 35*; *Suzuki-Tokio 20920*, Nov. 16, . 1940 under forest, alt. 650 m.

A widely dispersed species in the tropics and subtropics in Asia, Polynesia, Australia and New Zealand.

A very distinct species not easily to be confounded with any other Chinese species of the genus perhaps except *A. pubescens* Schkuhr., which has dichotomously forked fronds and densely pubescent stipe and lamina. The species is further differentiated from its relatives in having stipe broadly grooved on the upper side (not terete) and the sporangia borne on the veinlets as well as on the parenchymatic tissues between the veinlets.

There is a form in Malaysia having lamina devoid of the characteristic hairs on the under side and dry olive-green; it is called var. *affine* v. A. v. R. (l. c.), and is also known from Hainan and Fujian (*C. Wang 35202*, *35784* from Hainan; *Tang Siu Ging 16359* from Hokchiang, Fujian).

Plate 225. 1. Habit sketch (natural size). 2. Fertile pinna (× 3.). 3. Sterile pinna, showing dentate margin (× 3). 4. Cross section of the basal part of stipe (greatly enlarged). 5. Scale from the base of stipe (× 60). 6. Hairs from the under side of lamina (× 100). 7. Under side of indusium, showing the distribution of sporangia and hairs (× 10). 8-9. Sporangium with spores (× 100).

ADIANTUM DIAPHANUM Blume
长尾铁线蕨

图版 226　掌叶铁线蕨

　　根状茎粗大，厚木质，短匍匐状，斜出，有密生根丛，顶端有密生的棕色披针形鳞片；叶簇生，叶柄长 15—35 cm，光亮，栗褐色，基部以上光滑，叶体长 18—35 cm，宽与长相等或较宽，自叶柄顶端二叉平分为左右两半边的弯弓侧枝，从每个侧枝上边生有等距的约 4—6 条向上的一回羽状的线状披针形羽片，中间的一对羽片最发达，长 12—28 cm，宽 2.5—3.2 cm，直立；小羽片多数，箆齿状，有小叶柄，开展，长 1.3—2 cm，基部宽约 7 mm，对开式，下边直切，基部上方近叶轴处较宽，顶端钝形，有几个微钝齿牙，上边分裂，有时裂至 1/3 处；两边侧生羽片向外渐短，直至外顶端的一条，长仅 4—6 cm，也有同形的小羽片；叶为薄草质，上面为亮绿色或暗绿色；叶脉明显，二叉分枝；每小羽片上通常有 3—4 个子囊群，假囊群盖为肾形至横生长圆形。

　　分布：本种广布于北半球温带森林中，在我国见于东北，内蒙古，山西，陕西（秦岭），河南，甘肃，云南西北部，四川西南部高山，西康及青藏高原；喜马拉雅山区（海拔 2300—3000 m），日本，朝鲜，西伯利亚东部，苏联远东部分及美国北部也产。

　　图注：1. 本种全形（自然大）；2. 能育小羽片（放大 3 倍）；3. 不育小羽片，表示分裂的边缘（放大 3 倍）；4. 叶柄基部的鳞片（放大 20 倍）；5—6. 孢子囊及孢子（放大 100 倍）。

ADIANTUM PEDATUM Linné

ADIANTUM PEDATUM Linné Sp. Pl. **2**: 1095. 1753; Thunb. Fl. Jap. 339. 1781; Sw. Syn. Fil. 121. 1806; Hook. Sp. Fil. **2**: 28. 1851; Hook. & Bak. Syn. Fil. 125. 1867; Franch. Pl. David. **1**: 348. 1884; Diels in Engl. u. Prantl, Nat. Pflanzenfam. **1**: iv. 284. 1899; in Bot. Jahrb. **29**: 200. 1900 (pro parte); Kom. Fl. Mansh. **1**: 143. 1901; C. Chr. Ind. Fil. 31. 1905; in Bot. Gaz. **56**: 331. 1913; Ogata, Ic. Fil. Jap. **4**: t. 151. 1931; Kitagawa in Rep. 1st. Sci. Exped. Manch. **4**: ii. 87. 1935; Tagawa in Journ. Jap. Bot. **14**: 392. 1938; Kitagawa, Lineamenta Fl. Mansh. 26. 1939; Ching in Acta Phytotax. Sinica **6**: 322. 1957.

Adiantum pedatum L. var. *glaucinum* (non Christ) C. Chr. in Journ. Wash. Acad. Sci. **17**: 498. 1927.

　　Rhizome thick, woody, short-creeping, oblique, densely radicose, and clothed at the apex with brown, broadly lanceolate *scales*; *fronds* caespitose, a few together, stipe 15-35 cm long, polished, chestnut-brown, glabrous above the base, lamina 18-35 cm long, as broad or broader, forked from the top of stipe into two spreading curved main lateral branches, each having 4-6 equally spaced, simply pinnate linear-lanceolate pinna on the upper side; *central pinnae* are the most developed, 12-28 cm long, 2.5-3.2 cm broad, erect; *pinnules* numerous, pectinate, petiolulate, patent, 1.3-2 cm broad, about 7 mm broad at base, dimidiate, broader on the upper side nearest to the rachis, apex obtuse with a few blunt teeth, the upper margin lobed, sometimes one-third way down; the lateral pinna become gradually shorter and shorter as they go outwards, and the outermost are only 4-6 cm long with similar pinnules; *texture* thin herbaceous, light green or bluish green on the under side; *veins* distinct, dichotomous; *sori* usually 3-4 to each pinnule along the upper margin, indusium reniform to transversely oblong.

　　A widely distributed species in forests of the temperate zone in the Northern Hemisphere. In China specimens have been seen from the Northeastern Provinces, Chahar, Inner Mongolia, Shanxi, Shaanxi (Tsingling Range) Henan, Gansu, W. Yunnan, Sichuan, Sikang and Tibet.

　　Also in the Himalayas (alt. 2300-3000 m.), Japan, Korea, Eastern Sibiria, the USSR Far East and Northeastern America.

　　Plate 226. 1. Habit sketch (natural size). 2. Fertile pinnule (× 3). 3. Sterile pinnule, showing lobed upper margin (× 3). 4. Scale from the base of stipe (× 20). 5-6. Sporangium with spores (× 100).

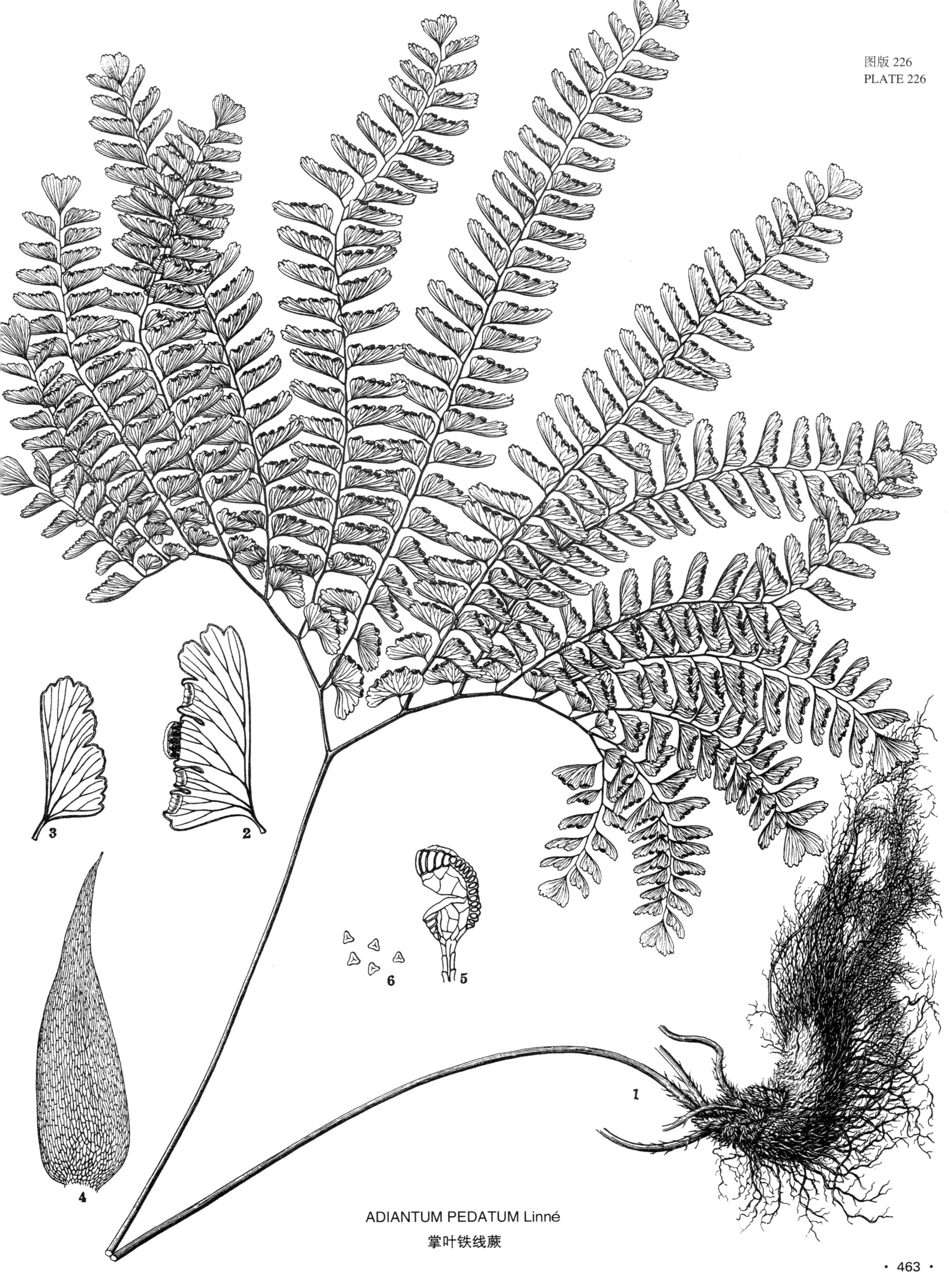

ADIANTUM PEDATUM Linné
掌叶铁线蕨

图版 227　灰背铁线蕨

　　根状茎厚而粗壮，木质，斜出，顶端密生深棕色的坚厚鳞片，披针形，渐尖头，全缘；叶簇生，直立，叶柄长 12—25 cm，粗壮，黑色，极光亮，圆形，基部以上光滑，叶体长 25—35 cm，宽与长相等，叶柄顶端以锐角二叉平分为左右两半边，从弯弓形而多曲折的两边侧枝上面生出等距的 3—6 对一回羽状的线状披针形直立羽片，中间的一对最长，长达 25—35 cm，向外逐渐顺次变短，最外面的羽片（顶生的）长仅 7—10 cm，中间的羽片宽 3—3.5 cm，有许多小羽片（约 45 对），有小叶柄，近生，篦齿状，水平开展，为三角形的对开式，长 1.2—1.6 cm，基部上方斜形或截形，宽 7 mm，下边直切，上边同样为直线形，有锯齿，并在亚急尖或钝形的顶端常有 3—5 个尖头三角形齿牙，在顶生小扇形的小羽片下，侧生小羽片较下部的逐渐变小，外缘的一条羽片上的小羽片也是同形，但较少，最外面的一条羽片约由 10—15 对小羽片组成；叶为草质，上面绿色，下面灰绿色，上下两面光滑；叶脉细而明显，到达每一齿牙的顶端；子囊群圆形或横生肾形，每片有 4—5 个，生于浅缺刻里，为凹口分开，假囊群暗色，有浅色膜质的宽边缘；孢子为尖四面体，浅棕色，透明而光滑。

　　分布：本种为我国西南部特有的蕨种，分布于四川（峨眉山），贵州，湖北西部及云南西北部亚高山，常见于海拔 1000—1500 m 的森林下，在峨眉山尤多。本种在一般形体上酷似北温带的掌叶铁线蕨（*Adiantum pedatum* L.），曾被以前学者认为同义词或变种看待，但因叶柄几乎为黑色，叶的上面光亮，小羽片显然为三角形，其急尖或钝形顶端有 3—6 个尖三角形齿牙，上边有密锯齿，每小羽片上边有略为圆形或较圆的假囊群盖和根状茎有深棕色光亮较狭又坚厚的鳞片等特征，故大有区别。在地理分布上，本种生长在低海拔或低纬度地方，这里掌叶铁线蕨已不能生存。

　　图注：1. 本种全形（自然大）；2. 能育羽片（放大 3 倍）；3. 不育羽片，表示上边缘的尖锯齿（放大 3 倍）；4. 叶柄基部的鳞片（放大 20 倍）；5—6. 孢子囊及孢子（放大 100 倍）。

ADIANTUM MYRIOSORUM Baker

ADIANTUM MYRIOSORUM Baker in Kew Bull. Misc. Inform. **1898**: 230; Ching in Bull. Fan Mem. Inst. Biol. Bot. Ser. **11**: 54. 1941 cum descr.; in Acta phytotax. Sinica **6**: 324. 1957.

Adiantum pedatum L. var. *myriosorum* Christ in Bull. Herb. Boiss. Ⅱ. **1903**: 511; in Bull. Soc. Bot. France **52**: Mém. Ⅰ. 111, 510. 1905.

Adiantum pedatum (non L.) Diels in Bot. Jahrb. **29**: 200. 1900; Christ in Bull. Soc. Bot. France **52**: Mém. Ⅰ. 62. 1905; in Bot. Gaz. **51**: 346. 1911; C. Chr. in Acta Hort. Gotheb. **1**: 95. 1924; Hand-Mzt. Symb. Sin. **6**: 38. 1929 (pro parte).

Adiantum pedatum L. var. *glaucinum* Christ in Bull. Herb. Boiss. **6**: 957. 1898; C. Chr. in Contr. U. S. Nat. Herb. **26**: 310. 1931.

Adiantum pedatum L. var. *protrusum* Christ in Bull. Acad. Géog. Bot. Mans. **1904**: 110.

Rhizome strong, thick, woody, oblique, densely clothed at growing apex in dark brown, firm, lanceolate, acuminate, entire *scales*; *fronds* caespitose, a few together, erect, stipe 12-25 cm long, stout, ebeneous, highly polished, terete, glabrous above the base, lamina 25-35 cm long and as broad, forked from the top of stipe under an acute angle into two spreading, curved, flexuose lateral branches, from the upper side of which spring 3-6 equally spaced, simply pinnate, linear-lanceolate upright *pinna* on each side of the two equal halves, the *two central pinnae* much the longest, 25-35 cm long, the others gradually shortened as they go outwards, with the outermost (the terminal) pinnae 7-10 cm long and spreading; the central pinnae 3-3.5 cm broad, with numerous (to 45 pairs), close, pectinate, horizontally patent, triangular, dimidiate and petio-lulate *pinnules*, 1.2-1.6 cm long, 7 mm broad at the upper-inner, straight or oblique base, the lower edge cut straight, the upper edge also straight and serrate under the acute apex, which is always provided with 3-5 rather sharply deltoid teeth, the upper pinnule gradually smaller below the small narrowly fan-shaped terminal one, the pinnules of the outer pinna being similar but decreased in number, and the outermost pinna consisting of only about 10-15 pinnules; *texture* herbaceous, green above, glaucescent below, glabrous in all parts; *veins* fine and distinct, one to each apical tooth; *sori* roundish or transversely reniform, attached to shallow sinuses, 4-5 along the upper edge of each pinnule, separated by a notch, indusium dark-colored with broad, pale-colored membranaceous margin; spores sharply tetraedric, pale-brown, transluscent and smooth.

　　A distinct and pretty fern endemic in Southwest China: Sichuan (Mt. Omei), Guizhou, Western Hubei and N. W. Yunnan, where the fern is locally abundant under forests at elevations from 1000-1500 m.

　　In general habit this species exactly resembles *Adiantum pedatum* L. of Northeast Asia and North America, to which it was previously reduced as a synonym or a variety, but it differs in the almost black stipe, glaucescent underside of fronds, decidedly triangular pinnules invariably provided with 3-6 rather sharply deltoid teeth at the acute or blunt apex and with serrate upper margin, the generally roundish and more numerous indusia to each pinnule and in the dark brown, glossy, narrower, firmer scales on rhizome. It grows at a lower elevation.

　　Plate 227. 1. Habit sketch (natural size). 2. Fertile pinna (× 3). 3. Sterile pinna, showing serrate upper margin (× 3). 4. Scale from the base of stipe (× 20). 5-6. Sporangium with spores (× 100).

ADIANTUM MYRIOSORUM Baker
灰背铁线蕨

图版 228　毛叶茯蕨

根状茎短而直立或斜生，有密的长根丛；叶簇生，柄长 20—30 cm 或较长，禾秆色，颇粗大，下半部有暗棕色阔披针形有刚毛的鳞片和开展的灰白色针状长刚毛混生，同样的刚毛向上分布到中轴全部，叶体长 20—35 cm，宽 10—20 cm，长圆形，渐尖头，向基部不变狭，一回羽状，顶端为羽状深裂；羽片 15—20 对，开展，下部为阔缺刻分开，向上较接近，下部 6—8 对分离，几无柄，上部各对向顶端愈来愈多合生，基部一对长与上方一对相等或有时稍长，水平开展，常为亚镰刀形，基部等形，变狭，或等阔，中部各对长 5—10 cm，宽 1—1.5 cm，线状披针形，渐尖头，基部等形，截形，羽状深裂到 1/2 深处，成许多斜出长圆形的圆头裂片；叶为草质，干后为淡棕绿色，中轴及羽片上下两面有开展的针状灰白长刚毛；叶脉在裂片 5—6 对，分离，斜出，稍弯弓，仅基部上方一脉达于缺刻底部；子囊形长形，沿生于下部几对叶脉，有刚毛，因每个子囊具有向上直立的 5—6 根针状刚毛所致。

分布：福建北部仙霞岭及台湾阿里山。

本种广布于印度南部，斯里兰卡，中国东部和日本，向北展伸到北海道及朝鲜半岛南部。在中国却少见，过去关于本种在中国西部的报导实为错误，而是：

云南茯蕨（*Leptogramma yunnanensis* Ching）新种。

叶体较狭长，一般宽 6—8 cm，羽片斜向上，较短较阔，长 4—5 cm，基部宽 1.2—1.7 cm，短渐尖头或亚急尖头，中轴及中肋和小肋下面有较短而较细软的毛。

分布：云南西北部，维西，澜沧怒江分水岭，冯国楣号码 4474，生林下沟中，海拔 2100 m；贡山，菖蒲桶，冯国楣号码 7366，生杂木林下，海拔 1900—2000 米。

图注：1. 本种全形（自然大）；2. 羽片的一部，表示叶脉及子囊群（放大 4 倍）；3. 叶柄基部的鳞片（放大 20 倍）；4. 叶柄上的刚毛（放大 40 倍）；5—6. 孢子囊及孢子（放大 100 倍）。

LEPTOGRAMMA MOLLISSIMA (Fisch.) Ching

LEPTOGRAMMA MOLLISSIMA (Fisch.) Ching in *Sinensia*. **7**: 102 t. 9. 1936.
Gymnogramma mollissima Fisch. ex Kunze in Linnaea **23**: 255, 310. 1850.
Gymnogramma totta Hook. Sp. Fil. **5**: 138, 1864; Hook. & Bak. Syn. Fil. 376. 1867 (pro parte).
Dryopteris africana C. Chr. Ind. Fil. 250. 1905; Suppl. Ⅲ. 80 1933 (pro parte); Ogata, Ic. Fil. Jap. **1**: t. 20. 1928.
Leptogramma loveii Nakai in Bot. Mag. Tokio **45**: 103. 1931 (non J. Sm. 1841).

Rhizome short, erect, or oblique, densely radicose; *fronds* caespitose, stipe 20-30 cm or longer, stramineous, rather strong, copiously clothed in dark brown, broad-lanceolate and setose *scales* and also spreading setose needle-like gray *hairs*, which extend upwards throughout the rachis, lamina 20-35 cm long, 10-20 cm broad, oblong, acuminate, base not narrowed, simply pinnate; *pinnae* 15-20 pairs under the pin-natifid apical part, patent, separated by broad sinuses below, closer upwards, the lower 6-8 pairs free and subsessile, the upper ones becoming more and more adnate upwards, the basal pair as long as the upper next one, or even longer, horizontally patent, often subfalcate and slightly broader above the contracted lower side of the base, the middle ones 5-10 cm long, 1-1.5 cm broad, linear-lanceolate, acuminate, base equal, truncate, pinnatisect to half way down into many oblong, rounded and oblique *lobes*; *texture* herbaceous, dry brownish-green, rachis and both sides with spreading, needle-like long gray hairs; *veins* in the lobes 5-6 pairs, oblique, curved, only the anterior basal vein runs to the sinus; *sori* elongate-oblong, along the several lower veins, setose, due to the setose sporangia.

Fujian: SihYaling on the border of southern Zhejiang, *Chiu Pei-chih 109*, August, 1953. Taiwan: Mt. Arisan.

Widely distributed from Southern India, Sri Lanka, East China to Japan as far north as Hokaido and Korea peninsula. In China however, it is a rare fern. Plants from West China previously referred to the present species should be: **Leptogramma yunnanensis** Ching, sp. nov.

Fronde anguste elongata, plerumque 6-8 cm lata, pinnis lateralibus obliquis, brevioribus latioribusque, 4-5 cm longis, 1.2-1.7 cm basi latis, breviter acuminatis aut subacutis, rachide et lamina subtus secus costam costulamque pilis brevioribus molliter obtectis.

N. Y. Yunnan: Weisih Hsien, Mekong-Salwin Divide, *K. M. Feng 4474* in wooded ravine, alt. 2100 m.; Kong Shan Hsien, Chonmutung, *K. M. Feng 7366*, in thickets, alt. 1900-2000 m.

The plants from Northern India referred to by Clarke and Hope may also belong here rather than to *L. mollissima* of S. India.

Plate 228. 1. Habit sketch (natural size). 2. Portion of pinna, showing venation and sori (× 4). 3. Scale from the base of stipe (× 20). 4. Hairs from the stipe (× 40). 5-6. Sporangium with spores (× 100).

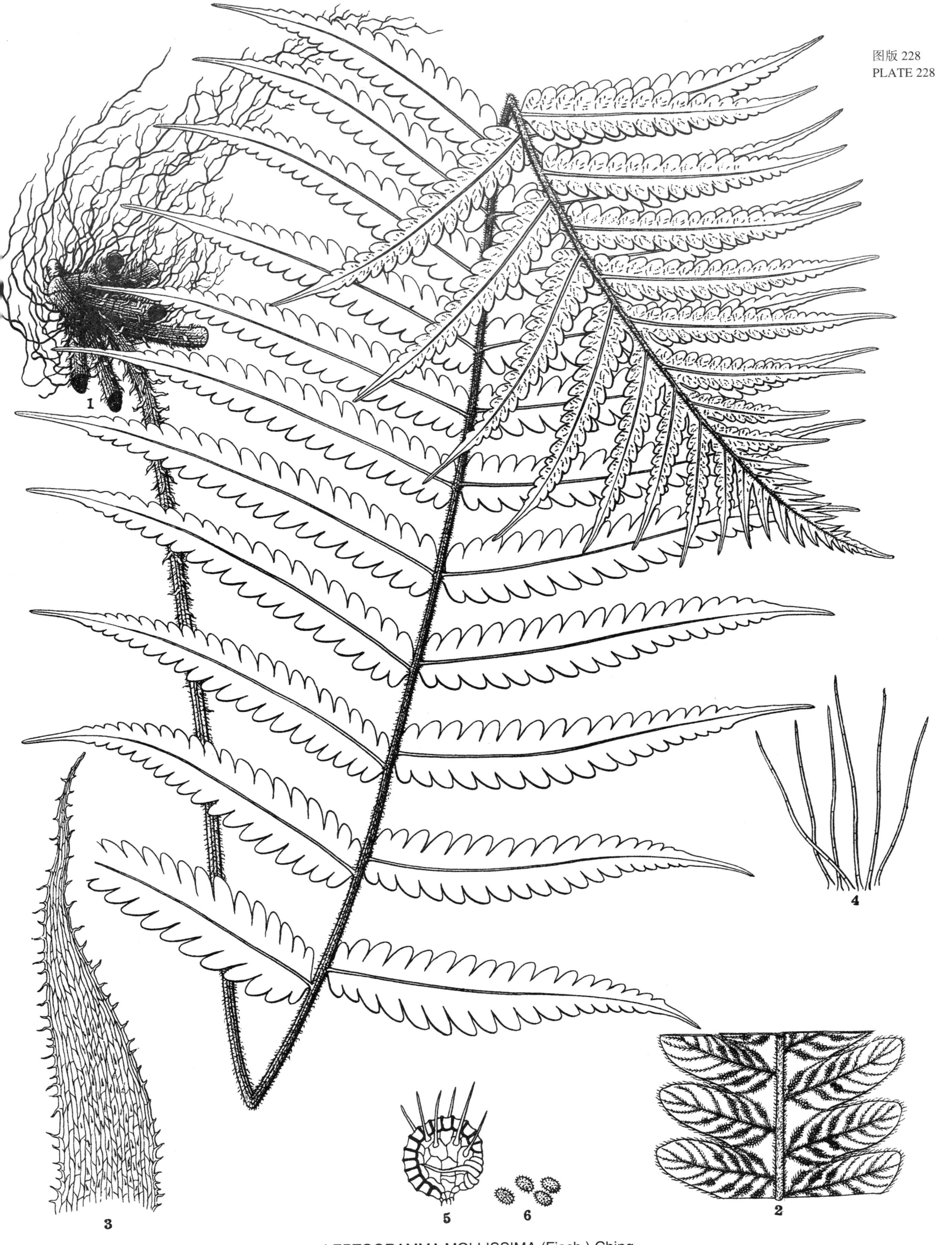

LEPTOGRAMMA MOLLISSIMA (Fisch.) Ching
毛叶茯蕨

根状茎短而直立或斜生,有密的根丛;叶簇生,柄长 6—12 cm,暗禾秆色,基部有适量的鳞片及密生的刚毛,鳞片棕色,披针形,有刚毛附生,刚毛灰白色,长针形,由单细胞组成,叶体通常长 12—20 cm,或较长,宽 7—10 cm,狭三角状长圆形,渐尖头,基部不变狭,间或稍狭,一回羽状;顶部为羽状深裂;羽片约为 10 对,斜向上,颇接近,下面几对分离,并有明显的短柄,向上部各对愈来愈多的合生,基部一对长同上面一对,很少有时较短,几为水平开展,中部羽片 4—5 cm,一般宽 6—8 mm,罕较宽,披针形,急尖头或亚渐尖头,基部相等,截形或圆形,羽状深裂到大约 1/2 深处,成 10 对的斜出三角卵形全缘的裂片;叶质粗糙,干后淡棕色或淡褐绿色,在中轴,中肋及小肋下面有适量的灰白色针状开展的刚毛,上面光滑,唯中肋有短毛;叶脉在裂片为 4 对,分离,斜出,弯弓,基部上方一脉达于缺刻底部,下方一脉达于叶边,下面有刚毛;子囊群长圆形,通直,每片 1—2 对,沿生于下二对脉上,有毛,因子囊顶部有 4 根直立针状刚毛。

分布:本种为长江南岸各省的特有种,自四川(峨眉山)向东经贵州,广西,湖南,江西(庐山)达于浙江及安徽南部,向南经云南达于越南北部,生于海拔 400—1000 m 的林下沟边。

图注:1. 本种全形(自然大);2. 羽片的一部,表示叶脉及子囊群(放大 4 倍);3. 叶柄基部的鳞片(放大 30 倍);4. 叶柄基部的针状刚毛(放大 60 倍);5—6. 孢子囊及孢子(放大 100 倍)。

LEPTOGRAMMA SCALLANI (Christ) Ching

LEPTOGRAMMA SCALLANI (Christ) Ching in *Sinensia* **7**: 101 t. 7. 1936; Tard. et C. Chr. in Fl. Gen. Ind. **7**: 372. fig. 44 (1). 1941.

Aspidium scallani Christ in Bull. Soc. Bot. Ital. **1901**: 296.

Dryopteris scallani C. Chr. Ind. Fil. 291. 1905; in Acta Hort. Gotheb. **1**: 55. 1924.

Rhizome short, erect or oblique, densely radicose; *fronds* caespitose, stipe 6-12 cm long, dark stramineous, basal part moderately scaly and densely setose hairy, *scales* brown, lanceolate, setose, *hairs* gray, needle-like, long and unicellular; lamina usually 12-20 cm long or longer, 7-10 cm broad, subdeltoid-oblong, acuminate, base not or sometimes slightly narrowed, simply pinnate under the pinnatifid apical part; *pinnae* about 10 pairs, oblique, rather close, the lower several pairs free and distinctly petiolate, the upper ones becoming more and more adnate upwards, the basal pair as long as the upper next ones and almost horizontally patent, or sometimes considerably shortened, the middle ones 4-5 cm long, generally 6-8 mm, or rarely broader, lanceolate, acute or subacuminate, base equal, truncate or rounded, pinnatisect about half way or less down into about 10 pairs of deltoid-ovate, oblique entire *lobes*; *texture* thickly herbaceous, dry brownish or blackish-green, rachis, costa and costules below copiously clothed in long spread-ing gray needle-like hairs, quite glabrous above except the tomentose costa; *veins* in lobes 4 pairs, oblique and curved, the anterior basal one runs to the sinus, the posterior basal one to the margin some way above the sinus, setose hairy underneath; *sori* oblong, straight, 1-2 pairs in each lobe along the lower veins, sporangia provided with 4 erect needle-like hairs.

A very distinct endemic species inhabiting deep ravines by stream side under forests in the Provinces of Sichuan (Mt. Omei), Guizhou, Yunnan, Guangxi, Hunan, Jiangxi (Lushan), Zhejiang, and southern part of Anhui, at elevations of 400-550 m. Also Vietnam, Indo-China.

Plate 229. 1. Habit sketch (natural size). 2. Portion of pinna, showing venation and sori (× 4). 3. Scale from the base of stipe (× 30). 4. Hairs from the lower part of stipe (× 60). 5-6. Sporangium with spores (× 100).

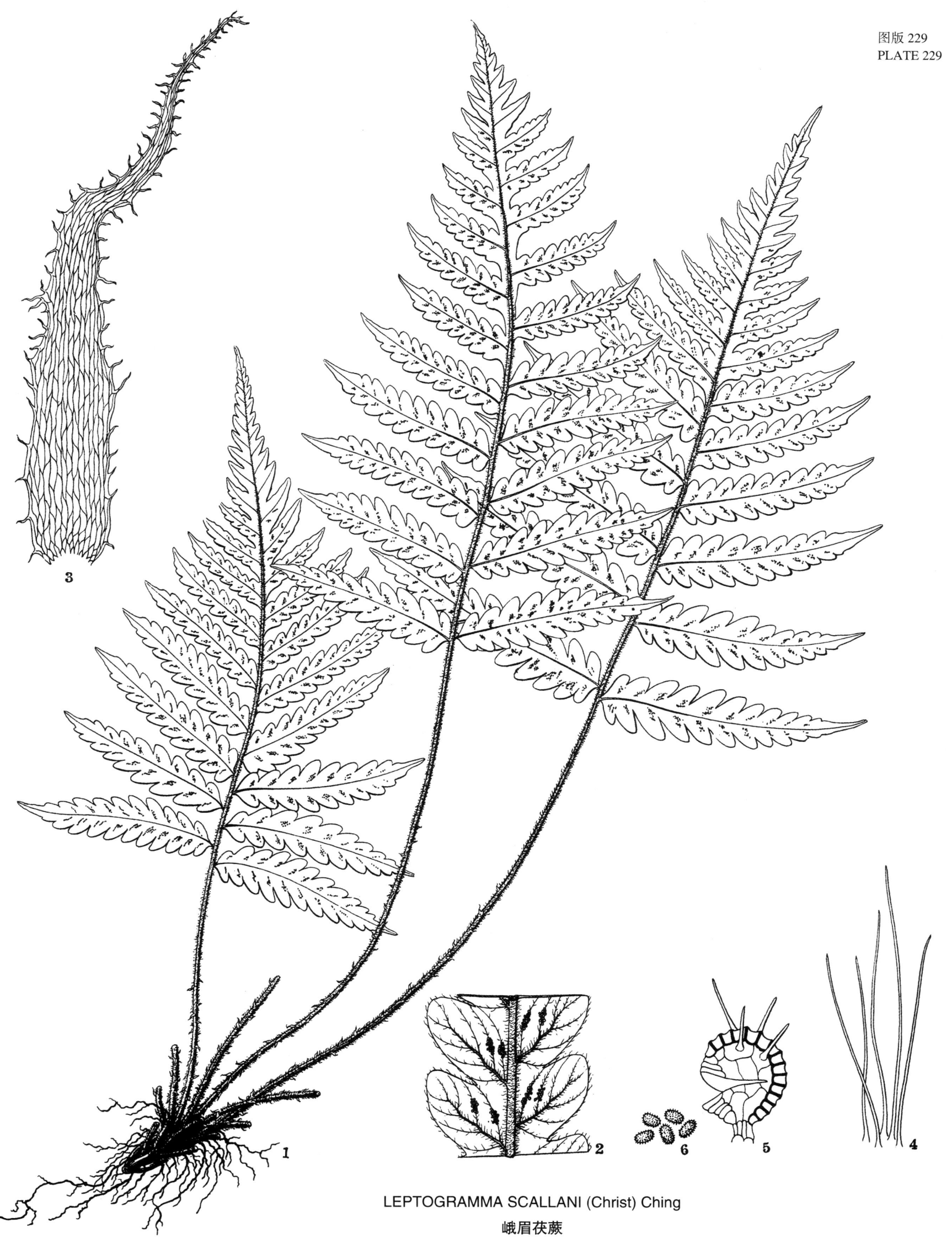

LEPTOGRAMMA SCALLANI (Christ) Ching
峨眉茯蕨

图版 230　尾叶茯蕨

根状茎斜出或为短匍匐状；叶簇生于生长顶端，柄细弱，全体被有适量的刚毛，但无鳞片，长 5—10 cm，叶体长三角状披针形，向顶部为长渐尖头，长 10—16 cm，戟状，基部以上宽 3—5 cm，一回羽状，上半部为深羽裂；分离羽片 3—5 对，无柄，向上各对逐渐愈来愈多的合生，基部一对羽片常最长，长 2—4 cm，宽 1 cm，阔披针形，水平开展或略向下，顶端急尖或亚钝头，基部不等，即下方斜切，上方亚截形，边缘浅羽裂为 4—6 对的斜出钝三角形的全缘裂片，基部以上的羽片远较短，接近，斜开展，和基部一对同形，但羽裂较浅，基部多少合生，亚圆形或亚截形，上下相等，顶端钝头或急尖头；叶为草质，上下两面被适量刚毛，边缘有睫毛；叶脉在裂片通常 3 对，不分叉，分离，都达于叶边，小肋弯曲；子囊群线形，沿生于下部的脉上，有刚毛，因子囊顶部有 4—5 根直立针状刚毛所致。

分布：福建福州及台湾阿里山特产。

本种是本属各种中最特出的一种，不同于其他各种的特征为其小而狭长的叶，基部有一对特长的水平开展羽片，全形如戟。

图注：1. 本种全形（自然大）；2. 中部羽片，表示叶脉及子囊群（放大 3 倍）；3. 叶柄上的刚毛（放大 30 倍）；4—5. 孢子囊及孢子（放大 100 倍）。

LEPTOGRAMMA CAUDATA Ching

LEPTOGRAMMA CAUDATA Ching in *Sinensia* **7**：98 t. 4. 1936.

Dryopteris africana Hayata, Ic. Pl. Form. **4**：187 t. 124, A-B. 1914 (non C. Chr. Ind. Fil. 1905).

Rhizome oblique, or short-creeping; *fronds* fasciculate at the growing apex of the rhizome, stipe gracile, stramineous, moderately hirsute throughout, 5-10 cm long, lamina elongate, deltoid-lanceolate, gradually long-acuminate, 10-16 cm long, 3-5 cm broad above the hastate base, simply pinnate below the broad and deeply pinnatifid upper half; *free pinnae* 3-5 pairs, sessile, the upper ones become more and more adnate as they go upwards, the basal pair of pinnae generally much the longest, 4-2 cm long, 1 cm broad, broadly lanceolate, acute, horizontally patent or somewhat deflexed, apex acute or obtusish, base unequal, i. e. the lower side cut oblique, the upper side as broad as the upper part, pinnatifid less than half way down into 6-4 oblique, deltoid, obtuse entire lobes; *pinnae* above the basal pair half as long, patent or oblique-patent, more or less adnate, of similar shape as the basal ones, less deeply lobato-incised, close, base roundish or subtruncate, equal, apex obtuse or acute; *texture* herbaceous, moderately hirsute on both sides, margin ciliate; *veins* in lobes usually 3-jugate, simple, free from flexuose costule, the basal pair of the contiguous groups reach the leaf margin a little way above the sinus; *sori* linear along the lower veins of each group, hirsute due to the presence near the top of the sporangia of some upright pale-colored needle-like unicellular hairs.

Fujian: Foochow, Lukiang, *F. P. Metcalf 5002C* (type), August 10, 1926. Taiwan: Mt. Arisan, *Hayata et Sasaki*, January 1912; *Faurie 529*, June, 1914.

A peculiarly distinct species, different from all other known species of the genus in the narrowly elongate fronds with much produced basal pair of pinna.

Plate 230. 1. Habit sketch (natural size). 2. A middle pinna, showing venation and sori ($\times$ 3). 3. Hairs from the stipe ($\times$ 30). 4-5. Sporangium with spores ($\times$ 100).

LEPTOGRAMMA CAUDATA Ching
尾叶茯蕨

根状茎斜出或为短匍匐状；叶簇生于生长顶端，柄细软，暗禾秆色，长 8—14 cm，基部有疏鳞片和刚毛混生，鳞片卵状披针形，渐尖头，暗棕色，有刚毛附生，刚毛淡灰色，针形，开展，由多个细胞组成，柄上部及中轴全部也有同样刚毛，叶体长 14—24 cm，宽 4—5 cm，阔披针形，渐尖头，向基部略变狭，一回羽状，顶部为深羽裂；羽片 7—10 对，多少合生，唯下面几对分离，无柄，比中部的略短，中部羽片为卵状披针形，长 2.5—4 cm，基部宽 1.3—1.5 cm，圆形或截形，急尖头或短渐尖头，基部一对羽片较短，长不及 2 cm，长圆卵形，开展或略向下，上部的羽片开展，为颇狭的缺刻分开，边缘亚全缘或者通常为波状圆齿形；叶为粗草质，干后为淡黄绿或淡褐色，在上下两面沿中肋，叶脉及小脉有适量的短刚毛；侧脉 8—10 对，明显，亚开展，小脉羽状，3—4 对，斜向上方，稍弯弓，其中通常只基部一对顶端交结，在中肋两旁成三角形的网眼，顶部延伸小脉向外达于浅缺刻；子囊群线形，沿生于下面几对小脉上，而基部一对较上方各对为长，有刚毛，因子囊群均有 5—6 根针状硬毛生于顶部及子囊柄上也有刚毛所致。

分布：四川西南部（峨眉山），贵州（梵净山）及云南北部，为本区特有种，生于海拔 1000—2400 m 的杂木林下沟中，在形体上颇似贯众的小羽片变型 Cyrtomium Fortunei J. Sm. f. *polypteron* (Diels) Ching。

图注：1. 本种全形（自然大）；2. 中部的羽片，表示叶脉，边缘及子囊群（放大 3 倍）；3. 叶柄基部的鳞片（放大 40 倍）；4. 叶柄上的针状刚毛（放大 30 倍）；5—6. 孢子囊及孢子（放大 100 倍）。

STEGNOGRAMMA CYRTOMIOIDES（C. Chr.）Ching

STEGNOGRAMMA CYRTOMIOIDES (C. Chr.) Ching in *Sinensia* **7**: 95 t. 3. 1936.

Dryopteris stegnogramma var. *cyrtomioides* C. Chr. in Acta Hort. Gotheb. **1**: 56. 1924.

Rhizome oblique or short-creeping; *fronds* caespitose at the growing apex of the rhizome, stipe slender, 8-14 cm long, dark stramineous, the basal part is both sparsely scaly and hirsute, *scales* dark brown, ovate-lanceolate, acuminate and setose, *hairs* pale-colored, needle-like, spreading, multicellular, the upper part of stipe and the entire rachis are similarly hirsute, lamina 14-24 cm long, 4-5 cm broad, broadly lanceolate, acuminate, slightly narrowed towards the base, simply pinnate under the deeply pinnatifid apical part; *pinnae* 7-10 pairs, more or less adnate, only several lower pairs free and sessile, slightly shorter than the middle ones, which are 2.5-4 cm long, 1.3-1.5 cm broad at the broader round-truncate base, ovate-lanceolate, acute or short-acuminate, the lowest pair less than 2 cm long, oblong-ovate, patent or often somewhat de-flexed, the upper ones patent, separated by rather narrow sinuses, margin subentire or more often undulate-crenate; *texture* thickly herbaceous, flavo-greenish, or blackish, both sides moderately and shortly hirsute on costa, veins and veinlets; *lateral veins* 8-10-jugate, distinct, subpatent, pinnate; veinlets 4-3 pairs, obliquely ascending, slightly curved, of which usually only the basal pair of opposite groups united into a triangular mesh along each side of the costa with excurrent veinlet running towards the margin as the upper veinlets; *sori* linear, along the lower several veinlets of each group, of which the lowest are much longer than those above, densely hirsute due to the setose sporangia.

Sichuan: Ta-shiang-ling, *Harry Smith 13499*, *10069*, *2104* (type), in thickets, alt. 2400 m.; Mt. Omei, *W. P. Fang 3007*, in thickets, alt. 1000-1100 m. Guizhou: Yinkiang, Vanching Shan, *Y. Tsiang 7953*, Dec. 30, 1930; *A. N. Steward 687*; Kweiyang, *Bodinier 2765* (1899); Tsingay, *Bodinier 2100*. Yunnan: Paipintien, *E. E. Maire* (1916), alt. 400 m.

This distinct species is endemic in West China. In habit, it resembles *Cyrtomium Fortunei* J. Sm. f. *polypteron* (Diels) Ching from the same region.

Plate 231. 1. Habit sketch (natural size). 2. One middle pinna, showing venation, margin and sori (× 3). 3. Scale from the base of stipe (× 40). 4. Hairs from the stipe (× 30). 5-6. Sporangium with spores (× 100).

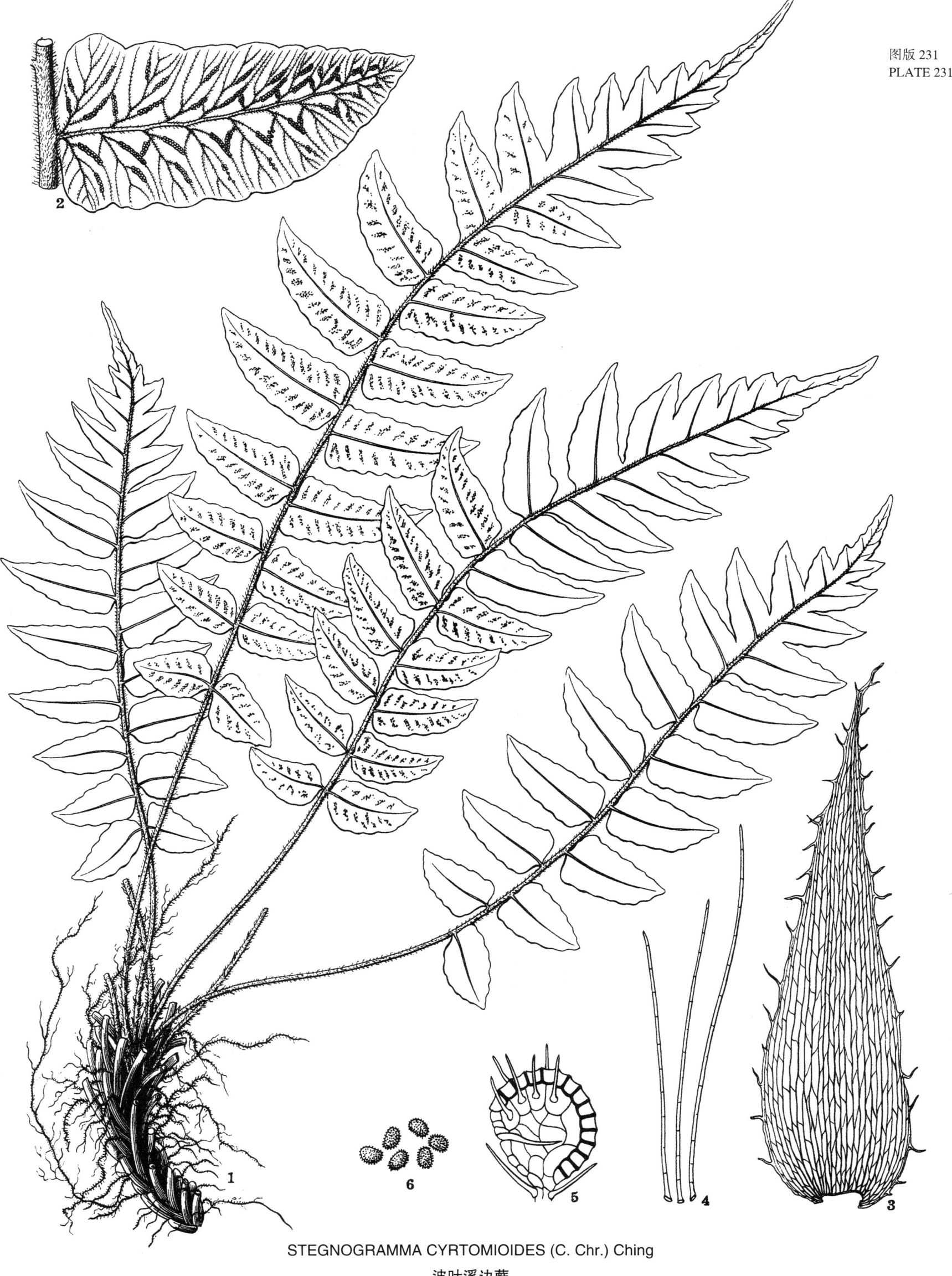

STEGNOGRAMMA CYRTOMIOIDES (C. Chr.) Ching
波叶溪边蕨

根状茎斜出向上或短匍匐状，为老的叶柄基部所覆盖；叶簇生，柄长 20 cm，或过之，暗禾秆色，下部有颇多的鳞片和开展的灰白色针状刚毛混生，鳞片为披针形，暗棕色，有刚毛附生，向上部亚光滑或有刚毛疏生，叶体长 40 cm 或过之，宽约 12 cm，长圆披针形，渐尖头，向基部几不变狭，在羽状深裂的顶部下面为一回羽状；羽片约为 10 对，其中下部 5—8 对分离而无柄，上部的多少合生，开展，彼此为较宽的缺刻所分开，互生，长 5—7 cm 或较长，基部一对略短或和上面的羽片同长，宽 1.4—2 cm，阔披针形，略为短渐尖头，基部上下方都略为斜切，亚圆形，羽状浅裂到 1/3 深处或更浅一些，成为 10—13 对的阔圆形的斜向小裂片，上方的羽片边缘只为圆齿状的浅裂；叶质粗糙，干后为黝绿色，中轴及上下两面都有针状刚毛；叶脉在每一小裂片为 4—6 对，单脉，斜向上方，稍弯弓，下面 2 对或有时只一对顶端结合，并有延伸顶脉直达缺刻；子囊群线形沿生于脉上，也稍为弯弓，下面一对顶端通常也结合，无盖，孢子囊顶部经常有 4 根或较多的上向针状单细胞刚毛；孢子两面形，长圆，有密细刺状突起。

分布：四川西部（大相岭）及云南西北部，向西经北缅甸至印度北部，生于海拔 2400 m 以上的次生杂木林沟内。

本种为本区内特有种，过去曾错误地被认为和南洋群岛的 *Stegnogramma aspidioides* Bl. 为同一种，作者曾加以研究，指出两种的差别（参阅 *Sinensia* 7 卷 94 页 1936 年）。

图注：1. 本种全形（自然大）；2. 羽片的一部，表示叶脉及子囊群（放大 3 倍）；3. 叶柄基部的鳞片（放大 40 倍）；4. 叶柄上的刚毛（放大 40 倍）；5—6. 孢子囊及孢子（放大 100 倍）。

STEGNOGRAMMA ASPLENIOIDES J. Smith

STEGNOGRAMMA ASPLENIOIDES J. Smith apud Ching, Gen. Stegn. and Lept. in *Sinensia* **7**: 94 t. 2. 1936.

Gymnogramma aspidioides Hook. Sp. Fil. **5**: 150. 1864; Hook. & Bak. Syn. Fil. 378. 1867; Clarke in Trans. Linn. Soc. II. Bot. **1**: 569. 1880 (pro parte).

Dryopteris stegnogramma C. Chr. Ind. Fil. 294. 1905 (pro parte).

Dryopteris stegnogramma var. *asplenioides* C. Chr. in Acta Hort. Gotheb. **1**: 56. 1924.

Rhizome obliquely ascending or short-creeping, densely clothed in the old basal parts of the stipes of previous years; *fronds* fasciculate, stipe 20 cm or longer, dark stramineous, copiously clothed in the lower part in both dark brown lanceolate setose *scales* and the spreading needle-like long gray *hairs*, almost sub-glabrous or sparsely hirsute upwards, lamina 40 cm or longer, about 12 cm broad, oblong-lanceolate, acuminate, hardly narrowed towards the base, simply pinnate under the pinnatifid apical part; *pinnae* about 10 pairs, of which the lower 5-8 pairs free and sessile, the upper ones slightly adnate, patent, alternate, separated by rather broad sinuses, 5-7 cm long, or longer, the basal ones not or slightly shortened, 1.4-2 cm broad, broad-lanceolate, rather short-acuminate, base equally oblique, roundish, pinnatisect about one-third way or less down into 10-13 pairs of broad, rounded, oblique lobes, the upper pinna only lobato-crenate; *texture* thickly herbaceous, opaque green, hairy along the rachis and on both sides of pinna with needlelike hairs; *veins* in lobes 4-6 pairs, simple, obliquely ascending and curved, the lower 2 pairs (or sometimes only 1 pair) of the opposite groups jointed below the sinus with excurrent veinlets running straight to the sinus; *sori* linear, curved along several pairs of lower veins with the lowest pair often jointed or even run up a short way along the excurrent veinlet, sporangia setose with 4 or more erect needle-like long unicellular hairs.

Sichuan: Tashiangling, *Harry Smith 2116*, in thickets, alt. 2400 m. Yunnan: Kongshan Hsien, Chonmutong, *K. M. Feng 7578*, alt. 2300 m., common; Yangpie Hsien, *R. C. Ching 22483*, in thickets.

Khasia: *Griffith* (type); *Hooker et Thomson*, 5-6000 ft. alt. Assam: Shillong Wood, *C. B. Clarke 37345*, February, 1885.

This species is very different from the Javanese *S. aspidioides* Bl. in many essential characters as was already pointed out by me long ago (l. c.). From that species it can be easily distinguished by much larger size, much longer and pinnatisect or lobato-crenate pinna, generally 2 pairs of anastomosed veins and by the costa below provided with spreading needle-like long hairs.

Plate 232. 1. Habit sketch (natural size). 2. Portion of pinna, showing venation and sori ($\times$ 3). 3. Scale from the basal part of stipe ($\times$ 40). 4. Hairs from stipe ($\times$ 40). 5-6. Sporangium with spores ($\times$ 100).

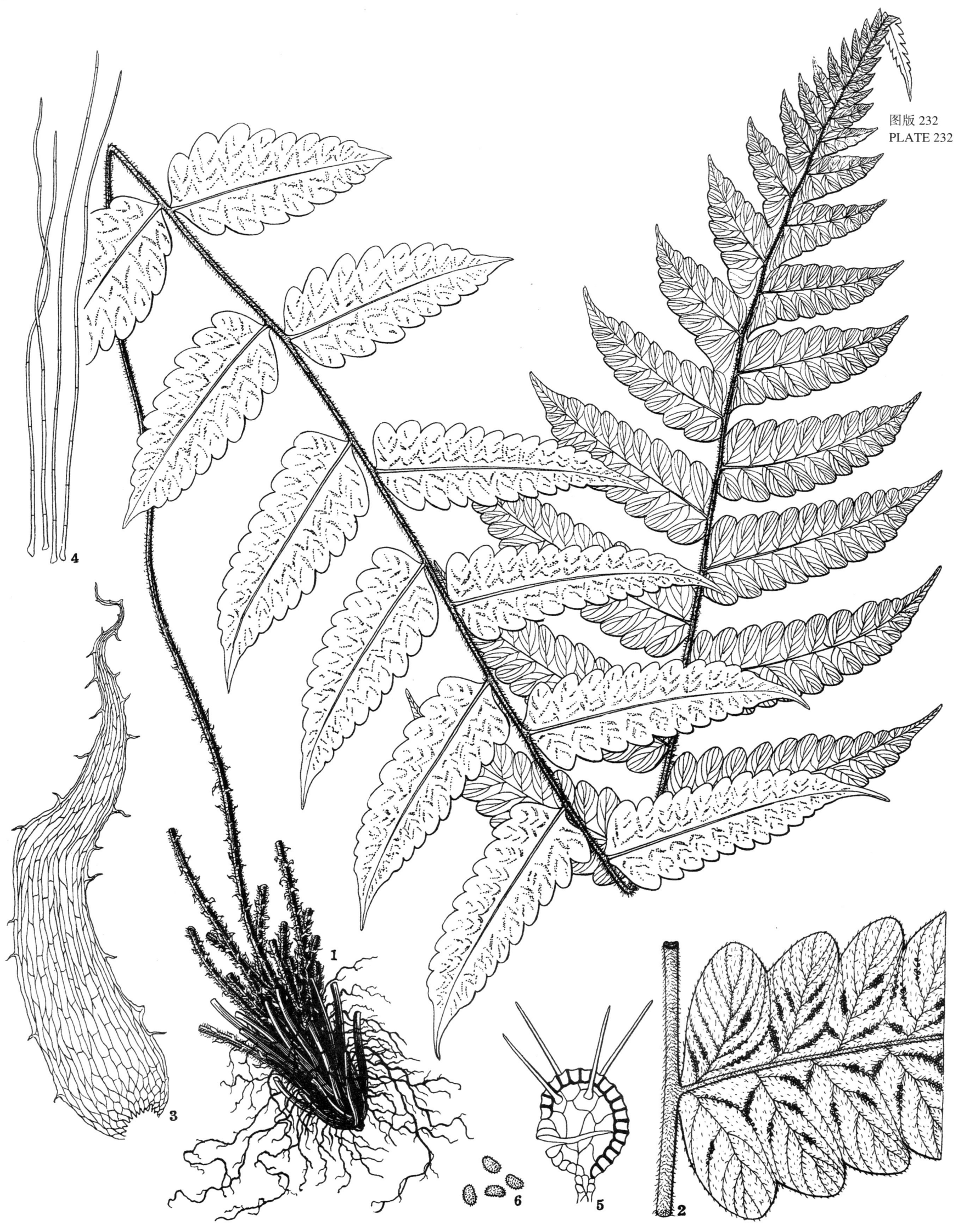

STEGNOGRAMMA ASPLENIOIDES J. Smith
浅裂溪边蕨

根状茎短而直立,有密的根丛,同叶柄一起,有暗棕色披针形渐尖头的鳞片密覆,边有锯齿,长达 5 mm;叶多数簇生,叶柄长 5—7 cm,棕禾秆色,具密鳞片,叶体长 15—25 cm,宽 2 cm,线状披针形,渐尖头,向基部极少变狭,一回羽状;羽片多数(40—60 对),栉比状瓦覆排列,几成直角开展,长约 1 cm,宽 5 mm,几无小柄,亚长方圆形,基部极不等,下边切成直线,基部上方截形,并有一个急尖头三角形耳片和中轴平行,上边缘为波状,向圆头有几个短而开展的尖锯齿,基部 1—2 对羽片略小,并多少下向;叶为草质,干后淡棕色,两面光滑,但中轴有鳞片疏生;叶脉分叉,而基部上方一脉为羽状;子囊群小,大都生于羽片的上半边,5—7 个,边内生,下边缺如或只有 1—2 个生于顶端下,囊群盖灰色,最后脱落。

分布:贵州平番县,卡弗雷里(Cavalerie)号码 892,3712;桐梓县,蒋英号码 5176。

本种为一明显的地方种,形体颇似锯齿耳蕨(Polystichum hecatopterum Diels),但羽片为圆钝头,不具长刺形的锯齿,下部羽片不为强度下向,故易区别。本种也近于对生耳蕨 P. deltodon (Bak.) Diels,但羽片较多,较密,为栉比状瓦覆排列,顶端为圆钝,也易于区别。

图注:1. 本种全形(自然大);2. 羽片,表示叶脉及子囊群位置(放大 4 倍);3. 叶柄基部的鳞片(放大 20 倍);4—5. 孢子囊及孢子(放大 100 倍)。

POLYSTICHUM DIELSII Christ

POLYSTICHUM DIELSII Christ in Bull. Acad. Géogr. Bot. Mans 16: 238. 1906; C. Chr. Ind. Fil. Suppl. I. 64. 1912.

Hemesteum dielsii Léveillé Fl. Kouy-tscheou 497. 1915.

Polystichum hecatopterum C. Chr. in Bull. Acad. Géogr. Bot. Mans. 14: 114. 1904 (non Diels).

Rhizome short, erect, densely radicose and together with stipes clothed in dense dark brown, lanceolate, acuminate, dentate *scales* over 5 mm long; *fronds* caespitose, 6-12 together, stipe 5-7 cm long, rufostramineous, copiously clothed in the similar but smaller scales than those at the base, lamina 15-25 cm long, 2 cm broad, linear-lanceolate, acuminate, hardly narrowed downward, pinnate; *pinnae* numerous (40-60-jugate), pectinately imbricate, recto-patent, about 1 cm long, 5 mm broad, subsessile, subrectangular-oblong, base strongly unequal, cut straight below, truncate above with an acute deltoid auricle parallel to the rachis, upper margin crenate, but with several sharp, short, spreading teeth towards the rounded apex, the lowest 1-2 pairs of pinna slightly smaller and more or less deflexed; *texture* herbaceous, dry brownish, both sides glabrous except the sparsely scaly rachis; *veins* forked but the anterior basal one pinnate; *sori* small, mostly confined to the upper half of the pinna, 5-7 and intramarginal, none or 1-2 on the lower side below the apex, indusiate, indusium peltate, gray, fallen off at last.

Guizhou: Pinfa, *Cavalerie 892* (type), 3712, on limestone cliff; Houang tsa-pa, *Cavalerie 7061*, 7647 (pro parte); Ta-ny, *Esquirol 3579*; Tungtze, *Y. Tsiang 5176* (typical).

A very distinct local species with a habit more like *P. hecatopterum* Diels (see Ching, Ic. Fil. Sin. 1: t. 12. 1930) man any other related species of the group of *P. deltodon* (Bak.) Diels, but differs in the margin of the pinna having only a few shorter spiny teeth towards the obtuse apex below which the upper margin is undulate or crenate and in the less deflexed lower pinna. From *P. deltodon* (Bak.) Diels (see Ching, l. c. t. 11.), our fern differs in having numerous, pectinately imbricate pinna with obtuse apex provided with a few longer teeth.

Plate 233. 1. Habit sketch (natural size). 2. Pinna, showing venation and sori ($\times$ 4). 3. Scale from the base of stipe ($\times$ 20). 4-5. Sporangium with spores ($\times$ 100).

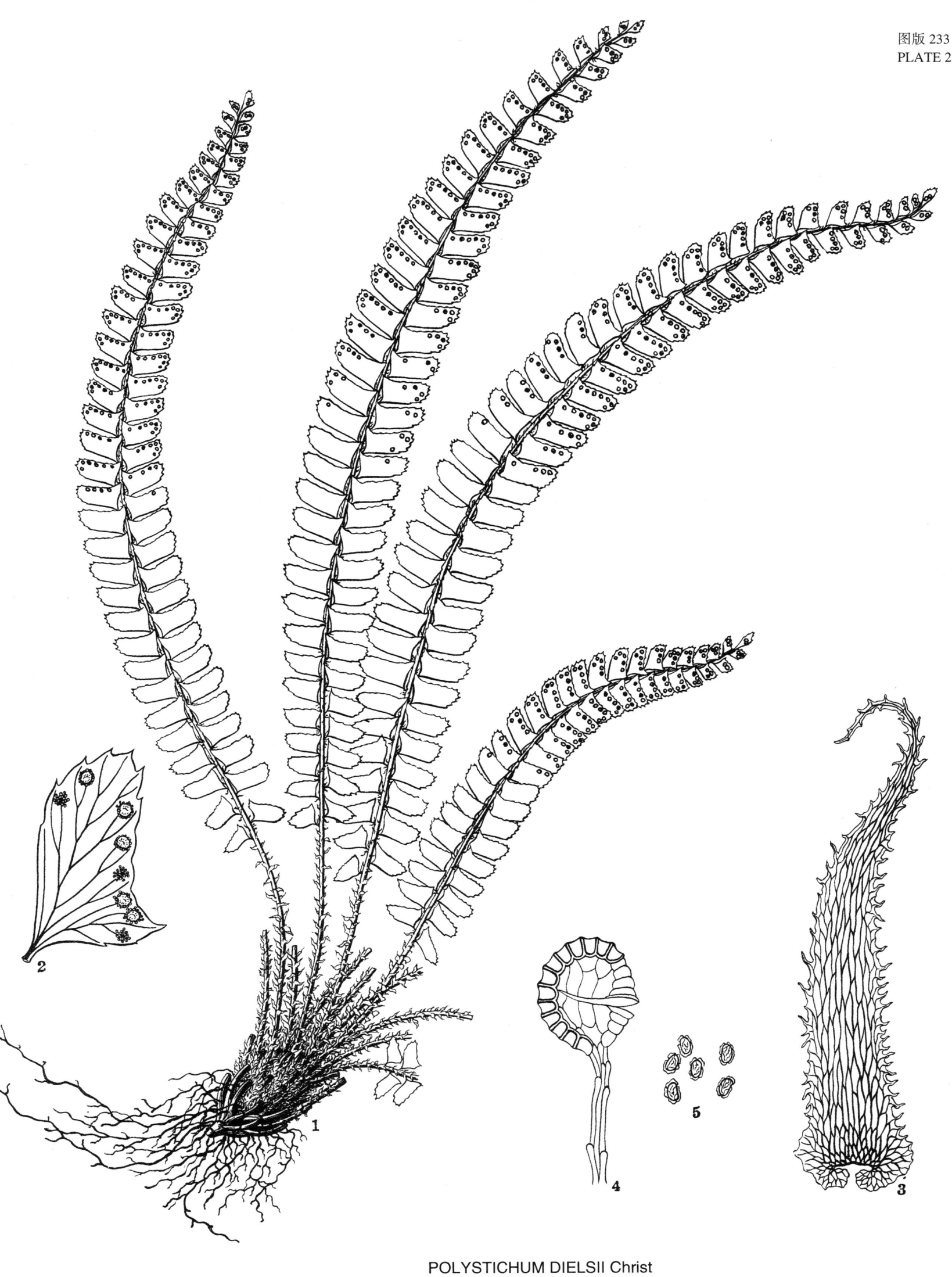

POLYSTICHUM DIELSII Christ
圆顶耳蕨

图版 234　尖顶耳蕨

　　根状茎短而直立,有密的根丛,同叶柄一起,有红棕色狭卵状披针形的鳞片密生,边具粗睫毛;叶多数簇生,叶柄长 7—12 cm,淡禾秆色,上部鳞片疏生,叶体长 20—25 cm,宽 3.2—4 cm,线披针形,渐尖头,向基部不变狭,一回羽状;羽片 20—28 对,有小柄,栉比状排列,斜出开展,向顶部逐渐缩小,但基部羽片不缩小,长 2 cm,基部宽 6 mm,亚镰刀状的斜方披针形,顶端为短头,具一微突头,基部不等,下边切成直线形,基部上方为斜截形,并具一小三角耳形突起,有一刺尖头,上边缘有波状短齿;叶为纸质,干后为淡绿色,中轴及下面有小鳞片疏生;侧脉约有 8 对,分叉,或基部上方一脉为羽状,两面不显明,子囊群中位,上边有 6—8 个,下边 4—6 个,顶生于每群脉的上方一短小脉;囊群盖大而圆,边缘有不规则的锯齿,最后脱落。

　　分布:云南东南部,屏边县,大围山,冯国楣号码 5173(模式标本),生林下石灰岩上,常见。贵州,精城,青龙山,邓世伟 90379;中国科学院植物所生态组号码 2001,海拔 1150 m;平番,卡弗雷里(Cavalerie)号码 741,4172(部分)。

　　本种为西南山区特有种,但过去常被与圆顶耳蕨(*Polystichum dielsii* Christ)和平番耳蕨(*Polystichum pinfaense* Christ)混同一起,其实本种与前二种相似而很不相同,应成为一个独立之种。

　　图注:1.本种全形(自然大);2.羽片,表示叶脉及子囊群位置(放大 4 倍);3.叶柄基部的鳞片(放大 20 倍);4.囊群盖(放大 20 倍);5—6.孢子囊及孢子(放大 100 倍)。

POLYSTICHUM EXCELLENS Ching

POLYSTICHUM EXCELLENS Ching, sp. nov.

　　Species gregis *Polystichi dielsii* Christ, lamina multo latiora, pinnis lateralibus longioribus, apice acutis, marginibus usque ad apicem leviter crenatis, sori supra costam plerumque 6-8, infra costam 4-6 insigniter diversa.

　　Rhizome short, erect, densely radicose and together with stipes densely clothed with rufo-brown narrowly ovate-lanceolate-fimbriate *scales*; *fronds* caespitose, often numerous together, stipe 7-12 cm long, pale stramineous, sparsely scaly upward, lamina 20-25 cm long, 3.2-4 cm broad, linear-lanceolate, acuminate, not narrowed towards base, pinnate; *pinnae* 20-28 pairs, petiolulate, pectinate, obliquely patent, gradually abbreviate towards apex, the basal ones not shortened, 2 cm long, 6 mm broad at the base, subfalcately rhombic-lanceolate, apex acute, mucronate, base uneqal, cut straight below, obliquely truncate above and with a low small deltoid mucronate auricle, margin undulato-crenate with a few low teeth; *texture* char-taceous, dry pale green, rachis and also trie under side sparsely scaly; *lateral veins* about 8 pairs, forked, or pinnate in the anterior base, indistinct; *sori* rnedial, 6-8 above the costa, 4-6 below, terminal on the anterior veinlet of each group, indusium gray, margin lacerate, fallen off at last.

　　Southeastern Yunnan: Pingpien Hsien, Tawei Shan, *K. M. Feng 5173* (type), *5170*, Oct. 26, 1954, on limestone cliff under forest, common; Mengtze, *Hancock 19* (pro parte). Guizhou: Tsingchen, Tsinglung Shan, *Tang Shi-wei 90397*, May 30, 1936; Ecological Survey Party of the Bot. Inst., Academia Sinica *No. 2001*, Oct. 18, 1956, alt. 1150 m.; Pin-fa, *Cavalerie 741*, 4172 (pro parte).

　　There has been a great confusion even by Christ himself as to the identity of *Polystichum Dielsii* Christ, *P. hecatopterum* Diels and *P. pinfaense* Christ. The distinction between the first two species has already been noted in the previous page, while the distinction between *P. pinfaense* Christ and our new species chiefly lies in the former having much narrower fronds (about 2 cm broad with smaller deltoid pinnae, 1 cm long, 5 mm broad at the base, margin with a few low teeth) and the sori confined mainly to the upper side of the costa, and usually none or only 1-2 below the costa.

　　Plate 234. 1. Habit sketch (natural size). 2. Pinna, showing venation and position of sori ($\times$ 4). 3. Scale from the base of stipe ($\times$ 20). 4. Indusium ($\times$ 20). 5-6. Sporangium with spores ($\times$ 100).

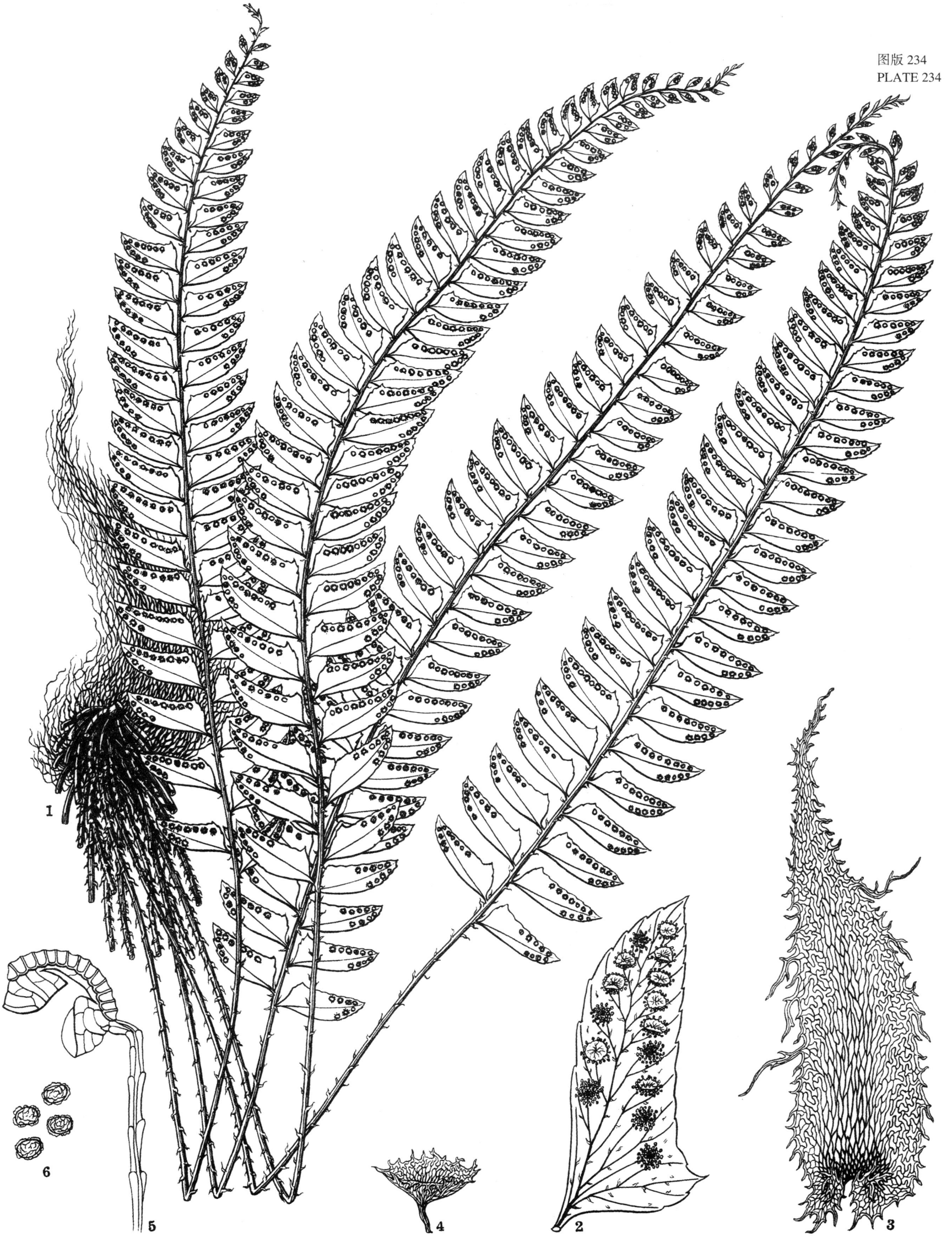

POLYSTICHUM EXCELLENS Ching
尖顶耳蕨

图版 235　倒叶耳蕨

根状茎短而直立，有密的根丛，同叶柄一起，被暗棕色阔卵形渐尖头的鳞片，边缘下半部全缘，上半部略具疏锯齿；叶簇生，叶柄长 7—9 cm，禾秆色，鳞片稀疏或上部几光滑，叶体长 10—30 cm，宽 3 cm，线披针形，渐尖头，向基部渐变狭，一回羽状；羽片 25—30 对，互生，彼此被宽缺刻分开，强度下向，向下部逐渐缩小，基部一对长宽各约 5 mm，三角形，中部羽片长 1—1.2 cm，基部宽 6—7 mm，斜方三角形，基部不等，下方切去，几成直线，基部上方为切形，并有一个小急尖耳形突起，顶端具一短刺头，上边缘有短锯齿疏生；叶为纸质，干后变淡棕色，中轴及上下两面颇光滑；叶脉下面明显，上面隐约难辨，分叉，唯基部上方一脉为羽状，基部下方一脉为单脉；子囊群小，中位，顶生于每群叶脉的上方一短小脉，限于羽片的外半部，上方 3—5 个，下方 1—3 个，有囊群盖。

分布：云南西北部，中甸县，俞德浚号码 11286，生石面，海拔 2200 m。

本种形体颇似芽胞耳蕨（*Polystichum stenophyllum* Christ），但叶端下部从不生芽胞，羽片为亚斜方形，强度下向，基部上方耳形突起不发达，彼此分离较远，顶端仅有一个短微尖头，下面光滑，不具紧贴叶面的鳞片，故易区别。

图注：1.本种全形（自然大）；2.羽片，表示叶脉及子囊群位置（放大 4 倍）；3.叶柄基部的鳞片（放大 30 倍）；4—5.孢子囊及孢子（放大 100 倍）。

POLYSTICHUM YUANUM Ching

POLYSTICHUM YUANUM Ching, sp. nov.

Species gregis *Polystichi stenophylli* Christ, differt fronde apice haud gemmifera radicanteque, pinnis subrhomboidis, valde deflexis, basi anteriore paulo auriculatis, nec pectinato-confertis sed spatio sat lato sese separatis, facie inferne subglabra et apice solomondo brevissime mucronatis.

Rhizome short, erect, densely radicose and together with stipe sparsely clothed in dark brown, broadly ovate, acuminate, denticulate *scales*; *fronds* caespitose, stipe 7-9 cm long, stramineous, sparsely scaly or subglabrous upward, lamina 20-30 cm long, 3 cm broad, linear-lanceolate, acuminate, gradually narrowed downward, pinnate; *pinnae* 25-30 pairs, alternate, separated by a broad sinus from each other, strongly de-flexed, lower ones gradually abbreviated, the lowest pair about 5 mm each way, triangular, the middle ones 1-1.2 cm long, 6-7 mm broad at the base, rhomboid-triangular, base unequal, cut straight below, truncate above with a small acute and mucronate auricle, margin crenato-serrate with short teeth; *texture* chartaceous, dry brownish, rachis and both sides quite glabrous; *veins* visible below, obscure above, forked except the anterior basal one which is pinnate, while the posterior basal one is simple; *sori* small, medial, terminal on the anterior veinlet of each group and generally confined to the outer part, 3-5 above and 1-3 below the costa, indusiate.

Northwestern Yunnan: Chungtien Hsien, Chiren, *T. T. Yü 11286*, on rock surface, alt. 2200 m., May 1st, 1937.

A very distinct species of the group of *P. deltodon* (Baker) Diels, especially related to *P. stenophyllum* Christ, from which it differs in characters diagnosed above.

Plate 235. 1. Habit sketch (natural size). 2. Pinna, showing venation and position of sori ($\times$ 4). 3. Scale from the base of stipe ($\times$ 30). 4-5. Sporangium with spores ($\times$ 100).

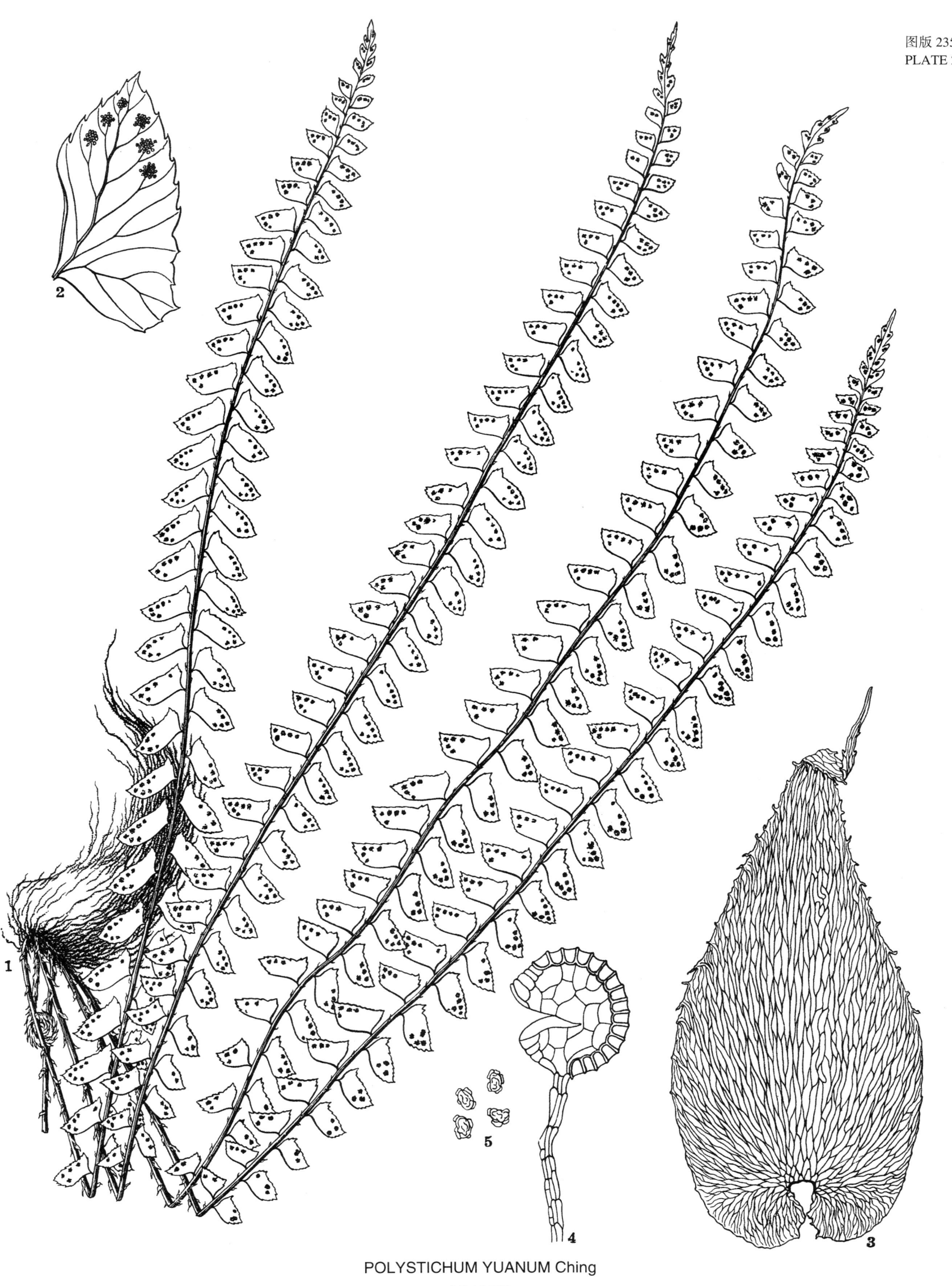

POLYSTICHUM YUANUM Ching
倒叶耳蕨

图版 236　广东耳蕨

　　根状茎短而直立,有密的根丛,同叶柄一起,有暗棕色宽披针形的鳞片疏生,边有睫毛;叶簇生,10 或更多成丛,叶柄长 10—12 cm,禾秆色,鳞片疏生,叶体长 20—30 cm,宽 2.5 cm,线披针形,渐尖头,向基部略变狭,一回羽状,羽叶 30—35 对,接近,向下反曲,长 1.2 cm,基部宽 6 mm,亚长方长圆形,下边切成直线,上基部为截形,且几不呈耳形,仅有一小尖角,近基部的羽片较小,长约 1 cm,上边缘为波状浅齿,向顶端亚钝头具一短细尖头;叶为纸质,干后淡棕绿色,两面颇光滑;叶脉分叉,但基部上方一脉为羽状,两面明显;子囊群小,通常上边有 8 个,下边在顶部以下有 2—3 个或缺如;囊群盖大而圆,久则脱落。

　　分布:广东北部,乳源县,生于西云寺旁的大石上。高锡朋号码 53790,1933 年 11 月 21 日。

　　本种为一个相当独特的蕨种,不易与其他相近之种混淆;与本种接近的一种当为云南西北部的倒叶耳蕨(*Polystichum Yuanum* Ching),但其叶柄上的鳞片较狭,羽片彼此紧接,为亚长方长圆形,顶端为亚钝头,边缘为波状浅齿,故易与云南种区别。

　　图注:1.本种全形(自然大);2.羽片,表示叶脉及子囊群(放大 4 倍);3.叶柄基部的鳞片(放大 40 倍);4—5.孢子囊及孢子(放大 100 倍)。

POLYSTICHUM KWANGTUNGENSE Ching

POLYSTICHUM KWANGTUNGENSE Ching, sp. nov.

　　Species *Polysticho Yuani* Ching affinis, differt squamis rhizomatis stipitisque angustioribus, pinnis con-fertis, subrectangulari-oblongis, marginibus crenatis, apice obtusis.

　　Rhizome short, erect, densely radicose, together with stipes sparsely clothed in dark brown, broadly lanceolate, fimbriate *scales*; *fronds* caespitose, 10 or more together, stipe 10-12 cm long, stramineous, sparsely scaly, lamina 20-30 cm long, 2.5 cm broad, linear-lanceolate, acuminate, slightly narrowed towards the base, pinnate; *pinnae* 30-35 pairs, confert, deflexed, 1.2 cm long, 6 mm broad at the base, subrectan-gular-oblong, cut straight below, truncate above and hardly auricled with an acute apex, the lowest ones smaller, about 1 cm long, margin undulato-crenate with a few short mucronate teeth towards the obtuse apex; *texture* chartaceous, dry brownish-green, quite glabrous; *veins* forked, but pinnate in the anterior basal group, visible on both sides; *sori* small, generally 8 above, 2-3 below towards the apex, indusiate.

　　Northern Guangdong: Chu-yuan Hsien, on the large rock by the Sieh Yuan Temple, *S. H. Ko* 53790, Nov. 21, 1933.

　　A very unique species not easily confounded with the related species by its subrectangular-oblong, de-flexed and crenate pinna with obtuse apex. South China is poor in the species of *Polystichum*, chiefly because of the mountains there being as a rule not high enough to provide suitable environments for the genus. The discovery of the present species there is quite interesting floristically.

　　Plate 236. 1. Habit sketch (natural size). 2. Pinna, showing venation and sori ($\times$ 4). 3. Scale from the base of stipe ($\times$ 40). 4-5. Sporangium with spores ($\times$ 100).

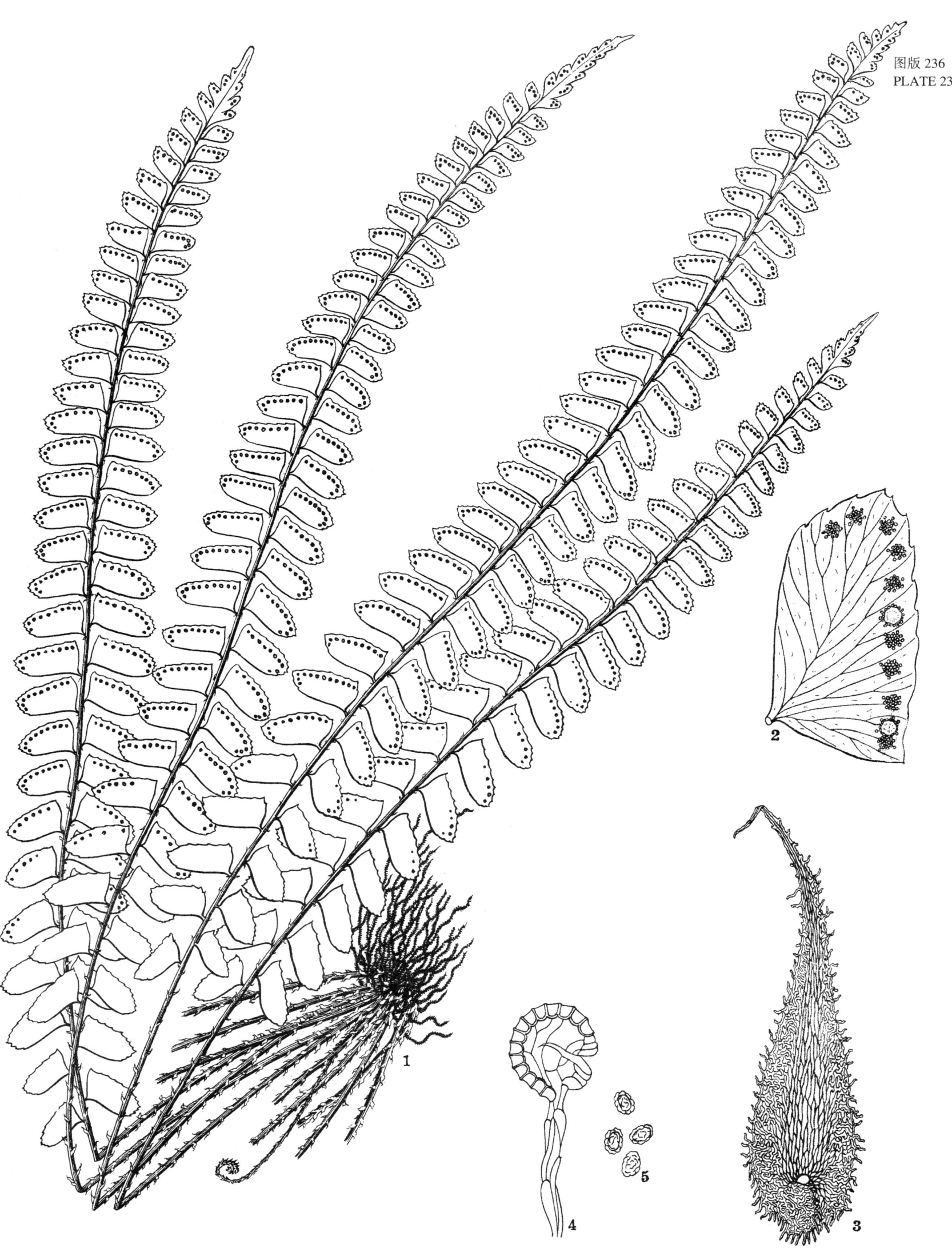

POLYSTICHUM KWANGTUNGENSE Ching
广东耳蕨

图版 237　刺叶耳蕨

根状茎短而直立,有密的根丛,同叶柄一样,起初有深棕色卵形渐尖头的鳞片疏生,后变成几乎光滑;叶多数簇生成丛,柄长 10—12 cm,灰禾秆色,叶体 20—28 cm,宽 3 cm,线状披针形,渐尖头,向基部略变狭,一回羽状;羽片 30—35 对,篦齿形的瓦覆排列,有小柄,向下反曲,长 1.5 cm,基部宽 7—8 mm,亚斜方三角形,短急尖头,或略为钝头,基部不等,下边切成直线形,上方内边截形,且略呈耳形,边缘疏生尖锯齿,有长刺头,最下的羽片略小,长不及 1 cm,下向;叶为草质,干后呈深棕色,两面光滑;叶脉不明显,分叉,基部上方一脉为羽状;子囊群小,边内生,上边约有 7—9 个,下边在顶端下 1—3 个,生于每脉的上方小脉顶端,囊群盖全缘。

分布:四川东南部,碚陵县,鸡子洞,侯学煜号码 495。

本种为一明显的种,属于对生耳蕨 P. deltodon (Bak.) Diels 组,叶柄基部以上光滑无鳞片,叶体细长,草质,羽片向下反曲,篦齿状瓦覆排列,并有疏生的长刺头锯齿,故易区别。

图注:1.本种全形(自然大);2.羽片全角,表示叶脉及子囊群(放大 4 倍);3.叶柄基部鳞片(放大 40 倍);4.囊群盖(放大 20 倍);5—6.孢子囊及孢子(放大 100 倍)。

POLYSTICHUM CONSIMILE Ching

POLYSTICHUM CONSIMILE Ching, sp. nov.

Species configuratione *Polystichi hecatopteridi* Diels proxime similis, differt stipitibus rachibusque squamis ovatis ab initio sparsis demun subevanescentibus conspersis, pinnis omnibus lateralibus deflexo-recurvatis, latioribus, basi anteriore paulo auriculatis, marginibus dentibus aristatis sparsis et brevioribus praeditis.

Rhizome short, erect, densely radicose, together with stipes at first sparsely clothed in dark-brown, ovate, acuminate *scales*, later become nearly naked; *fronds* caespitose, many together, stipe 10-12 cm long, of a dead-stramineous color, lamina 20-28 cm long, 3 cm broad, linear-lanceolate, acuminate, base slightly narrowed, pinnate; *pinnae* 30-35 pairs, pectinately imbricate, petiolulate, deflexo-recurvate, 1.5 cm long, 7-8 mm broad at the base, subrhombic-triangular, apex short-acute or bluntish, base unequal, cut straight below, truncate above and slightly auricled, margin sparsely aristate-serrate with long spreading spines, the lowest pinnae smaller, less than 1 cm long, deflexed; *texture* herbaceous, dry dark brown, glabrous on both sides; *veins* indistinct, forked, but the anterior basal one pinnate; *sori* small, intramarginal, about 7-9 on the upper side, 1-3 on the lower side below the apex, terminating anterior forking of each lateral vein, indusium entire.

Southeastern Sichuan: Peiling Hsien, Pégoupu, Gitzetung, *Hou Shou-Yu 495*.

A distinct species of the group of *P. deltodon* (Bak.) Diels with elongate glabrous fronds above the base of the stipe, of a herbaceous texture and the deflexly recurvate, pectinately imbricate pinnae with remote long spiny teeth.

Plate 237. 1. Habit sketch (natural size). 2. Pinna, showing venation and sori ($\times$ 4). 3. Scale from the base of stipe ($\times$ 40). 4. Indusium ($\times$ 20). 5-6. Sporangium with spores ($\times$ 100).

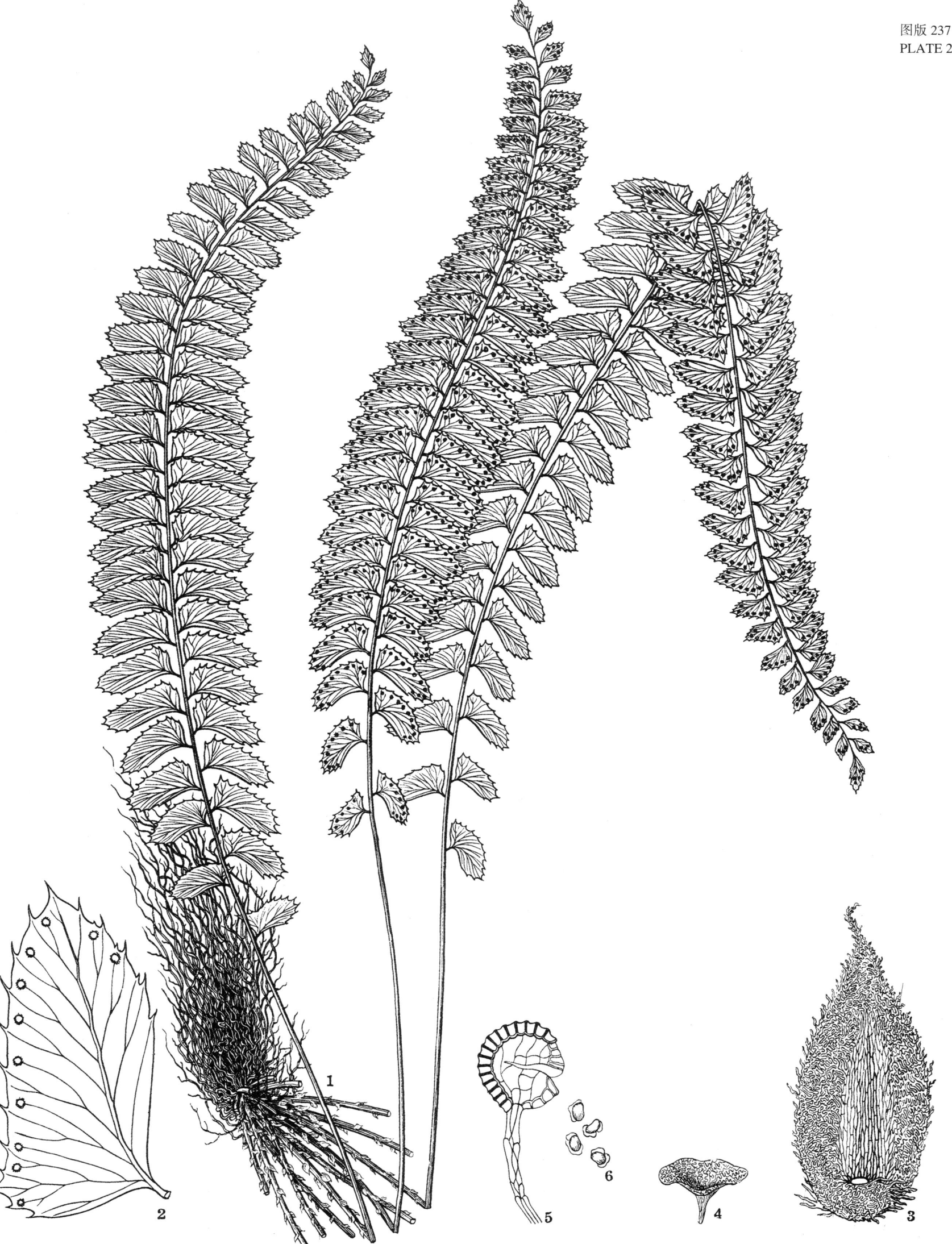

POLYSTICHUM CONSIMILE Ching
刺叶耳蕨

根状茎短而直立,有密的根丛,同叶柄一样,有红棕色宽卵形尾尖头的鳞片,边缘有睫毛;叶多数簇生,叶柄长 8—10 cm,棕禾秆色,具同样的宿存鳞片,与一些棕色披针形的钻状鳞片混生一起,叶端下约 4 cm 处有一棕色芽胞,能着地生根并发生幼株,一回羽状;羽片 40—50 对,篦齿状的瓦覆排列,有短柄,下部一些羽片向下反曲并略缩小,中部的羽片长 1.6 cm,基部宽 7 mm,三角形,有粗刺状的急尖头,基部下边为圆楔形,上方内边为截形,且具一大三角形的刺尖头耳形,叶边多少反卷,亚全缘而且有少数短硬的倒伏尖锯齿;叶为革质,干后为棕色,中轴有鳞片,下面有紧贴疏生的披针状鳞片,尤以幼时为多;叶脉紧密,分叉,但在基部上方为羽状,下面明显而凹陷,上面不明显;子囊群小,位于叶边与中肋之间,生于每组叶脉的上方一小脉,囊群盖圆形,全缘,最后脱落。

分布:四川西南部,湖北西部,云南西北部及台湾高山常见,生于海拔 2500—3300 m。日本的高山,北缅甸及印度的锡金也有分布。

本种为一极明显的种,最初发现于四川西南部宝兴县,在我国西南高山针叶森林带甚为普通,向西经北缅甸达于印度北部,在这里本种似常与 *Polystichum Atkinsoni* Bedd. 同生;台湾省和日本标本与中国内地的模式无异,仅有时较小而已。

图注:1.本种全形(自然大);2—3.二个羽片,表示叶脉,子囊群和叶脉上的鳞片(放大 4 倍);4.叶柄基部的鳞片(放大 10 倍);5—6.孢子囊及孢子(放大 100 倍)。

POLYSTICHUM STENOPHYLLUM Christ

POLYSTICHUM STENOPHYLLUM Christ in Bull. Soc. Bot. France **52**; Mém. I. 27. 1905; C. Chr. Ind. Fil. 587. 1905; in Contr. U. S. Nat. Herb. **26**: 284 t. 16. 1931.

Aspidium caespitosum Wall. var. *stenophyllum* Franch. MSS.

Polystichum levingei Hope. in herb.; Christ in Bull. Acad. Géogr. Bot. Mans. **10**: 260. 1902 (nom. nud.).

Polystichum deversum Christ in Bot. Gaz. **51**: 353. 1911; C. Chr. Ind. Fil. Suppl. I. 65. 1912; Hand.-Mzt. Symb. Sin. **6**: 26. 1929.

Polystichum pseudo-Maximowiczii Hayata, Ic. Pl. Form. **5**: 234. 1915; Ogata, Ic. Fil. Jap. **7**: t. 340. 1936.

Rhizome short, erect, densely radicose, together with the stipe densely clothed in rufo-brown broadly ovate, caudate and fimbriate *scales*; *fronds* caespitose, stipe 8-10 cm long, rufo-stramineous, clothed in similar and persistent scales intermixed with some lanceolate-subulate ones of a brown color, lamina 30-35 cm long, 2-3 cm broad, linear-lanceolate, gemmiferous and often radicant at about 4 cm below the apex, pinnate; *pinnae* 40-50 pairs, pectinately imbricate, subsessile, lower ones deflexed and slightly abbreviated, the middle ones 1.6 cm long, 7 mm broad at the base, triangular, aristately acute, base roundly cuneate below, and truncate above with a large deltoid aculeate-aristate auricle, margin repand, subentire with a few low stout decumbent teeth; *texture* coriaceous, dry brown, rachis scaly, under side with sparse, lanceolate appressed scales especially in the young state; *veins* confert, forked, but pinnate in the auricle, distinctly impressed below, obscure above; *sori* small, medial between the margin and costa, terminating the anterior veinlet of each group, indusium round, entire, fallen off at last.

Sichuan: Moupin, *David* (type), 1870; *E. H. Wilson 5404*; *W. P. Fang 6303, 1577*, May 27, 1930; *Scallan*; *Faber*. Hubei: Fang Hsien, *E. H. Wilson 2625* (type of P. *deversum* Christ). Yunnan: Hokin, Maeulshan, *Delavay*, May 1, 1890; Mekong, *H. Handel-Mazzetti 9754*; Pintscheoan, *Ducloux 156, 5414*; Weisih, Sihlung Shan, *K. M. Feng 4811*, alt. 2600-2800 m., on rocks under forest; *C. W. Wang 46037, 68698*, alt. 3600 m.; Chungtien, Chiao Tou, *K. M. Feng 3158*, alt. 2700 m.; Atuntze, *K. M. Feng 5497*, common, alt. 2900-3300 m., July 16, 1940; Tachiming Valley, *J. F. Rock 8629*, on shaded bank in bamboo forest. Tibet: Yatung, *H. E. Hobson* (1899). Taiwan: Mt. Arisan, common at 2500 m. altitude.

Also Japan and Sikkim: *J. D. Hooker* (1857) pro parte; Phuloot, *H. Levinge*, Oct. 15, 1880.

A very distinct and common fern in the wooded mountains in Southwestern China, westwardly to Upper Burma and Sikkim, where the species seems often to grow together with *Polystichum Atkinsoni* Bedd. The plant from Taiwan as figured by Ogata is typical of the species. The Japanese plant proves essentially the same, except smaller, the same form being also seen from West China.

Plate 238. 1. Habit sketch (natural size). 2-3. pinna, showing venation, sori and scales on veins (× 4). 4. Scale from the base of stipe (× 10). 5-6. Sporangium with spores (× 100).

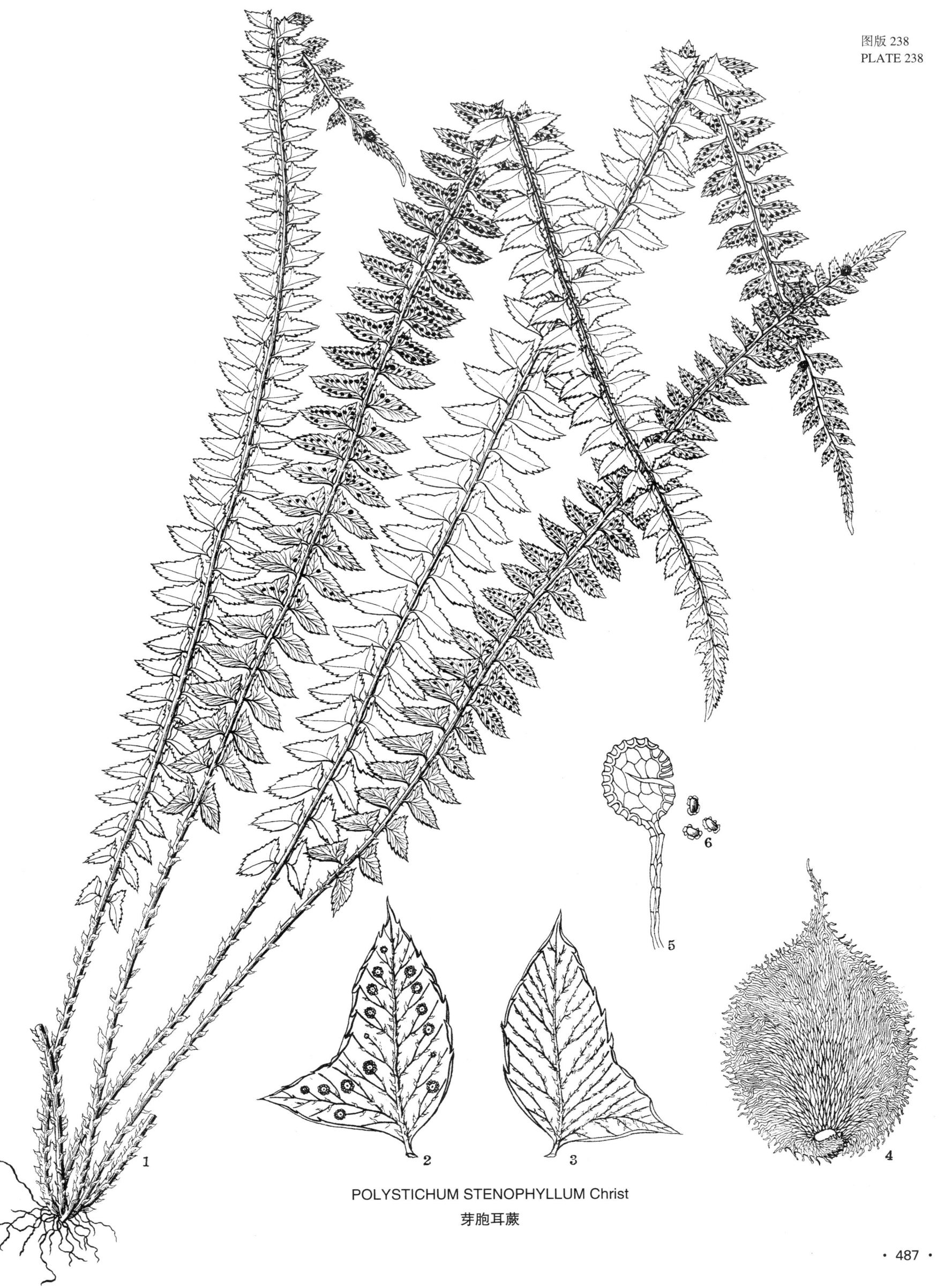

POLYSTICHUM STENOPHYLLUM Christ
芽胞耳蕨

图版 239 粗齿耳蕨

根状茎短而直立,有密的根丛,顶端有暗棕色卵状长渐尖的鳞片,具啮蚀状的边缘;叶多数簇生,柄长 8—14 cm,灰禾秆色,仅有少数同样的鳞片疏生,其余的光滑,叶体长 25—30 cm,中部宽 2.5 cm,基部宽 3 cm,线状披针形,向尾状顶部渐狭,向基部不变狭,一回羽状;羽片约为 40 对,有短柄,开展,近生,而向下部被宽缺刻分开,基部几对较大,长 1.5 cm,基部宽 1 cm,三角形,急尖而具短尖头,基部下方切成直线,上方为截形,上边缘及顶端以下有不规则的粗大三角形齿牙,具短尖头,中部的羽片长约 1.2 cm,由此向上逐渐缩小;叶为纸质,干后为淡棕绿色,中轴及两面均为光滑;叶脉下面明显,上面隐约不清,在基部上方为羽状,在大齿牙内为三叉,向上端小齿牙为二叉或单脉;子囊群大,边内生,接近,有盖。

分布:四川东南部,酉阳县,铜鼓潭,接近贵州边界,侯学煜号码 825。

本种为对生耳蕨 Polystichum deltodon (Bak.) Diels 组的另一个特出的种,有三角形的羽片,具粗大齿牙和向尾状叶端逐渐变狭的叶体。本种的形体颇似 Polystichum yaeyamense Makino 但远较大,而且叶端为渐狭尾状。

图注:1.本种全形(自然大);2.基部的羽片,表示叶脉及子囊群位置(放大 4 倍);3.中部的羽片(放大 4 倍);4.叶柄基部的鳞片(放大 40 倍);5—6.子囊群及孢子(放大 100 倍)。

POLYSTICHUM GROSSIDENTATUM Ching

POLYSTICHUM GROSSIDENTATUM Ching, sp. nov.

Species distinctissima ab omnibus affinibus adhuc cognitae in China differt frondis apicem versus grada-tim in cauda lineari longe attenuatis, pinnis lateralibus trigonis, marginibus superioribus grosse et irregulari-ter dentatis.

Rhizome short, erect, densely radicose, crowned at the apex with dense, dark brown, ovate, long-acuminate *scales* having erosed margin; *fronds* caespitose, many together, stipe 8-14 cm long, gray-stramineous, subglabrous with a few similar scales above the base, otherwise the entire plant glabrous, lamina 25-30 cm long, 2.5 cm broad at the middle, 3 cm broad at the base, linear-lanceolate, gradually long-attenuate towards the caudate apex, base not narrowed, pinnate; *pinnae* about 40 pairs, petiolulate, patent, the lower ones separated by broad sinuses, the basal pair slightly larger than those above, 1.5 cm long, 1 cm broad at the base, triangular, mucronately. acute, base cut straight below, truncate above, the upper and the outer margins coarsely and irregularly dentate with mucronate apex, the middle pinnae smaller, about 1.2 cm long, and thence gradually diminish upward; *texture* chartaceous, dry brownish green, rachis and both sides glabrous; *veins* distinctly raised below, obscure above, pinnate in the anterior base, triforked in the larger teeth, forked or simple in the upper smaller teeth; *sori* large, intramarginal, contiguous, indusiate.

Southeastern Sichuan, Youyang Hsien, Tungkootant, on the border of Guizhou, *Hou Shou Yu 825*.

This is one of the most distinct species of the group of *P. deltodon* (Baker) Diels by possessing triangular pinnae with coarsely and irregularly dentate margin and the gradually long-attenuate caudate apex of the fronds. In general habit, our new species resembles *Polystichum yaeyamense* Makino (see Ogata, Ic. Fil. Jap. **7**: t. 343. 1936), but differs in much larger size and caudate fronds.

Plate 239. 1. Habit sketch (natural size). 2. Basal pinna, showing venation and position of sori ($\times$ 4). 3. Middle pinna ($\times$ 4). 4. Scale from the base of stipe ($\times$ 40). 5-6. Sporangium with spores ($\times$ 100).

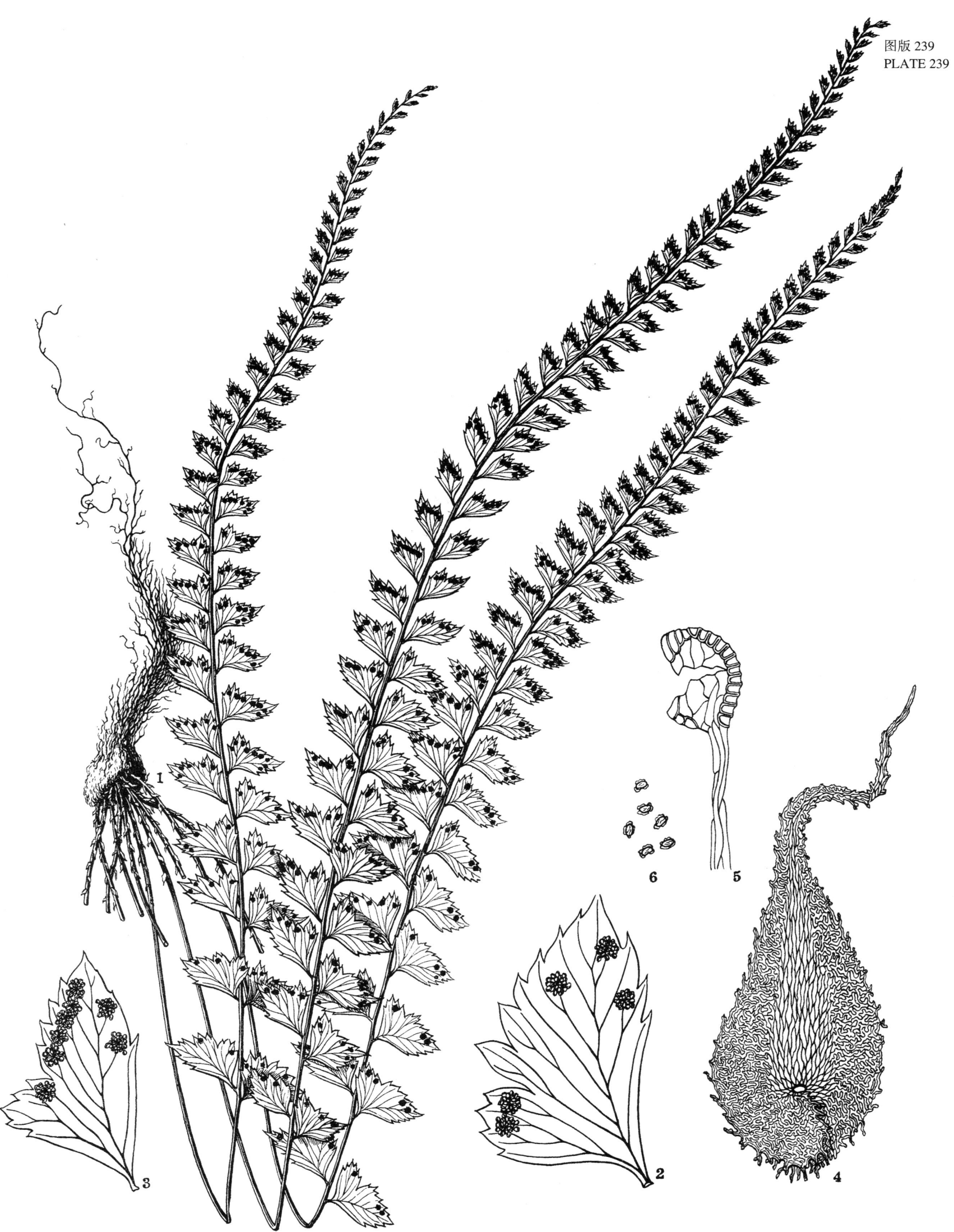

POLYSTICHUM GROSSIDENTATUM Ching
粗齿耳蕨

根状茎短而直立,有密的根丛,顶端有深棕色小卵形渐尖头的鳞片密生;叶簇生,开展,或生在石灰岩上常成莲座状叶丛,并以根系深入到石缝中,柄短,长 1—3 cm,棕禾秆色,有疏鳞片,叶体一般长 5—8 cm,宽 8—12 mm,线状披针形,渐尖头或往往为亚钝头,向基部为不显著的变狭,一回羽状;羽片 7—20 对,篦齿状或者紧接,有小柄,开展,亚四方形至斜方形,长 3—5 mm,宽 2—4 mm,下边全缘,切成直线,上边及外边有几个较大的三角的锯齿具张开的长硬刺头,有刺痛之感,基部上方不呈耳形;叶为亚革质,两面光滑,有光泽,尤以上面为显著,干后棕色或淡棕绿色;叶脉少数而稀疏,两面明显,单脉,仅基部上方的 1—2 脉为分叉,达于锯齿;子囊群小,少数,中肋上边 2—3 个,上中位,下边缺如,囊群盖大,圆形,全缘,不久脱落。

分布:湖北西部,四川西南部(峨眉山),贵州及湖南西部;广西可能也有。生石灰岩上,海拔 900—1600 m。

本种是对生耳蕨 Polystichum deltodon (Baker) Diels 组的最小的一种,为西南石灰岩地区的特有旱生种,形体变异极大,因此,过去有几个不同的名词,如 Polystichum nanum Christ,是本种生于干旱生境下的极端生态小型的代表。

图注:1. 本种全形(自然大);2. 羽片,表示叶脉及子囊群位置(放大 10 倍);3—4. 叶柄基部的鳞片(放大 50 倍);5—6. 孢子囊及孢子(放大 100 倍)。

POLYSTICHUM LANCEOLATUM (Baker) Diels

POLYSTICHUM LANCEOLATUM (Baker) Diels in Engl. Jahrb. **29**: 193. 1900; C. Chr. Ind. Fil. 583. 1905.

Aspidium lanceolatum Baker in Gard. Chron. n. s. **14**: 494. 1880.

Polystichum parvulum Christ in Bull. Acad. Géogr. Bot. Mans **14**: 114. cum fig. 1904.

Polystichum nanum Christ in Bull. Acad. Géogr. Bot. Mans **16**: 238. 1906.

Rhizome short, erect, very densely radicose, densely clothed at the apex with dark brown, ovate acuminate, small *scales*; *fronds* tufted, numerous together, often forming a rosette on dry rocks with roots penetrating deep into the crevices, stipe short, 1-3 cm long, rufo-stramineous, sparsely scaly, lamina generally 5-8 cm long, 8-12 mm broad, linear-lanceolate, acuminate or often obtusish at apex, hardly narrowed towards the base, pinnate; *pinnae* 7-20 pairs, pectinate or confert, petiolulate, patent, subquadrangular to rhombic, 3-5 mm long, 2-4 mm broad, entire and straight on the lower side, the upper and outer margins provided with a few relatively large, deltoid, spreading long-spiny and pungent teeth, anterior base hardly auricled; *texture* subcoriaceous, both sides glabrous and lustrous especially on the upper side, dry brown or brownish-green; *veins* few, simple except the anterior basal 1 or 2 which are forked, quite distinct on both sides; *sori* small, few, 2-3 on the upperside of the costa, supra-medial, indusium large, round, entire, soon falling off.

Western Hubei: Ichang Gorge, *Maries* (type); Kochen Hsien, *H. C. Chow 4067*, Oct. 20, 1935. Sichuan: Mt. Omei, on the way to Chouloutung from Hanchenping, *R. C. Ching 168*, *170*, on dry limestone cliff, alt. 900-1600 m.; *S. C. Chen 30045*; *E. H. Wilson 5384*; Pietanngtze, *S. C. Chen 30094*. Guizhou: Tsingay, *Cavalerie 1234* (type of *P. parvulum* Christ); Ganchow, *Bodinier 312*, *3797*, *4152*; Pinfa, *Cavalerie 1945* (type of *P. nanum* Christ), Nov. 2; Yunfanshan, Tuyun, *Y. Tsiang 5953*, on moist limestone cliff, July 20, 1930. Hunan: Yungsai Hsien, Ecological Survey Party of Bot. Inst., Academia Sinica No. 586, on limestone cliff, Aug. 18, 1953.

A very distinct small xerophytic fern endemic in the limestone region. *P. parvulum* Christ proves the same, while *P. nanum* Christ represents a still smaller form with fronds to 4 cm long, 9 mm broad, consisting of only 6-7 pairs of subimbricating pinnae, 4 mm long, 3.5 mm broad with a few spreading long spiny teeth.

One of the smallest species of the genus, not easily confounded with any other species except the less divided form of *P. Martinii* Christ, which differs, however, in the larger size and at least lobato-incised pinna with free anterior auricle. It often grows in the crevices of limestone cliff with little moisture from the rain, where the plant becomes stunted and spreads out in a rosette.

Plate 240. 1. Habit sketch (natural size). 2. Pinna, showing venation and sori (× 10). 3-4. Scales from the base of stipe (× 50). 5-6. Sporangium with spores (× 100).

POLYSTICHUM LANCEOLATUM (Baker) Diels
亮叶耳蕨

根状茎短而直立,有根丛;叶簇生,柄长 3—5 cm,深禾秆色,鳞片甚多,尤以基部为甚,红棕色,卵状长圆形,渐尖头,边为疏流苏状,向上部变为线状披针形或为纤维状,叶体为线状披针形,长 7—20 cm,宽 2—3 cm,向上逐渐变狭,顶端为线状尾形,向基部通常稍变狭,深二回羽裂或者二回羽状;羽片多数,开展,均匀排列,长 1—1.5 cm,基部宽 5—6 mm,亚斜方三角形,急尖头,基部不等或在下部羽片为亚不等,下边为楔形,上方内边为斜切,并有大而特出的锐裂耳片,羽状半裂到沿中肋两侧的狭翅,成为 4—5 对斜出的全缘或锐裂的线披针形并向内弯的渐尖裂片;叶为草质或坚草质,干后为淡黄绿色;中轴及中肋下面有纤维状鳞片疏生;叶脉在耳片为羽状,在其他裂片为分叉;子囊群小,在中肋两侧有 3—5 个,在耳片 2—3 个,中生,位于每脉的上方短小脉之顶,囊群盖大,膜质,边有圆波状齿,颇宿存。

分布:本种为亚高山种,广布于云南西北部及四川西南部(峨眉山)山地,海拔 1500—2800 m,生针叶林下石缝中。在喜马拉雅山区 2000—2800 m 处也极为常见。

本种为一秀丽喜阴的蕨种,其特点为线状披针形叶体,向上逐渐变狭,顶端有一长尾头,叶为草质,故易与相近之种区别。

图注:1. 本种全形(自然大);2. 羽片,表示叶脉及子囊群位置(放大 4 倍);3. 叶柄基部的鳞片(放大 20 倍);4—5. 孢子囊及孢子(放大 100 倍)。

POLYSTICHUM THOMSONI (Hooker fil.) Beddome

POLYSTICHUM THOMSONI (Hooker fil.) Beddome, Ferns Brit. Ind. t. 126. 1866; Handb. Ferns Brit. Ind. 206. 1883; Diels in Engl. u. Prantl, Nat. Pflanzenfam. **1**: iv. 191. 1899; C. Chr. Ind. Fil. 588. 1905; Christ in Bot. Gaz. **51**: 348. 1911; Hand.-Mzt. Symb. Sin. **6**: 28. 1929.

Aspidium thomsoni Hook. fil. in Hooker, 2nd. Cent. Ferns t. 25. 1860 (pro parte); Hook. & Bak. Syn. Fil. 251. 1864 (pro parte); Clarke, Ferns N. Ind. in Trans. Linn. Soc. II. Bot. **1**: 508. 1880.

Dryopteris thomsoni O. Ktze. Rev. Gen. Pl. **2**: 813. 1891.

Rhizome short, erect, radicose; *fronds* fasciculate, stipe 3-5 cm long, dark stramineous, copiously scaly especially near the base, *scales* rufo-brown, ovate-oblong, acuminate, sparsely fimbriate, becoming linear-lanceolate or fibrillose upward, lamina linear-lanceolate, 7-12 cm long, 2-3 cm broad, gradually long-attenuate and ended in a long linear cauda towards apex, not or usually only slightly narrowed towards the base, deeply bipinnatifid or subbipinnate; *pinnae* numerous, patent, regularly spaced, 1-1.5 cm long, 5-6 mm broad at the base, subrhombic-triangular, acute, base unequal or subequal in the lower pinnae, cuneate below, the upper side oblique with enlarged incised auricle, pinnatifid down to a narrow wing along the costa into 4-5 pairs of oblique, entire or cleft, lanceolate-acuminate, incurved segments; *texture* herbaceous or firmly so, rachis copiously and costa underneath sparsely fibrillose, dry brownish-green; *veins* pinnate in the auricle, forked in segments; *sori* small, 3-5 on each side of costa, 2-3 in the auricle, medial, terminal on short anterior veinlets, indusium large, membranaceous, crenate, quite persistent.

Yunnan: N. E. of Atuntze, *G. Forrest 20171*, Sept. 1921; Yungpeh mountains, *G. Forrest 15319*; Weisih, Mekong-Salwin Divide, *K. M. Feng 4475* on shaded rocks, alt. 2100 m.; Likiang, Yulung Shan, *H. Handel-Mazzetti 3587*; Youngpie Hsien, Leetzeping, *R. C. Ching 22422*; Chenkang Hsien, *C. W. Wang 72422* in rock crevices, alt. 2800 m.; Hokin, *R. C. Ching 23654* on shaded rocks. Sichuan: Mt. Omei, Sihshiangchu, *S. C. Chen 30114*, on stone wall, alt. 1500 m.; *E. H. Wilson 2607* (pro parte); *Faber* (pro parte).

N. India: The Himalayas, from Balti to Sikkim, alt. 2000—2800 m. very common.

A quite elegant and shade-loving fern, characterized by the long tail-like apex of the linear-lanceolate fronds and soft herbaceous texture.

Plate 241. 1. Habit sketch (natural size). 2. Pinna, showing venation and sori (× 4). 3. Scale from the base of stipe (× 20). 4-5. Sporangium with spores (× 100).

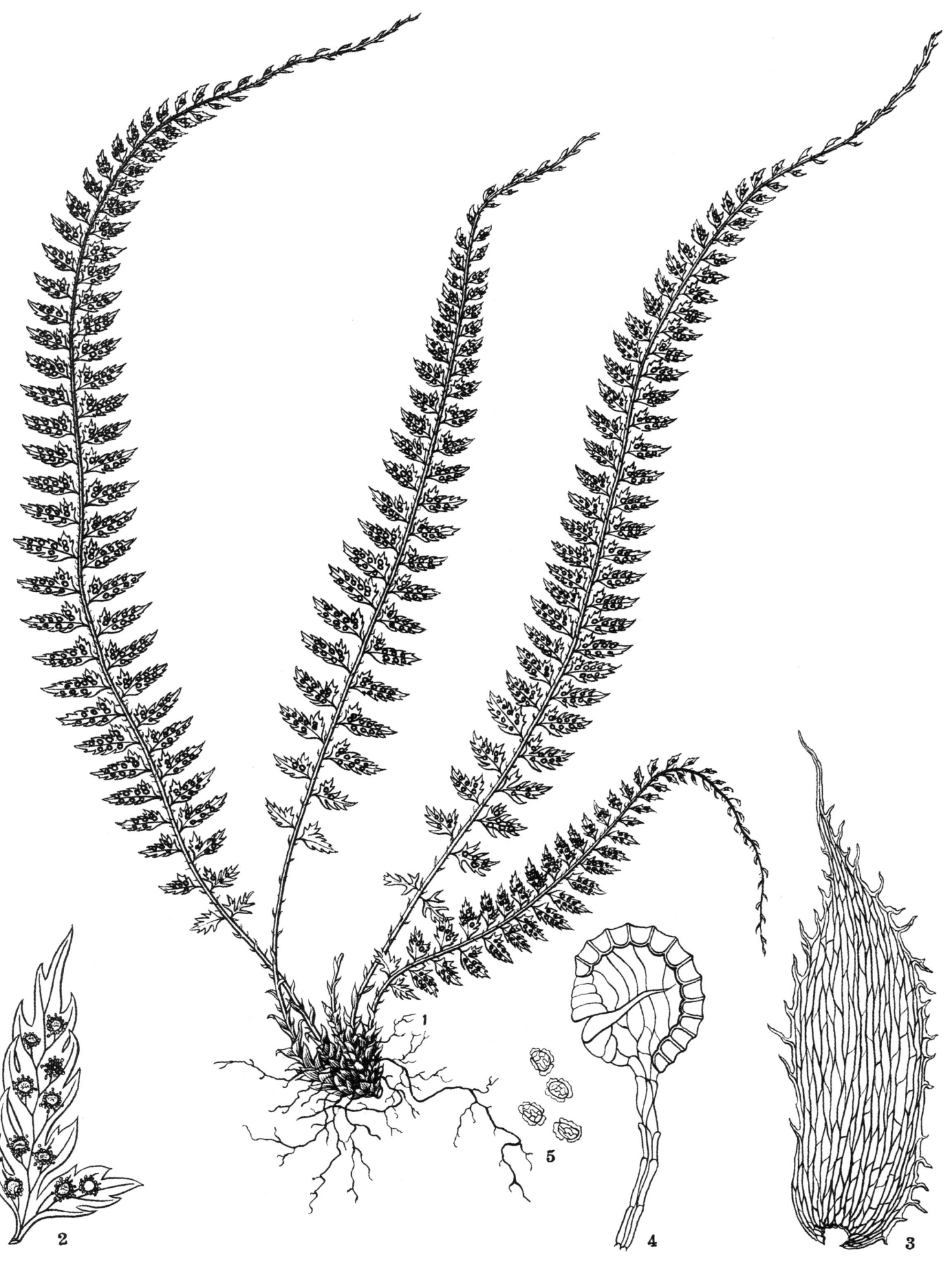

POLYSTICHUM THOMSONI (Hooker fil.) Beddome
尾叶耳蕨

图版 242　钳形耳蕨

根状茎短而直立，有密的根丛；叶簇生，柄长 13—15 cm，棕禾秆色，颇光滑，基部有卵状披针形渐尖头的鳞片，叶体长 22—30 cm，宽 4 cm，阔披针形，渐尖头，向基部不变狭，一回羽状；羽片 30—40 对，接近，开展，长 2 cm，基部以上宽 6 mm，亚镰刀形，急尖或亚急尖，有短尖头，基部极不等，下边为楔形，上方内边为截形，并且大而突出的三角有锐锯齿的耳片，上边为有规则地片状锐裂到 1/3—1/2 深处，成为 6—7 个通常有不等二裂的短尖头裂片，下边有 4—5 个较长的裂片，其中最下部两片为全缘，裂片均向内弯成镰刀形；叶为草质，干后为绿色；每羽片的叶脉 7—8 对，基部上方一条为羽状，其他为分叉，顶部及基部下方为单脉，两面明显；子囊群小，中生，每裂片有一个，但下部 4—5 片及上方耳片为不育，位于每脉的上方短小脉顶端，囊群盖大而圆，全缘，不久脱落。

分布：云南东南部，屏边县，大围山，蔡希陶号码 62760。生林下石上，海拔 1500 米。

本种为云南东南部特产，形体颇似贵州产的马氏耳蕨 (*Polystichum martinii* Christ)，但较大，叶柄几无鳞片，羽片较长较狭，镰刀形，基部上方有较大的耳片，裂片大都二裂成钳形，故易区别。

图注：1.本种全形(自然大)；2.羽片，表示叶脉及子囊群位置(放大 4 倍)；3.叶柄基部的鳞片(放大 30 倍)；4—5.孢子囊及孢子(放大 100 倍)。

POLYSTICHUM BIFIDUM Ching

POLYSTICHUM BIFIDUM Ching, sp. nov.

Species gregis *Polystichi martinii* Christ, differt majore, stipitibus squamis fere carentibus, pinnis lateralibus longioribus, augustioribusque, falcatis, basi anteriore auriculis majoribus, deltoideis rectis praeditis et segmentis pinnarum utraque costae latere plerumque bifidis.

Rhizome short, erect, densely radicose; *fronds* caespitose, stipe 13-15 cm long, stramineous-brown, quite glabrous except the base which is clothed with ovate-lanceolate, acuminate *scales*, lamina 22-30 cm long, 4 cm broad, broadly lanceolate, acuminate, not narrowed towards base, pinnate; *pinnae* 30-40 pairs, close, patent, 2 cm long, 6 mm broad above the base, subfalcate, apex acute, or acutish, mucronate, base very unequal, cuneate below, truncate above and provided with a prominent, deltoid, sharply serrate auricle regularly lobato-incised 1/3-1/2 down into 6-7 usually unequally bifid and mucronate segments on the upper side, 4-5 longer segments beneath, of which the lowest 2 entire and like the others incurved; *texture* herbaceous, dry green, glabrous; *veins* 7-8-jugate, the anterior basal one pinnate, the others forked, the uppermost and the posterior basal two simple, distinct below; *sori* small, medial, one to each segment except the lower 4-5 below, terminal on the anterior veinlet of each group, indusium large, round, entire, falling off soon.

Southeastern Yunnan: Pingpien Hsien, *H. T. Tsai 62760*, July 18, 1934, on rocks in forest, alt. 1500 m.

Plate 242. 1. Habit sketch (natural size). 2. Pinna, showing venation and sori ($\times$ 4). 3. Scale from the base of stipe ($\times$ 30). 4-5. Sporangium with spores ($\times$ 100).

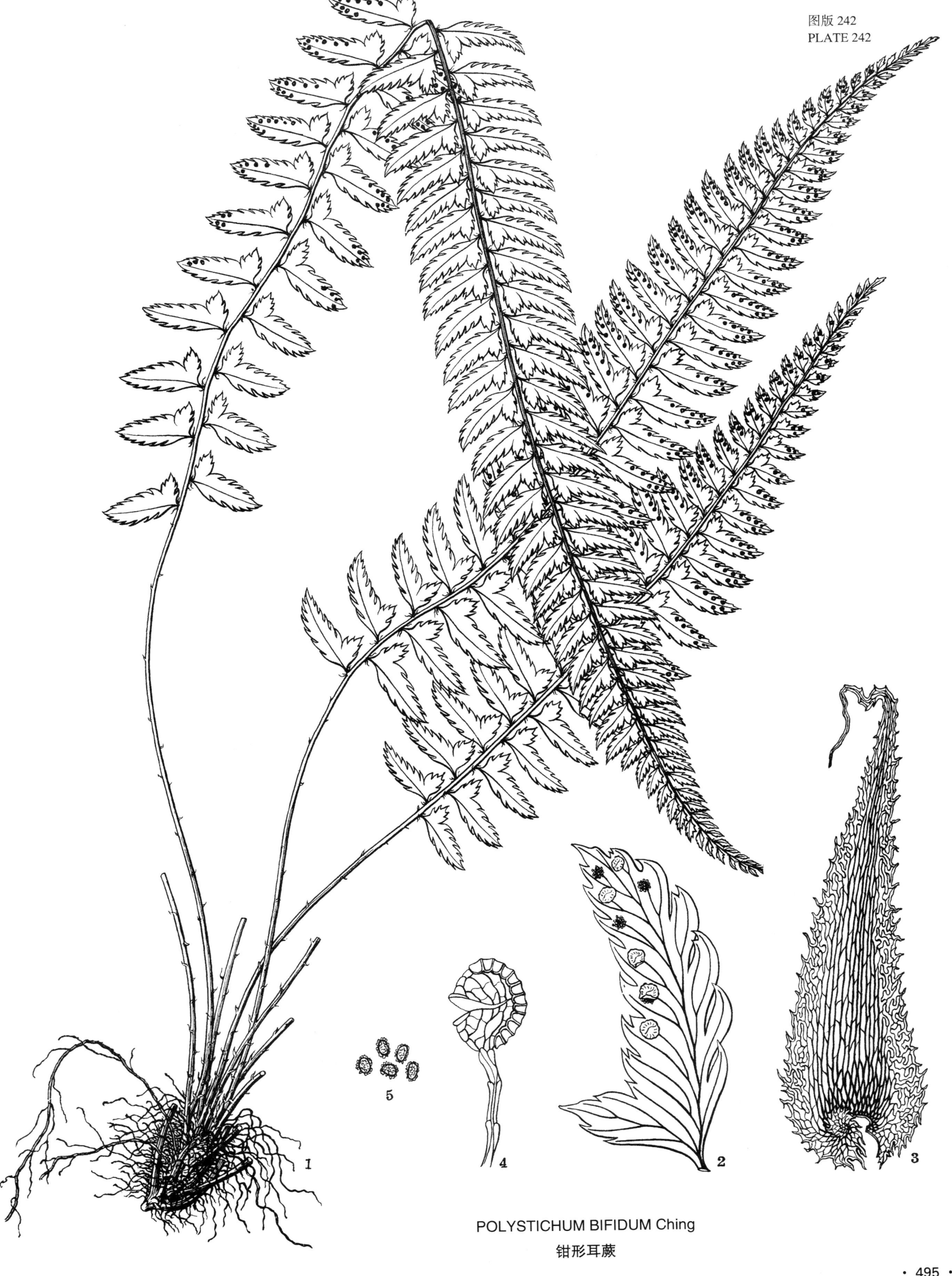

POLYSTICHUM BIFIDUM Ching
钳形耳蕨

图版 243　软骨耳蕨

根状茎短而直立，有密的根丛；叶多数簇生，柄长 12—15 cm，棕禾秆色，近基部为灰禾秆色，密被大而暗棕色或淡黑色有光泽的阔卵形渐尖头厚鳞片，饰以较淡的边缘，并混有红棕色线状钻形的鳞片，叶体长 25—30 cm，宽 5—6 cm，阔披针形，渐尖头，向基部略变狭，一回羽状；羽片约为 25 对，密接，有短柄，开展，长 2.5—3 cm，近基部宽 8 mm，亚镰刀形，短渐尖头或急尖头，基部不等，下方楔形，上方为明显的宽阔的急尖耳形，边缘有双重的软骨质急尖锯齿，无刺痛的感觉，基部的羽片较小，长不及 2 cm，不反曲向下，而为水平的开展；叶为革质，干后为棕色，中轴有稀疏的暗棕色阔卵形的鳞片，与一些线形棕色鳞片混生，上面光滑，下面有伏贴在叶体上的细鳞片，基部圆形，向上变成线状尖头；叶脉成羽状，以锐角由中肋斜出，上面为明显的隆起，下面常为凹陷；子囊群大形，两边各一列，接近中肋，在耳片常为二列，但在叶体为二回羽裂的变型(f. subbipinnata)，子囊群在中肋两边常为不规则的两列，生于每群叶脉的上方一小脉的背部，囊群盖大，膜质，深灰色，边有圆锯齿，颇能宿存。

分布：本种广布于云南西部及西北部，海拔 2700—3000 m 米的亚高山及四川西南部（峨眉山），海拔 1500 m，西藏东南部海拔 3000 m，台湾阿里山，海拔 2300—2600 m。印度，缅甸北部（海拔 2300—2660 m）及菲律宾普洛格山(Mt. Pulog)，海拔 2700 m 也有分布。

在叶的分裂度上，本种有时表现为二回羽裂(f. subbipinnata)。

图注：1. 本种全形（自然大）；2. 羽片，表示叶脉及子囊群位置（放大 3 倍）；3. 叶柄基部的鳞片（放大 20 倍）；4—5. 孢子囊及孢子（放大 100 倍）；6. 二回羽裂变型的羽片（放大 3 倍）。

POLYSTICHUM NEPALENSE (Sprengel) C. Christensen

POLYSTICHUM NEPALENSE (Sprengel) C. Christensen, Ind. Fil. 84. 1905. 585. 1906; in Contr. U. S. Nat. Herb. **26**: 284. 1931.

Aspidium nepalense Sprengel Syst. Veget. **4**: 97. 1827.

Aspidium marginatum Wall. List n. 366. 1828 (nom. nud.); Mett. Famgatt. Pheg. u. Asp. 39, n. 88. 1858; Hope in Journ. Bomb. Nat. Hist. Soc. **14**: 459 t. 17. 1902.

Polystichum marginatum Schott, Gen. Fil. ad t. 9. 1834.

Aspidium auriculatum Presl var. *marginatum* Hook. et Bak. Syn. Fil. 251. 1876.

Polystichum auriculatum var. *marginatum* Bedd. Ferns Brit. Ind. Suppl. t. 363. 1876; Handb. Ferns Brit. Ind. 204. 1883.

Aspidium auriculatum Don, Prodr. Fl. Nep. 3. 1825 (non Sw.).

Aspidium manmeiense Christ in Bull. Herb. Boiss. **6**: 765. 1898.

Polystichum atroviridissimum Hayata, Ic. Pl. Form. **4**: 190, f. 128. 1914.

Polystichum falcatipinnum Hayata, Ic. Pl. Form. **4**: 192 f. 130. 1914.

Rhizome short, thick, erect, densely radicose; *fronds* caespitose, many together, stipe 12-15 cm long, rufo-stramineous, or dead-stramineous near the base, densely clothed especially near the base in large, dark brown or blackish, shining broadly ovate-acuminate, thick *scales* often with lighter-colored margin, intermixed with rufo-brown linear-subulate scales; lamina 25-30 cm long, 5-6 cm broad, broadly lanceolate, acuminate, slightly narrowed towards the base, pinnate; *pinnae* about 25 pairs, close, patent, short-petiolate, 2.5-3 cm long, 8 mm broad near the base, subfalcate, short-acuminate or acute, base cuneate below, the upper with a distinct broad acute auricle, margin double-serrate with dense, white-callose, acute teeth which are not pungent to feeling, the basal pinnae smaller, less than 2 cm long, acute, not deflexed but horizontally patent; *texture* coriaceous, dry brown, rachis rather sparsely clothed in rufo-brown, large, ovate scales mixed with some brown linear ones, glabrous above, the under side copiously dotted on the veins with minute brown appressed scales with nearly orbicular base and long-cuspidate apex; *veins* in pinnate groups, slowly branching out from the costa under a sharp angle, distinctly raised above, visible and often impressed below; *sori* large in two costal rows and often an additional short row in the auricle (but in the *bipinnatifid* form, the sori often irregularly two-rowed on each side of costa), dorsal on the anterior veinlet of each group, indusium large, membranaceous, dark gray, margin crenate, quite persistent.

In China numerous specimens have been seen from Yunnan, S. W. Sichuan (Mt. Omei), Tibet (Yatung) and the Island Taiwan (Mt. Arisan) at elevations from 1500-3000 m. Also Northern India and the Philippines (Mt. Pulog, alt. 2750 m.).

A close relative of the Eurasian-American *Polystichum lonchitis* (L.) Roth and was considered as identical to each other by earlier botanists.

The degree of pinnation in a number of simply pinnate species of *Polystichum* of the group of *Lobatum* often varies considerably, as is examplified by the present and the following species, with which the present fern looks quite alike but differs in the large, broadly ovate dark brown shining scales on stipe mixed with brown fibrils and linear-subulate ones, in the biserrate, dense and white-callose teeth along the margin of the pinnae, the under side of which is copiously dotted with appressed orbicular-cuspidate minute scales and in the large costal sori with large brown membranaceous indusium.

Plate 243. 1. Habit sketch (natural). 2. Pinna, showing venation and sori (× 3). 3. Scale from the base of stipe (× 20). 4-5. Sporangium with spores (× 100). 6. Pinna of f. *subbipinnata* (C. Chr.) Ching (× 3).

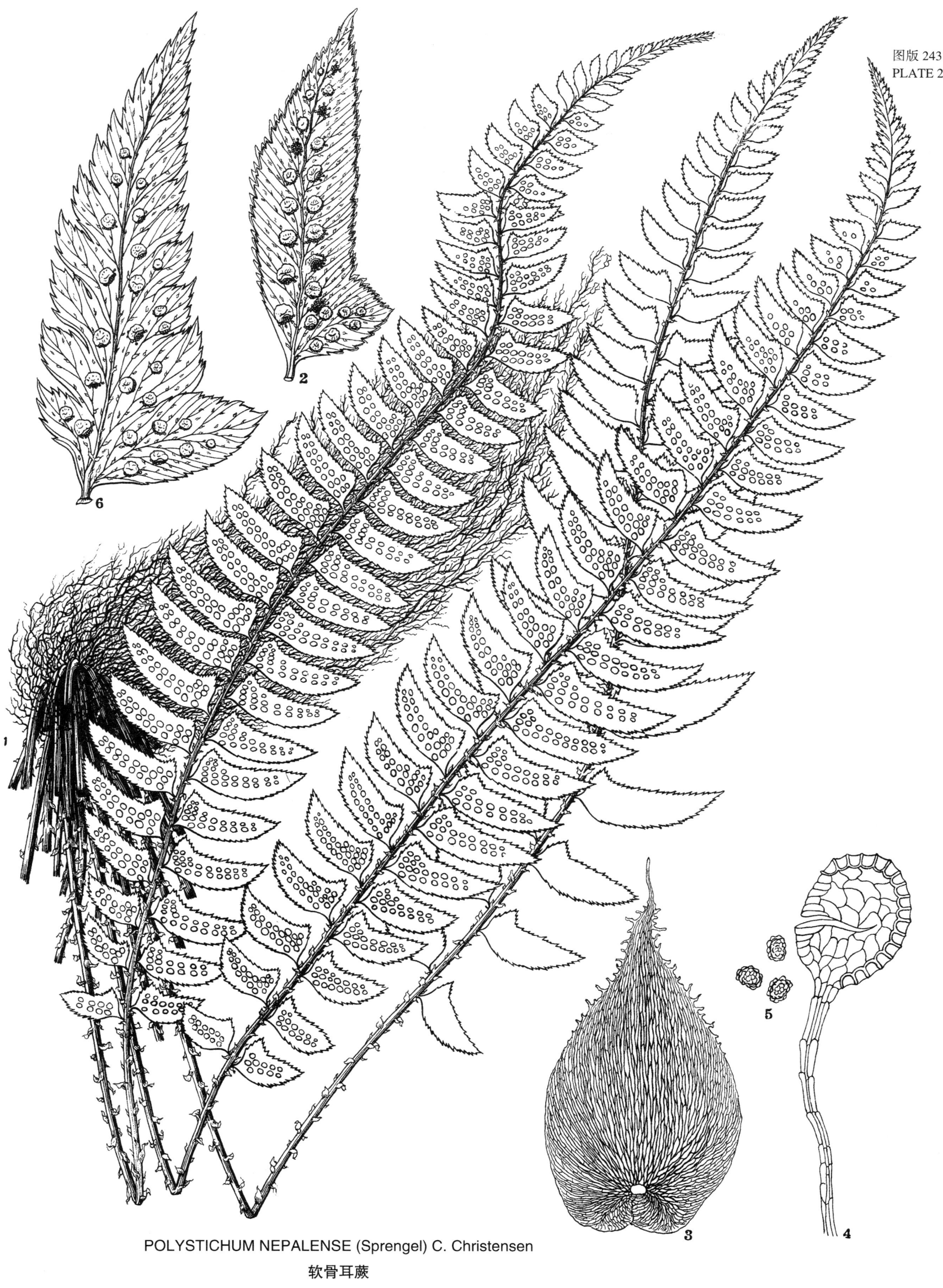

POLYSTICHUM NEPALENSE (Sprengel) C. Christensen
软骨耳蕨

根状茎短而直立,有宿存的老叶柄基部瓦覆盖着;叶簇生,柄长 20—30 cm,暗禾秆色,有密生的深棕色宽披针形厚鳞片,并夹入浅黑线状钻形的鳞片,饰以干膜质的边缘,脱落后留下由宿存的细疣状基部所形成粗糙叶柄,叶体长 30 cm 或更长一些,宽 10—15 cm,长圆披针形,渐尖头,基部不变狭,一回羽状;羽片约为 30 对,有短柄,开展,互生,下部的羽片为同宽的缺口所分开,长 7.5—10 cm,宽 1 cm,亚镰刀状披针形,渐尖头,基部不等,下方斜切,上方有一分离的或合生的卵形急尖头的耳片,边缘具有规则的锯齿或为波形浅裂片状,具一硬尖头;叶为革质,颇有光泽,中肋有黑色钻状鳞片密生,上下两面光滑,干后为淡棕色;叶脉不显明,羽状;子囊群在中肋两边各为一列,中生,但常为不规则的两列,囊群盖大,质硬,灰色,早落。

分布:四川西南部(峨眉山),东南部(南川,金佛山),贵州中部(贵阳)及云南北部。

本种为我国西南部的特有种,极近于大型的 *Polystichum otophorum* (Franch.),Bedd.,但除形体远较高大外,羽片为宽缺刻分开,缓渐尖头,而更特别的是边缘只有短尖头的锯齿,故易区别。在羽片的分裂度上,有 (1) 不裂的羽片变型 f. **integra**,(2) 半裂的羽片变型 f. **subbipinnatifida** 及 (3) 一回羽状的羽片变型 f. **bipinnata** 等三种变型,这是耳蕨属的许多种的叶之分裂度不稳定的又一证明。在鉴定种类时,应特别注意此点。

图注:1. 本种全形(自然大);2. 羽片(模式型)表示叶脉及子囊群位置(放大 2 倍);3. 叶柄基部的鳞片(放大 20 倍);4—5. 孢子囊及孢子(放大 100 倍);6. 羽片(全缘型),表示叶脉及子囊群位置(自然大);7. 羽片(二回羽叶型),表示叶脉及子囊群位置(自然大)。

POLYSTICHUM XIPHOPHYLLUM (Baker) Diels

POLYSTICHUM XIPHOPHYLLUM (Baker) Diels in Engl. u. Prantl, Nat. Pflanzenfam. **1**: iv. 189. 1899; C. Chr. Ind. Fil. 589. 1905.

Aspidium xiphophyllum Baker in Journ. Bot. **1888**: 227.

Polystichum praelongatum Christ in Bull. Acad. Géogr. Bot. Mans **12**: 260. 1902; C. Chr. Ind. Fil. 586. 1905.

Rhizome short, thick, erect, imbricately covered with the persistent bases of old stipes; *fronds* fascicled, several together, stipe 20-30 cm long, dark stramineous, densely clothed in dark brown, firm, broadly lanceolate *scales* intermixed with blackish linear-subulate ones with scarious margin and, upon falling, leaving the stipe roughened by the persistent minute warty bases, lamina 30 cm or longer, 10-15 cm broad, oblong-lanceolate, acuminate, not narrowed towards the base, pinnate; *pinnae* about 20 pairs, short-petiolate, spread, alternate, separated below by sinuses of the same width as the pinna, 7.5-10 cm long, 1 cm broad, subfalcate-lanceolate, acuminate, base unequal, obliquely cut away below, with a free, or adnate ovate acute auricle on the upper side, margin varies from regularly serrate, or lobulato-crenate with stout cuspidate teeth even to pinnate; *texture* coriaceous, rather lustrous, rachis densely clothed in black subulate scales, both sides glabrous, dry brownish; *veins* immersed, indistinct, pinnate; *sori* uniseriate, midway between the costa and margin, often irregularly biseriate, indusium large, firm, gray, deciduous.

Sichuan: Mt. Omei, *Faber 1040* (type); *Brown 137*; *S. C. Chen 30044*, *10068*, alt. 1300, August 28, 1953; *R. C. Ching 174*; *T. C. Peng 27*; *Z. C. Chen 322*; Nanchuan, *W. P. Feng 5725*. Guizhou: Pinfa, *Cavalerie 328*; Kweiyang, *Cavalerie 2641*. Yunnan: Hongpa, *Ducloux 5091*.

An endemic species in West China. In general habit and degree of pinnation of leaves, it most closely resembles the large form of *P. otophorum* (Franch.) Bedd., from which, however, it can be easily distinguished, besides the generally larger size, by the pinna being more widely separated by sinuses, the gradually acuminate apex and, above all, by the shortly cuspidate teeth along the margin. *P. otophorum* Franch. has, as a rule, fronds of much smaller size, shorter and contiguous pinna with short-acuminate apex and, above all, longer spiny teeth along the margin. As to the degree of pinnation of leaves, the two species vary in almost the same manner.

From ample material on hand, four closely related forms have been observed: 1, forma **integra**, having entire pinnae with regularly and finely cuspidate-serrate margin; 2, forma **typica**, having pinnae with a free ovate acute auricle at the anterior base; 3, forma **bipinnatifida**, having, besides a free anterior pinnule, the pinnae deeply pinnatifid, as represented by *Polystichum praelongatum* Christ, based on *Cavalerie No. 2641* from Guizhou and, lastly, forma **bipinnata**, a fully bipinnate form of larger size with many free pinnules, of which the anterior basal one much the largest, to 1.3 cm long, erect and appressed against the rachis and its anterior base may have an additional adnate lobule.

Plate 244. 1. Habit sketch of f. *typica* (natural size). 2. Pinna of f. *typica* (× 2). 3. Scale from the base of stipe (× 20). 4-5. Sporangium with spores (× 100). 6. Pinna of f. *integra* (natural size). 7. Pinna of f. *bipinnata* (natural size).

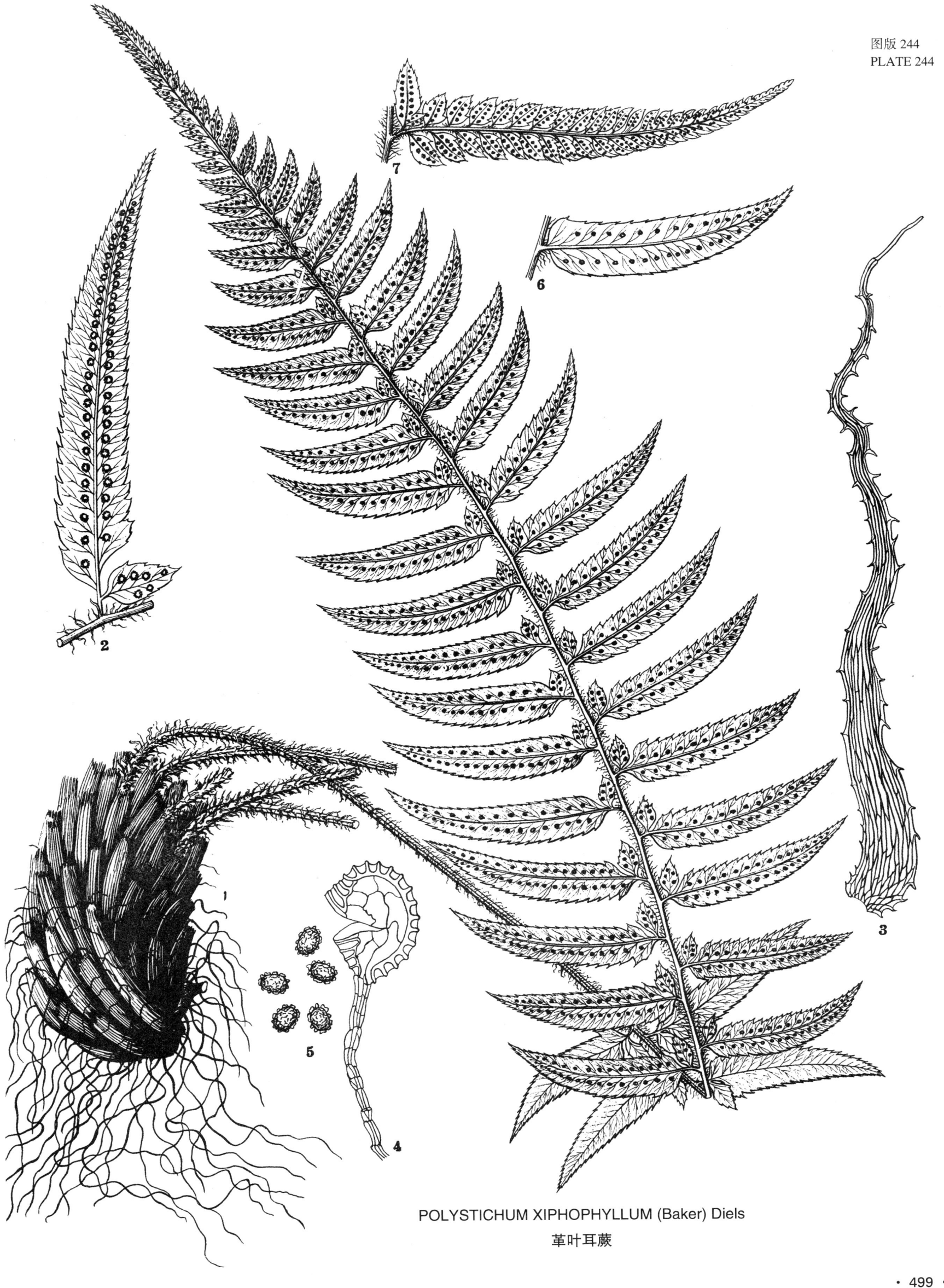

POLYSTICHUM XIPHOPHYLLUM (Baker) Diels
革叶耳蕨

根状茎短而直立，通常粗大而有老的叶柄基部覆盖着；叶簇生，柄长 12—25 cm，禾秆色，质颇软弱，基部有密生的，上部有疏生的深棕色宽披针形的鳞片，叶体长 22—25 cm，基部宽 5—7 cm，一回羽状，而基部一对羽片大大伸长张开，并且有如上部一样的羽状；基部一对以上的羽片 20—40 对，开展，长 2.5—3 cm，宽 9—12 mm，均匀排列，披针形，渐尖头，基部较阔，不等形，下方斜切，上方截形，并为耳形，耳片为卵形，急尖，有锯齿，上边缘为片状锐裂或为大锯齿，下边缘有几个倒伏的大锯齿；基部一对羽片伸长到 6—10 cm，宽不及 1 cm，有 6—10 对同形的而较小的小羽片，边缘有同样的锯齿；叶为草质，干后为淡绿色，中轴及叶之下面有淡棕色卵形薄鳞片；叶脉在耳片为羽状，其余为分叉，明显；子囊群小，一列，位于中肋与边缘中间，生于每脉的上方的一小脉之顶端，囊群盖大，膜质，有圆锯齿，宿存。

分布：本种在中国分布很广，从东北林区向南经河北，山东，河南，湖南，湖北，四川，贵州，达广西，广东，江西，浙江，江苏北部（海州）及安徽南部高山林下。日本及朝鲜半岛也产。

本种为耳蕨属一个最特出的种，由于它的基部一对羽片如两臂式的特别伸长张开并为羽状，它在形态上恰恰处于一回羽状叶的对生耳蕨 *P. deltodon* (Baker) Diels 和二回羽状的 *P. setiferum* (Forsk.) Woyar 之间，实为联结两组的桥梁，也为一回羽状种过渡到二回羽状种的中间形式。

同本种极相近的一种，或者可视为本种的一个地理型的是台湾耳蕨 *Polystichum hancockii* (Hance) Diels，它的形体较细瘦，基部一对羽片只比上部的稍为向两侧伸长，具少数几对较小的小羽片，或者有时很少伸长。早田藏文氏曾根据基部一对羽叶的不同长度，曾给以几个不同的名称，实是多余的事。

图注：1. 本种全形（自然大）；2. 羽片，表示叶脉及子囊群位置（放大 2 倍）；3. 叶柄基部的鳞片（放大 20 倍）；4. 囊群盖（放大 20 倍）；5—6. 孢子囊及孢子（放大 100 倍）。

POLYSTICHUM TRIPTERON (Kunze) Presl

POLYSTICHUM TRIPTERON (Kunze) Presl, Epim. Bot. 58. 1849; Diels in Engl. u. Prantl, Nat. Pflanzenfam. **1**: iv. 191. 1899; C. Chr. Ind. Fil. 589; Hand.-Mzt. Symb. Sin. **6**: 27. 1929; Ogata, Ic. Fil. Jap. **7**: t. 342. 1936.

Aspidium tripteron Kunze in Bot. Zeit. **1848**: 569; Mett. Farngatt. Pheg. u. Asp. n. 118. 1858; Hook., 2nd. Cent. Ferns t. 56. 1861; Hook. & Bak. Syn. Fil. 254. 1874; Christ, Farnkr. d. Erde 237. 1897.

Dryopteris triptera O. Ktze. Rev. Gen. Fil. **2**: 814. 1891.

Ptilopteris triptera Hayata in Bot. Mag. Tokio **41**: 706. 1927.

Aspidium tripteris Eaton in Perry, Narr. Exp. to China **2**: 330. 1856.

Rhizome short, erect, usually thick and covered with bases of old stipes; *fronds* caespitose, stipe 12-25 cm long, stramineous, rather soft, densely near the base and sparsely upward clothed in dark brown broadly lanceolate scales, lamina 22-35 cm long, 5-7 cm broad above the base, simply pinnate with the basal pair of pinnae always greatly lengthened out to 10 cm or more long and again similarly pinnate as the upper part; *pinnae* in the central main branch 20-40 pairs, patent, 2.5-3.5 cm long, 9-12 mm broad, regularly spaced, lanceolate and acuminate from the broader unequal base, which is cut straight or oblique below, truncate and auricled above, the auricle ovate, acute, serrate, the upper margin lobato-incised or coarsely serrate, while the lower margin with a few large linear, decumbent teeth; the *basal pair of pinnae* consists of 6-10 pairs of smaller pinnules of the same shape as the pinnae in the central part, 6-20 mm long, less than 1 cm broad, margin similarly serrate; *texture* herbaceous, dry light green, rachis and under side moderately clothed in light brown, ovate, thin scales; *veins* pinnate in the anterior side of the base, forked upward and below the costa, distinct; *sori* small, terminal on the anterior basal veinlet of each group, uniseriate between the costa and margin, indusium large, membranaceous, crenate and persistent.

A common fern in China, numerous specimens have been seen from Northeast China, Hebei, Shandong, Henan, Hunan, Sichuan, Guizhou, Guangxi, Guangdong, Jiangxi, Zhejiang, Northern Jiangsu and Southern Anhui Provinces.

Also known from Japan and Korea peninsula.

This is a peculiarly distinct species because of the lengthening out of the basal pair of pinnate pinnae, and, in the degree of pinnation, it stands exactly midway between the groups of simply pinnate *P. deltodon* (Bak.) Diels and of bipinnate *P. setiferum* (Forsk.) Woyar.

A closely related species, or perhaps be better considered as a geographic race is *Polystichum Hancockii* (Hance) Diels from Taiwan and also Guangdong, which is characterized by much slender habit with smaller pinnae, of which the basal pair is very short with only a few pairs of small pinnules. From Taiwan, forms with basal pair of pinnae remains practically the same as the upper ones, or only slightly lengthened out have also been observed and given different names by Hayata.

Plate 246. 1. Habit sketch (natural size). 2. Pinna, showing venation and sori ($\times$ 2). 3. Scale from the base of stipe ($\times$ 20). 4. Indusium ($\times$ 20). 5-6. Sporangium with spores ($\times$ 100).

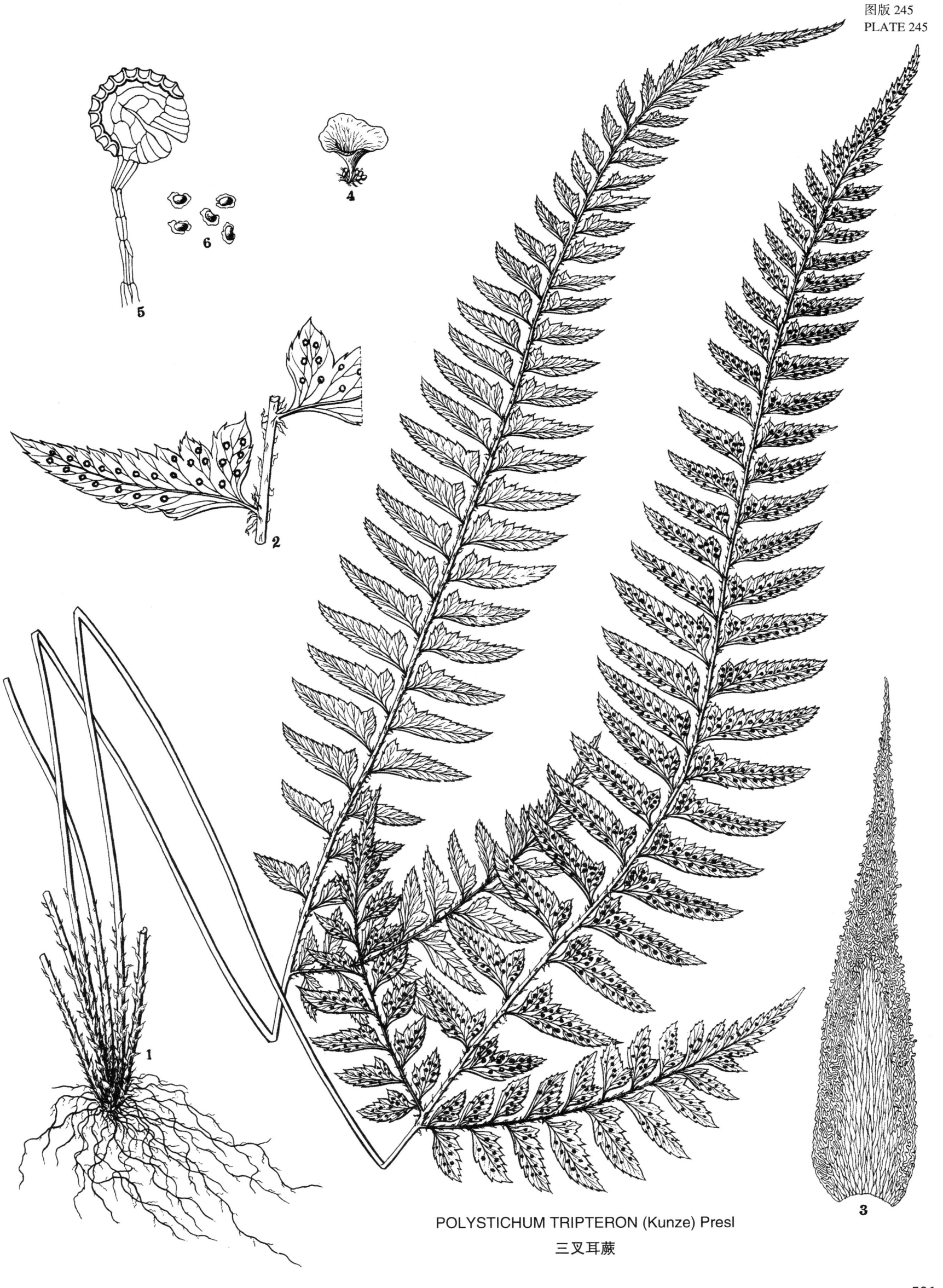

POLYSTICHUM TRIPTERON (Kunze) Presl 三叉耳蕨

根状茎短粗而直立，有老叶柄的基部瓦覆；叶簇生，柄长 15—25 cm，较为细长，深禾秆色，有大而深棕色的披针形鳞片，混以淡黑色的钻状鳞片，叶体长 20—30 cm，宽 10—15 cm，卵状长圆形，渐尖头，基部不变狭，二回羽状，顶部为一回羽状；羽片大约 15—20 对，彼此接近，亚镰刀形，渐尖头，长 5—9 cm，有柄，基部极不等，下方斜切，上方膨大并为截形，一回羽状，顶部为片状羽裂；小羽片接近或亚瓦覆状排列，椭圆长圆形，急尖头，有短尖头，全缘或通常有细尖头锯齿，基部上方的一个小羽片特大，靠近中轴，它的基部上方也为尖耳形，边缘有短尖头的锯齿或常为片状羽裂；基部一对羽片通常和它上面一对等长，亚水平开展；叶为亚革质，干后为淡棕色或淡绿色，中轴及中肋的下面有淡黑色或煤黑色钻状坚质鳞片密生，上面光滑，下面幼时有纤维状鳞片，后变为颇光滑；子囊群小，每小羽片有 2—4—6 对，中生，生于每群叶脉的上方一小脉的顶端，囊群盖圆形，灰色，质坚，全缘，最后脱落。

这是耳蕨属的一个普通的蕨种，长江以南各省包括云南，四川，贵州，湖北，湖南，江西，福建，浙江，安徽南部常见，向北分布至秦岭的太白山。日本，朝鲜半岛和越南北部也产。

本种的形体大小变异很大，模式标本采自对马岛，形体较小，但显然和中国产的形体较大的植物同属一种。中井猛之进氏（日本植物学杂志 39 卷 117 页）曾把此种和原产日本的 Polystichum polyblepharum Presl 混同起来，认为是同物异名，这是错误的，因为两种截然不同。

图注：1. 本种全形（自然大）；2. 羽片的一部，有小羽片着生，表示叶脉及子囊群位置（放大 4 倍）；3. 叶柄基部的鳞片（放大 30 倍）；4—5. 中轴上的鳞片（放大 10 倍）；6. 囊群盖（放大 20 倍）；7—8. 孢子囊及孢子（放大 100 倍）。

POLYSTICHUM TSUS-SIMENSE (Hooker) J. Smith

POLYSTICHUM TSUS-SIMENSE (Hooker) J. Smith, Hist. Fil. 219. 1875; Diels in Engl. u. Prantl, Nat. Pflanzenfam. **1**: iv. 191. 1899; Hand.-Mzt. Symb. Sin. **6**: 27. 1929; Ogata, Ic. Fil. Jap. **2**: t. 98. 1929; C. Chr. Ind. Fil. Supp. III. 165. 1933.

Aspidium tsus-simense Hooker, Sp. Fil. **4**: 16 t. 220. 1862; Miq. Prol. Fl. Jap. 340. 1867; Fr. et Sav. Enum. Pl. Jap. **2**: 231. 1876.

Polystichum lobatum var. *Tsus-simense* C. Chr. Ind. Fil. 583. 1905.

Aspidium monotis Christ in Bull. Bot. Soc. Ital. **1901**: 294.

Polystichum monotis C. Chr. Ind. Fil. 584. 1905.

Rhizome short, thick, erect, imbricately covered with the bases of old stipes; *fronds* caespitose, stipe 15-25 cm long, rather slender, dark stramineous, clothed with large, dark brown, lanceolate scales mixed with subulate blackish ones, lamina 20-30 cm long, 10-15 cm broad, ovate-oblong, gradually acuminate, base not narrowed, bipinnate below the simply pinnate apex; *pinnae* about 15-20 pairs, close, obliquely patent, sub-falcate, acuminate, 5-9 cm long, petiolate, base strongly unequal, oblique below, acroscopically produced and truncate above, pinnate under lobato-pinnatifid apex; *pinnules* contiguous or subimbricate, elliptic-oblong, acute, mucronate, entire or often with minute mucronate serrature, the anterior basal one much the largest, erect, close to the rachis, anterior base sharply auricled, margin cuspidato-serrate or often again lobato-pinnatifid, the basal pair of pinna usually as long as those next above, subhorizontally patent; *texture* subcoriaceous, dry brownish or greenish, rachis and the lower part of costa of pinnae beneath copiously clothed in firm, blackish, or coal-black subulate scales, glabrous above, under side fibrillose when young, and becoming quite glabrous at last; *veins* pinnate or forked in the pinnnule, indistinct on both sides; *sori* small, 2-4-6 pairs to each pinnule, medial, terminal on the anterior basal veinlet of each group, indusium orbicular, gray, firm, entire, falling off at last.

Another common fern in China, to the south of Yangtze; numerous specimens have been seen from Yunnan, Sichuan, Guizhou, Hubei, Hunan, Jiangxi, Fujian, Zhejiang, and southern part of Anhui, northwardly to the southern part of Shaanxi (Taipei Shan).

Also Japan, Korea peninsula and Vietnam (Chapa).

The type of the species, from Tsu Shima, represents rather a small form, while most of the specimens seen from China are generally much larger with the anterior basal pinnule of each pinna much larger and pinnatifid above the anterior basal auricle.

Nakai (Bot. Mag. Tokio **39**: 117) wrongly reduced the present species as a synonym of the Japanese *Polystickum polyblepharum* Presl, because he thought the two species are identical. I saw types of both species and found that they are very different, as was also pointed out by H. Handel-Mazzetti (Symb. Sin. **6**: 27). On the other hand, I am strongly inclined to consider the present species as being identical with the South African **Polystichum luctuosum** (*Aspidium luctuosum* Kze.) Ching, comb. nov., as was first proposed by Hope. (Journ. Bomb. Nat. Hist. Soc. **3**: 475).

Plate 246. 1. Habit sketch (natural size). 2. A part of pinna with pinnules, showing venation and sori ($\times$ 4). 3. Scale from the base of stipe ($\times$ 30). 4-5. Scales from the rachis ($\times$ 10). 6. Indusium ($\times$ 20). 7-8. Sporangium with spores ($\times$ 100).

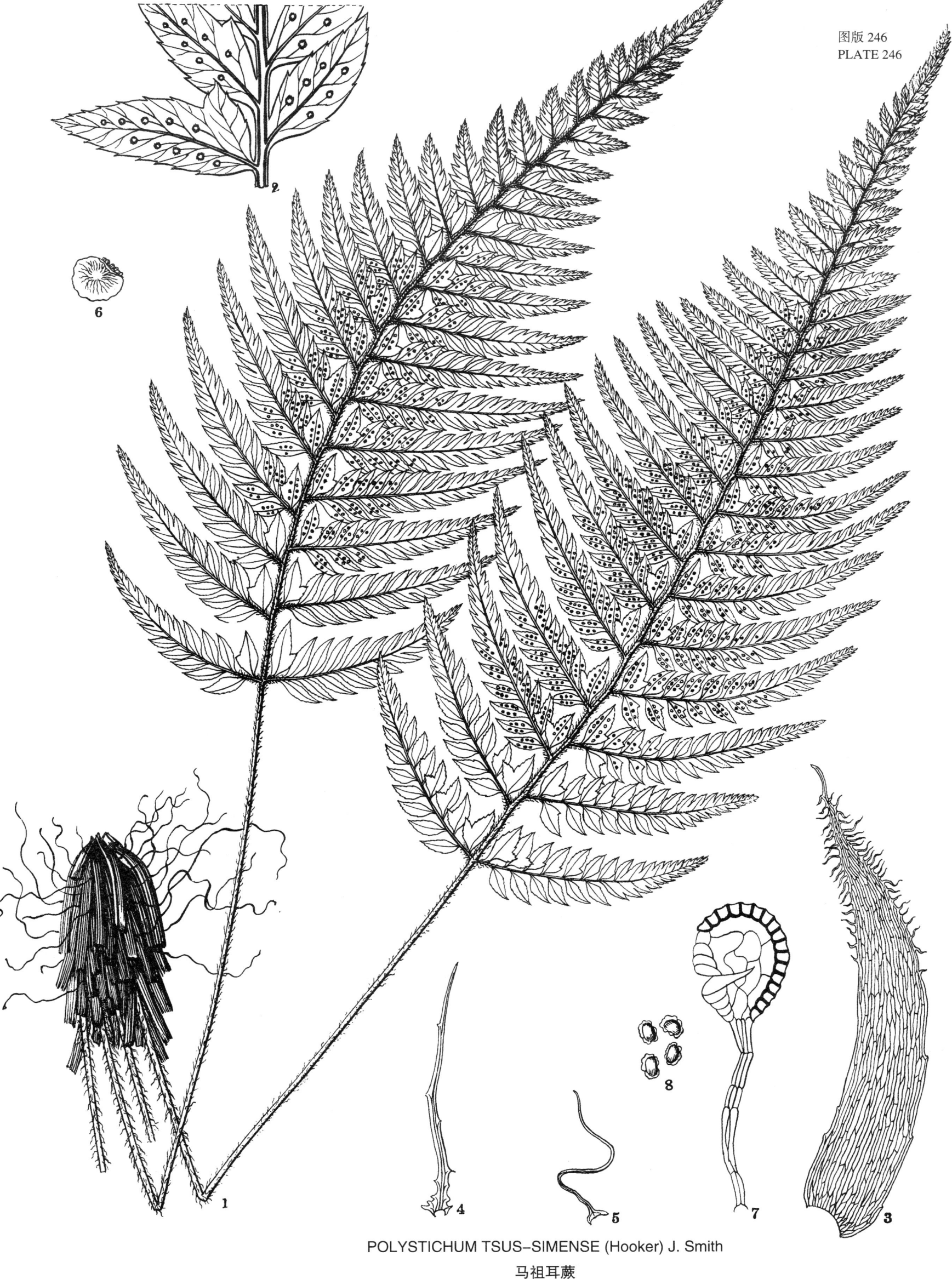

POLYSTICHUM TSUS-SIMENSE (Hooker) J. Smith
马祖耳蕨

根状茎短而直立,顶端密覆棕色卵形渐尖的大鳞片,边具睫毛;叶簇生,直立或张开,柄细长,长 8—14 cm,有同样和张开的鳞片;能育叶为线状披针形,长渐尖头,向基部几不变狭,长 15—20 cm,宽 2.5—3.5 cm,下部羽裂几达中轴,向上部羽裂较浅;裂片很多,接近,开展,下部的为长圆卵形,急尖头,全缘,长 1.2—1.4 cm,宽 7 mm,上方不为耳形突起,叶体顶部渐狭,全缘;亚革质,干后为棕色,上面光滑,下面连叶轴被有鳞片;叶脉不明显,羽状,每脉为二叉或少为三叉,小脉斜出,向上方都达于边缘;子囊群生中肋两边,各为一列,中央生,无囊群盖;不育叶的体形同于能育叶,通常羽裂较浅,有时为单叶,下部叶缘为波状,中肋顶端延长成一条无叶的鞭状匍匐茎,顶端向下着地,生出幼株。

分布:广东北江特有种。

这个独特的蕨种到现在为止,只见于广东北部,从现有材料看,是一个稀有的蕨种。

图注:1.本种全形(自然大);2.叶体的一部,表示叶脉及子囊群(放大 3 倍);3.叶柄基部的鳞片(放大 40 倍);4.叶下面的鳞片(放大 40 倍);5—6.孢子囊及孢子(放大 100 倍)。

CYRTOMIDICTYUM BASIPINNATUM (Baker) Ching

CYRTOMIDICTYUM BASIPINNATUM (Baker) Ching in Acta Phytotax. Sinica **6**: 262 t. 51. 1957.

Aspidium basipinnatum Baker in Journ. Bot. **1889**: 176; Christ in Warburg, Mons. 77. 1900.

Polystichum basipinnatum Diels in Engl. u. Prantl, Nat. Pflanzenfam. **1**: iv. 189. 1899; C. Chr. Ind. Fil. 579. 1905.

Rhizome short, erect, clothed at the growing tip in dense large brown and broadly ovate, acuminate, fimbriate *scales*; *fronds* tufted, upright or spreading and nodding, stipe slender, 8-14 cm long, copiously covered with similar spreading scales, *fertile lamina* linear-lanceolate, gradually acuminate, not narrowed towards base, 15-20 cm long, 2.5-3.5 cm broad, pinnatifid at the base almost down to the rachis but leaving a broad wing along the rachis upwards; *segments* numerous, close, patent, oblong-ovate, acute, entire, 1.2-1.4 cm long, 7 mm broad, not auricled in the anterior base, the apical part gradually tappering and entire; *texture* subcoriaceous, dry brown, glabrous above, scaly over the lower side and rachis; *veins* in segments obscure, close, all forked or rarely triforked with very obliquely ascending veinlets reaching the margin; *sori* medial and uniseriate on each side of costa, exindusiate. The *sterile lamina* similar to the fertile ones, usually less deeply pinnatifid or sometimes simple with wavy margin below, rachis lengthening out into a long whip-like nodding stolon and viviparous at the apex.

Guangdong: North River, ex Herb. Bot. Gard. Hong Kong *No. 103* (type); North River, ex Herb. Bot. Gard. Hong Kong *No. 209*; *C. G. Matthew*, January 4, 1904 and November 23, 1906; *Gerlach*; Wang Yuen District, Fan Shiu Shan, *S. K. Lau 2640*, October 7-30, 1933.

A very distinct and endemic species known so far only from the northern part of Guangdong, even where, judging from the herbarium material, it is a rare fern.

Plate 247. 1. Habit sketch (natural size). 2. Portion of lamina, showing venation and sori ($\times$ 3). 3. Scale from the base of stipe ($\times$ 40). 4. Scale from the under side of lamina ($\times$ 40). 5-6. Sporangium with spores ($\times$ 100).

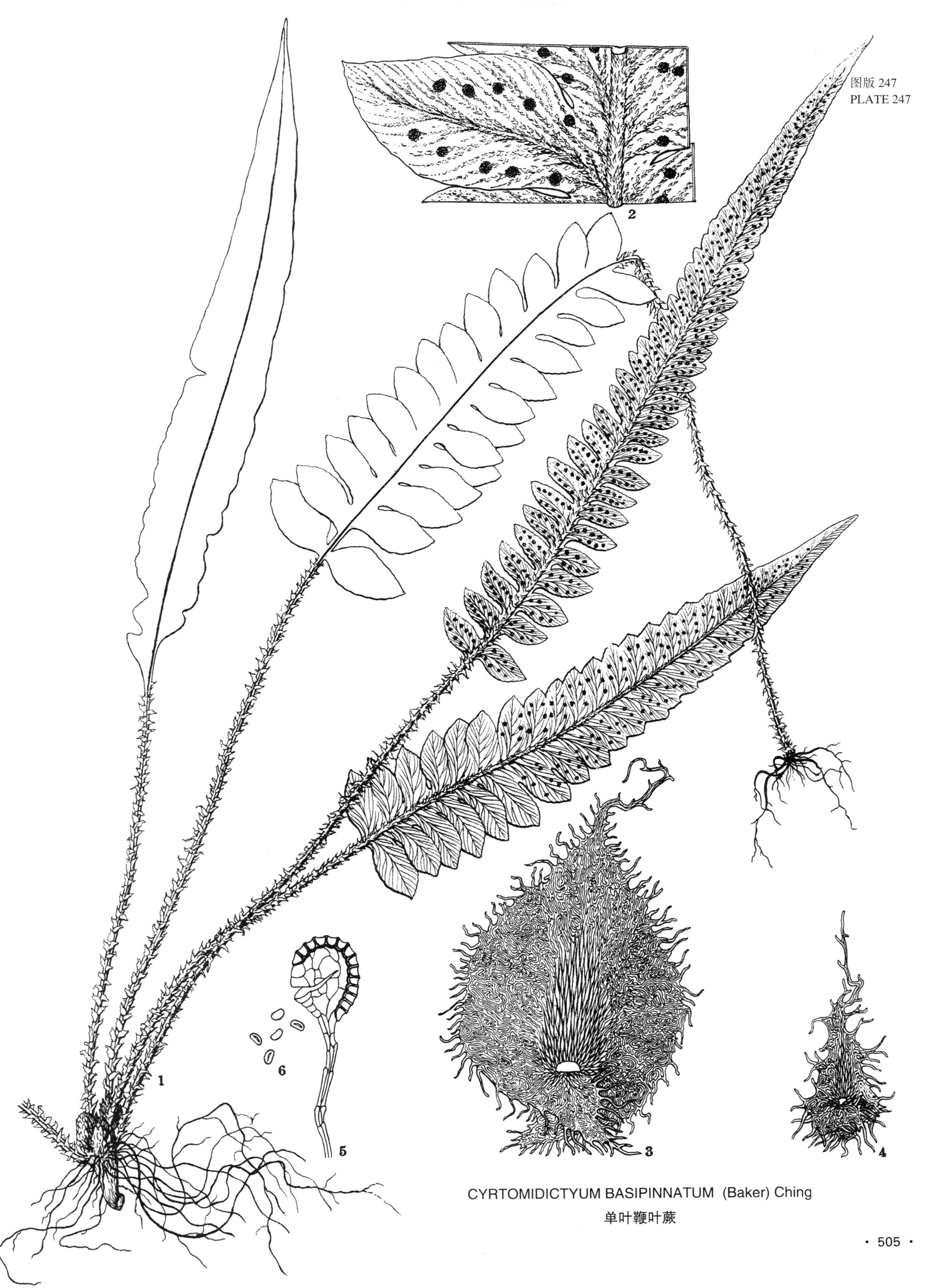

CYRTOMIDICTYUM BASIPINNATUM (Baker) Ching
单叶鞭叶蕨

根状茎短而直立,和叶柄都被有鳞片;叶簇生,柄长 15—20 cm,鳞片大,棕色,卵状披针形,长尾尖头,边有睫毛,宿存;能育叶体长 15—20 cm 或稍长,宽 4 cm,披针形,基部不变狭,向顶端为渐尖头,一回羽状;羽片很多(约 20 对),接近,开展,下部 5—7 对,与中轴分离,有短小叶柄,其余均或多或少合生,基部一对和上方各对等长或稍长,长 2—2.3 cm,长圆卵形,急尖头,不具短尖,基部下方斜形,上方呈圆耳形,全缘,多少呈反折波状;叶为革质,干后为棕色,叶轴密覆鳞片,羽片上面光滑,下面被小而紧贴的鳞片,边有睫毛,卵形尾尖头;叶脉不明显,分离,有时下部结合,羽状,或向顶端分叉,每组的基部上方一小脉向外不到达边缘;子囊群小而圆形,中央生,在中肋两侧各为 1 列,或在基部上方为 2 列;不育叶与能育叶同形,有较少的羽片,为宽缺刻分开,叶轴向顶端延长成鞭状有鳞片的匍匐茎。

分布:江西西北部,宜丰县,黄岗山麓特产。

本种形体颇似单叶鞭叶蕨 C. basipinnatum (Baker) Ching,但叶为羽状,基部上方为亚耳形,故易区别。

图注:1.本种全形(自然大);2.能育叶的羽片,表示叶脉及子囊群位置(放大 3 倍);3.不育叶的羽片(放大 3 倍);4.叶柄基部的鳞片(放大 30 倍);5.羽片下面的鳞片(放大 30 倍)。

CYRTOMIDICTYUM CONJUNCTUM Ching

CYRTOMIDICTYUM CONJUNCTUM Ching in Acta Phytotax. Sinica **6**: 263. t. 52. 1957.

Polystichum conjunctum Ching in herb.

Rhizome short, erect, densely radicose; *fronds* caespitose, spreading or nodding, stipe 15-20 cm long, clothed throughout in dense, large, brown, ovate-acuminate, ciliate scales with a long cauda, the *fertile lamina* 15-20 cm long or longer, 4 cm broad, lanceolate, not narrowed at base, gradually acuminate, simply pinnate; *pinnae* about 20 pairs, contiguous, patent, the lower 7-10 pairs free, petiolulate, the upper ones more or less adnate under the pinnatifid apical part, the basal pair as long as, or slightly longer than those next above, 2-2.3 cm long, 1 cm broad, ovate-oblong, acute, not mucronate, base subequal, round-cuneate below, rather bluntly auricled and truncate above, margin entire, or more or less repando-undulate; *texture* sub-coriaceous, dry brown, rachis copiously scaly, naked above and copiously clothed in minute, appressed, ovate-caudate, ciliate scales on the under side of pinnae; *veins* obscure, free or occasionally joint in the lower part, in pinnate groups, or forked towards apex, of which the anterior basal veinlet stops short far below the margin; *sori* small, rounded, medial and uniseriate on each side of costa, or biseriate near the base above, dorsal on the anterior basal veinlet of each group. The *sterile fronds* similar to the fertile with less pairs of pinnae separated from each other by broad sinuses, axis prolongated at apex into a long whip-like densely scaly stolon.

Northwestern Jiangxi: Ifeng Hsien, Hwangkan Shan, *Hsiung Yao-ko 6466*, Oct. 25, 1947, at the foot of cliff, rare.

A distinct local species, closely related to *C. basipinnatum* (Baker) Ching, differing in the pinnate fronds having free pinnae in the lower part, which are slightly auricled on the upper side of the base.

Plate 248. 1. Habit sketch (natural size). 2. Pinna from the fertile frond, showing venation and sori (× 3). 3. Pinna from the sterile frond (× 3). 4. Scale from the basal part of stipe (× 30). 5. Scale from the under side of pinna (× 30).

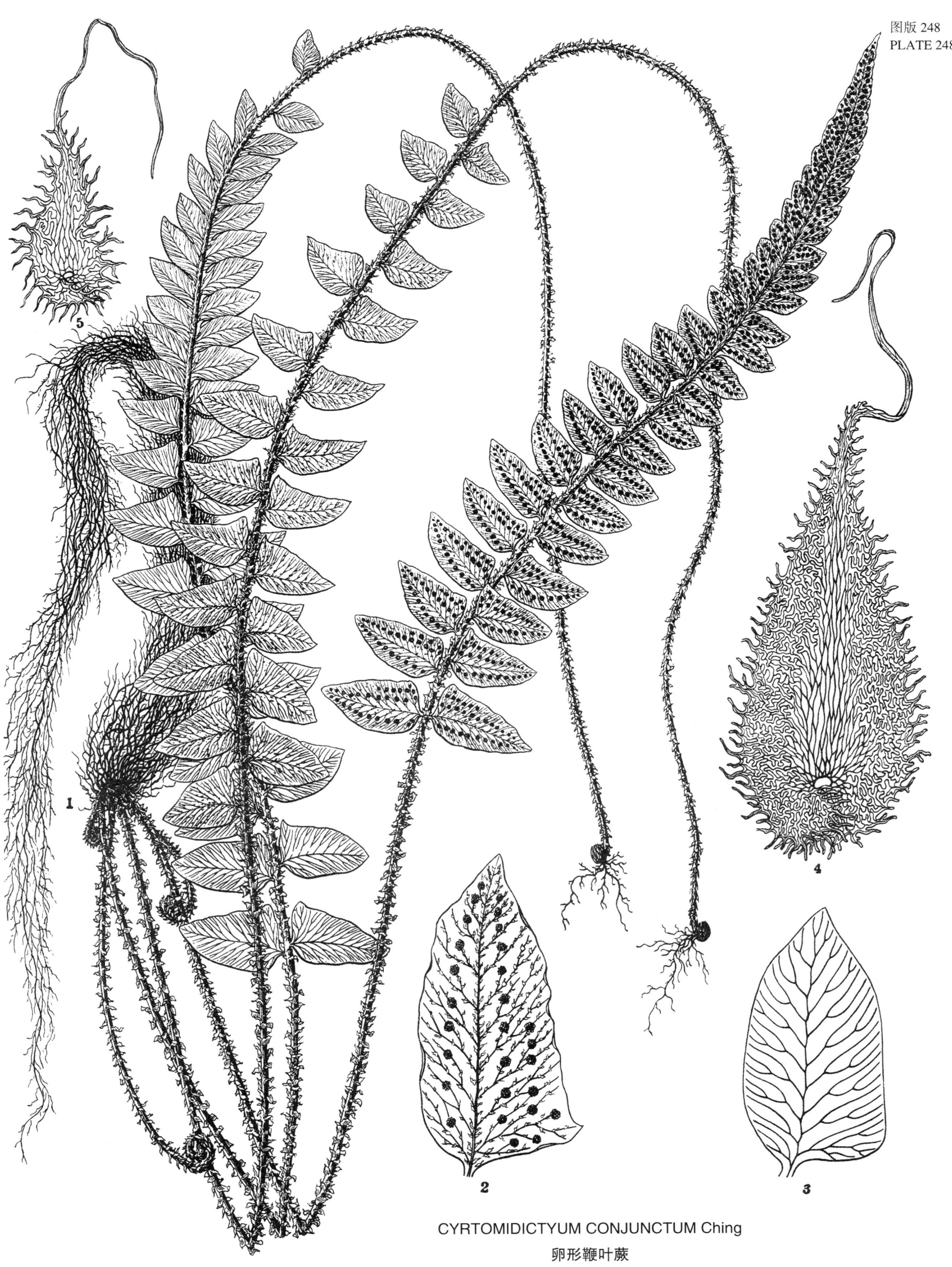

CYRTOMIDICTYUM CONJUNCTUM Ching
卵形鞭叶蕨

根状茎短而向上，有密的根丛；叶簇生，大都开展或伏地，亚二型，能育叶的柄长 15—30 cm，基部以上直径 2.5 mm，有大卵形的包着或亚覆瓦状的软质棕色鳞片密生，基部为心脏形，边缘有长而密的睫毛，叶体长同叶柄，宽 8—15 cm，长矩圆形到宽矩圆披针形，一回羽状，顶部为羽状深裂的渐尖头，向基部不变狭；羽片 8—15 对，具短小柄，或向上几无柄，开展，互生，有宽缺刻分开，基部一对等长或罕比上方的稍长，镰状披针形，一般长 5.5—10 cm，宽 1.2—1.8 cm，渐尖头，全缘，基部不等，下方斜切，上方为耳状尖三角形突起；叶为厚纸质或亚革质，干后棕色，叶轴与叶柄有同样的鳞片密生，羽片上部光滑，下面散生伏贴的薄质卵状披针形而边有长睫毛的鳞片，中肋明显，上面稍凹陷，下面隆起；叶脉不明显，斜出，小脉羽状成组，每组的基部上方一脉向外止于中途，其他的都到达边缘，分离，但有时下部叶脉联结；子囊群小而圆，背生于每组的 2—3 条小脉上，在中肋两边为 2 列，或在耳形突起处为不规则的三列；不育叶的柄远较细长，并有鳞片，叶体较能育叶为狭，羽片数也较少，形体也较小，彼此分离较远，中轴顶端延伸成一无叶而有鳞片的鞭状匍匐茎，顶端生一胞芽，着地后行无性繁殖。

分布：中国东部，包括福建，江西，湖南，安徽的南部以及接近浙江边境的江苏南部。日本与朝鲜半岛也产。过去日本学者曾从中国台湾报告此种，但据了解，不像本种，而似为下一种。

图注：1. 本种全形（自然大）；2. 羽片的一部，表示叶脉及子囊群位置（放大 3 倍）；3. 叶柄基部的鳞片（放大 20 倍）；4. 羽片下面的鳞片（放大 20 倍）；5—6. 孢子囊及孢子（放大 100 倍）。

CYRTOMIDICTYUM LEPIDOCAULON (Hooker) Ching

CYRTOMIDICTYUM LEPIDOCAULON (Hooker) Ching in Bull. Fan. Mem. Inst. Biol. Bot. Ser. **10**: 162. 1940; in Acta Phytotax. Sinica **6**: 264 t. 53. 1957.

Aspidium lepidocaulon Hooker, Sp. Fil. **4**: 12 t. 217. 1862; Hook. & Bak. Syn. Fil. 250. 1874; Fr. et Sav. Enum. Pl. Jap. **2**: 230. 1876; Palibin in Acta Hort. Bot. Petrop. **11**: 41. 1901.

Polystichum lepidocaulon J. Sm. Ferns Brit. & Fore. 286. 1866; Diels in Engl. u. Prantl, Nat. Pflanzenfam. **1**: iv. 189. 1899; C. Chr. Ind. Fil. 582. 1905; Ching in *Sinensia* **3**: 331. 1933; Ogata, Ic. Fil. Jap. **5**: t. 246. 1933; DeVol, Ferns East. China 78. 1945 (pro parte).

Dryopteris lepidocaula O. Ktze. Rev. Gen. Pl. **2**: 812. 1891.

Rhizome short, ascending, densely radicose; *fronds* tufted, spreading and nodding, stipe of the *fertile frond* 15-30 cm long, about 2.5 mm across above the base, copiously clothed in large, brown, ovate, clasping or subimbricating, ciliate *scales* with cordate base, lamina as long as the stipe, 8-15 cm broad, elongate-oblong to broadly oblong-lanceolate, apex acuminate and pinnatifid, base not narrowed, simply pinnate; *pinnae* 8-15 pairs, shortly petiolate, patent, alternate, separated by sinuses nearly as broad, the basal pair as long as those next above, or rarely somewhat longer, usually 5.5-10 cm long, 1.2-1.8 cm broad, lanceolate-falcate, gradually acuminate, entire throughout or nearly so, base unequal, round-cuneate below, truncate and with a triangular acute auricle above; *texture* thickly chartaceous or subcoriaceous, dry brownish, rachis similarly scaly as the stipe, glabrous above and quite copiously clothed below in scattered, appressed, thin and fimbriate scales, costa of pinnae slender, grooved above, and raised beneath; *veins* obscure, veinlets in pinnate group, of which the anterior basal one stops short about half way to the margin, occasionally the lower ones of the same group, or of the opposite groups joint, otherwise all free; *sori* small, dorsal on the lower 2-3 veinlets of each group, biseriate on each side of the costa, 3-seriate on the auricle. The *sterile frond* with much more slender and longer scaly stipe, lamina narrower than the fertile, usually consisting of less number of smaller pinnae, which are far apart from each other, the rachis is prolongated into a long whip-like scaly stolon, ended in a scaly bud as a means of vegetative reproduction.

At present this species has been reported from East China, including Fujian, Jiangxi, Hunan, the southern part of Anhui and Jiangsu on the border of Zhejiang. Also known from Japan and Korea peninsula: Tsu Shima, *Wilford 565* (type), May, 1857.

Like the other species of the genus, the present fern also prefers to grow in wooded ravines by stream side at low elevations and often forms a dense patch, throwing out long wiry stolons in all directions.

Plate 249. 1. Habit sketch (natural size). 2. Portion of soriferous pinna, showing venation and position of sori (× 3). 3. Scale from the basal part of stipe (× 20). 4. Scale from the under side of pinna (× 20). 5-6. Sporangium with spores (× 100).

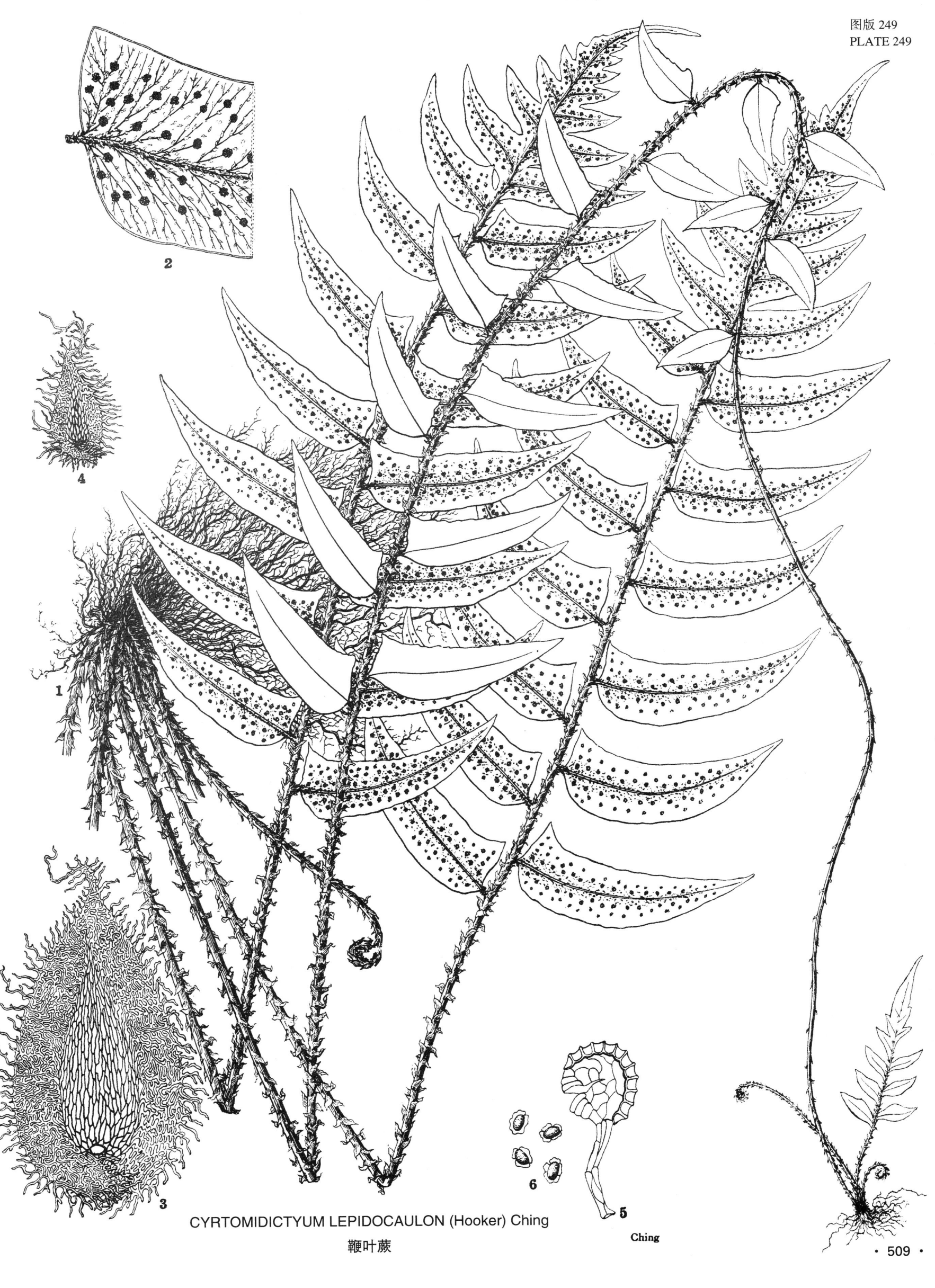

CYRTOMIDICTYUM LEPIDOCAULON (Hooker) Ching
鞭叶蕨

图版 250　普陀鞭叶蕨

在一般形体上,本种酷似鞭叶蕨 C. lepidocaulon（Hook.）Ching,但较小,能育叶全长仅 15—35 cm（包括柄长 10—20 cm）,叶体长 10—15 cm,宽 5.5—10 cm,分离羽片 5—8 对,通常长 4—5 cm,宽 0.8—1 cm,镰状披针形,基部上方也同样有耳状尖三角形突起,全缘；叶为革质；叶脉不明显,每组小脉 3—4 条,向外都达于边缘；子囊群中央生,在中肋两侧各为一列,仅在基部上方耳形突起处有时为亚二列；不育叶有较长的柄,羽片数较少,较小,疏生于长鞭状而有鳞片的叶轴下部。

本种极似前种,过去曾当作同一种看待,但羽片较短,子囊群在中肋两侧各为一列,并每组的小脉都到达边缘,故易区别。

分布：浙江普陀,黄岩,杭州,镇海及台湾,成块状生于常绿林下的溪沟边。

图注：1.本种全形（自然大）；2.能育叶的羽片,表示叶脉及子囊群位置（放大 2 倍）；3.叶柄基部的鳞片（放大 20 倍）；4—5.孢子囊及孢子（放大 100 倍）。

CYRTOMIDICTYUM FABERI（Baker）Ching

CYRTOMIDICTYUM FABERI (Baker) Ching in Acta Phytotax. Sinica **6**: 265 t. 54. 1957.

Nephrodium faberi Baker in Ann. Bot. **5**: 318. 1891.

Polystichum lepidocaulon C. Chr. Ind. Fil. 582. 1905; Ito, Illustr. Pl. Form. t. 64. 1931; DeVol, Ferns East. China 78. 1945 (pro parte).

Polystichum Faberi Ching in herb.

Very much similar to the preceeding species in general habit, but differs in smaller size, *fertile fronds* 15-35 cm long including stipes 10-20 cm long, lamina 10-15 cm long, 5.5-10 cm broad, free *pinnae* 5-8 pairs, 4-5 cm long, 8-10 mm broad, lanceolate-falcate, also with a triangular acute auricle on the anterior side of the base, entire; *texture* coriaceous; *veins* obscure, veinlets 3-4 in each group, *all* reaching the margin; *sori* medial and uniseriate on each side of the costa, dorsal only on the anterior basal veinlet of each group, or occasionally biseriate in thè auricle. The *sterile frond* with longer stipe consisting of a few pairs of smaller and widely separated pinnae along the wiry and scaly rachis, which is prolongated into a very long stolon.

Zhejiang: Ningpo, Puto Island, *E. Faber 205* (type); *J. Guckett, Forbes 739*; *H. Migo*, June 5, 1936; Whangyen, *N. T. Wang*, August 10, 1944; Hangchow, Lingyin Monastery, *R. C. Ching* (1919), by stream side under broad-leaved forests; *Mary S. Matthew 10218* (pro parte); Cheng-hai, *K. K. Tsoong 135*. Also Taiwan.

A distinct local species, distinguished from *C. lepidocaulon* (Hooker) Ching chiefly in the much shorter pinnae, the uniseriate sori on each side of costa and in that all the veinlets of each group reach the margin. According to the photograph by Ito (I. c.), the plants from Taiwan reported as *Polystichum lepidocaulon* (Hooker) J. Sm. seems to belong to the present species.

Plate 250. 1. Habit sketch (natural size). 2. Pinna, showing venation and sori ($\times$ 2). 3. Scale from the basal part of stipe ($\times$ 20). 4-5. Sporangium with spores ($\times$ 100).

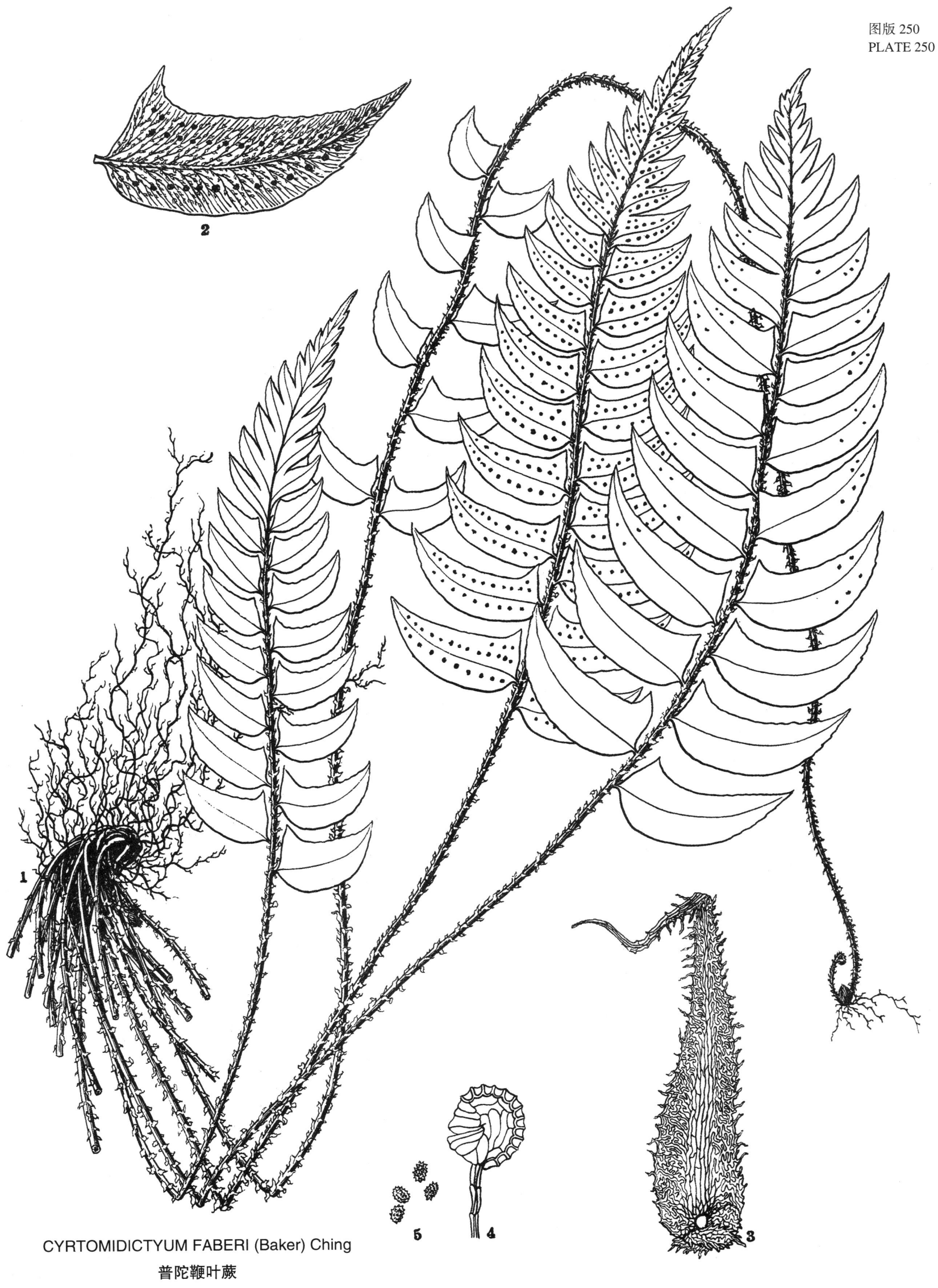

CYRTOMIDICTYUM FABERI (Baker) Ching
普陀鞭叶蕨

拉丁名索引 Index to Latin Names

A

Acrophorus stipellatus 362
Adiantum bonatianum 452
Adiantum capillus-junonis 436
Adiantum capillus-veneris 442
Adiantum caudatum 430
Adiantum chienii 326
Adiantum davidi 282
Adiantum diaphanum 460
Adiantum edentulum 444
Adiantum edgeworthii 284
Adiantum faberi 114
Adiantum fengianum 446
Adiantum flabellulatum 456
Adiantum gravesii 324
Adiantum greenii 62
Adiantum induratum 458
Adiantum juxtapositum 434
Adiantum muticum 454
Adiantum myriosorum 464
Adiantum nanum 62
Adiantum pedatum 462
Adiantum philippense 438
Adiantum refractum 280
Adiantum roborowskii 328
Adiantum sinicum 432
Adiantum smithianum 448
Adiantum soboliferum 440
Adiantum venustum 450
Angiopteris fokiensis 414
Antrophyum formosanum 296
Antrophyum petiolatum 80
Archangiopteris bipinnata 416
Archangiopteris caudata 426
Archangiopteris henryi 4
Archangiopteris hokouensis 418
Archangiopteris latipinna 424
Archangiopteris somai 420
Archangiopteris subrotundata 422
Archangiopteris tonkinensis 428
Arthromeris himalayensis 304
Arthromeris lungtauensis 306
Arthropteris obliterata 218
Aspidium ebenium 16
Aspidium longicrure 14
Asplenium adnatum 108

Asplenium crinicaule 234
Asplenium exiguum 356
Asplenium finlaysonianum 44
Asplenium fugax 354
Asplenium interjectum 360
Asplenium loriceum 358
Asplenium prolongatum 232
Asplenium sampsoni 230
Asplenium sarelii 228
Asplenium saxicola 236
Athyrium anisopterum 42
Athyrium goeringianum 226
Athyrium otophorum 224
Athyrium sheareri 222

B

Blechnum eburneum 56
Bolbitis heteroclita 244
Bolbitis subcordata 246

C

Camptosorus sibiricus 220
Cheilanthes chusana 276
Cheilanthes duclouxii 272
Cheilanthes hancockii 274
Cheilanthes trichophylla 278
Cheiropleuria bicuspis 100
Christopteris tricuspis 188
Colysis bonii 182
Colysis digitata 404
Colysis hemionitidea 398
Colysis hemitoma 402
Colysis henryi 184
Colysis longipes 186
Colysis morsei 408
Colysis pentaphylla 406
Colysis wrightii 400
Colysis wui 396
Coniogramme caudata 342
Coniogramme fraxinea 340
Coniogramme intermedia 292
Coniogramme procera 344
Cyclopeltis crenata 262
Cyclophorus calvatus 164
Cyrtomidictyum basipinnatum 504
Cyrtomidictyum conjunctum 506
Cyrtomidictyum faberi 510

Cyrtomidictyum lepidocaulon 508
Cyrtomium aequibasis 374
Cyrtomium falcatum 260
Cyrtomium fortunei 258
Cyrtomium fraxinellum 34
Cyrtomium hemionitis 28
Cyrtomium lonchitoides 32
Cyrtomium muticum 376
Cyrtomium nephrolepioides 30
Cystopteris moupinensis 10

D

Davallia mariesii 212
Davallia orientalis 214
Dictyocline griffithii 240
Diplazium macrophyllum 50
Diplazium pellucidum 48
Diplazium pullingeri 46
Dipteris chinensis 316
Drynaria fortunei 392
Drynaria sinica 394
Dryopteris championi 370
Dryopteris enneaphylla 12
Dryopteris liankwangensis 368
Dryopteris scottii 366
Dryopteris serrato-dentata 364

E

Elaphoglossum austro-sinicum 98
Elaphoglossum mcclurei 166

G

Gleichenia cantonensis 310
Gleichenia laevissima 312
Gleichenia splendida 314
Gymnocarpium remoti-pinnatum 352
Gymnopteris vestita 58

H

Helminthostachys zeylanica 2
Hemigramma decurrens 378
Humata assamica 216
Hypodematium crenatum 248
Hypodematium cystopteroides 252
Hypodematium fordii 250

L

Lemmaphyllum adnascens 158
Lemmaphyllum drymoglossoides 160
Lemmaphyllum microphyllum 162
Lemmaphyllum subrostratum 156
Lepisorus angustus 148

Lepisorus bicolor 132
Lepisorus clathratus 136
Lepisorus contortus 152
Lepisorus eilophyllus 118
Lepisorus heterolepis 146
Lepisorus kuchenensis 130
Lepisorus lewisii 116
Lepisorus loriformis 124
Lepisorus macrosphaerus 126
Lepisorus macrosphaerus var. asterolepis 128
Lepisorus obscure-venulosus 134
Lepisorus oligolepidus 142
Lepisorus pseudonudus 122
Lepisorus sordidus 138
Lepisorus subconfluens 144
Lepisorus sublinearis 120
Lepisorus suboligolepidus 140
Lepisorus thunbergianus 154
Lepisorus ussuriensis 150
Leptogramma caudata 470
Leptogramma mollissima 466
Leptogramma scallani 468
Leucostegia hookeri 382
Leucostegia immersa 380
Leucostegia multidentata 384
Lindsaya chienii 38
Lindsaya chinensis 40
Lindsaya decomposita 322
Lindsaya lobbiana 320
Loxogramme chinensis 94
Loxogramme ensiformis 390
Loxogramme grammitoides 386
Loxogramme salicifolia 388

M

Microlepia chrysocarpa 36
Microlepia tenera 204
Microsorium buergerianum 172
Microsorium fortuni 168
Microsorium hymenodes 170
Microsorium membranaceum 178
Microsorium punctatum 174
Microsorium zippelii 176

N

Neocheiropteris palmatopedata 96
Neocheiropteris phyllomanes 180

O

Oleandra cumingii 346
Oleandra undulata 350
Oleandra wallichii 104

Oleandra whangii 348
Onychium contiguum 330
Onychium ipii 336
Onychium moupinense 332
Onychium tenuifrons 334

P

Pellaea smithii 270
Phyllitis cardiophylla 54
Phyllitis delavayi 52
Phymatodes nigrovenia 202
Plagiogyria assurgens 318
Plagiogyria henryi 60
Pleurosoriopsis makinoi 338
Polypodium amoenum 200
Polypodium dareaeformioides 82
Polypodium dielseanum 196
Polypodium ellipticum 90
Polypodium lachnopus 192
Polypodium leveillei 92
Polypodium manmeiense 190
Polypodium mengtzeense 84
Polypodium microrhizoma 194
Polypodium niponicum 198
Polypodium oblongiosorum 86
Polypodium triglossum 88
Polystichum acanthophyllum 20
Polystichum bifidum 494
Polystichum chingae 372
Polystichum chunii 26
Polystichum consimile 484
Polystichum deltodon 22
Polystichum dielsii 476
Polystichum duthiei 18
Polystichum excellens 478
Polystichum fimbriatum 264
Polystichum grossidentatum 488
Polystichum hecatopterum 24
Polystichum kwangtungense 482
Polystichum lanceolatum 490
Polystichum nepalense 496
Polystichum omeiense 268
Polystichum otophorum 266
Polystichum stenophyllum 486

Polystichum thomsoni 492
Polystichum tripteron 500
Polystichum tsus-simense 502
Polystichum xiphophyllum 498
Polystichum yuanum 480
Pseudodrynaria coronans 412
Pteris actiniopteroides 70
Pteris dactylina 288
Pteris deltodon 64
Pteris dimorpha 68
Pteris excelsa 74
Pteris fauriei 286
Pteris hui 66
Pteris insignis 112
Pteris paupercula 72
Pyrrosia drakeana 302
Pyrrosia sheareri 300

Q

Quercifilix zeylanica 242

S

Saxiglossum taeniodes 298
Schizoloma ensifolium 290
Stegnogramma asplenioides 474
Stegnogramma cyrtomioides 472

T

Taenitis blechnoides 110
Tectaria macrodonta 256
Tectaria subtriphylla 254
Trichomanes tereticaulum 6

V

Vittaria forrestiana 294
Vittaria nana 78
Vittaria pauciareolata 76

W

Woodsia cinnamomea 8
Woodsia macrochlaena 208
Woodsia manchuriensis 210
Woodwardia harlandii 106
Woodwardia orientalis 238

中文名索引 Index to Chinese Names

矮铁线蕨　62
矮叶书带蕨　78
爱氏铁线蕨　284
白背铁线蕨　282
半月形铁线蕨　438
抱石莲　160
抱树莲　162
鞭叶蕨　508
鞭叶铁线蕨　430
波氏双盖蕨　46
波氏星蕨　172
波叶溪边蕨　472
苍山铁线蕨　432
长柄车前蕨　80
长柄茶蕨　350
长柄线蕨　186
长盖铁线蕨　448
长生铁角蕨　232
长瓦韦　122
长尾铁线蕨　460
长叶实蕨　244
车前蕨　296
陈氏耳叶蕨　26
赤色岩蕨　8
翅柄铁线蕨　440
川石莲　158
川水龙骨　196
刺耳叶蕨　20
刺叶耳蕨　484
粗齿耳蕨　488
大囊岩蕨　208
大瓦韦　126
大叶贯众　376
大叶双盖蕨　50
带瓦韦　124
单叶鞭叶蕨　504
单叶贯众　28
单叶扇蕨　180
倒叶耳蕨　480
低头贯众　30
低头铁角蕨　356
地耳蕨　242
滇耳蕨　372
滇贯众　374
滇冷蕨　10
滇水龙骨　190
滇瓦韦　120

滇线蕨　406
滇星蕨　170
东方狗脊　238
杜氏耳叶蕨　18
杜氏粉背蕨　272
断线蕨　398
对生耳叶蕨　22
峨眉茯蕨　468
峨眉铁线蕨　114
峨眉耳蕨　268
峨眉瘤足蕨　318
二回原始观音座莲　416
二尖燕尾蕨　100
二形凤尾蕨　68
冯氏铁线蕨　446
福建观音座莲　414
福氏星蕨　168
福氏肿足蕨　250
傅氏凤尾蕨　286
高山茶蕨　104
高山耳蕨　266
高山凤了蕨　344
高山鳞毛蕨　364
高山三叉蕨　256
高山瓦韦　118
高山乌蕨　330
高山阴石蕨　216
革叶耳蕨　498
格氏铁线蕨　62
骨牌蕨　156
贯众　258
光里白　312
光石韦　164
光蹄盖蕨　224
广东耳蕨　482
广东里白　310
过山蕨　220
哈氏狗脊　106
海南实蕨　246
海南铁线蕨　458
海州骨碎补　212
韩氏粉背蕨　274
合生铁角蕨　108
河口原始观音座莲　418
荷叶对开盖蕨　52
鹤庆铁线蕨　454
黑柄三叉蕨　16

黑鳞莩蕨　202
黑鳞瓦韦　138
亨利马蹄蕨　4
亨氏瘤足蕨　60
亨氏线蕨　184
胡氏凤尾蕨　66
槲蕨　392
华凤了蕨　292
华槲蕨　394
华假铁线蕨　40
华剑蕨　94
华南茶蕨　346
华南骨碎补　214
华南舌蕨　98
华中铁角蕨　228
槐叶贯众　34
黄胞鳞蕨　36
黄毛凤尾蕨　72
黄瓦韦　128
灰背铁线蕨　464
霍氏膜盖蕨　382
芨瓦韦　146
戟蕨　188
尖顶耳蕨　478
尖叶原始观音座莲　428
睫毛蕨　338
金毛裸蕨　58
锯齿耳叶蕨　24
柯氏蹄盖蕨　226
宽书带蕨　294
阔叶剑蕨　390
阔叶书带蕨　76
阔叶原始观音座莲　424
莱氏水龙骨　92
莱氏线蕨　400
濑水龙骨　192
栗柄水龙骨　194
连珠瓦韦　144
两广鳞毛蕨　368
两色瓦韦　132
亮叶耳蕨　490
鳞瓦韦　142
岭南铁角蕨　230
琉璃节肢蕨　304
柳叶剑蕨　388
陇铁线蕨　328
庐山石韦　300

庐山蹄盖蕨 222	全缘凤了蕨 340	狭叶乌蕨 334
庐山瓦苇 116	全缘凤尾蕨 112	仙霞铁线蕨 434
卵形鞭叶蕨 506	全缘贯众 260	象牙乌毛蕨 56
洛氏林蕨 320	软骨耳蕨 496	小叶剑蕨 386
马氏线蕨 408	三叉耳蕨 500	星蕨 174
马祖耳蕨 502	三叉蕨 254	芽胞耳蕨 486
满洲岩蕨 210	三叶水龙骨 88	崖姜蕨 412
毛粉背蕨 278	山东肿足蕨 252	岩凤尾蕨 64
毛膜盖蕨 384	扇叶铁线蕨 456	燕尾三叉蕨 14
毛铁角蕨 234	圣蕨 240	瑶山𦶟蕨 348
毛叶凤了蕨 342	史氏旱蕨 270	瑶山瓦苇 130
毛叶茯蕨 466	史氏鳞毛蕨 366	叶氏乌蕨 336
毛足铁线蕨 452	蜀铁线蕨 280	宜昌金星蕨 12
蒙自水龙骨 84	双扇蕨 316	阴地铁角蕨 354
膜盖蕨 380	水龙骨 198	友水龙骨 200
膜叶双盖蕨 48	硕里白 314	圆柄石衣蕨 6
膜叶星蕨 178	宿蹄盖蕨 42	圆顶耳蕨 476
木坪乌蕨 332	台湾原始观音座莲 420	圆基原始观音座莲 422
南海铁角蕨 358	藤蕨 218	月芽铁线蕨 444
嫩毛蕨 204	铁线蕨 442	粤节肢蕨 306
拟叉蕨 378	团叶铁线蕨 436	粤铁角蕨 236
拟凤尾蕨 290	椭圆水龙骨 90	粤铁线蕨 324
拟贯众 32, 262	瓦鳞耳蕨 264	粤瓦苇 134
拟鳞毛蕨 362	瓦苇 154	毡毛石韦 302
拟鳞瓦苇 140	网脉单盖蕨 44	张氏鳞毛蕨 370
拟石苇 298	网脉林蕨 322	掌凤尾蕨 288
扭瓦苇 152	网眼瓦苇 136	掌叶铁线蕨 462
彭氏线蕨 182	尾叶耳蕨 492	掌叶线蕨 404
普陀鞭叶蕨 510	尾叶茯蕨 470	掌状扇蕨 96
戚氏星蕨 176	尾叶原始观音座莲 426	肢节蕨 352
钱氏假铁线蕨 38	乌柄水龙骨 82	肿足蕨 248
钱氏铁线蕨 326	乌苏里瓦苇 150	舟山粉背蕨 276
钳形耳蕨 494	吴氏线蕨 396	胄叶线蕨 402
黔铁角蕨 360	溪凤尾蕨 74	珠带水龙骨 86
浅裂溪边蕨 474	锡兰七指蕨 2	猪鬣凤尾蕨 70
琼崖对开盖蕨 54	细叶铁线蕨 450	竹叶蕨 110
琼崖舌蕨 166	狭叶瓦苇 148	

分类和名称变化对照表
Taxonomic and Nomenclatural Changes

图谱原名 Name in *Icones*		图版号 Plate No	现接受名 Present accepted name	
Helminthostachys zeylanica	锡兰七指蕨	1	Helminthostachys zeylanica (L.) Hook.	七指蕨
Archangiopteris henryi	亨利马蹄蕨	2	Archangiopteris henryi Christ & Gies.	亨利原始观音座莲
Trichomanes tereticaulum	圆柄石衣蕨	3	Selenodesmium siamense (Christ) Ching & Chu H. Wang	广西长筒蕨
Woodsia cinnamomea	赤色岩蕨	4	Woodsia cinnamomea Christ	赤色岩蕨
Cystopteris moupinensis	滇冷蕨	5	Cystopteris moupinensis Franch.	宝兴冷蕨
Dryopteris enneaphylla	宜昌金星蕨	6	Dryopteris enneaphylla (Baker) C. Chr.	宜昌鳞毛蕨
Aspidium longicrure	燕尾三叉蕨	7	Tectaria simonsii (Baker) Ching	燕尾叉蕨
Aspidium ebenium	黑柄三叉蕨	8	Tectaria ebenina (C. Chr.) Ching	黑柄叉蕨
Polystichum duthiei	杜氏耳叶蕨	9	Sorolepidium glaciale Christ	玉龙蕨
Polystichum acanthophyllum	刺耳叶蕨	10	Polystichum stimulans (Kunze & Mett.) Bedd.	猫儿刺耳蕨
Polystichum deltodon	对生耳叶蕨	11	Polystichum deltodon (Baker) Diels	对生耳蕨
Polystichum hecatopterum	锯齿耳叶蕨	12	Polystichum hecatopterum Diels	芒齿耳蕨
Polystichum chunii	陈氏耳叶蕨	13	Polystichum chunii Ching	陈氏耳蕨
Cyrtomium hemionitis	单叶贯众	14	Cyrtomium hemionitis Christ	单叶贯众
Cyrtomium nephrolepioides	低头贯众	15	Cyrtomium nephrolepioides (Christ) Copel.	低头贯众
Cyrtomium lonchitoides	拟贯众	16	Cyrtomium lonchitoides (Christ) Christ	小羽贯众
Cyrtomium fraxinellum	槐叶贯众	17	Cyrtogonellum fraxinellum (Christ) Ching	柳叶蕨
Microlepia chrysocarpa	黄胞鳞蕨	18	Microlepia chrysocarpa Ching	金果鳞盖蕨
Lindsaya chienii	钱氏假铁线蕨	19	Lindsaea chienii Ching	钱氏鳞始蕨
Lindsaya chinensis	华假铁线蕨	20	Lindsaea chingii C. Chr.	碎叶鳞始蕨
Athyrium anisopterum	宿蹄盖蕨	21	Athyrium anisopterum Christ	宿蹄盖蕨
Asplenium finlaysonianum	网脉单盖蕨	22	Asplenium finlaysonianum Wall. ex Hook.	网脉铁角蕨
Diplazium pullingeri	波氏双盖蕨	23	Monomelangium pullingeri (Baker) Tagawa	毛轴线盖蕨
Diplazium pellucidum	膜叶双盖蕨	24	Allantodia hirtipes (Christ) Ching	鳞轴短肠蕨
Diplazium macrophyllum	大叶双盖蕨	25	Allantodia megaphylla (Baker) Ching	大羽短肠蕨
Phyllitis delavayi	荷叶对开盖蕨	26	Sinephropteris delavayi (Franch.) Mickel	水鳖蕨
Phyllitis cardiophylla	琼崖对开盖蕨	27	Boniniella cardiophylla (Hance) Tagawa	细辛蕨
Blechnum eburneum	象牙乌毛蕨	28	Struthiopteris eburnea (Christ) Ching	荚囊蕨
Gymnopteris vestita	金毛裸蕨	29	Paragymnopteris vestita (Wall. ex C. Presl) K. H. Shing	金毛裸蕨
Plagiogyria henryi	亨氏瘤足蕨	30	Plagiogyria stenoptera (Hance) Diels	耳形瘤足蕨
Adiantum greenii	格氏铁线蕨	31 (1-5)	Adiantum gravesii Hance	白垩铁线蕨
Adiantum nanum	矮铁线蕨	31 (6-9)	Adiantum mariesii Baker	小铁线蕨
Pteris deltodon	岩凤尾蕨	32	Pteris deltodon Baker	岩凤尾蕨
Pteris hui	胡氏凤尾蕨	33	Pteris hui Ching	胡氏凤尾蕨
Pteris dimorpha	二形凤尾蕨	34	Pteris cadieri Christ	条纹凤尾蕨
Pteris actiniopteroides	猪鬣凤尾蕨	35	Pteris actiniopteroides Christ	猪鬣凤尾蕨
Pteris paupercula	黄毛凤尾蕨	36	Pellaea paupercula (Christ) Ching	凤尾旱蕨
Pteris excelsa	溪凤尾蕨	37	Pteris excelsa Gaud.	溪边凤尾蕨
Vittaria pauciareolata	阔叶书带蕨	38	Haplopteris elongata (Sw.) E. H. Crane.	唇边书带蕨
Vittaria nana	矮叶书带蕨	39	Haplopteris flexuosa (Fée) E. H. Crane.	书带蕨
Antrophyum petiolatum	长柄车前蕨	40	Antrophyum obovatum Baker	长柄车前蕨
Polypodium dareaeformioides	乌柄水龙骨	41	Gymnogrammitis dareiformis (Hook.) Ching ex Tardieu & C. Chr.	雨蕨
Polypodium mengtzeense	蒙自水龙骨	42	Polypodiastrum argutum (Wall. ex Hook.) Ching	尖齿拟水龙骨
Polypodium oblongiosorum	珠带水龙骨	43	Lepisorus stenistos (C. B. Clarke) Y. X. Lin	狭带瓦韦
Polypodium triglossum	三叶水龙骨	44	Neocheiropteris triglossa (Baker) Ching	三叉扇蕨
Polypodium ellipticum	椭圆水龙骨	45	Colysis elliptica (Thunb.) Ching	线蕨

续表

图谱原名 Name in *Icones*		图版号 Plate No	现接受名 Present accepted name	
Polypodium leveillei	莱氏水龙骨	46	Colysis leveillei (Christ) Ching	绿叶线蕨
Loxogramme chinensis	华剑蕨	47	Loxogramme chinensis Ching	中华剑蕨
Neocheiropteris palmatopeda	掌状扇蕨	48	Neocheiropteris palmatopedata (Baker) Christ	扇蕨
Elaphoglossum austro-sinicum	华南舌蕨	49	Elaphoglossum yoshinagae (Yatabe) Makino	华南舌蕨
Cheiropleuria bicuspis	二尖燕尾蕨	50	Cheiropleuria integrifolia (D. C. Eaton ex Hook.) M. Kato, Y. Yatabe, Sahashi & N. Murak.	全缘燕尾蕨
Oleandra wallichii	高山茶蕨	51	Oleandra wallichii (Hook.) C. Presl	高山条蕨
Woodwardia harlandii	哈氏狗脊	52	Chieniopteris harlandii (Hook.) Ching	崇澍蕨
Asplenium adnatum	合生铁角蕨	53	Asplenium adnatum Copel.	合生铁角蕨
Taenitis blechnoides	竹叶蕨	54	Taenitis blechnoides (Willd.) Sw.	竹叶蕨
Pteris insignis	全缘凤尾蕨	55	Pteris insignis Mett. ex Kuhn	全缘凤尾蕨
Adiantum faberi	峨眉铁线蕨	56	Adiantum roborowskii Maxim. f. faberi (Baker) Y. X. Lin	峨眉铁线蕨
Lepisorus lewisii	庐山瓦韦	57	Lepisorus lewisii (Baker) Ching	庐山瓦韦
Lepisorus eilophyllus	高山瓦韦	58	Lepisorus eilophyllus (Diels) Ching	高山瓦韦
Lepisorus sublinearis	滇瓦韦	59	Lepisorus sublinearis (Baker) Ching	滇瓦韦
Lepisorus pseudonudus	长瓦韦	60	Lepisorus pseudonudus Ching	长瓦韦
Lepisorus loriformis	带瓦韦	61	Lepisorus loriformis (Wall. ex Mett.) Ching	带叶瓦韦
Lepisorus macrosphaerus	大瓦韦	62	Lepisorus macrosphaerus (Baker) Ching	大瓦韦
Lepisorus macrosphaerus var. asterolepis	黄瓦韦	63	Lepisorus asterolepis (Baker) Ching	黄瓦韦
Lepisorus kuchenensis	瑶山瓦韦	64	Lepisorus kuchenensis (Wu) Ching	瑶山瓦韦
Lepisorus bicolor	两色瓦韦	65	Lepisorus bicolor Ching	两色瓦韦
Lepisorus obscure-venulosus	粤瓦韦	66	Lepisorus obscure-venulosus (Hayata) Ching	粤瓦韦
Lepisorus clathratus	网眼瓦韦	67	Lepisorus clathratus (C. B. Clarke) Ching	网眼瓦韦
Lepisorus sordidus	黑鳞瓦韦	68	Lepisorus sordidus (C. Chr.) Ching	黑鳞瓦韦
Lepisorus suboligolepidus	拟鳞瓦韦	69	Lepisorus suboligolepidus Ching	拟鳞瓦韦
Lepisorus oligolepidus	鳞瓦韦	70	Lepisorus oligolepidus (Baker) Ching	鳞瓦韦
Lepisorus subconfluens	连珠瓦韦	71	Lepisorus subconfluens Ching	连珠瓦韦
Lepisorus heterolepis	芰瓦韦	72	Lepisorus heterolepis (Rosenst.) Ching	异叶瓦韦
Lepisorus angustus	狭叶瓦韦	73	Lepisorus angustus Ching	狭叶瓦韦
Lepisorus ussuriensis	乌苏里瓦韦	74	Lepisorus ussuriensis (Regel & Maack) Ching	乌苏里瓦韦
Lepisorus contortus	扭瓦韦	75	Lepisorus contortus (Christ) Ching	扭瓦韦
Lepisorus thunbergianus	瓦韦	76	Lepisorus thunbergianus (Kaulf.) Ching	瓦韦
Lemmaphyllum subrostratum	骨牌蕨	77	Lepidogrammitis rostrata (Bedd.) Ching	骨牌蕨
Lemmaphyllum adnascens	川石莲	78	Lepidogrammitis adnascens Ching	贴生骨牌蕨
Lemmaphyllum drymoglossoides	抱石莲	79	Lepidogrammitis drymoglossoides (Baker) Ching	抱石莲
Lemmaphyllum microphyllum	抱树莲	80	Lemmaphyllum microphyllum C. Presl	伏石蕨
Cyclophorus calvatus	光石韦	81	Pyrrosia calvata (Baker) Ching	光石韦
Elaphoglossum mcclurei	琼崖舌蕨	82	Elaphoglossum mcclurei Ching	琼崖舌蕨
Microsorium fortuni	福氏星蕨	83	Microsorum fortunei (T. Moore) Ching	江南星蕨
Microsorium hymenodes	滇星蕨	84	Lepidomicrosorium subhemionitideum (Christ) P. S. Wang	云南鳞果星蕨
Microsorium buergerianum	波氏星蕨	85	Lepidomicrosorium buergerianum (Miq.) Ching & K. H. Shing	鳞果星蕨
Microsorium punctatum	星蕨	86	Microsorum punctatum (L.) Copel.	星蕨
Microsorium zippelii	戚氏星蕨	87	Microsorum zippelii (Blume) Ching	显脉星蕨
Microsorium membranaceum	膜叶星蕨	88	Microsorum membranaceum (D. Don) Ching	膜叶星蕨
Neocheiropteris phyllomanes	单叶扇蕨	89	Neolepisorus ovatus (Bedd.) Ching	盾蕨
Colysis bonii	彭氏线蕨	90	Colysis pedunculata (Hook. & Grev.) Ching	长柄线蕨
Colysis henryi	亨氏线蕨	91	Colysis henryi (Baker) Ching	矩圆线蕨
Colysis longipes	长柄线蕨	92	Colysis elliptica (Thunb.) Ching var. longipes (Ching) L. Shi & X. C. Zhang	线蕨长柄变种
Christopteris tricuspis	戟蕨	93	Christiopteris tricuspis (Hook.) Christ	戟蕨
Polypodium manmeiense	滇水龙骨	94	Metapolypodium manmeiense (Christ) Ching	篦齿蕨

续表

图谱原名 Name in *Icones*		图版号 Plate No	现接受名 Present accepted name	
Polypodium lachnopus	濑水龙骨	95	Polypodiodes lachnopus (Wall. ex Hook.) Ching	濑水龙骨
Polypodium microrhizoma	栗柄水龙骨	96	Polypodiodes microrhizoma (C. B. Clarke ex Baker) Ching	栗柄水龙骨
Polypodium dielseanum	川水龙骨	97	Polypodiastrum dielseanum (C. Chr.) Ching	川拟水龙骨
Polypodium niponicum	水龙骨	98	Polypodiodes niponica (Mett.) Ching	日本水龙骨
Polypodium amoenum	友水龙骨	99	Polypodiodes amoena (Wall. ex Mett.) Ching	友水龙骨
Phymatodes nigrovenia	黑鳞莳蕨	100	Phymatopteris nigrovenia (Christ) Pic. Serm.	毛叶假瘤蕨
Microlepia tenera	嫩毛蕨	101 (Fas. 2)	Microlepia tenera Christ	薄叶鳞盖蕨
Woodsia macrochlaena	大囊岩蕨	101 (Fas. 3)	Woodsia macrochlaena Mett. ex Kuhn	大囊岩蕨
Woodsia manchuriensis	满洲岩蕨	102	Protowoodsia manchuriensis (Hook.) Ching	膀胱蕨
Davallia mariesii	海州骨碎补	103	Davallia mariesii T. Moore ex Baker	骨碎补
Davallia orientalis	华南骨碎补	104	Davallia divaricata Blume	大叶骨碎补
Humata assamica	高山阴石蕨	105	Humata assamica (Bedd.) C. Chr.	长叶阴石蕨
Arthropteris obliterata	藤蕨	106	Arthropteris palisotii (Desv.) Alston	爬树蕨
Camptosorus sibiricus	过山蕨	107	Camptosorus sibiricus Rupr.	过山蕨
Athyrium sheareri	庐山蹄盖蕨	108	Anisocampium sheareri (Baker) Ching ex Y. T. Hsieh	华东安蕨
Athyrium otophorum	光蹄盖蕨	109	Athyrium otophorum (Miq.) Koidz.	光蹄盖蕨
Athyrium goeringianum	柯氏蹄盖蕨	110	Athyrium iseanum Rosenst.	长江蹄盖蕨
Asplenium sarelii	华中铁角蕨	111	Asplenium sarelii Hook.	华中铁角蕨
Asplenium sampsoni	岭南铁角蕨	112	Asplenium sampsonii Hance	岭南铁角蕨
Asplenium prolongatum	长生铁角蕨	113	Asplenium prolongatum Hook.	长叶铁角蕨
Asplenium crinicaule	毛铁角蕨	114	Asplenium crinicaule Hance	毛轴铁角蕨
Asplenium saxicola	粤铁角蕨	115	Asplenium saxicola Rosenst.	石生铁角蕨
Woodwardia orientalis	东方狗脊	116	Woodwardia orientalis Sw.	东方狗脊
Dictyocline griffithii	圣蕨	117	Dictyocline griffithii T. Moore	圣蕨
Quercifilix zeylanica	地耳蕨	118	Quercifilix zeylanica (Houtt.) Copel.	地耳蕨
Bolbitis heteroclita	长叶实蕨	119	Bolbitis heteroclita (C. Presl) Ching	长叶实蕨
Bolbitis subcordata	海南实蕨	120	Bolbitis subcordata (Copel.) Ching	华南实蕨
Hypodematium crenatum	肿足蕨	121	Hypodematium crenatum (Forsk.) Kuhn	肿足蕨
Hypodematium fordii	福氏肿足蕨	122	Hypodematium fordii (Baker) Ching	福氏肿足蕨
Hypodematium cystopteroides	山东肿足蕨	123	Hypodematium sinense K. Iwats.	山东肿足蕨
Tectaria subtriphylla	三叉蕨	124	Tectaria subtriphylla (Hook. & Arn.) Copel.	三叉蕨
Tectaria macrodonta	高山三叉蕨	125	Tectaria coadunata (Wall. ex Hook. & Grev.) C. Chr.	大齿叉蕨
Cyrtomium fortunei	贯众	126	Cyrtomium fortunei J. Sm.	贯众
Cyrtomium falcatum	全缘贯众	127	Cyrtomium falcatum (L. f.) C. Presl	全缘贯众
Cyclopeltis crenata	拟贯众	128	Cyclopeltis crenata (Fée) C. Chr.	拟贯众
Polystichum fimbriatum	瓦鳞耳蕨	129	Polystichum fimbriatum Christ	瓦鳞耳蕨
Polystichum otophorum	高山耳蕨	130	Polystichum otophorum (Franch.) Bedd.	高山耳蕨
Polystichum omeiense	峨眉耳蕨	131	Polystichum omeiense C. Chr.	峨眉耳蕨
Pellaea smithii	史氏旱蕨	132	Pellaea smithii C. Chr.	西南旱蕨
Cheilanthes duclouxii	杜氏粉背蕨	133	Aleuritopteris duclouxii (Christ) Ching	裸叶粉背蕨
Cheilanthes hancockii	韩氏粉背蕨	134	Cheilosoria hancockii (Baker) Ching & K. H. Shing	大理碎米蕨
Cheilanthes chusana	舟山粉背蕨	135	Cheilosoria chusana (Hook.) Ching & K. H. Shing	毛轴碎米蕨
Cheilanthes trichophylla	毛粉背蕨	136	Pellaea trichophylla (Baker) Ching	毛旱蕨
Adiantum refractum	蜀铁线蕨	137	Adiantum edentulum Christ f. refractum (Christ) Y. X. Lin	蜀铁线蕨
Adiantum davidi	白背铁线蕨	138	Adiantum davidii Franch.	白背铁线蕨
Adiantum edgeworthii	爱氏铁线蕨	139	Adiantum edgeworthii Hook.	普通铁线蕨
Pteris faurei	傅氏凤尾蕨	140	Pteris faurei Hieron.	傅氏凤尾蕨
Pteris dactylina	掌凤尾蕨	141	Pteris dactylina Hook.	指叶凤尾蕨

续表

图谱原名 Name in *Icones*		图版号 Plate No	现接受名 Present accepted name	
Schizoloma ensifolium	拟凤尾蕨	142	Lindsaea ensifolia Sw.	剑叶鳞始蕨
Coniogramme intermedia	华凤了蕨	143	Coniogramme intermedia Hieron.	普通凤了蕨
Vittaria forrestiana	宽书带蕨	144	Haplopteris doniana (Mett. ex Hieron.) E. H. Crane	带状书带蕨
Antrophyum formosanum	车前蕨	145	Antrophyum formosanum Hieron.	台湾车前蕨
Saxiglossum taeniodes	拟石苇	146	Pyrrosia angustissima (Gies. ex Diels) Tagawa & K. Iwats.	石蕨
Pyrrosia sheareri	庐山石苇	147	Pyrrosia sheareri (Baker) Ching	庐山石韦
Pyrrosia drakeana	毡毛石苇	148	Pyrrosia drakeana (Franch.) Ching	毡毛石韦
Arthromeris himalayensis	琉璃节肢蕨	149	Arthromeris himalayensis (Hook.) Ching	琉璃节肢蕨
Arthromeris lungtauensis	粤节肢蕨	150	Arthromeris lehmannii (Mett.) Ching	节肢蕨
Gleichenia cantonensis	广东里白	151	Diplopterygium cantonense (Ching) Ching	粤里白
Gleichenia laevissima	光里白	152	Diplopterygium laevissimum (Christ) Nakai	光里白
Gleichenia splendida	硕里白	153	Dicranopteris splendid (Hand.-Mazz.) Tagawa	大羽芒萁
Dipteris chinensis	双扇蕨	154	Dipteris chinensis Christ	中华双扇蕨
Plagiogyria assurgens	峨眉瘤足蕨	155	Plagiogyria assurgens Christ	峨眉瘤足蕨
Lindsaya lobbiana	洛氏林蕨	156	Lindsaea lucida Blume	亮叶鳞始蕨
Lindsaya decomposita	网脉林蕨	157	Lindsaeahainaniana (K. U. Kramer) Lehtonen & Tuomisto	海南网脉鳞始蕨
Adiantum gravesii	粤铁线蕨	158	Adiantum lianxianense Ching & Y. X. Lin	粤铁线蕨
Adiantum chienii	钱氏铁线蕨	159	Adiantum chienii Ching	北江铁线蕨
Adiantum roborowskii	陇铁线蕨	160	Adiantum roborowskii Maxim.	陇南铁线蕨
Onychium contiguum	高山乌蕨	161	Onychium contiguum C. Hope	黑足金粉蕨
Onychium moupinense	木坪乌蕨	162	Onychium moupinense Ching	木坪金粉蕨
Onychium tenuifrons	狭叶乌蕨	163	Onychium tenuifrons Ching	蚀盖金粉蕨
Onychium ipii	叶氏乌蕨	164	Onychium moupinense Ching var. ipii (Ching) K. H. Shing	湖北金粉蕨
Pleurosoriopsis makinoi	睫毛蕨	165	Pleurosoriopsis makinoi (Maxim. ex Makino) Fomin	睫毛蕨
Coniogramme fraxinea	全缘凤了蕨	166	Coniogramme fraxinea (D. Don) Diels	全缘凤了蕨
Coniogramme caudata	毛叶凤了蕨	167	Coniogramme pubescens Hieron.	骨齿凤了蕨
Coniogramme procera	高山凤了蕨	168	Coniogramme procera Fée	直角凤了蕨
Oleandra cumingii	华南荼蕨	169	Oleandra cumingii J. Sm.	华南条蕨
Oleandra whangii	瑶山荼蕨	170	Oleandra musifolia (Blume) C. Presl	光叶条蕨
Oleandra undulata	长柄荼蕨	171	Oleandra undulata (Willd.) Ching	波边条蕨
Gymnocarpium remoti-pinnatum	肢节蕨	172	Gymnocarpium remote-pinnatum (Hayata) Ching	羽节蕨
Asplenium fugax	阴地铁角蕨	173	Asplenium capillipes Makino	阴地铁角蕨
Asplenium exiguum	低头铁角蕨	174	Asplenium yunnanense Franch.	云南铁角蕨
Asplenium loriceum	南海铁角蕨	175	Asplenium loriceum Christ	南海铁角蕨
Asplenium interjectum	黔铁角蕨	176	Asplenium interjectum Christ	贵阳铁角蕨
Acrophorus stipellatus	拟鳞毛蕨	177	Acrophoruspaleolatus Pic. Serm.	鱼鳞蕨
Dryopteris serrato-dentata	高山鳞毛蕨	178	Dryopteris serrato-dentata (Bedd.) Hayata	刺尖鳞毛蕨
Dryopteris scottii	史氏鳞毛蕨	179	Dryopteris scottii (Bedd.) Ching ex C. Chr.	无盖鳞毛蕨
Dryopteris liankwangensis	两广鳞毛蕨	180	Dryopteris liankwangensis Ching	两广鳞毛蕨
Dryopteris championi	张氏鳞毛蕨	181	Dryopteris championii (Benth.) C. Chr.	阔鳞鳞毛蕨
Polystichum chingae	滇耳蕨	182	Polystichum chingae Ching	滇耳蕨
Cyrtomium aequibasis	滇贯众	183	Cyrtomium aequibasis (C. Chr.) Ching	等基贯众
Cyrtomium muticum	大叶贯众	184	Cyrtomium macrophyllum (Makino) Tagawa	大叶贯众
Hemigramma decurrens	拟叉蕨	185	Hemigramma decurrens (Hook.) Copel.	沙皮蕨
Leucostegia immersa	膜盖蕨	186	Leucostegia immersa (Wall. ex Hook.) C. Presl	大膜盖蕨
Leucostegia hookeri	霍氏膜盖蕨	187	Araiostegia hookeri (T. Moore ex Bedd.) Ching	宿枝小膜盖蕨
Leucostegia multidentata	毛膜盖蕨	188	Paradavallodes multidentatum (Hook. & Baker) Ching	假钻毛蕨
Loxogramme grammitoides	小叶剑蕨	189	Loxogramme grammitoides (Baker) C. Chr.	匙叶剑蕨

续表

图谱原名 Name in *Icones*		图版号 Plate No	现接受名 Present accepted name	
Loxogramme salicifolia	柳叶剑蕨	190	Loxogramme salicifolia (Makino) Makino	柳叶剑蕨
Loxogramme ensiformis	阔叶剑蕨	191	Loxogramme formosana Nakai	台湾剑蕨
Drynaria fortunei	槲蕨	192	Drynaria roosii Nakaike	槲蕨
Drynaria sinica	华槲蕨	193	Drynaria sinica Diels	秦岭槲蕨
Colysis wui	吴氏线蕨	194	Colysis pedunculata (Hook. & Grev.) Ching	长柄线蕨
Colysis hemionitidea	断线蕨	195	Colysis hemionitidea (C. Presl) C. Presl	断线蕨
Colysis wrightii	莱氏线蕨	196	Colysis wrightii (Hook.) Ching	褐叶线蕨
Colysis hemitoma	胄叶线蕨	197	Colysis hemitoma (Hance) Ching	胄叶线蕨
Colysis digitata	掌叶线蕨	198	Colysis digitata (Baker) Ching	掌叶线蕨
Colysis pentaphylla	滇线蕨	199	Colysis elliptica (Thunb.) Ching var. pentaphylla (Baker) L. Shi & X. C. Zhang	滇浅蕨
Colysis morsei	马氏线蕨	200	Colysis elliptica (Thunb.) Ching	线蕨
Pseudodrynaria coronans	崖姜蕨	201	Pseudodrynaria coronans (Wall. ex Mett.) Ching	崖姜蕨
Angiopteris fokiensis	福建观音座莲	202	Angiopteris fokiensis Hieron.	福建观音座莲
Archangiopteris bipinnata	二回原始观音座莲	203	Archangiopteris bipinnata Ching	二回原始观音座莲
Archangiopteris hokouensis	河口原始观音座莲	204	Archangiopteris hokouensis Ching	河口原始观音座莲
Archangiopteris somai	台湾原始观音座莲	205	Archangiopteris henryi Christ & Gies.	亨利原始观音座莲
Archangiopteris subrotundata	圆基原始观音座莲	206	Archangiopteris henryi Christ & Gies.	亨利原始观音座莲
Archangiopteris latipinna	阔叶原始观音座莲	207	Archangiopteris henryi Christ & Gies.	亨利原始观音座莲
Archangiopteris caudata	尾叶原始观音座莲	208	Archangiopteris henryi Christ & Gies.	亨利原始观音座莲
Archangiopteris tonkinensis	尖叶原始观音座莲	209	Archangiopteris tonkinensis (Hayata) Ching	尖叶原始观音座莲
Adiantum caudatum	鞭叶铁线蕨	210	Adiantum malesianum Ghatak	假鞭叶铁线蕨
Adiantum sinicum	苍山铁线蕨	211	Adiantum sinicum Ching	苍山铁线蕨
Adiantum juxtapositum	仙霞铁线蕨	212	Adiantum juxtapositum Ching	仙霞铁线蕨
Adiantum capillus-junonis	团叶铁线蕨	213	Adiantum capillus-junonis Rupr.	团羽铁线蕨
Adiantum philippense	半月形铁线蕨	214	Adiantum philippense L.	半月形铁线蕨
Adiantum soboliferum	翅柄铁线蕨	215	Adiantum soboliferum Wall. ex Hook.	翅柄铁线蕨
Adiantum capillus-veneris	铁线蕨	216	Adiantum capillus-veneris L.	铁线蕨
Adiantum edentulum	月芽铁线蕨	217	Adiantum edentulum Christ	月芽铁线蕨
Adiantum fengianum	冯氏铁线蕨	218	Adiantum fengianum Ching	冯氏铁线蕨
Adiantum smithianum	长盖铁线蕨	219	Adiantum fimbriatum Christ	长盖铁线蕨
Adiantum venustum	细叶铁线蕨	220	Adiantum venustum D. Don	细叶铁线蕨
Adiantum bonatianum	毛足铁线蕨	221	Adiantum bonatianum Brause	毛足铁线蕨
Adiantum muticum	鹤庆铁线蕨	222	Adiantum edentulum Christ f. muticum (Ching) Y. X. Lin	鹤庆铁线蕨
Adiantum flabellulatum	扇叶铁线蕨	223	Adiantum flabellulatum L.	扇叶铁线蕨
Adiantum induratum	海南铁线蕨	224	Adiantum induratum Christ	圆柄铁线蕨
Adiantum diaphanum	长尾铁线蕨	225	Adiantum diaphanum Blume	长尾铁线蕨
Adiantum pedatum	掌叶铁线蕨	226	Adiantum pedatum L.	掌叶铁线蕨
Adiantum myriosorum	灰背铁线蕨	227	Adiantum myriosorum Baker	灰背铁线蕨
Leptogramma mollissima	毛叶茯蕨	228	Leptogramma pozoi (Lag.) Ching subsp. mollissima (Fischer ex Kunze) Nakaike	毛叶茯蕨
Leptogramma scallani	峨眉茯蕨	229	Leptogramma scallanii (Christ) Ching	峨眉茯蕨
Leptogramma caudata	尾叶茯蕨	230	Leptogramma tottoides H. Ito	小叶茯蕨
Stegnogramma cyrtomioides	波叶溪边蕨	231	Stegnogramma cyrtomioides (C. Chr.) Ching	贯众叶溪边蕨
Stegnogramma asplenioides	浅裂溪边蕨	232	Stegnogramma asplenioides (C. Chr.) J. Sm. ex Ching	浅裂溪边蕨
Polystichum dielsii	圆顶耳蕨	233	Polystichum dielsii Christ	圆顶耳蕨
Polystichum excellens	尖顶耳蕨	234	Polystichum excellens Ching	尖顶耳蕨
Polystichum yuanum	倒叶耳蕨	235	Polystichum yuanum Ching	倒叶耳蕨
Polystichum kwangtungense	广东耳蕨	236	Polystichum kwangtungense Ching	广东耳蕨
Polystichum consimile	刺叶耳蕨	237	Polystichum consimile Ching	涪陵耳蕨
Polystichum stenophyllum	芽胞耳蕨	238	Polystichum stenophyllum Christ	狭叶芽胞耳蕨
Polystichum grossidentatum	粗齿耳蕨	239	Polystichum subdeltodon Ching	粗齿耳蕨
Polystichum lanceolatum	亮叶耳蕨	240	Polystichum lanceolatum (Baker) Diels	亮叶耳蕨

续表

图谱原名 Name in *Icones*		图版号 Plate No	现接受名 Present accepted name	
Polystichum thomsoni	尾叶耳蕨	241	Polystichum thomsonii (Hook. f.) Bedd.	尾叶耳蕨
Polystichum bifidum	钳形耳蕨	242	Polystichum bifidum Ching	钳形耳蕨
Polystichum nepalense	软骨耳蕨	243	Polystichum nepalense (Spreng.) C. Chr.	尼泊尔耳蕨
Polystichum xiphophyllum	革叶耳蕨	244	Polystichum xiphophyllum (Baker) Diels	剑叶耳蕨
Polystichum tripteron	三叉耳蕨	245	Polystichum tripteron (Kunze) C. Presl	戟叶耳蕨
Polystichum tsus-simense	马祖耳蕨	246	Polystichum tsus-simense (Hook.) J. Sm.	对马耳蕨
Cyrtomidictyum basipinnatum	单叶鞭叶蕨	247	Cyrtomidictyum basipinnatum (Baker) Ching	单叶鞭叶蕨
Cyrtomidictyum conjunctum	卵形鞭叶蕨	248	Cyrtomidictyum conjunctum Ching	卵状鞭叶蕨
Cyrtomidictyum lepidocaulon	鞭叶蕨	249	Cyrtomidictyum lepidocaulon (Hook.) Ching	鞭叶蕨
Cyrtomidictyum faberi	普陀鞭叶蕨	250	Cyrtomidictyum lepidocaulon (Hook.) Ching	鞭叶蕨